COMPUTER INTEGRATED MACHINE DESIGN

Charles E. Wilson
New Jersey Institute of Technology

Prentice Hall
Upper Saddle River, New Jersey Columbus, Ohio

Library of Congress Cataloging-in-Publication Data

Wilson, Charles E.
Computer integrated machine design/Charles E. Wilson.
p. cm.
Includes bibliographical references and index.
ISBN 0-02-428390-8
1. Machine design—/Data processing. 2. Computer-aided design.
I. Title.
TJ233.W55 1997
621.8'15'0285—dc20 96-28563
CIP

Editor: Stephen Helba
Production and Design Coordination: bookworks/Karen Fortgang
Cover Design: Proof Positive/Farrowlyne Associates
Production Manager: Pamela D. Bennett
Illustrations: Graphic World
Marketing Manager: Frank Mortimer, Jr.

This book was set in Times Roman by Graphic World and was printed and bound by Quebecor Printing/Book Press. The cover was printed by Phoenix Color Corp.

Simon & Schuster/A Viacom Company
Upper Saddle River, New Jersey 07458

Printed in the United States of America

10 9 8 7 6 5 4 3 2 1

ISBN 0-02-428390-8

Prentice-Hall International (UK) Limited, *London*
Prentice-Hall of Australia Pty. Limited, *Sydney*
Prentice-Hall of Canada, Inc., *Toronto*
Prentice-Hall Hispanoamericana, S.A., *Mexico*
Prentice-Hall of India Private Limited, *New Delhi*
Prentice-Hall of Japan, Inc., *Tokyo*
Simon & Schuster Asia Pte. Ltd., *Singapore*
Editora Prentice-Hall do Brasil, Ltda., *Rio de Janeiro*

To Liz

Brief Contents

Contents

Preface

The Accreditation Board for Engineering and Technology (ABET)[1] describes engineering technology as a part of the technological field that requires the application of science and engineering methods combined with technical skills in the support of engineering activities. Subject matter in an engineering technology program has its roots in mathematics and basic science and carries knowledge further toward application. ABET makes the following statement on computer competency:

> Engineering technicians and technologists are dependent on the computer to effectively perform their job functions. It is therefore essential that students acquire a working knowledge of computer usage. Instruction in applications of software for solving technical problems and student practice within appropriate technical courses is required for all programs.

Machine design offers unlimited opportunities for computer usage. This text takes advantage of the general availability of computers. With the help of a computer, we are able to examine a design problem efficiently and thoroughly. We can consider alternative design options with a few keystrokes. Computers do not replace thinking and creative work; they reduce computational drudgery and allow us more time for important design tasks.

Machine designers face problems that have many possible solutions. These solutions may include unsafe or impractical designs as well as acceptable and optimum designs. Example problems throughout the text illustrate design methods and principles that simulate the real design environment. Most problems are solved with the aid of a computer, and the results

[1]*Criteria for Accrediting Programs in Engineering Technology—Effective for Evaluations during the 1995–96 Accreditation Cycle*, Baltimore, MD: Accreditation Board for Engineering and Technology.

are checked and discussed. If the assumptions, design decisions, and calculations do not lead to the desired result, appropriate changes are made. End-of-chapter design problems follow a similar pattern. Emphasis is placed on evaluation and interpretation of results and on development of design skills. Computer-generated graphics are used to add insight to design decisions.

Most of the chapters in the text are free-standing, allowing for topic selection to meet the goals of a particular curriculum. The free-standing chapters also allow flexibility for times when technology faculty want to enhance their teaching by introducing topics and design problems based on their industrial experience or consulting assignments.

Many of the homework problems can be easily modified for in-class tests. For example, if a computer is not available, it may take too long to design a countershaft with loads in two planes. The problem can be made to fit in a test by giving the value of moment and torque at various points on the shaft. Some problems appear difficult at first glance. However, students often respond enthusiastically to challenging problems that require decision making and evaluating results.

Design errors may have serious consequences, including loss of life. Responsible practicing designers check and recheck assumptions, design decisions, and calculations, using alternative procedures where available. They seek help from other professionals to ensure safe products. When practical, materials and products are tested under service conditions. Responsible practicing designers consider the consequences of foreseeable product use and even product misuse. Safety considerations should also be foremost to students, who will soon be practicing designers.

It is the reader's responsibility to assess methods and formulas to determine their applicability to a particular design situation. Although the reviewers, the editors, and the author have made every attempt to ensure accuracy, errors are certain to appear. Corrections and suggestions are welcome.

ACKNOWLEDGMENTS

I wish to thank all who helped with this book. The content was greatly influenced by the reviewers who include Thomas G. Boronkay, University of Cincinnati; Janak Dave, University of Cincinnati; William Ferraioli, Capital Community-Technical College; Frank S. Irlinger, Nashville State Technical Institute; Thomas F. Lukach, Stark Technical College; John G. Nee, Central Michigan University; and Jack Zecher, Indiana University/Purdue University. Thanks also to Steve Helba, Syl Huning, and others at Prentice Hall, my students and colleagues at New Jersey Institute of Technology who made helpful suggestions, and the companies that provided illustrations and other material.

Charles E. Wilson

CHAPTER 1

Creative Design

Symbols

a	acceleration	p	pressure
$a, b, c,$		P	load
h, t	dimensions	r, R	radius
A	area	S_W	working strength
C	cost function	S_{YP}	yield strength
d	diameter, distance to parallel axis	V	volume
$\mathbf{F}_I$	inertia-force	x, y, z	axes
I_M	mass-moment-of-inertia	Z_{CG}	distance to center-of-gravity
M	mass	α	angular acceleration
$\mathbf{M}_I$	inertia-moment	γ	mass density
N_{FS}	factor of safety	σ	stress

Units

acceleration	in/s^2 or m/s^2
angular acceleration	rad/s^2
dimensions and distances	in or m (often mm)
load	lb or N
pressure, stress, and strength	lb/in^2 or Pa (often MPa)
mass	lb · s^2/in or kg
mass density	lb · s^2/in^4 or kg/m^3
mass-moment-of-inertia	lb · s^2 · in or kg · m^2

Engineering design combines art and science, analysis and synthesis. Successful engineers are inventive and imaginative, while employing scientific principles to develop a satisfactory design. Consider, for example, the air bag crash restraint for automobiles. It was necessary to design a system to detect a crash and signal the air bag to inflate (in about 1/20 second), while avoiding the possibility of unintended inflation. The designers had to use the laws of physics as well as their ingenuity; they had to analyze many possible solutions in order to select the best one.

Knowledge from the entire engineering technology curriculum goes into formulating a design. Communication is as important as technology, whether a designer is working individually or as part of a team. Designers must be able to understand what is needed and describe a design graphically, verbally, and in written form.

COMPUTER-AIDED ENGINEERING (CAE)

The design process includes these steps:

- recognizing a need
- formulating specifications for a product to meet that need (e.g., applied static and dynamic forces, size and weight limitations, environmental requirements, and factor of safety)
- employing creative synthesis (inventive, imaginative combination of parts to form a new product)
- drafting
- analyzing (determining stresses and deflections, and comparing them with strength and acceptable deflection)
- redesigning as required
- manufacturing and testing

Figure 1.1 shows the interrelation between steps in the design process.

Computer-aided engineering (CAE) software can help perform some of the tasks in the design process, particularly the drafting and analysis steps. Using CAE software, a designer can easily answer such questions as "What if a given dimension is changed? How will that affect other components? What is the effect on stress and deflection?"

Concurrent Engineering (CE)

An Institute for Defense Analysis report (1986) defines CE as "a systematic approach to integrated, concurrent design of products and their related processes." The report adds, "This approach is intended to cause the developers, from the outset, to consider all elements of the product life cycle from concept through disposal, including quality, cost, schedule, and user requirements." CE requires teamwork and coordination involving research and development engineers, designers, manufacturing engineers and technicians, and personnel from other functions. The focus is on product-based teams rather than on departments. With the CE approach, many aspects of product design and development can be carried out simultaneously. Thus, CE is a parallel process, while the traditional process is a series approach. Problems sometimes result from lack of coordination in the traditional approach. CE attempts to

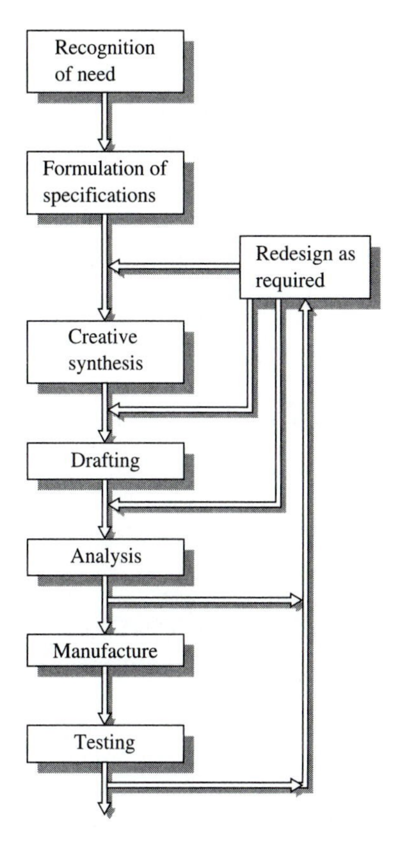

Figure 1.1 The design process.

eliminate such problems in the concept and design phase. Mechanical design automation (MDA) tools—including software for drafting, solid modeling, stress analysis, and manufacturing—facilitate integration of the design-through-manufacturing process. The solid model, which is discussed later in this chapter, serves as a means of communication for the CE team.

COMPUTER-AIDED DRAFTING (CAD)

Computer-aided drafting or design (CAD) software packages take advantage of the computer's ability to store, process, and retrieve information. A designer describes the proposed design and then displays it on the computer monitor. The system enables the designer to view a proposed machine part from various locations and to produce a hardcopy (print). Design changes can be made quickly and inexpensively at this point.

In addition, CAD can be used to create an accurate three-dimensional (3-D) geometry database, produce a bill of materials, and to eliminate the need for a prototype, or create a prototype via stereo-lithography (more on this in Chapter 19). Thus, CAD can reduce concept-to-production time, improving competitiveness.

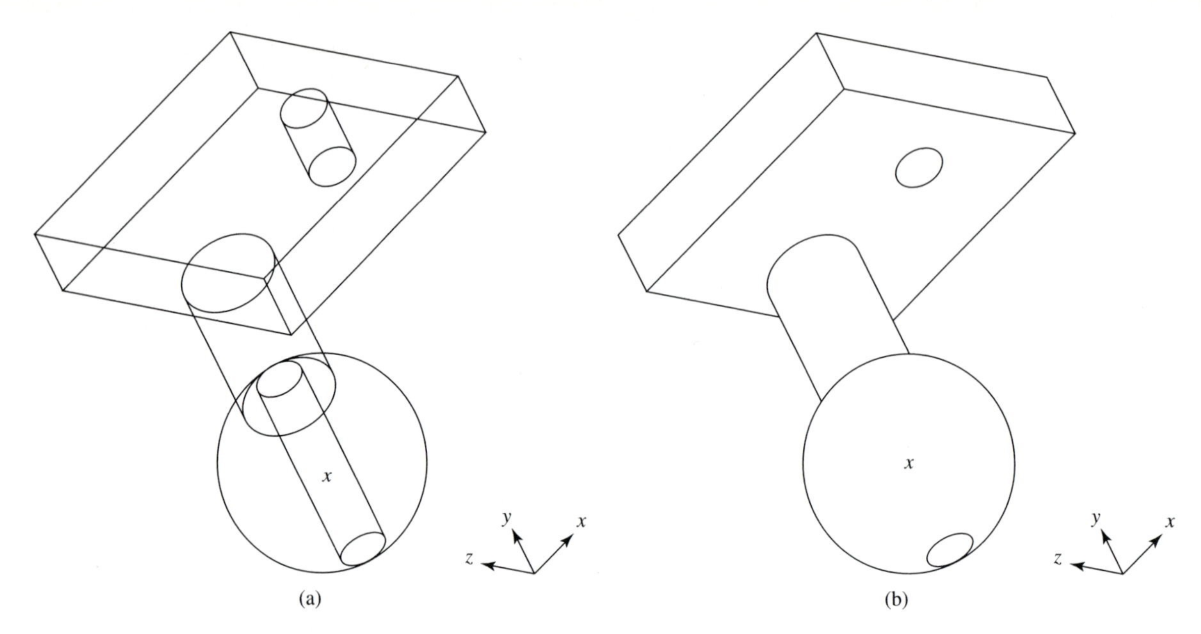

Figure 1.2 (a) Solid model; (b) solid model with hidden lines removed.

CAD data can be shared with other tasks, including finite element analysis (FEA) and computer-integrated manufacturing (CIM). CAD has a variety of capabilities, each designed to meet a different engineering need. Some of these capabilities are described in the paragraphs that follow.

Wire-Frame Models. Wire-frame models represent parts as lines indicating the edge of the part. Curved surfaces can be represented by contours. Wire-frame models are sometimes ambiguous if all the part edges are shown. Some clutter and ambiguity in wire-frame models can be remedied by removing hidden lines (i.e., lines that someone viewing the actual part would not see).

Creation of a Solid Model. The term **solid model** or **solid object** refers to a three-dimensional description of a part in terms of solid components. It can be created by combining **primitives** (i.e., familiar solid shapes such as right circular cylinders, spheres, cones, and blocks). The designer or draftsman gives the computer the dimensions and relationship of these solid shapes, rather than simply specifying boundaries as in a wire-frame model.

Figure 1.2 shows a solid model of part of a ball-joint. Part (a) of the figure shows the depth of the drilled holes, but the figure is ambiguous because of the presence of hidden lines. The CAD software was instructed to suppress hidden lines in Part (b).

Machine design is not limited to the design of shafts, gears, springs, and other machine parts. Designers apply their skills in many other areas. For example, one project in the medical rehabilitation field involves study and redesign of an orthosis. The orthosis is a device to aid people with a drop-foot problem, a loss of muscle control resulting from a stroke or head injury.

Figure 1.3 (a) shows a computer-aided design (CAD) model of an orthosis. The toe is at the top right of that figure. Part (b) of the figure shows another view, with the toe at the bottom right. The model is shaded in Part (c), and the hidden lines are removed. Then, in Part (d), a solid model of the soft tissue of the foot is inserted into the device. The bones and ligaments which form the inner core of the foot are modeled in Part (e). Rehabilitation devices of this type are ordinarily handcrafted. This project involves analysis leading to computer-aided-design and computer-aided-manufacture of the device.

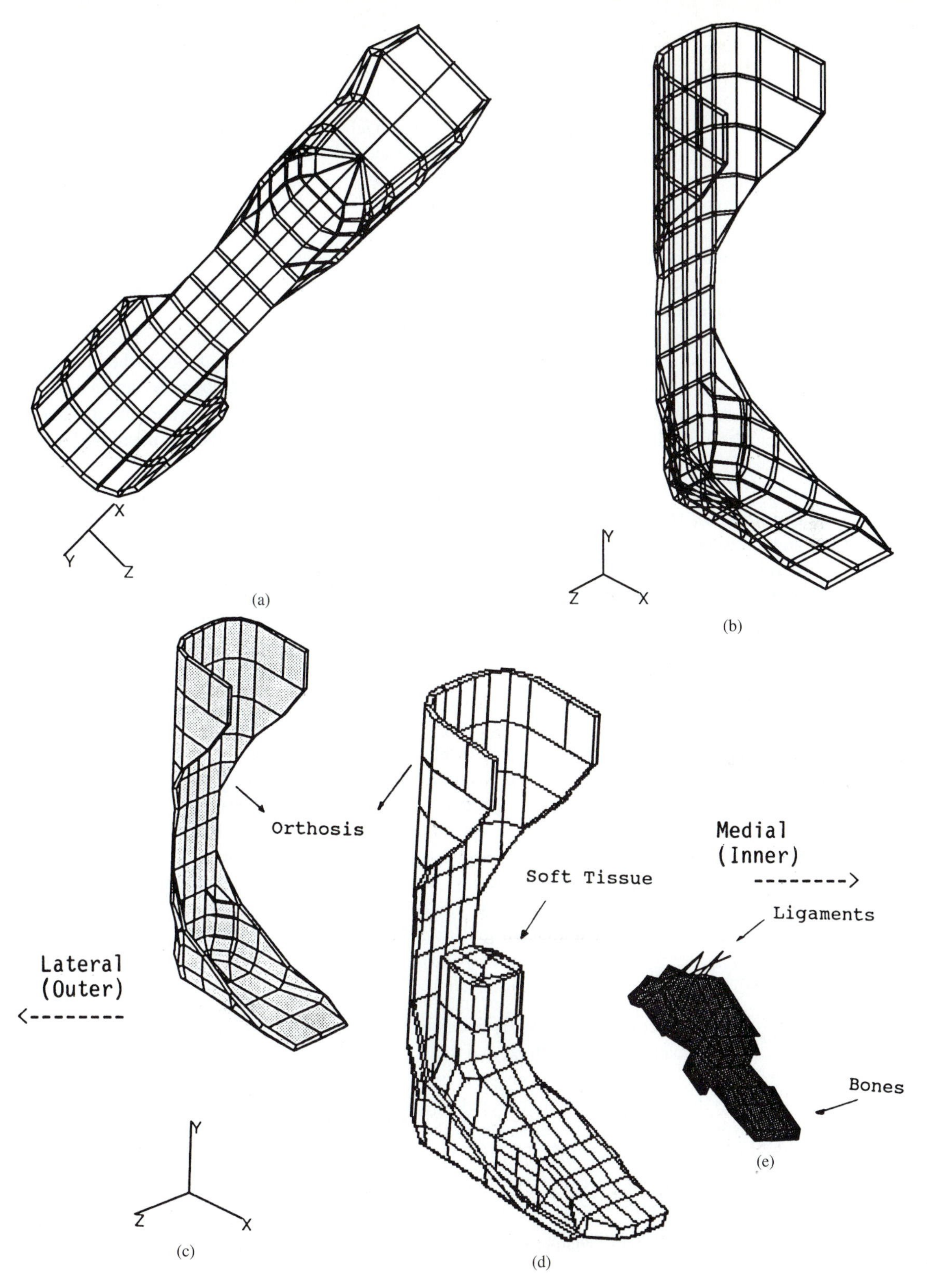

Figure 1.3 CAD model of an orthosis. (a) View of unloaded model; (b) view with toe at bottom right; (c) shaded view with hidden lines removed; (d) assembled model; (e) model of bones and ligaments (inner core of the foot). (Courtesy of Tai-Ming Chu)

Extrusion. An alternative method of solid model creation involves describing a two-dimensional profile and then moving it to generate a cylindrical surface. A circular profile generates a right circular cylinder. In general, the result resembles what an extrusion die would produce.

Revolving a Profile. A two-dimensional profile can be described and then revolved about an axis to form a surface of revolution. The resulting solid model might be equivalent to a part produced on a lathe.

Skinning. A blade of a jet engine or a steam turbine could not easily be described in terms of primitives or formed by extrusion or profile revolution. We could, however, define a set of profile cross sections and then "skin" a surface over the profiles to approximate the blade shape. Solid modeling techniques and software capabilities are described in detail by Lawry (1994).

Construction Operations (Boolean Operations). Construction operations include cut, join, and intersect. We can cut one solid object with another, join one solid object to another, or determine the intersection of two solid objects. We can also make plane cuts, add material at internal corners to form fillets, round off external corners by removing material, and drill blind or through-holes. If the thickness of a part is increased after a through-hole has been drilled, the hole will still penetrate the part. If a cylinder is used to cut the part, we could get an unintended result. If the part is made thicker, the hole could become a blind hole rather than a through-hole.

Display Options. CAD software can display a solid model from various viewpoints. Once the part description has been stored, a simple command can display front, side, top, oblique, and isometric views, as well as perspective views from arbitrary locations. The software can suppress hidden lines to simulate an opaque object, or it can show the lines as if the object were transparent. An object can be shaded, and assemblies can be displayed using different colors to distinguish various components. Some programs can automatically dimension the part and develop cross-sectional views.

DESIGN IMPLICATIONS

A designer's first configuration rarely becomes the final design. Ordinarily, the design process goes through many iterations: stress and deflection considerations, cost and weight optimization, and interaction among various parts in an assembly.

If part dimensions and form never changed, CAD would offer few advantages over manual drafting. However, since that is generally not the case, it is comforting to know that CAD systems allow a dimensional change to be made with a few keystrokes. The designer can investigate questions such as "What if I redesign the part to be symmetrical for ease of assembly? If I change a particular dimension, will the part interfere with other parts in the assembly?" When using integrated computer-aided drafting/finite element analysis (CAD/FEA)

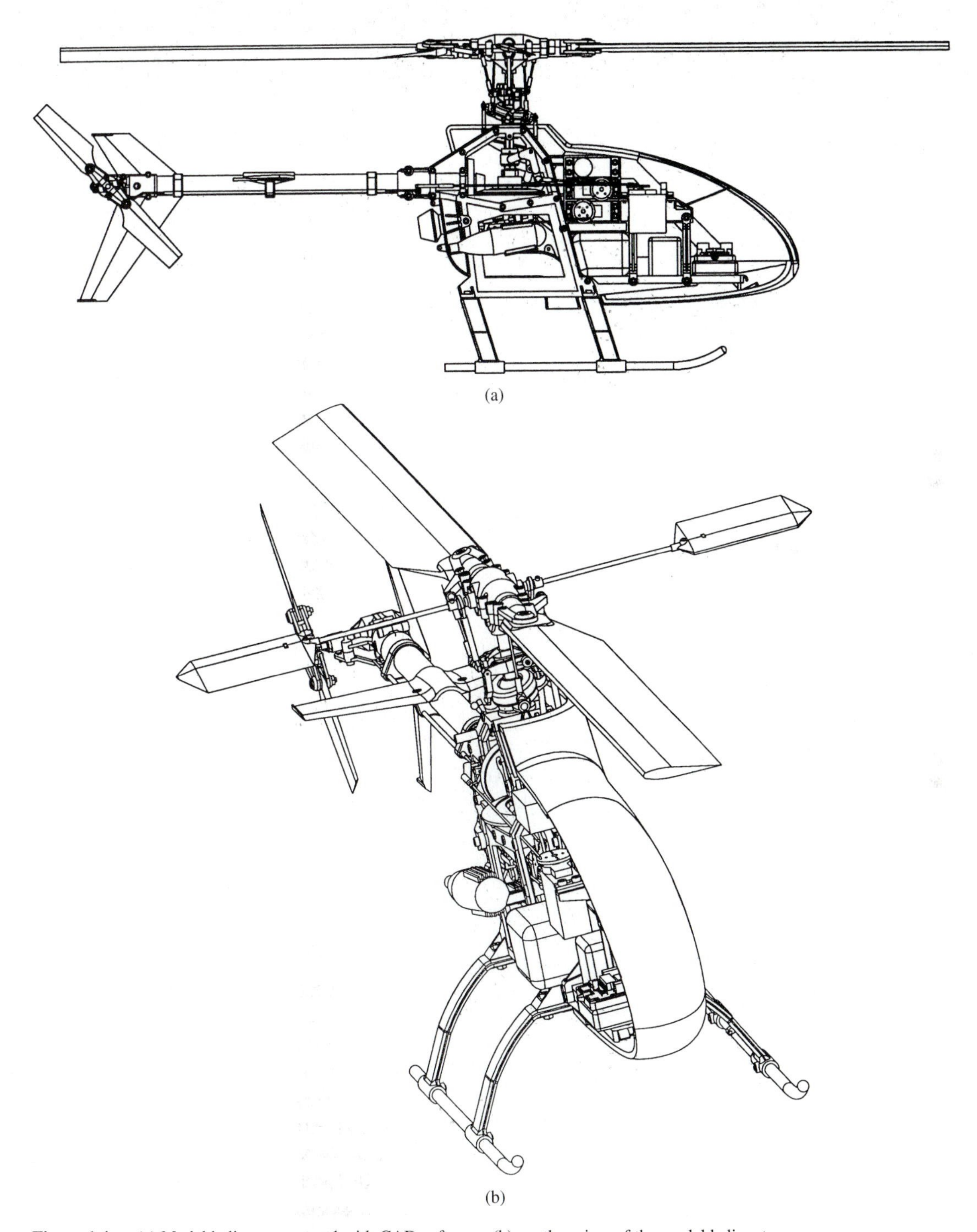

Figure 1.4 (a) Model helicopter created with CAD software; (b) another view of the model helicopter.

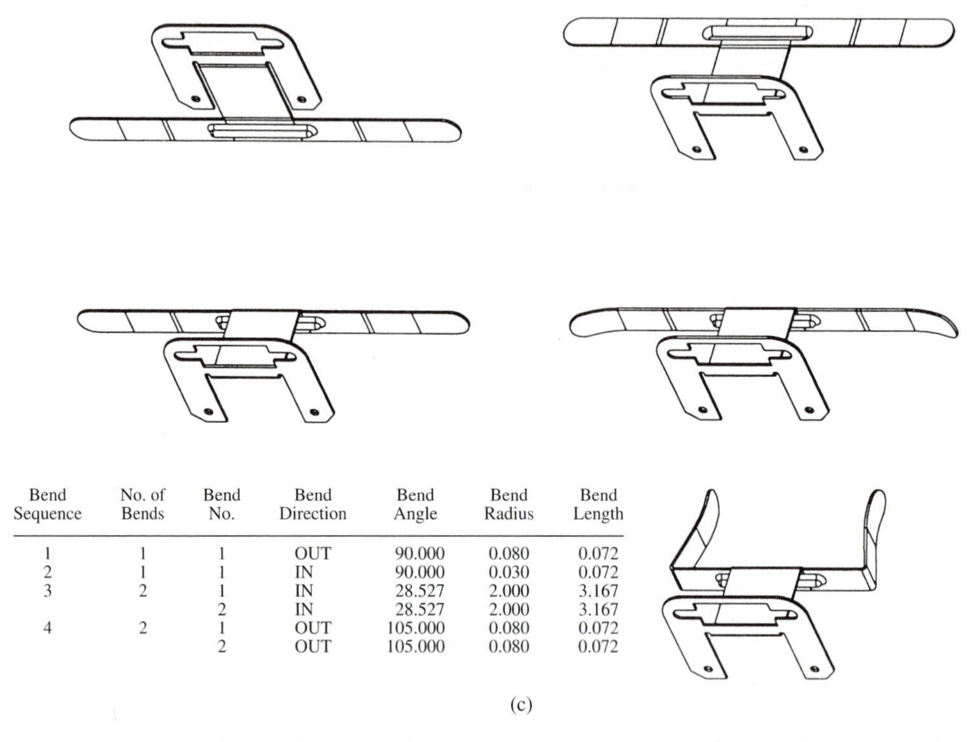

Bend Sequence	No. of Bends	Bend No.	Bend Direction	Bend Angle	Bend Radius	Bend Length
1	1	1	OUT	90.000	0.080	0.072
2	1	1	IN	90.000	0.030	0.072
3	2	1	IN	28.527	2.000	3.167
		2	IN	28.527	2.000	3.167
4	2	1	OUT	105.000	0.080	0.072
		2	OUT	105.000	0.080	0.072

(c)

Figure 1.4 *continued* (c) a progressive sheet-metal bending operation. (Courtesy of Parametric Technology Corporation)

systems, the designer may ask, "Will stresses be reduced if I increase the fillet radius at a particular location?" Computer-aided drafting/computer-aided manufacture (CAD/CAM) system users can also examine the implications of a dimension change on manufacturing time and cost.

A feature-based CAD system can create models as combinations of design features such as slots, ribs, and chamfers. A parametric CAD system assigns symbols to model dimensions, rather than fixed values in inches or millimeters. An engineer using a parametric system can relate one dimension or feature to another. When one parameter is changed, all related parameters are automatically updated. For example, four pins, each of nominal diameter

$$d_1 = 0.375 \text{ in}$$

are to fit into clearance holes of nominal diameter

$$d_2 = d_1 + 0.010$$

If stress analysis results lead the engineer to change d_1 to 0.500 in, then all of the hole diameters are automatically updated from $d_2 = 0.385$ to $d_2 = 0.510$ in. Parametric CAD systems make it easier for an engineer to improve a design because changes are rapidly incorporated.

Computer software modules are available for detailing, dimensioning, and tolerancing of CAD-generated parts. Other modules manage assembly of CAD-generated parts. Manufacturing modules are available to generate manufacturing process plans and tool programs for drilling, turning, and milling. These modules can also provide manufacturing time estimates and cost estimates from CAD models.

Figure 1.4(a) and (b) shows two views of a miniature helicopter assembly model created with the aid of *Pro/ENGINEER*™ CAD/CAM modules. The fuselage was created using a surface module; sheet-metal parts were bent using a sheet-metal module; and component machining was simulated with a manufacturing module. Part (c) of the figure shows progressive bending operations used to form a clip to hold servos on the helicopter. A sheet-metal module was used to model the operations that form the clip. Figure 1.5 shows an exploded view of a heat-gun assembly model created with the aid of an assembly module.

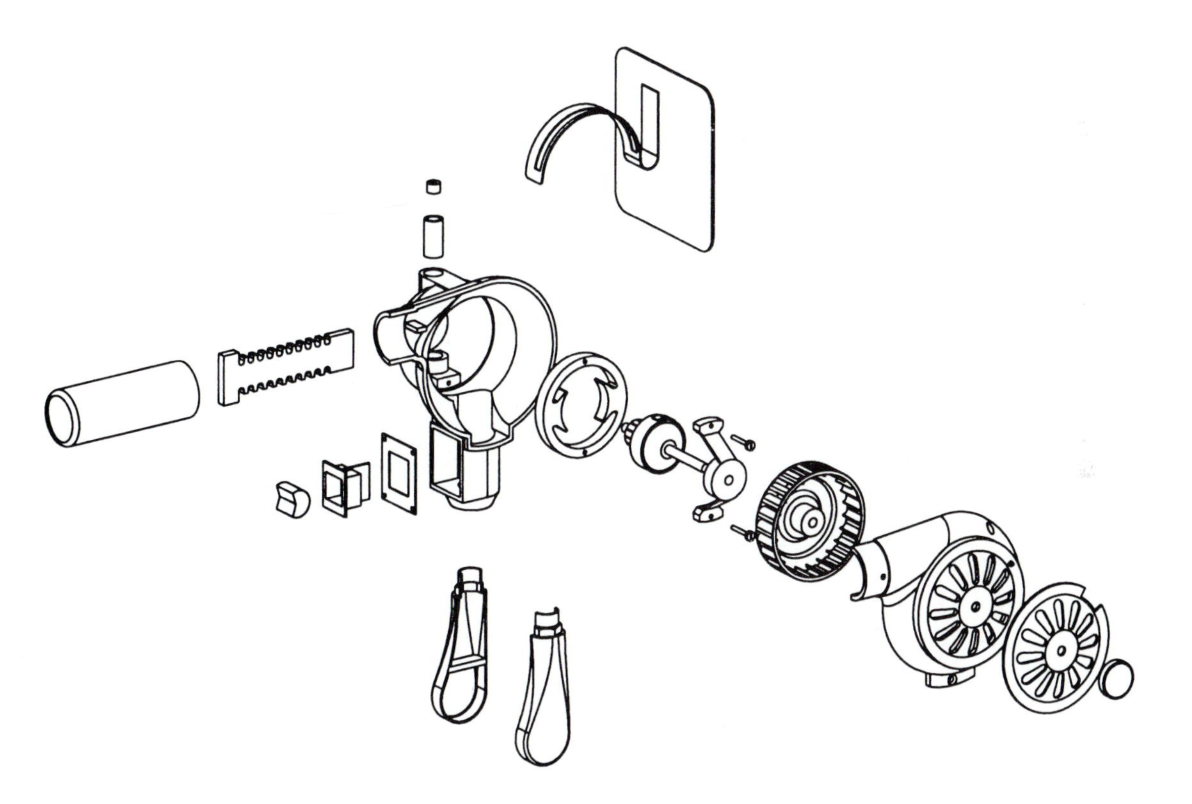

Figure 1.5 Exploded view of a heat-gun assembly. (Courtesy of Parametric Technology Corporation)

EXAMPLE PROBLEM Computer-Aided Drafting

A connecting rod is to have a 15 mm diameter bore and a 25 mm diameter bore at a distance of 150 mm center-to-center. Make a drawing of the connecting rod.

Design Decisions: For the preliminary design of the connecting rod, outer diameters (ODs) will be twice inner diameters (IDs), and bearing surfaces will be 25 mm long. Center section thickness will be 20 mm. These arbitrary decisions are our judgment of reasonable proportions. Dimensions will be changed if the preliminary design is found unacceptable owing to stress or other considerations.

Solution: As shown in Figure 1.6, we draw a 30 mm diameter by 25 mm long (right circular) cylinder with its center at the origin and its axis in the y-direction. A 50 mm diameter by 25 mm long cylinder is drawn and moved so that its center lies at 150,0,0 (x, y, z coordinates in mm). We now form the center section joining the two cylinders by defining four cross sections. Cross sections a, b, c, and d are located, respectively, at $x = 0$, 50, 100, and 150 mm. They are defined by the coordinates $0,\pm 10,\pm 15$; $50,\pm 10,\pm 15$; $100,\pm 10,\pm 18$; and $150,\pm 10,\pm 25$. That is, cross section a is a rectangle defined by the coordinates $0, 10, 15$; $0, 10, -15$; $0, -10, -15$; and $0, -10, 15$. The skinning procedure forms a smooth center section with the four cross sections. We now define a 15 mm diameter cylinder and use it to cut (drill) the 30 mm diameter cylinder. A 25 mm diameter cylinder is defined and moved and used to cut

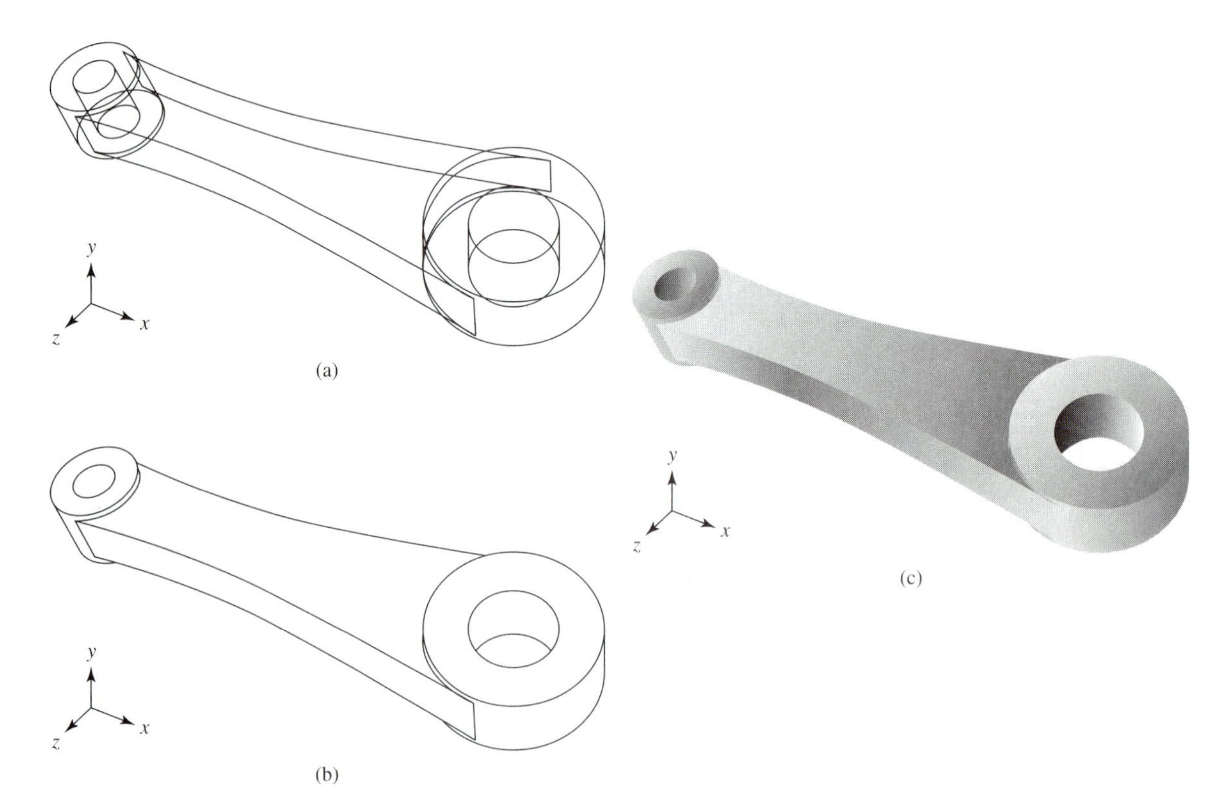

Figure 1.6 Computer-aided drafting example. (a) Solid model of connecting rod before removal of hidden lines; (b) solid model of connecting rod with hidden lines removed; (c) shaded view of solid model of connecting rod.

the 50 mm cylinder. The "drills" should be somewhat longer than 25 mm to avoid the possibility of drilling a blind hole.

The solid model can be displayed in various positions and with different simulated lighting conditions. Figure 1.6(a) shows the connecting rod before hidden lines have been eliminated. Part (b) of the figure shows the same view without hidden lines. Part (c) is a shaded view.

Solution Alternatives: In this example, we can drill the 50 mm cylinder before moving it. However, it is often safer to do the drilling operation after joining operations, since the joining of two solid objects may partially fill a drilled hole. We can form the center section of the connecting rod by defining a surface made up of lines and arcs. The surface can then be extruded to form a solid object. The extrusion command applies only to formation of the solid model. Extrusion is a process of shaping by forcing a material through a die. It is not an appropriate process for forming an actual connecting rod.

■■

PROPERTIES

Various properties of a part are important in the planning of its manufacture. For example, if a part is to be cast, we want to know the volume of material required. Part mass is important in the aerospace industry and even in the automotive industry, where there is emphasis on weight reduction. Part mass and surface area must be determined before we can analyze heat transfer characteristics during manufacture and during end-use. Moving parts require a kinematic and dynamic analysis to ensure that they withstand inertia forces and torques. Part mass, center-of-gravity, and mass-moment-of-inertia are essential elements in dynamic analysis.

All of these properties can be determined by hand calculation. However, it is more efficient to use CAD software to determine properties of solid objects, including surface area, volume, mass, center-of-gravity, mass-moment-of-inertia, and principal axes. Either customary U.S. units (inch–pound–second) or SI units (meter–kilogram–second) are recommended. Density can be specified in kg/m^3 or $lb \cdot s^2/in^4$. To obtain the latter units, divide weight density (lb/in^3) by the acceleration of gravity ($386\ in/s^2$). Mass will then be determined in kg or $lb \cdot s^2/in$ units.

Center-of-Gravity

The center-of-gravity or center-of-mass is obviously the geometric center in a sphere, a right circular cylinder, or a cube or other rectangular parallelepiped, if the density is uniform. In general, the distance of the center-of-gravity from some arbitrary plane is given by

$$z_{CG} = \frac{\int z\, dM}{\int dM} \tag{1.1}$$

where integration takes place over the entire body

z_{CG} = distance from the selected plane to the center-of-gravity

z = distance from that plane to an element

dM = mass element

For a composite body made up of n parts, Equation (1.1) is equivalent to

$$z_{CG} = \frac{\sum_{i=1}^{n} z_i M_i}{\sum_{i=1}^{n} M_i} \quad \textbf{(1.2)}$$

where M_i = mass of the ith part

z_i = distance from the center-of-gravity of the ith part to the selected plane

Mass-Moment-of-Inertia

The mass-moment-of-inertia of a solid body about a given axis is the second moment-of-mass about that axis.

$$I_M = \int r^2\, dM \quad \textbf{(1.3)}$$

where the expression is integrated over the entire body

I_M = mass-moment-of-inertia (lb · s² · in or kg · m²)

r = distance to axis (in or m)

Table 1.1 Properties of a few selected solids.

Solid	Volume	Mass-moment-of-inertia	Axis
Sphere, radius R	$4\pi R^3/3$	$0.4MR^2$	Through CG
Disk, radius R, thickness c	$c\pi R^2$	$0.5MR^2$	x, through CG
Rod, radius R, length c	$c\pi R^2$	$M(R^2/4 + c^2/12)$	y, through CG
Rectangular parallelepiped, a by b by c	abc	$M(b^2 + c^2)/12$	y, through CG

$dM = \gamma\, dV$ = mass element ($\text{lb} \cdot \text{s}^2/\text{in}$ or kg)

γ = mass density ($\text{lb} \cdot \text{s}^2/\text{in}^4$ or kg/m^3)

dV = volume element (in^3 or m^3)

If we know the mass-moment-of-inertia about an axis through the center-of-gravity, we can find the mass-moment-of-inertia about a parallel axis.

$$I_M = I_{MCG} + Md^2 \tag{1.4}$$

where I_{MCG} = mass-moment-of-inertia about an axis through the center-of-gravity ($\text{lb} \cdot \text{s}^2 \cdot \text{in}$ or $\text{kg} \cdot \text{m}^2$)

I_M = mass-moment-of-inertia about a parallel axis ($\text{lb} \cdot \text{s}^2 \cdot \text{in}$ or $\text{kg} \cdot \text{m}^2$)

M = mass ($\text{lb} \cdot \text{s}^2/\text{in}$ or kg)

d = distance between the parallel axes (in or m)

The mass-moment-of-inertia and volume of a few solid shapes are given in Table 1.1. The disk and circular rod are equivalent. That is, the mass-moment-of-inertia of the circular rod about the long axis (x-axis) is given by

$$I_{Mx} = 0.5MR^2$$

For a slender rod with rectangular or circular cross section ($R << c$ or $a << c$ and $b << c$), the mass-moment-of-inertia about the y-axis through the center-of-gravity is approximated by

$$I_{My} = Mc^2/12$$

EXAMPLE PROBLEM Mass-Moment-of-Inertia

A rectangular-cross-section, reinforced plastic bar is pivoted about one end. It is necessary to find the mass-moment-of-inertia to perform a dynamic analysis.

Design Decision: We will select glass-reinforced, liquid-crystal polymer for the rod (specific gravity $sg = 1.7$).

Solution: Suppose that the rectangular bar in Table 1.1 rotates about an axis y' through the center of one end and parallel to the y-axis. The y'-axis lies a distance $d = c/2$ from the y-axis through the center-of-gravity. Using Equation (1.4), the parallel axis equation, we find

$$\begin{aligned} I_{My'} = I_{MyCG} + Md^2 &= M(b^2 + c^2)/12 + M(c/2)^2 \\ &= M(b^2/12 + c^2/3) \end{aligned}$$

We find the weight density of the plastic by multiplying the specific gravity by the density of water ($0.03613\ \text{lb/in}^3$). If dimensions a, b, and c are in inches, we find the mass of the bar by multiplying volume (abc) by weight density and dividing by the acceleration of gravity ($386\ \text{in/s}^2$).

$$M = 1.7 \cdot 0.03613abc/386 = 1.59 \cdot 10^{-4}abc\ \text{lb} \cdot \text{s}^2/\text{in}$$

If metric (SI) units are used, with a, b, and c in meters, the calculation is a bit easier since the mass density of water is $1000\ \text{kg/m}^3$. The mass of the bar is

$$M = 1.7 \cdot 1000abc = 1700abc\ \text{kg}$$

EXAMPLE PROBLEM Connecting Rod—Properties Obtained with Solid Modeling Software and an Approximate Check of Results

Consider the connecting rod modeled in the previous example "Computer-Aided Drafting." Determine properties that could be of use in kinematic analysis and manufacture of the connecting rod.

Design Decisions: The connecting rod will be made of aluminum. This decision should result in low weight and low inertia-forces.

Solution Summary: The mass density of aluminum is about 2770 kg/m³. *I-DEAS*™ software was used to model the connecting rod and obtained the following results:

$$\text{Total surface area} = 0.0248 \text{ m}^2$$
$$\text{Volume} = 0.000129 \text{ m}^3$$
$$\text{Mass} = 0.357 \text{ kg}$$
$$\text{Center-of-gravity: } 0.0884 \text{ m from left bearing center}$$
$$\text{Moment-of-inertia about left bearing centerline} = 0.003793 \text{ kg} \cdot \text{m}^2$$
$$\text{Moment-of-enertia about CG} = 0.001000 \text{ kg} \cdot \text{m}^2$$

Approximate Verification: The rod was broken into three parts: the right bearing, the left bearing, and the center section. The center section was (crudely) approximated as a rectangular parallelepiped extending from bearing center to bearing center. The solution is based on the properties of a disk and rectangular parallelepiped as given in Table 1.1. For a disk of radius R_O with a central hole of radius R_I, mass-moment-of-inertia is given by

$$I_M = 0.5\gamma\pi c(R_O^4 - R_I^4)$$

where I_M = mass-moment-of-inertia about the central axis (through the disk center-of-gravity)

c = disk thickness

γ = mass density

Equation (1.4), the parallel axis equation, is used to find the mass-moment-of-inertia about another parallel axis. The results are given in the detailed solution which follows.

Detailed Solution:

Density (kg/m³): $\gamma = 2770$

Dimensions:

Left End / *Right End*

$R_{o1} = 0.015$ $\quad R_{I1} = 0.0075$ $\quad R_{I2} = 0.0125$ $\quad R_{O2} = 0.025$ $\quad c_2 = 0.025$

$c_1 = 0.025$ $\quad d_1 = 0$ $\quad d_2 = 0.150$

Center Section

$a = 0.020$ $\quad b = 0.030$ $\quad c_3 = 0.150$ $\quad d_3 = 0.075$

Volume:

$$V_1 = \pi \cdot (R_{O1}^2 - R_{I1}^2) \cdot c_1 \qquad V_2 = \pi \cdot (R_{O2}^2 - R_{I2}^2) \cdot c_2$$
$$V_3 = a \cdot b \cdot c_3$$
$$V_1 = 1.32536 \cdot 10^{-5} \qquad V_2 = 3.68155 \cdot 10^{-5} \qquad V_3 = 9 \cdot 10^{-5}$$
$$V = V_1 + V_2 + V_3 \qquad V = 1.40069 \cdot 10^{-4}$$

Mass (kg):

$$M_1 = V_1 \cdot \gamma \qquad M_2 = V_2 \cdot \gamma \qquad M_3 = V_3 \cdot \gamma$$
$$M_1 = 0.03671 \qquad M_2 = 0.10198 \qquad M_3 = 0.2493$$
$$M = M_1 + M_2 + M_3 \qquad M = 0.38799$$

Center-of-Gravity (x = 0 at Left Bearing Center):

$$x_{CG} = \frac{M_1 \cdot d_1 + M_2 \cdot d_2 + M_3 \cdot d_3}{M} \qquad x_{CG} = 0.08762$$

Mass-Moment-of-Inertia about Axis of Left Bearing (kg/m²):

$$I_{M1} = 0.5 \cdot \gamma \cdot \pi \cdot c_1 \cdot (R_{O1}{}^4 - R_{I1}{}^4) + M_1 \cdot d_1{}^2 \qquad I_{M2} = 0.5 \cdot \gamma \cdot \pi \cdot c_2 \cdot (R_{O2}{}^4 - R_{I2}{}^4) + M_2 \cdot d_2{}^2$$

$$I_{M3} = M_3 \cdot \frac{b^2 + c^2}{12} + M_3 d_3{}^2$$

$$I_{M1} = 5.16269 \cdot 10^{-6} \qquad I_{M2} = 0.00233 \qquad I_{M3} = 0.00189$$
$$I_{M0} = I_{M1} + I_{M2} + I_{M3} \qquad I_{M0} = 0.00423$$

Mass-Moment-of-Inertia about CG:

$$I_{MCG} = I_{M0} - M \cdot x_{CG}{}^2 \qquad I_{MCG} = 0.00125$$

■■

In the preceding example problem, the properties of the connecting rod obtained with solid modeling software were approximately verified by the rough check. We could have obtained greater accuracy by using a more accurate portrayal of the actual geometry and by breaking the connecting rod into more elements.

MathCAD™ was used for the above calculations, but other mathematics software would have given the same results. The user of *MathCAD* types a colon when defining variables and an equal sign when calculating and displaying results. In this text, the equal sign is used both for defining variables and for showing calculated results.

SELECTING SOFTWARE: CHOOSING AMONG PROGRAMMING LANGUAGES, SPREADSHEETS, AND EQUATION SOLVERS (MATHEMATICS SOFTWARE)

A pocket calculator may be the best tool for solving simple, single-answer, classroom-type problems. Design problems involving finite element analysis (FEA) require a great many complicated calculations, suggesting the use of specialized programs. As noted earlier, computer-aided drafting/computer-aided manufacturing (CAD/CAM) programs are useful tools for relieving the designer of tedious work. Many machine design problems are not so complicated that they require specialized dedicated programs, but they may be too complicated for a pocket calculator. We can then choose between programming languages (e.g., BASIC, FORTRAN, C, and C++), spreadsheets, and mathematics software (e.g., *MathCAD*™ and *TK Solver*™).

There is not one best software package for all types of machine design problems. Every year, major advancements are made in both hardware and software. The program that provides the most efficient solutions today may be overtaken by another within a few months.

However, the purchaser can consider some of the following items when making software decisions:

- **Available hardware.** Software packages will not work well if computer speed, random access memory, and available hard disk space are inadequate. When practical, one should choose the software before purchasing the hardware. Personal computer magazines report scores for computers they test, comparing their ability to run actual off-the-shelf software.
- **Educational goals.** If the secondary goal of a course is to teach FORTRAN programming, then that goal would override other considerations.
- **Experience with software.** If a user is already familiar with a particular spreadsheet, then maximum efficiency in problem solving might be obtained by using that familiar tool.
- **Equation form.** Mathematics software that shows built-up equations, upper- and lower-case Roman and Greek letters, and subscripts and superscripts can make equation writing and debugging easier. This form may be more familiar to the reader than the typical programming language form.
- **Computational features.** Mathematics software features that machine designers use include equation solvers, derivatives, integrals, summations, products, iteration, statistical functions, curve fitting, and matrix operations.
- **Symbolic equations.** Convenient features include symbolic equation solvers; integration, differentiation, and simplification of symbolic equations; and symbolic matrix manipulation.
- **Presentation form.** Machine designers must present their work in a readable form. Useful features include word processor-like text capability, the ability to mix calculations with comments and explanations, and the ability to cut and paste sections of a document.
- **Plotting.** Machine designers use rectangular plots, contour plots, or three-dimensional surface plots to gain insights into a problem and to optimize a design. They may want several traces per plot, identified with legends.

NEWTON'S LAWS AND RELATED PRINCIPLES

Isaac Newton (1642–1727) recognized and published some of the key principles of mechanics in 1686. The principles, which we call **Newton's laws,** are as important today as they were in the seventeenth century. Briefly stated, they are as follows:

- **Newton's first law.** When a body is at rest or moving at constant velocity, the resultant of all forces acting on the body is zero.
- **Newton's second law.** The acceleration of a body is proportional to (and in the direction of) the resultant force on the body and inversely proportional to the mass of the body.
- **Newton's third law.** When one body exerts a force on a second body, the second body exerts an equal and opposite force on the first.

D'Alembert's Principle

Jean d'Alembert (1717–83) combined Newton's first and second laws by defining an inertia-force (reverse-effective-force) equal to the mass times the acceleration. The direction of the inertia-force is opposite the acceleration direction. The inertia-force is

$$\mathbf{F}_I = -M\mathbf{a}$$

where $\mathbf{F}_I$ = inertia-force (lb or N)

M = mass (lb · s²/in or kg)

$\mathbf{a}$ = (vector) acceleration (in/s² or m/s²)

Using customary U.S. units,

$$M = w/g$$

where w = weight (lb)

g = acceleration of gravity = 386 in/s²

We can then state that the resultant of all forces and effective forces on a body equals zero.

Moment and Torque Equilibrium

Newton's laws and d'Alembert's principle can be applied to moments (including torsional moments) as well. The inertia-moment (reverse-effective-moment) is given by

$$\mathbf{M}_I = -I_M\boldsymbol{\alpha}$$

where $\mathbf{M}_I$ = inertia-moment (lb·in or N·m)

I_M = mass-moment-of-inertia (lb·s²·in or kg·m²)

$\boldsymbol{\alpha}$ = angular acceleration (rad/s²)

We can now be more general and state that all forces and moments on a body (including enertia effects) must be in equilibrium.

Free-Body Diagrams

The free-body diagram is the key to using Newton's laws and related principles. Usually, the free-body represents a single part. All of the forces and moments (including inertial effects) acting *on* a given body are shown.

When we study an assembly, an exploded view is convenient. Using an exploded view, we can show the force (or moment) that one part of an assembly exerts on a second part and the equal and opposite force (or moment) that the second part exerts on the first.

An assembly consisting of two or more parts can also be represented by a free-body diagram. So can a portion of one part. In each case, we must show the inertia effects and the forces and moments acting on the free body. If a portion of a part is represented, we represent stresses by resultant forces and moments on the cut section.

STRESS AND STRAIN

Consider a solid body (e.g., a machine part) deformed by external forces. Imagine that the body is made up of a large number of particles or elements. The particles in the body rearrange themselves so that forces are in equilibrium. **Normal strain** is the change in length of an element of the body divided by the original length (the length of the element before loading). **Shear strain** is a change in angle between faces of an element. **Stress** is the force on a face of that element divided by the area of the face. If the force is parallel to the element face, the stress is a **shear stress.** If the force is perpendicular to the face, the stress is a **normal stress** (tensile or compressive). In mechanical engineering, tensile stresses are identified as positive. Figure 1.7 shows an element subject to normal (tensile) stress and an element subject to shear stress.

DETERMINATION OF MATERIAL PROPERTIES

Engineers design products to function properly and safely. Thus, engineering analysis considers stresses, deflections, and failure theories. A designer needs to estimate the loads that a proposed design will encounter and to make tests and assumptions regarding material behavior—properties such as strength, homogeneity, linear elasticity, and isotropy.

Homogeneity implies a consistent, uniform structure and composition throughout the part. **Linear elasticity** refers to a constant ratio of strain to stress (deflection to load) in the part and complete recovery to the original shape when the load is removed. **Isotropy** is the possession of equal properties in all directions. These properties are idealizations; that is, they are mathematical representations that only approximate the more complicated composition and behavior of actual materials such as metals and plastics.

Steel, under typical loading (below the proportional limit), is nearly linearly elastic. The

Figure 1.7 Stress and deformation. (a) Normal stress; (b) shear stress.

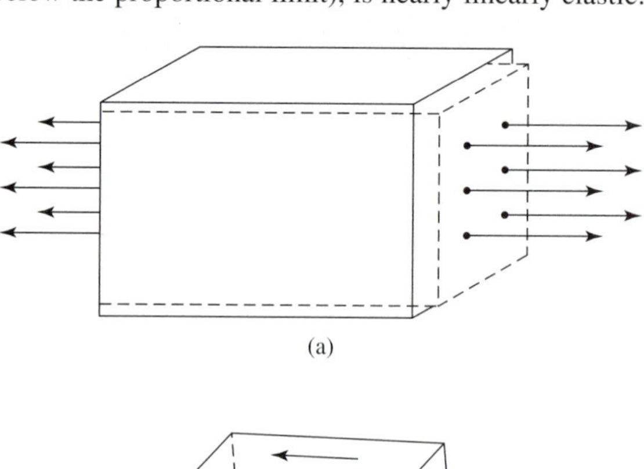

(a)

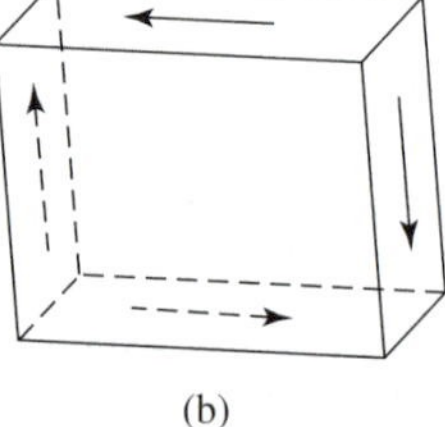

(b)

load-to-deflection ratio for most types of rubber is not constant; that is, rubber exhibits significant nonlinear properties. Metals more closely approximate homogeneity than reinforced concrete. Fiber-reinforced materials are essentially isotropic if the fibers are randomly oriented but anisotropic (not isotropic) if the fibers all lie in one direction.

To ensure efficient, cost-effective designs, engineers and technicians must test and evaluate materials. The **tensile test** is one means of measuring material response to loading. Figure 1.8(a) shows a system for tensile testing. Part (b) of the figure, a printer output from the

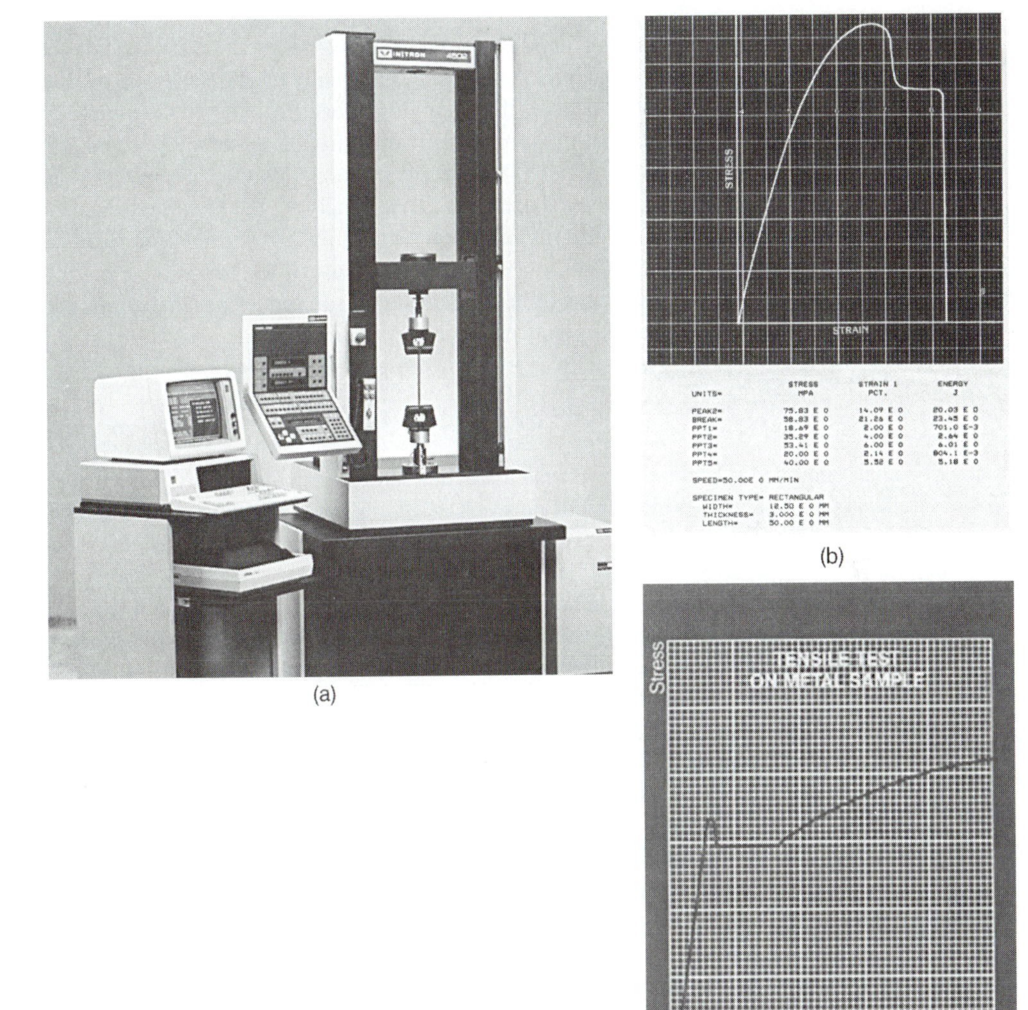

(a) (b) (c)

Figure 1.8 The tensile test. (a) A digital control tensile-testing system; (b) and (c) printer output showing stress versus strain produced by the tensile-testing system. (Courtesy of Instron Corporation)

system, shows a plot of stress versus strain. This test of a nonmetal specimen shows a nonlinear stress-versus-strain relationship. Stress peaks at about 76 MPa (10,900 psi). Note that the material then continues to deform (plastically) at reduced stress. The strain is about 0.14 at peak stress and about 0.21 at fracture. Part (c) shows the results of a tensile test on a metal sample. Note the linear stress-versus-strain relationship during the first part of this test.

ANALYTICAL METHODS FOR FINDING STRESSES AND DEFLECTIONS

Analytical methods of determining stresses in a body subject to given forces include the theory of elasticity and strength of materials methods. The **theory of elasticity** is a method that yields a stress distribution which satisfies the equations of equilibrium and the boundary conditions and is compatible with continuous displacements throughout the body. Thus, a theory of elasticity solution is sometimes called the "exact solution." Stress distributions for relatively simple part configurations have been found analytically by the theory of elasticity.

Strength of materials methods are analytical techniques and formulas for stress analysis based on empirical (experimental) and theoretical studies of the behavior of engineering materials. Designers using strength of materials methods sometimes assume stress distributions that do not satisfy the requirements of the theory of elasticity.

FINITE ELEMENT ANALYSIS (FEA)

The analytical methods noted above treat a machine part as a continuum, a body made up of infinitesimal elements. **Finite element analysis (FEA)** involves dividing a part into elements of finite size. A set of simultaneous equations relates the displacements and forces on each element.

Many applications result in tentative part designs that are too complicated to solve analytically by the theory of elasticity but for which we cannot assume a stress distribution. The finite element method may provide the best solution to such problems.

Although finite element concepts were used in stress analysis long before the general availability of computers, elaborate FEA computer programs now can analyze complicated part geometries. To determine stresses and deflections using the finite element method, we begin by modeling a tentative part design as if the part consisted of a number of finite elements. These elements can appear as a mesh of triangles, rectangles, or other quadrilaterals on the part surface. The user can define element boundaries, or the FEA program automatically generates them. Ordinarily, the user will specify the element density in various regions of a model so that the mesh is finer at critical locations.

The user then specifies material properties. Known loads and inertia-forces on the part are represented as surface forces and body forces, and restraints or boundary conditions are identified as zero-displacement of zero-rotation conditions on the boundary of the model. If part geometry and loading are symmetrical about the part midline, then only half of the part need be modeled. The symmetry condition is then represented by boundary conditions on the midline cut.

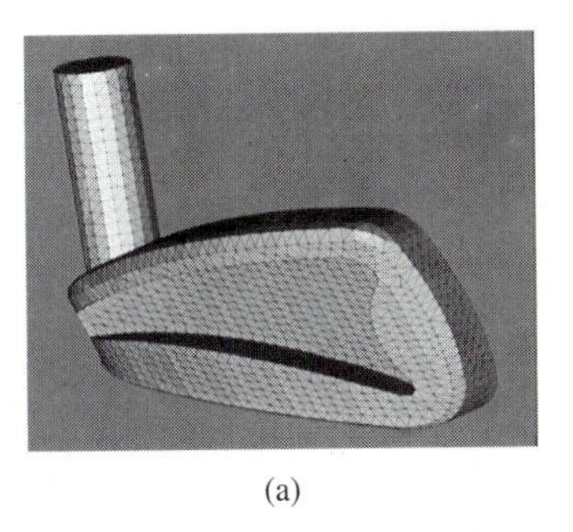

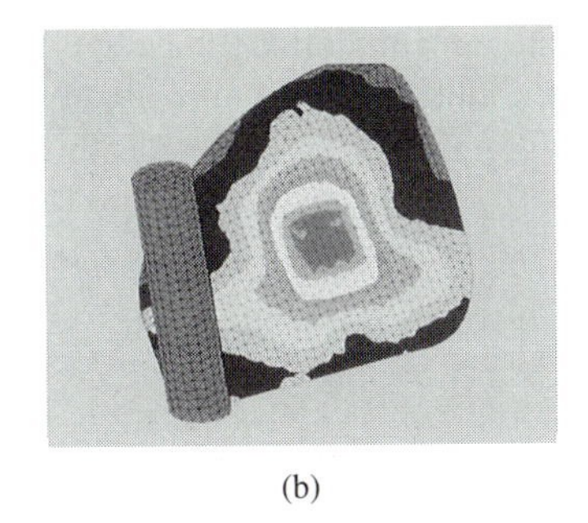

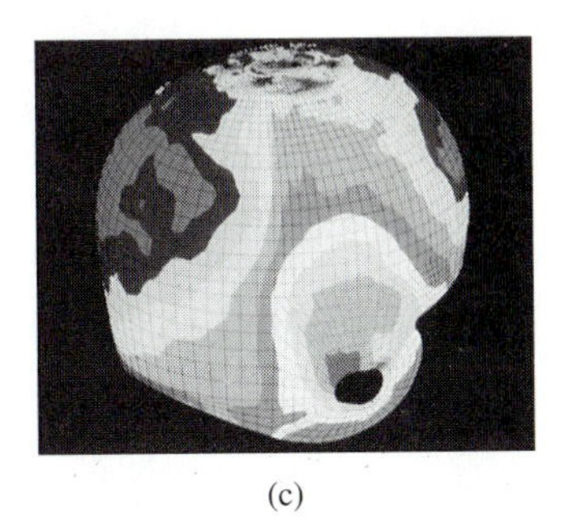

Figure 1.9 Models for finite element analysis. (a) Model of golf club head showing finite element mesh; (b) stress pattern on golf club head; (c) stress pattern on a helmet. (Courtesy of the MacNeil-Schwendler Corporation)

The FEA program constructs stiffness matrices to represent the elasticity equations governing the problem. The user then directs the program to solve for displacements and stresses resulting from applied static and dynamic loads. Some designs must consider thermal stresses as well. Stresses and displacements can be shown as contour plots overlying the part sketch. A sketch of the shape of the part after loading is an alternative form of presentation. Deformation of the part is greatly exaggerated in this type of presentation, because actual deflections are usually imperceptible.

Figure 1.9 shows models used to analyze a golf club and a helmet. Part (a) shows a finite element mesh on the golf club head. Loads and constraints were applied to the model, and Part (b) shows the resulting stress pattern. Part (c) shows a stress pattern on a helmet. Although the software generated color contours, the stress contours are shown shaded in Figure 1.9.

Figure 1.10 (a) and (b) shows two views of deflections of an orthosis, a rehabilitation devise described earlier in this chapter. Results are based on finite element analysis. Solid lines show the model after deflection due to dorsi-flexion (bending at the ankle). The unloaded

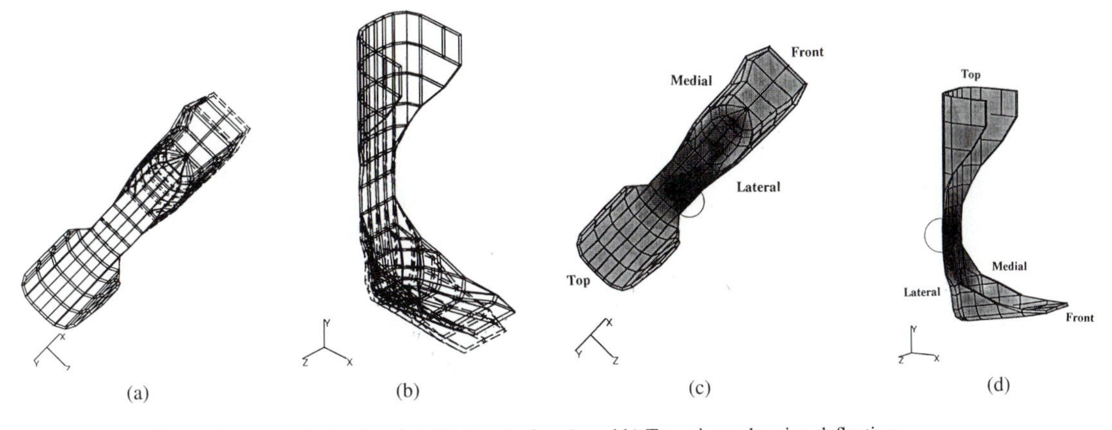

Figure 1.10 Finite element analysis of a rehabilitation device. (a and b) Two views showing deflection; (c and d) two views showing stress. (Courtesy of Tai-Ming Chu)

position of the model is shown with dotted lines. Parts (c) and (d) of the figure show stresses on the model in two views. The region of maximum stress is circled. These results provide information for redesign for greater strength and/or lower flexibility of the neck region of the orthosis.

Some designers may be interested only in whether a design is safe; that is, they are concerned only with the maximum and minimum stress. A complete FEA solution can provide much more information. The stress plot shows regions of maximum stress, possibly suggesting changes in fillet radii or other design refinements. These results may also indicate a need for a reanalysis with a finer mesh at critical locations to improve solution accuracy. If the deflection plot is inconsistent with the expected deformed shape, it may point to an error in specifying loading or boundary conditions. If the designer models only one-half (or one-quarter) of the part, to take advantage of symmetry, then the deformed shape of the model should be consistent with the symmetry condition. For example, at a plane of symmetry, any displacement perpendicular to that plane indicates an error in applying boundary conditions.

FEA software is designed to aid in engineering design, from mesh generation through solution of equations and presentation of results. However, the results must not be accepted blindly. It is still the engineer's responsibility to properly define the problem, guide the solution, and interpret the results.

FACTOR OF SAFETY

Part dimensions are often governed by the strength of the material. **Factor of safety** (N_{FS}), where $N_{FS} > 1$, is the ratio of material strength (usually ultimate strength or yield point) to actual or calculated stress. Alternatively, factor of safety can be defined as the ratio of load at failure to actual or calculated load. The factor of safety provides a margin of safety to account for uncertainties such as errors in predicted loading of a part, variations in material properties, and differences between the ideal model and actual material behavior.

Safety factors are sometimes specified in codes and contracts. When a part failure could result in injury or death, the design requires a higher safety factor than if failure would result only in inconvenience. If the loading is uncertain, and if material properties cannot be closely controlled, then the safety factor will be higher than if these parameters are known with reasonable certainty.

A safety factor of 1.5 or 2.0 would be adequate for a drive component in a household product such as a vacuum cleaner. Then that component would be able to withstand 1.5 or 2.0 times the expected load. A critical component in a passenger elevator, however, would require a higher factor of safety. Material properties tend to vary from one sample to another, even when the samples are supposed to be identical, and we can never be absolutely certain about the loading that a design will encounter. Add to this the possibility of manufacturing errors. Safety factors take these uncertainties into account. Greater uncertainty calls for higher safety factors.

Selecting a factor of safety is sometimes complicated by conflicting demands, for example, performance requirements that depend on low weight versus human safety considerations. The aircraft industry tries to balance such demands by using materials with well-documented properties and by obtaining accurate worst-case loading estimates. When weight, safety, and performance are all important, designers may resort to redundant load paths at critical locations. Nondestructive testing procedures are used to verify design calculations. As a last resort, data from past failures help designers avoid repeating design errors. Detailed analyses of aircraft accidents and incidents, for example, are disseminated to operators and designers.

OPTIMUM DESIGN

As noted above, a good design meets performance, cost, safety, and robustness requirements. An **optimum design** is the best obtainable solution to a design problem within given constraints. Of course, the final design depends on a seemingly unlimited number of variables. Even if we were able to optimize the design of the lowly paper clip, different designers would find a number of "optimum" designs employing various configurations, materials, and manufacturing processes.

Problems with Only One Variable

When faced with a design problem with hundreds of possible choices, a designer may make a number of design decisions based on experience, reducing the problem to a single variable. A solution to obtain the optimum result is often straightforward in such a case.

EXAMPLE PROBLEM Supporting a Static Load

We are required to support a static load *(P)* of 2500 lb with four equal tension members of given length.

Design Decisions: We will select a material with a tensile strength S_U of 18,000 psi and use a safety factor N_{FS} of 3 based on the uncertainty of loading and the potential consequences of failure. We will use a circular cross section based on availability of material.

Solution: The working strength of the material S_W is defined as the failure strength divided by the factor of safety. Thus, we calculate the working strength in tension.

$$S_W = \frac{S_U}{N_{FS}} = \frac{18{,}000}{3} = 6000 \text{ psi}$$

Suppose that we select four $^3/_4$ in diameter rods. The tensile stress in the rods is

$$\sigma_{\text{tensile}} = \frac{\text{load}}{\text{cross-sectional area}} = \frac{P}{4\pi D^2/4}$$

$$= \frac{2500}{4\pi 0.75^2/4} = 1415 \text{ psi}$$

Our decision produces a satisfactory design since

$$\sigma_{\text{tensile}} = 1415 \text{ psi} < S_W = 6000 \text{ psi}$$

This is not necessarily the lowest-cost solution, however. We can determine the optimum rod diameter by setting

$$\sigma_{\text{tensile}} = S_W$$

$$2500/(4\pi D^2/4) = 6000$$

from which

$$D = [2500/(4\pi 6000/4)]^{1/2} = 0.3642 \text{ in}$$

We would actually use the next larger commercially available rod size. Note that this simple illustration ignores the connections at the ends of the rods. A more complete analysis might change the results.

■■

Cost Function (Objective Function)

The **cost function**—also called the objective function, the criterion function, the payoff function, and the figure of merit—is the function that we wish to optimize in designing a part or device. In many cases, our goal is simply minimizing cost, including the cost of materials and manufacture. In some situations, we select weight as the cost function.

Constraints include limitations on variables—the size of the members or maximum allowable stress, for example. In the preceding example problem, the optimum design was governed by maximum allowable stress. It resulted in the smallest diameter supports corresponding to minimum weight, and probably minimum cost. In another situation, the cost function could be a weighted combination of cost, weight, and reliability. This combined cost function might take the form

$$C = K_1 \times \text{material cost} + K_2 \times \text{manufacturing cost} + K_3 \times \text{product weight} + K_4 \times (1 - \text{reliability})$$

where C = cost function to be minimized

$K_1, \ldots, K_4$ = constants chosen according to the relative importance of cost, weight, reliability, and so on

Sometimes the cost function consists of positive attributes which we wish to maximize.

Minimizing a Cost Function. Some design problems involve several variables related by geometric conditions, strength considerations, and other constraints. Optimizing these designs may involve both analytical and numerical methods.

EXAMPLE PROBLEM Optimum Design of a Pressure Vessel—Computer Solution

Three cubic meters of gas are to be confined at a gage pressure p of 700,000 Pa (1 Pa = 1 N/m^2). Design a pressure vessel for this application.

Design Decisions: We will select steel with a yield point S_{YP} of 296 MPa (1 MPa = 10^6 Pa = 1 N/mm^2) and a safety factor N_{FS} of 5. The vessel will be cylindrical with wall thickness t_1 (to be determined). The thickness of the heads will be $t_2 = 1.8t_1$. Assume that material and manufacturing costs are related to the area A and thickness t of the material by the cost function

$$C = K_1 A_1 t_1^{0.5} + K_2 A_2 t_2^{0.5}$$

where $K_1 = 25$

$K_2 = 20$

Subscripts 1 and 2 refer to the cylindrical wall and the heads, respectively.

The above cost function is for illustrative purposes only. A valid cost function could be based on a study of material and manufacturing costs for similar products made in the same plant.

Solution: Since gas volume (V) is specified, cylinder height (h) and radius (r) are not independent but are related by

$$V = \pi r^2 h \qquad \textbf{(a)}$$

from which

$$h = V/(\pi r^2) \qquad \textbf{(b)}$$

A safe cylinder wall thickness can be determined from thin-wall shell theory, which relates tangential stress to internal pressure. The working strength (S_W) of the material (i.e., the allowable stress) will be based on the yield strength.

$$S_W = S_{YP}/N_{FS} \qquad \textbf{(c)}$$

Equating tangential stress to working strength, we have

$$S_{YP}/N_{FS} = \sigma_\theta = pr/t_1 \qquad \textbf{(d)}$$

from which

$$t_1 = N_{FS}\,pr/S_{YP} \qquad \textbf{(e)}$$

The area of the steel making up the cylindrical part of the shell is

$$A_1 = 2\pi rh \qquad \textbf{(f)}$$

and the total area of the steel making up the two heads is

$$A_2 = 2\pi r^2 \qquad \textbf{(g)}$$

Table 1.2 Computer solution for optimum design of a pressure vessel.

Radius r (m)	Height h (m)	Wall Thickness t_1 (mm)	Head Thickness t_2 (mm)	Cost Function C
0.05	381.972	0.591	1.064	72.955
0.1	95.493	1.182	2.128	51.638
0.15	42.441	1.774	3.193	42.274
0.2	23.873	2.365	4.257	36.8
0.25	15.279	2.956	5.321	33.195
0.3	10.61	3.547	6.385	30.683
0.35	7.795	4.139	7.449	28.899
0.4	5.968	4.73	8.514	27.645
0.45	4.716	5.321	9.578	26.805
0.5	3.82	5.912	10.642	26.308
0.55	3.157	6.503	11.706	26.107
0.6	2.653	7.095	12.77	26.17
0.65	2.26	7.686	13.834	26.476
0.7	1.949	8.277	14.899	27.011
0.75	1.698	8.868	15.963	27.765
0.8	1.492	9.459	17.027	28.731
0.85	1.322	10.051	18.091	29.904
0.9	1.179	10.642	19.155	31.281
0.95	1.058	11.233	20.22	32.861
1	0.955	11.824	21.284	34.644

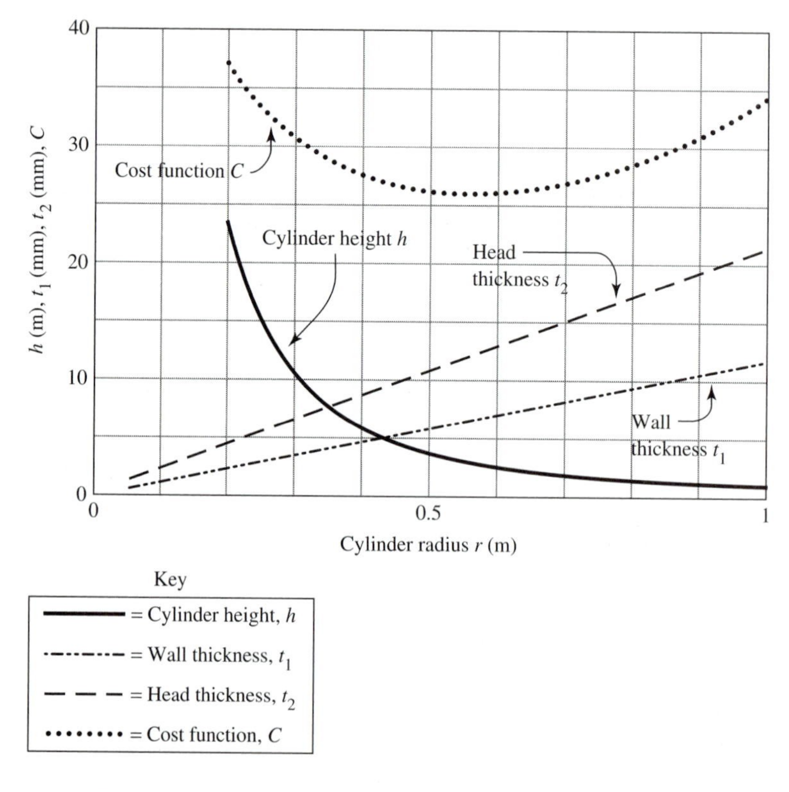

Figure 1.11 Computer solution of optimization of a pressure vessel.

We can use Equations (b), (e), (f), and (g) to evaluate the cost function for various values of r where $t_2 = 1.8t_1$. Table 1.2 and Figure 1.11 show the results that *MathCAD*™ mathematics software produces for $0.05 \leq r \leq 1$ m in 0.05 m increments. (Other software or a user-written program could have been used instead.) Thicknesses are shown in millimeters, while radius and height are in meters.

Table 1.2 and Figure 1.11 show that the optimum cylinder radius is about 0.55 m, based on minimum cost function. This value corresponds to a height of 3.157 m and a thickness of 6.503 mm in the cylindrical portion. Thin-wall shell theory is generally considered valid for $t < 0.1\,r$; our results are well within this range. There is stress concentration where the cylindrical part of the shell joins the heads, but our analysis neglected this effect.

This example uses a pressure vessel problem to introduce a simple optimization procedure. The actual design of unfired pressure vessels is ordinarily governed by codes.

Using Elementary Calculus to Aid in Optimum Design[1]

We see in the pressure vessel problem that minimum cost function corresponds to zero slope in the plot of C versus r. This classic elementary calculus problem of locating an extremum is illustrated in the following example problem.

EXAMPLE PROBLEM Optimum Design—Calculus Solution

Design a pressure vessel of volume V to confine gas at pressure p.

(a) Solve the problem in general terms.
(b) Using the specifications and cost function from the preceding example problem, find optimum dimensions of the pressure vessel.

Design Decisions: We will select a cylindrical form and attempt to optimize the design. Assume that a cost function in the form

$$C = K_1 A_1 t_1^{0.5} + K_2 A_2 t_2^{0.5}$$

is valid for this problem where subscripts 1 and 2 refer, respectively, to the cylindrical wall and heads of the shell, and

$$t_2 = K_3 t_1$$

Solution (a): We use the equations of the preceding example to write cost function in terms of shell radius (r). After simplifying, we have

$$C = (pN_{FS}/S_{YP})^{0.5}(2K_1 V r^{-0.5} + 2K_2 \pi K_3^{0.5} r^{2.5})$$

---

[1]This section can be omitted without loss of continuity. The results of a calculus solution to the pressure vessel problem are essentially the same as those of the preceding computer solution.

We can now calculate the derivative

$$dC/dr = (pN_{FS}/S_{YP})^{0.5}(-K_1Vr^{-1.5} + 5K_2\pi K_3^{0.5}r^{1.5})$$

and set $dC/dr = 0$ to obtain

$$-K_1Vr^{-1.5} + 5K_2\pi K_3^{0.5}r^{1.5} = 0$$

from which

$$r = [K_1V/(5\pi K_2K_3^{0.5})]^{1/3} = r_{\text{extremum}}$$

for a cost function extremum. If the derivative dC/dr is negative for $r < r_{\text{extremum}}$ and positive for $r > r_{\text{extremum}}$, then we have located the minimum cost function. If dC/dr is positive for $r < r_{\text{extremum}}$ and negative for $r > r_{\text{extremum}}$, then the corresponding cost function is a maximum.[2] If the derivative dC/dr does not change sign, then we have located an inflection point in the C-versus-r plot.

Solution (b): We use the equation for r_{extremum} from Part (a) of this example and the equations from the previous example. We find that $r = 0.562$ m, $h = 3.018$ m, $t_1 = 6.651$ mm, and $t_2 = 11.971$ mm. These produce a cost function (C) of 26.098, slightly better than 26.107, the value obtained in the preceding example. Ordinarily, we would test to ensure that the extremum that we calculated results in a **minimum** cost function. In this case, it is obvious from the table and plots obtained previously that we do have a minimum. We might now adjust material specifications to commercially available thicknesses. If the data are already entered into a computer program, it is convenient to continue using the computer as we recheck stresses and other variables.

Practical Considerations and Constraints

Mathematical models are limited in their application because of the assumptions upon which they are based. The designer must be alert to practical considerations, avoiding blind faith in mathematical results, whether these results are produced by commercially available programs or the designer's own calculations. Cost functions in a practical design may be more complicated than the cost functions used in the above examples.

Practical considerations would suggest limits on thickness and on the ratio of shell height to radius. For example, let us attempt to design a minimum weight thin shell to the specifications of the last example problem. The cost function becomes shell weight:

$$C = K_1A_1t_1 + K_2A_2t_2$$

where $K_1 = K_2 =$ density

Trying values of radius $0.05 \le r \le 1$, we find that the minimum cost function corresponds to $r = 0.05$ m, $h = 382$ m, and $t_1 = 0.591$ mm, obviously absurd proportions. A calculus solution suggests an optimum design for r/h approaching zero. If we add constraints on radius and

[2] In some cases, the derivative of a cost function may be zero at two or more values of the independent variable. In such cases, we can locate a relative extremum, rather than the desired overall minimum cost function. This difficulty is more likely to occur when we are solving complicated optimization problems using numerical methods. While there is no certain way of avoiding errors of this type, it helps to check several regions of a problem, particularly in a multivariable problem.

thickness, the optimum solution (minimum weight in this case) would correspond to minimum allowable thickness or minimum allowable radius.

Multivariable Optimization. Design decisions and constraints are used to reduce the number of independent variables in an optimization problem. However, it is not necessary to reduce the number of independent variables to only one. By permitting two or more independent design variables, we can improve the quality of the final design, but the tradeoff is a more difficult optimization procedure. Multivariable optimization procedures will be considered in a later section.

ATTRIBUTES OF GOOD DESIGN

It is generally assumed that a good design meets performance and cost goals. Another attribute of a good design is **robustness,** a resistance to quality loss or deviation from desired performance. A robust design permits production process variability—close tolerances and narrow specifications are not required. **Noise factors** are the uncontrollable, undesirable elements that reduce quality in manufacturing and use. Production machinery wear, corrosion, and variability of product use patterns are noise factors. A robust design is relatively insensitive to noise factors. General techniques of good design and design for manufacture are discussed by Shina (1991).

DESIGN RESEARCH AND COMPUTER-BASED MODELS OF DESIGN PROCESSES

Design research refers to studies of the thought processes, strategies, and problem-solving methods that go into producing creative engineering designs. If we can understand how designers design, then it may be possible to teach the skills and strategies used by designers and develop better computer-aided design tools (software).

The **scientific method** of research involves identifying a problem, gathering data, formulating a hypothesis, and testing the hypothesis empirically. The last of these steps, hypothesis testing through experimentation, is particularly difficult to replicate because it involves human thought processes. For this reason, design research may be in a pretheory stage, with more computer-based studies required to discover design strategies and develop theoretical foundations.

Finger and Dixon (1989) reviewed and summarized research in mechanical engineering design, including computer-based models of the design process. Their review paper discusses design research tools, including the following:

- techniques borrowed from artificial intelligence studies such as **protocol analysis** which records the actions of the designer as the design evolves
- proposed computer-based **cognitive models** which describe and simulate human problem-solving skills
- case studies of the design process

- **morphological analysis,** the generation and selection of design alternatives, which involves dividing a problem into subproblems, solving the subproblems, and combining the solutions to form a global solution
- **prescriptive models** which describe either the process by which a design should be generated or the attributes required of the final product

A FEW SUGGESTIONS FOR WRITING COMPUTER PROGRAMS

Engineering practice involves substantial computer use. Designers and engineers may find the following suggestions helpful when writing programs and using computer software:

- Identify the actual problem. A computer cannot do this step for you. Problems encountered in engineering practice are often ill-defined, requiring many assumptions. There is no point in solving the problem correctly if you have not solved the correct problem.
- Begin with a simple program and build on it as necessary.
- Identify the variables and include comments. You may forget important details and limitations if you use the program at a later date.
- Output the results of intermediate calculations to check for technical errors and to aid in debugging.
- Verify results. Make a few hand (calculator) calculations to see if the results are reasonable.
- The computer is your servant, not vice versa. Do not spend hours making unessential refinements in a program. The quality of the design and analysis and the interpretation of results are far more important.

WORKING "SMARTER" AND DESIGNING "SMARTER"

Working harder is not always the most effective means of accomplishing a task. Improvements in productivity have resulted from application of the principles of industrial engineering, including knowledge derived from task study or time-and-motion study. Early working-smarter studies investigated a task and determined how to do that task most efficiently. It was taken for granted that the task had to be done. Drucker (1992) suggests asking, "What is the task? What do we try to accomplish? Why do it at all?" Increases in productivity in "knowledge and service work" may result from redefining the task and eliminating unnecessary tasks.

This principle can be applied to engineering design. Possible working-smarter questions include the following: "What does a particular design task accomplish? Can this task be eliminated or combined with another task? Can it be automated?" Possible answers include implementing a concurrent engineering (CE) program. With CE, the CAD-software-generated solid model can replace detailed layout drawings in the early design stages. Some "downstream" processes, including numerically controlled (NC) machining and testing, can be automated. Finally, detailed drawings can be produced (by the CAD system) for use by suppliers and manufacturing.

Going one step further, we can extend the working-smarter principle to designing smarter. Possible designing-smarter questions include the following: "What does a particular component or part attribute do? Can we eliminate the component or attribute without sacrificing performance or safety?" For example, a design may call for a close tolerance on one part and a finished surface on another because "it was always done this way." Can we combine the component or attribute with another to simplify the design? Can we simplify or revise the design to speed assembly? For example, the base of an assembly may be molded with integral fasteners. A cleverly designed base may serve as a tray for the other parts of the assembly.

References and Bibliography

Carter, D. E., and B. S. Baker. *Concurrent Engineering.* Reading, MA: Addison-Wesley, 1992.

Drucker, Peter F. *Managing for the Future—The 1990s and Beyond.* New York: Truman Talley Books/Dutton, 1992.

Finger, S., and J. R. Dixon. "A Review of Research in Mechanical Engineering Design. Part I: Descriptive, Prescriptive, and Computer-Based Models of Design Processes." *Research in Engineering Design* (1989):51–67.

Institute for Defense Analysis. *IDA Report R-338,* 1986.

Lawry, M. H. *I-DEAS Master Series Student Guide.* 2d ed. Milford, OH: Structural Dynamics Research Corporation, 1994.

Shina, S. G. *Concurrent Engineering and Design for Manufacture of Electronic Products.* New York: Van Nostrand Reinhold, 1991.

Design Problems

1.1. A connecting rod is to have a 20 mm diameter bore and a 30 mm diameter bore at a distance of 240 mm center-to-center. Draw the connecting rod in a plan view, a view from a different angle showing hidden lines, a view suppressing hidden lines, and a shaded representation. Use solid modeling software if available.

Design Decisions. Outer diameters (ODs) will be twice inner diameters (IDs), and bearing surfaces will be 35 mm long. The center section will have a 35 mm by 40 mm rectangular cross section.

1.2. A link (part of a spatial linkage) is to have the male part of a spherical pair (ball-joint) at each end, at a center-to-center distance of 75 mm. The radius of the sphere at the left end is to be 10 mm, and the radius of the sphere at the right end 15 mm. Use a circular-cross-section center section with 8 mm radius. Make a drawing of the link. Use solid modeling software if available.

1.3. A connecting rod is to have a 20 mm diameter bore and a 30 mm diameter bore at a distance of 240 mm center-to-center.

Design Decisions. Outer diameters (ODs) will be twice inner diameters (IDs), and bearing surfaces will be 35 mm long. The center section will have a 35 mm by 40 mm rectangular cross section. The part will be made of steel. The origin of coordinates will be located midway between the bearings. Let the x-axis lie along the line joining the bearing centers, and let the z-axis be parallel to the bearing centers. Find surface area, volume, mass, center-of-gravity, moments-of-inertia about the origin, and principal moments-of-inertia about the center-of-gravity. Use solid modeling software.

1.4. Consider the steel connecting rod in Problem 1.3. Calculate approximate volume, mass, center-of-gravity, moment-of-inertia about the z-axis through the origin, and moment-of-inertia about the z-axis through the center-of-gravity.

1.5. A steel link (part of a spatial linkage) is to have the male part of a spherical pair (ball-joint) at each end, at a center-to-center distance of 75 mm. The radius of the sphere at the left end is to be 10 mm, and the radius of the sphere at the right end 15 mm. Use a circular-cross-section center section with 8 mm radius. Find surface area, volume, mass, center-of-gravity, moments-of-inertia about the origin, and principal moments-of-inertia about the center-of-gravity. Use solid modeling software. Let the x-axis lie along the line joining the centers of the spheres, with the origin at the center of the 10 mm radius sphere.

1.6. Consider the steel link in Problem 1.5. Calculate approximate volume, mass, center-of-gravity, moment-of-inertia about the z-axis through the origin, and moment-of-inertia about the z-axis through the center-of-gravity.

1.7. Design a pressure vessel for 4 m^3 of gas at a gage pressure of 400,000 Pa.

Design Decisions. Use a cylindrical form with cylinder wall thickness t_1 and head thickness $t_2 = K_3t_1$. Select steel with a yield point of 320 MPa, and use a safety factor of 4. Optimize the design using a cost function in the form

$$C = K_1A_1t^{0.5} + K_2A_2t_2^{0.5}$$

where

$$K_1 = 30$$
$$K_2 = 25$$
$$K_3 = 1.5$$

Use a computer to tabulate and plot cylinder height, wall thickness, and cost function against radius for $0.05 \leq r \leq 1$.

***1.8.** Use methods of elementary calculus to determine the optimum design for a pressure vessel to meet the requirements in Problem 1.7. Use the same design decisions.

1.9. Design a pressure vessel for 4 m^3 of gas at a gage pressure of 800,000 Pa.

Design Decisions. Use a cylindrical form with cylinder wall thickness t_1 and head thickness $t_2 = K_3t_1$. Select steel with a yield point of 350 MPa, and use a safety factor of 6. Optimize the design using a cost function in the form

$$C = K_1A_1t_1^{0.5} + K_2A_2t_2^{0.5}$$

where

$$K_1 = 20$$
$$K_2 = 15$$
$$K_3 = 1.3$$

Use a computer to tabulate and plot cylinder height, wall thickness, and cost function against radius for $0.05 \leq r \leq 1$.

***1.10.** Use methods of elementary calculus to determine the optimum design for a pressure vessel to meet the requirements in Problem 1.9. Use the same design decisions.

1.11. Design a pressure vessel for 2 m^3 of gas at a pressure of 600,000 Pa.

Design Decisions. Use a cylindrical form with cylinder wall thickness t_1 and head thickness $t_2 = K_3t_1$. Select steel with a yield point of 280 MPa, and use a safety factor of 3.5. Optimize the design using a cost function in the form

$$C = K_1A_1t_1^{0.5} + K_2A_2t_2^{0.5}$$

where

$$K_1 = 35$$
$$K_2 = 35$$
$$K_3 = 1.1$$

Use a computer to tabulate and plot cylinder height, wall thickness, and cost function against radius for $0.05 \leq r \leq 1$.

***1.12.** Use methods of elementary calculus to determine the optimum design for a pressure vessel to meet the requirements in Problem 1.11. Use the same design decisions.

Partial answers to selected problems are given following the last chapter of the text. In addition, suggested machine design projects are also provided.

*An asterisk indicates a more difficult problem.

CHAPTER 2

Materials for Machine Design

Symbols

A	original cross-sectional area	$S_{U(c)}$	ultimate strength (compression)
E	modulus of elasticity (tension)	S_{YP}	yield point (tension)
E_B	modulus of elasticity (bending)	$S_{YP(c)}$	yield point (compression)
EL	elastic limit	U_R	modulus of resilience
F	force in the tensile or bending test	U_T	toughness
L	gage length, unsupported length	δL	change in gage length
N	number of cycles to failure	ϵ	strain
S'_N	endurance limit	ϵ_U	strain at fracture
S_N	fatigue strength	σ	stress
S_U	ultimate strength (tension)		

Units

area	in^2 or mm^2
force	lb or N
length	in or mm
strain	dimensionless
stress, strength, modulus of elasticity, modulus of toughness, and modulus of resilience	psi or MPa

Mechanical engineering designs include ferrous and nonferrous metals, plastics, composites, and other materials. Material selection is based on strength, hardness, elasticity, ductility, machinability, density, response to heat treatment, resistance to environmental degradation, cost, and many other factors. Material properties of a few important engineering materials will be discussed in this chapter. When selecting materials, design engineers utilize materials science textbooks, materials handbooks, and catalogs from manufacturers and suppliers for comprehensive information. Useful references include DeGarmo, Black, and Kohser (1988); *Machine Design* (periodical); *Metals Handbook* (1985); *Thomas Register of American Manufacturers* (1992) and *Thomas Register Catalog File* (1992); and *Sweet's Catalog File* (1991).

MATERIAL PROPERTIES REQUIRED IN DESIGN

Designers are interested in both physical properties and mechanical properties of materials. Physical properties include density, melting point, and coefficient of thermal expansion. Mechanical properties include strength, ductility, toughness, and modulus of elasticity.

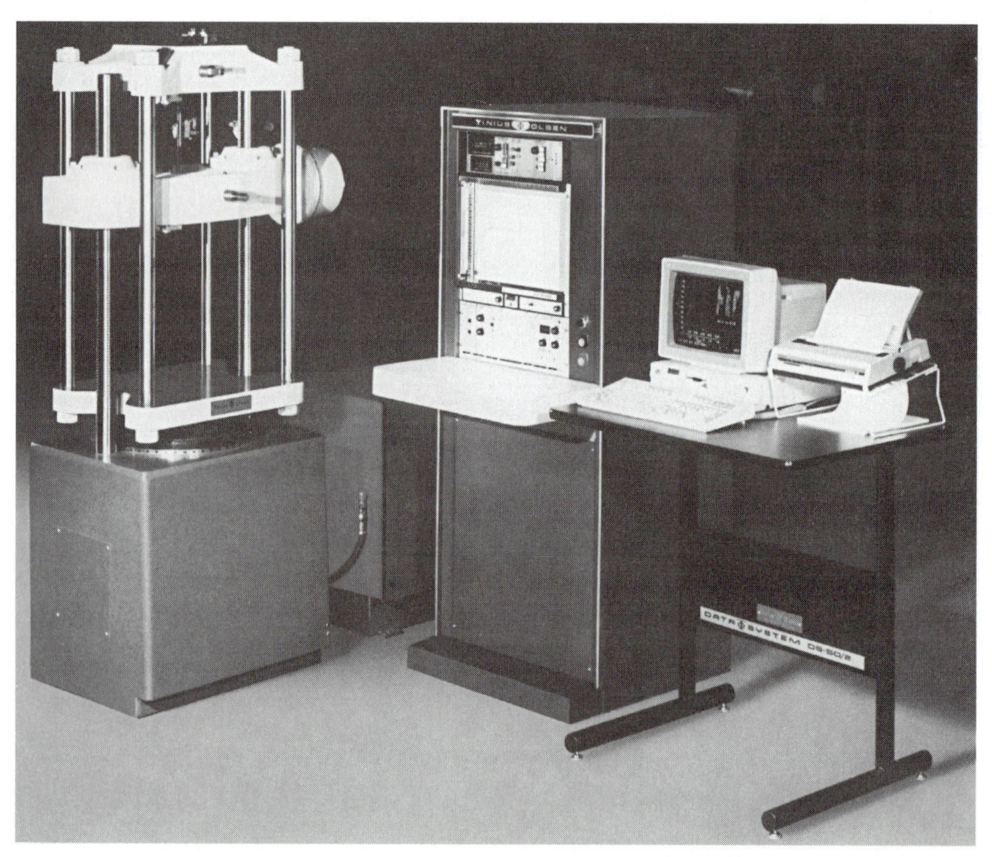

Figure 2.1 A hydraulic testing machine with 120,000 lb capacity load frame shown with servo controls and data system. (Photograph courtesy of Tinius Olsen Testing Machine Co., Inc., Willow Grove, PA)

The Tensile Test

Tensile testing usually involves loading a test specimen in the axial direction and measuring load-versus-deformation relationships. Specimen size, loading rate, and other test details are standardized for a given material, so that results are repeatable. A hydraulic testing machine with 120,000 lb capacity load frame is shown with servo controls and data system in Figure 2.1. The results of tensile tests are often shown as stress–strain diagrams as in Figure 2.2.

Stress, the load per unit area, is given by

$$\sigma = F/A \tag{2.1}$$

where, for the tensile test, σ = axial stress (psi or MPa)

F = axial force (lb or N)

A = original cross-sectional area of test specimen (in^2 or mm^2)

The cross-sectional area of a ductile material is reduced considerably during plastic deformation, but engineers use original area A to calculate strength.

Strain, the unit change in length, is given by

$$\epsilon = \delta L/L \tag{2.2}$$

where ϵ = axial strain

L = original gage length, the axial distance between two points on the specimen (in or mm; 2 in is commonly used)

δL = change in gage length

A material is considered **elastic** if it deforms when loaded and returns to its original shape as the load is removed. **Plastic deformation** is a change in shape that remains after a load is removed. Many metals and other materials deform elastically for stresses below a limiting value, the **elastic limit,** and deform plastically at higher stresses.

In most low-carbon steels, the stress-versus-strain curve has a horizontal tangent at some load. The corresponding stress is called the **yield point** or yield strength (S_{YP}) as identified in Figure 2.2(a). Typical hot-rolled structural steel shapes have yield points of 36,000 to 65,000 psi (248 to 448 MPa), while some high-strength alloy steels are available with yield points above 200,000 psi (above 1400 MPa).

Yield point is determined by the offset method for materials that do not have a zero-slope yield, as in Figure 2.2(b). In most cases, the **offset-method yield point** is defined as the stress that will produce 0.2% plastic deformation. The resulting stress is identified as $S_{YP(0.2\%\ offset)}$, or simply S_{YP}.

The **ultimate strength** (S_U) is the nominal stress (load/original cross section) in the test specimen when it fractures (see Figure 2.2). The ultimate strength of typical rolled structural steel shapes ranges from 58,000 to 80,000 psi (400 to 552 MPa), while alloy steels are available with much higher ultimate strengths.

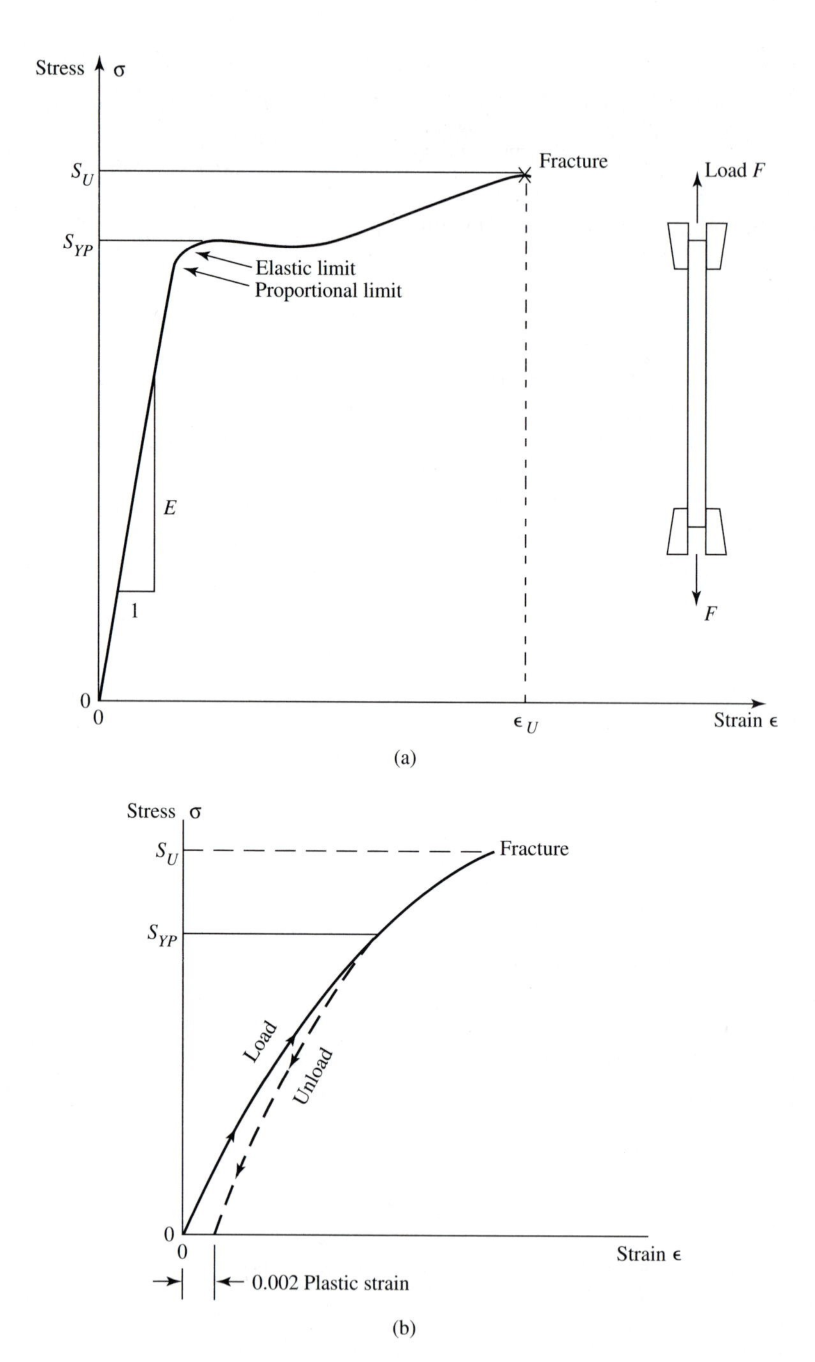

Figure 2.2 (a) Stress–strain diagram typical of some steels; (b) stress–strain diagram with yield point obtained by 0.2% offset method.

Steel and some other materials have a stress–strain curve with a fairly straight portion for stresses between zero and the **proportional limit,** a stress somewhat below the yield point. The slope of that straight section is the **elastic modulus** (E). The elastic modulus of steel is about $30 \cdot 10^6$ psi (207,000 MPa). The elastic modulus is also called Young's modulus. Figure 2.2(a) shows slope E; the relative values of proportional limit and elastic limit; yield point; and ultimate strength for a steel specimen. While yield point and ultimate strength have been reported for most engineering materials, elastic limit and proportional limit are seldom reported.

If there is little or no plastic deformation before fracture, the material is considered **brittle.** A **ductile** material will undergo significant plastic deformation before fracture. A ductile material tends to neck down (decrease in cross section) as it is tested in tension and to finally fracture at the narrowest section. The cross-sectional area of the specimen at the fracture is compared with the original cross-sectional area. **Reduction in area** refers to the percent decrease in cross section. We calculate **percent elongation** by comparing the plastic deformation in the region of necking down with the original gage length. For hot-rolled structural steel shapes, 15% to 23% elongation in a 2 in gage length is typical. Gray cast iron exhibits less than 1% elongation.

Resilience

Resilience is the capacity of a material to absorb energy without permanent deformation. The **modulus of resilience** (U_R) is the energy that can be absorbed elastically by a unit volume of material. It is given by the area under the stress–strain curve up to the elastic limit in the tensile test.

$$U_R = \int_0^{\mathrm{EL}} \sigma \, d\,\epsilon \approx \mathrm{EL}^2/(2E) \approx S_{\mathrm{YP}}^2/(2E) \tag{2.3}$$

where U_R = modulus of resilience (psi or MPa)

EL = elastic limit (psi or MPa)

σ = stress in tensile test (psi or MPa)

ϵ = strain in tensile test

E = modulus of elasticity (psi or MPa)

S_{YP} = yield point (psi or MPa)

Toughness

The capacity of a unit volume of material to absorb energy without fracture is called **toughness.** Toughness is given by the total area under the stress–strain curve in the tensile test.

$$U_T = \int_0^{S_U} \sigma \, d\,\epsilon \approx (S_{\mathrm{YP}} + S_U)\epsilon_U/2 \tag{2.4}$$

where U_T = toughness (psi or MPa)

S_{YP} = yield point (psi or MPa)

S_U = ultimate strength (psi or MPa)

ϵ_U = strain at fracture in tensile test

Compressive Strength

Compression tests may utilize the same testing equipment as tensile tests. A large cross-sectional area test specimen is ordinarily used because a slender specimen could fail in elastic stability (i.e., buckle). The yield point and the ultimate strength in compression can be identified as $S_{YP(c)}$ and $S_{U(c)}$, respectively. The tensile strength of cast iron is substantially lower than the compressive strength. The tensile strength of concrete is so low that engineers specify steel reinforcing rods in concrete when tensile loads are likely. For static loading, steel behaves about the same in tension and compression. Therefore, tables of properties for steel do not report compressive strengths; we assume that $S_{YP(c)} = S_{YP}$ and $S_{U(c)} = S_U$ for steel.

Hardness

Hardness has been defined as resistance to wear, resistance to scratching, and resistance to denting. Scratch tests can help in identifying minerals, but they are not precise enough to classify steel grades. Wear tests provide useful information to engineers but require substantial time to conduct. Indentation tests are popular for engineering materials because they are quick and repeatable.

In the **Brinell hardness test,** a 1 cm steel ball is pressed into the material surface by a load of 500, 1500, or 3000 kg. Figure 2.3(a) shows a production tester designed for automatic determination of Brinell hardness values. The **Brinell hardness number (BHN)** is defined by

$$\text{BHN} = F/A_{\text{SURFACE}} \tag{2.5}$$

where BHN = Brinell hardness number (It is customary to report BHN without identifying units.)

F = load (kg)

A_{SURFACE} = surface area of the indentation (mm²)

The **Rockwell™ hardness test,** like the Brinell test, is based on indenting the specimen. In the Rockwell test, a minor load is used to seat a hardened ball or diamond penetrator. A major load is then applied and released. The additional penetration due to the major load is automatically indicated. Rockwell hardness scales A, B, C, D, F, and G specify different loads and indenters. The results of Rockwell tests should also include the scale. Figure 2.3(b) shows a Rockwell hardness tester with a robotic manipulator for automatic loading, unloading, and sorting. Software and computer-controlled hardware are also used to aid in production testing of materials. Part (c) of the figure shows a Rockwell hardness tester with a personal computer for use with specialized software.

Vickers and **Knoop hardness testers** use diamond penetrators. The pyramid-shaped indenter used in the Vickers test produces a square-shaped indentation, while the elongated

(a)

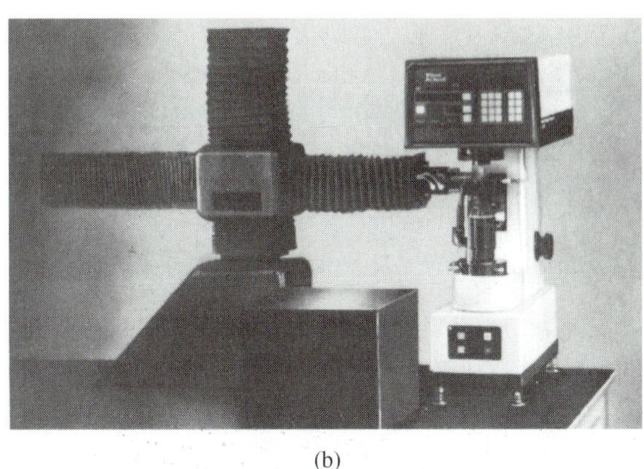

(b)

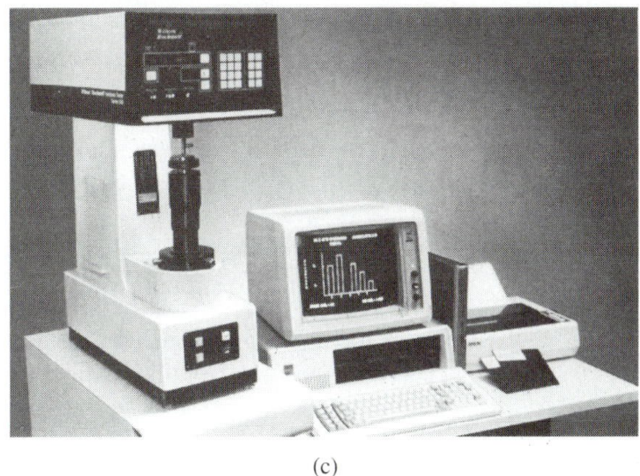

(c)

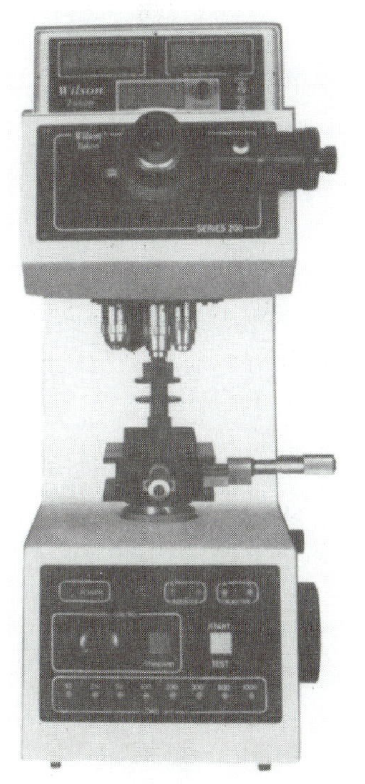

(d)

Figure 2.3 Hardness testers. (a) Digital Brinell hardness tester; (b) Rockwell™ hardness tester designed for automatic loading, unloading, and sorting by a robot; (c) Rockwell™ hardness tester with personal computer; (d) microhardness tester designed to calculate and display Knoop or Vickers hardness values. (Courtesy of Wilson Division, Instron Corporation)

penetrator in the Knoop test produces a diamond-shaped indentation. The indentations can be measured with a microscope. The hardness number of the material is defined as a function of load and indentation measurement. Figure 2.3(d) shows a microhardness tester with a digital display package. The display indicates the indent length in microns. Knoop or Vickers hardness number is automatically calculated and displayed and then converted to equivalent Rockwell B and C hardness numbers.

We can examine the effect of heat treatment on a steel bar by cutting the bar and measuring hardness at various depths (i.e., at various locations on the cut section). Hardness test results can be related approximately to one another and to ultimate strength. For plain carbon steels and low-alloy steels, for example,

$$\text{BHN} \approx \text{VHN} \tag{2.6}$$

$$\text{BHN} \approx 10 \cdot R_C \quad \text{for } R_C > 20 \tag{2.7}$$

$$S_U \text{ (psi)} \approx 500 \cdot \text{BHN} \quad \text{for BHN} \leq 600 \tag{2.8}$$

$$S_U \text{ (MPa)} \approx 3.45 \cdot \text{BHN} \quad \text{for BHN} \leq 600 \tag{2.9}$$

where BHN = Brinell hardness number

VHN = Vickers hardness number

R_C = Rockwell scale C hardness number

S_U = ultimate strength

All four of the above relationships are approximate because each test uses a different procedure. There are no precise relationships between BHN, VHN, R_C, and S_U.

EXAMPLE PROBLEM Hardness and Strength

A hardness test on a plain carbon steel rod indicated a Brinell hardness number BHN of 180. Estimate the following:

(a) the Rockwell (C scale) hardness number

(b) the Vickers hardness number

(c) the ultimate strength

Solution (a): $\text{BHN} \approx 10 \cdot \text{R}_C$

from which

$$R_C \approx \text{BHN}/10 \approx 180/10 \approx 18$$

Solution (b): $\text{VHN} \approx \text{BHN} \approx 180$

Solution (c): For BHN ≤ 600,

$$S_U \approx 500 \cdot \text{BHN} \approx 500 \cdot 180 \approx 90{,}000 \text{ psi}$$

or

$$S_U \approx 3.45 \cdot \text{BHN} \approx 3.45 \cdot 180 \approx 621 \text{ MPa}$$

Impact Strength

Impact testing involves subjecting a specimen to suddenly applied load. In the **Charpy test,** the specimen has a V-notch, U-notch, or keyhole-shaped notch at midlength. The specimen is supported at the ends and is struck with a pendulumlike hammer on the side opposite the notch. In an **Izod test,** a notched specimen is clamped near the notch and is struck at the free end. The energy lost by the hammer as it breaks the material is reported as the **impact strength,** usually in ft·lb.

The head (hammer) of a universal impact tester is designed to accommodate the tooling for Charpy, Izod, and tension impact tests. Although the specimen is held in the machine base in most impact testers, some machines are designed to carry the specimen in the head. Machines of varying capacities and configurations are available for testing metals, plastics, adhesives, and other materials. Figure 2.4 shows two models of impact testers. Many engineering materials exhibit poor impact resistance at low temperatures. Temperature test chambers are available to aid in evaluating impact resistance at various temperatures.

Standardized test procedures must be strictly followed to give valid repeatable results. Additional test details are given by DeGarmo et al. (p. 51 ff., 1988). Impact test results cannot predict the energy that could be absorbed by an actual machine part. However, engineers compare impact strengths when selecting materials for applications where sudden loads are likely.

(a) (b)

Figure 2.4 Impact testing machines. (Photographs courtesy of Tinius Olsen Testing Machine Co., Inc., Willow Grove, PA)

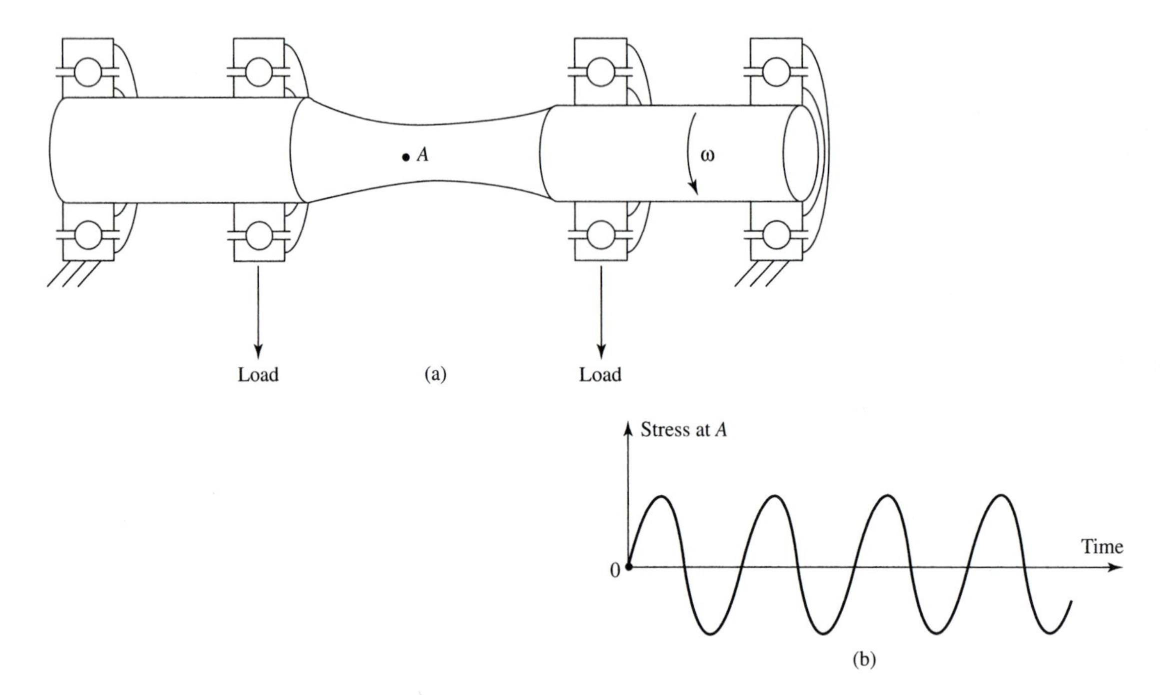

Figure 2.5 Fatigue (reversed stress).

Fatigue Strength and Endurance Limit

Fatigue refers to fracture resulting from repeated loading. It is one of the most important, but least understood, aspects of engineering design. As a result, fatigue failure is probably the most common mode of catastrophic failure.

Figure 2.5 shows a rotating shaft with a lateral load. Tensile stress due to bending will be greatest on the surface where the shaft is narrowest. Due to the rotation, we have **reversed stress,** which is stress varying continuously from positive to negative (i.e., from tension to compression). Reversed stress is common in machinery, particularly in shafts in gear, chain, and belt drives.

Fatigue tests are designed to produce loading similar to that of Figure 2.5. A beam or shaft is loaded to produce a given reversed stress amplitude, the test is run until the specimen fails, and the number of cycles to failure is recorded.

A rotating-beam fatigue testing machine is shown in Figure 2.6(a). This machine has an adjustable-speed spindle, operating at speeds from 500 to 10,000 rpm. It has a calibrated beam which can apply a moment up to 200 in·lb to the cantilevered end of the specimen. Part (b) of the figure identifies the testing machine components.

Fatigue test results are sensitive to test specimen configuration, surface finish, and other factors. Therefore, test specimens must be standardized so that valid comparisons of materials and heat treatments are possible. Figure 2.6(c) shows suggested configurations and surface finish details for rotating-beam specimens.

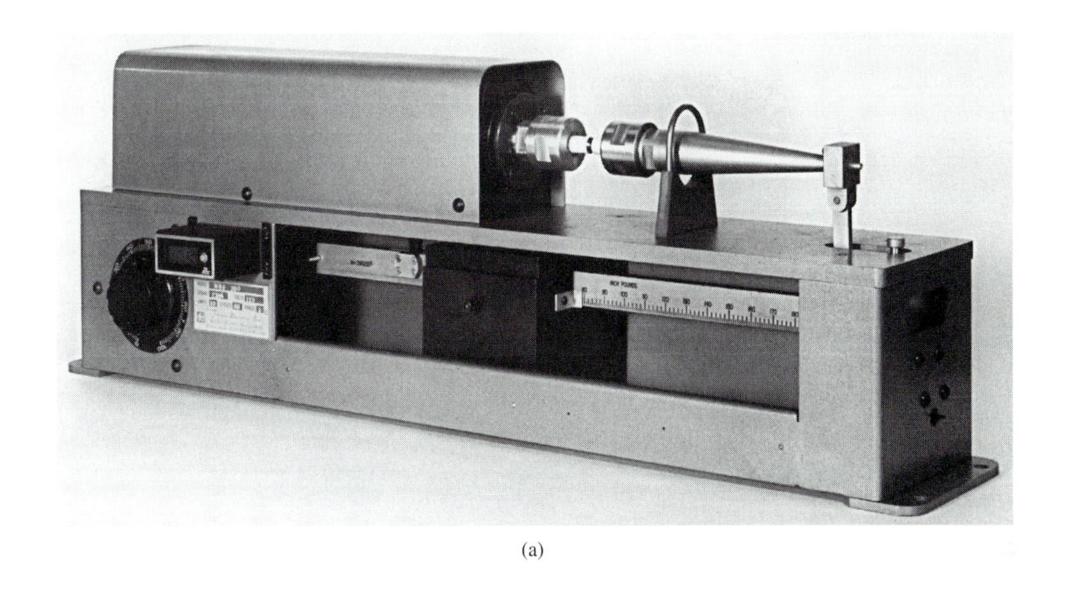

(a)

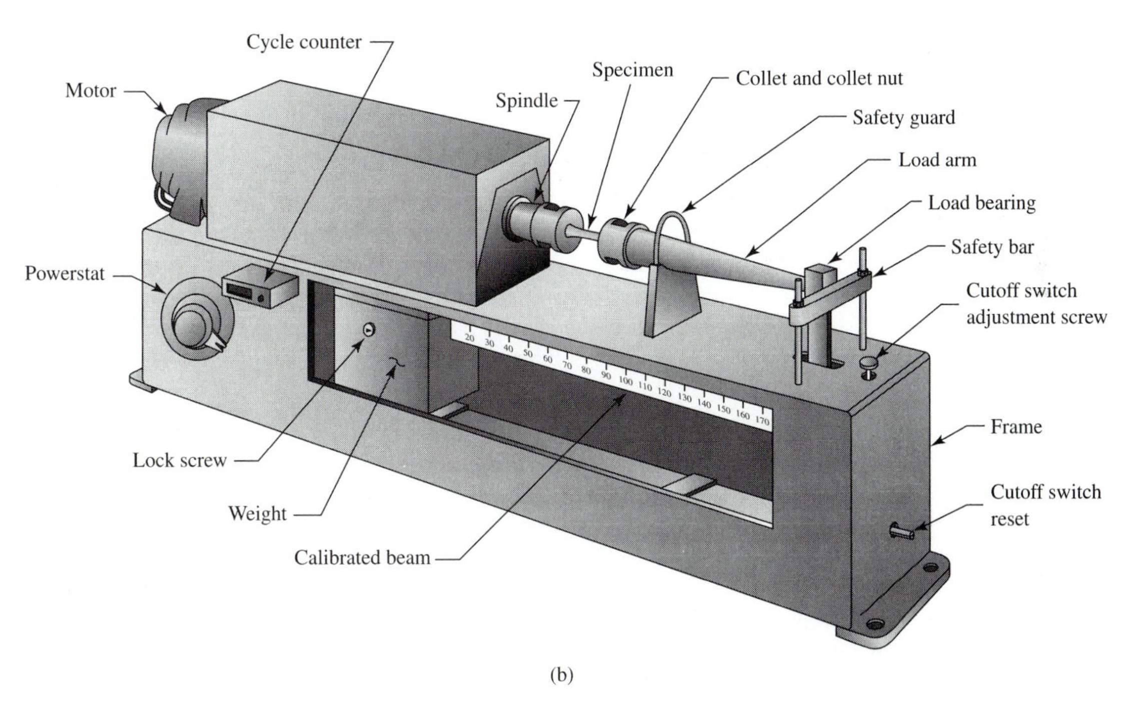

(b)

Figure 2.6 (a) Rotating-beam fatigue testing machine; (b) components of the rotating-beam fatigue testing machine;

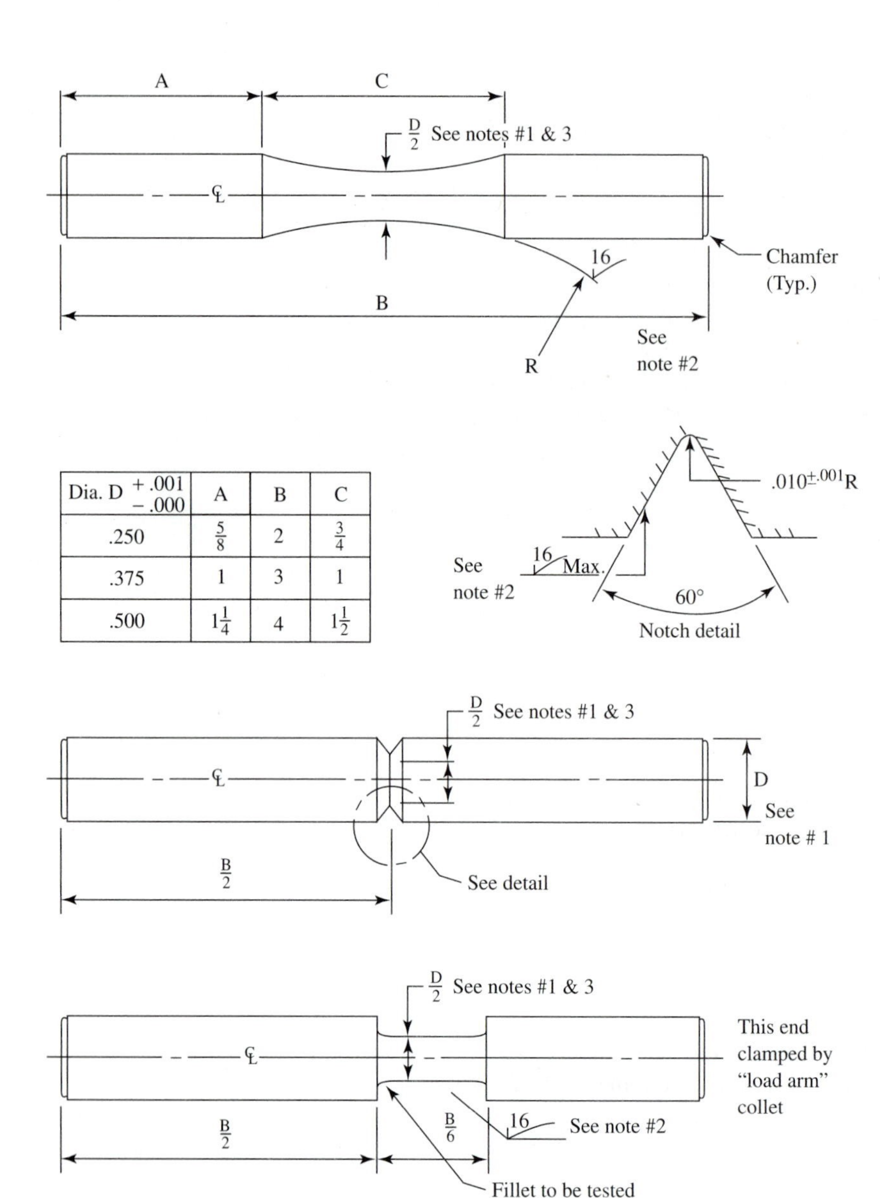

Dia. D $^{+.001}_{-.000}$	A	B	C
.250	$\frac{5}{8}$	2	$\frac{3}{4}$
.375	1	3	1
.500	$1\frac{1}{4}$	4	$1\frac{1}{2}$

Notes:

1. Diameters "D" and "$\frac{D}{2}$" to be concentric within 0.001.
2. Test section finish to be free of nicks, dents, scratches, and circumferential tool marks. Polish longitudinally, progressing through 0,00 and 000 emery paper. Do not buff.
3. Adjust dimensions $\frac{D}{2}$ to identity through all specimens ± .0002.

Figure 2.6 *continued* (c) suggested test specimen configuration for the rotating-beam fatigue testing machine. (Courtesy of Fatigue Dynamics, Inc.)

A series of tests may produce an *S–N* diagram similar to Figure 2.7, where a logarithmic scale is used for number of stress cycles to failure. The upper curve is typical of steel, and the lower curve is typical of a nonferrous metal. The solid lines are a curve fit to mean values, and the dashed lines indicate the scatter of data.

S–N curves for steel specimens tend to "level off"; that is, there appears to be a stress level, the **endurance limit** (S'_N), at which the specimen will last indefinitely. Most materials seem not to have an endurance limit. For materials without an endurance limit, test results may be reported as **fatigue strength** (S_N), the failure stress for a specified number of cycles (say, 10^6 or 10^7 cycles). Fatigue strength and endurance limit data are available for only a limited number of materials because fatigue tests are expensive and time-consuming. Endurance limit and fatigue strength test results cannot be applied directly to most engineering design problems. It is usually necessary to adjust these "ideal conditions" data for actual design conditions. Adjustments may be necessary for surface finish, stress concentration, reliability, and size.

Rotating-beam fatigue testing machines produce a reversed stress at a point on the surface of the test specimen, with zero mean load (see Figure 2.5). A sheet- and plate-bending (flexural-fatigue) testing machine can produce a fluctuating stress with a nonzero mean. The flexural-fatigue testing machine shown in Figure 2.8 has a 40-lb load capacity and can be adjusted for a stroke up to 2 in. Mean load is changed by adjusting the position of the vise which retains the specimen or load cell. Test speed is adjustable from 200 to 2400 cycles per minute.

The word *fatigue,* as used by engineers, usually refers to failures after thousands or millions of stress cycles, where maximum stress is *below the yield point.* When a soft metal is deformed by stresses *exceeding the yield point,* it may **work harden;** that is, its yield point may increase as a result of the deformation. For example, you may bend a wire paper clip repeatedly, causing it to work harden in the area of the bend. Then the hardened area will not yield, and the wire will break at further attempts to bend it.

Homogeneity and Isotropy

Homogeneous materials are essentially uniform throughout. When molten metals solidify, they actually assume crystalline structures. For mechanical engineering purposes, however, it is convenient to idealize steel and other metals as homogeneous.

If a material exhibits the same mechanical properties (strength, elastic modulus, etc.) regardless of the loading direction, it is **isotropic.** Homogeneous cast materials are usually considered isotropic. Materials lacking this property are **anisotropic** (i.e., not isotropic).

Orthotropic materials (a special class of anisotropic materials) can be described by giving their properties in three perpendicular directions. Wood may be considered orthotropic; it has one set of properties "with the grain" and another set of properties in either of the other two mutually perpendicular directions. A practical design utilizing wood can cause the higher tensile stresses to be in the direction of the grain, with lower tensile stresses in the cross-grain directions. Some materials become orthotropic due to cold-working such as rolling and drawing. Composites such as glass-fiber-reinforced plastics and carbon-fiber-reinforced plastics are also orthotropic if the fibers are oriented in a particular direction.

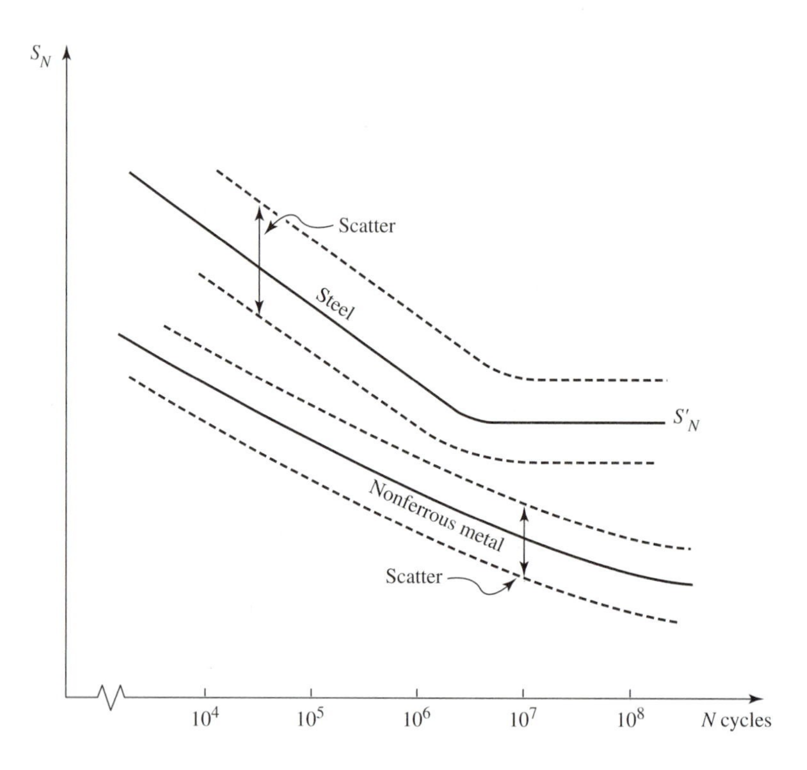

Figure 2.7 S–N diagram for reversed stress.

Figure 2.8 A 40-lb capacity sheet- and plate-bending (flexural-fatigue) testing machine. (Courtesy of Fatigue Dynamics, Inc.)

Engineering Design Implications

With so many different materials available and so many ways of judging them, rational material selection may seem nearly impossible. The examples that follow are intended to help in relating test data to engineering needs.

(a) Need: Rigidity

Properties: Elastic modulus and yield point.

Example: Camshaft deflection can result in unreliable valve behavior. We would select a material with a high yield point and a high modulus of elasticity. (Steel has a high modulus of elasticity. High-strength steel has a high yield point as well.)

(b) Need: Strength

The material must not deform plastically under static load.

Property: Yield point.

Example: A measuring instrument shaft has bent due to higher-than-predicted loads. Redesign, using material with a higher yield point.

(c) Need: Wear Resistance

Property: Hardness.

Example: Gear teeth in an air-operated wrench are wearing out. Redesign, using harder steel (higher BHN, VHN, or R_C).

(d) Need: Strength

The material must be able to withstand occasional overload without fracture.

Properties: Toughness and impact resistance.

Example: Farm equipment and power shovels sometimes hit large rocks. A ductile steel with moderate strength will dent or bend, absorbing energy (acceptable behavior for this application.)

(e) Need: Reliability and Safety

Heavily loaded rotating machinery must be reliable and safe.

Properties: Endurance limit and yield point.

Example: Gear, pulley, and chain drive shafts. Select a material with a high endurance limit and a high yield point. Consider stress concentration and surface finish in the design.

Specific Materials

Tensile strength, elastic modulus, and other properties are used to aid in selection of the best material for a particular application. Materials properties data are collected from a number of sources, and testing methods vary. Descriptive terms are relative; for example, a high-strength plastic may not be as strong as a mild steel. Thus, comparisons between and among materials in different groups may not be valid. Data contained herein is for illustrative purposes; actual design data must be obtained from material suppliers.

PLASTICS

The noun **plastic** refers to a variety of organic compounds, most produced by polymerization. Plastics are usually fabricated while in a plastic (an adjective here) state. They can be extruded or cast into various shapes. Engineers sometimes select plastics to replace parts traditionally made of metal to take advantage of the ease of fabrication, corrosion resistance, and light weight of plastics.

The two main groups of plastics are thermoplastics and thermosets. Thermoplastics soften at elevated temperatures and become harder and stronger when cooled. The process is repeatable, allowing for recycling of thermoplastics. Thermosets are heated to promote an irreversible reaction which hardens them. Thermosetting plastics cannot be reheated and reworked.

As an alternative, polymers can be classified by structure: amorphous, crystalline, and liquid crystalline. Other classes include copolymers, alloys, and elastomers. Glass or mineral reinforcement is often used to improve the mechanical properties of plastics.

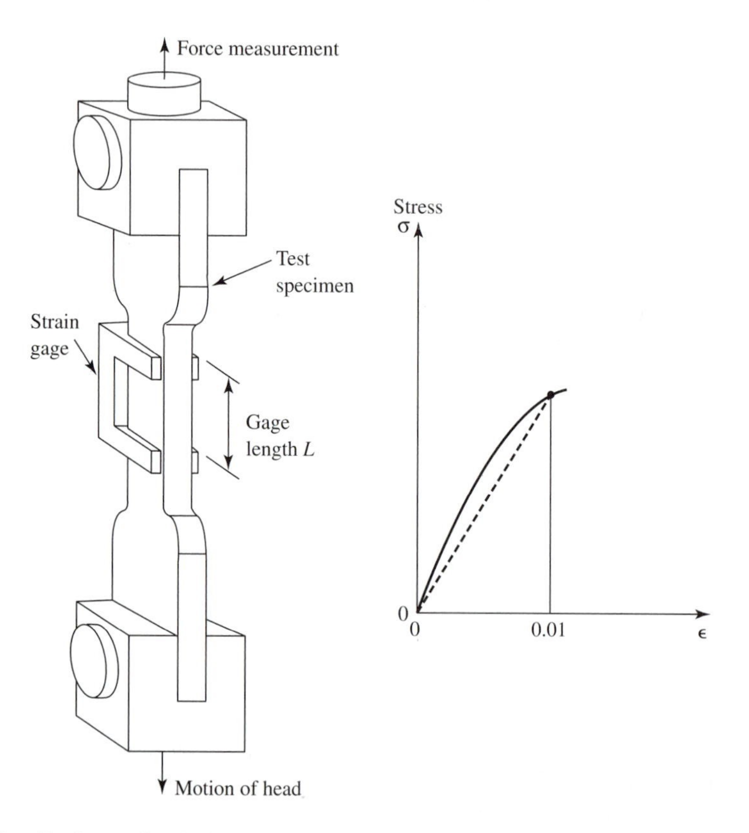

Figure 2.9 Tension test for plastics.

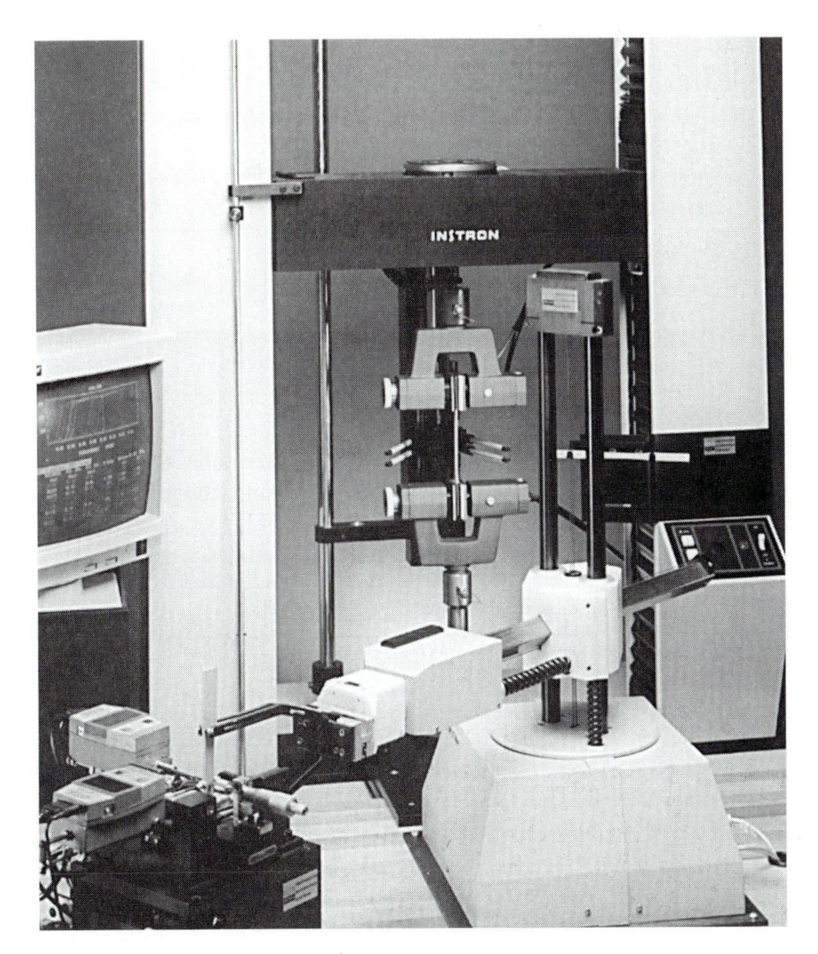

Figure 2.10 Production tension testing of plastics with the aid of robotics. (Courtesy of Instron Corporation)

Tension Test

A 1/8 in thick test specimen is used in the typical tension test for plastics. Specimen width narrows to 1/2 in where a strain gage extensometer is attached, as shown in Figure 2.9. Most plastics have a **nonlinear** stress–strain relationship; the ratio of stress to strain varies.

Figure 2.10 shows a system designed for testing more than 300 plastic specimens per shift with the aid of robotics. Desirable features in a laboratory robot for materials testing include the following:

- a servo-controlled arm capable of precision sample placement
- self-changing grippers with tactile feedback to limit pressure on the sample and to allow testing of delicate materials
- integration with a computer to collect data from the testing machine

Tensile Modulus

In the plastics industry, the modulus of elasticity determined from the tensile test is called the **tensile modulus.** It is calculated from test results as follows:

$$E = (F/A)/(\delta L/L) = FL/(A\ \delta L) \tag{2.10}$$

where E = tensile modulus (psi or MPa)

F = tensile load (lb or N)

A = cross-sectional area at gage (in^2 or mm^2)

= 1/8 in. by 1/2 in. in a typical test

L = gage length (in or mm)

δL = change in gage length (in or mm)

Caution: Some manufacturers report the **initial tensile modulus,** the slope of the stress–strain curve near 0,0. Others report the **tensile secant modulus,** the slope of the stress–strain curve at 1% deformation. Note that the "slope of the stress–strain curve" is measured in psi/(in/in) or MPa/(mm/mm). The tensile secant modulus is given by

$$E_{1\%\text{STRAIN}} = \sigma_{1\%\text{STRAIN}}/0.01 = 100\sigma_{1\%\text{STRAIN}} \tag{2.11}$$

where $\sigma_{1\%\text{STRAIN}} = F/A$ at the instant that $\delta L/L = 1\%$

Bending

Stress is approximately proportional to strain almost up to the yield point for steel and some other materials; that is, such materials are **linearly elastic.** For linearly elastic materials, we obtain the same value of elastic modulus *(E)* whether determined by a tensile test or a bending test. As a result of the *nonlinear* stress–strain behavior of most plastics, the elastic modulus in bending differs from the tensile modulus.

For a centrally loaded, simply supported beam (Figure 2.11), the deflection under the load is given by

$$w = FL^3/(48EI) \tag{2.12}$$

where $w = w(L/2)$, the deflection at the center (in or mm)

L = length between supports (in or mm)

E = elastic modulus, called the bending or flexure modulus, E_B, for nonlinear materials (psi or MPa)

I = moment-of-inertia (in^4 or mm^4)

We can determine **bending modulus** by using a known load to deform a beam as in Figure 2.11. The moment-of-inertia for a rectangular beam oriented as in the figure is given by

$$I = bh^3/12 \tag{2.13}$$

If deflection w is accurately measured, we obtain

$$E_B = FL^3/(4bh^3w) \tag{2.14}$$

where E_B = bending modulus (psi or MPa)

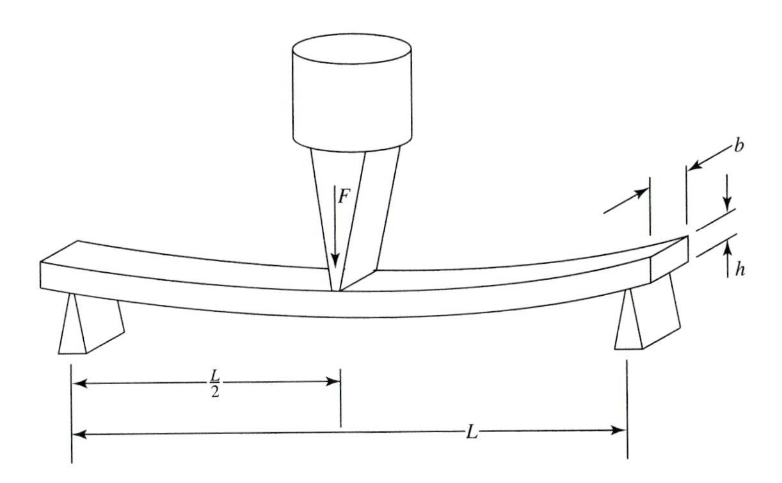

Figure 2.11 Bending test.

Bending tests of plastics can also be automated. Figure 2.12 shows a testing machine set up for high-volume automatic testing of plastic specimens in bending.

Bending Stress

The maximum tensile stress due to the bending of a linearly elastic material is given by

$$\sigma = Mc/I$$

where σ = maximum tensile stress due to bending (psi or MPa)

M = maximum bending moment (lb·in or N·mm)

c = distance from the neutral axis to the extreme fiber (in or mm)

I = moment-of-inertia (in^4 or mm^4)

For the centrally loaded, simply supported beam used in the bending test (Figure 2.11),

F = load applied at center (lb or N)

L = length between supports (in or mm)

b = width (in or mm)

h = thickness (in or mm)

For a rectangular cross section loaded as in Figure 2.11,

$$c = h/2 \text{ (in or mm)}$$
$$I = bh^3/12 \text{ (in}^4 \text{ or mm}^4\text{)}$$

resulting in a bending stress of

$$\sigma = 1.5FL/(bh^2)$$

If the rectangular beam is replaced with one of circular cross section with diameter d, then

$$c = d/2$$
$$I = \pi d^4/64$$

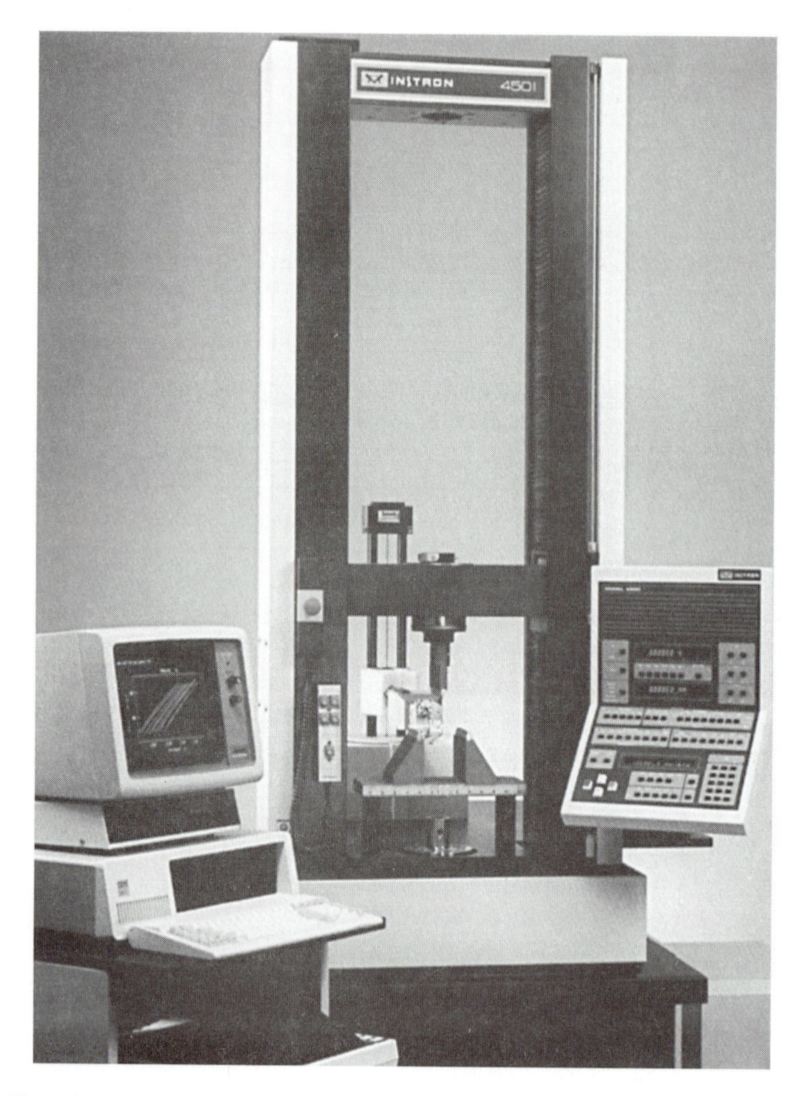

Figure 2.12 Automated bending test of plastic specimens. (Courtesy of Instron Corporation)

Table 2.1 Safety factors for preliminary part design.

	Failure Noncritical	Failure Critical
Steady or intermittent (nonfatigue) loading	2–4	4–10
Fluctuating (fatigue) loading	4–10	10–20

Maximum tensile stress due to bending a centrally loaded beam with circular cross section is given by

$$\sigma = 8FL/(\pi d^3)$$

The assumptions used to derive the above equations do not strictly apply to plastics and other nonlinear materials. However, these equations are used to define the bending strengths of plastics and thus can be used for design.

Safety Factor

As noted in Chapter 1, safety factor (N_{FS}) provides a margin of safety to account for uncertainties in loading and material properties and for errors in formulating a model of behavior. Obviously, we cannot express uncertainties as precise quantities; thus, selection of a safety factor requires considerable judgment. First, we must evaluate the consequences of a part failure. We must then describe the loading pattern. Hoechst (*Designing with Plastic,* 1990) suggests design strengths (working strengths) of 5% to 50% of strength values published in marketing data sheets. If we convert these recommendations into safety factors, the results are as given in Table 2.1.

The values in this table are to be used for preliminary design analysis. Before a product is marketed, the part or assembly should be tested at maximum end-use load and temperature in the presence of any chemicals that the part may encounter. It may be necessary to consider loading, chemical exposure, and temperature before end-use (e.g., during assembly, in a paint oven, etc.). Impact tests should be performed at the lowest temperatures expected during product use or shipping.

SELECTED TYPES OF PLASTICS

There is a seemingly countless array of types of plastics, with many different grades available within each type. New and modified formulations are developed every year. A few of the engineering plastics are described below.

Nylon

Nylon is often selected when toughness, strength, and abrasion resistance are important. Nylon also has good **lubricity,** that is, a low coefficient of friction. Glass reinforcement substantially increases both strength and stiffness. Impact-modified nylon is also available for applications requiring high impact strength. Applications of engineering-grade nylon include bearings, gears, under-the-hood automotive parts, and tool housings.

Liquid-Crystal Polymer Plastics

Plastics in this group have high strength and elastic modulus, good dimensional stability, and little mold shrinkage. Applications include components for aircraft, aerospace, automotive, medical, and dental use. Part complexity and miniaturization are possible due to the low melt-viscocities of liquid-crystal polymers.

Polyphenylene Sulfide (PPS)

PPS is a polymer with good thermal characteristics (as compared with other polymers). It is used for valves, gears, fittings, and other applications where high-temperature performance and chemical resistance are critical.

Injection-Moldable Polyethylene Terephalate (PET)

This thermoplastic polyester is glass reinforced to provide good physical properties. It has high-quality surface finish and good heat resistance. It is used for tools and electrical applications.

Long-Fiber-Reinforced Thermoplastics

Long-fiber-reinforced thermoplastics have greater strength and impact resistance than short-fiber-reinforced thermoplastics. Applications include metal part replacement, for example, automotive parts, plumbing fittings, mechanical handling equipment, hydraulic parts, appliance parts, and tools.

Polyarylate

This plastic has good impact strength, ultraviolet light stability, and flame retardance. It is available transparent or opaque and is used for lenses, safety panels, shields on industrial machinery, solar collection panels, and other products where resistance to sunlight is required.

Phenolics

Phenolic resins are thermosetting materials, ordinarily compounded with fibers and fillers for dimensional stability and moldability. Applications of various phenolic compounds include automatic transmission rings, power-brake system components, valve components in automatic transmissions, and pump housings.

Polycarbonates

Polycarbonates are high-molecular-weight engineering thermoplastics with high impact strength over a wide range of temperatures. The fatigue strength (for 10 million load cycles) is 1000 psi for general-purpose polycarbonate and 5000 psi for 20% glass-reinforced polycarbonate. Applications include computers and business machines, vacuum cleaner impellers, kidney dialyzers, blood oxygenators, and solar cell covers.

Table 2.2 Properties of selected engineering-grade plastics.

Type of Plastic	Tensile Strength (psi)	Tensile Elongation (%)	Flexural Modulus (10^6 psi)	Flexural Strength (psi)	Specific Gravity
Unfilled nylon	8000–13,000	15–80	0.27–0.45	15,000–17,000	1.04–1.14
Glass-reinforced nylon	17,000–32,000	2–4	0.75–1.6	27,500–40,000	1.23–1.47
Glass-reinforced liquid-crystal polymer	23,000–37,000	1–3	1.6–2.9	30,000–43,500	1.50–1.89
Glass-reinforced polyphenylene sulfide	10,000–24,000	0.5–1.5	1.9–2.2	14,800–34,000	1.64–2.00
Glass-reinforced thermoplastic polyester	16,500–27,500	1–3	1.1–2.2	24,000–42,400	1.40–2.10
Long-fiber-reinforced thermo-plastics	10,000–42,000	1.1–3	0.8–3	18,000–67,000	1.12–1.88
Unfilled polyarylate	10,000	50	0.33	14,500	1.21
Glass-reinforced polyarylate	13,000–24,000	24	0.47–1.43	22,000–34,000	1.28–1.53
Phenolic compounds	5000–9000	—	1.0–2.5	9000–11,000	1.35–1.84
Glass-reinforced polycarbonate	16,000–18,500	4–6	0.8–1.2	19,000	1.35–1.43

Source: This information has been compiled and adapted from a number of sources, including *Designing with Plastic—The Fundamentals* (1990) and *Machine Design Component Design Reference Issue* (1990). The data are intended for use in academic problems only. Manufacturers' data and specification sheets must be consulted for actual design purposes.

Typical properties of some selected plastics are given in Table 2.2. Note that the flexural strength of plastics generally exceeds the tensile strength. Also, reinforced plastics have higher tensile and flexural strengths than unfilled plastics of the same type, but reinforcing fibers reduce the percent elongation. New and improved plastic formulations are constantly being developed. As a result, the properties of some plastics may fall outside the range of data given in the table.

EXAMPLE PROBLEM Direct Tension

Design a part to support a steady tension load F of 190 lb. Failure of the part would be critical.

Design Decisions: Unfilled nylon will be selected. The part cross section will be circular.

Solution: Referring to the table of properties of engineering-grade plastics (Table 2.2), we will use the lower value of tensile strength for unfilled nylon (8000 psi). The table of safety factors (Table 2.1) for preliminary part design suggests safety factors from 4 to 10 for steady loads when failure is critical. We will use a safety factor of 5, which results in a working strength of

$$S_W = 8000/5 = 1600 \text{ psi}$$

As in the tensile test, axial stress is given by

$$\sigma = F/A$$

Setting the tensile stress equal to the working strength, we obtain

$$S_W = \sigma = F/A$$

from which we find the cross-sectional area.

$$A = F/S_W = 190/1600 = 0.11875 \text{ in}^2$$

For a circular section,

$$A = \pi d^2/4$$

The required part diameter is given by

$$d = (4A/\pi)^{0.5} = (4 \cdot 0.11875/\pi)^{0.5} = 0.389 \text{ in}$$

Other Considerations: The properties of thermoplastics are temperature dependent. If the part might be subject to high temperature, the material selection should be reconsidered.

■■

■■ EXAMPLE PROBLEM Bending of a Plastic Part

A part is to span a distance of 95 mm between supports, carrying a static bending load of 3000 N at the center. Failure of the part would not endanger human life or cause undue hardship. Design the part.

Design Decisions: The part cross section will be rectangular, with an aspect ratio (width to thickness) of $r_A = b/h = 2.75$. We decide to specify a grade of nylon with a bending strength S_B of 117 MPa. We use a safety factor N_{FS} of 3.

Solution: The working strength is given by

$$S_W = S_B/N_{FS} = 117/3 = 39 \text{ MPa}$$

Assuming simple supports, the maximum bending stress in a centrally loaded rectangular beam is

$$\sigma = 1.5FL/(bh^2)$$

Setting bending stress σ equal to working stress S_W and substituting $r_A h$ for width b, we obtain

$$S_W = 1.5FL/(r_A h^3)$$

Solving for thickness h, we have

$$\begin{aligned} h &= [1.5FL/r_A S_W)]^{1/3} \\ &= [1.5 \cdot 3000 \cdot 95/(2.75 \cdot 39)]^{1/3} = 15.86 \text{ mm} \end{aligned}$$

The part could be 16 mm thick by 44 mm wide.

■■

STEEL

Steel is formed from iron, carbon, and other alloying elements. It can be cast to a desired shape. More commonly, molten steel is allowed to solidify into ingots or is continuously cast. The ingot or continuous casting can then be hot-rolled to form I-beams, plates, angle-sections, and other mill forms. Finished parts are then formed by cold-rolling, drawing, forging, stamping, and machining. Steel that has undergone mechanical working (hot- or cold-rolling, forging, or drawing) is called **wrought steel.**

Carbon steel composition is limited to 1.65% manganese, 0.60% silicon, and 0.60% copper. Plain carbon steels with fewer than 30 points of carbon are designated as low-carbon steels (where points of carbon represent hundredths of one percent). Medium-carbon steels have 30 to 80 points of carbon, and high-carbon steels have more than 80 points of carbon. The combined effects of steel composition, heat treatment, and mechanical working during processing determine the properties of finished steel.

Alloy steels are iron–carbon alloys that contain either higher percentages of manganese, silicon, or copper than specified for carbon steel or significant amounts of other alloying elements (particularly nickel, chromium, molybdenum, and vanadium).

Alloying Elements

Some of the effects of alloying elements are as follows:

- **Carbon.** Affects response to heat treatment. Steels are limited to about 2% dissolved carbon, the maximum that can solidify as a single-phase alloy. Most steels have less than 1% carbon. Increased carbon content results in increased hardness and tensile strength but reduced ductility and weldability.
- **Vanadium.** Improves hardenability, toughness, and strength.
- **Chromium.** Improves wear resistance, hardenability, corrosion resistance, and elevated-temperature behavior.
- **Molybdenum.** Improves hardenability and increases high-temperature strength and creep strength.
- **Nickel.** Improves toughness and ductility at low temperatures; improves response to heat treatment, reducing the tendency of steel to crack during quenching; and improves corrosion resistance.

Steel Designations

The American Iron and Steel Institute (AISI) uses a numbering system for steel to identify type, points of carbon, and other alloying elements. The first two digits of a four-digit AISI number identify the steel type (principal alloys, etc.). The last two digits give the nominal-points of carbon (where 100 points = 1%). A few of the designations are as follows:

AISI Number	*Steel Type*
10xx	Plain carbon
11xx	Free-cutting
13xx	Manganese
2xxx	Nickel
3xxx	Nickel–chromium
4xxx	Molybdenum
5xxx	Chromium
6xxx	Chromium–vanadium
8xxx	Chromium–nickel–molybdenum

For example, a plain carbon steel with about 30 points of carbon would be designated as AISI 1030. Five digits are used for steels with 100 or more points of carbon.

The Unified Numbering System (UNS) utilizes five-digit numbers with the following letter prefixes: G for plain carbon and alloy steel, A for aluminum, C for copper-base alloys, and S for stainless steel. For most types of steel, the first four digits of UNS designations correspond to the AISI numbers.

Steel types are also designated by steel mills and by American Society for Testing and Materials (ASTM) standards. A few characteristics and applications of these steel types are described below.

Carbon Steel. These structural steels have the lowest cost per pound. The following properties are given in the as-rolled condition without heat treatment:

> **ASTM A 36.** The basic structural steel for construction. The minimum yield point is 36,000 psi except for plates over 8 in thick. The minimum yield point for plates over 8 in thick is 32,000 psi.
>
> **ASTM A 283.** Structural steel for general application. The minimum yield point is 24,000 to 33,000 psi, depending on grade.

High-Strength Low-Alloy Steel. These steels have moderate amounts of alloying elements other than carbon. Use of these steels instead of ASTM A 36 can result in a weight saving since parts can be made with smaller dimensions. Properties are given in the as-rolled condition:

> **ASTM A 441.** Steel with higher strength obtained by addition of vanadium. The minimum yield point is 40,000 to 50,000 psi.
>
> **ASTM A 572.** Steel recommended for riveted, bolted, or welded construction of bridges, buildings, and other structures, depending on grade. The minimum yield point is 42,000 to 65,000 psi.

Quenched and Tempered Alloy Steel.

> **ASTM A 514.** Heat-treated alloy steel with high strength-to-weight ratios and good notch-toughness. The minimum yield strength is specified as 100,000 psi for thicknesses up to 2.5 in, and as 90,000 psi for thicknesses over 2.5 in and up to 6 in. Roller-quenched grades are used for welded bridges, high-rise buildings, television towers, missile-transporting and -erecting equipment, farm and construction machinery, trucks, trailers, and storage tanks.

Ultrahigh-Strength Steel. Steels with yield points over 180,000 psi are considered to be in the ultrahigh-strength category. This group includes certain modified medium-carbon low-alloy steels; modified chrome–molybdenum–vanadium tool steels; high-nickel maraging steels; chrome–nickel–molybdenum steels; chrome–molybdenum–cobalt steels; and nickel-cobalt steels. After heat treatment, yield points of steels in this group range to 300,000 psi and above. Some ultrahigh-strength modified medium-carbon low-alloy steels will elongate up to

10% (for 2 in gage length) before fracture. Ultrahigh-strength steels meet the demands of the aerospace industry for high strength-to-weight ratios. Applications include aircraft fuselage frames, airframe fasteners and components, wear plates, highway grader blades and buckets, axles, shafts, and gears.

Properties Typical of Steel

As noted above, the yield point of engineering steels can range from about 24,000 to over 300,000 psi (165 to over 2070 MPa). Elongation in a 2-in gage length may range from near zero to about 30%. These properties depend on composition (carbon and alloy content), heat treatment, and specimen size. The elastic modulus, however, is about $30 \cdot 10^6$ psi for all carbon steels and alloy steels. The compressive yield point and ultimate compressive strength are ordinarily assumed to equal the tensile yield point and ultimate tensile strength, respectively. However, slender parts in compression may fail in buckling. The yield point in shear and the ultimate shear strength are ordinarily assumed to equal one-half of the corresponding values in tension. The endurance limit for polished specimens in reversed bending is about 40% to 60% of the ultimate strength, with 50% the customary assumption. The endurance limit is greatly reduced by surface roughness and stress concentration due to abrupt changes in part dimensions. Steel density is about 0.283 lb/in^3 (7830 kg/m^3).

Heat Treatment of Steel

Heat treatment has a major effect on the hardness and toughness of steel. Heat treatments for hardening steel include quenching, tempering, austempering, martempering, precipitation hardening, and freezing. Case-hardening processes include gas carburizing, nitriding, and induction-hardening. Heat treatments for grain refinement and uniformity include annealing, normalizing, and stress relieving. A few of these processes are described below. Detailed discussions of heat treatment are found in metallurgy and materials texts. See, for example, *Metals Handbook* (1985), DeGarmo et al. (1988), Neely and Kibbe (1987), and Doyle et al. (1985).

Quenching involves heating (to about 1650°F or 900°C for steel) and immersing in a cold medium in order to increase hardness and strength. Iced brine and iced water provide very rapid cooling because of the favorable heat transfer properties and high specific heat of water. Colder media (e.g., liquid nitrogen) cannot transfer heat away from the hot metal as well. If slower cooling is desired, oil, molten salt, molten metal, or air quenching may be specified.

The effect of heat treatment depends on steel composition and size. At a given temperature, the heat content of a steel bar is proportional to the volume. Cooling, which depends on heat transfer at the surface, is proportional to surface area. For example, a long 2 in diameter bar has four times the volume of a 1 in diameter bar, but only about twice the surface area. The smaller bar can be cooled more rapidly, resulting in higher strength and hardness. Severe quenching, however, results in a nonuniform grain structure, brittleness, and residual stresses.

Tempering is a reheating process to improve structural uniformity and remove residual stresses. Tempering temperatures for steel range from 200°F to 1300°F (93°C to 700°C). Tempering usually follows immediately after quenching to avoid the possibility of cracking due to residual stress. The effect of the quench-and-temper process is improved toughness, hardness, wear resistance, and strength. The process is most effective with alloy and plain carbon steels having more than 35 points of carbon.

Austempering and **martempering** involve interrupted quenching, holding steel in a high-temperature bath to produce a particular microstructure. These processes produce strong, tough, hard, wear-resistant steel with minimum distortion.

Annealing is a heat treatment consisting of heating and then slow cooling to decrease hardness. When materials are hardened by cold-working (cold-forming, -bending, and -drawing), they may be annealed so that further cold-forming or machining is possible. Annealing is also used to reduce residual stress in castings, forgings, and welded assemblies.

Case-hardening processes are intended to produce a hard surface to reduce wear, while leaving a more ductile interior. **Induction-hardening** is a case-hardening process in which the surface of a steel part is heated by a high-frequency current, and then the part is quenched. Another case-hardening process, **nitriding,** involves heating a steel part in a bath of molten salts or in an ammonia-containing atmosphere. Nitriding is used to surface-harden bearings, gears, and other parts requiring high reliability and minimal distortion.

CAST IRON

Cast iron is an alloy of iron, carbon, silicon, and manganese. The carbon content of cast iron ranges from 2% to 4%. Due to its high carbon content, molten iron is very fluid, making intricate castings possible. Cast iron types include gray iron, ductile or nodular iron, white iron, malleable iron, and high-alloy irons.

Gray iron is used for gears, automotive blocks, brake disks and drums, and other machine parts. During solidification of the casting, some of the carbon precipitates out of the supersaturated solution in the form of graphite flakes, counteracting shrinkage. Gray iron has good wear resistance, even when lubrication is marginal. Gray iron castings also have the ability to damp vibration, probably owing to the graphite flakes dispersed through the casting.

Gray cast iron is specified by tensile strength in thousands of pounds per square inch (kpsi). For example, Class 20 and Class 60 gray irons have minimum tensile strengths of 20,000 and 60,000 psi (about 138 and 414 MPa), respectively. Tensile strength values refer to ultimate tensile strength of a test bar; gray cast iron has no distinct yield point. An engineer must consider the size of the casting at locations where stress is critical, because the strength of cast iron is sensitive to cross-sectional size. Small cross sections cool rapidly, reaching higher unit strengths than large sections. A large cross section may fail at a lower stress than the test bar.

Gray cast iron is stronger in compression than in tension. Typical compressive strengths of Class 20 and Class 60 gray irons are 95,000 and 170,000 psi (655 and 1170 MPa), respectively. Where practical, machine parts made of cast iron are designed so that tensile stresses are lower than compressive stresses. The density of cast iron ranges from 0.25 to 0.26 lb/in^3 (6920 to 7200 kg/m^3).

Ductile or **nodular iron** is a type of cast iron alloyed with magnesium which precipitates out carbon as small spheres. It can be specified by tensile strength, yield strength, and percent elongation. Ductile iron is used for heavily loaded gears and other parts requiring higher strength and better impact resistance than typical gray cast irons.

White iron is a cast iron produced by a process that forms iron carbide, a very hard material. White iron is used for rolling mills, crushers, and other applications where wear and abrasion resistance are critical. Unless specially formulated and heat treated, white iron tends to be brittle.

White iron can be converted to **malleable iron** by a special heat-treatment process. Malleable iron has good wear and impact resistance and good machinability. It is used for cast heavy-duty bearing surfaces on farm machinery, highway vehicles, and railroad rolling stock.

High-alloy irons are cast irons containing more than 3% nickel, chromium, or zinc. These alloying metals are added to improve hardness, wear and abrasion resistance, or corrosion resistance.

STAINLESS STEEL

Stainless steels are iron–carbon alloys with high percentages of chromium. Corrosion resistance is imparted by formation of chromium oxide on the surface of the steel. Wrought stainless steels are generally categorized by metallurgical structure: austenitic, ferritic, martensitic, and precipitation-hardening. Cast stainless steels are generally designated according to whether their principal application requires heat resistance or corrosion resistance.

The American Iron and Steel Institute (AISI) designates stainless steels according to alloying elements (in addition to iron and carbon) and metallurgical structure. The AISI 200 series represents alloys of chromium, nickel, and manganese, having an austenitic structure. AISI 300 series alloys contain chromium and nickel and also have an austenitic structure. AISI 400 series stainless steels have a ferritic or martensitic structure, and the only alloying element is chromium. The AISI 500 series designates martensitic stainless steels with less than 12% chromium.

Austenitic wrought stainless steels have excellent corrosion resistance, but they are hardenable only by cold-working. Ferritic wrought stainless steels are more corrosion resistant than martensitic stainless steels, and they can be moderately hardened by cold-working. Martensitic wrought stainless steels can be hardened by heat treatment. Hardened and tempered martensitic stainless steels have high strength and toughness. They are usually formed in the annealed condition. Precipitation-hardening wrought stainless steels are hardened by a special lower-temperature heat treatment. This process is designed to avoid distortion of precision parts.

ALUMINUM AND ALUMINUM ALLOYS

Aluminum and aluminum alloys are light in weight and easy to form. When aluminum is exposed to air, an oxide forms on the surface, protecting the base metal from corrosion in most environments. The ultimate strength of pure untempered aluminum is about 13,000 psi (90 MPa), and the yield strength is 5000 psi (34 MPa). The ultimate strength of common aluminum alloys ranges from about 19,000 to 83,000 psi (130 to 570 MPa), and the yield strength ranges from about 8000 to 73,000 psi (55 to 500 MPa). The higher values are attained by heat treatment. Aluminum and aluminum alloys are also strengthened by cold-working.

The density of common aluminum alloys ranges from 0.095 to 0.101 lb/in^3 (2630 to 2800 kg/m^3), and the modulus of elasticity ranges from $10 \cdot 10^6$ to $10.6 \cdot 10^6$ psi (69,000 to 73,000 MPa). Note that both the density and the elastic modulus of aluminum are about one-third that of steel. Designers consider these properties along with strength and cost when deciding whether to choose aluminum, steel, or high-strength plastics for a particular application.

OTHER NONFERROUS ENGINEERING METALS

Magnesium, beryllium, copper, nickel, zinc, tungsten, tin, tantalum, columbium, molybdenum, and titanium are also used in engineering design. Some of these metals are relatively expensive but have unique properties. Beryllium, for example, meets the severe requirements imposed on space reentry vehicles. Others, including bronze, an alloy of copper and tin, have been used since ancient times. The Bronze Age, a period of human culture between the Stone Age and the Iron Age, was named for the bronze implements and weapons used during that period.

Copper and its alloys are noted for good thermal and electrical conductivity and corrosion resistance. They are easily formed by hot- or cold-working. However, copper and its alloys have relatively low strength-to-weight ratios and low strength at elevated temperatures.

Magnesium, with a typical density of 0.064 lb/in^3 (1770 kg/m^3), is the lightest of the structural metals. Magnesium alloys are easily machined and can be formed by most metal-working processes. Magnesium alloys can be joined by welding or adhesives. Stress concentration and galvanic corrosion should be considered when mechanical fasteners are used. Magnesium die casting alloys are competitive with plastics for many applications. Magnesium die castings can be held to tolerances as close as 0.0002 in (0.005 mm). In some designs, walls may be cast as thin as 0.025 in (less than 1 mm). Draft angle, the slope of the die sides, may be 0° or near zero for magnesium die castings, while dies for plastics usually require greater draft angles. Applications for magnesium die castings include electronic ignition module covers, computer printer carriages, and four-wheel-drive vehicle transaxle housings.

Tin and zinc are commonly used as corrosion-resistant coatings for steel (tin-plating and -galvanizing). Zinc and its alloys are ideal for die-casting because of their low melting point and low cost. Zinc die-castings are generally suitable for low-stress applications.

Refractory metals include columbium, molybdenum, tantalum, and tungsten. They have high melting points and greater strength at high temperatures than common structural metals. Protective coatings may be required on these metals to prevent oxidation at high temperatures.

POWDER METALLURGY

Powder metallurgy (P/M) is a manufacturing process in which metal powders are compacted in a die and then sintered or heated in a controlled-atmosphere furnace, metallurgically bonding the particles. The two main groups of P/M engineering applications are as follows:

- Parts that are difficult or impractical to make by other methods, including parts made of difficult-to-fabricate materials such as tungsten, tungsten carbide, and molybdenum. Porous bearings and filters are examples of parts that cannot be efficiently made by processes other than P/M.
- Part designs for which P/M is a cost-effective alternative to machining, casting, and forging. The P/M process can maintain close dimensional tolerances, minimize or eliminate the need for machining, and facilitate the manufacture of complex shapes.

Common P/M materials include iron, carbon steel, low-alloy steel, ferromagnetic materials, copper, bronze, brass, titanium, and aluminum alloys. Metal powders can be produced by atomization, electrolysis, or chemical or oxide reduction. Plain carbon steel can be produced in the P/M process by mixing iron and graphite powders and sintering to form alloys of up to 75 points of carbon. Other alloying elements can be added as powders. As an alternative, prealloyed compositions such as low-alloy steel, stainless steel, or brass can be sintered.

The mechanical properties of P/M parts are related to final density, the density of the manufactured part. Density can be given as a percentage of theoretical density, the ratio of part density to the density of its wrought metal counterpart. If the density of a P/M part is below 75% of theoretical density, the part is considered **low density;** if 75% to 90% of theoretical density, the part is **medium density;** and if over 90% of theoretical density, the part is **high density.** Structural and mechanical parts typically have "dry" densities from 80% to above 95%. Self-lubricating bearings may have "wet" densities of about 75% (after oil impregnation).

Part density, determined by weighing a P/M part in air and weighing it when immersed in water, is given by

$$\gamma = \gamma_W W_A/(W_{IA} - W_{IW} + W_T) \qquad \textbf{(2.15)}$$

where γ = part density

γ_W = density of water

W_A = weight of unimpregnated part in air

W_{IA} = weight of oil-impregnated part in air

W_{IW} = weight of oil-impregnated part in water

W_T = tare weight (weight of suspending wire or basket)

Structural and mechanical P/M part samples are impregnated with oil to prevent water absorption when determining displacement. Weights can be expressed (consistently) in any units. Part density will be in the same units as the units used for water density. The density of water is about 0.997 g/cm^3 (997 kg/m^3 or 0.036 lb/in^3) at 68°F (20°C).

P/M specimens are described in terms of yield strength (by the 0.2% offset method), ultimate tensile strength, elongation, impact strength, fatigue strength, elastic modulus, and hardness. In addition, a bending strength (transverse rupture strength) is determined by applying a vertical load to the center of a specimen that is simply supported near the ends as in Figure 2.11. Bending strength is given by

$$S_B = 3FL/(2bh^2) \tag{2.16}$$

where S_B = bending strength at fracture (psi or MPa)

F = load at fracture (lb or N)

L = distance between supports (in or mm)

b = cross-sectional width (in or mm)

h = cross-sectional depth (in or mm)

Table 2.3 Typical properties of selected P/M materials. Strength is given in kpsi (thousands of psi), and modulus in millions of psi.

Material	Ultimate Strength S_U	Yield Strength S_{YP}*	Bending Strength S_B	Fatigue Strength S_N†	Young's Modulus E
Unalloyed iron	18–38	13–25	36–95	7–14	14–20.5
Carbon steel (0.3%–0.6% carbon)	24–38	18–28	48–76	9–14	14–18
Heat-treated carbon steel (0.3%–0.6% carbon)	60–80		105–140	32–30	16.5–19
Carbon steel (0.6%–0.9% carbon)	29–57	25–40	51–100	11–22	12–19
Heat-treated carbon steel (0.6%–0.9% carbon)	65–95		100–145	25–36	15–19.5
Heat-treated low-alloy steel	85–155		130–230	32–59	16.5–25
Stainless steel	39–60	17–42	80–127		
Heat-treated stainless steel	105		113		

Source: Adapted principally from MPIF Standard 35 (1990–91). Additional data will be found in this and subsequent editions of the standard.

Notes: (a) 1 MPa ≈ 145 psi; 1 kpsi ≈ 6.895 MPa.

(b) For heat-treated P/M materials, yield strengths approximate ultimate strengths

(c) The above typical values are for comparison purposes and academic design problems only. Minimum strengths are lower. Strengths depend on manufacturing method and part size.

(d) Stress concentration must be considered, particularly when stresses vary with time.

*kpsi by 0.2% offset method

†kpsi at 10^7 cycles of reversed loading on smooth, unnotched specimens.

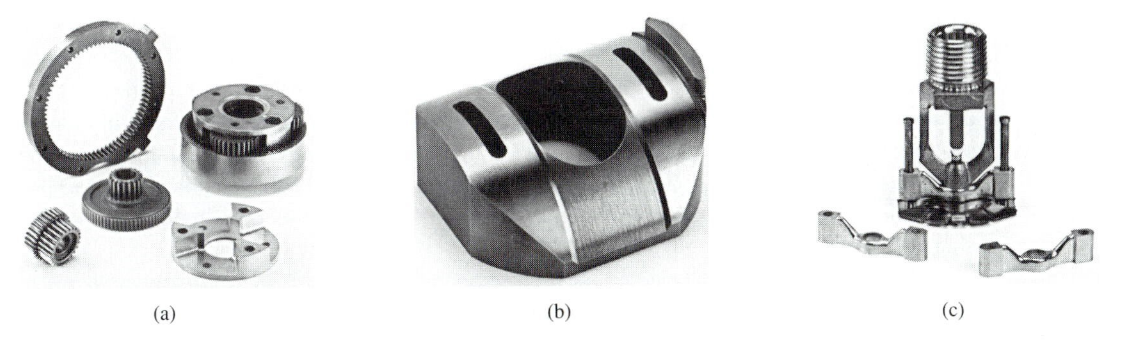

(a) (b) (c)

Figure 2.13 Examples of parts produced by powder metallurgy. (a) A compound planetary gear drive containing P/M nickel–steel parts; (b) a variable-displacement cam made of copper-infiltrated steel; (c) a brass yoke used in a commercial sprinkler system. (Courtesy of American Powder Metallurgy Institute)

For loading in pounds and dimensions in inches, bending strength is given in lb/in^2. If the load is given in newtons and the dimensions in millimeters, bending strength is given in N/mm^2 or MPa (1 MPa = 1 N/mm^2).

The typical fatigue strength of P/M materials, based on 10^7 cycles of reversed loading, is about 38% of the typical ultimate strength. The stress that wrought carbon and alloy steels can sustain for 10^7 cycles is considered the endurance limit (S_E), sustainable indefinitely. This assumption is applied to P/M carbon and alloy steel parts as well. For P/M parts made of other materials, lower fatigue strengths would apply when the stress cycles exceeded 10^7. Typical properties of some P/M materials are given in Table 2.3.

Figure 2.13 shows examples of parts produced by powder metallurgy. Part (a) of the figure shows a compound planetary gear drive containing P/M nickel steel parts; the drive is used in an automatic door. Part (b) shows a variable-displacement cam made of copper-infiltrated steel; the cam is used in a hydraulic pump. Part (c) is a brass yoke used in a commercial sprinkler system; yoke assemblies must pass corrosion tests as well as tests under fire conditions.

P/M Part Design Considerations

P/M parts are usually limited to about 3 in (76 mm) in height because the pressing action is only from the top and bottom. Height, of course, refers to the pressing direction in the die, not to orientation of the part as it is used. The most suitable part shape for P/M processing is one with uniform dimensions in the pressing direction. Ideal P/M designs include rectangular and circular cross sections, cams, and spur gears. More complicated shapes can be repressed after sintering or formed with multiple-motion tooling. Holes in the pressing direction can be produced by core-rods which extend through the tools. Guidelines for design are given in the *P/M Design Guidebook* (1983) and the *Powder Metallurgy Design Manual* (1989).

STRUCTURAL CERAMICS

Ceramics are hard, heat-resistant materials made from oxides, carbides, and nitrides of metals and nonmetals including aluminum, silicon, boron, and magnesium. Manfacturing methods for clay-based ceramics have been known for thousands of years. Many ceramics are brittle and have low tensile strength. Structural ceramics, those useful for mechanical engineering applications, are formulated to improve mechanical properties while retaining heat-resistance characteristics.

Structural ceramics are selected for high-temperature applications (often above the melting point of steel) and for situations where abrasion resistance is critical. Applications for structural ceramics include jet-engine turbine blades and other components, exhaust valves and turbochargers for reciprocating engines, cutting tools, armor plates, pump components, and pressure-blasting nozzles.

When ductile steel is loaded statically, high localized stresses (stress concentration) may be redistributed by local yielding. Most ceramics do not exhibit a yield point; they tend to fracture without measurable yielding. Thus, engineering designs for ceramics should avoid stress concentration by avoiding abrupt changes in cross section at critical locations and by attempting to distribute loads over a large area. Mechanical and thermal shock should be avoided whenever possible. Prototype testing, computer simulation, and nondestructive testing and evaluation are recommended for critical applications.

COMPOSITE MATERIALS

Sometimes optimum properties can be obtained by combining dissimilar materials. Adobe, a sun-dried brick of clay and straw, is an example of an early composite. The tensile strength of thin glass fibers (measured in psi or MPa) is much greater than the tensile strength of glass specimens with larger cross sections. Thus, engineers use high-strength glass fibers to reinforce thermoplastics intended for high-stress applications. Compare, for example, the tensile and flexural strengths of glass-reinforced nylon with those properties for unfilled nylon as given in Table 2.2. Optimum performance-to-cost ratios are typically obtained with 20% to 40% glass in thermoplastic molding compounds, but over 60% glass is sometimes used.

Carbon fibers are also used for reinforcing nylon, polyetheramide, polysulfone, polyester, polyphenylene sulfide, and other thermoplastics. Usually, the composite contains 10% to 40% carbon fibers. Carbon-fiber-reinforced plastics are usually superior to glass-reinforced plastics in terms of strength-to-weight ratio and wear resistance. Typically, carbon-reinforced plastics have a lower coefficient of thermal expansion and lower mold shrinkage than their glass-reinforced counterparts.

Strengths of typical carbon-reinforcing fibers range from 275,000 to 820,000 psi (1900 to 5650 MPa), and the elastic moduli range from $33 \cdot 10^6$ to $110 \cdot 10^6$ psi (230,000 to 760,000 MPa). When reinforcing fibers are aligned in a particular direction, the composite is orthotropic. Strength and elastic modulus of the composite are greater in the longitudinal direction (parallel to the fibers) than in the transverse direction (perpendicular to the fibers). Typical carbon-fiber-reinforced composites have tensile strengths ranging from 110,000 to 470,000 psi (760 to 3240 MPa) in the longitudinal direction, and only 3000 to 10,000 psi (20 to 70 MPa) in the transverse direction. For these composites, the elastic moduli are between $20 \cdot 10^6$ and $72 \cdot 10^6$ psi (140,000 to 500,000 MPa) in the longitudinal direction. Typical moduli in the transverse direction range from 10^6 to $4 \cdot 10^6$ psi (7000 to 28,000 MPa).

EXAMPLE PROBLEM Determination of Required Part Size and Weight Comparisons

A 4 in long bar (measured between simple supports) will be subject to a static bending load F of 100 lb at the center. Failure will not endanger human life or cause undue expense. A bar of minimum weight is desirable. Specify the bar.

Design Decisions: A circular-cross-section bar will be selected, using a safety factor N_{FS} of 3. We will compare unfilled nylon, glass-reinforced nylon, and glass-reinforced liquid-crystal polymer. Approximate weight will be based on the unsupported length.

Solution: We equate bending stress in the bar to working strength of the material, obtaining

$$S_W = \frac{S}{N_{FS}} \geq \sigma = \frac{Mc}{I} = \frac{FL/4}{\pi d^3/32} = \frac{8FL}{\pi d^3}$$

from which the required bar diameter is given by

$$d = \sqrt[3]{\frac{8FLN_{FS}}{\pi S}}$$

We find the weight density of the material by multiplying specific gravity sg by 0.03613, the density of water. The approximate weight of the bar is

$$W = 0.03613 \cdot sg \cdot L \cdot \pi d^2/4$$

If we use mathematics software to avoid repetitive hand calculations, we write d and W as functions of the strength and specific gravity of the material; that is,

$$d(S) = \sqrt[3]{\frac{8FLN_{FS}}{\pi S}}$$

$$W(S,sg) = 0.03613 \cdot sg \cdot L \cdot \pi d(S)^2/4$$

Using average values of strength and specific gravity, the computations yield the following values:

Unfilled Nylon

$S = 16{,}000 \qquad sg = 1.09$
$d(S) = 0.576 \qquad W(S,sg) = 0.041$

Glass-Reinforced Nylon

$S = 33{,}750 \qquad sg = 1.35$
$d(S) = 0.449 \qquad W(S, sg) = 0.031$

Glass-Reinforced Liquid-Crystal Polymer

$S = 36{,}750 \qquad sg = 1.70$
$d(S) = 0.436 \qquad W(S,sg) = 0.37$

Glass-reinforced nylon is the obvious choice among the three materials considered. A 0.449 in diameter glass-reinforced nylon bar meets the requirements.

EXAMPLE PROBLEM Strength-to-Weight Ratio

Compare the ratios of allowable bending load to weight for various materials.

Decisions: We will arbitrarily select a 6 in long, 1/2 in diameter bar with the load at the center and simple supports at the ends. A safety factor of 3 will be used. Unfilled nylon, glass-reinforced nylon, glass-reinforced liquid-crystal polymer, heat-treated carbon steel P/M (60 to 90 points of carbon, 82% of theoretical density), and aluminum alloy will be considered.

Solution Summary: Bending stress in the bar is set equal to working strength S/N_{FS}, and the equation is rewritten to obtain allowable bending load F at the center of the bar. Mathematics software was used to obtain the comparative values of ratios of bending load to weight given in the detailed solution which follows.

Detailed Solution:

Circular-Cross-Section Bar Supporting a Central Load:

Basis: median bending (flexural) strength and specific gravity.

Load

F at center

Length

$L = 6$ in

Safety Factor

$N_{FS} = 3$

Diameter

$d = 0.5$ in

Allowable Load (lb)

$$F(S) = \pi \cdot d^3 \cdot \frac{S}{8 \cdot L \cdot N_{FS}}$$

Weight (lb)

$$W(sg) = \pi \cdot d^2 \cdot L \cdot sg \cdot \frac{0.03613}{4}$$

Unfilled Nylon:

$$S = 16{,}000 \qquad sg = 1.09$$

$$F(S) = 43.633 \qquad W(sg) = 0.046$$

$$\frac{F(S)}{W(sg)} = 940$$

Glass-Reinforced Nylon:

$$S = 33{,}750 \qquad sg = 1.35$$

$$F(S) = 92.039 \qquad W(sg) = 0.057$$

$$\frac{F(S)}{W(sg)} = 1{,}602$$

Glass-Reinforced Liquid-Crystal Polymer:

$$S = 36{,}750 \qquad sg = 1.70$$

$$F(S) = 100.22 \qquad W(sg) = 0.072$$

$$\frac{F(S)}{W(sg)} = 1{,}385$$

Heat-Treated Carbon Steel P/M (60 to 90 Points of Carbon):

Assume 82% theoretical density.

$$S = 80{,}000 \qquad sg = 0.82 \cdot \frac{0.283}{0.03613}$$

$$F(S) = 218.166 \qquad W(sg) = 0.273$$

$$\frac{F(S)}{W(sg)} = 798$$

Aluminum Alloy:

$$S = 51{,}000 \qquad sg = \frac{0.1}{0.03613}$$

$$F(S) = 139.081 \qquad W(sg) = 0.118$$

$$\frac{F(S)}{W(sg)} = 1{,}181$$

ELASTOMERS

Natural rubber, synthetic rubbers, and rubberlike materials are called **elastomers.** Elastomers, which include both thermosetting and thermoplastic compounds, have in common the ability to undergo large deformations and then recover after removal of the force causing the deformation. Elongation at failure can range from 100% to 1350%. Elastomers are used in the automotive, aircraft, aerospace, paper, chemical, and petroleum industries, as well as many others.

A few of the many engineering applications include crankshaft and oil filter bushing seals, valve cover and valve stem seals, rollover safety check valves, transaxle transfer support seals, vacuum tubing, tires, belting, wellhead seals, O-rings, V-rings, gaskets, flue duct coatings, and expansion joints. Typical properties of some elastomers are given in Table 2.4.

Table 2.4 Typical properties of selected elastomers.

Elastomer Type	Specific Gravity (psi)	Tensile Strength	Elongation (%)
Polyurethane	1.1–1.24	4000–9000	225–570
Copolyester	1.15–1.25	4500–7600	250–800
Olefin	0.9–1.0	650–4000	180–600
Fluoroelastomers	1.8	1500–2500	100–400
Silicones	1.1–1.6	700–1800	100–800
Acrylic	1.1	2000	100–400
Nitrile	1.0	1000–3000	400–600

Source: Based partially on data supplied by 3M Industrial Chemical Products Division.

DATABASES AND EXPERT SYSTEMS

The information presented in this chapter provides only a brief introduction to comparative material properties. Engineers engaged in design obtain physical and performance characteristics of materials from manufacturers, suppliers, and standards agencies. Manufacturers publish this information (or database) in catalogs and product data sheets, and they file it in hard-copy form or on microfiche, tape, or disk.

A **database management system** (called simply a **database**) is an electronic information-handling software system designed to increase the efficiency of data storage and retrieval. An empty database serves as a template with tools for making data entry thorough and efficient. A typical company uses a database to store product information, as well as other information related to sales, customers, and suppliers. The database must also provide easy access to data. It must enable a user to retrieve necessary data in the most useful form for making design decisions, writing reports, controlling inventory, and making other engineering, manufacturing, and business decisions.

Artificial intelligence (AI) is an area of computer science research into understanding and simulating human intelligence. Goals of AI include development of computer programs that can solve problems, interpret visual scenes and natural langauge, and learn from experience.

AI research has led to development of **expert systems,** computer programs that attempt to duplicate the knowledge and judgment of a human expert. **Knowledge engineers** consult libraries and interview experts in a given field to compile a base of knowledge and rules. They then devise an expert system for solving problems in that field. The expert system contains a large database of knowledge and if–then rules for making decisions. A human user consults with the expert system, describing a problem, answering questions about the problem, and gathering test data if necessary. The expert system applies the rule set to the facts, using inference to provide recommendations about and/or solutions to the problem. The expert system may even describe and explain the reasoning process that it used to solve the problem. Expert systems designed for solving a limited range of small problems are sometimes called **knowledge systems.**

A number of expert systems have been developed for medical diagnostic tasks. One such system, EEG Analysis System, accepts an electroencephalogram signal directly and classifies it as normal or abnormal based on interpretation of a fast-Fourier-transform analysis. An expert system called MECHO solves problems in mechanics involving, for example, mechanical elements and their moments-of-inertia, accepting English text input. A mathematical expert system, MACSYMA, performs symbolic integration and algebraic manipulation as well as numerical solutions. Other examples of expert systems include a program that troubleshoots problems with diesel-electric locomotives and a program to help design noise barriers. A variety of expert systems are described by Smith (1989) in a dictionary that includes over 2000 definitions related to AI and expert systems. A system to relate materials, geometry, and manufacturing in the design of mechanical components is described by Hadley, Langrana, and Steinberg (1989).

A Database/Expert System: Interactive Software for Material Selection

Some manufacturers supply database/expert system software to their customers to aid in product selection. One such example is the *3M Elastomer Selection and Design Guide* (1989). This guide is supplied on a disk for use with a PC. The user is asked a series of interactive questions, and a product recommendation is made on the basis of the user's needs. The use of interactive software is described in the following example.

EXAMPLE PROBLEM Elastomer Selection Using Interactive Software

An elastomer is needed in a particular design. An engineering technician has determined the required mechanical properties and environmental conditions.

Design Decisions: The technician will use the 3M software to evaluate the applicability of various elastomers.

Solution: The questions posed by the interactive software and the technician's replies are as follows:

Q: What is your upper temperature requirement in °C or °F?

A: 250C (482F)

Q: What is your upper continuous temperature requirement in °C or °F?

A: 200C (292F)

Q: What is your low-end temperature requirement in °C or °F?

A: −20C (−4F)

Q: What are your maximum tensile strength requirements? Up to:

A: 2500 psi

Q: What is your Shore A hardness requirement?

A: 50±5

Q: What is your elongation requirement in %?

A: 300

Q: What is your compression set requirement at 23°C (73°F)?

A: 10

Q: What is your compression set requirement at 175°C (347°F)?

A: 10

Q: What is your compression set requirement at 200°C (392°F)?

A: 15

The software used its knowledge base and if–then rules based on American Society for Testing and Materials (ASTM) standard tests to conclude that a particular proprietary fluoroelastomer was suitable. Table 2.5 shows the test performance of the recommended product, as well as the test performance of two other products.

At the technician's request, the software tabulated properties of the recommended elastomer (Fluorel™) beside properties of a silicone elastomer; the results are shown in Table 2.6. The technician then requested information on strength and percent elongation of the recommended elastomer when exposed to high temperatures for long periods. The software responded with a plot of tensile strength versus temperature and exposure time and a plot of percent elongation versus temperature and exposure time. Both plots are based on 60-day tests of the material at temperatures from 400°F to 500°F; they are given in Figures 2.14 and 2.15.

Table 2.5 Product selection using interactive software.

Criteria	Fluorel™	Aflas™	Sylon™	
Upper temp	Yes	Yes	No	Based on the answers to your questions,
Upper cont. temp	Yes	Yes	Yes	we recommend the following products:
Low-end temp	Yes	No	Yes	
Tensile strength	Yes	Yes	No	
Shore A hardness	Yes	No	Yes	
Elongation (%)	Yes	Yes	Yes	Fluorel
Modulus @ 100% Elong	Yes	Yes	Yes	
Compression set @:				Press any key to continue
23°C = 73°F	Yes	No	Yes	
175°C = 347°F	Yes	No	Yes	
200°C = 392°F	Yes	No	No	

Source: 3M Industrial Chemical Products Division.

Table 2.6 Comparative product information provided by interactive software.

Criteria	Fluorel™	Silicone
ASTM D-2000 designation	HK	FC, FE, GE
Specific gravity	1.8	1.1–1.6
Hardness "Durometer" range, Shore A	50–90	20–90
Tensile strength range, psi	1000–2500	750–1500
Elongation range %	100–500	100–800
Compression set	Excellent	Good
Electrical resistance	Good	Excellent
Stress relaxation	Good	Good
Impact strength	Good	Excellent
Abrasion resistance	Good	Good
Tear resistance	Good	Fair
Heat aging at 212°F	Excellent	Excellent
Heat aging at 177°F	Excellent	Excellent

Source: 3M Industrial Chemical Products Division.

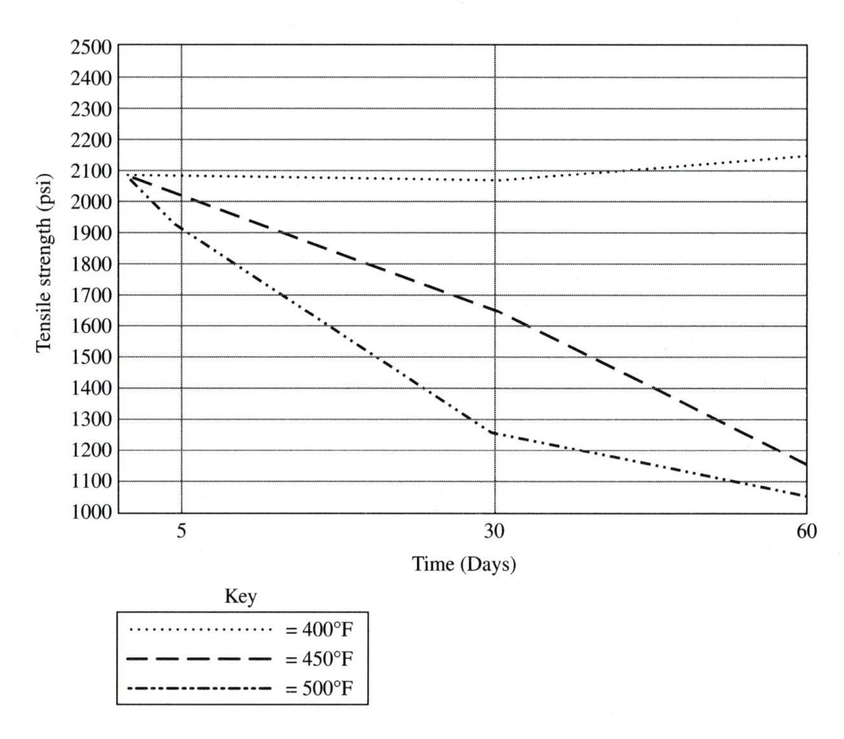

Figure 2.14 Tensile strength versus exposure time at 400°F to 500°F. (Courtesy of 3M Industrial Chemical Products Division)

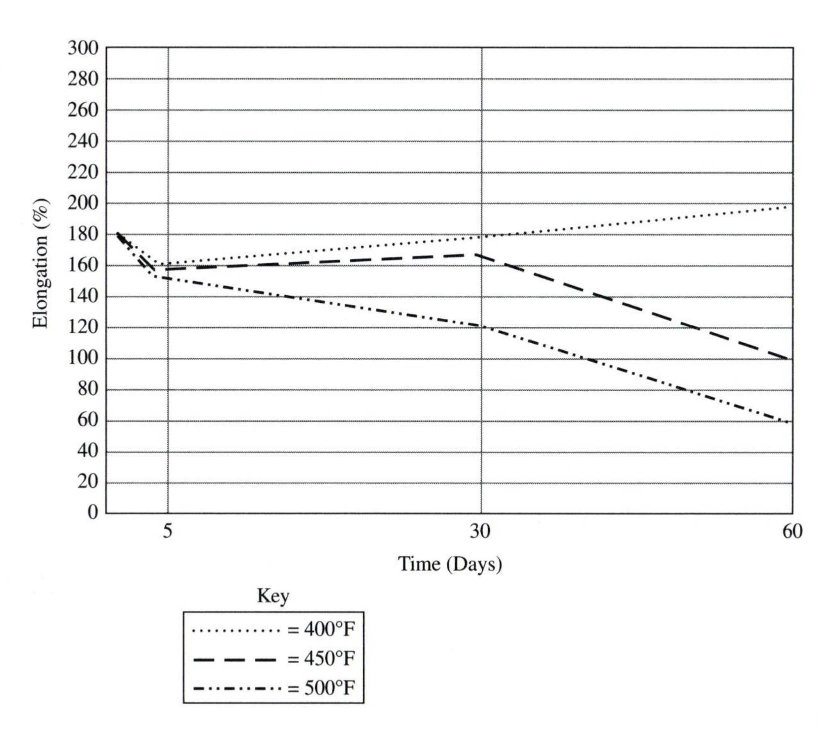

Figure 2.15 Percent elongation (tensile test) versus exposure time at 400°F to 500°F. (Courtesy of 3M Industrial Chemical Products Division)

References and Bibliography

DeGarmo, E. P., J. T. Black, and R. A. Kohser. *Materials and Processes in Manufacturing.* 7th ed. New York: Macmillan, 1988.

Designing with Plastic—The Fundamentals. Design Manual TDM-1. Chatham, NJ: Hoechst Celanese Corporation, 1990.

Doyle, L. E., C. Keyser, J. Leach, G. Schrader, and M. Singer. *Manufacturing Processes and Materials for Engineers.* 3d ed. Englewood Cliffs, NJ: Prentice-Hall, 1985.

Hadley, A. L., N. A. Langrana, and L. Steinberg. "Design Using Features for Mechanical Components." *Preprints · NSF Engineering Design Research Conference.* Amherst, MA (June 11–14, 1989): 81–96.

Machine Design. Cleveland, OH: Penton/IPC, Inc. A biweekly periodical that publishes an annual materials reference issue.

Machine Design Component Design Reference Issue. Cleveland, OH: Penton/IPC, Inc., October 1990.

Metals Handbook. Metals Park, OH: American Society for Metals, 1985.

MPIF Standard 35. *Materials Standards for P/M Structural Parts.* Princeton, NJ: Metal Powder Industries Federation, 1990–91.

Neely, J. E., and R. R. Kibbe. *Modern Materials and Manufacturing Processes.* New York: John Wiley, 1987.

P/M Design Guidebook. Princeton, NJ: Metal Powder Industries Federation, 1983.

Powder Metallurgy Design Manual. Princeton, NJ: Metal Powder Industries Federation, 1989.

Product Line Performance Guide. PL-1. Chatham, NJ: Hoechst Celanese Corporation, 1990.

Smith, Raoul, ed. *Dictionary of Artificial Intelligence.* New York: Facts on File, 1989.

Structural Steel Data for Architectural and Engineering Students. Bethlehem, PA: Bethlehem Steel Corporation, 1978.

Sweet's Catalog File. New York: Sweet's Division, McGraw-Hill Information Systems, 1991.

Thomas Register of American Manufacturers and *Thomas Register Catalog File.* New York: Thomas Publishing Co., 1992.

3M Elastomer Selection and Design Guide (software). St. Paul, MN: 3M Industrial Chemical Products Division, 1989.

Design Problems

2.1. A plain carbon steel rod has a Brinell hardness number BHN of 150.
Estimate the following:
(a) Rockwell (C scale) hardness number
(b) Vickers hardness number
(c) ultimate strength

2.2. A plain carbon steel rod has a Brinell hardness number BHN of 130.
Estimate the following:
(a) Rockwell (C scale) hardness number
(b) Vickers hardness number
(c) ultimate strength

2.3. Design a part to support a steady tension load F of 250 lb. Failure of the part would be critical.

Design Decisions: Glass-reinforced nylon will be selected. The part cross section will be circular. Refer to Table 2.2 and use the lower value of tensile strength. Use the median value from the appropriate range in Table 2.1.

2.4. Design a part to support a steady tension load F of 275 lb. Failure of the part would not be critical.

Design Decisions: Glass-reinforced liquid-crystal polymer will be selected. The part cross section will be rectangular, with width twice the thickness. Refer to Table 2.2 and use the lower value of tensile strength. Use the median value from the appropriate range in Table 2.1.

2.5. A part is to span a distance of 3.5 in between supports, carrying a steady bending load of 440 lb at the center. Failure of the part could endanger human life or cause undue hardship. Design the part.

Design Decisions: The part cross section will be rectangular, specifying an aspect ratio (width to thickness) of $R_A = b/h = 1.5$. It is decided to specify a phenolic compound. Refer to Table 2.2 and use the median value of bending strength. Use the median value from the appropriate range in Table 2.1.

2.6. A part is to span a distance of 5 in between supports, carrying a fluctuating bending load of 400 lb at the center. Failure of the part could endanger human life or cause undue hardship. Design the part.

Design Decisions: The part cross section will be rectangular, specifying an aspect ratio (width to thickness) $r_A = b/h = 3$. It is decided to specify a phenolic compound. Refer to Table 2.2 and use the median value of bending strength. Use the median value from the appropriate range in Table 2.1.

2.7. A part is to span a distance of 5 in between supports, carrying a fluctuating bending load of 400 lb at the center. Failure of the part could endanger human life or cause undue hardship. Design the part.

Design Decisions: The part cross section will be rectangular, specifying an aspect ratio (width to thickness) of $r_A = b/h = 3$. It is decided to specify a glass-reinforced polyester compound. Refer to Table 2.2 and use the minimum value of bending strength. Use the median value from the appropriate range in Table 2.1.

2.8. A part is to span a distance of 100 mm between supports, carrying a fluctuating bending load of 4000 N at the center. Failure of the part would not endanger human life or cause undue hardship. Design the part.

Design Decisions: The part cross section will be rectangular, specifying an aspect ratio (width to thickness) of $r_A = b/h = 3.25$. It is decided to specify a glass-reinforced nylon compound with a bending strength of 190 MPa. Use the median value from the appropriate range in Table 2.1.

2.9. A 5 in long bar (measured between supports) will be subject to a fluctuating load F of 200 lb at the center. Failure will not endanger human life or cause undue expense. A bar of minimum weight is desirable. Specify the bar.

Design Decisions: A circular-cross-section bar will be selected, using a safety factor N_{FS} of 7. We will compare glass-reinforced polyphenylene sulfide, glass-reinforced thermoplastic polyester, and a long-fiber-reinforced thermoplastic.

2.10. A 6-in long bar (measured between supports) will be subject to a static load F of 150 lb at the center. Failure could endanger human life or cause undue expense. A bar of minimum weight is desirable. Specify the bar.

Design Decisions: A circular-cross-section bar will be selected, using a safety factor N_{FS} of 7. We will compare unfilled polyarylate, glass-reinforced polyarylate, and glass-reinforced polycarbonate.

2.11. Compare the ratios of allowable bending load to weight for glass-reinforced polyphenylene sulfide, glass-reinforced thermoplastic polyester, heat-treated low-alloy steel P/M (82% theoretical density), and untempered aluminum.

Design Decisions: Select a 10-in long, 3/4 in diameter bar with simple supports at the ends; use a safety factor of 7. Determine allowable load at the center and load-to-weight ratio. Base calculations on median strength and specific gravity values.

2.12 Compare the ratios of allowable bending load to weight for the following:

(a) unfilled polyarylate
(b) glass-reinforced polyarylate
(c) Class 60 gray cast iron

Design Decisions: Select a 1-in long, 3/8-in diameter bar with simple supports at the ends; use a safety factor of 7. Base calculations on median strength and specific gravity values. Determine allowable load at the center and load-to-weight ratio.

CHAPTER 3

Static Stresses in Machine Members and Failure Theories for Static Stress

Symbols

A	cross-sectional area	S_{YP}	yield point
b, h	dimensions	S_U	ultimate strength
c	distance from neutral axis to extreme fiber	S_{SW}, S_W	working strength in shear and in tension, respectively
D	diameter, determinant	T	torque
I	moment-of-inertia	V	shear force
J	polar-moment-of-inertia	z	distance from neutral axis
M	bending moment	σ, τ	tensile and shear stress
N_{FS}	factor of safety		
P	load		

Units

bending moment and torque	in · lb or N · mm
dimensions and distances	in or mm
load and shear force	lb or N
moment-of-inertia and polar-moment-of-inertia	in^4 or mm^4
stress and strength	psi or MPa

STATIC AND TIME-VARYING STRESSES

Stresses in a machine member can be time varying or steady. Steady bending loads on rotating parts result in time-varying stresses. Design for time-varying stress, whether caused by time-varying loads or part rotation, is based on fatigue failure theories. In this chapter we will consider only static stresses, the stresses due to constant loads on stationary parts.

WORKING STRENGTH AND FACTOR OF SAFETY

Consider a heavily loaded machine part. If the load per unit area (the stress) exceeds the strength of the material, then the part will fail. To avoid failure, we select a **factor of safety** N_{FS} greater than 1 based on the consequences of a failure and the degree of uncertainty of loading conditions and material properties. If failure could endanger human life, then a high factor of safety might be chosen. In actual design situations, determining loads, stresses, and material strength and selecting a failure theory may be the most difficult part of the design process.

Allowable tensile stress or **working strength** is given by

$$S_W = S_{YP}/N_{FS} \tag{3.1}$$

or

$$S_W = S_U/N_{FS} \tag{3.2}$$

where S_W = working strength (psi or MPa)

S_{YP} = yield point strength (psi or MPa)

S_U = ultimate tensile strength (psi or MPa)

The elastic limit of a typical material, that is, the limit of elastic behavior, falls somewhat below the yield point strength. When stresses exceed the yield point strength, then significant plastic (inelastic) deformation is likely. Since plastic deformation of machine parts cannot ordinarily be tolerated, designers usually base working strength on yield point strength.

UNIFORM TENSION

Material specimens in the tensile test (discussed in Chapters 1 and 2) and some machine parts are subject to uniform tension in a single direction. We arbitrarily place the x-axis in that direction. Then the tensile stress is given by

$$\sigma_x = P/A \tag{3.3}$$

and the design criterion is

$$S_W \geq \sigma_x \tag{3.4}$$

where σ_x = tensile stress in the x-direction (psi or MPa)

P = load in the x-direction (lb or N)

A = area of a critical cross section, perpendicular to the x-axis (in^2 or mm^2)

If part dimensions are given in meters, and loads in newtons, then stress and strength are expressed in pascals.

STRESS DUE TO BENDING

Actual loading conditions seldom lead to uniform tension throughout a part. Bending loads result in a tensile stress proportional to the bending moment and the distance from the neutral axis, and inversely proportional to the cross-sectional moment-of-inertia.

$$\sigma_x = Mz/I \tag{3.5}$$

where σ_x = tensile stress in the x-direction (psi or MPa)
M = bending moment at the location of interest, usually the maximum bending moment (in · lb or N · m)
z = distance from the neutral axis (in or mm)
I = moment-of-inertia about the neutral axis (in^4 or mm^4)

Moment-of-inertia, bending moment, and deflection are considered in detail in Chapter 5. We will consider only a few simple bending problems at this time.

Figure 3.1 shows a rectangular cantilever beam subject to bending. Part (a) shows the loading, Part (b) the stress distribution, and Part (c) the location of the neutral axis on a cross section. At the extreme fiber (the bottom of the beam), $z = c$ and

$$\sigma_x = Mc/I \tag{3.6}$$

For the rectangular beam loaded as in Figure 3.1, the neutral axis in bending is the central axis in the y-direction (i.e., $c = h/2$). The moment-of-inertia about that axis is

$$I_{\text{RECTANGLE}} = bh^3/12 \tag{3.7}$$

from which the maximum bending stress at any cross section in the rectangular beam is given by

$$\sigma_{x(\text{RECTANGLE})} = 6M/(bh^2) \tag{3.8}$$

If the rectangular beam in Figure 3.1(a) is replaced by a solid circular shaft or beam of diameter D, the neutral axis of any cross section is a horizontal diameter, $c = D/2$, and the moment-of-inertia of the circular section about the neutral axis is

$$I_{\text{CIRCLE}} = \pi D^4/64 \tag{3.9}$$

from which the maximum bending stress at any cross section in the circular beam is

$$\sigma_{x(\text{circle})} = \frac{32M}{\pi D^3} \tag{3.10}$$

If the beam shown in Figure 3.1(a) is clamped at the left end ($x = 0$), and if there is a concentrated load P at $x = x_1$, then bending moment is given by

$$M = P(x_1 - x) \text{ for } x \leq x_1$$

Maximum moment occurs at the clamped end and is given by

$$M_{\text{MAX}} = Px_1$$

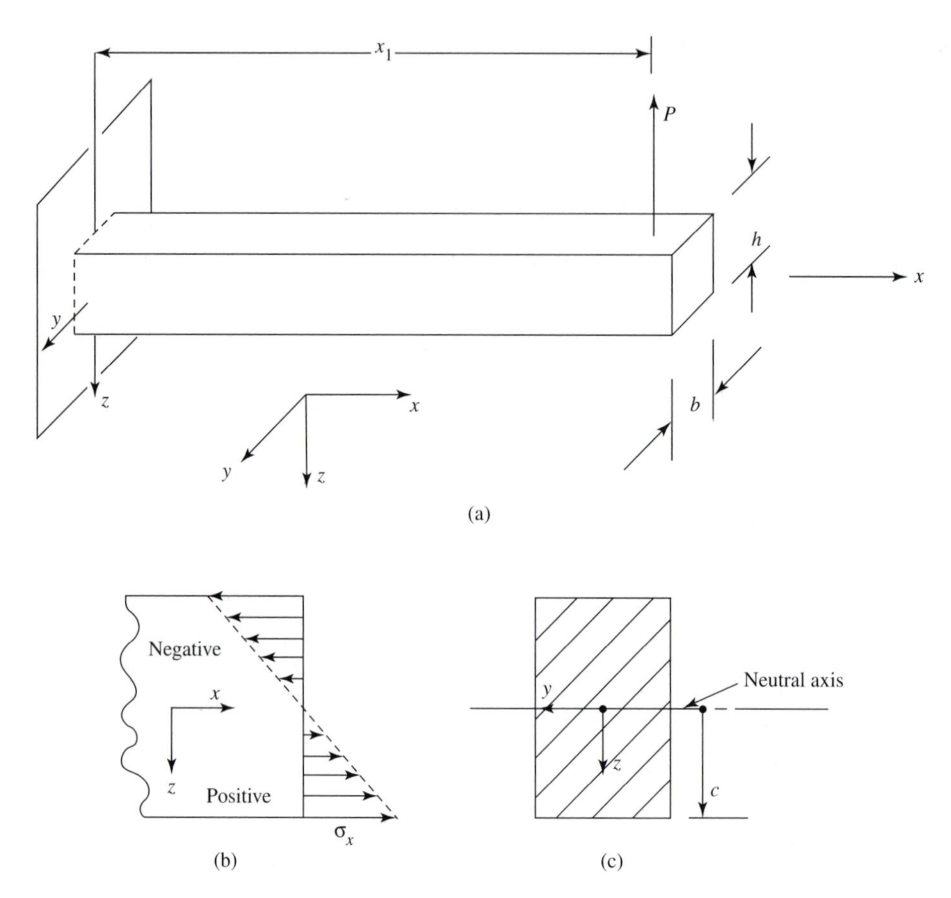

Figure 3.1 Bending of a rectangular beam. (a) Loading; (b) stress distribution; (c) locations of neutral axis on a cross section.

Note that the beam shown in Figure 3.1 is **statically determinate** if the loading is given. Reaction forces and bending moment can be calculated from equilibrium considerations alone. That is, we need to consider only equilibrium of forces and moments. If a beam is overconstrained, then it is **statically indeterminate,** and we must consider deflections to calculate reactions and moments. Statically indeterminate beams are more difficult to solve. More will be said on this topic in Chapter 5, "Bending of Machine Members."

EXAMPLE PROBLEM Statically Loaded Cantilever Beam

A machine part is to be fixed at the left end ($x = 0$) and is to carry a 500 N static load at $x = 130$ mm. Design the part.

Design Decisions: The part will be made of steel with a yield point S_{YP} of 700 MPa. The safety factor N_{FS} will be 3. A rectangular cross section 25 mm wide will be used.

Solution: We substitute maximum bending moment

$$M_{\text{MAX}} = Px_1$$

into the stress equation

$$\sigma_{x(\text{RECTANGLE})} = 6M/(bh^2)$$

and use the design criterion relating stress and working strength

$$S_W \geq \sigma_x$$

where
$$S_W = S_{\text{YP}}/N_{\text{FS}}$$

to obtain

$$S_{\text{YP}}/N_{\text{FS}} \geq 6Px_1/(bh^2)$$

from which

$$h^2 \geq 6Px_1N_{\text{FS}}/(bS_{\text{YP}}) = 6 \cdot 500 \cdot 130 \cdot 3/(25 \cdot 700) = 66.86$$
$$h \geq 8.18 \text{ mm}$$

The part will be made with a constant cross section 25 mm wide by at least 8.18 m thick. We might specify a thickness of 10 mm.

■■

SHEAR STRESS DUE TO BENDING

Transverse shear stress results from shear force in a beam. For a rectangular beam loaded as in Figure 3.1,

$$\tau_{xz(\text{MAX})} = 1.5\ V/A \tag{3.11}$$

where $\tau_{xz(\text{MAX})}$ = maximum transverse shear stress (psi or MPa)
V = shear force (lb or N)
$A = bh$ = cross-sectional area (in^2 or mm^2)

For the cantilever beam loaded at the end as in Figure 3.1, shear force magnitude equals the load.

$$V = -P \quad \text{for } 0 < x < x_1$$

Transverse shear stress is maximum at the neutral axis and zero at the top and bottom of a vertically loaded beam.

■■ **EXAMPLE PROBLEM Design for Minimum Weight**

(a) Design a part to meet the requirements in the preceding example problem. Weight is to be minimized.
(b) Evaluate the design.

Design Decisions: The design decisions will be unchanged, but variable thickness will be considered.

Solution (a): For a cantilever beam with a single concentrated load, bending moment is proportional to distance from the load

$$M = P(x_1 - x)$$

and bending stress equals

$$\sigma_x = 6P(x_1 - x)/(bh^2)$$

For more efficient use of material, we can design the beam so that tensile stress σ_x at the extreme fiber $z = c = h/2$ is constant over the length of the beam. We now use the design criterion relating stress and working strength

$$S_W \geq \sigma_x$$

where $S_W = S_{YP}/N_{FS}$

to obtain

$$S_{YP}/N_{FS} \geq 6P(x_1 - x)/(bh^2)$$

from which

$$\begin{aligned} h^2 &\geq 6P(x_1 - x)N_{FS}/(bS_{YP}) \\ &\geq 6 \cdot 500(130 - x)3/(25 \cdot 700) \\ h &\geq 0.7171[130 - x]^{1/2} \end{aligned}$$

Thickness will vary according to the equation

$$h = 0.7171[130 - x]^{1/2}$$

Thus, $h = 8.18$ mm at the clamped end.

Although the equation calls for zero thickness at the load location, we must specify some value of minimum thickness to accommodate the load and avoid failure in shear. The "concentrated" load must actually be spread over a finite area of the beam. If we decide to use a minimum thickness of 1 mm, then

$$\begin{aligned} \tau_{xz(\text{MAX})} &= 1.5V/A = -1.5P/(bh) = -1.5 \cdot 500/(25 \cdot 1) \\ &= 30 \text{ MPa} \end{aligned}$$

in the thinner section. A thickness of 1 mm near the right end is more than adequate with regard to transverse shear stress.

Solution (b): The design decisions could be modified to reduce manufacturing cost. For example, we could consider a beam of constant depth and varying width.

■■

TORSION

Torsion loads cause warping in parts with noncircular cross sections, resulting in complicated stress distributions. Circular cross sections are easier to analyze and design. Torsion of solid circular shafts and hollow circular shafts results in a shear stress proportional to radius.

$$\tau = Tr/J \tag{3.12}$$

with a maximum value at the outer surface of the shaft

$$\tau = TD/(2J) \tag{3.13}$$

where τ = shear stress due to torsion (psi or MPa)
r = radius, a variable (in or mm)
D = shaft diameter (in or mm)
J = polar-moment-of-inertia (in^4 or mm^4)

For a solid circular shaft,

$$J = \pi D^4/32 \tag{3.14}$$

and, for a hollow shaft,

$$J = \pi(D^4 - D_i^4)/32 \tag{3.15}$$

where D_i = inside diameter

BENDING AND DIRECT TENSION

When a beam supports a bending load and an axial load as well, the stresses combine easily. Consider a rectangular beam with a positive moment M and a positive direct tension load P. The bending load results in a tensile stress in the x-direction which varies from

$$\sigma_x = 6M/(bh^2)$$

on the beam surface in tension to

$$\sigma_x = -6M/(bh^2)$$

on the beam surface in compression. The direct tension load causes a stress

$$\sigma_x = P/(bh)$$

which is simply added to the bending stress to produce a maximum stress

$$\sigma_x = 6M/(bh^2) + P/(bh)$$

The situation is more complicated when stresses are in different directions and when combining tensile and shear stresses.

STRESS AT A POINT

A stress can be described according to its direction and the plane on which it acts. Imagine a small piece "cut" from the surface or the interior of a machine part. Figure 3.2 is a representation of an infinitesimal element at some point in a solid body. For this sketch, the positive directions of the x-, y-, and z-axes are chosen to be, respectively, to the right, upward, and toward the viewer. A plane of the element perpendicular to the x-axis on the right side of the element is identified as a $+x$-face; the plane on the opposite side of the element is a $-x$-face; and so on. The figure also shows normal and shear stresses on each face. These stresses depend on the loading on the solid body.

Normal Stress

A stress is identified by the face on which it acts and by its direction. Normal stresses σ_x, σ_y, and σ_z act on the $+x$-, $+y$-, and $+z$-faces, respectively, in the $+x$-, $+y$-, and $+z$-directions (and on the $-x$-, $-y$-, and $-z$-faces, respectively, in the $-x$-, $-y$-, and $-z$-directions).

Shear Stress

Stresses parallel to the face on which they act are called **shear stresses.** A shear stress may be identified by a two-letter subscript, the first letter identifying the face, and the second the direction. Thus, τ_{xy} is a shear stress on the $+x$-face in the $+y$-direction. The far side of the element would show a shear stress on the $-x$-face in the $-y$-direction, also identified as τ_{xy}.

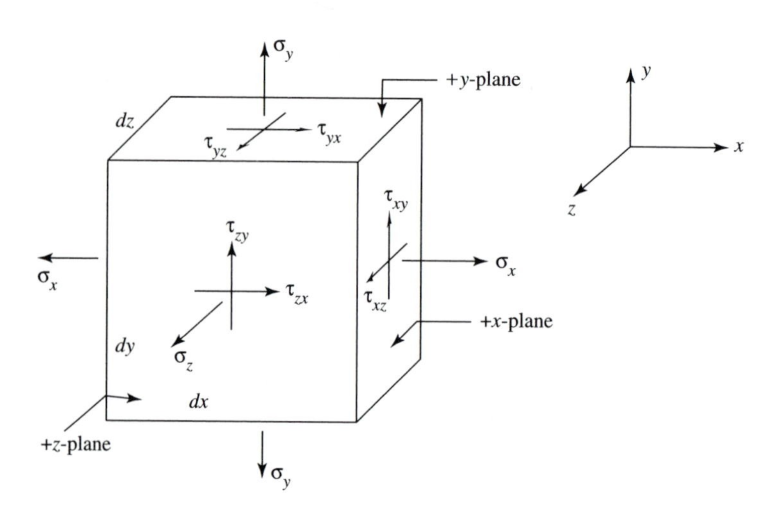

Figure 3.2 Stress at a point.

Combined Stress

A normal stress and two shear stresses are shown on each face of the element in Figure 3.2. However, the stresses are not all independent. Stress magnitudes are related as follows:

$$|\tau_{yx}| = |\tau_{xy}| \qquad |\tau_{zy}| = |\tau_{yz}| \quad \text{and} \quad |\tau_{zx}| = |\tau_{xz}|$$

Thus, in general, we have three normal stresses, σ_x, σ_y, and σ_z, and three independent shear stresses, τ_{xy}, τ_{yz}, and τ_{xz}.

Instead of considering all of these stresses separately, we can combine them to form a single-number descriptor of stress at a point. Single-number descriptors in common use are maximum normal stress, maximum shear stress, and von Mises stress. These descriptors can be obtained by hand or computer calculation, and they can be displayed as the result of finite element analysis.

PRINCIPAL STRESSES

For any state of stress at a point, there exist three mutually perpendicular planes which are free of shear stress. These planes are called the **principal planes.** The normal stresses on the principal planes are called the **principal stresses** σ_1, σ_2, and σ_3 (see Figure 3.3). The three principal stresses include the maximum and minimum normal stresses.

The *x*-, *y*-, and *z*-axes can be arbitrary; that is, we can choose these directions for our convenience. At a point in a solid body, for a given set of loads, the principal stresses and the principal planes are invariant. There is only one set of principal stresses and principal planes at a point. If the loading on a body and the selection of axes result in the case where

$$\tau_{xy} = \tau_{yz} = \tau_{xz} = 0$$

then stresses σ_x, σ_y, and σ_z are the principal stresses.

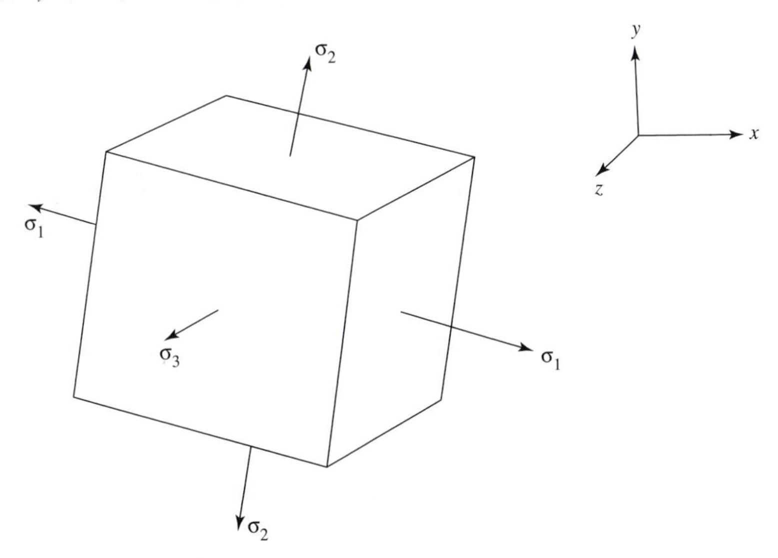

Figure 3.3 Principal stresses.

Principal Stresses: Plane Stress Case

It is rare to find all six stress components present at a point in a machine part. In actual practice, we may find only one, two, or three stress components with significant magnitude. **Plane stress** refers to the situation where one of the principal stresses is zero. For example, if

$$\sigma_z = \tau_{xz} = \tau_{yz} = 0$$

at a given point, then

$$\sigma_3 = 0$$

The other two principal stresses are given by

$$\sigma_1 = \frac{\sigma_x + \sigma_y}{2} - \sqrt{\left(\frac{\sigma_x - \sigma_y}{2}\right)^2 + \tau_{xy}^{\ 2}} \tag{3.16}$$

and

$$\sigma_2 = \frac{\sigma_x + \sigma_y}{2} - \sqrt{\left(\frac{\sigma_x - \sigma_y}{2}\right)^2 + \tau_{xy}^{\ 2}} \tag{3.17}$$

for this state of plane stress.

EXAMPLE PROBLEM Principal Stresses

The following state of stress exists at the most severely loaded point in a machine part:

$$\sigma_x = 36{,}500 \qquad \sigma_y = 2500 \qquad \sigma_z = 0$$
$$\tau_{xy} = 13{,}000 \qquad \tau_{yz} = 0 \qquad \tau_{xz} = 0$$

where all values are in psi.

Find the principal stresses at that point.

Solution: Note that the above stress pattern corresponds to plane stress. Thus, principal stress

$$\sigma_3 = 0$$

Using mathematical software to evaluate Equations (3.16) and (3.17), we find the other two principal stresses at the point.

$$\sigma_1 = 40{,}436 \quad \text{and} \quad \sigma_2 = -6436 \text{ psi}$$

Note that the numbering order of the principal stresses is arbitrary.

THE MAXIMUM NORMAL STRESS THEORY

A design based on the **maximum normal stress theory** requires that

$$S_W \geq \sigma_{\text{MAX}} \tag{3.18}$$

where $S_W = S_{YP}/N_{FS}$ or S_U/N_{FS}, the working strength

σ_{MAX} = the maximum normal stress, the greatest of σ_1, σ_2, and σ_3

Cast iron and some other materials are substantially stronger in compression than in tension. When practical, such materials are used in designs where compressive stress exceeds tensile stress. A satisfactory design must satisfy Equation (3.18) at the location where positive stresses are greatest and the following equation where negative stresses are greatest:

$$S_{W(c)} \geq \sigma_{MAX,c} \tag{3.19}$$

where $S_{W(c)} = S_{U(c)}/N_{FS}$, the working strength in compression

$\sigma_{MAX,c}$ = negative stress of greatest magnitude

$S_{U(c)}$ = ultimate strength in compression

EXAMPLE PROBLEM The Maximum Normal Stress Theory

Determine the minimum material strength requirement for the state of stress described in the preceding example problem.

Design Decisions: Use the maximum normal stress theory. Select a safety factor of 2, and base the design on yielding.

Solution: The criterion is

$$S_W \geq \sigma_{MAX}$$

where $S_W = S_{YP}/N_{FS}$

$\sigma_{MAX} = \sigma_1 = 40{,}436$ psi as found above

Thus,

$$S_{YP}/2 \geq 40{,}436$$

and the yield point requirement for the material is

$$S_{YP} \geq 80{,}872 \text{ psi}$$

Ordinarily, we do not design parts as described in this example. The designer usually selects a material first and then determines the required part dimensions.

MAXIMUM SHEAR STRESS: PLANE STRESS CASE

Suppose that the loading of a machine part results in plane stress. Then

$$\tau_{MAX} = \text{the largest of} \begin{cases} |(\sigma_1 - \sigma_2)/2| \\ |\sigma_1/2| \\ |\sigma_2/2| \end{cases} \tag{3.20}$$

where τ_{MAX} = maximum shear stress

The nonzero principal stresses are identified as σ_1 and σ_2.

If principal stresses σ_1 and σ_2 have opposite signs, then the first expression in Equation (3.20) applies, and

$$\tau_{MAX} = \frac{|\sigma_1 - \sigma_2|}{2} = \sqrt{\left(\frac{\sigma_x - \sigma_y}{2}\right)^2 + \tau_{xy}^2} \qquad \textbf{(3.21)}$$

Caution: Equation (3.21), which is found in some references and is incorporated in some finite element analysis software, calculates τ_{MAX} in the xy-plane. If principal stresses σ_1 and σ_2 have the same sign, then use of Equation (3.21) could result in an unsafe design. Thus, it is best to first calculate the principal stresses and then use Equation (3.20).

EXAMPLE PROBLEM Maximum Shear Stress

Stress components at a critical point in a machine member are calculated to be

$$\sigma_x = 650 \qquad \sigma_y = -250 \quad \text{and} \quad \tau_{xy} = 300$$

where all values are in MPa.

The other stress components are zero. Find maximum shear stress at that point.

Solution: This is a plane stress situation, and we can take $\sigma_3 = 0$. Equations (3.16) and (3.17), which were already coded into mathematics software for a previous example, yield principal stresses $\sigma_1 = 741$ and $\sigma_2 = -341$ MPa. Then Equation (3.20) gives maximum shear stress at the point $\tau_{MAX} = 541$ MPa.

MOHR CIRCLES FOR PLANE STRESS

Equations (3.16), (3.17), and (3.20) can be expressed in graphical form as shown in Figure 3.4(a). The procedure is as follows:

1. Draw the normal stress axis horizontal and the shear stress axis vertical.
2. Plot points (σ_x, τ_{xy}) and $(\sigma_y, -\tau_{xy})$.
3. These points are the diameter of the Mohr circle in the xy-plane. The center is on the normal stress axis at $(\sigma_x + \sigma_y)/2$. Draw the Mohr circle.
4. Note principal stresses σ_1 and σ_2 where the circle crosses the normal stress axis.
5. For plane stress, $\sigma_3 = 0$. Mark principal stress σ_3 at the origin. Complete the diagram by drawing the Mohr circles in the 1–3-plane and the 2–3-plane. Their diameters are, respectively, $\sigma_1 - \sigma_3$ and $\sigma_2 - \sigma_3$.
6. The radius of the largest circle gives the maximum shear stress τ_{MAX}. Note the previous caution in the paragraph following Equation (3.21).

A common special case of plane stress occurs when the only nonzero stress components are σ_x and τ_{xy}. The Mohr circle construction for this case is illustrated in Figure 3.4(b). The procedure is the same as outlined above except that $\sigma_y = 0$.

Graphical constructions have diminished in importance with the general availability of computers. However, there is still some interest in the Mohr circles, which give information

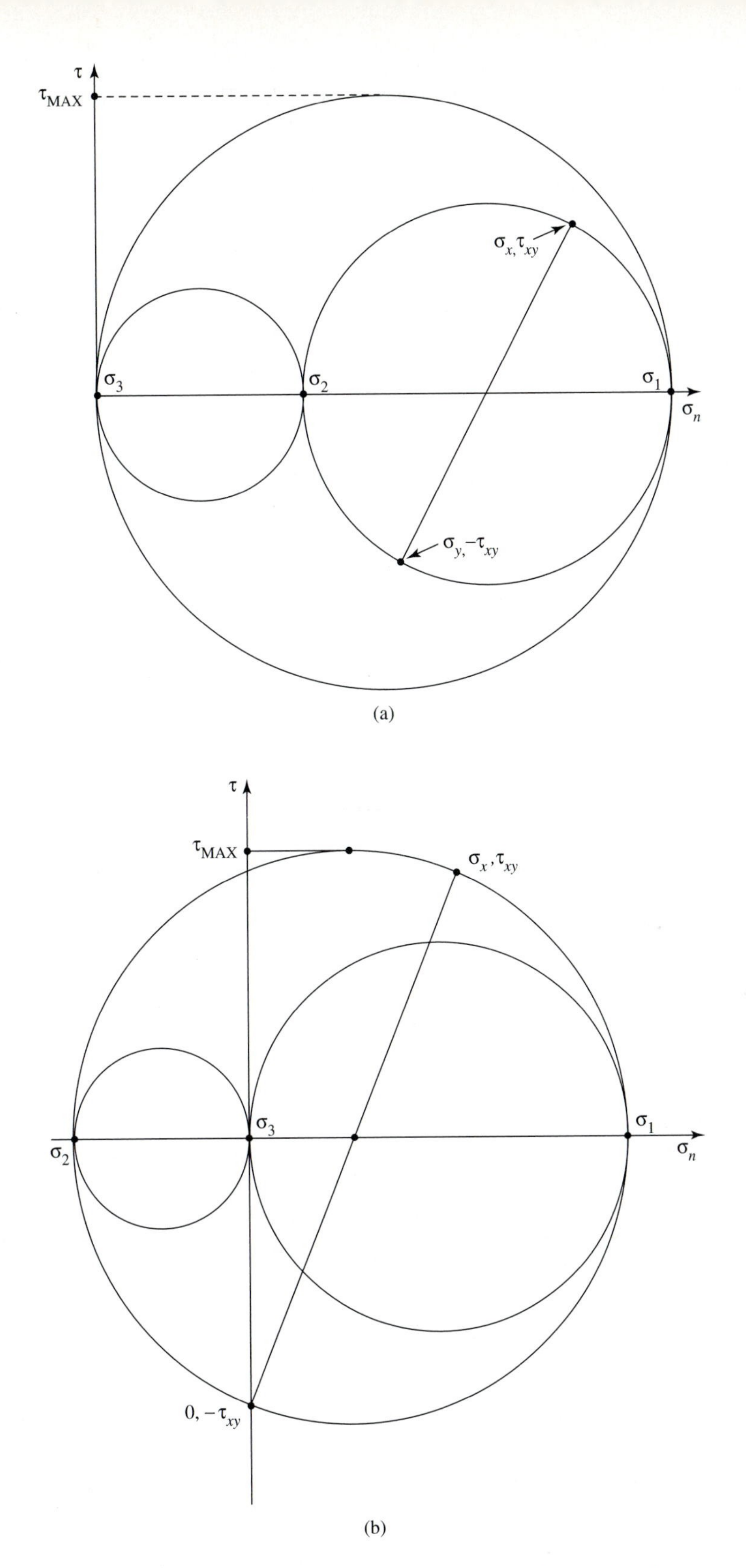

Figure 3.4 (a) Mohr circles for plane stress; (b) special case of plane stress (σ_x and τ_{xy} are the only nonzero stress components)

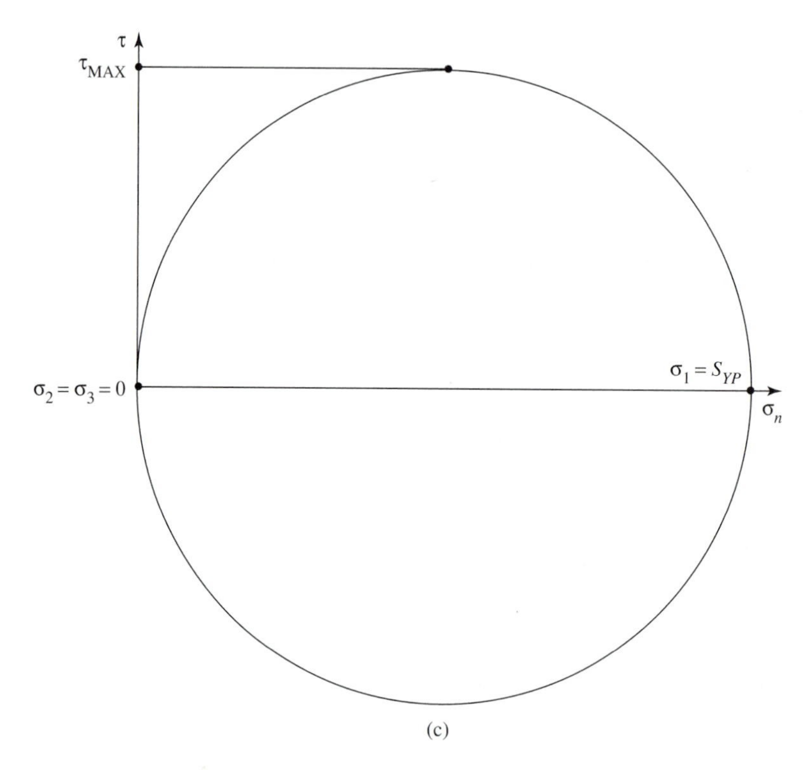

Figure 3.4 *continued* (c) Mohr circle for a tension test.

on the location of the principal planes and the plane of maximum shear stress, and which provide added insight into combined stress. See, for example, Timoshenko (1955), Deutschman, Michels, and Wilson (1975), or Logan (1991).

THE MAXIMUM SHEAR STRESS THEORY

Consider a test specimen of cross-sectional area A subject to a tension load P in the x-direction. Let the load be increased until the specimen begins to yield. In the absence of other loading,

$$\sigma_1 = \sigma_x = P/A = S_{YP}$$

where S_{YP} = the reported yield point for the specimen

A Mohr circle for the tension test is shown in Figure 3.4(c). In the tension test, maximum shear stress is given by

$$\tau_{MAX} = \sigma_x/2 = S_{YP}/2$$

Metallurgical examination has shown that some materials fail due to shear stress during a tension test. Rearranging the above equation and introducing a safety factor N_{FS}, we have the basis for the maximum shear stress theory.

$$\frac{S_{YP}}{2\,N_{FS}} \geq \tau_{MAX} \tag{3.22}$$

Combining this relationship with Equation (3.20), we have the **maximum shear stress theory for plane stress.**

$$S_{YP}/N_{FS} \geq \text{the largest of} \begin{cases} |\sigma_1 - \sigma_2| \\ |\sigma_1| \\ |\sigma_2| \end{cases} \tag{3.23}$$

THE VON MISES THEORY

The **von Mises theory** is based on the assumption that failure occurs when the distortion energy at some point in a member equals the distortion energy at failure (of the same material) in a tension test. It is assumed that the energy due to uniform tension or compression in all directions does not contribute to failure. The von Mises theory is also called the **distortion energy theory.**

For a member subject to plane stress due to static loads, the von Mises theory leads to the following design criterion:

$$S_{YP}/N_{FS} \geq \sigma_{VM} = (\sigma_1^2 + \sigma_2^2 - \sigma_1\sigma_2)^{1/2} \tag{3.24}$$

where σ_{VM} is called the **von Mises stress.**

Another failure theory based on octahedral shear stress results in the same design criterion.

EXAMPLE PROBLEM Design Based on the Maximum Normal Stress Theory, the Maximum Shear Stress Theory, and the von Mises Theory

The part sketched in Figure 3.5 must support a static load P of 900 lb. The dimensions are $a = 4$ in and $b = 5$ in. What is the minimum safe diameter for the circular section?

Design Decisions: Steel with a yield point of 68,000 psi will be selected for the part. Stress concentration will be neglected, but a safety factor of 6 will be used because of uncertainty concerning the loading. The required diameters and weights obtained by different failure theories will be compared.

Solution Summary: Shear stress due to torsion will be maximum anywhere on the surface of the circular section. Tension (or compression) due to bending moment will be greatest at the top (or bottom) of the circular section at the fixed end. We will base the design on combined stress at point A, arbitrarily locating the xyz-coordinate system as shown in Figure 3.5.

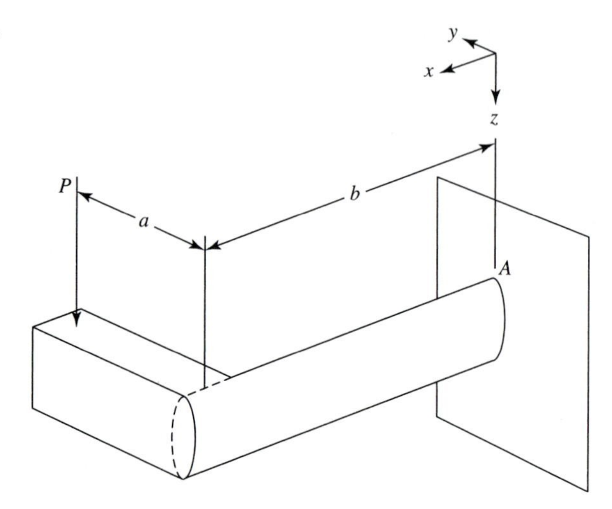

Figure 3.5 Design for a static load.

At point A, bending moment M and torque T, given by

$$M = bP \qquad \text{and} \qquad T = aP,$$

result in stresses

$$\sigma_x = Mc/I = 32M/(\pi D^3)$$

and

$$\tau_{xy} = Tr/J = 16T/(\pi D^3)$$

We cannot evaluate these equations numerically since diameter D is unknown. To get around this difficulty, let us multiply both sides of the above equations by D^3 to obtain

$$\sigma_{xn} \equiv \sigma_x D^3 = 32M/\pi$$

and

$$\tau_{xyn} \equiv \tau_{xy} D^3 = 16T/\pi$$

The subscript n indicates that stresses are multiplied by D^3.

We used mathematical software to calculate principal stresses, maximum shear stress, and von Mises stress; the calculations are given in the detailed solution which follows. Again, these stress terms containing the unknown D^3 factor are indicated with a subscript n.

The maximum normal stress theory requires that

$$S_{\text{YP}}/N_{\text{FS}} \geq \sigma_{\text{MAX}}$$

or, in this case,

$$S_{\text{YP}}/N_{\text{FS}} \geq \sigma_{1n}/D^3$$

We can rearrange this equation to find minimum safe diameter based on the maximum normal stress theory.

$$D_{MIN} = (\sigma_{1n} N_{FS} / S_{YP})^{1/3} = 1.6645 \text{ in}$$

Minimum safe diameters found on the basis of the maximum shear stress theory and von Mises theory are 1.7302 in and 1.7008 in, respectively. The resulting weights of the circular section range from about 0.6158 to 0.6654 lb/in of length.

The detailed solution follows.

Detailed Solution: The following values are given in inches and pounds:

$$P = 900 \qquad a = 4 \qquad b = 5$$
$$S_{YP} = 68{,}000 \qquad N_{FS} = 6$$
$$T = a \cdot P \qquad T = 3600$$
$$M = b \cdot P \qquad M = 4500$$

Subscript n: Divide these values by diameter cubed to obtain stress.

$$\sigma_{xn} = 32 \cdot \frac{M}{\pi} \qquad \sigma_{xn} = 45{,}837$$

$$\tau_{xyn} = 16 \cdot \frac{T}{\pi} \qquad \tau_{xyn} = 18{,}335$$

$$\sigma_y = 0 \qquad \sigma_z = 0$$
$$\tau_{yz} = 0 \qquad \tau_{xz} = 0$$

Principal Stresses:

$$\sigma_3 = 0$$

$$\sigma_{1n} = \frac{\sigma_{xn}}{2} + \sqrt{\left(\frac{\sigma_{xn}}{2}\right)^2 + \tau_{xyn}{}^2} \qquad \sigma_{1n} = 52{,}268$$

$$\sigma_{2n} = \frac{\sigma_{xn}}{2} - \sqrt{\left(\frac{\sigma_{xn}}{2}\right)^2 + \tau_{xyn}{}^2} \qquad \sigma_{2n} = -6431$$

Maximum Normal Stress Theory:

$$D_{MIN} = \left(\sigma_{1n} \cdot \frac{N_{FS}}{S_{YP}}\right)^{1/3} \qquad D_{MIN} = 1.6645$$

$$I = \pi \cdot \frac{D_{MIN}{}^4}{64} \qquad I = 0.376827$$

Weight of Circular Section (lb/in):

$$w = 0.283 \cdot \pi \cdot \frac{D_{MIN}{}^2}{4} \qquad w = 0.6158$$

Maximum Shear Stress Theory:

$$\tau_n = \begin{bmatrix} \dfrac{|\sigma_{1n} - \sigma_{2n}|}{2} \\ \dfrac{|\sigma_{2n}|}{2} \\ \dfrac{|\sigma_{1n}|}{2} \end{bmatrix}$$

$$\tau_{MAXn} = \max(\tau_n) \qquad \tau_{MAXn} = 29{,}350 \qquad 2 \cdot \tau_{MAXn} = 58{,}700$$

$$D_{MIN} = \left(2 \cdot \tau_{MAXn} \cdot \frac{N_{FS}}{S_{YP}}\right)^{1/3} \qquad D_{MIN} = 1.7302$$

$$w = 0.283 \cdot \pi \cdot \frac{D_{MIN}^2}{4} \qquad w = 0.6654$$

von Mises Theory:

$$\sigma_{VMn} = (\sigma_{1n}^2 + \sigma_{2n}^2 - \sigma_{1n} \cdot \sigma_{2n})^{1/2} \qquad \sigma_{VMn} = 55{,}763$$

$$D_{MIN} = \left(\sigma_{VMn} \cdot \frac{N_{FS}}{S_{YP}}\right)^{1/3} \qquad D_{MIN} = 1.7008$$

$$w = 0.283 \cdot \pi \cdot \frac{D_{MIN}^2}{4} \qquad w = 0.643$$

■■

The careful reader will note that transverse shear stress was not mentioned in the preceding example. Transverse shear stress is zero at point *A* and reaches a maximum at the neutral axis (a horizontal diameter of the circular cross section). However, tensile stress due to bending is zero at the neutral axis. As with most cases combining bending and torque, the design is governed by the combined stress at the point where tension stress due to bending is maximum (point *A* in the preceding example).

WHICH FAILURE THEORY IS THE BEST?

If the above question had a simple answer, then only the best failure theory would have been discussed in this text. Designers using cast iron and other brittle materials usually favor the maximum normal stress theory, based on ultimate strengths in tension and compression. For ductile steel and other ductile materials, the von Mises theory is a popular procedure. The maximum shear stress theory has the advantage of being more conservative; that is, it often results in a safer design. However, note the caution that applies when principal stresses have the same sign.

For unidirectional stress problems, that is, those involving simple bending and tension loads, the maximum normal stress theory, the maximum shear stress theory, and the von Mises theory will give the same result. In many problems involving combined stress, the results of these three theories vary only slightly. When varying loads are present or when rotating members are considered, a fatigue failure theory must be used.

PRINCIPAL STRESSES: THE THREE-DIMENSIONAL CASE[1]

The principal stresses at a point are given by the solution to the following equation, given in determinant form:

$$\begin{vmatrix} \sigma_x - \sigma & \tau_{xy} & \tau_{xz} \\ \tau_{xy} & \sigma_y - \sigma & \tau_{yz} \\ \tau_{xz} & \tau_{yz} & \sigma_z - \sigma \end{vmatrix} = 0 \qquad (3.25)$$

The determinant can be evaluated by hand calculations or by mathematics software with symbolic capability. After collecting terms by powers of σ, we have

$$\begin{aligned} &-\sigma^3 + (\sigma_x + \sigma_z + \sigma_y) \cdot \sigma^2 + (\tau_{xy}^2 - \sigma_x \cdot \sigma_y - \sigma_x \cdot \sigma_z - \sigma_y \cdot \sigma_z + \tau_{xz}^2 + \tau_{yz}^2) \\ &\cdot \sigma + \sigma_x \cdot \sigma_y \cdot \sigma_z - \tau_{xz}^2 \cdot \sigma_y - \tau_{xy}^2 \cdot \sigma_z - \sigma_x \cdot \tau_{yz}^2 + 2 \cdot \tau_{xy} \cdot \tau_{xz} \cdot \tau_{yz} \\ &= 0 \end{aligned} \qquad (3.26)$$

A symbolic solution to cubic Equation (3.26) is long and complicated. It is usually more convenient to use mathematics software to evaluate the determinant numerically. The following procedure can be used:

1. Determine the normal and shear stresses on the x-, y-, and z-planes.
2. Arrange the stresses in an array as in Equation (3.25).
3. Using mathematics software, find the three roots of the determinant of the array. A numerical solution can begin with a rough estimate of each root.
4. Check to see that the determinant is actually zero when each of the roots is substituted into it. Allow a tolerance for rounding off of values.
5. Identify the three roots as principal stresses σ_1, σ_2, and σ_3.

EXAMPLE PROBLEM Principal Stresses—Three-Dimensional Case

The following normal and shear stresses (psi) exist at a given point in a body:

$$\sigma_x = 1000 \qquad \sigma_y = 13{,}000 \qquad \sigma_z = 5500$$
$$\tau_{xy} = 11{,}500 \qquad \tau_{yz} = 7000 \qquad \tau_{xz} = 7340$$

Find the three principal stresses.

Summary Solution: We will solve the stress determinant numerically, using mathematics software. We must give the program starting values; that is, we must make three estimates of the principal stresses. We will make a high guess, a low guess, and an inbetween guess. If the same answer comes up twice, we will try another guess. The principal stresses will have the same units as the input values: lb/in^2 or MPa. In certain special cases, two, or even all three, principal stresses are identical, but that does not appear likely in this problem. The detailed solution follows.

Detailed Solution: The stresses in an array are

$$M(\sigma) = \begin{bmatrix} \sigma_x - \sigma & \tau_{xy} & \tau_{xz} \\ \tau_{xy} & \sigma_y - \sigma & \tau_{yz} \\ \tau_{xz} & \tau_{yz} & \sigma_z - \sigma \end{bmatrix}$$

[1]Three-dimensional design problems are less common and more difficult than plane stress design problems. The sections involving the three-dimensional case can be omitted without loss of continuity.

The first estimate of principal stress is

$$\sigma = 35{,}000$$

The stress determinant is

$$D(\sigma) = |M(\sigma)|$$

One principal stress found by setting the determinant equal to zero is

$$\sigma_{\text{prin}} = \text{root}(D(\sigma),\sigma)$$

First Root

$$\sigma_{\text{prin}} = 24{,}956$$
$$\sigma_1 = \sigma_{\text{prin}}$$

Check: Does the solution satisfy the condition that the determinant equal zero?

$$D(\sigma_{\text{prin}}) = 0$$

The second estimate is

$$\sigma = 0$$
$$\sigma_{\text{prin}} = \text{root}(\text{D}(\sigma),\sigma)$$

Second Root

$$\sigma_{\text{prin}} = 1321$$
$$\sigma_2 = \sigma_{\text{prin}}$$

Check:

$$D(\sigma_{\text{prin}}) = 0$$

The third estimate is

$$\sigma = -14{,}000$$
$$\sigma_{\text{prin}} = \text{root}(D(\sigma),\sigma)$$

Third Root

$$\sigma_{\text{prin}} = -6778$$
$$\sigma_3 = \sigma_{\text{prin}}$$

Check:

$$D(\sigma_{\text{prin}}) = 0$$

FAILURE THEORIES FOR THREE-DIMENSIONAL STATIC STRESS FIELDS

The **maximum normal stress theory** has the same form in two- and three-dimensional stress fields. After determining the three principal stresses, we compare the greatest principal stress with the working strength.

$$S_{\text{YP}}/N_{\text{FS}} \geq \sigma_{\text{MAX}} \tag{3.27}$$

Or we use an equivalent expression involving ultimate tensile or compressive strength.

Maximum shear stress in a three-dimensional stress field is given by

$$\tau_{\text{MAX}} = \text{the largest of} \begin{cases} |\sigma_1 - \sigma_2|/2 \\ |\sigma_1 - \sigma_3|/2 \\ |\sigma_2 - \sigma_3|/2 \end{cases} \quad \textbf{(3.28)}$$

and the **maximum shear stress theory** takes the form

$$S_{\text{YP}}/N_{\text{FS}} \geq \text{the largest of} \begin{cases} |\sigma_1 - \sigma_2| \\ |\sigma_1 - \sigma_2| \\ |\sigma_2 - \sigma_3| \end{cases} \quad \textbf{(3.29)}$$

The **von Mises theory** in three dimensions requires that

$$S_{\text{YP}}/N_{\text{FS}} \geq (\sigma_1^2 + \sigma_2^2 + \sigma_3^2 - \sigma_1\sigma_2 - \sigma_2\sigma_3 - \sigma_3\sigma_1)^{1/2} \quad \textbf{(3.30)}$$

■■ EXAMPLE PROBLEM Design Criteria for a Three-Dimensional Stress Field

Consider the stresses at a point as given in the preceding example problem. Determine required working strength.

Solution: The principal stresses found previously are substituted in Equations (3.22), (3.28), (3.29), and (3.30). The largest principal stress is $\sigma_1 = 24{,}956$. Thus, the maximum normal stress theory requires that

$$S_{\text{YP}}/N_{\text{FS}} \geq 24{,}956 \text{ psi}$$

Maximum shear stress is $\tau_{\text{MAX}} = 15{,}867$. Thus, the maximum shear stress theory requires that

$$S_{\text{YP}}/N_{\text{FS}} \geq 31{,}734 \text{ psi}$$

The von Mises stress is 28,559, resulting in the following criterion:

$$S_{\text{YP}}/N_{\text{FS}} \geq 28{,}559 \text{ psi}$$

■■

References and Bibliography

Deutschman, A. D., W. J. Michels, and C. E. Wilson. *Machine Design—Theory and Practice.* New York: Macmillan, 1975.

Logan, D. L. *Mechanics of Materials.* New York: HarperCollins, 1991.

Timoshenko, S. *Strength of Materials.* Part I. Princeton, NJ: Van Nostrand, 1955.

Design Problems

3.1. A machine part in the form of a cantilever beam is fixed at the left end ($x = 0$). The part must support a 150 lb static load located 6 in from the fixed end.

Design Decisions: The part will be made of 1/2 in thick aluminum alloy with a yield strength of 40,000 psi. A safety factor of 1.5 will be used.

(a) Find the required width of the part if width is to be constant.

(b) Design the part with variable width for minimum weight.

3.2. A machine part in the form of a cantilever beam is fixed at the left end ($x = 0$). The part must support a 75 lb static load located 4 in from the fixed end.

Design Decisions: The part will be made of 0.375 in thick steel with a yield strength of 60,000 psi. A safety factor of 2.5 will be used.

(a) Find the required width of the part if width is to be constant.
(b) Design the part with variable width for minimum weight.

3.3. A machine part in the form of a cantilever beam is fixed at the left end ($x = 0$). The part must support a 95 lb static load located 4.5 in from the fixed end.

Design Decisions: The part will be made of 1/4 in thick steel with a yield strength of 58,000 psi. A safety factor of 2.25 will be used.

(a) Find the required width of the part if width is to be constant.
(b) Design the part with variable width for minimum weight.

3.4. A shaft must resist 1200 in · lb pure torsion (i.e., no bending). Find the required shaft diameter.

Design Decisions: Allow a working strength of 20,000 psi in shear. Use a solid shaft.

3.5. A hollow shaft must resist 1200 in · lb pure torsion (i.e., no bending). Find the required outer diameter for the shaft.

Design Decisions: Allow a working strength of 20,000 psi in shear. Let the inner diameter be 60% of the outer diameter.

3.6. A hollow shaft must resist 1200 in · lb pure torsion (i.e., no bending). Find the required outer diameter for the shaft.

Design Decisions: Allow a working strength of 20,000 psi in shear. Let the inner diameter be 0.5 in.

3.7. The following stresses exist at a point in a body:

$$\sigma_x = 53{,}000 \qquad \sigma_y = -4000 \qquad \tau_{xy} = 17{,}500$$

where all values are in psi.

Other stress components are zero. Find the following:

(a) principal stresses
(b) maximum shear stress
(c) von Mises stress

3.8. The following stresses exist at a point in a body:

$$\sigma_x = 55 \qquad \sigma_y = -25 \qquad \tau_{xy} = 47$$

where all values are in MPa.

Other stress components are zero. Find the following:

(a) principal stresses
(b) maximum shear stress
(c) von Mises stress

3.9. The part sketched in Figure 3.5 must support a static load P of 400 lb. The dimensions are $a = 7$ in and $b = 3$ in. What is the minimum safe diameter for the circular section?

Design Decisions: Steel with a yield point of 60,000 psi will be selected for the part. A safety factor of 3 will be used. Find the minimum required diameter and weight based on the following:

(a) maximum normal stress theory
(b) maximum shear stress theory
(c) von Mises theory

3.10. The part sketched in Figure 3.5 must support a static load P of 400 lb. The dimensions are $a =$ 3.5 in and $b = 4.5$ in. What is the minimum safe diameter for the circular section?

Design Decisions: Steel with a yield point of 70,000 psi will be selected for the part. A safety factor of 3 will be used. Find the minimum required diameter and weight based on the following:

(a) maximum normal stress theory
(b) maximum shear stress theory
(c) von Mises theory

3.11. The part sketched in Figure 3.5 must support a static load P of 2000 N. The dimensions are $a =$ 150 mm and $b = 175$ mm. What is the minimum safe diameter for the circular section?

Design Decisions: Material with a yield point of 400 MPa will be selected for the part. A safety factor of 2.5 will be used. Base your results on the following:

(a) maximum normal stress theory

(b) maximum shear stress theory
(c) von Mises theory

3.12. The part sketched in Figure 3.5 must support a static load P of 3500 N. The dimensions are a = 180 mm and b = 155 mm. What is the minimum safe diameter for the circular section?

Design Decisions: Material with a yield point of 470 MPa will be selected for the part. A safety factor of 2.0 will be used. Base your results on the following:

(a) maximum normal stress theory
(b) maximum shear stress theory
(c) von Mises theory

***3.13.** The following stresses exist at a point:

$\sigma_x = 300 \quad \sigma_y = 200 \quad \sigma_z = 150$
$\tau_{xy} = 250 \quad \tau_{yz} = 100 \quad \tau_{wxz} = 50$

where all values are in MPa.

(a) Find the three principal stresses.
(b) Evaluate the stress determinant for each solution. The determinant should equal zero (±tolerance for rounding).
(c) Find maximum shear stress.
(d) Find von Mises stress.

***3.14.** The following stresses exist at a point:

$\sigma_x = 220 \quad \sigma_y = 135 \quad \sigma_z = 55$
$\tau_{xy} = 115 \quad \tau_{wyz} = 70 \quad \tau_{wxz} = 40$

where all values are in MPa.

(a) Find the three principal stresses.
(b) Evaluate the stress determinant for each solution. The determinant should equal zero (±tolerance for rounding).
(c) Find maximum shear stress.

***3.15.** The following stresses exist at a point:

$\sigma_x = 4800 \quad \sigma_y = -13{,}000 \quad \sigma_z = 5500$
$\tau_{xy} = 11{,}500 \quad \tau_{yz} = 7000 \quad \tau_{xz} = 7340$

where all values are in psi.

(a) Find the three principal stresses.
(b) Evaluate the stress determinant for each solution. The determinant should equal zero (±tolerance for rounding).
(c) Find maximum shear stress.
(d) Find von Mises stress.

***3.16.** The following stresses exist at a point:

$\sigma_x = 13{,}000 \quad \sigma_y = -9000 \quad \sigma_z = 8000$
$\tau_{xy} = 17{,}500 \quad \tau_{yz} = 0 \quad \tau_{xz} = 0$

where all values are in psi.

(a) Find the three principal stresses.
(b) Evaluate the stress determinant for each solution. The determinant should equal zero (±tolerance for rounding).
(c) Find maximum shear stress.
(d) Find von Mises stress.

*Problems involving three-dimensional stress are marked with an asterisk. Solutions tend to be difficult and time-consuming.

CHAPTER 4

Dynamic Loading of Machine Members (Design for Endurance)

Symbols

A	cross-sectional area
C_F	surface-finish correction factor
C_R	reliability factor
C_S	size factor
d, D	diameter
e	base of natural logarithms
$F(X)$	reliability
K_F	fatigue stress concentration factor
K_T	theoretical stress concentration factor
ln	natural logarithm
M	bending moment
N	number of cycles to failure
N_{FS}	factor of safety
p	pressure
P	load
$P(E)$	probability of an event
q	notch sensitivity index
Q	samples-per-failure ratio
r	radius
R	reliability
S_E	endurance limit (corrected for surface finish, etc.)
S_N	fatigue strength for N cycles of reversed stress
S'_N	endurance limit of test specimen
S_U	ultimate strength
S_{YP}	yield point
SD	standard deviation
t	thickness; time
X	number of standard deviations from the mean
σ_M	mean stress
σ_R	range stress

Units

area	in^2 or mm^2
force	lb or N
length, diameter	in or mm
moment	in·lb or N·mm
stress, strength, pressure	psi or MPa

FAILURE THEORIES FOR FATIGUE LOADING

Fracture resulting from repeated loading is called **fatigue failure.** As noted in Chapter 2, fatigue is a frequent cause of catastrophic failure, possibly because fatigue failure is poorly understood and difficult to predict.

REVERSED STRESS

Reversed stress refers to a stress–time history where stress at a point varies repeatedly from tension to compression, with zero mean. Figure 4.1 shows rotating-beam fatigue-testing machines which produce reversed stress. Part (a) of the figure shows a machine for room-temperature tests. The machine in Part (b) is designed to perform high-temperature tests. Specimen details and the testing procedure are similar to the rotating-beam test shown in Chapter 2.

If a specimen fails at N cycles of reversed stress, the stress amplitude is designated as S_N, the **fatigue strength,** with the number of cycles indicated. A plot of stress versus number of stress cycles (S–N plot) shows the results of a series of fatigue tests (see Chapter 2). This test series is expensive because it requires many carefully prepared specimens and substantial testing time. The S–N plot for a typical steel specimen has a horizontal region (i.e., it "levels off") after a high number of reversed stress cycles. This failure stress is called the **endurance limit for the test specimen,** S'_N. Many engineering materials do not level off; they apparently do not have an endurance limit.

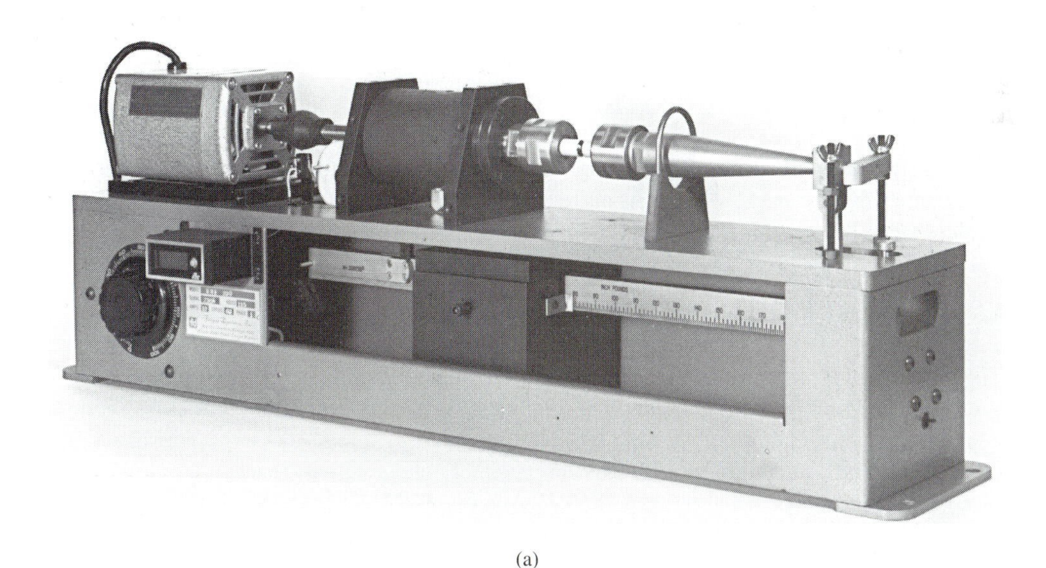

(a)

Figure 4.1 Rotating-beam fatigue-testing machines. (a) Room-temperature test model; (Courtesy of Fatigue Dynamics, Inc.)

(b)

Figure 4.1 *continued* (b) high-temperature test model. (Courtesy of Fatigue Dynamics, Inc.)

APPROXIMATING FATIGUE STRENGTH AND ENDURANCE LIMIT

Actual test values are preferable to approximations of fatigue strength and endurance limit based on ultimate strength. However, in the absence of test data, the approximations given in Table 4.1 can be used for preliminary design calculations.

ENDURANCE LIMIT: CORRECTIONS FOR ACTUAL DESIGN CONDITIONS

The capacity of an actual machine part to endure reversed loading depends on material characteristics and on heat treatment, surface roughness, stress concentration, part cross section, and temperature. In addition, we can consider welds in high-stress regions and desired reliability of the part. We could test each part and assembly under actual service conditions to ensure a safe design. The usual practice, however, is to refer to results of the rotating-beam fatigue test and make appropriate corrections. The ways in which surface finish and other variables influence fatigue strength are discussed in several references, including Deutschman, Michels, and Wilson (1975); DeGarmo, Black, and Kohser (1988); Lipson and Juvinall (1963); and Logan (1991).

Table 4.1 Approximations for preliminary design calculations.

Material	Ultimate Strength S_U	Approximation of S_N or S'_N	Number of Cycles N
Wrought steel	≤200,000 psi (≤1380 MPa)	$S'_N = 0.5S_U$	—
Wrought steel	>200,000 psi (>1380 MPa)	$S'_N = 100{,}000$ psi $S'_N = 690$ MPa	— —
Cast steel	≤88,000 psi (≤607 MPa)	$0.4S_U \leq S'_N \leq 0.45S_U$	—
Cast steel	>88,000 psi (>607 MPa)	$S'_N = 40{,}000$ psi $S'_N = 275$ MPa	—
Cast iron	—	$0.4S_U \leq S'_N \leq 0.45S_U$	—
Magnesium alloys	—	$0.35S_U \leq S_N \leq 0.38S_U$	10^6 to 10^8
Wrought aluminum alloys	≤40,000 psi (≤275 MPa)	$0.38S_U \leq S_N \leq 0.4S_U$	$5 \cdot 10^8$
Cast aluminum alloys	≤50,000 psi (≤345 MPa)	$0.16S_U \leq S_N \leq 0.3S_U$	$5 \cdot 10^8$
Titanium alloys	—	$0.45S_U \leq S'_N \leq 0.65S_U$	—
P/M* carbon and alloy steels	—	$S'_N = 0.38S_U$	—
Other P/M* materials	—	$S_N = 0.38S_U$	10^7
Some engineering plastics†	—	$0.4S_U \leq S_N \leq 0.5S_U$	†

*Powder metallurgy.

†Plastics vary widely in properties. Many are unsuitable for fatigue loading. Parts and assemblies should be tested at maximum end-use load and temperature.

We will examine the correction factors needed to adjust test specimen endurance limit in the paragraphs that follow. We will then be able to calculate corrected endurance limit for practical applications.

Surface Roughness

Surface finish has a major influence on part life under fatigue loading. The effect on machine members can be accounted for by a **surface-finish correction factor,** C_F. Some typical values for ferrous metals and alloys are as follows:

Surface Finish	*Surface-Finish Correction Factor* C_F
Polished test specimen	1.0
Ground	0.84–0.90
Machined	0.64–0.80
Hot-rolled	0.28–0.73
As forged	0.18–0.57

The lowest values in the range for the as-forged condition apply to high-tensile-strength steel. Steels with low tensile strength are less sensitive to surface roughness.

Size Effects

The standard fatigue test specimen diameter at the failure section is 0.30 in, as shown in Chapter 2. Larger diameter parts tend to have lower fatigue strengths. The following expression was obtained from regression analysis on some fatigue data:

$$C_S = 1 - 0.1265 \ln(D/0.3) = 0.8477 - 0.1265 \ln D \quad \textbf{(4.1)}$$

where C_S = size factor

D = diameter (in) for $D \geq 0.3$ in

ln = natural logarithm

Regression techniques are discussed later in this chapter.

Converting the size factor equation to metric units, we obtain

$$C_S = 1 - 0.1265 \ln(D/7.62) = 1.2569 - 0.1265 \ln D \quad \textbf{(4.2)}$$

where D = diameter (mm) for $D \geq 7.6$ mm

For $D \leq 0.3$ in ($D \leq 7.6$ mm), a size factor $C_S = 1$ is suggested. For a noncircular section subject to reversed bending, the term D can be used to refer to some equivalent dimension at a critical cross section.

STRESS CONCENTRATION

Consider a plate with a tensile load (see Figure 4.2). If a hole is drilled in the plate, the tensile stress at points A and B will be substantially higher than the average stress at the net cross section. This condition is called **stress concentration.** The **stress concentration factor** is defined by

$$K_F \text{ or } K_T = \sigma_{\text{MAX}}/\sigma_{\text{NOM}} \quad \textbf{(4.3)}$$

where K_F = fatigue stress concentration factor

K_T = theoretical stress concentration factor

σ_{MAX} = maximum stress, which occurs at points A and B (psi or MPa)

$\sigma_{\text{NOM}} = P/A_{\text{NET}} = \sigma_o(a - r)/a$ = nominal stress, the average stress at the net cross section (psi or MPa)

$A_{\text{NET}} = 2(a - r)t$ = net cross-sectional area (in^2 or mm^2)

$P = 2at\sigma_o$ = total tensile force on each end of the plate (lb or N)

σ_o = tensile stress at each end (psi or MPa)

When σ_{MAX} is based on analysis using the theory of elasticity or finite element analysis, the result is identified as K_T, the **theoretical stress concentration factor.** For example, if the hole shown in Figure 4.2 is very small compared with the other plate dimensions, then based on the theory of elasticity,

$$K_T = 3 \quad \text{for } r << a \quad \text{and} \quad r << b$$

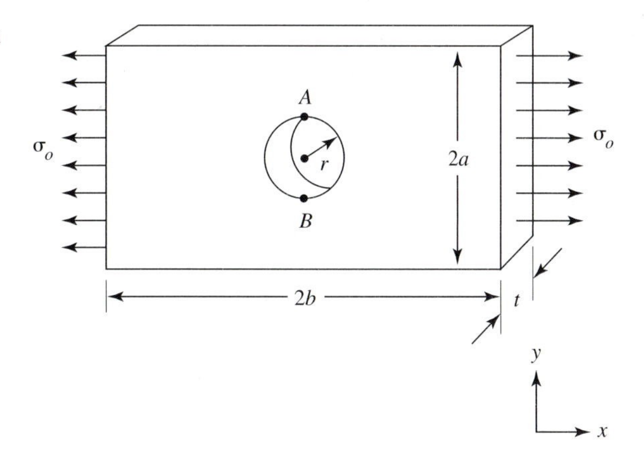

Figure 4.2 Stress concentration, plate in tension with central hole.

When the stress concentration factor is based on the results of fatigue tests, it is designated K_F, the **fatigue stress concentration factor.** For a typical sled runner keyway in a quenched and drawn shaft (Figure 4.3),

$$K_F = 1.6$$

based on the results of reversed bending tests.

The fatigue stress concentration factor K_F can be estimated if the notch sensitivity of the material is known. In the absence of this information, stress calculations can be based on K_T, the theoretical stress concentration factor.

A Note on Notch Sensitivity and Stress Concentration

Consider a machine part with a notch (a stress raiser such as a fillet, a hole, or some other change in cross section). We plan to subject the part to reversed stress (fatigue loading). As noted above, theoretical stress concentration factor K_T is based on the theory of elasticity or finite element analysis results. Values of K_T depend on the shape of a part made of an idealized

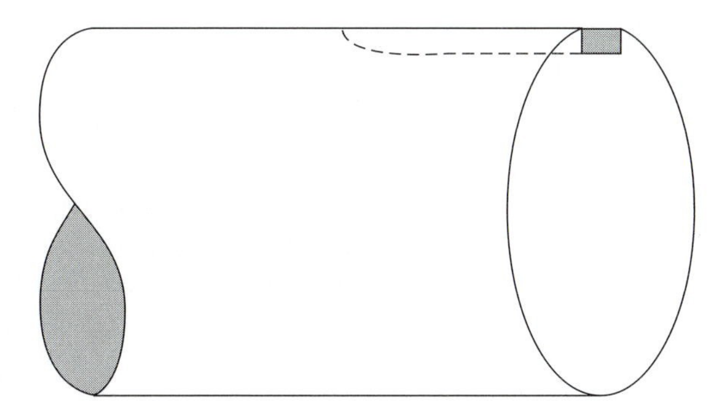

Figure 4.3 Sled runner keyway.

material which is assumed to be perfectly elastic, homogeneous, and isotropic. We can rearrange the definition of K_T in this form:

$$K_T = \frac{\text{strength of unnotched part}}{\text{theoretical strength of notched part}}$$

It is assumed that the cross-sectional area of the unnotched part is reduced to the net cross-sectional area of the notched part. Theoretical stress concentration factor sometimes overestimates the effect of stress raisers; that is, it leads us to underestimate part strength.

Fatigue stress concentration factor K_F is based on the results of tests using real materials. It depends on material, geometry, and type of loading. Fatigue stress concentration factor can be defined as follows:

$$K_F = \frac{\text{strength of unnotched part}}{\text{actual strength of notched part}}$$

Strength refers to failure load after many cycles of reversed stress. Again, the cross-sectional area of the unnotched part should equal the net cross-sectional area of the notched part.

The stress concentration factors are related by the **notch sensitivity index** as follows:

$$K_F = 1 + q(K_T - 1)$$

where q = notch sensitivity index, which has a range $0 \le q \le 1$.

Values of q for several materials and a range of notch radii are given by Juvinall (1967).

Collins (1981) notes that q approaches unity for some finer-grained materials, such as quenched and tempered steels, and for some coarser-grained materials, such as annealed and normalized aluminum alloys. Notch sensitivity factor also approaches unity for large notch radii, say, $r > 1/4$ in or $r > 7$ mm. In those cases, $K_F \approx K_T$; there is no need to distinguish between fatigue and theoretical stress concentration factor.

Cast iron, however, has a lower notch sensitivity. Test results show a typical value: $q \approx 0.2$. Holes, fillets, and other stress raisers have a smaller effect than they have in high-strength steel, possibly because of the composition of cast iron which includes precipitated carbon. These irregularities reduce the strength of an unnotched specimen. As a result, the additional strength loss due to stress raisers is not as great as predicted by the theoretical stress concentration factor.

Comparing the Effects of Static and Fatigue Loading

Let the plate in Figure 4.2 be made of a ductile material. If the load is increased slowly, plastic deformation (local yielding) will occur at points A and B when the stresses reach the yield point. Thus, the stresses will be redistributed, possibly averting failure until the average stress at the net cross section reaches the yield point. For this reason, we often neglect stress concentration when considering static loads. If the same plate is subject to fatigue loading, stress concentration must be considered. Test results show that fatigue failures occur at reversed stress amplitudes which are less than the yield point.

THEORETICAL STRESS CONCENTRATION FACTORS FOR DESIGN

Stress concentration factors can be determined analytically in some cases. This rather difficult process is described in standard references on the theory of elasticity, including those by Timoshenko and Goodier (1951), Wang (1953), and Ugural and Fenster (1987). Books with extensive graphs showing stress concentration factors include Peterson (1974) and Deutschman et al. (1975). Stress concentration factors for some cases of particular importance to machine design are given in this section. In most cases, they are given in equation form for use in computer-assisted design.

Plate in Tension with a Central Hole

This situation, shown in Figure 4.2, is a classical elasticity problem. The theoretical stress concentration factor is given by

$$K_T = 1.923 - 0.475 \cdot \ln(r/a + 0.1) \quad \text{for } 0 < r/a \leq 0.6 \quad \text{and} \quad r << b \qquad \textbf{(4.4)}$$

where $K_T = \sigma_{\text{MAX}}/\sigma_{\text{NOM}}$ = stress concentration factor based on regression analysis of theoretical results

r = hole radius

a = half-width of plate

b = half-length of plate

σ_{MAX} = maximum tensile stress, which occurs at points A and B

$\sigma_{\text{NOM}} = P/A_{\text{NET}} = \sigma_o(a - r)/a$ = nominal stress, the average stress at the net cross section

$A_{\text{NET}} = 2(a - r)t$ = net cross-sectional area

$P = 2at\sigma_o$ = total tensile force on each end of the plate

σ_o = tensile stress at each end

Plate in Bending with a Central Hole

For a small hole in a thick plate, as shown in Figure 4.4, the results are about the same as with the direct tension problem.

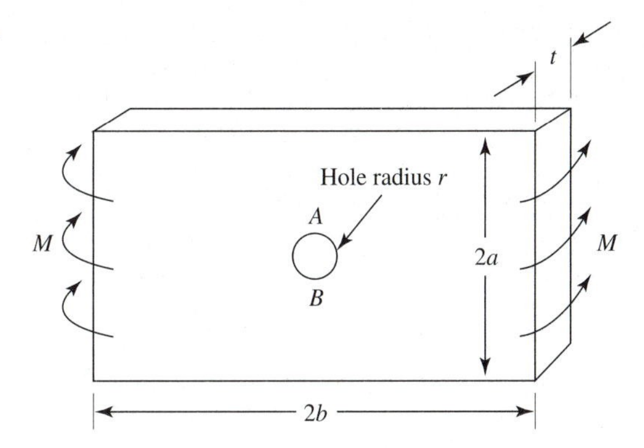

Figure 4.4 Stress concentration in bending.

$$K_T = 1.923 - 0.475 \cdot \ln(r/a + 0.1) \quad \text{for } 0 < r/a \leq 0.6 \quad r << b \quad \text{and} \quad r << t \tag{4.5}$$

where $K_T = \sigma_{\text{MAX}}/\sigma_{\text{NOM}}$

$= $ approximate stress concentration factor for $r/a < 0.1$

K_T decreases with increasing values of r/a.

$\sigma_{\text{NOM}} = Mc/I = 3M/[(a - r)t^2]$

$= $ nominal bending stress on the surface of the net section

Axisymmetric Case

Consider a circular plate with radial stress σ_o at the outer radius (see Figure 4.5). If there is a hole in the center of the plate, the maximum tensile stress is the tangential stress at the hole, and

$$K_T = 2r_o^2/(r_o^2 - r_i^2) \tag{4.6}$$

where $K_T = \sigma_{\text{MAX}}/\sigma_o = $ stress concentration factor

$\sigma_{\text{MAX}} = \sigma_\theta(r_i)$

$= $ tangential stress at the surface of the hole

$\sigma_o = $ radical stress at the outer radius

$r_o = $ outer radius

$r_i = $ hole radius

Note that stress concentration factor is defined in terms of σ_o in this case, not in terms of nominal stress on a net area.

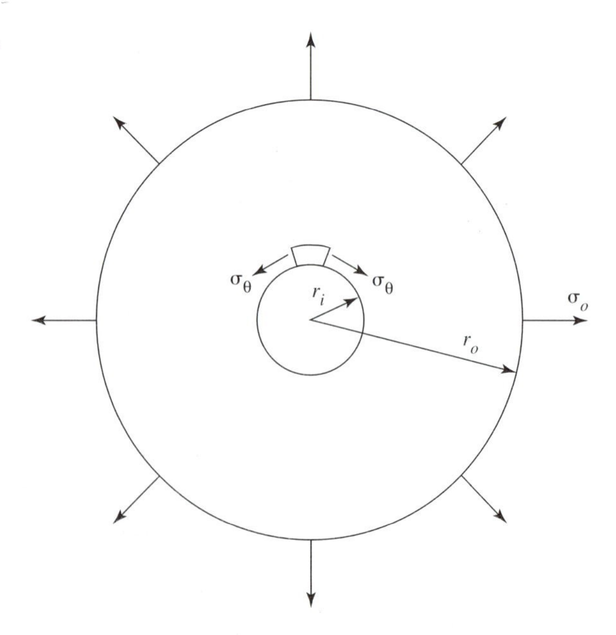

Figure 4.5 Stress concentration, axisymmetric case.

The axisymmetric case approximates the situation where a small hole is drilled in a large spherical pressure vessel. Assume that the hole is plugged to prevent escape of gas. For a thin-wall vessel,

$$\sigma_r = pR/(2t) \qquad \textbf{(4.7)}$$

where σ_r = radial stress at a large distance from the hole (e.g., a distance of 10 times the hole radius)

p = internal pressure

R = radius of the spherical vessel

t = wall thickness of the vessel ($t << R$)

For $r_i << R$, Equation (4.6) can be used with $r_i << r_o$. Stress concentration factor K_T approaches 2, and maximum stress approaches pR/t.

Biaxial Tension

Consider a large square plate with a small hole approximately in the center (see Figure 4.6). If tensile stresses $\sigma_x = \sigma_y = \sigma_o$ on the edges of the plate, then the stress concentration factor is approximately the same as for the axisymmetric case. That is,

$$K_T = 2a^2\,(a^2 - r_i^2) \qquad \textbf{(4.8)}$$

Figure 4.6 Biaxial tension.

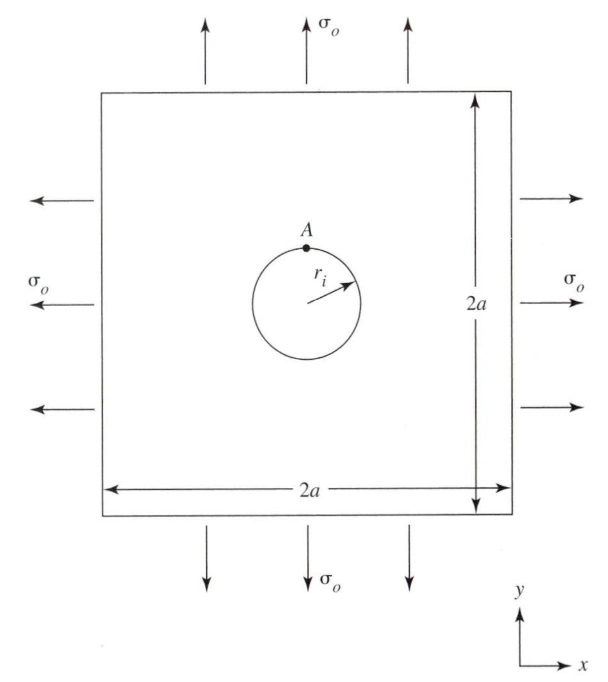

where $K_T = \sigma_{MAX}/\sigma_o$ = approximate stress concentration factor

σ_o = stress at both outer edges

Results should be reasonable if $a > 5r_i$; K_T approaches 2 for $r_i << a$.

When the results of our analysis disagree with our intuition, we usually suspect an error. If a plate with a small central hole is loaded only by an x-direction stress σ_o, then maximum stress approaches

$$\sigma_{MAX} = 3\sigma_o$$

If we add an equal y-direction stress σ_o to the x-direction stress, then maximum stress approaches

$$\sigma_{MAX} = 2\sigma_o$$

where plate size is many times hole size in both cases. In this case, the paradoxical result can be verified by the theory of elasticity and by finite element analysis.

Biaxial Stresses with Opposite Sign

Figure 4.7 shows a square plate with x-direction stress σ_o and y-direction stress $-\sigma_o$ (compression) at the edges. The result is the same as if we had put shear stress on the diagonal planes. In this case,

$$K_T \text{ approaches } 4 \quad \text{for } r << a$$

where $K_T = \sigma_{MAX}/\sigma_o$ = stress concentration factor

r = hole radius

$2a$ = length of a side of the square

Elliptical Holes

Figure 4.8(a) shows a plate with x-direction tensile stress σ_o at the right and left edges. There is an elliptical hole with axes $2r_x$ and $2r_y$ in the x- and y-directions, respectively. If hole size is very small compared to plate size, then

$$K_T = 1 + 2r_y/r_x \tag{4.9}$$

where $K_T = \sigma_{MAX}/\sigma_o$ = stress concentration factor

Equation (4.9) predicts very high stress concentration for a narrow elliptical hole with its long axis perpendicular to the load direction. Figure 4.8(b) shows an elliptical hole with its long axis parallel to the load. Predicted stresses would be lower in this case.

Stepped Shafts

A stepped shaft subject to bending is shown in Figure 4.9. The stress concentration factor is given by

$$K_T = 0.945(r/d)^{-0.229} \quad \text{for } D/d = 1.1 \tag{4.10}$$

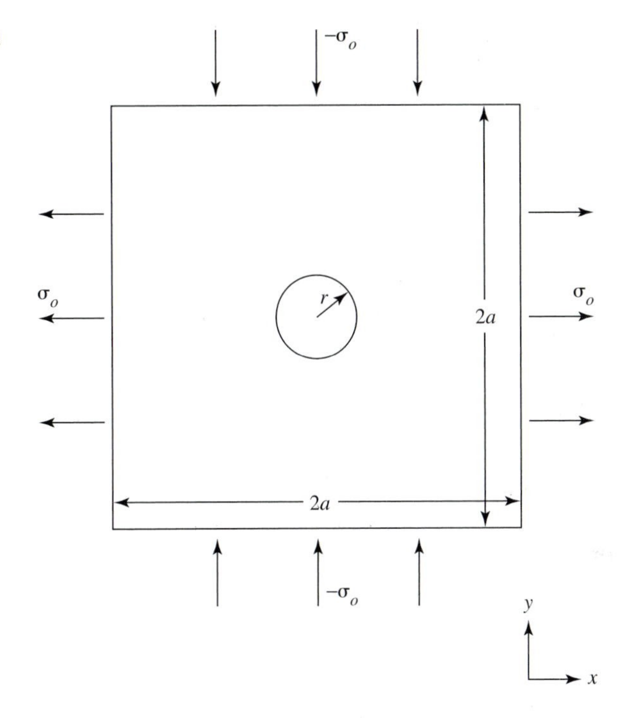

Figure 4.7 Stress concentration, biaxial stresses with opposite sign.

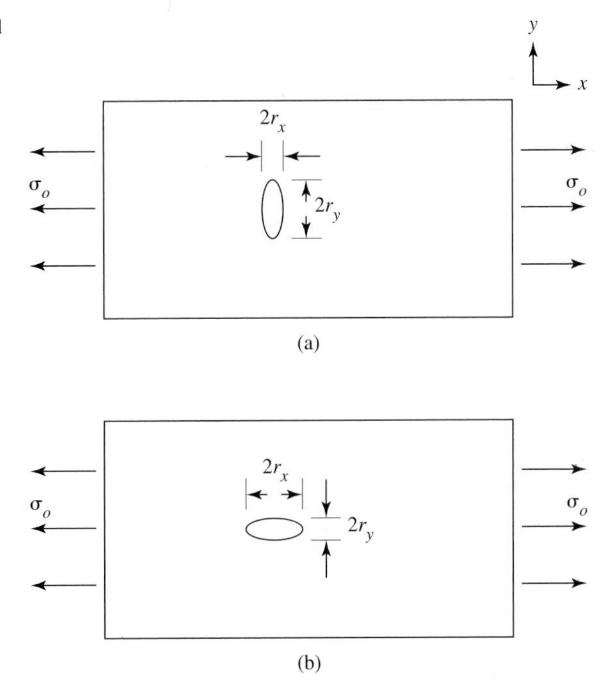

Figure 4.8 Stress concentration, elliptical hole.

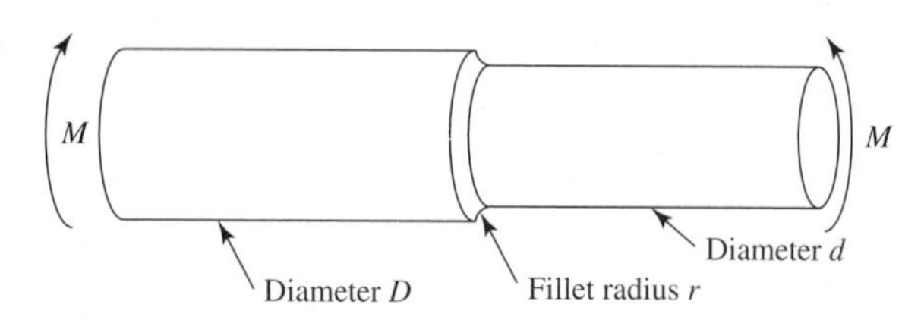

Figure 4.9 Stepped shaft in bending.

where $K_T = \sigma_{\text{MAX}}/\sigma_{\text{NOM}}$

$= $ stress concentration factor approximated by regression analysis

$\sigma_{\text{NOM}} = Mc/I = 32M/(\pi d^3)$

$= $ nominal stress in the smaller shaft section near the step

$M = $ bending moment at the step

$r = $ fillet radius

$d = $ smaller shaft diameter

$D = $ larger shaft diameter

Larger D/d ratios result in greater stress concentration.

REDUCING STRESS CONCENTRATION

Good design practice includes reduction of stress concentration wherever practical, particularly in parts subject to fatigue loading (including rotating shafts). Plans should be reviewed for the possibility of eliminating holes and section changes in high-stress regions. The stress concentration equation for elliptical holes indicates that stress is reduced if the long axis of the ellipse lies in the load direction, as in Figure 4.8(b). This result suggests the configuration shown in Figure 4.10(a), where the large hole could be a required access hole. Adding the two

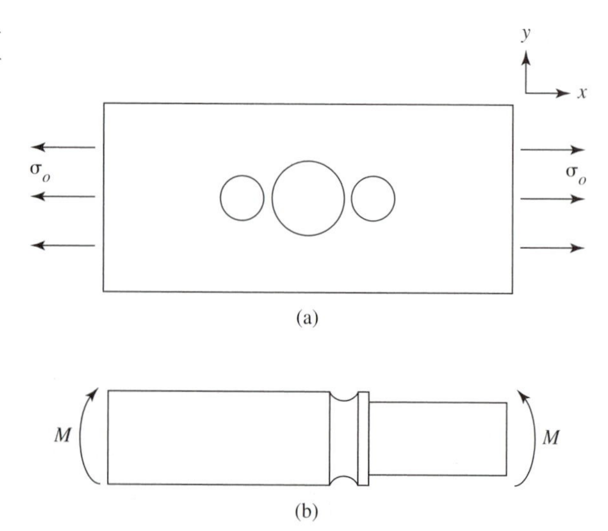

Figure 4.10 Reducing stress concentration. (a) Plate in tension; (b) shaft in bending.

smaller holes can reduce the maximum stress. It is suggested that finite element analysis be used to determine optimum hole size and spacing.

Sharp internal corners and small-radius fillets should be replaced by fillets with large radii. Sometimes a small-radius fillet is specified so that a gear or pulley hub can be precisely located against a step in a shaft. Figure 4.10(b) shows one method of reducing stress concentration. In this case, a large-radius groove is cut into the larger shaft section near the step. The surface of the shaft between the step and the groove is essentially stress-free, and the approximate stress concentration factor calculation can be based on the radius of the groove.

STATISTICAL DISTRIBUTION OF DATA: RELIABILITY

Reported fatigue strength S_N and endurance limit S'_N are usually based on the average of several test specimens. The approximate equations based on ultimate strength also give average fatigue strength or endurance limit. Figure 4.11(a) shows an *S–N* curve with the mean values of stress plotted against stress cycles to failure for a typical steel. The scatter of the data about the mean could be represented by the probability distribution curve shown at the end of the *S–N* curve. Statistical methods are used to predict failure rates and design for an acceptable level of reliability.

Normal Distributions

Of the many possible statistical models, we will assume that the data behave as a **normal distribution** about the mean. Figure 4.11(b) shows our **normal curve** assumption. The probability distribution is rotated so that the stress at fracture is shown horizontally, and the probability of fracture at that stress is the height of the curve. The unshaded tail of the normal curve represents the weaker specimens which fail. The **reliability** represented by the shaded region is the fraction of specimens that survive. At the mean ($X = 0$), the reliability is 0.50. We must lower the stress to get higher reliability.

The **cumulative normal function** represents the area under a normal distribution curve. If software with the cumulative normal function is available, we can use it to predict

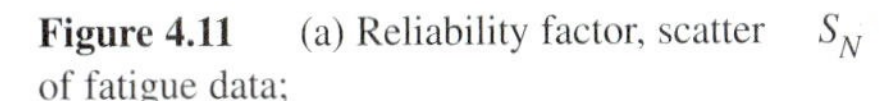

Figure 4.11 (a) Reliability factor, scatter of fatigue data;

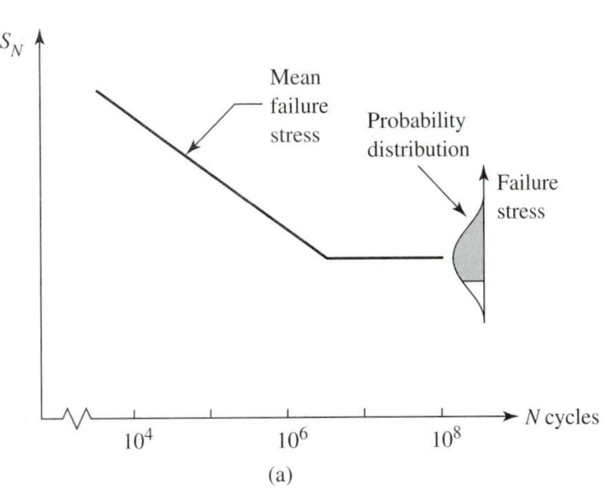

Figure 4.11 *continued* (b) normal distribution assumption.

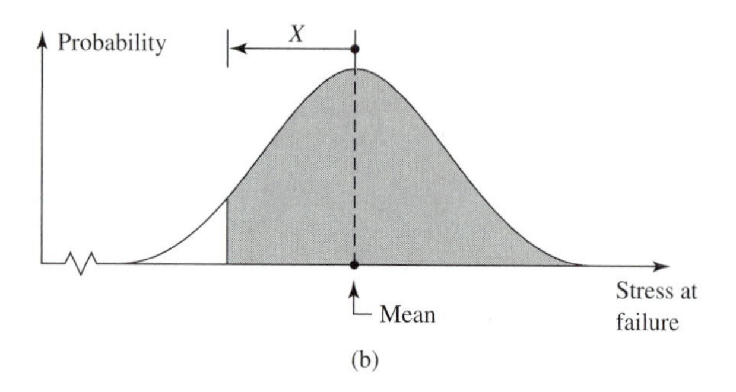

reliability. Otherwise, we can predict reliability by evaluating the complicated integral that follows:

$$F(X) = \int_0^X \frac{1}{(2 \cdot \pi)^{0.5}} \cdot e^{(-x^2/2)}\, dx + 0.5 \quad \textbf{(4.11)}$$

where $F(X)$ = reliability or cumulative normal function
X = number of standard deviations from the mean
x = a dummy variable

The **standard deviation** is a measure of the amount of scatter of the data. If we have collected a set of data, the standard deviation can be computed by mathematics software, or with a scientific calculator that "knows" the standard deviation formula, or from a user-written program. If we have only an estimate of S_N or S'_N based on ultimate strength, then we must estimate the standard deviation as well. A review of some test information suggests the following:

$$SD = 0.08S_N \quad \text{or} \quad SD = 0.08S'_N \quad \textbf{(4.12)}$$

where SD = approximate standard deviation of fatigue test data

To avoid confusing standard deviation with stress, the symbol σ (sigma), which is commonly used for standard deviation, will not be used here.

Reliability Factor

The **reliability factor** reduces the allowable stress on a part to obtain better than 50% survival probability. Using the standard deviation estimate of Equation (4.12), we have

$$C_R = 1 - 0.08X \quad \textbf{(4.13)}$$

where C_R = reliability factor
X = reduction in allowable stress (measured in standard deviations)

Failure Rate and Samples-per-Failure Ratio

The **failure rate** is given by $1 - F(X)$. If we take the reciprocal of the failure rate, we obtain the **samples-per-failure ratio**

$$Q = 1/[1 - F(X)] \quad \textbf{(4.14)}$$

where Q = ratio of total samples to predicted failures, the samples-per-failure ratio

Table 4.2 shows X, the reduction in allowable stress measured in standard deviations; $C_R(X)$, the reliability factor; $F(X)$, the reliability; and $Q(X)$, the samples-per-failure ratio. This table was generated by mathematics software, which solved for $F(X)$ for various values of X. With some mathematics software it is necessary only to call for the preprogrammed cumulative normal distribution function to calculate reliability.

The reliability factor assumes that 1 standard deviation corresponds to an 8% reduction in allowable stress. For a reliability factor of $C_R = 0.76$, corresponding to $X = 3$ standard deviations, the predicted reliability is almost 0.999 (i.e., almost 99.9%). Some manufacturers have set $X = 6$ standard deviations (called the *six sigma level*) as a reliability goal. The six sigma level is a reliability better than 0.999999999, or a failure rate of less than 1 in 1 billion. The samples-per-failure ratio is

$$Q(X) = 1.13 \text{ billion}$$

as shown in Table 4.2. With the standard deviation that we have assumed, the six sigma level is obtained with $C_R = 0.52$.

Table 4.2 Reliability.

$$F(X) = \int_0^X \frac{1}{(2 \cdot \pi)^{0.5}} \cdot e^{(-x^2/2)}\, dx + 0.5$$

$$Q(X) = \frac{1}{1 - F(X)}$$

$$C_R(X) = 1 - 0.08 \cdot X$$

X	$C_R(X)$	$F(X)$	$Q(X)$	X	$C_R(X)$	$F(X)$	$Q(X)$
1	0.92	0.841	6	3	0.76	0.9986501021	741
1.1	0.912	0.864	7	3.2	0.744	0.9993128627	1455
1.2	0.904	0.885	9	3.4	0.728	0.9996630725	2968
1.3	0.896	0.903	10	3.6	0.712	0.9998408953	6285
1.4	0.888	0.919	12	3.8	0.696	0.9999276594	13,823
1.5	0.88	0.933	15	4	0.68	0.9999683416	31,587
1.6	0.872	0.945	18	4.2	0.664	0.9999866748	75,046
1.7	0.864	0.955	22	4.4	0.648	0.9999946181	185,809
1.8	0.856	0.964	28	4.6	0.632	0.999997931	483,321
1.9	0.848	0.971	35	4.8	0.616	0.9999992654	1,361,265
2	0.84	0.977	44	5	0.6	0.9999997133	3,487,703
2.1	0.832	0.982	56	5.2	0.584	0.9999999003	10,027,604
2.2	0.824	0.986	72	5.4	0.568	0.9999999666	29,938,655
2.3	0.816	0.989	93	5.6	0.552	0.9999999892	92,767,210
2.4	0.808	0.992	122	5.8	0.536	0.9999999967	300,888,349
2.5	0.8	0.994	161	6	0.52	0.9999999991	1,132,372,833
2.6	0.792	0.995	215				
2.7	0.784	0.997	288				
2.8	0.776	0.997	391				
2.9	0.768	0.998	536				
3	0.76	0.999	741				

The relationship between X (standard deviations from the mean), $F(X)$ (reliability), and Q (samples-per-failure ratio) shown in Table 4.2 can be applied to other problems as well. They apply to any data which we assume to be normally distributed, and for which failure is represented by the lower tail of the normal distribution curve. Values of reliability factor C_R apply only to fatigue data.

EXAMPLE PROBLEM Design for Given Levels of Reliability

Find the design criteria for the following failure rates:

1/2 1/10 1/20 1/50 1/100 1/1000 1/10,000 1/100,000 1/1,000,000

Assumptions: We will assume that the material strengths are normally distributed and that the standard deviation of the fatigue strength is 8% of the mean value.

Solution: A failure rate of 1 in 2 represents a reliability $F(X)$ of 0.50, corresponding to $X = 0$ standard deviations from the mean, and a reliability factor C_R of 1. A failure rate of 1 in 10 corresponds to the reliability

$$F(X) = \int_0^X \frac{1}{(2 \cdot \pi)^{0.5}} \cdot e^{(-x^2/2)}\, dx + 0.5 = 0.9$$

This equation was solved for X with the aid of mathematics software. An estimate, $X = 1.3$, was used as a starting point for the calculations. The result is $X = 1.282$ standard deviations. The reliability factor is

$$C_R = 1 - 0.08X = 0.897$$

The solutions for failure rates of 1 in 20 to 1 in 1 million are listed below.

Reliability F(X)	*Standard Deviations* X	*Reliability Factor* C_R
0.95	1.645	0.868
0.98	2.054	0.836
0.99	2.326	0.814
0.999	3.09	0.753
0.9999	3.719	0.702
0.99999	4.265	0.659
0.999999	4.753	0.62

It would be difficult to solve the above equation for X by hand calculations. However, if a computer is unavailable, the values of X, C_R, and Q can be approximated with sufficient accuracy by interpolating Table 4.2.

CORRECTED ENDURANCE LIMIT

As indicated by the preceding paragraphs, design of parts for fatigue loading requires effort and judgment. Failure to account for fatigue loading and endurance limit correction factors sometimes results in catastrophic failures.

When the correction factors is applied to test data, the endurance limit for use in design of an actual part is given by

$$S_E = S'_N C_F C_R C_S / K_F \tag{4.15}$$

where S_E = corrected endurance limit, the strength of a material subject to reversed tension, after corrections for surface finish, reliability, size, and stress concentration

S'_N = endurance limit of a specially prepared test specimen

S_N = fatigue strength of a specially prepared test specimen subject to N cycles of reversed stress (*Note:* If the *S–N* diagram for the test does not level off, and/or if the part is to be designed for finite life, S_N is used instead of S'_N.)

C_F = surface-finish correction factor

C_R = reliability correction factor

C_S = size correction factor

K_F = fatigue stress concentration factor

K_T = theoretical stress concentration factor

In most cases, K_F is unavailable; thus, K_T is used instead.

EXAMPLE PROBLEM Design of a Part for Fatigue Loading (with Stress Concentration)

Design a part similar to the link shown in Figure 4.12 to carry a load varying from 3900 lb tension to 3900 lb compression. There must be a 0.7 in diameter hole at the center of the part.

Design Decisions: The part will be made of steel having an ultimate strength of 144,000 psi. The surface of the hole will be equivalent to an average machined surface. The part will be designed for a failure rate of 1 in 20 million, and a safety factor of 1.2 will be chosen to account for uncertainties in material, dimensions, and so on. Total part width will be $2a$ = 1.9 in at the hole. Thickness h must be adequate to avoid fatigue failure.

Solution: The endurance limit of a steel test specimen can be estimated from the following relationship:

$$S'_N = \text{if} \quad (S_U < 200000, 0.5 \cdot S_U, 100000)$$

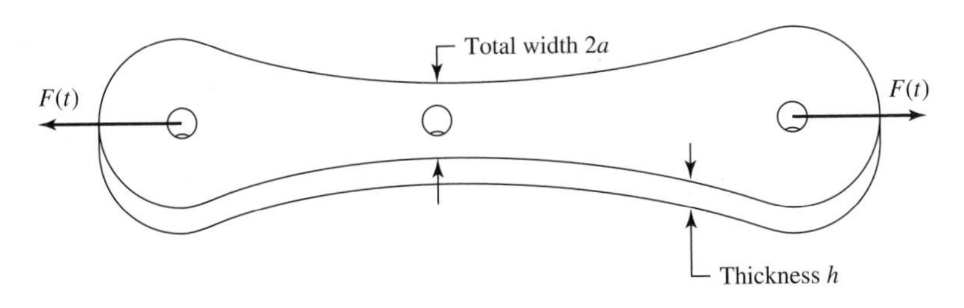

Figure 4.12 Fatigue and stress concentration.

The expression after the first comma applies if the inequality is true; the value after the second comma applies if the inequality is false. For $S_U = 144{,}000$ we find $S'_N = 72{,}000$ psi for a specially prepared, polished test specimen. This value must be corrected for size, surface finish, reliability, and stress concentration. Taking the average of the range of surface-finish correction factors for a machined surface, $C_F = 0.72$. The stress concentration factor is given by

$$K_T = 1.923 - 0.475 \cdot \ln[r/a + 0.1] = 2.283$$

The reliability is given by

$$F(X) = \int_0^X \frac{1}{(2 \cdot \pi)^{0.5}} \cdot e^{(-x^2/2)}\, dx + 0.5 = 1 - 1/Q$$

where Q (the samples-per-failure ratio) $= 20 \cdot 10^6$

Using mathematics software to solve the above equation numerically, beginning with the estimate $X = 5$, we find that $X = 5.327$ standard deviations. Assuming 1 standard deviation equal to 8% of the endurance limit, the reliability factor is

$$C_R = 1 - 0.08X = 0.574$$

Alternatively, X and C_R could have been estimated by interpolating the table of reliability factors.

The first estimate of size factor will assume a cross-sectional area equal to that of a 1 in diameter bar:

$$C_S = 0.8477 - 0.1265 \cdot \ln D = 0.8477$$

We can now calculate corrected endurance limit.

$$S_E = S'_N C_F C_R C_S / K_T = 11{,}045 \text{ psi}$$

Stress amplitude at the cross section through the hole should not exceed the corrected endurance limit divided by the factor of safety, from which

$$F/A_{\text{NET}} = S_E/N_{\text{FS}}$$

Rearranging, we find

$$A_{\text{NET}} = F \cdot N_{\text{FS}}/S_E = 0.424 \text{ in}^2$$

where $A_{\text{NET}} = h(2a - 2r)$

from which we find required plate thickness

$$h = A_{\text{NET}}/(2a - 2r) = 0.353 \text{ in}$$

A Check of the Assumptions: The assumptions seem reasonable, except that the size factor assumption can be reevaluated. The area of a circular cross section is

$$A = \pi D^2/4$$

Equating this value to A_{NET}, we find an equivalent diameter

$$D = 2(A_{\text{NET}}/\pi)^{1/2} = 0.735$$

This new value of D results in higher values of size factor and corrected endurance limit, and a smaller A_{NET}, resulting in a required thickness of $h = 0.338$ in. This fine-tuning results in a thickness decrease of about 4%, which might be of interest where weight is critical.

■■

ROTATING SHAFTS SUBJECT TO BENDING

Consider a point on the surface of a rotating solid circular shaft subject to bending. With each rotation, tension stress at the point is given by

$$\sigma(t) = \sigma_R \sin(\omega t) \quad \textbf{(4.16)}$$

where $\sigma(t)$ = instantaneous tensile stress which varies from $-\sigma_R$ to $+\sigma_R$

t = time

$\sigma_R = 32M/(\pi D^3)$

= range stress, the amplitude of the stress variation (psi or MPa) (*Note:* Range stress is also called *alternating stress.*)

M = bending moment (lb · in or N · mm)

D = shaft diameter (in or mm)

Figure 4.13(a) illustrates the stress variation at a point on the surface of a shaft rotating at an angular velocity of 400 rad/s, where σ_R = 1000 psi. The stress–time relationship is sinusoidal, and the mean stress is zero. The stress pattern of a rotating shaft subject to bending alone (i.e., free of torsion and thrust) is similar to the rotating-beam fatigue test. Thus, the design criterion is

$$S_E/N_{FS} \geq \sigma_R \quad \textbf{(4.17)}$$

where S_E = corrected endurance limit

N_{FS} = safety factor

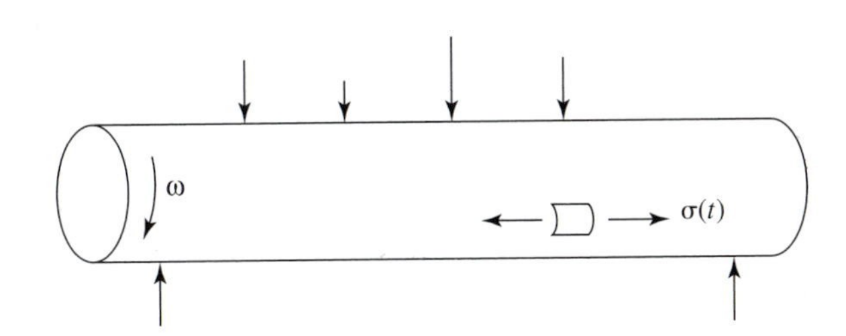

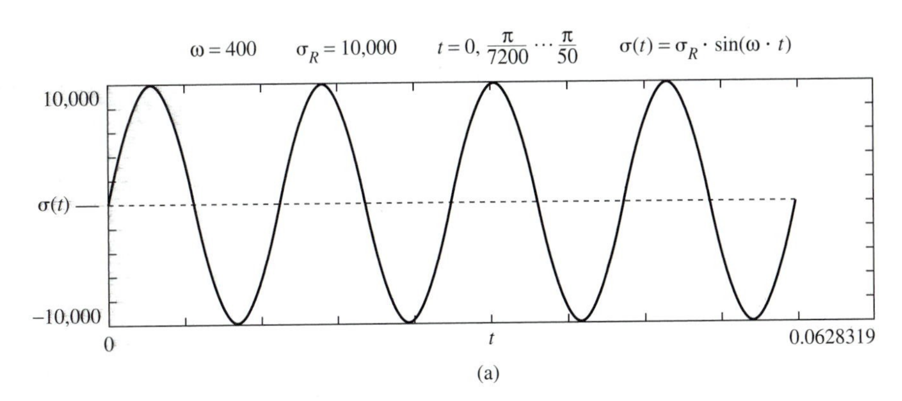

Figure 4.13 Fatigue loading with (a) zero mean stress;

DESIGN FOR FATIGUE LOADING WITH MEAN AND RANGE STRESS

Figure 4.13(b) illustrates a stress–time pattern in which the stress at a point varies continuously from zero to a maximum value. This situation can arise from one-way bending, where bending moment ranges from zero to a maximum value. In this case, range stress equals mean

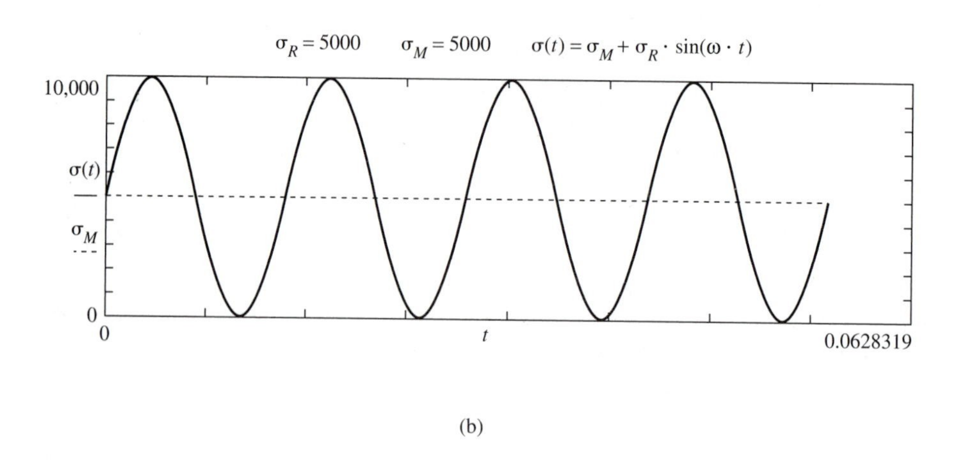

(b)

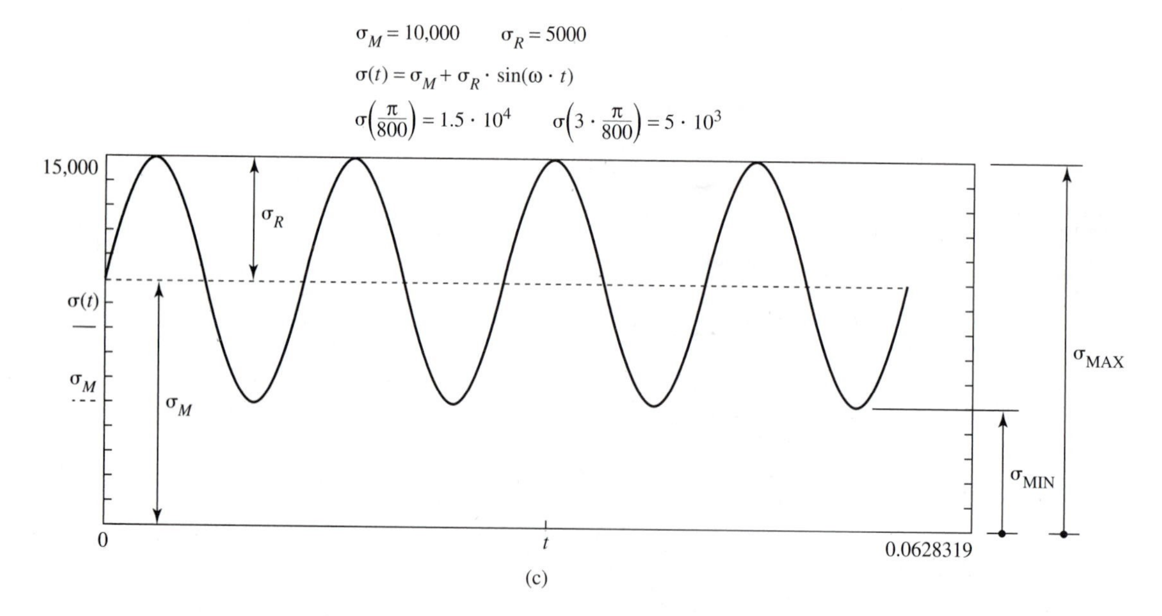

(c)

Figure 4.13 *continued* (b) one-way bending; (c) nonzero mean stress.

stress. Figure 4.13(c) shows a stress pattern with mean stress unequal to zero and unequal to range stress. If the actual stress–time pattern is irregular, we simply define

$$\sigma_M = (\sigma_{\text{MAX}} + \sigma_{\text{MIN}})/2 \tag{4.18}$$

and

$$\sigma_R = (\sigma_{\text{MAX}} - \sigma_{\text{MIN}})/2 \tag{4.19}$$

where σ_{MAX} and σ_{MIN} = maximum and minimum stress at a point, respectively

Figure 4.14 shows a fatigue-testing machine for plate and sheet specimens. The testing machine is designed so that the user can specify both mean and range loading.

When end-use loading fatigue tests of a material or prototype part are not practical, we use a theoretical fatigue behavior model. The idea is to establish a boundary limiting mean and range stress. A combination of mean and range stress within the boundary is safe, outside the boundary unsafe, and just on the boundary most economical (i.e., optimum). Not everyone agrees where the boundary should be.

Figure 4.15 shows safe-stress lines plotted with mean stress σ_M horizontal and range stress σ_R vertical. Since yielding usually constitutes machine failure, we will begin with the **yield criterion,** given by

$$S_{\text{YP}}/N_{\text{FS}} \geq |\sigma|_{\text{MAX}}$$

or

$$S_{\text{YP}}/N_{\text{FS}} \geq |\sigma_M| + \sigma_R$$

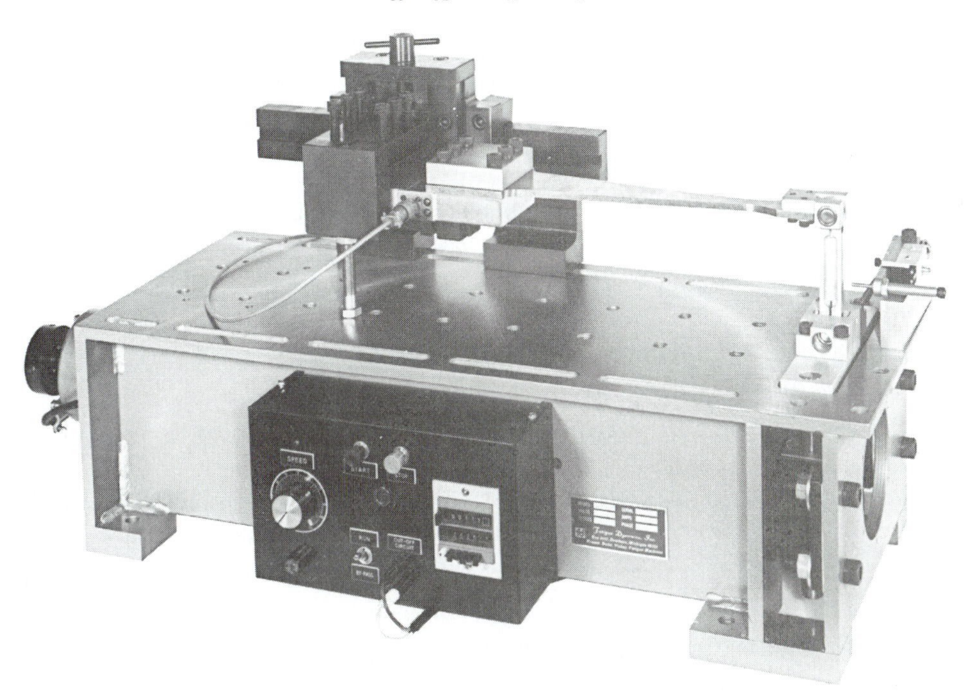

Figure 4.14 A 150-lb capacity sheet- and plate-bending (flexural-fatigue) testing machine. (Courtesy of Fatigue Dynamics, Inc.)

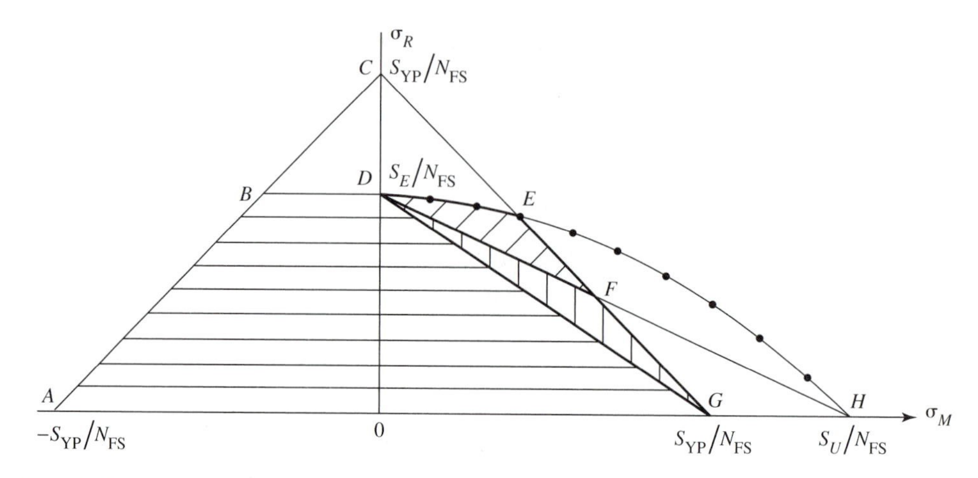

Figure 4.15 Safe-stress limits based on the Soderberg, Goodman, and Gerber yield criteria.

Line segments *AC* and *CG* on Figure 4.15 represent the yield criterion. If we assume that compressive mean stresses below the yield point do not contribute to failure, then we are limited by the endurance limit in this region. Thus, we have the **range stress limit** given by Inequality (4.17)

$$S_E/N_{FS} \geq \sigma_R$$

and represented by line segment *BD* in Figure 4.15.

SODERBERG CRITERION

The **Soderberg criterion** is represented on Figure 4.15 by straight line *DG* between S_E/N_{FS} on the range stress axis and S_{YP}/N_{FS} on the mean stress axis. The Soderberg line can be represented by the following equation:

$$\frac{1}{N_{FS}} = \frac{\sigma_M}{S_{YP}} + \frac{\sigma_R}{S_E}$$

The inequality describing the Soderberg safe-stress region is

$$S_{YP}/N_{FS} \geq \sigma_M + \sigma_R S_{YP}/S_E \tag{4.20}$$

If the inequality is not satisfied, we must redesign to reduce stresses.

We may be required to design a part for known mean and range loads. In that case, Inequality (4.20) can be rearranged to find the limiting value of mean stress as follows:

$$\sigma_M = \frac{S_{YP}/N_{FS}}{1 + (\sigma_R/\sigma_M)(S_{YP}/S_E)} \tag{4.21}$$

where σ_M = the maximum allowable mean stress for the given loading conditions

$\sigma_R/\sigma_M = P_R/P_M$

= the ratio of range to mean stress or the ratio of range to mean load

Part dimensions are then determined by relating σ_M found by Equation (4.21) to mean load P_M.

EXAMPLE PROBLEM Design for Range and Mean Loading Using the Soderberg Criterion

The load at the center of a simply supported bar varies from 55 to 175 lb. The supports are 36 in apart. Design the bar.

Design Decisions: Use steel with an ultimate strength of 91,000 psi and a yield point of 58,000 psi. The surface will be as-machined. Base the design on infinite life with 99.99% reliability. Estimate a size equivalent to $D = 1$ in. Let the width of the cross section be four times the depth. Use a safety factor of 1.5. Use S_{N1} to represent test specimen endurance limit approximation.

Solution: The corrected endurance limit is determined as follows:

$$C_F = \frac{0.64 + 0.80}{2} \qquad C_F = 0.72 \qquad D = 1$$

$$C_S = 0.8477 - 0.1265 \cdot \ln(D) \qquad C_S = 0.848 \qquad K_T = 1$$

$$C_R = 0.702 \qquad S_U = 91{,}000 \qquad S_{N1} = 0.5 \cdot S_U$$

$$S_E = S_{N1} \cdot C_F \cdot C_R \cdot \frac{C_S}{K_T} \qquad S_E = 19{,}495$$

Mean and Range Loading:

$$P_{\text{MAX}} = 175 \qquad P_{\text{MIN}} = 55 \qquad N_{\text{FS}} = 1.5$$

$$P_R = \frac{P_{\text{MAX}} - P_{\text{MIN}}}{2} \qquad P_R = 60 \qquad S_{\text{YP}} = 58{,}000$$

$$P_M = \frac{P_{\text{MAX}} + P_{\text{MIN}}}{2} \qquad P_M = 115$$

Ratio of Range to Mean Stress:

$$r_{RM} = \frac{P_R}{P_M}$$

Maximum Allowable Mean Stress from Equation (4.21):

$$\sigma_M = \frac{S_{\text{YP}}/N_{\text{FS}}}{1 + r_{RM} \cdot (S_{\text{YP}}/S_E)} \qquad \sigma_M = 15{,}150$$

Moment:

$$L = 6$$

$$M_M = P_M \cdot \frac{L}{4} \qquad M_M = 172.5$$

Aspect Ratio:

$$a = b/h \qquad a = 4$$

Depth *Width*

$$h = \left(\frac{6 \cdot M_M}{a \cdot \sigma_M}\right)^{1/3} \qquad h = 0.258 \qquad b = a \cdot h \qquad b = 1.03$$

Revised Size Factor:

$$D = 2 \cdot \left(\frac{b \cdot h}{\pi}\right)^{1/2} \qquad D = 0.581$$

$$C_S = 0.8477 - 0.1265 \cdot \ln(D) \qquad C_S = 0.916$$

$$S_E = S_{N1} \cdot C_F \cdot C_R \cdot \frac{C_S}{K_T} \qquad S_E = 21{,}074$$

$$\sigma_M = \frac{S_{YP}/N_{FS}}{1 + r_{RM} \cdot (S_{YP}/S_E)} \qquad \sigma_M = 15{,}873$$

Corrected Dimensions:

Depth — *Width*

$$h = \left(\frac{6 \cdot M_M}{a \cdot \sigma_M}\right)^{1/3} \qquad h = 0.254 \qquad b = a \cdot h \qquad b = 1.014$$

The dimensions change slightly due to the corrected size factor. Errors are possible when typing the equations.

Partial Check: We will see if Equation (4.20) is satisfied.

$$M_M = P_M \cdot \frac{L}{4}$$

$$\sigma_M = 6 \cdot \frac{M_M}{b \cdot h^2} \qquad \sigma_M = 15{,}873.369$$

$$\sigma_R = \sigma_M \cdot \frac{P_R}{P_M}$$

$$\frac{S_{YP}}{N_{FS}} = 38{,}666.667 \qquad \sigma_M + \sigma_R \cdot \frac{S_{YP}}{S_E} = 38{,}666.667$$

Equation (4.20) is satisfied.

It is often most cost-effective to use standard, "off-the-shelf" material sizes. To do so, we round off the calculated values to the next larger commercial size. For example, in the preceding example problem, the calculated depth is 0.254 in. Thus, we could specify 5/16 in thick steel or 7 mm thick steel.

A Note on Design for Fatigue Loading

Note that the Soderberg criterion is applied when mean stress is greater than zero. The yield criterion and the range stress limit apply when mean stress is negative.

Fatigue design problems can be solved graphically. Stress is described by the point (σ_M, σ_R). If the point falls on or below the Soderberg safe-stress line (line *DG* on Figure 4.15), then the design is safe. The most economical dimensions result if the point (σ_M, σ_R) falls on the Soderberg safe-stress line. Computer solutions are usually preferred because evaluation of design changes and "what-if" analysis require only a few keystrokes. A graphical solution serves as a check.

THE GOODMAN AND GERBER CRITERIA FOR FATIGUE LOADING[1]

The Goodman Criterion

The **Goodman safe-stress line** is a straight line joining points S_E/N_{FS} and S_U/N_{FS} on a σ_R versus σ_M plot (line *DH* on Figure 4.15). The Goodman safe-stress region is described by

$$S_U/N_{FS} \geq \sigma_M + \sigma_R S_U/S_E \quad \textbf{(4.22)}$$

The Goodman-Yield Criterion

Yielding is unacceptable for most machine design applications. To avoid the possibility that a part will yield, designers can use the **Goodman-yield criterion.** The Goodman-yield safe-stress region is bounded by the broken line *DFG* in Figure 4.15, defined by the following inequalities:

$$S_U/N_{FS} \geq \sigma_M + \sigma_R S_U/S_E \quad \text{and} \quad S_{YP}/N_{FS} \geq \sigma_M + \sigma_R \quad \textbf{(4.23)}$$

The Goodman-yield criterion permits the designer higher stress combinations than the Soderberg criterion; it is less conservative.

The Gerber-Yield Criterion

The Gerber-yield safe-stress region is bounded by curve *DE* and yield line *EG* on Figure 4.15. The following inequalities apply:

$$1/N_{FS} \geq (\sigma_M/S_U)^2 + \sigma_R/S_E \quad \text{and} \quad S_{YP}/N_{FS} \geq \sigma_M + \sigma_R \quad \textbf{(4.24)}$$

The Gerber-yield criterion allows higher stress combinations than the Goodman-yield and Soderberg criteria.

EXAMPLE PROBLEM A Program to Plot Fatigue Design Criteria

Sketch a flowchart describing a computer program to plot the fatigue design criteria introduced above.

Solution: The above equations are rewritten so that mean stress is given in terms of the other quantities. At several steps we must determine which curve applies by tests such as "Does the Goodman line fall below the yield line at this point?" One possible flowchart is shown in Figure 4.16. Other forms are also acceptable.

---

[1] This section describes alternatives to the Soderberg criterion. This section and the flowchart example problem that follows can be omitted without loss of continuity.

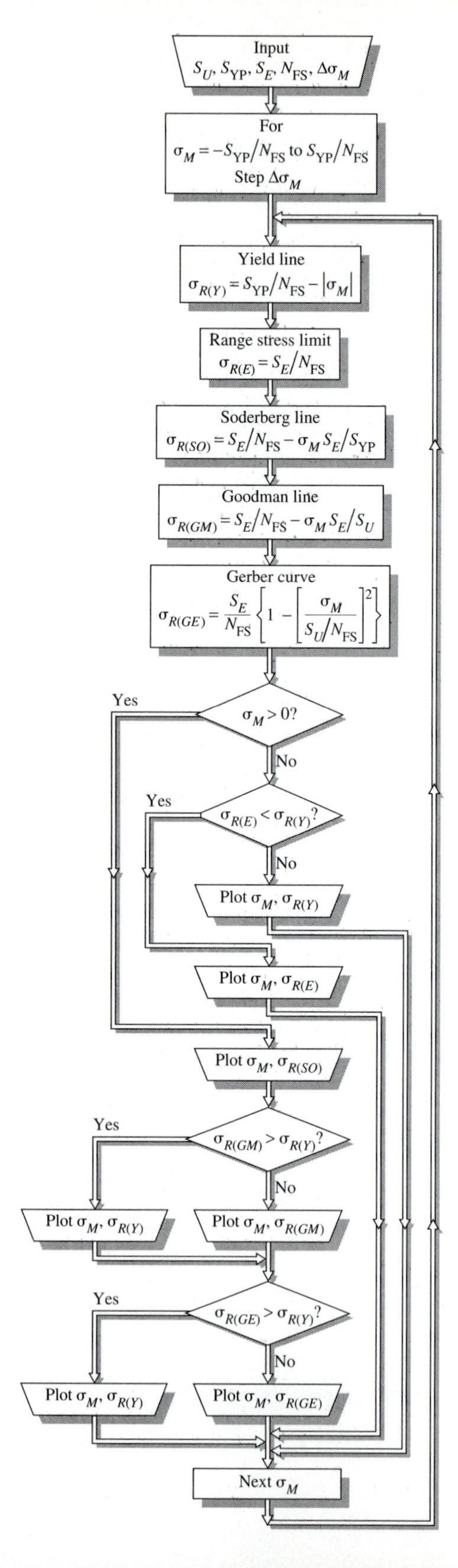

Figure 4.16 Flowchart.

PROBABILITY THEORY WITH POTENTIAL APPLICATIONS TO MACHINE DESIGN

As noted in Chapter 1, an important design attribute is robustness, a resistance to quality loss or deviation from desired performance. Quality loss can occur from batch to batch during the manufacturing process, or it can occur during end-use. Machine parts subject to fatigue loading are a particular challenge to the designer. We try to build in reliability, but we recognize that perfection cannot be achieved. Thus, we select a reliability factor as discussed in the section on endurance limit. The machine designer uses probability theory concepts to establish part and system reliability.

The probability of an event can fall between zero (it cannot happen) and unity (it is certain). We could call a machine failure an event; we would also call survival of the machine an event. Some events are independent of one another, and some are related. If two events are **mutually exclusive,** then one or the other can occur, but not both.

Independent Events

If events E_1 and E_2 are **independent,** then the probability of *both* occurring is

$$P(E_1 \text{ and } E_2) = P(E_1)P(E_2) \qquad \textbf{(4.25)}$$

where $P(E_1)$ = probability of event E_1, and so on

The probability of occurrence of *all* events in a group of independent events is

$$P(E_1 \text{ and } E_2 \text{ and } \cdots \text{ and } E_N) = P(E_1)P(E_2)\cdots P(E_N) \qquad \textbf{(4.26)}$$

where $E_1, E_2, \cdots, E_N$ are N independent events

The probability of occurrence of *either* of two independent events is

$$P(E_1 \text{ or } E_2) = P(E_1) + P(E_2) - P(E_1)P(E_2) \qquad \textbf{(4.27)}$$

Mutually Exclusive Events

The probability of occurrence of either of two mutually exclusive events is

$$P(E_1 \text{ or } E_2) = P(E_1) + P(E_2) \qquad \textbf{(4.28)}$$

Obviously, mutually exclusive events are not independent, and

$$P(E_1 \text{ and } E_2) = 0 \qquad \textbf{(4.29)}$$

where events E_1 and E_2 are mutually exclusive

SYSTEM RELIABILITY (RELIABILITY WHEN SEVERAL COMPONENTS ARE INVOLVED)

We can predict the reliability of a system if we know the reliability and arrangement of the individual components.

Redundant Components (Parallel Systems)

Suppose we design a system so that it does not fail unless two independent components (or two independent subsystems) fail. Then system reliability is given by

$$R = 1 - P(E_1)P(E_2) \tag{4.30}$$

where R = system reliability $0 < R < 1$

$R = 1$ is perfect

$P(E_1) = 1 - R_1$, the probability of failure of component 1

R_1 = reliability of component 1

$P(E_2) = 1 - R_2$, the proability of failure of component 2

R_2 = reliability of component 2

Components 1 and 2 are independent.

Brake systems and aircraft ignition systems can be designed as parallel systems.

Caution

1. Some parallel systems operate at reduced capacity when one component fails.
2. The redundant components may not be truly independent. Failure of one component may affect operation of the other, or both may be damaged simultaneously by another event or accident.
3. Where possible, a sensor and an indicator should be installed to warn the user of a partial failure.

Series Systems

If one component (or subsystem) in a series system fails, then the system fails. Failure in a series system is analogous to failure of one link in a chain. Most machine systems are series systems. In a series system, **nonfailure** (i.e., reliability) of a component is an independent event. The reliability of a series system is

$$R = R_1R_2R_3 \cdots R_N \tag{4.31}$$

where R = system reliability

R_1 = reliability of component or subsystem 1

R_N = reliability of component or subsystem N

EXAMPLE PROBLEM Reliability of a Parallel System

A conveyor is designed with two separate and independent braking systems. The reliabilities of the two braking systems (for the life of the conveyor) are

$$R_1 = 0.99 \quad \text{and} \quad R_2 = 0.995$$

(a) Find the probability of total failure of the combined braking system.
(b) Find the reliability of the combined braking system.

Decision: The system will be treated as a parallel system.

Solution (a): The failure probabilities are

$$P_{E1} = 1 - R_1 \qquad P_{E2} = 1 - R_2$$
$$P_{E1 \text{ and } E2} = P_{E1} \cdot P_{E2} \qquad P_{E1 \text{ and } E2} = 5 \cdot 10^{-5}$$

Solution (b): The reliability is

$$R = 1 - P_{E1} \cdot P_{E2} \qquad R = 0.99995$$

Note that actual reliability is affected by the frequency of system inspection and maintenance. ■■

EXAMPLE PROBLEM Reliability of a Series System

A machine is made up of 10 critical components. Failure of any single component could cause the machine to fail.

(a) Find the reliability of the machine if each component has a reliability of 0.999.
(b) Find the required reliability of each component if the machine reliability must be 0.9999.

Decision: The system will be treated as a series system.

Solution (a): The reliability of the machine is

$$R_i = 0.999 \qquad N = 10$$
$$R = R_i^N \qquad R = 0.99$$

Solution (b): The required reliability is

$$R = 0.9999$$
$$R_i = R^{1/N} \qquad R_i = 0.9999899995$$

Check:

$$R_i^N = 0.9999$$

■■

HINTS FOR USE OF REGRESSION ANALYSIS TO OBTAIN DESIGN EQUATIONS[2]

Some designs are based directly on simple formulas which can be derived analytically. In other cases, we may have to rely on a series of tests or use limited data from complicated analytical studies and finite element analysis (FEA). **Regression analysis** is used to generate an equation that approximates the relationship between sets of variables. For example, we may have FEA-generated stress concentration factors for various values of r/a, the ratio of hole

[2]The material in this section can be omitted without loss of continuity.

radius to half-plate-width. To use these values in computer-aided design studies, we can enter them into a database and then have the computer look them up and interpolate as necessary. Or we can calculate a regression equation relating stress concentration factor to r/a.

Linear regression formulas fit a straight line to the data, resulting in an equation in the form

$$y = A + Bx \tag{4.32}$$

where y = the dependent variable
x = the independent variable
A = the y-intercept of the line (the constant)
B = the slope of the line (the x-coefficient)

Linear regression formulas are preprogrammed in mathematics software, in spreadsheets, and in scientific calculators. A scientific calculator may require that the x- and y-values be entered in pairs. A spreadsheet will deal with ranges of values in column form (where the x- and y-ranges are of equal length). When using mathematics software, we enter the x- and y-values as vectors (with the same number of elements). If preprogrammed formulas are not available, refer to Chapter 10, the section entitled "Regression Analysis: User-Written Programs." In each case, the regression formulas give the values of A and B.

In general, the regression equation will be approximate; for a given value of x, the value of $A + Bx$ will not exactly equal the value of y in the original data. If an xy-plot of the raw data does not approximate a straight line, the regression equation will be a "poor fit" to the data.

In case of a poor fit to the raw data, we can try to fit the linear regression equation to various functions of the raw data. In each case, we can plot or tabulate the data points beside the results of the regression equation results to see if we have obtained a better fit. It may require some patience to get a good fit.

Logarithmic Regression

If a linear regression of the raw data does not produce a satisfactory result, we can pair values of the natural logarithm of x with y. The preprogrammed or user-written linear regression formulas will produce constants A and B. The result is a logarithmic regression equation in the form

$$y = A + B \ln x \tag{4.33}$$

where ln = natural logarithm

If the values of x are formed into a vector, we must take the natural logarithm of each term of the vector. When using mathematics software, we must instruct the software to take the logarithm term-by-term; we can accomplish this by "vectorizing" the calculation.

Exponential Regression

As another alternative, we can pair values of x and the natural logarithm of y, obtaining the regression equation

$$\ln y = \ln A + Bx \tag{4.34}$$

or the equivalent expression

$$y = Ae^{Bx} \tag{4.35}$$

The latter form [Equation (4.35)] is called an **exponential regression.** The y-intercept produced by the linear regression formulas is ln A in this case. To find constant A, we use

$$A = e^{(y\text{-intercept})}$$

Power Regression

Pairing values of ln x and ln y results in the regression equation

$$\ln y = \ln A + B \ln x \tag{4.36}$$

or the equivalent power regression form

$$y = Ax^B \tag{4.37}$$

The y-intercept produced by the linear regression formulas is again ln A.

EXAMPLE PROBLEM A Regression Equation for Stress Concentration Data

The data listed below apply to the plate with a central hole (Figure 4.2). Theoretical stress concentration factor K has been determined for seven ratios of hole size to plate width. The first value applies to a very small hole (r/a approaches zero). If the hole is eliminated, there is no stress concentration ($K_T = 1$).

Diameter-to-Width Ratio R = r/a	*Theoretical Stress Concentration Factor* K
0	3.00
0.1	2.71
0.2	2.51
0.3	2.36
0.4	2.24
0.5	2.15
0.6	2.10

Represent these data in the form of an equation.

Decision: Use preprogrammed linear regression formulas.

Solution: The relationship is obviously nonlinear. We can try logarithmic regression. To avoid the logarithm of zero, we define

$$R_1 = R + 0.1 = r/a + 0.1$$

and calculate

$$\ln(R_1) \quad \text{(term-by-term)}$$

The linear regression formulas are applied, yielding

$$A = \text{intercept}[\ln(R_1), K] = 1.9232$$

and

$$B = \text{slope}[\ln(R_1), K] = -0.475$$

The theoretical stress concentration factor is then approximated by

$$K = A + B \ln(R_1 + 0.1)$$

or

$$K_T = 1.923 - 0.475 \ln(r/a + 0.1)$$

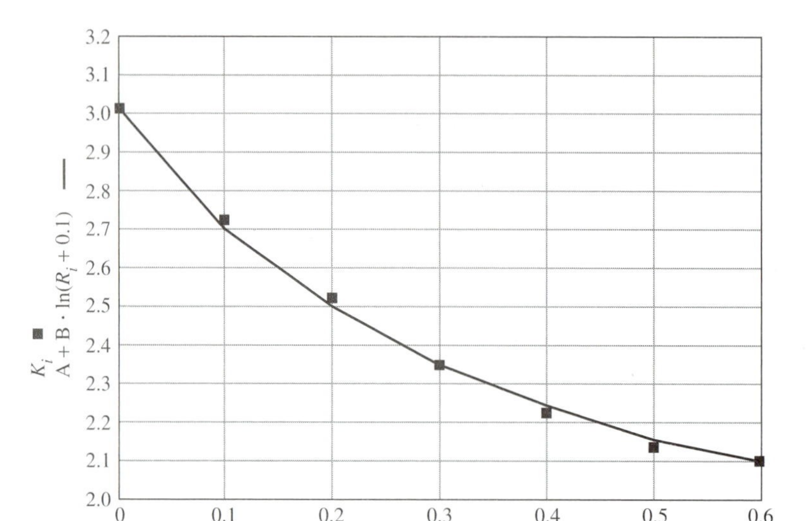

Figure 4.17 Stress concentration, plate with central hole in tension.

Figure 4.17 shows the regression formula plotted as a curved line; the given values of stress concentration factor are shown as square symbols. The regression formula is a close, but not precise, fit to the data.

IMPROVING RESISTANCE TO FATIGUE

Design for fatigue loading is particularly challenging. We can avoid fatigue failures by designing with an adequate safety factor, selecting appropriate materials, improving surface finish, and reducing stress concentration. The significance of these variables was discussed in earlier sections. Preloading of bolts to avoid fatigue failures will be covered in Chapter 15, "Fasteners." Shot-peening is another method of improving fatigue resistance.

Shot-Peening

If a part is bombarded with small, spherical pellets, a zone of compressive stress is formed at the surface of the part. This cold-working process is called **shot-peening.**

We will illustrate an application of shot-peening by considering a part with a time-varying bending load, as shown in Figure 4.18(a). For the loading shown, stress is distributed over a cross section, as shown in Figure 4.18(b). Maximum stress occurs at the bottom of the part and varies from zero to

$$\sigma_{MAX} = \frac{Mc}{I}$$

where σ_{MAX} = maximum tensile stress due to bending
M = maximum bending moment
c = half of part depth
I = moment-of-inertia of cross section

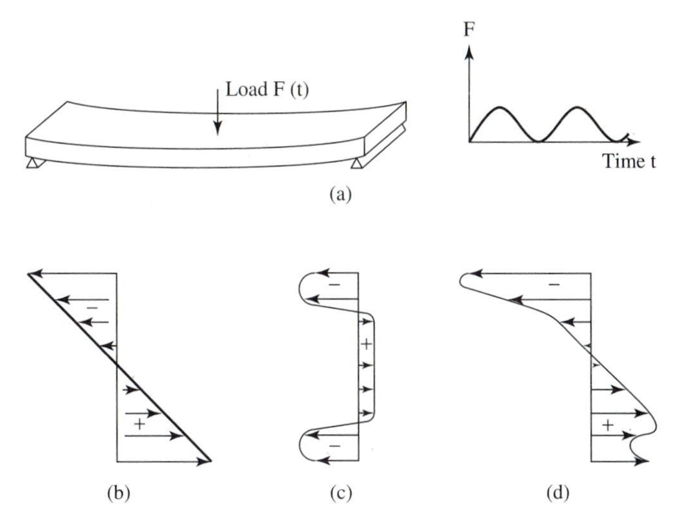

Figure 4.18 Shot-peening to improve fatigue resistance (not to scale). (a) Time-varying bending load on a part; (b) stress distribution at a cross section; (c) residual stresses due to shot-peening; (d) stress distribution due to loading of the shot-peened part.

Suppose that the top and bottom surfaces are shot-peened, producing residual stresses as shown in Figure 4.18(c). The maximum compressive stress due to peening may exceed one-half the yield point of the material. The middle region of the part must be in tension to balance the compressive stresses at the surface. Then the maximum stress due to bending, illustrated in Figure 4-18(d), is obtained by combining the stress patterns. Note that the shot-peening has reduced the maximum tensile stress. Shot-peening has increased the

Figure 4.19 Stress corrosion cracking in a steam turbine rotor. (Courtesy of Metal Improvement Company, Inc.)

maximum compressive stress on the top surface. However, fatigue cracks are not known to initiate or propagate in a region in compression. Thus, fatigue resistance should be improved.

In the above application, it would seem that there is no reason to shot-peen the top surface of the part. However, shot-peening only one side could cause distortion. Peen-forming takes advantage of this distortion, using it to shape parts. One of the first applications was the peen-forming of aircraft wing skins.

Equipment is available for shot-peening using metal, glass, or ceramic pellets, utilizing computer controls and monitors to document the process. The use of shot-peening to increase resistance to fatigue failures, corrosion fatigue, and stress corrosion cracking is considered in detail in a booklet produced by Metal Improvement Company (n.d.). Figure 4.19 shows typical stress corrosion cracking in the steeples holding the turbine blades in place on a steam turbine rotor. This problem occurs in both fossil fuel and nuclear power plants. Shot-peening has been used to reduce the probability of stress corrosion failures of this type.

References and Bibilography

Collins, J. A. *Failure of Materials in Mechanical Design.* New York: Wiley Interscience, 1981.

DeGarmo, E. P., J. T. Black, R. A. Kohser. *Materials and Processes in Manufacturing.* 7th ed. New York: Macmillan, 1988.

Deutschman, A. D., W. Michels, and C. Wilson. *Machine Design—Theory and Practice.* New York: Macmillan, 1975.

Juvinall, R. C. *Stress, Strain, and Strength.* New York: McGraw-Hill, 1967.

Lipson, C., and R. Juvinall. *Handbook of Stress and Strength.* New York: Macmillan, 1963.

Logan, D. L. *Mechanics of Materials.* New York: HarperCollins, 1991.

Peterson, R. E. *Stress Concentration Factors.* New York: John Wiley, 1974.

Shot-Peening Applications. 7th ed. Paramus, NJ: Metal Improvement Company, Inc., n.d.

Timoshenko, S., and J. N. Goodier. *Theory of Elasticity.* 2d ed. New York: McGraw-Hill, 1951.

Ugural, A. C., and S. K. Fenster. *Advanced Strength and Applied Elasticity.* Second SI Edition. New York: Elsevier, 1987.

Wang, C-T. *Applied Elasticity.* New York: McGraw-Hill, 1953.

Design Problems

4.1. Suppose that the scatter of fatigue test data for a particular material showed a standard deviation of 10% of the mean (instead of 8% as assumed previously).

(a) Tabulate the reliability factor for values of x ranging from 1 to 6 standard deviations.

(b) Tabulate the reliability factor for values of reliability $F(X)$ ranging from 0.90 to 0.9999.

4.2. Design a part similar to the link shown in Figure 4.12 to carry a load varying from 1500 lb tension to 1500 lb compression. There must be a 0.8 in diameter hole at the center of the part.

Design Decisions: The part will be made of steel having an ultimate strength of 147,000 psi. The surface of the hole will be equivalent to an average machined surface. The part will be designed for a failure rate of 1 in 5 million, and a safety factor of 1.5 will be chosen to account for uncertainties in material, dimensions, and so on. Total part width will be $2a = 1.6$ in at the hole. Thickness h must be adequate to avoid fatigue failure.

4.3. Design a part similar to the link shown in Figure 4.12 to carry a load varying from 5500 lb tension to 5500 lb compression. There must be 1 in diameter hole at the center of the part.

Design Decisions: The part will be made of steel having an ultimate strength of 96,000 psi. The surface of the hole will be equivalent to an average machined surface. The part will be designed for a failure rate of 1 in 50 million, and a safety factor of 1.2 will be chosen to account for uncertainties in material, dimensions, and so on. Total part width will be $2a = 2.2$ in at the hole. Thickness h must be adequate to avoid fatigue failure.

4.4. Design a part similar to the link shown in Figure 4.12 to carry a load varying from 4500 lb tension to 4500 lb compression. There must be a 0.84 in diameter hole at the center of the part.

Design Decisions: The part will be made of steel having an ultimate strength of 147,000 psi. The surface of the hole will be equivalent to an average machined surface. The part will be designed for a failure rate of 1 in 100 million, and a safety factor of 1.5 will be chosen to account for uncertainties in material, dimensions, and so on. Total part width will be $2a = 1.4$ in at the hole. Thickness h must be adequate to avoid fatigue failure.

4.5. A simply supported bar is to be designed to withstand a time-varying load at the center.
- **(a)** Find the corrected edurance limit based on an estimate of final dimensions.
- **(b)** Find allowable mean stress.
- **(c)** Find the required dimensions of the bar.
- **(d)** Revise the size factor; recalculate dimensions if necessary.
- **(e)** Using the design criterion in another form, check results.

Design Decisions: The following design requirements and design decisions will apply. The bar will be designed for infinite life, using the Soderberg criterion.

Load Range (lb)		*Length between Supports (In)*
$P_{MAX} = 200$	$P_{MIN} = -20$	$L = 7.5$
Reliability Factor		*Surface-Finish Factor*
$C_R = 0.76$		$C_F = 0.64$
Material Properties (psi)		*Safety Factor*
$S_U = 100{,}000$	$S_{YP} = 65{,}000$	$N_{FS} = 1.4$
Estimate of Equivalent Size		*Aspect Ratio* $a = b/h$
$D = 1$		$a = 3$

4.6. A simply supported bar is to be designed to withstand a time-varying load at the center.
- **(a)** Find the corrected endurance limit based on an estimate of final dimensions.
- **(b)** Find allowable mean stress.
- **(c)** Find the required dimensions of the bar.
- **(d)** Revise the size factor; recalculate dimensions if necessary.
- **(e)** Using the design criterion in another form, check results.

Design Decisions: The following design requirements and design decisions will apply. The bar will be designed for infinite life, using the Soderberg criterion.

Load Range (lb)		*Length between Supports (In)*
$P_{MAX} = 400$	$P_{MIN} = 40$	$L = 10$
Reliability Factor		*Surface-Finish Factor*
$C_R = 0.52$		$C_F = 0.80$
Material Properties (psi)		*Safety Factor*
$S_U = 70{,}000$	$S_{YP} = 48{,}000$	$N_{FS} = 1.25$
Estimate of Equivalent Size		*Aspect Ratio* $a = b/h$
$D = 1$		$a = 5$

4.7. A simply supported bar is to be designed to withstand a time-varying load at the center.
- **(a)** Find the corrected endurance limit based on an estimate of final dimensions.
- **(b)** Find allowable mean stress.
- **(c)** Find the required dimensions of the bar.
- **(d)** Revise the size factor; recalculate dimensions if necessary.
- **(e)** Using the design criterion in another form, check results.

Design Decisions: The following design requirements and design decisions will apply. The bar will be designed for infinite life, using the Soderberg criterion.

Load Range (lb)		*Length between Supports (In)*
$P_{MAX} = 800$	$P_{MIN} = 0$	$L = 8.5$
Reliability Factor		*Surface-Finish Factor*
$C_R = 0.6$		$C_F = 0.70$

Material Properties (psi)		*Safety Factor*
$S_U = 110{,}000$	$S_{YP} = 80{,}000$	$N_{FS} = 1.2$

Estimate of Equivalent Size	*Aspect Ratio* $a = b/h$
$D = 1$	$a = 3.5$

4.8. Write a computer program to plot safe-stress lines for the yield, Soderberg, Goodman-yield, and Gerber-yield criteria. Select reasonable values for material properties and factor of safety. Test the program with the values that you selected.

4.9. A machine is designed with two separate and independent braking systems. The reliabilities of the two braking systems are

$$R_1 = 0.999 \quad \text{and} \quad R_2 = 0.998$$

(a) Find the probability of total failure of the combined braking system.
(b) Find the reliability of the combined braking system.

Design Decisions: The system will be treated as a parallel system.

4.10. A machine is made up of 8 critical components. Failure of any single component could cause the machine to fail.
(a) Find the reliability of the machine if each component has a reliability of 0.99.
(b) Find the required reliability of each component if the machine reliability must be 0.99995.

Design Decisions: The system will be treated as a series system.

4.11. A machine is made up of 20 critical components. Failure of any single component could cause the machine to fail.
(a) Find the reliability of the machine if each component has a reliability of 0.95.
(b) Find the required reliability of each component if the machine reliability must be 0.999999.

Design Decisions: The system will be treated as a series system.

4.12. Use the data in the last example problem "A regression Equation for Stress Concentration Data."
(a) Try to fit the data to an equation.
(b) Evaluate your result by plotting the data points and the equation.

Design Decisions: Use power regression. Try an equation in the form

$$K = A(R + 1)^B$$

4.13. Use the data in the last example problem "A Regression Equation for Stress Concentration Data."
(a) Try to fit the data to an equation.
(b) Evaluate your result by plotting the data points and the equation.

Design Decisions: Use power regression. Try an equation in the form

$$K = A(R + 0.1)^B$$

4.14. Refer to Figure 4.9, a stepped shaft. Let $D/d = 1.1$.
(a) Try to fit a regression equation to the following stress concentration data:

r/d	K
0.01	2.7
0.05	1.9
0.1	1.6
0.15	1.46
0.2	1.35
0.25	1.3
0.3	1.25

(b) Evaluate your result by plotting the data points and the regression equation.

Design Decisions: Use power regression. Try an equation in the form

$$K = A(r/d)^B$$

CHAPTER 5

Bending of Machine Members

Symbols

A	area	q	distributed load
b	width	R	radius, reaction
c	greatest distance from the neutral axis (distance from neutral axis to extreme fiber)	$S(x, a, n)$	singularity function
		S_E	corrected endurance limit
		S_U	ultimate strength
GE	greater than or equal to	S_W	working strength
h	height (dimension perpendicular to the neutral axis)	S_{YP}	yield point strength
		t	thickness
		V	shear force
I	moment-of-inertia	w	deflection
IF–THEN–ELSE	logic statement	$w^{‡‡}$	weight per unit length
K	curvature	x, y, z	distances, coordinates
L	length, load	θ	slope
M	bending moment	Σ	summation
N_{FS}	factor of safety	σ	stress
P	concentrated load		

Units

bending moment	in · lb or N · m
dimensions	in or mm
load, shear force	lb or N
moment-of-inertia	in^4 or mm^4
stress, strength, modulus of elasticity	psi or MPa

Members subject to bending (flexure) due to lateral loads are called **beams.** Shafts, leaf springs, gear teeth, and many other structural members and machine elements are subject to bending. Thus, the concepts developed for beams can be applied to any machine member where bending stress and deflection govern the design.

BEAMS WITH CONCENTRATED AND DISTRIBUTED LOADS

Figure 5.1 illustrates beams with various load and support conditions. Assume in each case that the loads are applied in a plane of symmetry (an xz-plane) through the longitudinal axis of the beam (the x-axis). Loads applied over a small area of the beam are idealized as concentrated loads, that is, loads applied at a point. Gear and sheave loads on shafts are ordinarily treated as concentrated loads. The beams in Parts (a), (c), (d), and (e) of Figure 5.1 are subject to concentrated loads. A distributed load is a force that is applied over a substantial portion of a beam. The beam in Figure 5.1(b) is subject to a distributed load expressed in force per unit length. The weight of a beam itself is a distributed load. If the weight of a machine member is very small compared with other loads, the effect of weight can be neglected.

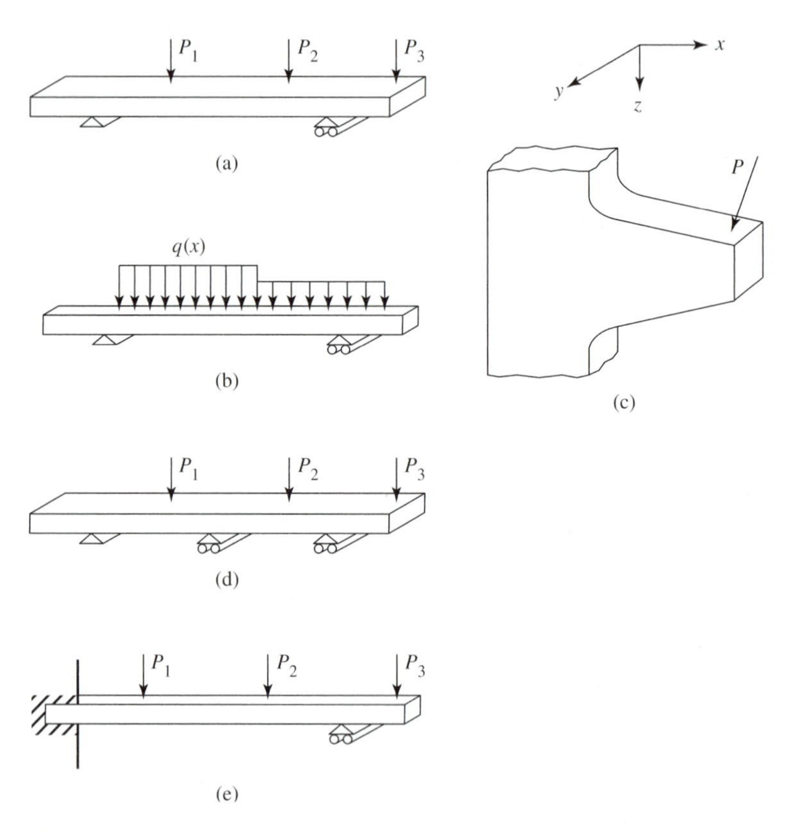

Figure 5.1 Bending of beams with concentrated and distributed loads.

SUPPORT REACTIONS

Actual machine member supports are usually idealized to simplify analysis. A **simple support** consists of a concentrated reaction force. Simple supports are used to represent ball bearings, roller bearings, journal bearings, and other supports that permit some rotation *in the plane of the bending deflection.* The beam supports shown symbolically in Parts (a), (b), and (d) of Figure 5.1 represent simple supports. Rotation is permitted about the y-axis. If beam supports do not permit relative motion along the beam axis, axial forces in the beam could result in additional restraint. It is customary to neglect this effect when designing machines.

Clamped or **built-in supports** provide a reaction moment and a reaction force, permitting no rotation in the plane of bending deflection. The slope of a bending deflection curve is zero at a clamped support. Clamped supports are used to represent supports that are very rigid. A clamped support is shown symbolically at the left end of the beam in Figure 5.1(e). A gear tooth [e.g., the rack tooth in Figure 5.1(c)] would ordinarily be considered clamped.

EQUILIBRIUM OF STATICALLY DETERMINATE BEAMS

If the support reactions of a beam can be determined from the equations of statics (moment and force equilibrium), then the problem is **statically determinate.** The beam illustrated in Figure 5.1(a) is statically determinate, assuming that the loads are known. Taking the sum of the moments about one reaction permits us to calculate the other reaction. We can calculate both reaction forces in this manner and check our values by considering the sum of the forces on the beam. Figure 5.1(b) and (c) also represents statically determinate problems. If the number of unknown reaction forces and moments exceeds the number of independent equations of statics that describe the problem, then the problem is **statically indeterminate.** The beams represented by Figure 5.1(d) and (e) are statically indeterminate.

SHEAR FORCE AND BENDING MOMENT

Shafts, beams, and other machine members have internal forces and moments. Suppose that the beam in Figure 5.2(a) is cut perpendicular to the x-axis at an arbitrary location. In general, it would be necessary to apply a force V (called **shear** or **shear force**) and a couple M (called **moment** or **bending moment**) to the cut face to maintain equilibrium of forces and moments, as shown in Figure 5.2(b). Part (c) shows the customary sign convention for concentrated loads, distributed loads, shear force, and bending moment. Engineers plot shear force and bending moment versus location on a beam to aid them in design.

EXAMPLE PROBLEM Support Reactions, Shear Force, and Bending Moment

Find the shear and moment (i.e., shear force and bending moment) in an idler gear shaft.

Design Decisions: The idler gear is to be mounted between bearings as in Figure 5.3(a).

Solution: *Modeling the Actual Problem:* The bearings will be treated as simple supports, and the total resultant load will be treated as a point load on the shaft, as shown in Figure 5.3(b).

Figure 5.2 Shear force and bending moment. (a) Loading; (b) cut section; (c) sign convention.

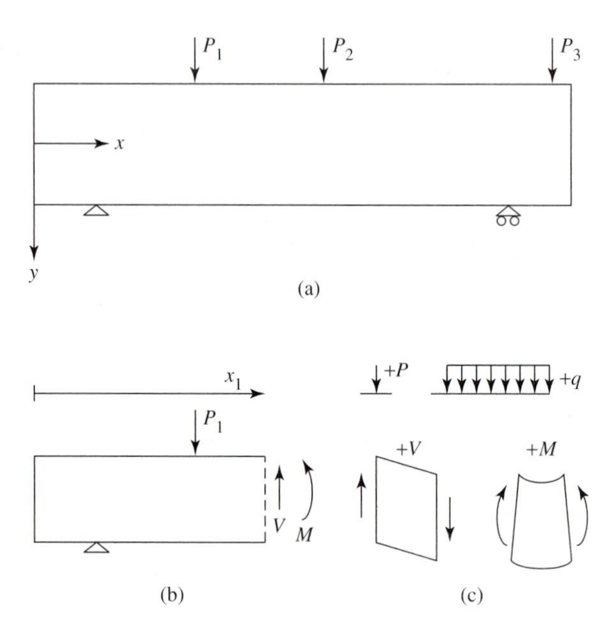

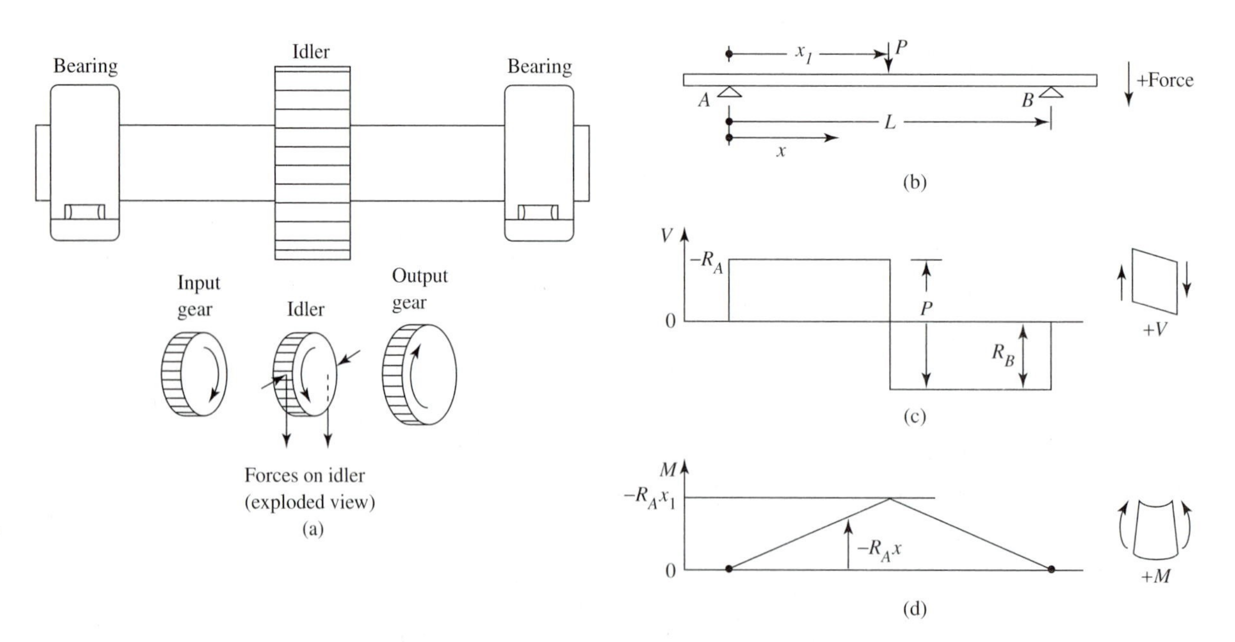

Figure 5.3 Idler gear shaft.

Reactions: We must first find the reactions at the bearings. If we sum the moments about one reaction, only one unknown is left. Taking the moments of forces about point A, we have

$$Px_1 + R_B L = 0$$

from which the reaction force at B is found to be

$$R_B = -Px_1/L$$

where downward forces will be considered positive. Moment equilibrium about point B leads to the reaction at A.

$$R_A = -P(L - x_1)/L$$

Adding the forces and reactions, we see that the forces on the shaft balance.

$$R_A + P + R_B = 0$$

Shear: Positive shear is defined as in Figure 5.3(c). The shear changes by the negative of the value of each reaction and load. The result is

$$V = -R_A \qquad 0 < x < x_1$$

and

$$V = -R_A - P = R_B \qquad x_1 < x < L$$

as plotted in Figure 5.3.

Moment: Moment at any location is equal to the integral of the shear with respect to x.

$$M = \int V\, dx \tag{5.1}$$

If there are no concentrated moments, then Equation (5.1) is equivalent to the area under the shear plot.

$$\begin{aligned} M &= -R_A x && 0 \le x \le x_1 \\ M &= -R_B(L - x) && x_1 \le x \le L \\ M_{\text{MAX}} &= -R_A x_1 = Px_1(L - x_1)/L && x = x_1 \end{aligned}$$

as shown in Figure 5.3(d). If the load is centered between the bearings, then

$$M_{\text{MAX}} = PL/4 \quad x = x_1 = L/2$$

Design Implications: In order to minimize bending moment, the engineer must make the design compact, using the smallest practical space between bearings.

■■

SINGULARITY FUNCTIONS APPLIED TO MACHINE DESIGN

Problems like the preceding example can be solved quickly by hand calculations or by graphical or semigraphical methods. However, design studies, and problems involving multiple loads or loads in two planes, are more efficiently handled by computer. Singularity functions can be used to reduce the programming effort.

Singularity functions are a family of functions based on the unit step function (also called **Heaviside's unit function**). Singularity functions can be defined as follows. For $n \ge 0$,

$$S(x,a,n) = \begin{cases} 0 & x < a \\ (x - a)^n & x \ge a \end{cases} \tag{5.2}$$

Figure 5.4 illustrates singularity functions.

The singularity function with $n = -2$, that is,

$$S(x,a,-2) = \begin{cases} \pm\infty & x = a \\ 0 & \text{elsewhere} \end{cases} \tag{5.3}$$

is equivalent to a unit concentrated moment at $x = a$.

Setting $n = -1$, we have

$$S(x,a,-1) = \begin{cases} \infty & x = a \\ 0 & \text{elsewhere} \end{cases} \tag{5.4}$$

The function $S(x,a,-1)$, equivalent to the Dirac delta function, can be used to represent a unit concentrated load at $x = a$.

The unit step function, that is,

$$S(x,a,0) = \begin{cases} 0 & x < a \\ (x-a)^0 = 1 & x \geq a \end{cases} \tag{5.5}$$

can be used to represent a distributed load over all or part of a machine member.

Figure 5.4 Singularity functions.

n	Application	Plot
−2	Concentrated moment	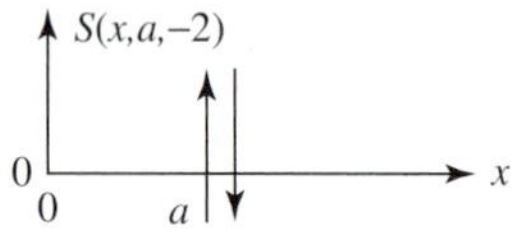 $S(x,a,-2)$; 0, 0, a, x
−1	Concentrated force*	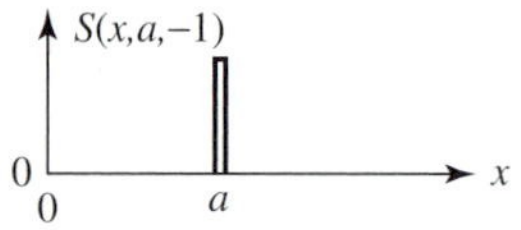 $S(x,a,-1)$; 0, 0, a, x
0	Distributed load*	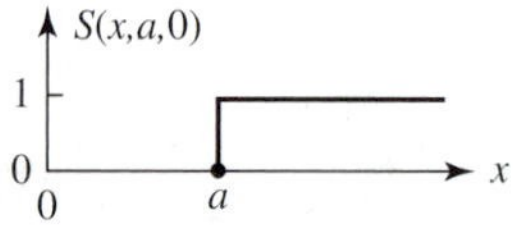 $S(x,a,0)$; 1, 0, 0, a, x
1	Ramp*	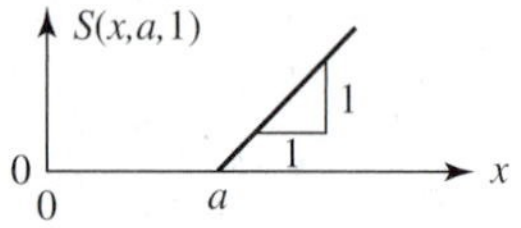 $S(x,a,1)$; 1, 1, 0, 0, a, x
2	Parabola*	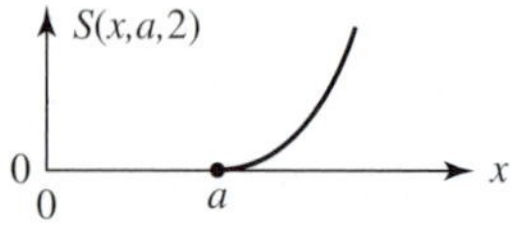 $S(x,a,2)$; 0, 0, a, x

*These functions can also be used to represent shear, moment, etc.

The ramp function, that is,

$$S(x,a,1) = \begin{cases} 0 & x < a \\ (x-a)^1 = x - a & x \geq a \end{cases} \tag{5.6}$$

could be used to represent a linearly increasing load on a beam. However, this and other functions like the parabolic function, which is

$$S(x,a,2) = \begin{cases} 0 & x < a \\ (x-a)^2 & x \geq a \end{cases} \tag{5.7}$$

appear as we represent shear, moment, slope, and deflection of machine members.

The singularity function $S(x,a,n)$ can be written in the form $\langle x - a\rangle^n$, which is suitable for hand calculation. However, the form $S(x,a,n)$ is more easily "understood" by a computer.

EXAMPLE PROBLEM Representing Loading with Singularity Functions

A machine member is loaded as follows.

Concentrated moment: 500 in lb clockwise at $x = 1.25$ in

Concentrated load: 200 lb downward at $x = 5.5$ in

Distributed load: 15 lb/in downward between $x = 3.5$ in and $x = 8$ in

Bearing reactions R_A and R_B are located at $x = A$ and $x = B$, respectively.

Represent the loading with singularity functions.

Design Decisions: Bearing locations A and B must be selected.

Assumptions: The bearings will be assumed to act as simple supports (concentrated reactions). The sign convention is as follows: Downward concentrated loads, distributed loads, and reactions are positive, upward negative. The clockwise moment is negative in the loading expression; the sign changes in the shear expression. The result is a positive jump in a moment plot for a clockwise concentrated moment.

Solution: The loading and reactions can be described as follows:

$$\text{Load } L(x) = -500S(x,1.25,-2) + 200S(x,5.5,-1) + R_AS(x,A,-1) + R_BS(x,B,-1) + 15[S(x,3.5,0) - S(x,8,0)]$$

This equation should follow an expression for calculating R_A and R_B. Note that the step function $S(x,3.5,0)$ "turns on" the distributed load and $S(x,8,0)$ turns it off.

Design Implications: This equation leads to similar equations for shear force and bending moment. Bending moment is needed to design the part on the basis of stress.

Singularity Functions in a Computer Program

Equation 5.2, which defines the singularity function for $n \geq 0$, is equivalent to the logical IF–THEN–ELSE statement in BASIC:

$$\text{IF } x >= a \text{ THEN } S(x,a,n) = (x - a)^n \text{ ELSE } S(x,a,n) = 0 \tag{5.8}$$

or the FORTRAN block IF statement (for $n \geq 0$):

$$\begin{aligned} &\text{IF } (x.\text{GE}.a)\ S(x,a,n) = (x - a)^n \\ &\text{ELSE } S(x,a,n) = 0 \\ &\text{END IF} \end{aligned} \tag{5.9}$$

Using *MathCAD*™, the following simple definition can be used (for $n \geq 0$):

$$S(x,a,n) = \text{IF}[x \geq a,\ (x - a)^n,\ 0] \tag{5.10}$$

Singularity functions can also be defined in terms of an IF expression in *Lotus 1–2–3* and other spreadsheets.

Singularity Functions Applied to Shear and Moment in Machine Members: Integration of Singularity Functions

Analysis of beams, shafts, and other machine members requires integration. Shear force is given by

$$V(x) = -\int L\ dx \tag{5.11}$$

and bending moment by

$$M(x) = \int V\ dx \tag{5.12}$$

Integration of singularity functions follows simple rules.

$$\int_0^x S(x,a,-2)\ dx = S(x,a,-1) \tag{5.13}$$

$$\int_0^x S(x,a,-1)\ dx = S(x,a,0) \tag{5.14}$$

and

$$\int_0^x S(x,a,n)\ dx = S(x,a,n+1)/(n+1) \quad \text{for } n \geq 0 \tag{5.15}$$

That is,

$$\int_0^x S(x,a,0)\ dx = S(x,a,1)$$

$$\int_0^x S(x,a,1)\ dx = S(x,a,2)/2$$

and

$$\int_0^x S(x,a,2)\ dx = S(x,a,3)/3 \quad \text{etc.}$$

The integration limits could have been shown as $-\infty$ to x. However, they are given as 0 to x since we usually orient the coordinate axes so that a machine member lies in the $+x$ region.

EXAMPLE PROBLEM Analysis of a Beam with Several Loads

A 160 mm long beam is loaded as follows:

F_i (N)	X_i (mm from left end)
1000	10
−500	60
−300	80
500	105
1200	120
850	135

In addition, there is a distributed load q of 20 N/mm over the entire length. Supports identified as reactions R_A and R_B are located, respectively, at $A = 40$ mm and $B = 150$ mm from the left end, as shown in Figure 5.5(a). Find the reactions, and plot shear force and bending moment.

Decisions: The reactions will be treated as simple supports. Note that the six concentrated loads and the two reactions divide the beam into nine regions which could be represented by nine equations. However, a solution in that form could be long and tedious. Instead, we will utilize mathematical software and the singularity function. The sign convention is that downward distributed loads, concentrated loads, and reactions are positive.

Solution: *Reactions:* We must find the two unknown reactions. Taking moments about point A, only one unknown, R_B, appears in the equation

$$M_{\text{ABOUTA}} = 0$$

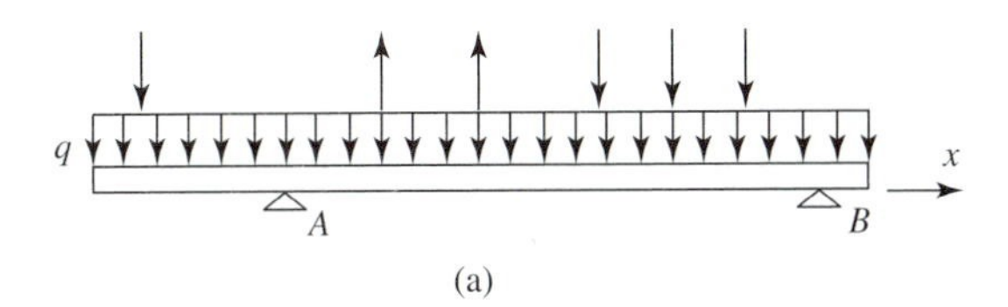

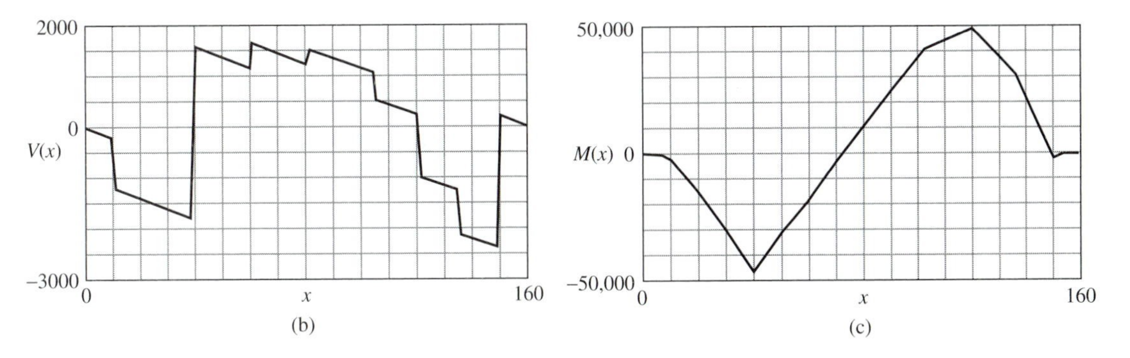

Figure 5.5 Analysis of a beam with several loads.

The contributions of the concentrated loads to the moment about A are $F_0(X_0 - A)$, $F_1(X_1 - A)$, and so on. We let the computer sum these contributions by typing

$$\sum_i [F_i(X_i - A)]$$

where $i = 0, \ldots, 5$

When finding the reactions, we can replace the distributed load by its resultant, the total distributed load qL located at the beam center $L/2$. Its contribution to moment about A is

$$qL(L/2 - A)$$

Adding all the contributions yields

$$M_{\text{ABOUTA}} = \sum_i [F_i(X_i - A)] + qL(L/2 - A) + R_B(B - A) = 0$$

from which

$$R_B = \{-1/(B - A)\}\left\{\sum_i [F_i(X_i - A)] + qL(L/2 - A)\right\} = -3357 \text{ N}$$

Similarly, the sum of the moments about B leads to

$$R_A = \{1/(B - A)\}\left\{\sum_i [F_i(X_i - B)] + qL(L/2 - B)\right\} = -2593 \text{ N}$$

To check the reactions, we type

$$\sum F + R_A + R_B + q \cdot L =$$

The result is zero; the beam is in equilibrium.

Beam Loading: Loading is given by the following single equation:

$$L(x) = R_A S(x,A,-1) + R_B S(x,B,-1) + qS(x,0,0) + \sum_i [F_i S(x,X_i,-1)]$$

In this case, the actual distributed load $qS(x,0,0)$ is used; if we used the resultant, then the shear and moment calculations would be incorrect. To avoid the difficulties in defining the singularity function $S(x,a,n)$ for negative values of n, we will not include $L(x)$ in a computer program.

We now program the definition of the singularity function for $n \geq 0$ in a form appropriate for the chosen language or software. For example,

$$S(x,a,n) = \text{IF}[x \geq a, (x - a)^n, 0] \quad \textbf{(5.10)}$$

which yields $(x - a)^n$ if $x \geq a$ is true, and zero if false.

Shear: Shear force is given by

$$V(x) = -\int L(x)\,dx$$

Using the integration rules for the singularity functions, we obtain

$$V(x) = -\left\{R_A S(x,A,0) + R_B S(x,B,0) + qS(x,0,1) + \sum_i [F_i S(x,X_i,0)]\right\}$$

Setting $x = 0, 0.01L, \ldots, L$, we request a plot of $V(x)$ versus x which appears in Figure 5.5(b). As expected, shear force is zero at both ends of the beam since there are no concentrated loads or reactions at the ends.

Moment: Bending moment is obtained by integrating the shear expression.

$$\begin{aligned} M(x) &= V(x)\,dx \\ &= -\{R_A S(x,A,1) + R_B S(x,B,1) \\ &\quad + qS(x,0,2)/2 + \textstyle\sum_i [F_i S(x,X_i,1)]\} \end{aligned}$$

This expression is plotted in Figure 5.5(c). As expected, moment is zero at the ends of the beam since the ends are not clamped. The greatest negative moment occurs at reaction *A*, and the greatest positive moment at the 1200 N load.

■■

BENDING STRESS

Tensile stress is of greatest consequence in the design of most beams. The tensile stress distribution due to bending loads in the *z*-direction is linear across a section of the beam, as illustrated in Figure 5.6. That is,

$$\sigma_x = Mz/I \qquad \textbf{(5.16)}$$

where σ_x = tensile stress (in the *x*-direction)
M = bending moment
z = distance from the neutral axis
I = moment-of-inertia, the second moment of area about the neutral axis

The **neutral axis** of a cross section of a beam is a line through the center-of-gravity of the section, perpendicular to the loading direction. The center-of-gravity of a rectangular cross section is the geometric center of the section. If the beam cross section is symmetric about the *y*-axis, and if loads are in the *z*-direction, then the *y*-axis is the neutral axis. The neutral axes of all cross sections form the **neutral surface.**

Assumptions and Limitations

To avoid twisting, load resultants must lie in a plane of symmetry of the beam. It is assumed that plane sections originally perpendicular to the beam axis remain plane after the beam is loaded. Stresses must be somewhat below the yield point of the material so that the proportional limit is not exceeded.

Tensile stress due to bending is zero at the neutral axis. It is greatest at the extreme fiber, the greatest distance from the neutral axis. Maximum tensile stress due to bending alone is given by

$$\sigma_{x(\mathrm{MAX})} = |Mc/I| \qquad \textbf{(5.17)}$$

where $c = |z_{\mathrm{MAX}}|$, the greatest distance from the neutral axis

We are particularly interested in the stress at locations of greatest bending moment and at points of stress concentration.

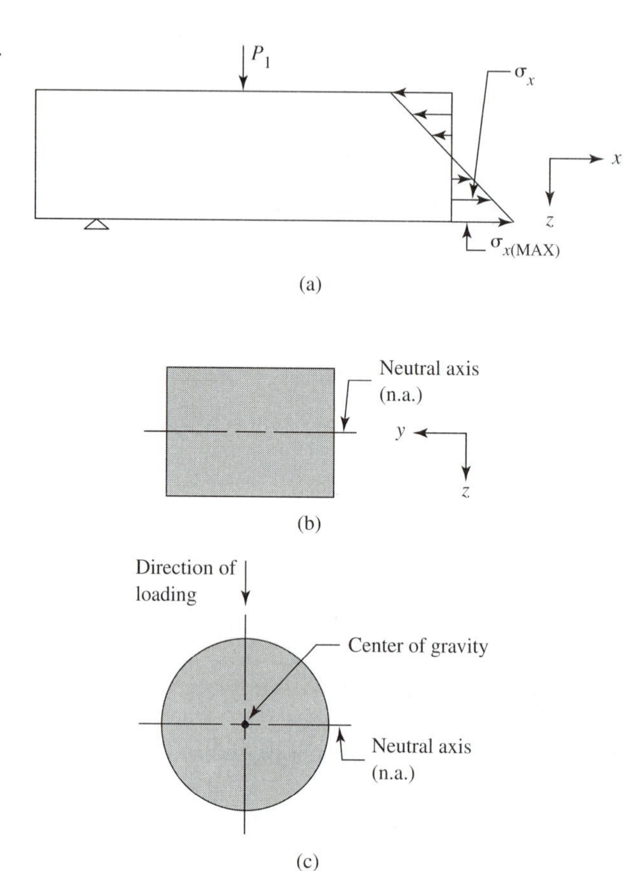

Figure 5.6 Bending stress.

LOCATING THE NEUTRAL AXIS

Consider a beam subject to loads in the z-direction. As noted above, if the beam is symmetric about the y-axis, then the y-axis is the neutral axis. The neutral axis of a circular shaft with vertical loads is the horizontal diameter of a cross section, as in Figure 5.6(c). In general, the neutral axis runs through the center-of-gravity which is located by

$$z_1 = \left[\int_{\text{CROSS SECTION}} v \, dA\right] / A \tag{5.18}$$

EXAMPLE PROBLEM Locating the Center-of-Gravity and the Neutral Axis of a Semicircular Section

Suppose that a beam of semicircular cross section is loaded as in Figure 5.7. Locate the center-of-gravity and the neutral axis of a cross section of the beam.

Solution: In general, the area of a cross section is given by

$$A = \int_{\text{CROSS SECTION}} dA \tag{5.19}$$

Figure 5.7 Locating the neutral axis.

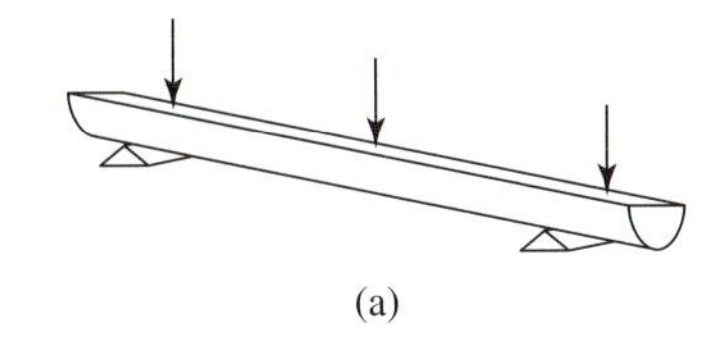

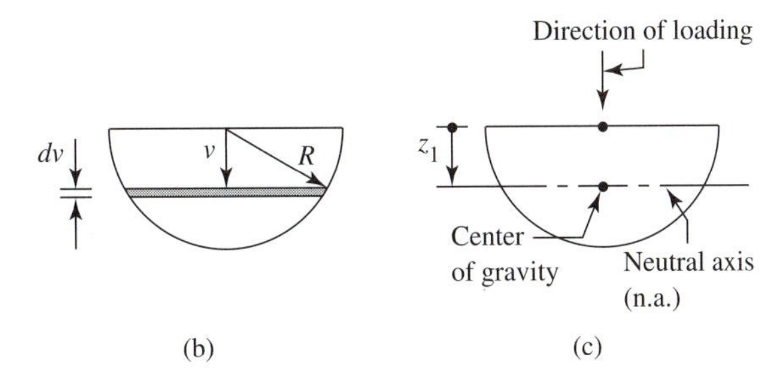

We need not integrate this expression, because we know that

$$A = \pi R^2/2 \quad \text{for a semicircular section}$$

where R = radius

The center-of-gravity is obviously located on a vertical line of symmetry of the cross section. To locate the center-of-gravity with reference to the flat surface, we note that

$$dA = 2(R^2 - v^2)^{1/2}\, dv \quad \text{for the semicircular section}$$

where v = distance from the flat surface as shown in Figure 5.7(b)

$$z_1 = \left[\int_{\text{CROSS SECTION}} v\, dA\right]/A$$

$$= \left[\int_0^R 2v(R^2 - v^2)^{1/2}\, dv\right]/A$$

We have a few choices in evaluating the integral: Use the rules of calculus, use a table of integrals, or use mathematics software or an integration routine on a calculator. All lead to the same answer.

$$z_1 = 4R/(3\pi) = 0.4244R$$

Since the loading is in a vertical plane, the neutral axis is horizontal, as shown in Figure 5.7(c).

Design Implications: Bending stress is proportional to distance from the neutral axis. For this section, bending stress magnitude is greater at the bottom of the section than at the flat surface.

■■

The center-of-gravity of a symmetrical I-beam is obviously located at the geometric center. However, the location of the center-of-gravity of T-sections, angle-sections, and channels is not obvious. Instead of using an integral expression, it may be more convenient to compute

$$z_T = \left[\sum_{i=1}^{n} A_i z_i\right] / \left[\sum_{i=1}^{n} A_i\right] \qquad \textbf{(5.20)}$$

where the section is broken up into n elements as follows:

A_i = area of ith element

z_i = distance of center-of-gravity of ith element from some arbitrary reference (say, the top of the section)

z_T = distance of center-of-gravity of the section from the same reference

EXAMPLE PROBLEM The Center-of-Gravity and the Neutral Axis of a T-Section

Locate the center-of-gravity and the neutral axis of a T-section. Referring to Figure 5.8, the dimensions in inches are

$$d = 2.95 \qquad b_f = 3.94 \qquad t_f = 0.215 \qquad t_w = 0.17$$

Design Decisions: The section will be oriented as the letter T, and the resultant loads and reactions will be in the plane of the center of the web.

Solution: The T-section is arbitrarily divided into two regions: (1) the flange and (2) the web below the flange. Areas and distances are

$$A_1 = b_f t_f \qquad z_1 = t_f/2$$
$$A_2 = (d - t_f)t_W \qquad z_2 = (d + t_f)/2$$

Mathematics software was used to compute the following:

$$A = \sum_{i=1}^{2} A_i = 1.312 \text{ in}^2$$
$$z_T = \left[\sum_{i=1}^{2} A_i z_i\right] / \left[\sum_{i=1}^{2} A_i\right] = 0.63$$

The center-of-gravity lies on a vertical line through the center of the web, 0.63 in below the top of the flange. For vertical loading, the neutral axis is a horizontal line through the center-of-gravity.

Design Implications: Most designers try to incorporate standard components to reduce cost. Standard structural T's are produced from wide-flange I-beams by shearing or flame-cutting. Rolled sections, including I-beams, angles, and channels, are available in various grades of steel conforming with the American Society for Testing and Materials (ASTM) standards. The standard American Iron and Steel Institute nomenclature for a structural T-section with the dimensions given in this example is WT3×4.5. The nominal section depth is 3 in, and the approximate weight is 4.5 lb/ft based on steel density of 490

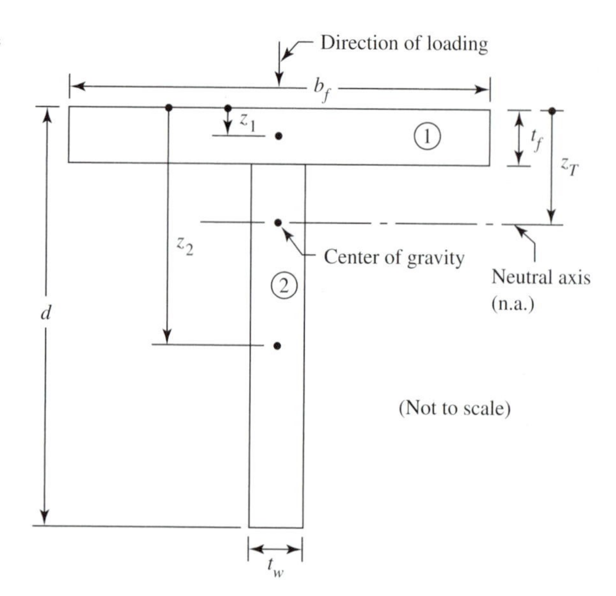

Figure 5.8 The center-of-gravity and the neutral axis of a T-section.

lb/ft³. If fillets and rounded edges are considered in computing area and center-of-gravity location, the results are $A = 1.34$ in² and $z_T = 0.623$ in (Bethlehem Steel Corporation, 1983).

CALCULATING MOMENT-OF-INERTIA

Strength and stiffness of a beam or shaft subject to bending are related to moment-of-inertia. Unless indicated otherwise, the term **moment-of-inertia** will refer to moment-of-inertia about the neutral axis. In general,

$$I = \int_{\text{CROSS SECTION}} z^2 \, dA \tag{5.21}$$

where I = moment-of-inertia

z = distance from the neutral axis

The moment-of-inertia of a beam with a rectangular cross section, as shown in Figure 5.9(a), is given by

$$I = \int_{-h/2}^{h/2} z^2 \cdot b \, dz = bz^3/3 \Big|_{-h/2}^{h/2} = bh^3/12 \tag{5.22}$$

where h = dimension of the section measured parallel to the load direction

b = dimension measured perpendicular to the load

n.a. = neutral axis

For a solid circular shaft [Figure 5.9(b)],

$$I = \pi D^4/64 \tag{5.23}$$

where D = diameter

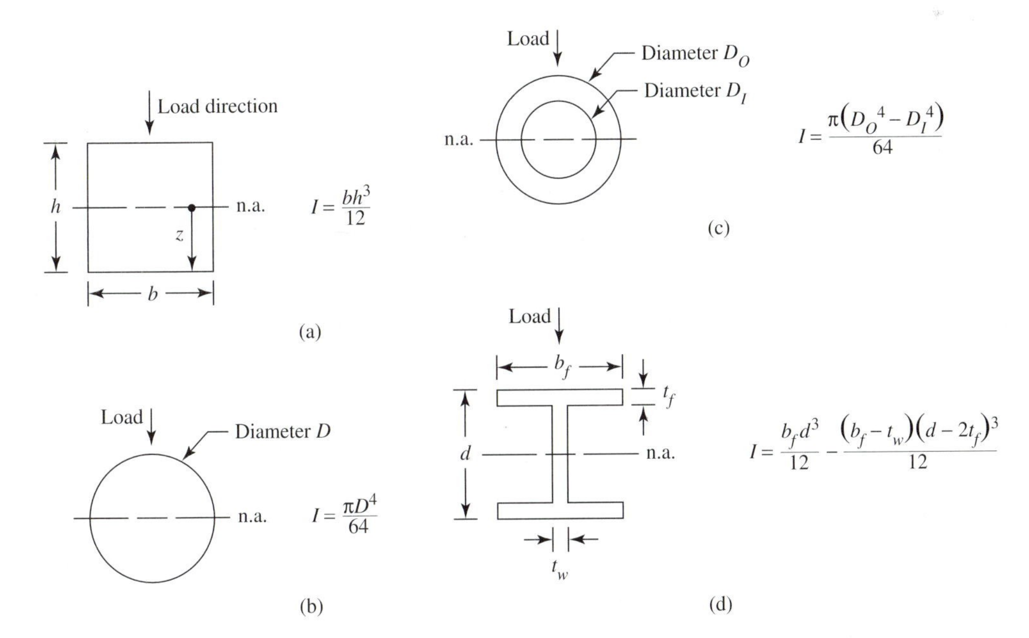

Figure 5.9 Moment-of-inertia. (a) Rectangle; (b) solid shaft; (c) hollow shaft; (d) I-beam.

We find the moment-of-inertia of a hollow shaft or hollow circular beam [Figure 5.4(c)] by subtracting the moment-of-inertia of the void.

$$I = \pi(D_O^4 - D_I^4)/64 \tag{5.24}$$

where D_O and D_I = outer and inner diameters, respectively

Similarly, for a symmetric I-beam [Figure 5.9(d)], we consider the enclosing rectangle and the void to obtain

$$I = b_f d^3/12 - (b_f - t_W)(d - 2t_f)^3/12 \tag{5.25}$$

where the dimensions are defined as in Figure 5.9(d)

EXAMPLE PROBLEM Moment-of-Inertia of an I-Beam and Comparison with Other Cross Sections

(a) An I-beam has the following cross-sectional dimensions:

$$d = 5.9 \qquad b_f = 3.94$$
$$t_f = 0.215 \qquad t_W = 0.17$$

where d = depth (in)

b_f = flange width (in)

t_f = flange thickness (in)

t_W = web thickness (in)

The beam is loaded as in Figure 5.9(d). Find the moment-of-inertia.

(b) A design calls for a beam 100 in long, supported at both ends. There is to be a concentrated load at the center. It is decided that the working strength ($S_W = S_{YP}/N_{FS}$) will be limited to 15,000 psi. Find maximum allowable load.

(c) Is it "good design" to select an I-beam to resist bending loads? Compare some other cross sections.

Solution (a): Using the given dimensions in Equation (5.25), we have

$$I = 3.94 \cdot 5.9^3/12 - (3.94 - 0.17)(5.9 - 2 \cdot 0.215)^3/12$$
$$= 16.01 \text{ in}^4$$

These dimensions correspond to a standard wide flange W6×9 I-beam with moment-of-inertia I = 16.4 in^4 and cross-sectional area A = 2.68 in^2 when fillets and rounded edges are considered (Bethlehem Steel Corporation, 1983). Standard sections can be identified by their shape, depth, and weight. For the W6×9 section, "W" indicates a wide-flange I-section, "6" the depth, and "9" the weight per foot. Depth and weight are usually rounded to whole numbers.

Solution (b): As noted earlier, maximum moment is given by

$$M_{MAX} = PL/4$$

at the center of a beam with a single concentrated load at the center. In this case,

$$M_{MAX} = 100P/4 = 25P$$

Maximum tensile stress due to bending alone is given by Equation (5.17).

$$\sigma_{x(MAX)} = Mc/I$$

Using the standard I-beam,

$$I = 16.4 \text{ in}^4$$

and

$$c = d/2 = 5.9/2 = 2.95 \text{ in}$$

Maximum tensile stress should not exceed working stress; that is,

$$\sigma_{x(\text{MAX})} = 25P \cdot 2.95/16.4 \leq S_W = 15{,}000$$

from which

$$P \leq 15{,}000 \cdot 16.4/(25 \cdot 2.95) = 3335$$

The load must not exceed 3335. By going back through the equations, we find that the units check if P is expressed in pounds. If the units on the left side of an equation are not the same as (or equivalent to) those on the right, an error has probably occurred.

Solution (c): Design Decision: We will evaluate sections on the basis of weight per unit length. The beam will be made as short as practical, subject to other design considerations.

Assumptions: Assume that the same safety factor applies and that the density and the yield point of the material are the same in all cases. (The latter assumption may not be strictly correct since rolling and other forming processes and size have some effect on strength.)

With these assumptions, we see from the above calculations that capacity of a beam to resist bending is proportional to I/c where I/c is called the **elastic section modulus** or simply the **section modulus.** Structural steel weighs about

$$490 \text{ lb/ft}^3 \quad \text{or} \quad 490/12^3 = 0.284 \text{ lb/in}^3$$

Beam weight per unit length is given by

$$w^{\ddagger\ddagger} = 0.284A \text{ lb/in} \quad \text{or} \quad 3.40A \text{ lb/ft}$$

where A = cross-sectional area (in^2)

Standard W6×9 Wide-Flange I-Beam:

Section Modulus	*Cross-sectional Area*
$I/c = 16.4/2.95 = 5.56 \text{ in}^3$	$A = 2.68 \text{ in}^2$

Weight per Unit Length

$w^{\ddagger\ddagger} = 0.284 \cdot 2.68 = 0.761$ lb/in or $3.4 \cdot 2.68 = 9.1$ lb/ft

Solid Rectangular Beam:

$$I = bh^3/12 \qquad c = h/2$$
$$I/c = bh^2/6 \qquad A = bh$$

For a square $a \times a$ cross-sectional beam,

$$I/c = a^3/6 \quad \text{and} \quad A = a^2$$

Suppose that we need a square section as strong as a W6×9 I-beam. Equating section modulus I/c of the square section to I/c for the I-beam, we have

$$a^3/6 = 5.56$$

from which

$$a = (6 \cdot 5.56)^{1/3} = 3.22$$
$$A = a^2 = 10.36 \text{ in}^2$$
$$w^{\ddagger\ddagger} = 0.284 \cdot 10.36 = 2.94 \text{ lb/in}$$

A square-cross-section beam appears to be a poor choice on the basis of weight. It weighs about 3.86 times as much as the W6×9 I-beam with the same bending strength.

Solid Circular Section:

$$I = \pi D^4/64 \qquad c = D/2 \qquad I/c = \pi D^3/32$$

Again, for equal strength we equate section modulus, obtaining

$$I/c = \pi D^3/32 = 5.56 \text{ in}^3$$

from which

$$D = (5.56 \cdot 32/\pi)^{1/3} = 3.84 \text{ in}$$
$$A = \pi D^2/4 = 11.58 \text{ in}^2$$
$$w^{\ddagger\ddagger} = 0.284 \cdot 11.58 = 3.29 \text{ lb/in}$$

which is even heavier than the square section.

By examining the equations for moment-of-inertia and section modulus, and by comparing the I-beam results with the solid circular and rectangular sections, we conclude that removal of material near the neutral axis of a section is likely to improve the strength-to-weight ratio. Let us test this conclusion on a hollow circular section where the ratio of inside diameter to outside diameter is

$$D_I/D_O = 0.8$$

Then

$$\begin{aligned} I &= \pi(D_O^4 - D_I^4)/64 \\ &= \pi[D_O^4 - (0.8D_O)^4]/64 \\ &= 0.02898D_O^4 \\ I/c &= 0.02898D_O^4/(D_O/2) = 0.05796D_O^3 \end{aligned}$$

If the hollow shaft is to have the same section modulus as the W6×9 I-beam, then

$$0.05796D_O^3 = 5.56 \text{ in}^3$$

from which

$$D_O = (5.56/0.05796)^{1/3} = 4.578 \text{ in}$$

The cross-sectional area is

$$\begin{aligned} A &= \pi(D_O^2 - D_I^2)/4 \\ &= \pi[4.578^2 - (0.8 \cdot 4.578)^2]/4 = 5.925 \text{ in}^2 \end{aligned}$$

and

$$w^{\ddagger\ddagger} = 0.284 \cdot 5.925 = 1.683 \text{ lb/in}$$

This hollow circular section is about one-half the weight of a solid circular section of equal strength.

■■

Weight is an important consideration in the design of machine elements, but it is not the only consideration. Gear and belt drive shafts are subject to bending loads, but hollow shafts are seldom used because manufacturing costs would be higher and larger bearings would be required. On the basis of section modulus alone, it would appear that bending members could be made of very thin tubing in order to save weight. However, there is a problem with local buckling or crippling when thin tubes are subject to bending. A similar problem occurs with other thin sections. If an I-section is too thin, the web may fail in shear, or the flanges may buckle. The mechanism of local buckling is complicated. However, we can use the proportions of commercially available rolled sections (I-beams, angles, channels, T's, etc.) as a guide to the design of machine elements subject to bending.

Moment-of-Inertia of Sections That Are Not Symmetrical about the Neutral Axis

The general equation for moment-of-inertia

$$I = \int_{\text{CROSS SECTION}} z^2 \, dA$$

is still valid, but the calculations for asymmetric sections are more difficult than for symmetric sections. Mathematics software may be a time saver when dealing with asymmetric sections.

The **parallel axis theorem** is helpful in some cases. It relates moment-of-inertia about one axis to moment-of-inertia about another axis as follows:

$$I_1 = I_0 + Az_s^2 \qquad \textbf{(5.26)}$$

where I_0 = moment-of-inertia about the neutral axis of a cross section

I_1 = moment-of-inertia about axis 1 which is parallel to the neutral axis

z_s = distance between the axes

A = area of the cross section

EXAMPLE PROBLEM Moment-of-Inertia of a Semicircular Cross-Section Beam

(a) Find the moment-of-inertia of a 10 mm radius semicircular beam loaded as in Figure 5.7.
(b) Check the result by another method.

Solution (a): Referring to Figure 5.7(b), we see that distance from the neutral axis is given by

$$z = v - z_1$$

and

$$dA = 2(R^2 - v^2)^{1/2} \, dv$$

as in the example problem "Locating the Center-of-Gravity and the Neutral Axis of a Semicircular Section." In that problem, we located the center-of-gravity of this section at

$$z_1 = 4R/(3\pi) = 0.4244R = 4.244 \text{ mm}$$

The moment-of-inertia equation becomes

$$I = \int_0^R (v - z_1)^2 \cdot 2(R^2 - v^2)^{1/2} \, dv$$

Using mathematics software or an integrating calculator to calculate this integral, we obtain

$$I = 1097 \text{ mm}^4$$

Solution (b): We know that the moment-of-inertia of a solid circular section about its neutral axis is

$$I_{\text{CIRCULAR SECTION}} = \pi D^4/64 = \pi R^4/4$$

The moment-of-inertia of the semicircular shaft about the top surface is one-half that, or

$$I_1 = \pi R^4/8$$

Using the parallel axis theorem, we find the moment-of-inertia of the semicircular section about its own neutral axis as follows:

$$\begin{aligned} I = I_0 &= I_1 - Az_s^2 \\ &= \pi R^4/8 - [\pi R^2/2][4R/(3\pi)]^2 = 1098 \text{ mm}^4 \end{aligned}$$

This value is essentially the same as that obtained above. ■■

In some cases, we can determine the moment-of-inertia of asymmetric sections by summing the contributions of two or more parts of the section, while utilizing the parallel axis theorem. T-sections, angles, and channels are usually analyzed this way. We can rewrite the parallel axis theorem in summation form as follows:

$$I_T = \sum_{i=1}^{n} [I_0 + Az_s^2]_i \qquad \textbf{(5.27)}$$

where the section is broken into n parts and

I_0 = moment-of-inertia of part of the section about its own central axis (parallel to the section neutral axis)

z_s = distance from the part central axis to the neutral axis of the section

If the section does not have a horizontal plane of symmetry, then, in general, maximum tensile and compressive stresses will not be equal.

■■ EXAMPLE PROBLEM Moment-of-Inertia and Stress in a T-Section

(a) Find the moment-of-inertia of the T-section described in the example problem "The Center-of-Gravity and the Neutral Axis of a T-Section."

(b) Find maximum allowable bending moment if tensile stress cannot exceed 18,000 psi.

(c) A cast iron T-section is to be simply supported near each end and subject to vertical downward loads between the supports. How should the section be oriented?

Solution (a): We break the T-section into two parts: (1) the flange and (2) the web below the flange. For the flange,

$$\begin{aligned} I_{01} &= b_f t_f^3/12 \\ z_{S1} &= t_f/2 - z_T \end{aligned}$$

and for the web,

$$I_{02} = t_W(d - t_f)^3/12$$

$$z_{S2} = (d + t_f)/2 - z_T$$

Using the parallel axis theorem and the values computed for the T-section in the earlier example, we obtain

$$I = 0.946 \text{ in}^4$$

If fillets and rounded edges are considered, the result is

$$I = 0.950 \text{ in}^4$$

for a standard WT3×4.5 structural T (Bethlehem Steel Corporation, 1983).

Solution (b): Based on the standard WT3×4.5 section, $z_T = 0.623$ in measured from the top of the flange. The greatest distance from the neutral axis to the extreme fiber (the bottom of the web) is

$$c = d - z_T = 2.95 - 0.623 = 2.327 \text{ in}$$

Bending stress is given by

$$\sigma_{x(\text{MAX})} = |Mc/I|$$

from which

$$M_{\text{MAX}} = \sigma_{x(\text{MAX})}\, I/c = 18{,}000 \cdot 0.950/2.237 = 7644 \text{ in} \cdot \text{lb}$$

Solution (c): Bending moment will be positive, resulting in tension at the bottom of the beam and compression at the top. If the section is oriented as an upside-down T, the extreme fiber of the flange will be in tension, and the end of the web in compression. Compression stress will greatly exceed tensile stress because the extreme fiber of the web is farther from the neutral axis than the extreme fiber of the flange. Cast iron is much stronger in compression than in tension. Therefore, the upside-down T orientation will best take advantage of the properties of cast iron.

LOCATING BEARINGS AND OTHER SUPPORTS

Bending stress is one criterion that can be used to determine the location of shaft bearings or beam supports. Other determining factors include deflection and vibration characteristics, including critical speed. Support and bearing locations are often forced upon us by other design features such as the location of gears in a transmission. If stress is selected as the criterion for locating supports, we note that maximum bending stress is directly proportional to the absolute value of bending moment in a shaft or beam of uniform cross section. Then the ideal support locations are those which result in the smallest absolute value of maximum bending moment for the given loading. Ordinarily, a designer will minimize bending stress by making a beam or shaft as short as practical. When loads are concentrated, supports are located as close as possible to loading points.

EXAMPLE PROBLEM Optimum Support Locations for a Uniformly Loaded Beam

(a) A 100 in-long beam with 10 lb/in uniform load is to be designed with two simple supports. Determine the optimum support locations.

(b) Shelves that are to be 9 in wide are to be designed for the above application, using a material with a tensile strength S_U of 12,800 psi. Find the required shelf thickness.

Solution (a): Design Decision: Support location will be based on bending stress. Thus, we will attempt to minimize the absolute value of bending moment.

For a beam of length L and vertical load per unit length q, the resultant load is qL at the midpoint $x = L/2$. Using moment equilibrium, we find the following support reactions:

$$\begin{aligned} R_A &= qL(L/2 - B)/(B - A) \quad \text{at } x = A \\ R_B &= -qL(L/2 - A)/(B - A) \quad \text{at } x = B \end{aligned} \tag{5.28}$$

We then verify equilibrium of vertical forces.

$$R_A + R_B + qL = 0$$

Shear force is given by

$$V = -[R_A S(x,A,0) + R_B S(x,B,0) + qS(x,0,1)] \tag{5.29}$$

where $S(x,a,n)$ = the singularity function

Bending moment is given by

$$M = -[R_A S(x,A,1) + R_B S(x,B,1) + qS(x,0,2)/2] \tag{5.30}$$

Since the loading is uniform, we will place the reactions equidistant from the center. Then

$$R_A = R_B = -qL/2$$

First Guess: Try reactions at the ends: $A = 0$ and $B = L$. Maximum moment is at the center.

$$\begin{aligned} |M|_{\text{MAX}} &= M(L/2) \\ &= -[R_A S(L/2,0,1) + R_B S(L/2,L,1) + qS(L/2,0,2)/2] \\ &= -[(-qL/2)(L/2 - 0)^1 + q(L/2 - 0)^2/2] \\ &= qL^2/8 \end{aligned}$$

We will try again since $|M|_{\text{MAX}}$ appears high.

Second Guess: Put both reactions at the center: $A = B = L/2$. Then

$$\begin{aligned} |M|_{\text{MAX}} &= M(L/2) \\ &= -[R_A S(L/2,L/2,1) + R_B S(L/2,L/2,1) + qS(L/2,0,2)/2] \\ &= -q(L/2 - 0)^2/2 = -qL^2/8 \end{aligned}$$

which is no improvement.

Third Guess: Try the one-third points: $A = L/3$ and $B = 2L/3$. Equations (5.29) and (5.30) were coded on mathematics software to calculate and plot shear force and bending moment for $L = 100$ in and $q = 10$ lb/in. The results are shown in Figure 5.10.

Maximum moment, which occurs at the supports, is given by

$$\begin{aligned} |M|_{\text{MAX}} &= M(L/3) \\ &= -[R_A S(L/3,L/3,1) + R_B S(2L/3,L/3,1) + qS(L/3,0,2)/2] \\ &= -[(q(L/3 - 0)^2/2] = -qL^2/18 = -5556 \text{ in} \cdot \text{lb} \end{aligned}$$

Figure 5.10 Beam with distributed load. Supports are at one-third points. (a) Shear force; (b) bending moment.

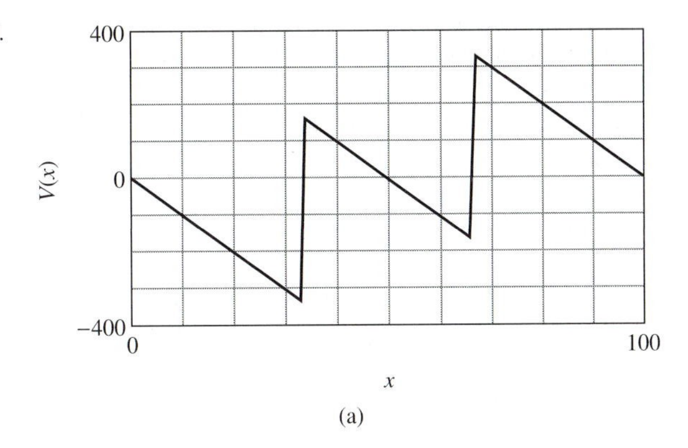

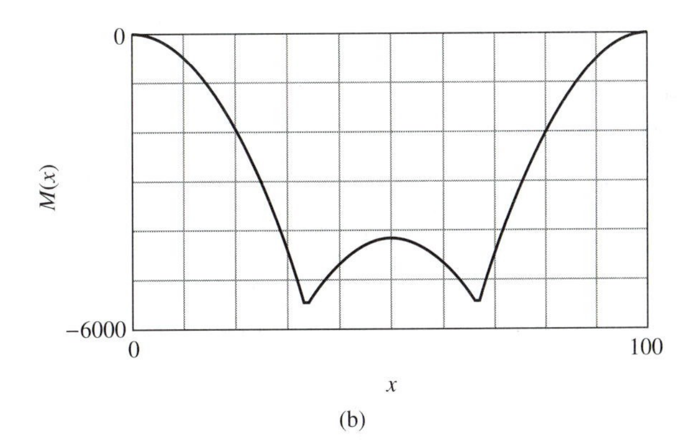

Although $|M|_{MAX}$ is less than one-half that obtained above, it is not the optimum value.

Fourth and Final Determination of Support Locations: Instead of continuing to guess, it is time to investigate the form of the shear and moment plots. Shear is negative to the left of support location A and positive to the right of A. If the supports are far apart, $|M|_{MAX}$ is greatest at $L/2$. If the supports are close together, $|M|_{MAX}$ is greatest at the supports. To minimize $|M|_{MAX}$, we adjust location A so that

$$|M(A)| = |M(L/2)|$$

Recall that

$$M = \int V \ dx$$

and note that this represents the area under the shear plot. For us to obtain $|M(A)| = |M(L/2)|$, the negative shear area to the left of A should equal one-half the positive area between A and $L/2$. The result is

$$qA \cdot A/2 = q(L - A)(L - A)/4$$

from which

$$A^2 + AL - L^2/4 = 0$$

The positive root is

$$A = L(-1 + 2^{1/2})/2 = 0.2071\, L$$

Figure 5.11 Optimum location of supports. (a) Shear force; (b) bending moment.

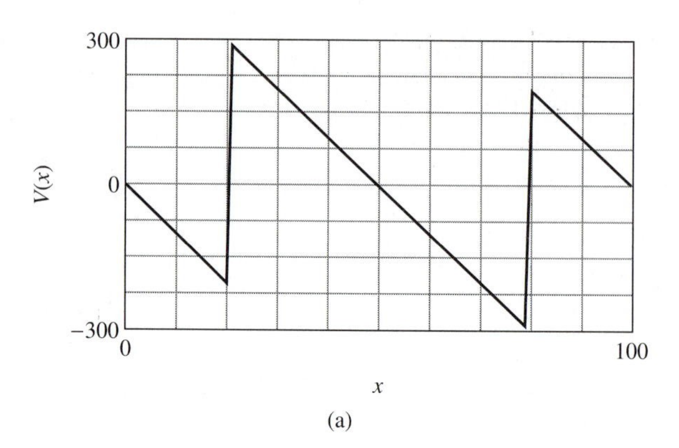

(a)

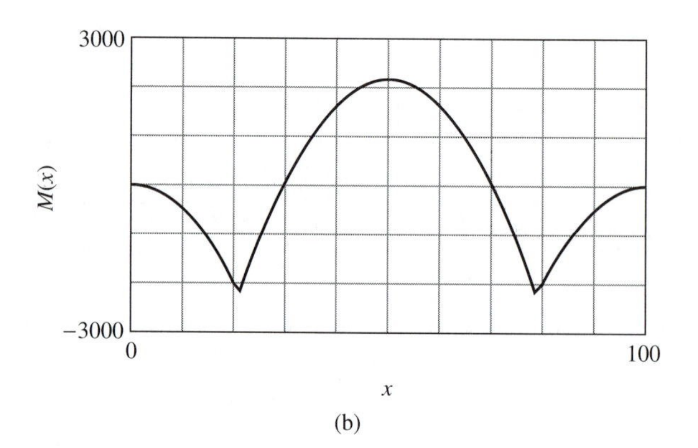

(b)

Using this value and $B = L - A$ for the support locations, we compute and plot the shear and moment with mathematics software. The results are shown in Figure 5.11, where

$$M(A) = M(B) = -2145 \quad \text{and} \quad M(L/2) = 2145 \text{ in} \cdot \text{lb}$$

Solution (b): Design Decisions: A safety factor of 5 will be used because of uncertainty regarding loading. The results of the fourth and final determination of support locations will be used. The shelf will be fastened at the supports in case a downward reaction force is required owing to partial loading. For a rectangular section,

$$I = bh^3/12 \quad \text{and} \quad c = h/2$$

We equate the working strength to the bending stress to obtain

$$S_W = S_U/N_{FS} = Mc/I = 6M/(bh^2)$$

from which

$$h^2 = 6MN_{FS}/(bS_U) = 6 \cdot 2145 \cdot 5/(9 \cdot 12{,}800)$$
$$h = 0.7474$$

We will make the shelf 0.75 in thick.

Table 5.1 Relationship between and among loading, shear force, bending moment, curvature, slope, and deflection.

Quantity	Equation	Sign Convention
Loading	L	$+L$ downward
Shear force	$V = -\int L\,dx$	$+V$
Bending moment	$M = \int V\,dx$	$+M$
Curvature	$K = d^2w/dx^2 = -M/[EI]$	$+K$
Slope	$\theta = dw/dx = \int K\,dx$	$+\theta$
Deflection	$w = \int \theta\,dx$	$+w$ downward

BEAM DEFLECTION

The slope and deflection of a beam can be determined from bending moment. Loading, shear force, bending moment, curvature, slope, and deflection are related as shown in Table 5.1.

There are many analytical, numerical, graphical, and semigraphical methods for determining the deflection of beams. The singularity function can be used to design and analyze beams with constant cross section as well as stepped beams and shafts.

EXAMPLE PROBLEM Deflection of a Cantilever Beam

A beam is clamped at the left end ($x = 0$) and free at the right end ($x = L$). It has a distributed downward load q over its entire length.

(a) Find shear force, bending moment, curvature, slope, and deflection. Use the singularity function.
(b) Rewrite the equations used to solve Part(a) without using the singularity function.
(c) Find slope and deflection at the right end.

Solution (a): Equilibrium of forces and moments requires a force reaction

$$R_0 = -qL$$

and a moment reaction

$$M_0 = -qL^2/2$$

both at $x = 0$. The sign conventions are as follows: downward distributed loads, concentrated loads, and reactions positive. The counterclockwise moment is entered as positive in the loading expression; its sign changes in the shear expression.

Written in singularity function form, the loading is

$$L = qS(x,0,0) - qLS(x,0,-1) + qL^2S(x,0,-2)/2$$

Using the above equations and the rules governing the singularity function, we obtain the following.

Shear Force:

$$V = -qS(x,0,1) + qLS(x,0,0) - qL^2S(x,0,-1)/2$$

Bending Moment:

$$M = -qS(x,0,2)/2 + qLS(x,0,1) - qL^2S(x,0,0)/2$$

Curvature:

$$K = [qS(x,0,2)/2 - qLS(x,0,1) + qL^2S(x,0,0)/2]/[EI]$$

Slope:

$$\theta = [qS(x,0,3)/6 - qLS(x,0,2)/2 + qL^2S(x,0,1)/2]/[EI]$$

Deflection:

$$w = [qS(x,0,4)/24 - qLS(x,0,3)/6 + qL^2S(x,0,2)/4]/[EI]$$

The slope and deflection equations ordinarily contain integration constants which depend on support conditions. Since, for the cantilever,

$$\theta(0) = w(0) = 0$$

the integration constants are zero.

Solution (b): If there were several concentrated loads, the beam would have to be divided into regions, resulting in several equations each for V, M, K, θ, and w. For the distributed loading in this problem, it is easy to evaluate the singularity functions. The results are as follows:

$$L = q$$
$$V = -qx + qL$$
$$M = -qx^2/2 + qLx - qL^2/2$$
$$K = [qx^2/2 - qLx + qL^2/2]/[EI]$$
$$\theta = [qx^3/6 - qLx^2/2 + qL^2x/2]/[EI]$$
$$w = [qx^4/24 - qLx^3/6 + qL^2x^2/4]/[EI]$$

Solution (c): The slope is

$$\theta(L) = [qL^3/6 - qLL^2/2 + qL^2L/2]/[EI] = qL^3/[6EI]$$

and the deflection is

$$w(L) = [qL^4/24 - qLL^3/6 + qL^2L^2/4]/[EI] = qL^4/[8EI]$$

STEPPED SHAFTS AND BEAMS

Stepped shafts are used when one shaft diameter is needed to accommodate gears or sheaves, and another to accommodate bearings. Problems involving slope and deflection are difficult because changes in cross section of a shaft or beam require special treatment. Shear and moment are not affected, but curvature, slope, and deflection are. Plotting of results helps us to verify that the boundary conditions (bearing restraints, etc.) are met, and it gives us insight into the results. Graphical methods were traditionally favored for analysis of stepped shafts. Currently, finite element analysis programs are commonly used. The example that follows illustrates the use of the singularity function to write a computer program for determining deflection of a stepped shaft. The results may lead us to change shaft diameter, load locations, bearing locations, or other parameters. Even though the initial programming effort is substantial, design changes can be investigated with little effort.

EXAMPLE PROBLEM Analysis of a Stepped Shaft

A steel shaft (see Figure 5.12) has bearings at locations x_1 and x_4. Shaft diameter is D_1 for $0 < x < x_2$ and D_2 for $x_2 < x < x_5$. There are vertical loads P_3 at x_3 and P_5 at x_5.

(a) Find the reactions. Express shear, moment, curvature, slope, and displacement in terms of singularity functions.

(b) Let the shaft length be 7.5 with bearings located at $x_1 = 0$ and $x_4 = 5$. Shaft diameter is $D_1 = 1.25$ for $0 < x < 2$ and $D_2 = 1.625$ for $2 < x < 7.5$. Loads are $P_3 = 250$ at $x_3 = 3.5$ and $P_5 = 400$ at $x_5 = 7.5$. Dimensions are given in inches and loads in pounds (positive downward). Calculate the reactions. Plot shear, moment, curvature, slope, and displacement versus shaft location. Calculate slope and displacement at the right end of the shaft.

Solution (a): The bearings will be treated as simple supports. Using moment equilibrium about the bearing at x_1, we find the bearing reaction at x_4.

$$R_4 = (-P_3x_3 - P_5x_5)/(x_4 - x_1)$$

The bearing reaction at x_1 is found from equilibrium of forces.

$$R_1 = -P_3 - R_4 - P_5$$

Shear force is found as in previous examples.

$$V(x) = -R_1S(x,x_1,0) - P_3S(x,x_3,0) - R_4S(x,x_4,0)$$

Recalling that

$$\int_0^x S(x,a,0)\ dx = S(x,a,1)$$

We integrate the shear force equation to obtain bending moment.

$$\begin{aligned} M(x) &= \int_0^x V\,dx \\ &= -R_1S(x,x_1,1) - P_3S(x,x_3,1) - R_4S(x,x_4,1) \end{aligned}$$

The curvature expression is complicated by the change in shaft diameter.

$$\begin{aligned} K(x) = {} & [R_1/(EI_1)][S(x,x_1,1) - S(x,x_2,1)] \\ & + [R_1/(EI_2)]S(x,x_2,1) \\ & - [M(x_2)/E][1/I_2 - 1/I_1]S(x,x_2,0) \\ & + [P_3/(EI_2)]S(x,x_3,1) \\ & + [R_4/(EI_2)]S(x,x_4,1) \end{aligned}$$

The first term in the curvature equation "turns on" the effect of reaction R_1 at x_1 and turns it off at the change in diameter. The effect of R_1 is then turned on again with the new value of moment-of-inertia. At x_2, the step function $S(x,x_2,0)$ adjusts the value of $-M/(EI)$ for the diameter change.

We find slope by integrating the curvature equation, following the rules for singularity functions. Recalling that

$$\int_0^x S(x,a,n)\ dx = S(x,a,n + 1)/(n + 1)$$

we find the change in slope from the left end.

$$\begin{aligned}\theta_A = &[R_1/(2EI_1)][S(x,x_1,2) - S(x,x_2,2)]\\ &+ [R_1/(2EI_2)]S(x,x_2,2)\\ &- [M(x_2)/E][1/I_2 - 1/I_1]S(x,x_2,1)\\ &+ [P_3/(2EI_2)]S(x,x_3,2)\\ &+ [R_4/(2EI_2)]S(x,x_4,2)\end{aligned}$$

where actual slope $\theta = \theta_A + C_2$

Integration constant C_2 will be found in a later step.

Integrating once more, we find the deflection variable

$$\begin{aligned}w_A = &[R_1/(6EI_1)][S(x,x_1,3) - S(x,x_2,3)]\\ &+ [R_1/(6EI_2)]S(x,x_2,3)\\ &- [M(x_2)/2E][1/I_2 - 1/I_1]S(x,x_2,2)\\ &+ [P_3/(6EI_2)]S(x,x_3,3)\\ &+ [R_4/(6EI_2)]S(x,x_4,3)\end{aligned}$$

where actual shaft deflection $w = w_A + C_1 + C_2x$

We find constants C_1 and C_2 by setting deflection $w = 0$ at reaction locations x_1 and x_4. At x_1,

$$0 = w_A(x_1) + C_1 + C_2x_1$$

and at x_4,

$$0 = w_A(x_4) + C_1 + C_2x_4$$

A simultaneous solution of the last two equations yields constants C_1 and C_2.

Solution (b): The problem was solved with the aid of mathematics software. Using the given data in the equations of Part (a), we find that $R_4 = -775$ and $R_1 = 125$ lb (positive upward). It is necessary only to define the singularity function once for $n > -1$, using

$$S(x,a,n) = \text{IF}[x \geq a, (x - a)^n, 0] \qquad \textbf{(5.31)}$$

or an equivalent statement. The equations of Part (a) are evaluated to plot the curves shown in Figure 5.12(b) through (g). Shear force and bending moment diagrams are shown in Figure 5.12(b) and (c).

Figure 5.12 Stepped shaft. (a) Loading.

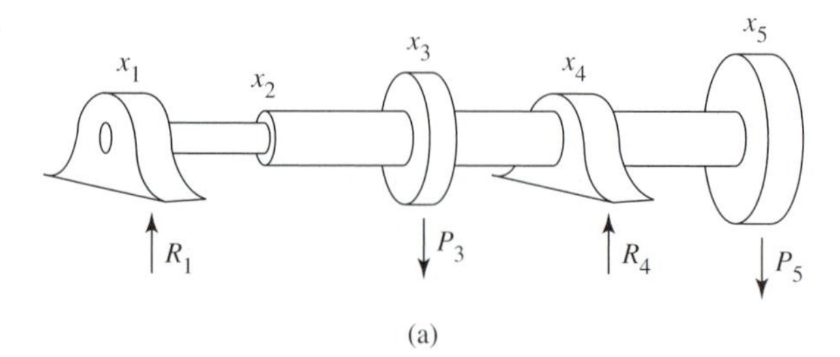

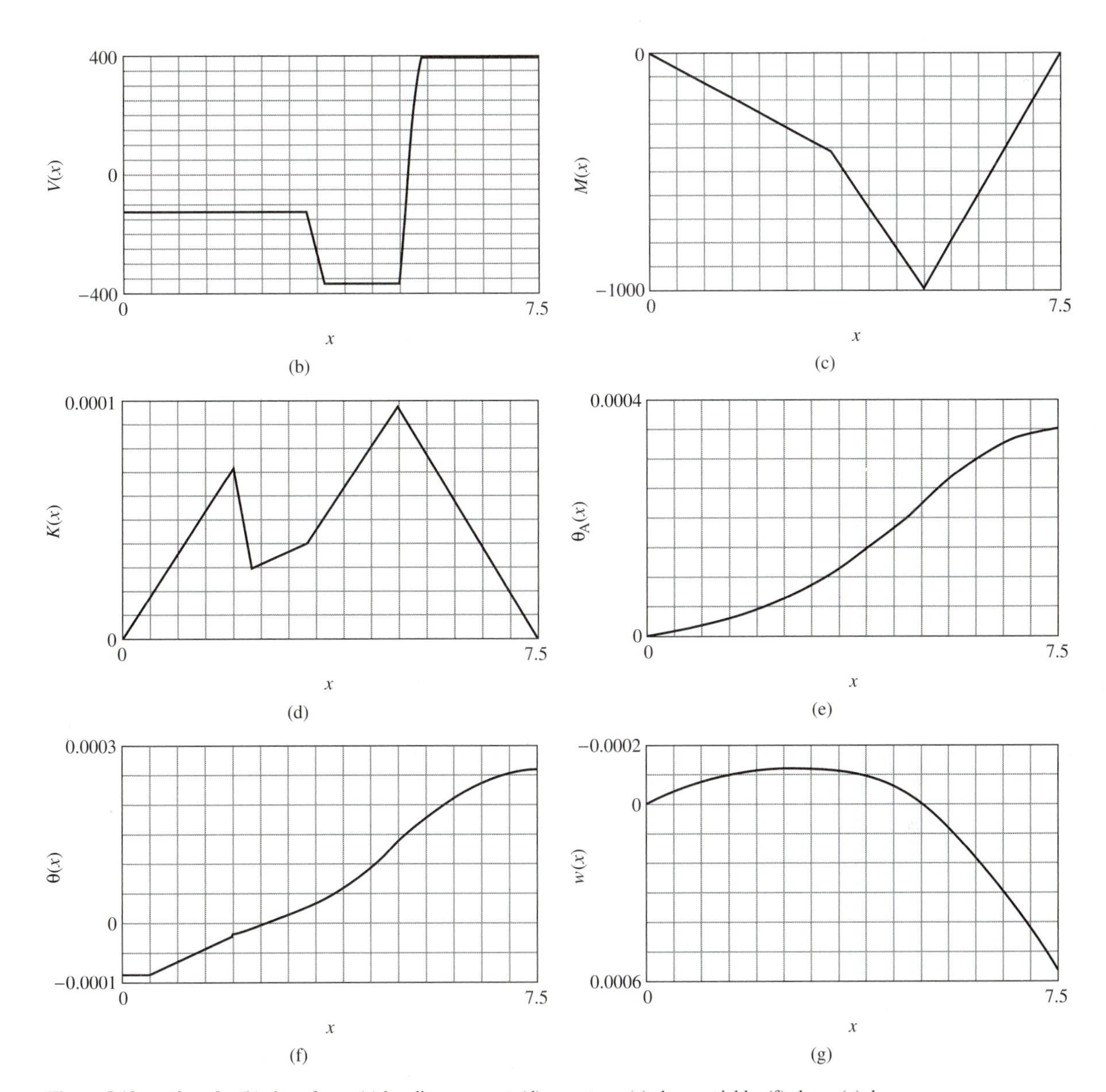

Figure 5.12 continued (b) shear force; (c) bending moment; (d) curvature; (e) slope variable; (f) slope; (g) deflection.

Using $E = 30 \cdot 10^6$ psi and $I = \pi D^4/64$, we plot curvature K. Note the jump in curvature at the diameter change [Figure 5.12(d)]; slope variable θ_A is plotted in Part (e). Noting that the left bearing is located at $x = 0$, we have deflection

$$w(x_1) = w(0) = 0 = w_A(0) + C_1 + C_2 \cdot 0$$

from which

$$C_1 = 0$$

The deflection is also zero at the right bearing, from which

$$C_2 = -w_A(x_4)/x_4 = -8.66 \cdot 10^{-5}$$

We now plot actual slope

$$\theta = \theta_A + C_2 \quad \text{Figure 5.12(f)}$$

and actual deflection

$$w = w_A + C_1 + C_2 x \quad \text{Figure 5.12(g)}$$

At the overhanging load, the slope is

$$\theta(7.5) = 2.599 \cdot 10^{-4} \text{ rad}$$

and the deflection is

$$w(7.5) = 5.482 \cdot 10^{-4} \text{ in}$$

■■

DESIGN OF BENDING MEMBERS

The design of shafts and other machine elements is considered in detail in other chapters. In this section we will deal briefly with bending members subject to static and fatigue loads.

In many cases, the design of beams is governed by tensile stress at the location of maximum bending moment. Maximum tensile stress at any cross section is given by

$$\sigma_x = \frac{Mc}{I} \tag{5.32}$$

where the beam axis is in the x-direction

σ_x = tensile stress in the x-direction (psi or MPa)

M = bending moment (lb·in or N·mm)

c = distance from the neutral axis to beam surface (in or mm)

I = moment-of-inertia about the neutral axis (in^4 or mm^4)

For static loading, the usual criterion for steel beams is

$$S_W = \frac{S_{YP}}{N_{FS}} \geq \sigma_X \tag{5.33}$$

where S_W = working stress (psi or MPa)

S_{YP} = yield point strength (psi or MPa)

N_{FS} = safety factor

Fatigue loading occurs when moment varies with time. A similar situation applies to rotating shafts. A point on the shaft "sees" a varying moment. If the stress is reversed, that is, if the stress at a point continuously ranges from tension to compression with zero mean stress, then the design criterion is

$$S_{RW} = \frac{S_E}{N_{FS}} \geq \sigma_R \tag{5.34}$$

where subscript R refers to range (reversed) stress

S_E = endurance limit corrected for surface finish, stress concentration, etc. (psi or MPa)

Maximum bending moment can be determined by the methods outlined earlier in this chapter. For shafts and other solid circular cross sections, moment-of-inertia and maximum distance from the neutral axis to shaft surface are given by

$$I = \frac{\pi D^4}{64}$$

$$c = \frac{D}{2}$$

where D = diameter

For rectangular beams, the moment-of-inertia and the distance from the neutral axis to the beam surface are given by

$$I = \frac{bh^3}{12}$$

$$c = \frac{h}{2}$$

Consider a rectangular beam, oriented so that the top surface is horizontal. For vertical loading, the neutral axis is the horizontal axis through the center of the beam cross section; h is measured in the vertical direction; and b is measured in the horizontal direction. Note that I–cross sections use material more efficiently and are preferred for large beams. However, use of rectangular cross-section members may result in a more compact design that is easier to fabricate.

Most bending member problems require the careful analysis outlined earlier in this chapter. However, the following simple case sometimes occurs.

Consider a beam or shaft of length L measured between simple supports. For a vertical load P halfway between the supports, maximum moment is given by

$$M_{\text{MAX}} = \frac{PL}{4}$$

where M_{MAX} = maximum bending moment for this special case

EXAMPLE PROBLEM Design of a Bending Member for Static Loads

A beam is to be loaded as in the example problem "Analysis of a Beam with Several Loads." Design the beam.

Design Decisions: The beam will have a width-to-depth ratio b/h of 1.5. We will use steel with a yield point of 450 MPa and a safety factor of 2.5.

Solution Summary: We use the results of the example problem, Equations (5.32) and (5.33), and the equations for I and c, rearranging to solve for beam depth. We then make a partial check of the values. The dimensions can be rounded to a beam cross section 11 mm deep by 16 mm wide. The calculations are given in the detailed solution which follows.

Detailed Solution:

Maximum Moment (N·mm) (Absolute Value):

$$M_{\mathrm{MAX}} = 50{,}000$$

Yield Point Strength (MPa):

$$S_{\mathrm{YP}} = 450$$

Width-to-Depth Ratio (b/h):

$$r_{bh} = 1.5$$

Safety Factor:

$$N_{\mathrm{FS}} = 2.5$$

Required Depth (mm):

$$h = \left(\frac{6 \cdot M_{\mathrm{MAX}} \cdot N_{\mathrm{FS}}}{r_{bh} \cdot S_{\mathrm{YP}}}\right)^{1/3} \qquad h = 10.357$$

Width (mm):

$$b = r_{bh} \cdot h \qquad b = 15.536$$

Dimensions can be rounded upward to convenient sizes.

Check:

Moment-of-Inertia:

$$I = \frac{b \cdot h^3}{12}$$

Stress:

$$\sigma_x = \frac{M_{\mathrm{MAX}} \cdot \dfrac{h}{2}}{I} \qquad \sigma_x = 180$$

Working Strength:

$$S_W = \frac{S_{\mathrm{YP}}}{N_{\mathrm{FS}}} \qquad S_W = 180$$

EXAMPLE PROBLEM Design for Reversed Loading

An idler gear has a resultant load of 3000 N. There is no thrust load. Design the idler shaft.

Design Decisions: The idler will be centered on a rotating shaft, with shaft bearings 100 mm apart center-to-center. The shaft will be made of steel with a corrected endurance limit of 190 MPa, and a safety factor of 3 will be used.

Solution Summary: Equations (5.32) and (5.34) will be used, along with the equations for I and c of a circular section. Rearranging, we find that $D = 22.93$ mm which can be rounded to 23 or 24 mm or 1 in. The results are partially checked in the detailed solution which follows.

Detailed Solution:

Vector Sum of Forces on Idler Gear (N):

$$P = 3000$$

Shaft Length between Bearings (mm):

$$L = 100$$

Maximum Bending Moment (N·mm)

$$M_{MAX} = \frac{P \cdot L}{4} \qquad M_{MAX} = 7.5 \cdot 10^4$$

Corrected Endurance Limit (MPa):

$$S_E = 190$$

Safety Factor:

$$N_{FS} = 3$$

Required Shaft Diameter (mm):

$$D = \left(\frac{32 \cdot M_{MAX} \cdot N_{FS}}{\pi \cdot S_E} \right)^{1/3} \qquad D = 22.934$$

Dimensions can be rounded upward to convenient sizes.

Check:

Moment-of-Inertia:

$$I = \frac{\pi \cdot D^4}{64}$$

Fatigue Stress (MPa):

$$\sigma_R = \frac{M_{MAX} \cdot \frac{D}{2}}{I} \qquad \sigma_R = 63.333$$

Allowable Fatigue Stress:

$$S_{RW} = \frac{S_E}{N_{FS}} \qquad S_{RW} = 63.333$$

■■

References and Bibliography

Anderson R. B. *The Student Edition of MathCAD* Version 2.0. Reading, MA: Addison-Wesley, 1989.

Bethlehem Steel Corporation. *Structural Shapes.* Catalog 3277C. Bethlehem, PA, May 1983.

MathCAD Plus 5.0 User's Guide. Cambridge, MA: Mathsoft, Inc., 1994.

Timoshenko, S. *Strength of Materials.* Part I. 3d ed. Princeton, NJ: Van Nostrand, 1955.

Timoshenko, S. *Strength of Materials.* Part II. 3d. ed. Princeton, NJ: Van Nostrand, 1956.

Design Problems

5.1. Consider the beam shown in Figure 5.5(a) and described in the example problem *"Analysis of a Beam with Several Loads,"* except that the distributed load is not present.
(a) Find the reactions.
(b) Plot shear force versus beam location.
(c) Plot bending moment.

5.2. An 8 in long beam is subject to concentrated loads as follows:

Load (lb)	*Location (in)*
200	1
400	3
550	4.5
−80	5
280	6

Supports are located at $x = 2$ in and $x = 7$ in.
(a) Find the reactions.
(b) Plot shear force versus beam location.
(c) Plot bending moment.

5.3. An 8 in long shaft is subject to concentrated loads as follows:

Load (lb)	*Location (in)*
100	2
−200	3
150	4
50	5
250	7

Bearings are located at $x = 1$ in and $x = 6$ in.
(a) Find the reactions.
(b) Plot shear force versus beam location.
(c) Plot bending moment.

5.4. A 16 in long beam is subject to concentrated loads as follows:

Load (lb)	*Location (in)*
300	5
280	7.5
145	8.5
330	10
−100	11.5
155	15.5

Supports are located at $x = 3$ in and $x = 13$ in.
(a) Find the reactions.
(b) Plot shear force versus beam location.
(c) Plot bending moment.

5.5. A uniform load of 10 N/mm is applied to a 160 mm long beam. Supports are located at $x = 30$ mm and $x = 130$ mm.
(a) Find the reactions.
(b) Plot shear force versus beam location.
(c) Plot bending moment.

5.6. A uniform load of 25 N/mm is applied to a 160 mm long beam. In addition, there are six concentrated loads of 1000 N each, at $x = 10, 50, 70, 90, 110$, and 150 mm. Supports are located at $x = 30$ mm and $x = 130$ mm.
(a) Find the reactions.
(b) Plot shear force versus beam location.
(c) Plot bending moment.

5.7. An I-beam has the following dimensions:

Depth = 8.14 in Flange width = 5.25 in

Flange thickness = 0.33 in Web thickness = 0.23 in

Loads and reactions will be perpendicular to the flanges.

(a) Find cross-sectional area.
(b) Find moment-of-inertia.
(c) Compare your results with published values for a standard I-beam, if available.

5.8. A T-section is cut from the I-beam described in Problem 5.7. Loads and reactions will be perpendicular to the flange.

(a) Find center-of-gravity and neutral axis in bending.
(b) Find cross-sectional area.
(c) Find moment-of-inertia.
(d) Compare your results with published values for a standard T-section, if available.

5.9. A 10 in long shaft has an elastic modulus of $E = 30 \cdot 10^6$ psi. Diameter $D_1 = 1.325$ to the left of a step at $x = 2$, and $D_2 = 1.5$ to the right of the step. A concentrated load of $P_3 = 400$ lb is located at $x_3 = 6$. Bearings are located at $x_1 = 0$ and $x_4 = 10$. (Dimensions are given in inches.)

Plot all of the following versus shaft location:

(a) shear
(b) moment
(c) curvature
(d) slope
(e) displacement
(f) Give slope and displacement at the load (location x_3).

5.10. A 10 in long shaft has an elastic modulus of $E = 30 \cdot 10^6$ psi. Diameter $D_1 = 1.0$ to the left of a step at $x_2 = 4$, and $D_2 = 1.25$ to the right of the step. A concentrated load of $P_3 = 800$ lb is located at $x_3 = 6$. Bearings are located at $x_1 = 0$ and $x_4 = 10$. (Dimensions are given in inches.)

Plot all of the following versus shaft location:

(a) shear
(b) moment
(c) curvature
(d) slope
(e) displacement
(f) Give slope and displacement at the step.

5.11. A 10 in long shaft has an elastic modulus of $E = 30 \cdot 10^6$ psi. Diameter $D_1 = 1.125$ to the left of a step at $x_2 = 3$, and $D_2 = 1.25$ to the right of the step. A concentrated load of $P_3 = 1400$ lb is located at $x_3 = 6$. Bearings are located at $x_1 = 0$ and $x_4 = 10$. (Dimensions are given in inches.)

Plot all of the following versus shaft location:

(a) shear
(b) moment
(c) curvature
(d) slope
(e) displacement
(f) Give displacement at the load.

5.12. A 5 in long shaft has an elastic modulus of $E = 30 \cdot 10^6$ psi. Diameter $D_1 = 1.25$ to the left of a step at $x_2 = 2$, and $D_2 = 1.625$ to the right of the step. A concentrated load of $P_3 = 250$ lb is located at $x_3 = 3.5$. Bearings are located at $x_1 = 0$ and $x_4 = 5$. (Dimensions are given in inches.)

Plot all of the following versus shaft location:

(a) shear
(b) moment
(c) curvature
(d) slope
(e) displacement
(f) Indicate slope at the left end and displacement at the load.

5.13. A 20 in long beam has an elastic modulus of $E = 10 \cdot 10^6$ psi. The moment-of-inertia $I_1 = 1$ in^4 to the left of a step at $x = 10$, and $I_2 = 2$ in^4 to the right of the step. A concentrated load of $P = 1000$ lb is located at $x = 10$. Bearings are located at $x_1 = 0$ and $x_4 = 20$. (Dimensions are given in inches.)

Plot all of the following versus shaft location:

(a) shear
(b) moment
(c) curvature
(d) slope
(e) displacement
(f) Indicate displacement at the load.

5.14. A beam is to be loaded as in Problem 5.5. Design the beam, and check your results.

Design Decisions: The beam will have a square cross section. We will use steel with a yield point of 400 MPa and a safety factor of 5.

5.15. A beam is to be loaded as in Problem 5.6. Design the beam, and check your results.

Design Decisions: The beam will have a width-to-depth ratio of 2.5. We will use steel with a yield point of 350 MPa and a safety factor of 6.

5.16. A beam is to be loaded as in Problem 5.6. Design the beam, and check your results.

Design Decisions: The beam will have a width-to-depth ratio of 3. We will use steel with a yield point of 375 MPa and a safety factor of 4.

5.17. An idler gear has a resultant load of 1800 N. There is no thrust load. Design the idler shaft.

Design Decisions: The idler will be centered on a rotating shaft, with shaft bearings 125 mm apart center-to-center. The shaft will be made of steel with a corrected endurance limit of 205 MPa, and a safety factor of 3.5 will be used.

5.18. An idler pulley has a resultant load of 6500 N. Design the idler shaft.

Design Decisions: The idler will be centered on a rotating shaft, with shaft bearings 85 mm apart center-to-center. The shaft will be made of steel with a corrected limit of 180 MPa, and a safety factor of 2 will be used.

5.19. An idler gear has a resultant load of 670 lb. There is no thrust load. Design the idler shaft.

Design Decisions: The idler will be centered on a rotating shaft, with shaft bearings 4 in apart center-to-center. The shaft will be made of steel with a corrected endurance limit of 27,500 psi, and a safety factor of 3 will be used.

5.20. An idler pulley has a resultant load of 1900 lb. Design the idler shaft.

Design Decisions: The idler will be centered on a rotating shaft, with shaft bearings 3.8 in apart center-to-center. The shaft will be made of steel with a corrected endurance limit of 30,000 psi, and we will use a safety factor of 2.5.

CHAPTER 6

Finite Element Analysis

Symbols

A	cross-sectional area	u, v, w	displacements
$\{D\}$	nodal displacements	δ	total displacement
E	elastic modulus	ν	Poisson's ratio
f	free (boundary condition)	$\sigma_1, \sigma_2, \sigma_3$	principal stresses
$\{F\}$	nodal forces	σ_{MAX}	maximum normal stress (maximum principal stress)
$[K]$	stiffness matrix		
K_T	theoretical stress concentration factor	σ_{NOM}	nominal stress
L	length	σ_R	range stress
N_{FS}	factor of safety	σ_x	x-direction normal stress
P	total load	σ_{VM}	von Mises stress
P_x	force per unit length	τ_{xy}	shear stress
S_E	corrected endurance limit	ω_x	rotation about the x-axis

Units

dimensions	in, m, or mm
forces	lb or N
stress, pressure, and elastic modulus	psi, Pa, or MPa

Beams, circular shafts, and simple tension members can be designed by using the methods of "strength of materials." Noncircular shafts and simple stress concentration problems can be analyzed by the theory of elasticity. However, the theory of elasticity requires us to satisfy the governing equations throughout a part as well as the boundary conditions, a difficult task for complicated structures and machine parts.

Finite element analysis (FEA) is a method that divides a structure or a machine part into small subregions called **finite elements.** A finite element boundary must be compatible with each of its neighbors. That is, forces must be transmitted from element to element, and the finite elements must fit together as in the real part.

Efficient FEA software, along with computers with adequate speed and memory, allows us to solve problems that could not have been solved in the past. FEA methods are not limited to stress analysis. They are used to model both static and dynamic problems, and they are used in heat transfer, fluid mechanics, and other fields, as well as in machine design. This chapter will expand on some FEA concepts that were mentioned in Chapter 1.

MODELING A REAL PROBLEM: LOADING

Most textbook problems are clearly defined, with the given data and assumptions leading to a single, unambiguous answer. However, typical real-world machine design problems are poorly defined. They require the designer to use considerable judgment and to make numerous assumptions. Real problems may have many correct (acceptable) solutions. Also, there may be many solutions that are useless, even though they are theoretically valid.

When designing a machine member, the designer usually first asks, "What will the loads be on the member?" In some cases, the designer can answer the question by adding a design feature to limit the loading. For example, if we must design a pressure vessel, it may be possible to incorporate a relief valve to limit pressure. Materials-handling equipment may incorporate a load cell to prevent overload. The loading on power equipment components may be limited by the maximum input power. In most cases, we must use judgment and experience to estimate the loading, trying to imagine the worst-case scenario if failure could have severe consequences.

Saint Venant's Principle

Saint Venant's principle makes it possible to model real design problems in simple form. In essence, it works as follows: Let the load on a body be replaced by a statically equivalent load. The stress pattern at a reasonable distance from the load will not be substantially changed. According to the same concept, a small change in the shape of a body in one region does not greatly affect the stress distribution at a reasonable distance from that region.

We frequently apply these principles to machine design problems. Consider a shaft carrying two pulleys and supported by journal bearings near each end. The belt loads are transmitted to the shaft as distributed loads beneath the pulleys. Using traditional methods of analysis (since we do not know the actual load distribution anyway), we replace the distributed load at each pulley by a concentrated force. The bearing reactions are actually distributed over an area beneath each bearing. However, when calculating the bearing reactions, we also treat them as concentrated forces. If there is a change in shaft diameter a reasonable distance from one of the pulleys, the resulting disturbance in the stress pattern does not substantially affect the stress pattern at that pulley.

When using Saint Venant's principle, we consider a "reasonable" distance to be any distance greater than the dimension of the distributed load or reaction. In this example, if the distributed load at a pulley is replaced by a statically equivalent, concentrated load, the stress distribution will be about the same, except within a distance of about one pulley-hub width from the disturbance. For the bearing, a reasonable distance is about one bearing length. Application of Saint Venant's principle in other design situations is discussed by Den Hartog (1952).

Caution: Saint Venant's principle is used to develop a model for analysis. If "point forces" were actually applied to machine members, infinite stresses would result. For example, a support reaction of 500 lb applied at a point is statically equivalent to an average bearing support pressure of 500 psi applied on an area of 1 in^2. If we actually attempted to use a knife-edge support instead of a bearing, stresses at the contact point would be extremely high, resulting in an early failure. Note also that a change in shaft diameter, a keyway, or other stress concentration will have an important local effect. The stress at the location of the stress concentration may be the greatest stress encountered; it may govern the design of the machine member.

When using FEA to study machine members, we can model loading as distributed forces. For two-dimensional problems, we use force per unit length. For three-dimensional problems, we use pressure, where tensile force per unit area becomes negative pressure. Valid results depend on using the best possible estimate of the force magnitude and the area over which it acts. If a load or reaction were assumed to act at a single point, the output of the FEA program could indicate unreasonably high stresses in that region. As a result, important stress information at critical locations could be obscured.

MODEL GEOMETRY

The behavior of a model should give the designer information about the behavior of a proposed machine element. It may not be necessary to make the model identical to the part in appearance. Most FEA solutions involve arrays of large matrix equations which require substantial computing time. In order to reduce programming effort and computing time, model geometry should be as simple as possible yet include necessary detail in critical areas. If critical areas on a part are not known in advance, a test run with a simple model may help the designer formulate a better model. In the early stages of design, a test run with a simple model may even provide data for redesign of a proposed machine element.

THREE-DIMENSIONAL STRESS FIELDS AND PLANE STRESS

In many cases, it is clear at the outset that we are dealing with **plane stress,** where the stress field can be described in terms of only two coordinate directions. For example, if stresses σ_x, σ_y, and τ_{xy} are the only nonzero stress components, then we have a plane stress case. If out-of-plane stresses are an order of magnitude less than in-plane stresses, then a plane stress assumption is likely to give reasonable results. This simplifies the problem, saving time for the designer and saving computing time, since a two-dimensional FEA model is adequate.

Table 6.1 Symmetry boundary conditions.

Plane	Displacements and Rotations					
	u	v	w	ω_x	ω_y	ω_z
yz $(x = x_1)$	0	f	f	f	0	0
xz $(y = y_1)$	f	0	f	0	f	0
xy $(z = z_1)$	f	f	0	0	0	f

Note: u, v, and w = displacements in the x-, y-, and z-directions, respectively. ω_x, ω_y, and ω_z are rotations in the x-, y-, and z-directions, respectively. f = free; 0 = restrained.

If out-of-plane stresses are significant, then a three-dimensional model must be used. In a general three-dimensional stress field, there are three normal stress components and three shear stress components.

SYMMETRY

A **plane of symmetry** exists in a body if the half of the body on one side of the symmetry plane is the mirror image of the other half, with the same condition applying to the loading and restraints. If there is a plane of symmetry of the body, the loading, and the restraints, then it is necessary to model only half of the body. If there are two planes of symmetry, then a one-quarter model may be used. We must then add appropriate restraints to the model boundaries that lie on the plane or planes of symmetry. By taking advantage of symmetry conditions, we can improve the efficiency of FEA in our design process.

Symmetry Boundary Conditions

If there is a plane of symmetry, then a point on that plane cannot move out of the plane. For example, if an xz-plane defined by $y = 0$ is a plane of symmetry, then one symmetry boundary condition is

$$v = 0 \quad \text{on the } y = 0 \text{ plane}$$

where v = displacement in the y-direction

Rotations about axes in the plane of symmetry are also impossible. Thus, if the xz-plane $y = 0$ is a plane of symmetry, then $\omega_x = 0$ and $\omega_z = 0$ on that plane, where ω_x and ω_z are rotations about the x-axis and z-axis, respectively. Symmetry boundary conditions are summarized in Table 6.1.

EXAMPLE PROBLEM Symmetry Boundary Conditions for FEA Modeling

Figure 6.1 shows a plate with a central hole and a tensile load at the left and right edges. Two additional holes were drilled in the plate in an attempt to reduce stress concentration. The designer wants to know if the additional holes actually help. Examine symmetry boundary conditions in preparation for finite element analysis modeling.

Assumptions: We will assume that the loading is a uniform tensile stress on the left and right edges of the plate. Considering the loading and the geometry, we then have a two-dimensional (plane stress) problem. We expect that the critical areas in the plate will be near the holes.

Figure 6.1 Symmetry boundary conditions.

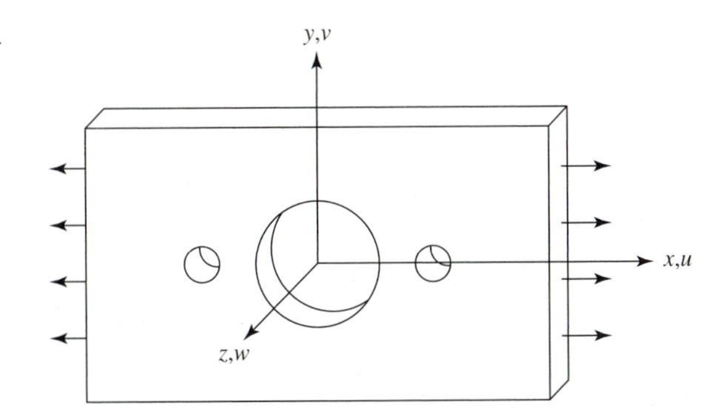

Solution: The x-, y-, and z-axes are established with the origin at the center of the large hole. Both geometry and loading are symmetrical about the x- and y-axes. Therefore, we need model only one-quarter of the plate, say, the upper-right quarter. A uniform load (lb/in or N/mm) will be applied to the right edge of the model. Symmetry requires that the edge along the y-axis above the hole cannot move in the x-direction (i.e., $u = 0$ on the $x = 0$ edge). However, the vertical dimension of the plate can change along that edge (i.e., v is free on the $x = 0$ edge). Along the $x = 0$ edge, there is no rotation in the xy-plane ($\omega_z = 0$). The boundary conditions for both symmetry planes are summarized below.

Symmetry Boundary Conditions:

Displacements and Rotations			
Edge	u	v	ω_z
$x = 0$	0	f	0
$y = 0$	f	0	0

■■

AXISYMMETRIC PROBLEMS

Axisymmetry implies geometric symmetry and loading symmetry about a single axis. For example, a horizontal circular plate is symmetric about a vertical axis through the center. The same is true if the plate has a round hole in the center. Let radius r be measured from the symmetry axis. If there is a constant vertical load or a load that depends only on r, then we have axisymmetry. Plate stress and deflection will depend only on r.

FINITE ELEMENT TYPES

Commercial finite element computer programs offer a wide variety of finite element configurations. The selection (the finite element library) varies, depending on the program. Some of the following elements may be included in a finite element library:

Line Elements	*Surface Elements*
Rod	Quadrilateral
Tube	Triangle
Bar	
Beam	
Curved beam	
Curved pipe	

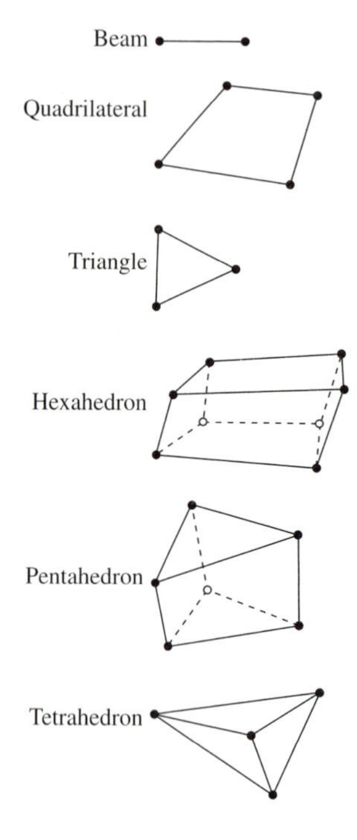

Figure 6.2 Finite elements.

Solid Elements	*Special Elements*
Hexahedron (brick)	Elements representing axisymmetric solids
Pentahedron (wedge)	Joint
Tetrahedron	Spring
	Crack
	Concentrated mass

Some of these element shapes are shown in Figure 6.2.

We usually select the simplest element configuration that can give us the information needed for a design. For example, beam elements "know" how actual beams behave under load. They are used to describe a structure made of one or more beams if we need to find stresses and displacements. However, if the beams have holes in them or changes in cross section, then there might be stress concentrations that could make our design unsafe. We would use surface elements or solid elements to find the local stresses in regions of stress concentration.

NODES

Nodes are points on a finite element model. The FEA model behaves as if internal and external forces are transmitted at the nodes. Elements that make up the model are connected at the nodes, and stresses and displacements are determined at the nodes. The large dots on the elements in Figure 6.2 indicate the nodes.

The elements shown in Figure 6.2 are linear or first-order elements. There are two nodes on each edge, and first-order equations are used to interpolate strain between the nodes. Second-order (parabolic) elements have three nodes per edge, and second-order equations are used to interpolate strain. Third-order (cubic) elements have four nodes per edge, and third-order equations are required to interpolate strain. Note that the terms used to identify the order of the elements do not refer to element shape.

FINITE ELEMENT MESH GENERATION

A model can be divided up into a grid of finite elements by the program user, or the program can be directed to automatically generate a finite element mesh. Most machine designers prefer a compromise, that is, semiautomatic mesh generation. For this option, the designer specifies the number of nodes along each boundary but lets the program complete the meshing task automatically.

Element Size

Holes, small-radius fillets, and other "disturbances" in stress patterns in a part cause sharp increases in stress in a small region. Stress concentration is particularly important when a machine part is subject to fatigue loading. A finite element mesh pattern made up of large elements may miss important local effects due to stress averaging within the elements. However, the use of small elements throughout a model can substantially increase solution time. The usual approach involves using experience, intuition, and preliminary runs with coarse meshing to determine critical locations on a model. A fine mesh is then prescribed for those critical locations, and a coarse mesh is used elsewhere.

Suppose, for instance, that a designer must determine the required thickness of the part sketched in Figure 6.3. The designer also wishes to find out if removing material by cutting an arc at *BC* will reduce stress at the small-radius fillet at *EF*. The actual part will be subject to fatigue loading. The designer creates a model by entering the required data in a finite element analysis program. An arbitrary thickness is selected for the model. Since stress will not

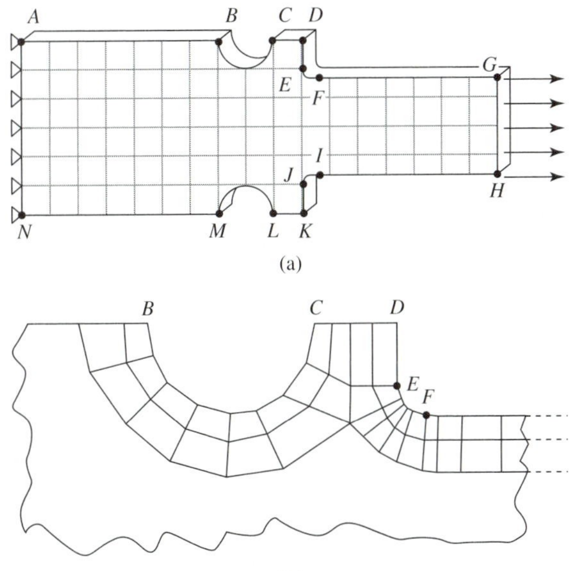

Figure 6.3 Modeling with finite elements. (a) First attempt at meshing; (b) revised mesh.

vary through the thickness, surface elements are used. As a first attempt at meshing, square elements are used.

If the preliminary model is processed, the results will indicate that stresses are symmetric about a horizontal plane through the part. The designer might have observed the symmetry of geometry and loading at the start. In addition, the results will show high stresses in the vicinity of arc *BC* and fillet *EF* (and the corresponding arcs at the bottom of the model). However, the stress will change abruptly from one element to the next in this region. Thus, the accuracy of the preliminary model is in doubt.

A revised model takes advantage of symmetry by including only half of the part above the symmetry axis. The program is directed to generate a finer mesh in the critical locations by specifying more nodes along arc *BC* and fillet *EF*. Figure 6.3(b) shows a possible mesh pattern in this vicinity.

Aspect Ratio

If one dimension of a finite element differs greatly from another, the program may produce poor results. When long, thin elements are generated in a critical region, it is advisable to change the number of specified nodes in an attempt to produce elements with approximately equal dimensions.

PREPROCESSING

Preparation of a model for finite element analysis is called **preprocessing.** Preprocessing tasks include describing a proposed design in terms understandable to a computer and computer processing up to the solution phase. Most of the preprocessing steps are discussed above.

THE SOLUTION PHASE

The element equation has the form

$$[K]\{D\} = \{F\} \tag{6.1}$$

where $\{F\}$ = nodal forces (lb or N)

$[K]$ = stiffness matrix

$\{D\}$ = node point displacement components (in, m, or mm)

Consider, for example, a pin-ended rod in tension. Let one end of the rod be fixed, and the other end loaded with a force F. The rod behaves like a simple spring, with spring rate K. For the rod,

$$K = AE/L \tag{6.2}$$

where A = cross-sectional area (in^2, m^2, or mm^2)

E = elastic modulus (psi, Pa, or MPa)

L = length (in, mm, or mm)

FEA programs are used to solve problems far more complicated than simple springs and rods. A typical element will have several node points, and each node point will have two displacement components in two-dimensional models and three displacement components in three-dimensional models. Each node will also have two or three force components in two- or three-dimensional models. The stiffness matrix, then, becomes complicated. A finite element analysis program may deal with hundreds of finite elements, all of which must fit together. Displacements and forces at common node points must agree, element to element. All practical FEA programs must be designed for efficient processing of data.

NASTRAN™, the National Aeronautics and Space Administration–sponsored computer program for structural analysis by the finite element method, was the first large-scale FEA program of its type. It became the basis for many other FEA programs. Some FEA programs are very powerful, very fast, and versatile. Others are user-friendly, sometimes at the expense of speed and generality. Most practical FEA programs represent hundreds of person-years of development. Thus, designers seldom find it practical to write their own FEA programs.

In the solution phase, we select the solution type. We then direct the FEA program to solve for displacements and stresses and to save these solution data sets.

POSTPROCESSING

The compiling and displaying of FEA results constitute **postprocessing.** We may be required to answer only the question, "Will maximum normal stress exceed the working strength at any point in the design?" Nevertheless, we direct the FEA program to show a deflection plot, with the deflected model shape overlaid on the unloaded model.

Finite element boundaries can be retained in the unloaded model and deflected model shape as an aid to visualizing the deflection at any location. The deflection plot can be used to verify that the boundary conditions are actually satisfied. Deflections are exaggerated since actual deflections in most engineering materials are not visible to the naked eye.

We will also direct the FEA program to plot maximum normal stress, maximum shear stress, or von Mises stress. The plot can be in the form of contour lines of equal stress overlaid on the model. As an alternative, stress ranges can be indicated by bands of colors. The colors can be continuous, blending into one another, or stepped with sharp boundaries. Stepped color contours define the boundaries of the stress regions more precisely.

We can direct the program to plot specific stress components (σ_x, σ_y, τ_{xy}, etc.). Although failure theories do not refer to these stress components explicitly, the plots can be used to verify that loading has been correctly applied to a boundary and that the model actually represents the proposed part design.

Deflection plots are also available as contour lines, continuous color bands, or stepped color bands. These options present numerical values of deflection throughout the model. However, a plot of deflected shape overlaid on the unloaded model shape is easier to read. By checking that plot, we can often tell intuitively whether the results are reasonable.

EXAMPLE PROBLEM FEA of a Plate with Stress Concentration

A proposed design includes a 200 mm wide by 50 mm thick plate. Two semicircular notches of 50 mm radius reduce the width to 100 mm, as shown in Figure 6.4(a). The plate carries a cyclic load that varies continuously from -2×10^6 N to $+2 \times 10^6$ N in the x-direction.

(a) Determine plate deflections.
(b) Determine maximum normal stresses.
(c) Determine von Mises stresses.
(d) Interpret your results, and evaluate the design.

Figure 6.4 FEA of a plate with stress concentration.

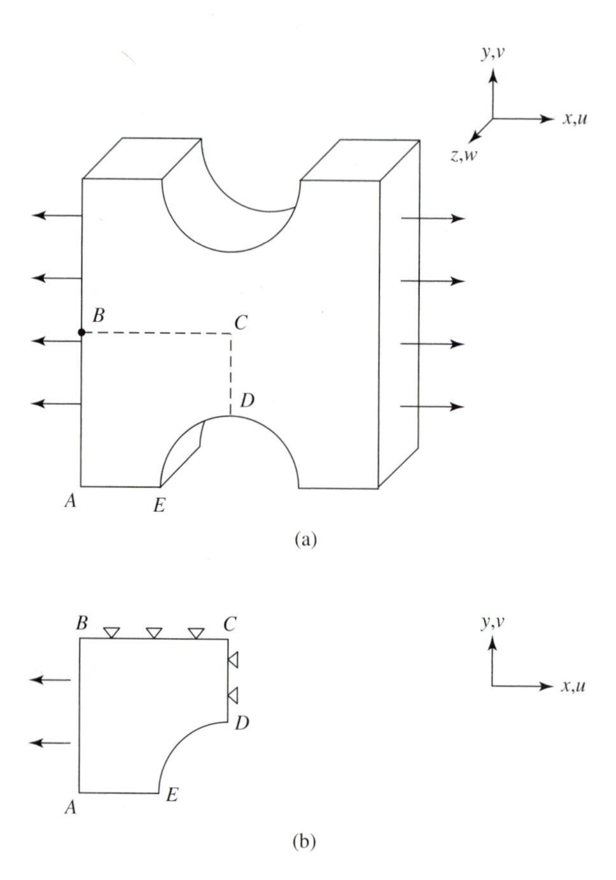

Assumption: We will assume that the load is uniformly distributed along the left and right edges.

Decisions: The plate will be made of steel. Since the load varies continuously, high stresses at the notches cannot be relieved by local yielding; we must consider stress concentration. The finite element method will be used. There are many FEA programs available for solving problems of this type. The user is likely to select a program from the software that is on hand.

Solution (a): Dividing the load amplitude by the area of the edge, we obtain

$$\sigma_x = 2 \cdot 10^6/(200 \cdot 50) = 200 \text{ MPa} = 200 \cdot 10^6 \text{ Pa}$$

or, dividing by the length of the edge, we have

$$P_x = 2 \cdot 10^6/200 = 10{,}000 \text{ N/mm} = 10 \cdot 10^6 \text{ N/m}$$

Loading and plate geometry are symmetric about both a horizontal axis and a vertical axis. We take advantage of symmetry by modeling only the lower-left quarter of the plate, as shown in Figure 6.4(b). We note also that this is a plane stress problem. A force per unit length of $10 \cdot 10^6$ N/m is applied to edge AB of the quarter-plate model. Edge BC lies in a symmetry plane. Considering symmetry and plane stress, the only nonzero displacement of edge BC is in the x-direction.

Displacements and rotations u, v, w, ω_x, ω_y, and ω_z are, respectively, f, 0, 0, 0, 0, and 0 on edge BC, where f = free. The only nonzero displacement on edge CD is in the y-direction. On that edge, the restraints are 0, f, 0, 0, 0, and 0. Point C is thus completely restrained.

Next, we divide up the surface of our model into finite elements as shown in Figure 6.5. Since we expect that stresses somewhere along the curved surface will be critical, we specify a large number of elements along that curve. We specify fewer nodes along the left and top edges. We specify proportional spacing of nodes along the edges joining the curve to the left and top edges. This produces elements with a good aspect ratio; most of them are roughly square. The program is directed to automatically generate the interior elements. Figure 6.5 shows the completed finite element mesh, with the nodes shown as asterisks.

We must specify the material properties. In this case, we will enter the elastic modulus and Poisson's ratio, respectively, as

$$E = 2.068 \cdot 10^{11} \text{ Pa} = 2.068 \cdot 10^{5} \text{ MPa}$$
$$\nu = 0.29$$

This step completes the preprocessing. We next direct the program to solve the problem for displacements and stresses and to save the solution. We will select postprocessing steps to obtain essential information (e.g., maximum stresses) and to check the solution validity (e.g., displacement form).

Figure 6.6(a) shows the form of the displacement of the plate (greatly exaggerated). The upper boundary and right-hand boundary of the quarter-plate model are straight. Since these lines are the symmetry axes of the entire plate, it appears that the symmetry boundary conditions are satisfied.

Displacement contours are shown in Figure 6.6(b). Displacements calculated by FEA range from zero at point C to

$$\delta_{\text{MAX}} = 3.20 \cdot 10^{-4} \text{ m (0.32 mm)} \quad \text{at point } A$$

Figure 6.5 Finite element mesh.

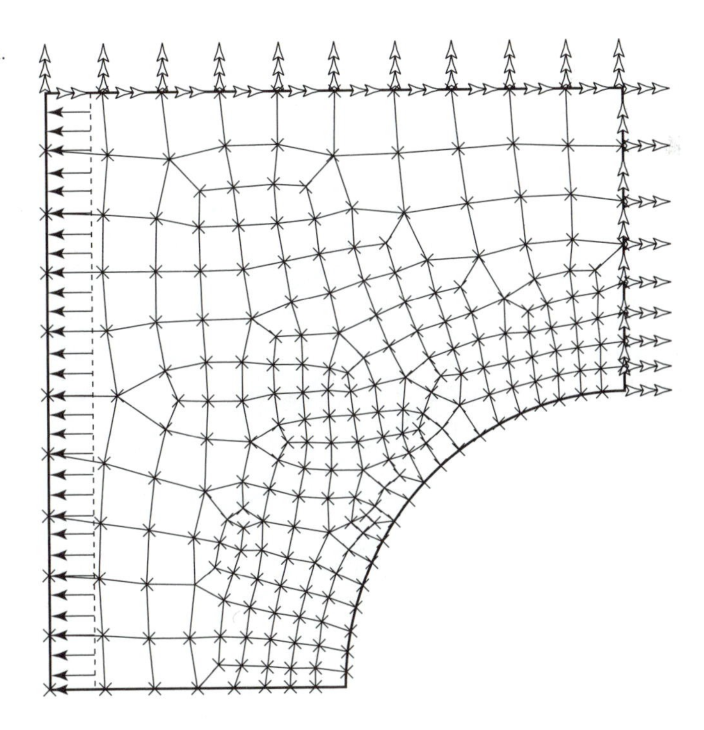

Figure 6.6 (a) Displacement. Unloaded shape is shown with dashed lines. (b) Displacement contours.

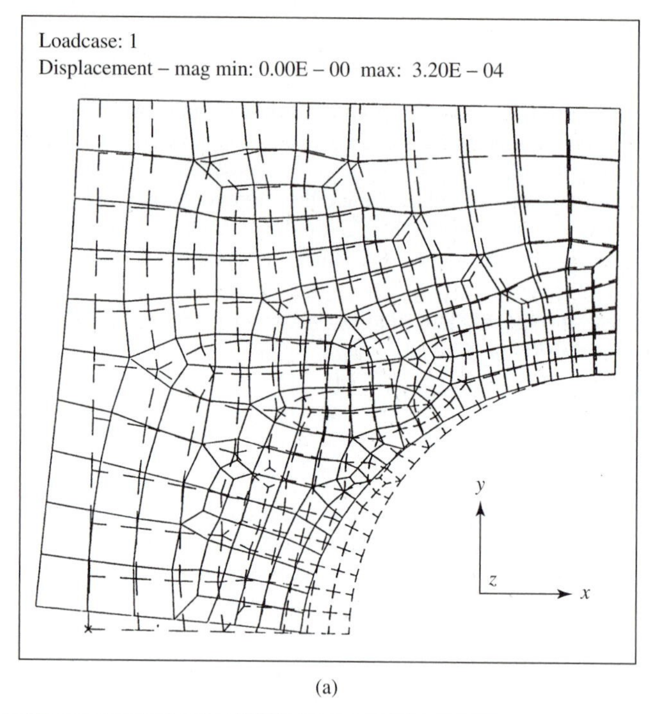

(a)

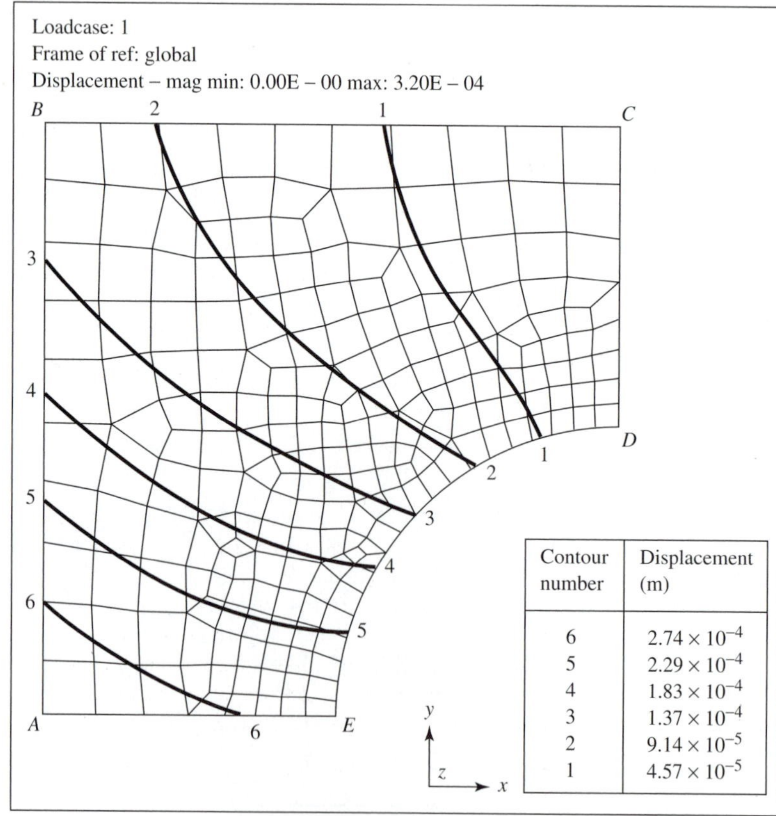

Contour number	Displacement (m)
6	2.74×10^{-4}
5	2.29×10^{-4}
4	1.83×10^{-4}
3	1.37×10^{-4}
2	9.14×10^{-5}
1	4.57×10^{-5}

(b)

The true displacement at point C is zero; it is fixed in the model. Point B on the horizontal symmetry axis lies between contours 2 and 3 produced by the FEA solution. Interpolating, we find that its displacement is

$$\delta_B \approx 0.114 \text{ mm} \quad \text{(approximately, based on FEA results)}$$

Computers sometimes make mistakes. Usually mistakes are the user's fault, but a software or hardware problem can cause errors. We can at least find out if our results are reasonable. Let us make a rough check of the deflection results. We will compare the deflection of the plate centerline with approximate results based on simple tension.

The entire plate is subject to a load

$$P = 2 \cdot 10^6 \text{ N}$$

The plate length is

$$L = 200 \text{ mm}$$

and it has an average cross-sectional area of roughly

$$A = 170 \cdot 50 = 8500 \text{ mm}^2$$

Total plate deflection along the horizontal symmetry axis, or twice the deflection of point B relative to fixed point C, can be estimated from

$$\begin{aligned} \delta = 2\delta_B &= PL/(AE) \\ &= 2 \cdot 10^6 \cdot 200/(8500 \cdot 2.068 \cdot 10^5) = 0.2276 \text{ mm} \end{aligned}$$

The result is a partial verification of the FEA results.

Note that interpolation between contour lines is unnecessary. If we need values of displacement or stress at a point, we can request a list of displacements and stresses at nodal points and a map of node locations.

Solution (b): The maximum normal stress σ_{MAX} is the greatest of σ_1, σ_2, or σ_3 where σ_1, σ_2, and σ_3 are the principal stresses at a point. Maximum normal stress can also be called **maximum principal stress.** Figure 6.7(a) shows maximum normal stress contours plotted over the finite element mesh. Plotted stress values are given in pascals (N/m^2). The greatest maximum normal stress is 710 MPa at point D.

Figure 6.7(b) shows the same data represented by stepped contours. The stress ranges can be represented by different colors selected from a palette.

Solution (c): For the plane stress case, von Mises stress is given by

$$\sigma_{\text{VM}} = (\sigma_1^2 + \sigma_2^2 - \sigma_1\sigma_2)^{1/2}$$

Figure 6.8(a) shows von Mises stress contours plotted over the finite element mesh. Values range from about zero at point E to a maximum of 691 MPa at point D.

Part (b) of Figure 6.8 shows von Mises stress as stepped contour bands. The original contour plot shows the bands in color. Part (c) is similar except that the von Mises stress contours are continuous. The colors in the original plot are "blended" at the boundaries. Stepped color contour bands are usually the most readable way to present stress patterns in a part. That form helps the designer locate critical points in the design and propose design changes where necessary.

Solution (d): The finite element analysis is based on a load of $2 \cdot 10^6$ N (the maximum load) over an area of 50 by 200 mm. Thus, we have an x-direction stress of

$$\sigma_x = 2 \cdot 10^6/(200 \cdot 50) = 200 \text{ MPa}$$

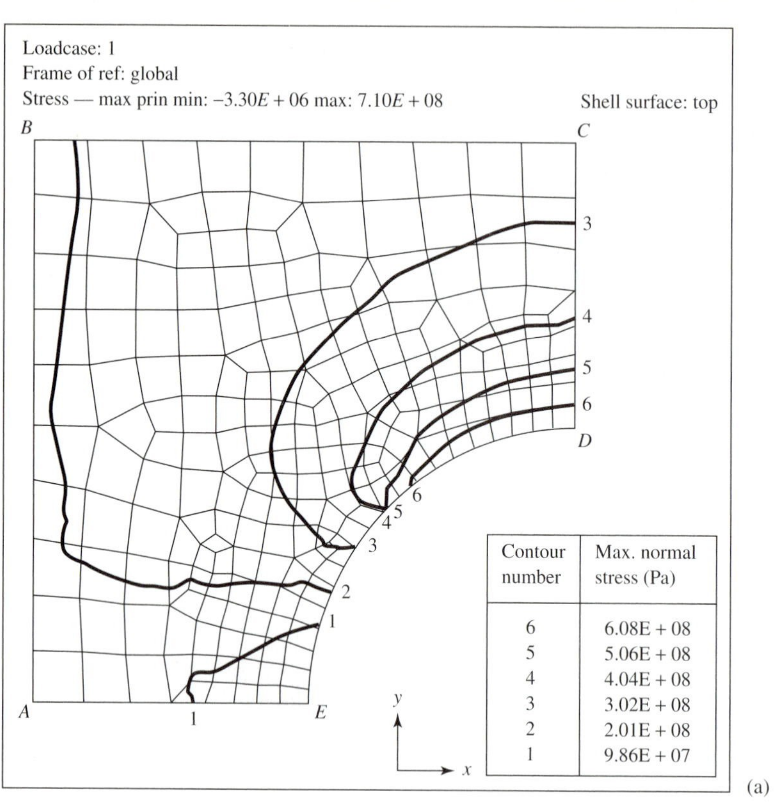

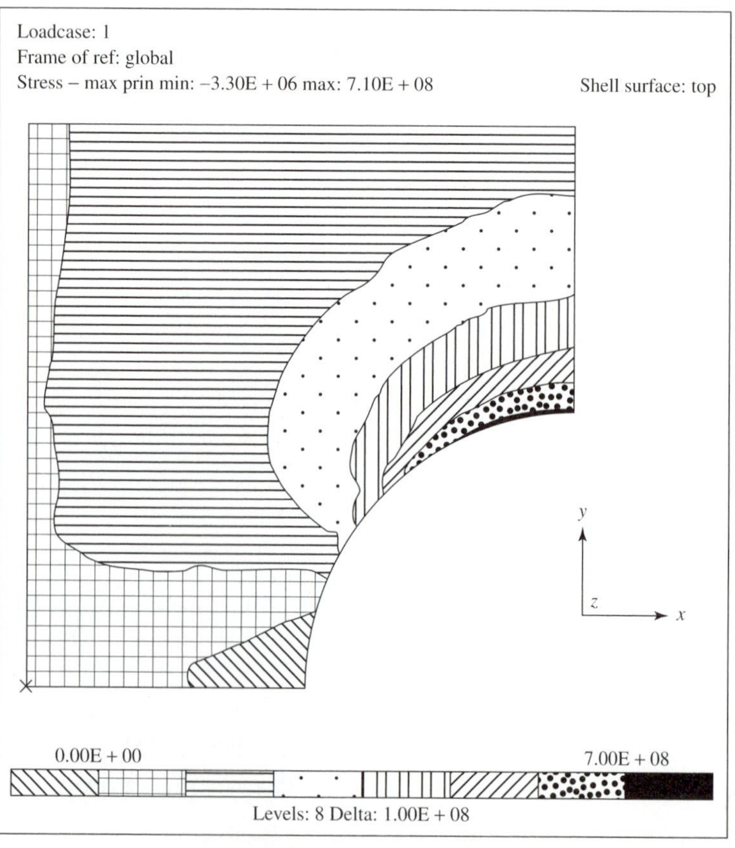

Figure 6.7 (a) Maximum normal stress contours overlaid on the finite element mesh; (b) maximum normal stress ranges represented by stepped contours.

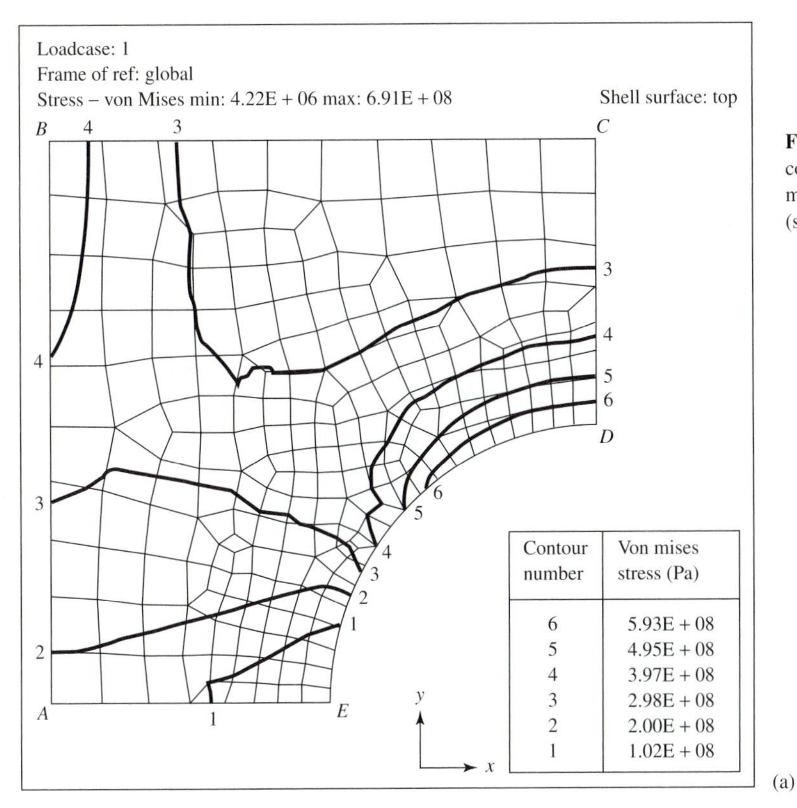

Figure 6.8 (a) Von Mises stress contours overlaid on the finite element mesh; (b) von Mises stress (stepped contour bands)

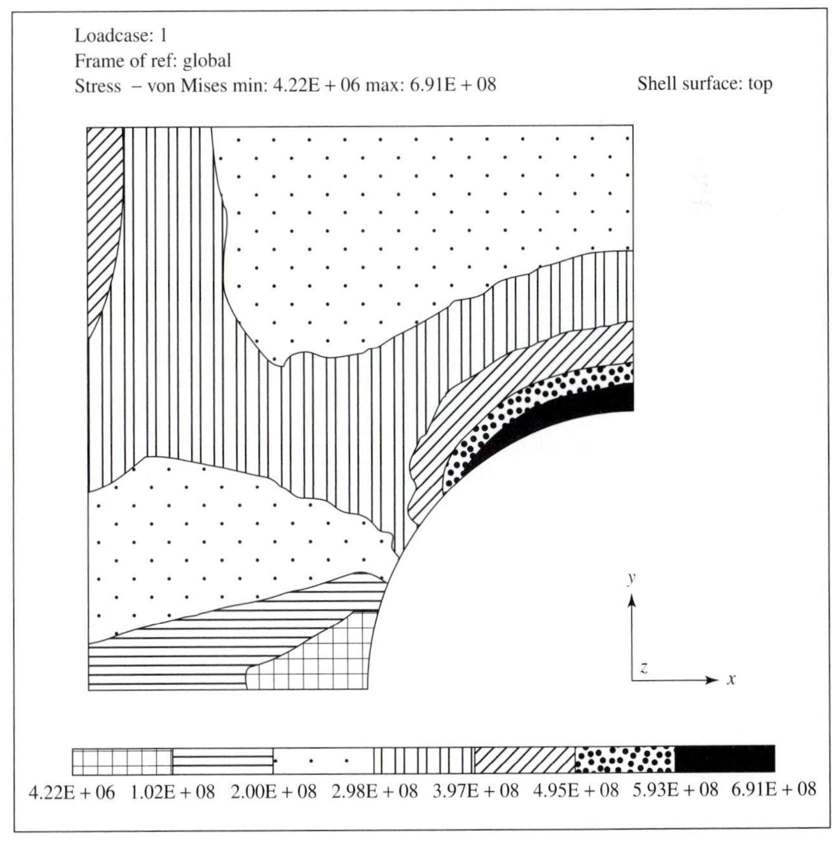

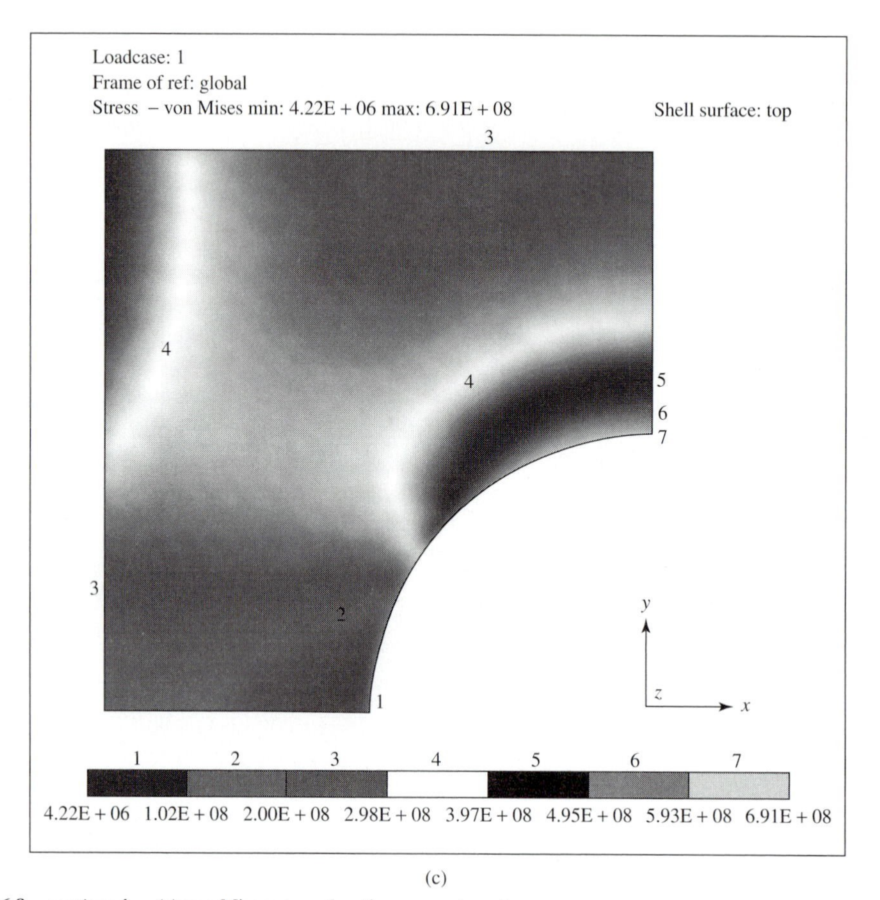

(c)

Figure 6.8 *continued* (c) von Mises stress (continuous contours).

on edge *AB*, the left edge of the plate. Near that edge, maximum normal stress should have about the same value. Checking the FEA plots, we see that the maximum normal stress plots agree roughly with our expectations.

The nominal stress is the average x-direction stress over the net cross section. It is

$$\sigma_{\text{NOM}} = \sigma_{x(\text{AVG})} = 2 \cdot 10^6/(100 \cdot 50) = 400 \text{ MPa}$$

on edge *CD*, a symmetry plane. Stress $\sigma_{x(\text{AVG})}$ on *CD* is double the value of σ_x on the left edge. Maximum normal stress should also equal x-direction stress on edge *CD*. The plotted FEA maximum normal stress contours on edge *CD* are around 400 MPa, as we expected.

The FEA plots show a maximum normal stress

$$\sigma_{\text{MAX}} = 710 \text{ MPa}$$

at point *D*. The stress concentration factor is defined by

$$K_T = \sigma_{\text{MAX}}/\sigma_{\text{NOM}} = 710/400 = 1.78$$

The maximum value of von Mises stress is

$$\sigma_{\text{VM}} = 691 \text{ MPa}$$

at point D. Since there is no loading normal to the notch surface, we would expect that

$$\sigma_2 = \sigma_3 = 0$$

and

$$\sigma_{VM} = \sigma_{MAX} = \sigma_1 = 710 \text{ MPa}$$

We can assume that the discrepancy (less than 3%) is due to the size of the finite element mesh in the vicinity of point D. Stresses σ_{VM} and σ_{MAX} at point D should converge as mesh size is reduced.

Overall, it appears that the analysis was a success, but the proposed design may be unsatisfactory. The design criterion for reversed tensile stress is

$$S_E/N_{FS} \geq \sigma_R$$

where S_E = corrected endurance limit

N_{FS} = factor of safety

σ_R = range stress

The value of σ_{VM} or σ_{MAX} is the range stress in this case. It is unlikely that we could find any material at reasonable cost to satisfy the design criterion. A redesign is suggested. Possibilities include (1) eliminating stress concentration by making the part with uniform cross section, (2) increasing part size, and (3) reducing load on the part.

■■

Users of FEA software are advised to test the program with a problem for which the solution is well known. If the answers disagree, then the user may have made a mistake in the preprocessing stage. A misapplication of the program could be responsible. More study of the manuals and more practice are needed before a design study that cannot be checked is attempted. Or the discrepancy may be due to a fault in the software.

■■ EXAMPLE PROBLEM FEA of an Axisymmetric Problem

A 5 mm thick, 40 mm radius circular aluminum plate has a 10 mm radius hole in the center. There is a 500 MPa radial stress on the outer radius.

(a) Using FEA, find displacements and stresses.
(b) Verify the results.

Solution (a): One-quarter of the plate is modeled. For aluminum, we will use an elastic modulus E of 69,000 MPa and Poisson's ratio ν of 0.3. We direct the program to make a fine mesh near the inner radius and a coarser mesh toward the outer radius.

Displacements and rotations u, v, w, ω_x, ω_y, and ω_z are, respectively, f, 0, 0, 0, 0, and 0 on the $y = 0$ edge where f = free and the coordinate origin is at the center of the full plate. The only nonzero displacement on the $x = 0$ edge is in the y-direction. On that edge, the restraints are 0, f, 0, 0, 0, and 0. The finite element mesh, the node points, and the restraints are shown in Figure 6.9.

The outer edge of the plate is given an edge pressure of

$$(-500 \cdot 10^6 \text{ Pa})(0.005 \text{ m}) = -2.5 \cdot 10^6 \text{ N/m}$$

where negative pressure corresponds to positive (tensile) stress. The preprocessing is completed and the solution phase is begun.

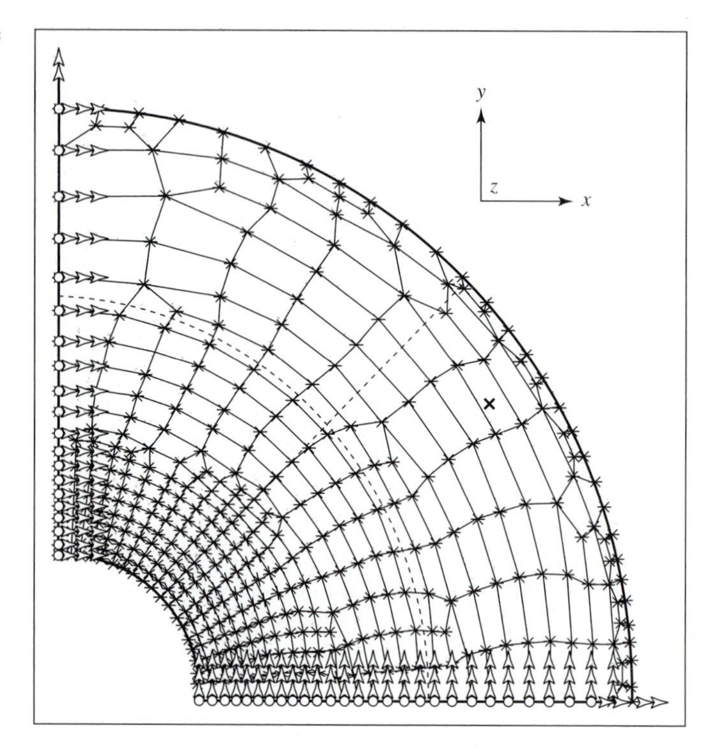

Figure 6.9 FEA model of axisymmetric problem.

Figure 6.10(a) shows the results of the first trial. The maximum normal stress contours obviously lack symmetry. Part (b) of the figure shows a displacement plot with the unloaded model shown by dashed lines. The displacement at the right-hand corner of the model is inconsistent with the symmetry boundary conditions required on the $y = 0$ edge. A restraint is missing on one or more nodes.

After applying the missing restraints, we repeat the solution phase. Figure 6.11(a) shows the displacement form of the corrected model. We have now achieved the required symmetry. Displacement contours are plotted in Part (b) of the figure. Based on the FEA results, minimum displacement is 0.147 mm (at the surface of the hole); maximum displacement is 0.241 mm (at the outer radius). There are 10 displacement contours in the figure. They range from 0.156 mm (contour 1) to 0.233 mm (contour 10) in steps of $8.55 \cdot 10^{-3}$ mm.

Maximum normal stress and von Mises stress are shown in Figure 6.12(a) and (b), respectively. The FEA results indicate that maximum normal stress (maximum principal stress) ranges from 565 MPa at the outer radius to 1080 MPa at the surface of the hole.

We can define a stress concentration factor as

$$K_T = \sigma_{\text{MAX}}/\sigma_{\text{NOM}} = 1080/500 = 2.16$$

where σ_{NOM} = stress at the outer radius

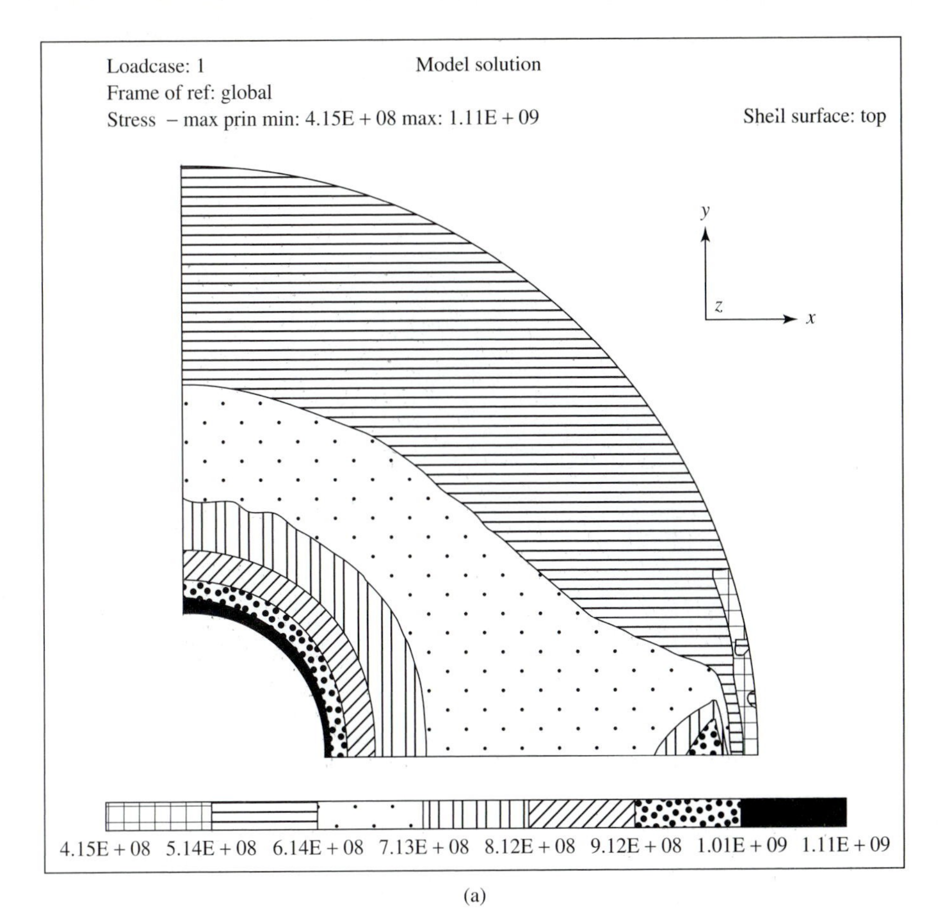

(a)

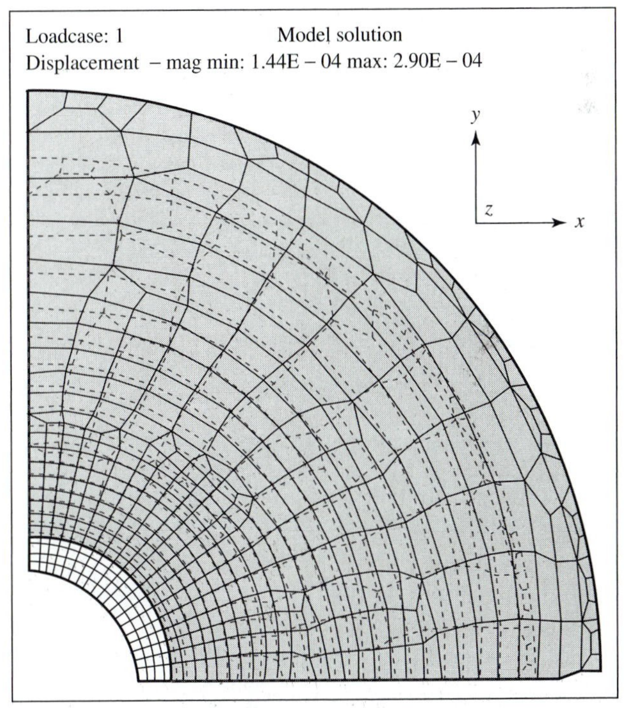

(b)

Figure 6.10 Maximum normal stress contours. Note the lack of symmetry. (b) Displacements. Results indicate a missing restraint.

Figure 6.11 Displacements (corrected model); (b) displacement contours.

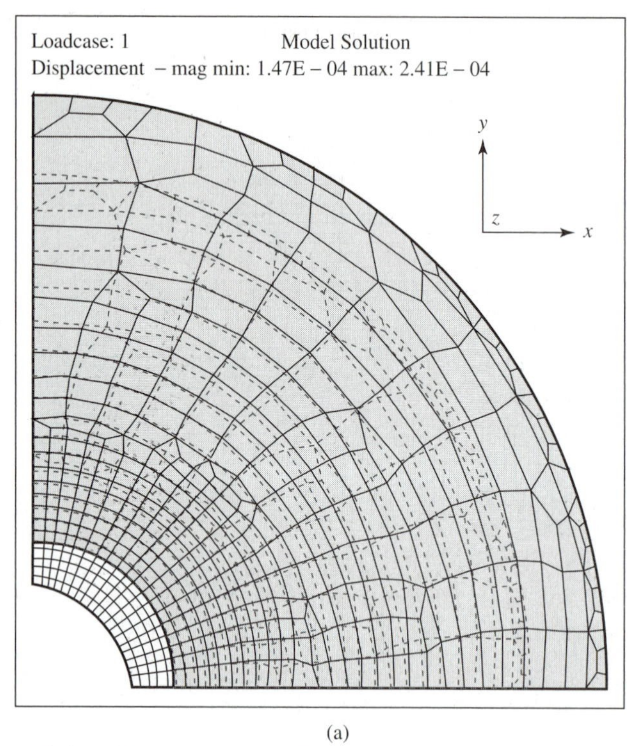

(a)

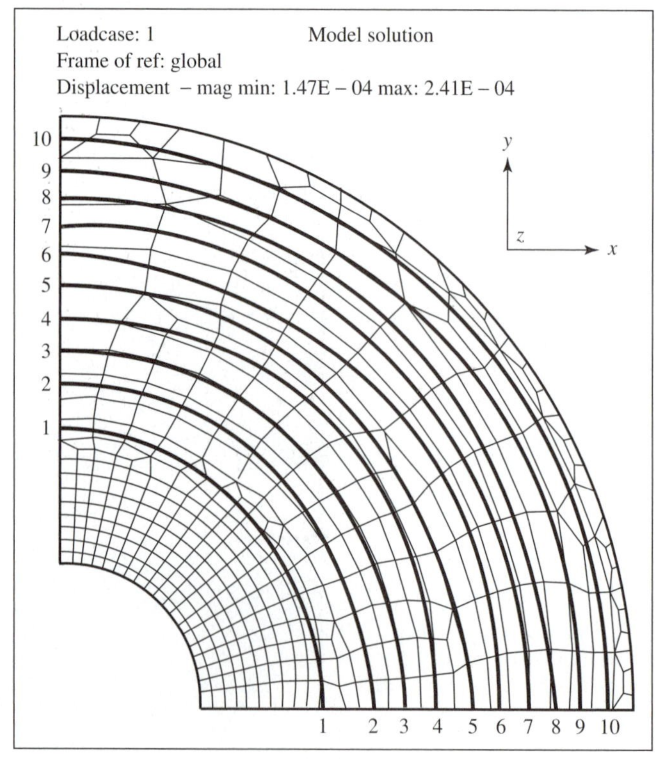

(b)

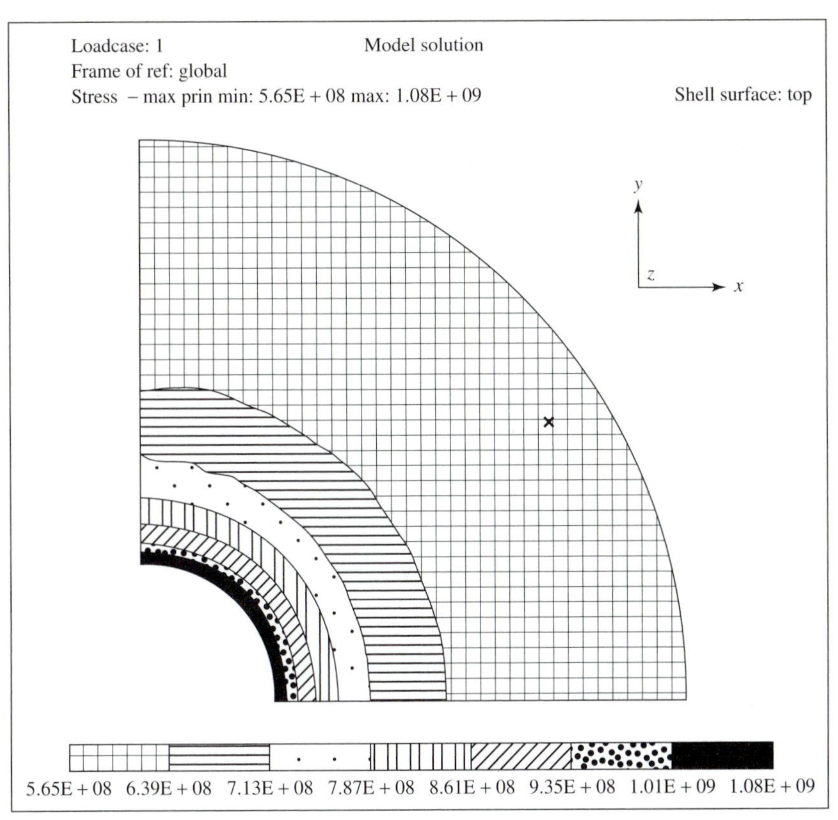

(a)

Figure 6.12 (a) Maximum normal stress in an axisymmetric plate; (b) von Mises stress in an axisymmetric plate.

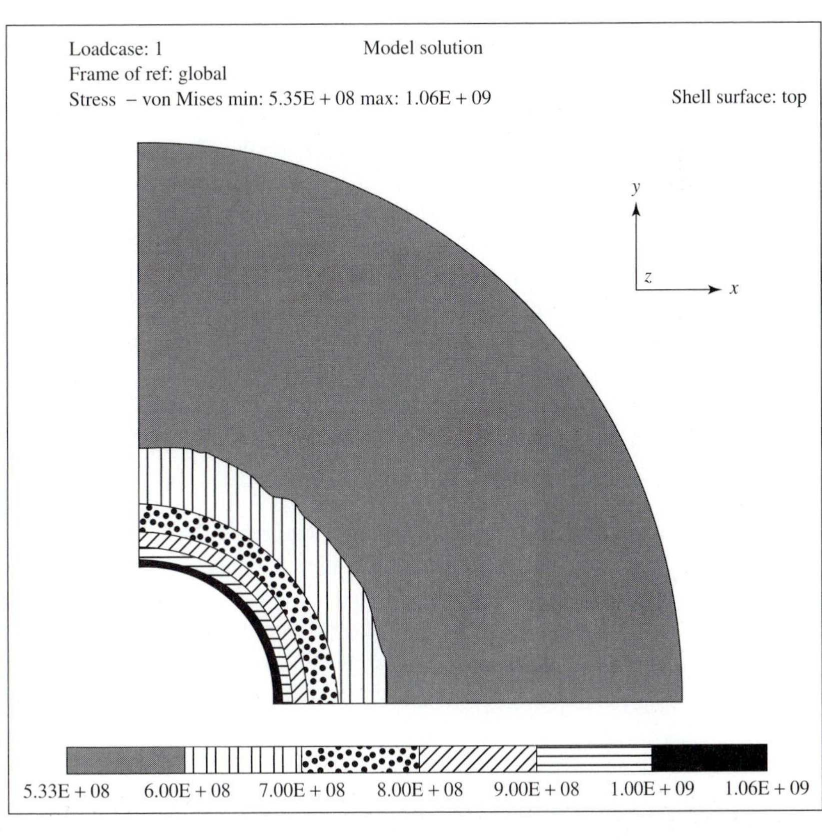

(b)

Solution (b): The tangential stress at radius r in a circular plate with pressure q at the outer radius is given by the following equation (Deutschman, Michels, and Wilson, 1975):

$$\sigma_\theta = -b^2q\left[1 + \frac{a^2}{r^2}\right]/[b^2 - a^2]$$

where a = inner radius = 10 mm

b = outer radius = 40 mm

$q = -\sigma_r(b) = -500$ MPa

Substituting the plate dimensions and stress at b, and setting $r = a$, we find

$$\sigma_{\text{MAX}} = \sigma_\theta(a) = -40^2[-500]\left[1 + \frac{10^2}{10^2}\right]/[40^2 - 10^2]$$
$$= 1067 \text{ MPa}$$

The values of stress obtained by FEA and those from the above equation differ by a little more than 1%.

■■

MODELING ACTUAL MACHINE DESIGN PROBLEMS—INTERPRETATION AND VERIFICATION OF RESULTS

After creativity, modeling and interpretation may be the most difficult tasks of a machine designer. While finite element analysis is a powerful tool, total reliance on computer-generated results can be treacherous. Some possible sources of FEA error are noted below.

1. **Hardware errors.** Hardware problems can sometimes cause data to be lost or misplaced. Fortunately, most computer hardware failures are obvious to the user and are unlikely to result in undetected errors.
2. **Software errors.** Some individuals who construct FEA and related programs are more skilled in the art of writing software than in the technical applications of the software, such as stress analysis and machine design. For example, in the preceding axisymmetric plate problem, the principal stresses are σ_θ, σ_r, and σ_z, where σ_θ is the largest principal stress and $\sigma_z = 0$. In general, maximum shear stress is the greatest value obtained by taking one-half the difference between principal stresses in any pair of planes. In a plane stress problem, maximum shear stress cannot be less than one-half the largest principal stress.

   At least one commercial FEA program computes maximum shear stress for the axisymmetric plate in the preceding problem as

   $$(\sigma_\theta - \sigma_r)/2$$

   The FEA program has found the greatest value of shear stress in the $r\theta$-plane, not the actual maximum shear. In some instances, an error of this type can lead to an unsafe design.
3. **Modeling errors.** Modeling of a real problem involves simplifying the problem to make a solution possible. Ideally, unimportant details are neglected, but critical items are modeled. If a problem is poorly modeled, the FEA results are meaningless. Mod-

eling skills can be improved by experience. Sometimes it is instructive to solve a problem by making a very simple model. The results from the first model are used to develop a second, improved model.

4. **Errors in boundary conditions.** This problem is a common one, but there is an easy solution. A careful study of the displacement shape and the stress contours will help the designer avoid boundary condition errors. Does a symmetry plane behave improperly? Does the stress pattern have unreasonable characteristics?
5. **Errors in units.** Units should be recorded. Use consistent units (lb, in, psi; N, m, Pa; or N, mm, MPa).

Verification

Errors can appear in software, just as errors are likely to be present in this text. Theory is subject to limitations that may not be apparent to the user. The user is advised to verify software reliability by solving a problem that can be checked with a standard reference. A successful solution will also verify the user's understanding of the software.

Of course, we use FEA software to solve problems that are not found in references. Solutions should be checked against hand-calculation solutions of the same problem in simplified form. Large discrepancies may indicate a need for reevaluating the assumptions and the solution.

Interpretation

Our results may lead to questions such as the following:

- Why did we solve this problem?
- Will the part fail?
- Is it overdesigned (not economical, not competitive)? Should we redesign?
- Do the FEA displacement and stress plots help us determine the sensitivity to certain design features (stress concentration, location of loads, etc.)?
- Do the FEA plots give us redesign hints?

Without interpretation, the FEA solution is a waste of time.

References and Bibliography

ANSYS™ Introduction to Finite Element Methods. Houston, PA: Swanson Analysis Systems, Inc., n.d.

Chandrupatla, T. R., and A. D. Belegundu. *Introduction to Finite Elements in Engineering.* Englewood Cliffs, NJ: Prentice Hall, 1993.

Den Hartog, J. P. *Advanced Strength of Materials.* New York: McGraw-Hill, 1952.

Deutschman, A., W. J. Michels, and C. Wilson. *Machine Design: Theory and Practice.* New York: Macmillan, 1975.

Grandin, H. *Fundamentals of the Finite Element Method.* New York: Macmillan, 1986.

Knight, C. E. *The Finite Element Method in Mechanical Design.* Boston: PWS-Kent Publishing Co., 1993.

Lawry, M. H. *I-DEAS Master Series™.* 2d ed. Milford, OH: Structural Dynamics Research Corporation, 1994.

MacNeal, R. H., ed. *MSC/NASTRAN Handbook for Linear Static Analysis.* Los Angeles: MacNeal-Schwendler Corp., 1981.

Zahavi, E. *The Finite Element Method in Machine Design.* Englewood Cliffs, NJ: Prentice Hall, 1991.

Design Problems

6.1. A 50 mm thick by 200 mm long by 200 mm wide steel plate has a 50 mm radius central hole, as shown in Figure 6.13. The plate will be subject to fatigue loading, with a maximum tensile stress of 200 MPa on the left and right edges as shown in the figure.

- **(a)** Model the problem. Apply loading and boundary conditions for solution by a finite element analysis program.
- **(b)** Form a finite element mesh.
- **(c)** Solve for the deflected form of the plate.
- **(d)** Solve for the stress distribution in the plate.
- **(e)** Make rough calculations to verify the validity of the results. Evaluate the results.

6.2. A 0.25 in thick by 3 in wide aluminum plate has a 0.5 in radius central hole (see Figure 6.13). The plate will be subject to fatigue loading, with a maximum tensile stress of 9000 psi on the left and right edges as shown in the figure.

- **(a)** Model the problem. Apply loading and boundary conditions for solution by a finite element analysis program.
- **(b)** Form a finite element mesh.
- **(c)** Solve for the deflected form of the plate.
- **(d)** Solve for the stress distribution in the plate.
- **(e)** Make rough calculations to verify the validity of the results. Evaluate the results.

Figure 6.13 Stress concentration.

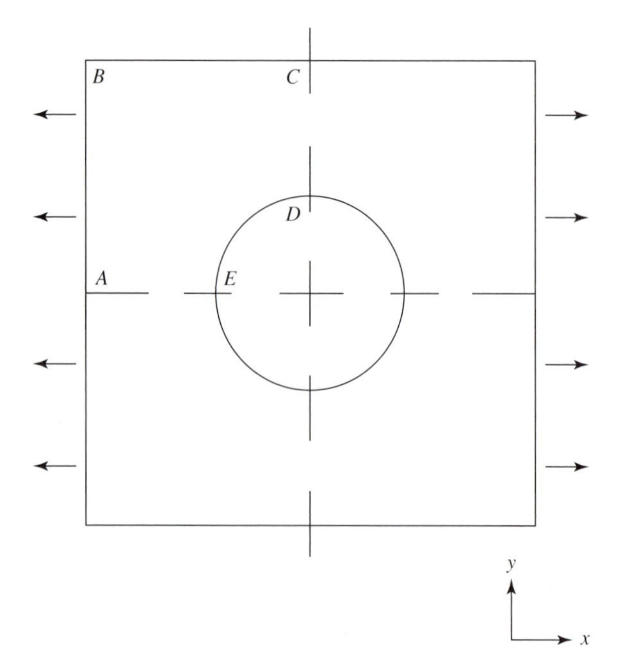

6.3. A 0.10 in thick by 2.5 in wide steel plate has a 0.25 in radius central hole (see Figure 6.13). The plate will be subject to fatigue loading, with a maximum tensile stress of 15,000 psi on the left and right edges as shown in the figure.

(a) Model the problem. Apply loading and boundary conditions for solution by a finite element analysis program.

(b) Form a finite element mesh.

(c) Solve for the deflected form of the plate.

(d) Solve for the stress distribution in the plate.

(e) Make rough calculations to verify the validity of the results. Evaluate the results.

6.4. A 0.125 in thick, 7 in diameter circular steel plate has a 1.5 in diameter hole in the center. The radial stress at the outer diameter is 10,000 psi.

(a) Use an FEA program to determine the stress distribution in the plate.

(b) Check the values by using a closed-form solution.

6.5. A 0.25 in thick, 5 in diameter circular steel plate has a 2 in diameter hole in the center. There is a tensile load of 5000 lb/in of circumference on the outer edge.

(a) Use an FEA program to determine the stress distribution in the plate.

(b) Check the values by using a closed-form solution.

CHAPTER 7

Elastic Stability (Column Buckling)

Symbols

A	cross-sectional area
E	modulus of elasticity (Young's modulus)
g	acceleration of gravity
I	least-moment-of-inertia
M	mass
P	load
P_{CR}	critical load
L	unsupported column length
N_{FS}	factor of safety
r	radius of gyration
S_C	compressive strength
S_{YP}	yield point strength
w	deflection in the y-direction
x	axial coordinate
σ	direct stress

Units

dimensions, radius of gyration	in or mm
load, critical load	lb or N
moment-of-inertia	in^4 or mm^4
stress, strength, modulus of elasticity	psi or MPa

A long, slender member that is loaded in compression may bend, and ultimately fail, even though stresses are initially below the yield point of the material. Failure under such circumstances is called a **buckling failure,** an **elastic stability failure,** or a **slender column failure.** This problem is encountered in machine design as well as in building design. A "column" need not be vertical to fail in buckling; if the load is compressive, we should check our design to rule out the possibility of an elastic stability failure.

THE EULER COLUMN

Figure 7.1 shows a pinned–pinned column with lateral restraint at the ends, also called an **Euler column.** The ideal Euler column is assumed to be perfectly straight before loading, and compressive load P is assumed to be axial. The differential equation describing the column is

$$\frac{d^2w}{dx^2} + \frac{Pw}{EI} = 0 \tag{7.1}$$

where x = the axial coordinate (vertical in the figure)

w = deflection of the column in the y-direction, at any location x

Deflection is measured from the x-axis (the centerline of the unloaded column).

P = axial load

E = elastic modulus

I = least-moment-of-inertia

We need not be proficient in the solution of differential equations to find the general solution to this one. The solution to equations of this form is well known and is found in

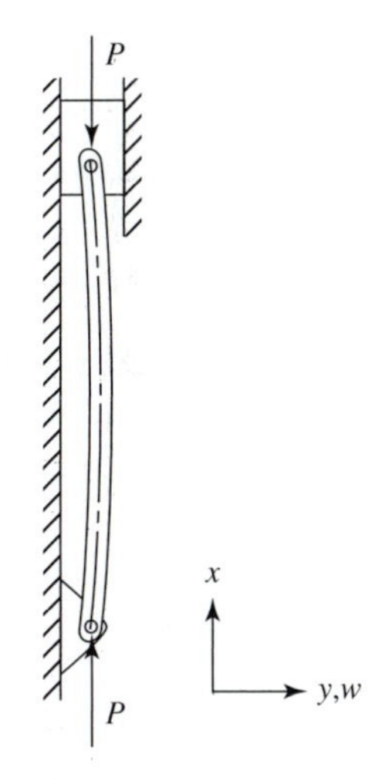

Figure 7.1 The Euler column.

scores of texts. Adapting the solution for the load and physical properties of the column problem, we have

$$w = C_1 \cos\left(\sqrt{\frac{P}{EI}}x\right) + C_2 \sin\left(\sqrt{\frac{P}{EI}}x\right) \tag{7.2}$$

where arbitrary constants C_1 and C_2 depend on end conditions; that is, the way the ends of the member are held in place.

For the Euler column, the restraints shown in Figure 7.1 require that deflection w be zero at both ends; that is,

$$w(0) = 0 \quad \text{and} \quad w(L) = 0 \tag{7.3}$$

The first end condition is satisfied if $C_1 = 0$. The second requires that

$$C_2 \sin\left(\sqrt{\frac{P}{EI}}L\right) = 0 \tag{7.4}$$

If we take $C_2 = 0$, then deflection $w(x) = 0$ for all values of x. Thus, the alternative is what we want; that is,

$$\sin\left(\sqrt{\frac{P}{EI}}L\right) = 0$$

Rejecting zero values of load P or length L, or infinite values of elastic modulus E or moment-of-inertia I, we have

$$\sqrt{\frac{P}{EI}}L = \pi$$

from which

$$P = P_{CR} = \frac{\pi^2 EI}{L^2} \tag{7.5}$$

where P_{CR} = critical load for the Euler column

If we apply a compressive load of P_{CR}, the ideal Euler column will fail by buckling. Other solutions are mathematically possible. For example,

$$\sin(2\pi) = 0, \sin(3\pi) = 0, \sin(4\pi) = 0, \ldots$$

However, these solutions result in higher values of critical load. Higher values of critical load apply only if we restrain the column somewhere along its length. The $w(x) = 0$ solution applies to the ideal column if the load is less than P_{CR}. Thus, if a member can be modeled as an Euler column, the design criterion is

$$\frac{P_{CR}}{N_{FS}} = \frac{\pi^2 EI}{L^2 N_{FS}} \geq P_{\text{COMPRESSIVE}} \tag{7.6}$$

This criterion applies only to **slender columns** and to certain end restraints. **Short columns** may fail when the compressive strength of the material is exceeded. The identification of slender columns and short columns will be treated in a later section.

EXAMPLE PROBLEM A Support Bracket

A bracket is required to support a load F of 300 lb as shown in Figure 7.2(a), where $L = 16$ in and $\theta = 40°$. Design the bracket.

Design Decisions: One-quarter in thick steel with a yield point S_{YP} of 60,000 psi will be used. The bracket parts will be bolted together and bolted to the large, vertical I-section column at the left. A safety factor N_{FS} of 4 will be used. Bar BC will be treated as a simply supported member (an Euler column) since the ends are not rigidly clamped.

Solution: We set the sum of the forces at point B equal to zero, as shown in Part (b) Figure 7.2. We find that a tensile force in horizontal member AB is

$$F_{AB} = F \tan \theta = 300 \tan 40° = 252 \text{ lb}$$

There is a compressive force

$$F_{CB} = F/\cos \theta = 300/\cos 40° = 392 \text{ lb}$$

in diagonal member CB.

Bar CB must be designed so that it does not fail in buckling. The design criterion is

$$\frac{P_{CR}}{N_{FS}} = \frac{\pi^2 EI}{L^2 N_{FS}} \geq P_{\text{COMPRESSIVE}}$$

from which we find the minimum moment-of-inertia.

$$I_{\text{MIN}} \geq \frac{L^2 N_{FS} F_{CB}}{\pi^2 E} = \frac{16^2 \cdot 4 \cdot 392}{30 \cdot 10^6 \cdot \pi^2} = 1.356 \cdot 10^{-3}$$

Moment-of-inertia is expressed in terms of bracket dimensions by

$$I = \frac{bh^3}{12}$$

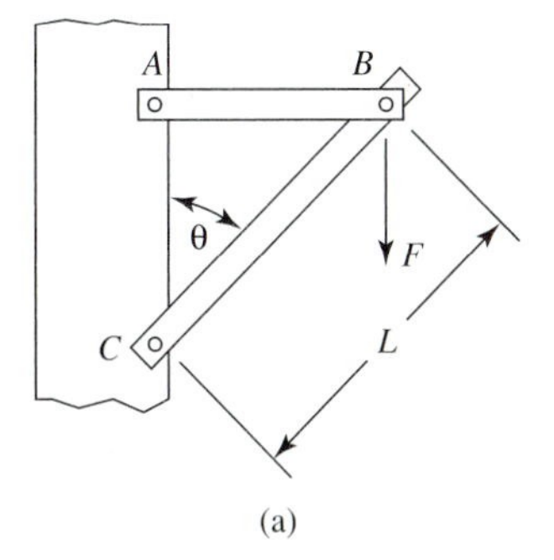

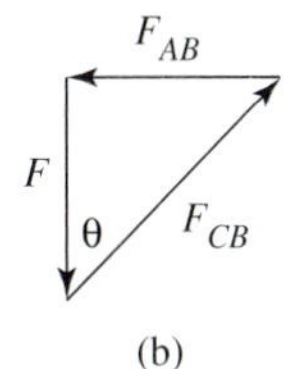

Figure 7.2 Support bracket.

from which we determine the minimum acceptable width for bar *CB*.

$$b = \frac{12I}{h^3} = \frac{12 \cdot 1.356 \cdot 10^{-3}}{0.25^3} = 1.041 \text{ in}$$

The direct stress in bar *CB* is

$$\sigma = \frac{F_{CB}}{bh} = \frac{392}{1.041 \cdot 0.25} = 1506 \text{ psi}$$

which is far below S_{YP}/N_{FS}. The buckling criterion governs the design of member *CB* of the bracket. For convenience in manufacture, the same dimensions can be used for bar *AB*. The designer must then select fasteners that are adequate for the loads at *A*, *B*, and *C*.

MOMENT-OF-INERTIA AND RADIUS OF GYRATION

Location of the neutral axis in bending and calculation of moment-of-inertia are discussed in detail in Chapter 5, "Bending of Machine Members." The properties of standard steel sections can be obtained from steel producers (e.g., Bethlehem Steel Corporation, 1978).

When a structural steel member is used as a beam (i.e., loaded in bending), the section is usually oriented to obtain the greatest moment-of-inertia. Take, for example, a W16×26 wide-flange I-beam. This section has a depth of about 16 in and a flange width of 5.5 in, and it weighs about 26 lb/ft. Ordinarily, we orient this beam so that the web is vertical, support the beam at two or more places on the bottom flange, and apply vertical loads on the top flange. The neutral axis in bending, then, is a horizontal line through the center of the web, axis *x*–*x* in Figure 7.3. The applicable moment-of-inertia I_x is 301 in^4. This orientation is optimum; that is, it produces the least stress and the least deflection in the beam.

If we orient an I-beam so that the web is horizontal, and if we apply vertical loads, then the applicable neutral axis is the web axis, axis *y*–*y* in Figure 7.3. The applicable moment-of-inertia is then $I_y = 9.59$ in^4. It would be inefficient to use a W16×26 wide-flange I-beam in this manner.

Figure 7.3 I-beam.

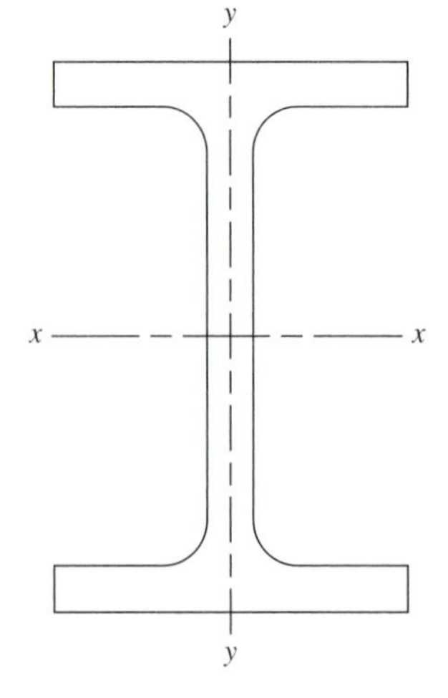

When a long, slender member is subject to compressive loads, we cannot base our design on the maximum moment-of-inertia. Buckling forces "look for" the axis that represents the least stiffness. The design must be based on minimum moment-of-inertia. For an I-beam similar to the one in the figure, I_y is the minimum moment-of-inertia. The minimum moment-of-inertia of an angle section is computed about a skew axis through the center-of-gravity.

Radius of Gyration

The **radius of gyration** is defined by

$$r = \sqrt{\frac{I}{A}} \qquad \textbf{(7.7)}$$

where r = radius of gyration (in or mm)

A = cross-sectional area (in^2 or mm^2)

I = moment-of-inertia (in^4 or mm^4)

I and r refer to the same axis (which should be identified).

We are usually interested in the least radius of gyration when considering elastic stability.

EXAMPLE PROBLEM An Angle-Section in Compression

Designers have decided to use a 60 in long L2×2×1/8 equal-leg steel angle-section as a stiffener in a machine support structure. The stiffener will be in compression. Find the maximum safe compressive load in the stiffener.

Design Decisions: The ends of the stiffener will be welded to other parts of the support structure. Although the welded ends may provide somewhat more restraint than simple supports, the ends are not rigidly clamped. A safety factor N_{FS} of 2.5 will be used.

Solution: We use a reference on standard steel sections (e.g., Bethlem Steel Corporation, 1978) to find the following properties for an L2×2×1/8 angle:

$$A = 0.484 \text{ in}^2 \qquad r_x = r_y = 0.626 \text{ in} \qquad r_z = 0.398 \text{ in}$$

The x-, y-, and z-axes are identified in Figure 7.4. The corresponding moments-of-inertia, given by

$$I = Ar^2$$

are

$$I_x = I_y = 0.190 \text{ in}^4 \quad \text{and} \quad I_z = 0.0767 \text{ in}^4$$

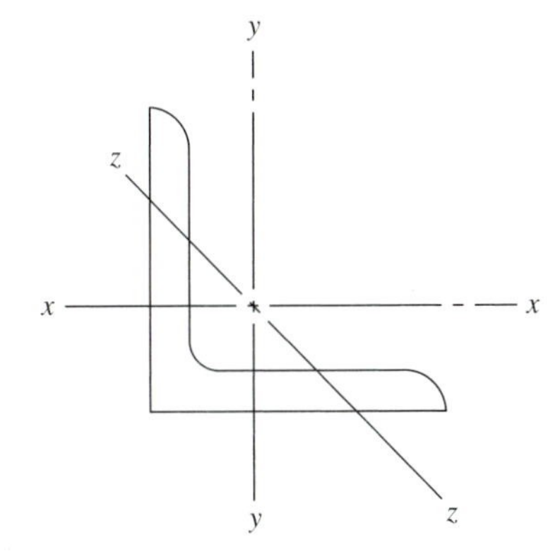

Figure 7.4 Angle-section.

Owing to the lack of rigidity at the ends, we will use the Euler column model. The result is

$$\frac{P_{CR}}{N_{FS}} = \frac{\pi^2 EI}{L^2 N_{FS}} = \frac{\pi^2 \cdot 30 \cdot 10^6 \cdot 0.0767}{60^2 \cdot 2.5} = 2523$$

Dividing the allowable force by the cross-sectional area, we calculate a direct stress of only 5213 psi. Thus, elastic stability governs; the compressive load in the stiffener is limited to 2523 lb.

END CONDITIONS

Although end conditions can easily be specified in textbook problems, it is sometimes difficult to determine the appropriate end conditions when actually designing machines. A careless assumption can lead to an unsafe design.

CLAMPED–FREE COLUMNS

The Euler column is restrained against lateral motion. The column shown in Figure 7.5 has clamped–free end conditions. If we take the coordinate origin at the top, then Equation (7.1), the same differential equation that describes the Euler column, applies. The general solution, Equation (7.2), applies as well, but the end conditions are different. The end conditions are

$$w(0) = 0 \quad \text{and} \quad \frac{dw}{dx}(L) = 0 \tag{7.8}$$

where x is measured from the top of the column

$w = w(x)$ = deflection relative to the top end

dw/dx = slope of the column

A procedure similar to that used with the Euler column leads to

$$P_{CR} = \frac{\pi^2 EI}{4L^2} \tag{7.9}$$

where P_{CR} = critical load for the clamped–free column

I = least-moment-of-inertia of the column cross section

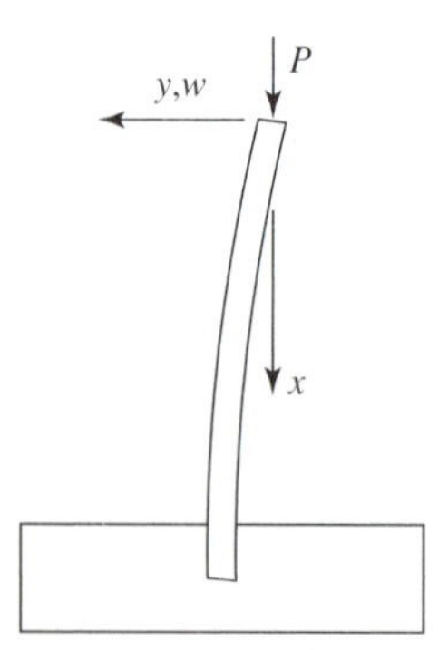

Figure 7.5 Clamped–free column.

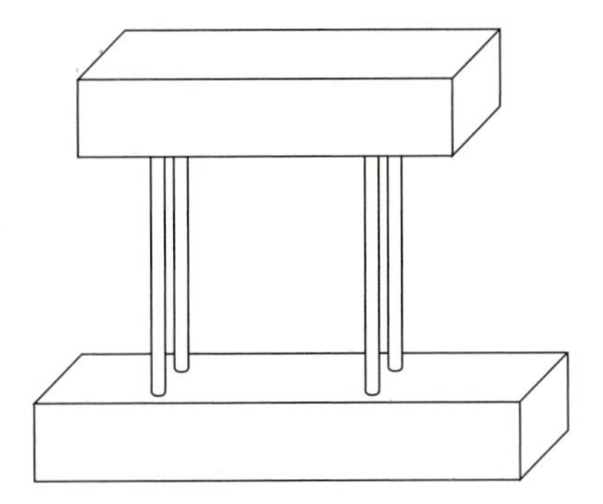

Figure 7.6 Clamped–clamped columns.

Clamped–free boundary conditions would occur if we wished to mount a heavy instrument on a slender, flagpole-like structure. Note that a slender clamped–free column can take only one-quarter the load of a similar Euler column.

CLAMPED–CLAMPED COLUMNS

Suppose that we support a heavy machine on four columns as in Figure 7.6. If both the support platform and the base are drilled to accommodate the columns, or if we weld both ends of each column to rigid braces, then the columns have clamped–clamped end conditions. However, the platform is free to move laterally (to the front or back, to the right or left). The critical load for a single column is given by

$$P_{CR} = \frac{\pi^2 EI}{L^2} \tag{7.10}$$

where P_{CR} = critical load of a single clamped–clamped column that is part of a group arranged as in Figure 7.6

Note that the above result is the same as for the Euler column. If the support contained only one or two columns, then the slope at the top would not be restricted, and clamped–free end conditions would apply. The per-column capacity would be only one-quarter as great.

CLAMPED–PINNED COLUMNS WITH LATERAL RESTRAINT

Figure 7.7 shows a column clamped at the bottom and pinned at the top. If the column deflects, a lateral force P_y prevents the top end from moving laterally. This lateral force appears in the differential equation describing the column and in the general solution. When we apply the boundary conditions, we obtain

$$\tan\left(L\sqrt{\frac{P_x}{EI}}\right) = L\sqrt{\frac{P_x}{EI}} \tag{7.11}$$

Equation (7.11) is written in the form

$$\tan(a) = a \tag{7.12}$$

where $a = L(P_x/EI)^{1/2}$

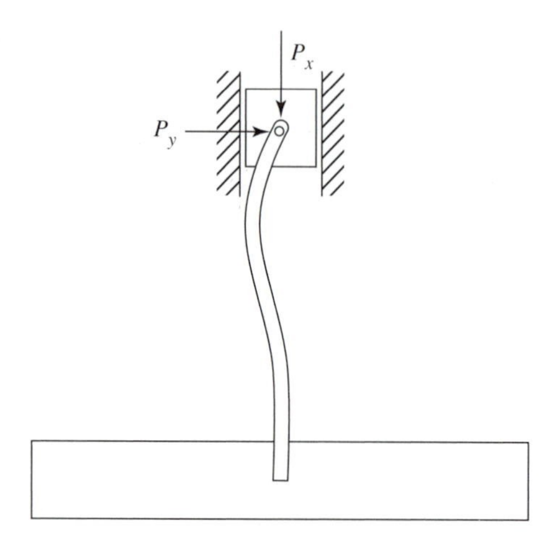

Figure 7.7 Clamped–pinned column with lateral restraint.

It can be solved by a user-written program or with the help of mathematics software. An obvious solution is $a = 0$. This is a trivial (i.e., useless) root. To avoid the trivial root, we begin iteration at, say, $a = 3$. The root-finding program yields

$$a = 4.49341$$

Identifying P_x as P_{CR}, we have

$$P_{CR} = \frac{a^2 EI}{L^2} = \frac{20.191 \cdot EI}{L^2} \tag{7.13}$$

where P_{CR} = critical load for the clamped–pinned column with lateral restraint

CLAMPED–CLAMPED COLUMNS WITH LATERAL RESTRAINT

Figure 7.8 shows a clamped–clamped column with lateral restraint. The additional restraint substantially increases the critical load. In this case, the critical load is

$$P_{CR} = \frac{4\,\pi^2 EI}{L^2} \tag{7.14}$$

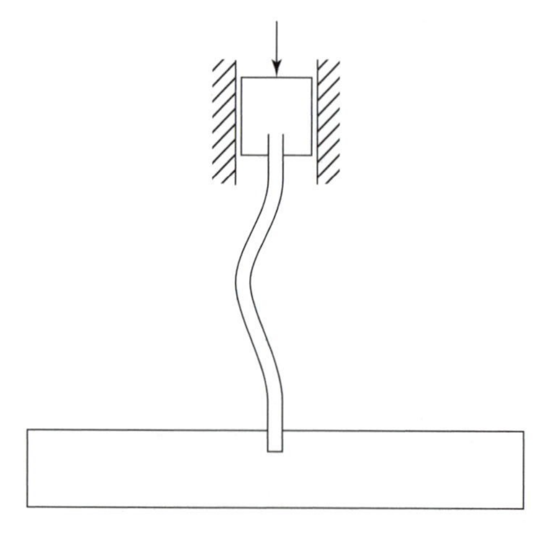

Figure 7.8 Clamped–clamped column with lateral restraint.

IDENTIFYING SLENDER COLUMNS AND SHORT COLUMNS

Columns are sometimes identified by their **slenderness ratio,** given by

$$\text{Slenderness ratio} = L/r$$

where L = unsupported column length
r = least radius of gyration

Slenderness ratio is a valid criterion for comparing columns made of the same material and having the same end conditions. However, designers of machines use many different materials, and end conditions vary. Thus, we must distinguish columns as follows:

$$\textbf{Slender columns:} \quad P_{CR}/A \leq S_C \tag{7.15}$$

$$\textbf{Short columns:} \quad S_C < P_{CR}/A \tag{7.16}$$

where P_{CR} = appropriate critical load considering material and end conditions
A = cross-sectional area
S_C = compressive strength of the material

Slender columns are designed on the basis of critical load. Short columns are designed on the basis of compressive stress. In either case, an appropriate factor of safety should be used. For steel, it is suggested that the yield point be used as the compressive strength of the material. When $P_{CR}/A \approx S_C$, the column can be classified as an intermediate column. The failure mode of an intermediate column may depend on its initial straightness.

EXAMPLE PROBLEM Design of a Two-Column Support, Considering Elastic Stability

Design a two-column support for a mass. The mass will rest on a small platform 200 mm above a rigid base, as shown in Figure 7.9. The total supported mass (including the platform) is 112.2 kg.

(a) Consider the possible modes of failure.
(b) Calculate the required moments-of-inertia.
(c) Find column dimensions for optimum design.
(d) Check the results.

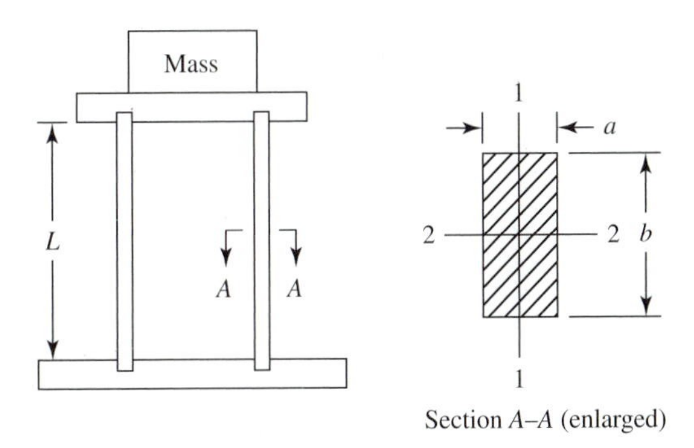

Figure 7.9 Two-column support.

Design Decisions: The mass will be supported on two rectangular aluminum bars. The bars will be recessed in the base and the platform, with an unsupported length of 200 mm. A safety factor of 3 will be used. We will try to use a minimum of material.

Mathematics software was used to aid in the design. Calculation details are shown later.

Solution (a): The total force on the pair of bars is

$$P = M \cdot g = 112.2 \text{ kg} \times 9.807 \text{ m/s}^2 = 1100 \text{ N}$$

The bars may fail in buckling by bending to the right or left. In that case, the platform will restrain end rotation of the bars. Each bar will have a vertical tangent at the top. Alternatively, the bars may fail in buckling by bending forward or backward. In that failure mode, the top ends of the bars will be free to rotate. Finally, we must check to see that the direct compressive stress on the bars is not excessive.

Solution (b): If the bars fail in buckling by bending about the 1–1 axis, then clamped–clamped end conditions apply, with lateral motion (left or right) possible. The critical load for a single bar is

$$P_{CR1} = \pi^2 EI_1/L^2$$

We multiply applied load P by the factor of safety and set the result equal to twice the critical load for one bar. Rearranging the equation and solving, we find the required moment-of-inertia of one bar about the 1–1 axis as follows:

$$I_1 = 96.92 \text{ mm}^4$$

If the bars fail in buckling by bending about the 2–2 axis, then clamped–free end conditions apply. In that case, the critical load for a single bar is

$$P_{CR2} = \pi^2 EI_2/(4L^2)$$

Solving as we did above, we find that a greater moment-of-inertia is required about the 2–2 axis.

$$I_2 = 387.66 \text{ mm}^4$$

Solution (c): The moments-of-inertia about the 1–1 and 2–2 axes are, respectively,

$$I_1 = a^3b/12 \quad \text{and} \quad I_2 = ab^3/12$$

Solving these equations simultaneously, we find dimensions a and b. Rounding the dimensions upward, we specify 5 mm by 10 mm bars oriented as in Figure 7.9.

Solution (d): Checking the results, we find that the critical load of a single bar is 1650 N about the 1–1 axis. The same value applies about the 2–2 axis. Thus, we have used material efficiently and have minimized weight. Since each bar takes one-half the load, we multiply one-half the load by the safety factor, obtaining again 1650 N. Dividing the load per bar by the cross-sectional area, we find the compressive stress. The compressive stress is low, indicating that buckling is the critical failure mode.

The calculation details follow.

Calculation Details:

Elastic Stability:

Two Rectangular-Cross-Section Aluminum Supports (Units: N, mm, MPa)

$$L = 200 \qquad P = 1100 \qquad E = 69{,}000$$

$$N_{FS} = 3 \qquad N_{FS} \cdot P = \frac{\pi^2 \cdot E \cdot I_1}{L^2} \cdot 2$$

1–1 Axis

$$I_1 = \frac{1}{2} \cdot N_{FS} \cdot \frac{P}{(\pi^2 \cdot E)} \cdot L^2 \qquad I_1 = 96.91591$$

$$N_{FS} \cdot P = \frac{\pi^2 \cdot E \cdot I_2}{L^2 \cdot 4} \cdot 2$$

2–2 Axis

$$I_2 = 2 \cdot N_{FS} \cdot \frac{P}{(\pi^2 \cdot E)} \cdot L^2 \qquad I_2 = 387.66366$$

Estimate $a = 1$, $b = 1$. Given

$$a^3 \cdot \frac{b}{12} = I_1$$

$$b^3 \cdot \frac{a}{12} = I_2$$

$$\begin{pmatrix} a \\ b \end{pmatrix} = \text{find}(a,b)$$

$$a = 4.911$$

$$b = 9.821$$

Check of Single Column about 1–1 Axis:

$$P_{CR1} = \frac{\pi^2 \cdot E \cdot a^3 \cdot b}{12 \cdot L^2} \qquad P_{CR1} = 1.65 \cdot 10^3$$

$$N_{FS} \cdot \frac{P}{2} = 1.65 \cdot 10^3$$

Check of Single Column about 2–2 Axis:

$$P_{CR2} = \frac{(\pi^2 \cdot E \cdot a \cdot b^3)}{12 \cdot 4 \cdot L^2} \qquad P_{CR2} = 1.65 \cdot 10^3$$

$$\sigma_C = \frac{\frac{P}{2}}{a \cdot b} \qquad \sigma_C = 11.404$$

■■

SPECIAL CONSIDERATIONS IN THE DESIGN OF COMPRESSION MEMBERS

Compression members sometimes fail because designers fail to recognize the possibility of buckling. For example, a shaft or a power screw may be heavily loaded in compression. A few special considerations and warnings are noted below.

- Examine designs carefully so that the possibility of buckling is not overlooked.
- Columns tend to buckle in a manner that takes the least energy. Look for the simplest buckling mode represented by the lowest nonzero solution to the critical load equation.

- Use the least-moment-of-inertia. An exception is the case when different end conditions apply to different buckling directions, as in the preceding example problem.
- End conditions have an important effect on critical load. Evaluate end conditions carefully. If you are uncertain about the rigidity of one end of a compression member, analyze the member as if it had a less rigid restraint. For example, few, if any, machine members could safely be represented as "clamped–clamped with lateral restraint."
- Use a factor of safety consistent with the uncertainty associated with the problem and the consequences of failure.

There are many forms of buckling that were not covered above.

- When a deep, narrow beam is subject to bending, the part of the beam in compression may buckle locally. This local buckling is sometimes called **crippling.**
- Thin shells subject to compressive loads may buckle. For example, a thin, cylindrical shell may be subject to an axial compressive load. The critical load equations given above cannot be used to predict the safe load on a thin shell.
- Buckling sometimes occurs as a result of torsional loads.
- Problems of this type are treated in a number of books and technical papers. See, for example, Timoshenko and Gere (1961), a standard reference that discusses various buckling problems.

References and Bibliography

Spiegel, L., and G. F. Limbrunner. *Advanced Statics and Strength of Materials.* New York: Merrill/Macmillan, 1991.

Structural Steel Data for Architectural and Engineering Students. Bethlehem, PA: Bethlehem Steel Corporation, 2d ed. New York: 1978.

Timoshenko, S. P., and J. M. Gere. *Theory of Elastic Stability.* 2d ed. New York: McGraw-Hill, 1961.

Design Problems

7.1. A bracket similar to the one in Figure 7.2 is required to support a vertical load F of 100 lb. Design the bracket.

Design Decisions: Let $L = 10$ in and $\theta = 45°$. Use 1/4 in thick aluminum with a yield point of 40,000 psi. Select a safety factor of 3, and treat bar CB as an Euler column.

7.2. A bracket similar to the one in Figure 7.2 is required to support a vertical load F of 150 lb. Design the bracket.

Design Decisions: Let $L = 12$ in and $\theta = 45°$. Use 3/8 in thick aluminum with a yield point of 40,000 psi. Select a safety factor of 2.5, and treat bar CB as an Euler column.

7.3. A bracket similar to the one in Figure 7.2 is required to support a vertical load F of 1500 N. Design the bracket.

Design Decisions: Let $L = 370$ mm and $\theta = 45°$. Use 6 mm thick steel with a yield point of 400 MPa. Select a safety factor of 2.5, and treat bar CB as an Euler column.

7.4. Design a two-column support for a mass. The mass will rest on a small platform 250 mm above a rigid base, as shown in Figure 7.9. The total supported load (including the platform) is 1800 N.

- Consider the possible modes of failure.
- Calculate the required moments-of-inertia.
- Find column dimensions for optimum design.
- Check the results.

Design Decisions: The mass will be supported on two rectangular steel bars. The bars will be recessed in the base and the platform. A safety factor of 3.5 will be used. Try to use a minimum of material.

7.5. Design a two-column support for a mass. The mass will rest on a small platform 12 in above a rigid base, as shown in Figure 7.9. The total supported load (including the platform) is 350 lb.

- Consider the possible modes of failure.
- Calculate the required moments-of-inertia.
- Find column dimensions for optimum design.
- Check the results.

Design Decisions: The mass will be supported on two rectangular aluminum bars. The bars will be recessed in the base and the platform. A safety factor of 2 will be used. Try to use a minimum of material.

7.6. Design a two-column support for a mass. The mass will rest on a small platform 8 in above a rigid base, as shown in Figure 7.9. The total supported load (including the platform) is 250 lb.

- Consider the possible modes of failure.
- Calculate the required moments-of-inertia.
- Find column dimensions for optimum design.
- Check the results.

Design Decisions: The mass will be supported on two rectangular aluminum bars. The bars will be recessed in the base and the platform. A safety factor of 3 will be used. Try to use a minimum of material.

CHAPTER 8

Shaft Design

Symbols

D	shaft diameter
I	moment-of-inertia
J	polar-moment-of-inertia
M	bending moment
N_{FS}	factor of safety
P_{hp}	horsepower
P_{kW}	power (kW)
R	shaft radius
S_E	endurance limit
S_{YP}	yield point strength
$S(x,a,n)$	singularity function
T	torque
V	shear force
x	axial coordinate
σ	normal stress
τ	shear stress
ω	angular velocity

Units

angular velocity	rad/s
bending moment, torque	in · lb or N · mm
dimensions	in or mm
load, shear force	lb or N
moment-of-inertia	in^4 or mm^4
power	hp or kW
stress, strength	psi or MPa

Many of the necessary techniques for shaft design have already been covered. Chapter 4, "Dynamic Loading of Machine Members," and Chapter 5, "Bending of Machine Members," are particularly relevant. We will now apply these tools, enabling us to design rotating shafts with bending loads in one or two planes and torsion loads.

NORMAL AND SHEAR STRESS IN ROTATING SHAFTS

Figure 8.1 shows a small part of a rotating shaft subject to bending and torsion. The loading is typical of shafts carrying gears and/or pulleys.

Torsional Stress

Consider an element A on the surface of the shaft. The shear stress on the element due to torsion is given by

$$\tau_{x\theta} = \frac{TR}{J} = \frac{16T}{\pi D^3} \tag{8.1}$$

where $\tau_{x\theta}$ = shear stress due to torsion (psi or MPa)

x = axial coordinate

θ = tangential coordinate attached to the rotating shaft

T = torque (in · lb or N · mm)

R = shaft radius (in or mm)

J = polar-moment-of-inertia of the shaft cross section

$= \pi D^3/32$ (in^4 or mm^4)

D = shaft diameter (in or mm)

Since radius R does not vary with time, the shear stress at point A due to torsion is constant with time for constant torque.

Figure 8.1 Stress in a rotating shaft. (a) Cross section; (b) stress on an element

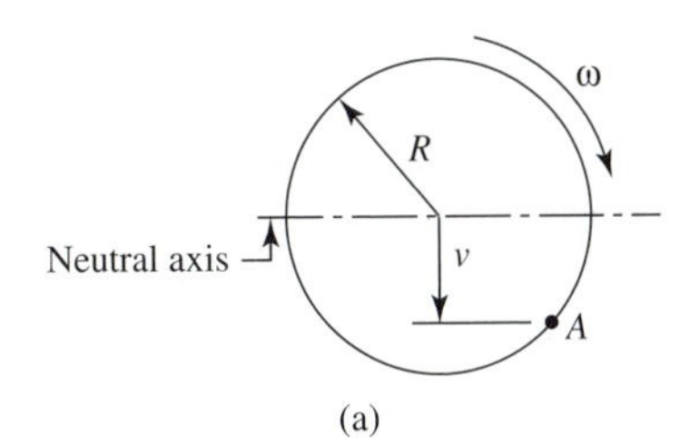

(a)

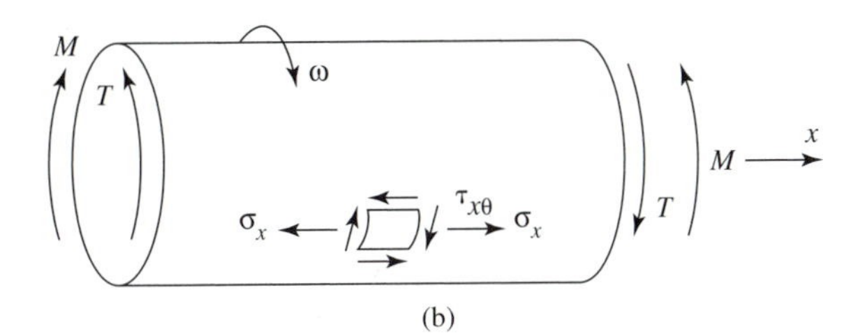

(b)

Figure 8.1 *Continued* (c) stress versus time.

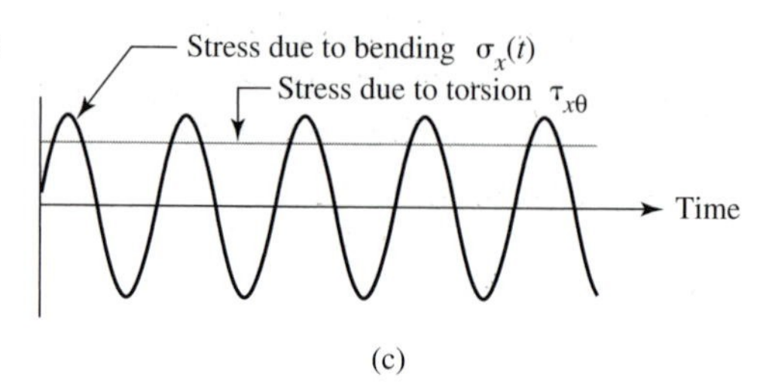

Bending Stress

Bending stress is the most significant effect of lateral loads (gear and pulley forces) on a shaft. For a solid circular shaft, the bending stress at point A [see Figure 8.1(a)] is given by

$$\sigma_x(t) = \frac{Mv}{I} \tag{8.2}$$

where $\sigma_x(t)$ = time-varying axial tensile stress due to bending (psi or MPa)
M = bending moment (lb · lb or N · mm)
v = distance from the neutral axis (in or mm)
I = moment-of-inertia about the neutral axis (in^4 or mm^4)

As the shaft rotates, point A on the surface of the shaft moves, so that the distance from the neutral axis has the form

$$v = c\sin(\omega t) \tag{8.3}$$

where $c = R = D/2$
ω = angular velocity (rad/s)
t = time (s)

The stress amplitude is

$$\sigma_x(\text{amplitude}) = \frac{Mc}{I} = \frac{32M}{\pi D^3} \tag{8.4}$$

and point A "feels" a stress σ_x that varies from

$$-\frac{Mc}{I} \quad \text{to} \quad \frac{Mc}{I}$$

Figure 8.1(c) shows the time variation of axial stress.

SHAFTS SUBJECT ONLY TO TORSION

In some cases, a shaft carries a torsion load without significant bending. Typical examples include shaft sections between flexible couplings or universal joints, which are free of lateral forces.

Principal Stresses for Pure Torsion

Referring to Chapter 3, "Static Stresses in Machine Members," we determine the principal stresses for the pure torsion case as follows:

$$\sigma_1 = \tau_{x\theta} \qquad \sigma_2 = -\tau_{x\theta} \qquad \sigma_3 = 0 \tag{8.5}$$

where stresses are given for a point on the surface of the shaft

$$\tau_{x\theta} = TR/J = 16T/(\pi D^3)$$

Maximum Normal Stress Theory

The maximum normal stress (the maximum principal stress) for pure torsion is

$$\sigma_{\text{MAX}} = \tau_{x\theta} = TR/J = 16T/(\pi D^3) \tag{8.6}$$

and the design criterion is

$$S_{\text{YP}}/N_{\text{FS}} \geq \sigma_{\text{MAX}} = TR/J = 16T/(\pi D^3) \tag{8.7}$$

The maximum normal stress theory is not generally considered a safe theory for the design of steel members subject to torsion.

Maximum Shear Stress

For the pure torsion case, the maximum shear stress is given by

$$\tau_{\text{MAX}} = \text{the maximum of} \begin{Bmatrix} \left|\dfrac{\sigma_1}{2}\right| \\ \left|\dfrac{\sigma_2}{2}\right| \\ \left|\dfrac{\sigma_1 - \sigma_2}{2}\right| \end{Bmatrix} = \left|\frac{\sigma_1 - \sigma_2}{2}\right|$$

$$= \tau_{x\theta} = \frac{TR}{J} = \frac{16T}{\pi D^3} \tag{8.8}$$

Maximum Shear Stress Theory

The maximum shear stress theory assumes that failure occurs when the maximum shear stress in a part equals the maximum shear stress at failure in a tensile test; that is,

$$\tau_{\text{MAX}}(\text{at failure}) = \frac{S_{\text{YP}}}{2} \tag{8.9}$$

This leads to the design criterion

$$\frac{S_{\text{YP}}}{N_{\text{FS}}} \geq \text{the maximum of} \begin{Bmatrix} |\sigma_1| \\ |\sigma_2| \\ |\sigma_1 - \sigma_2| \end{Bmatrix} \tag{8.10}$$

for a two-dimensional stress field. For the special case of pure torsion of a circular shaft, the design criterion becomes

$$\frac{S_{YP}}{N_{FS}} \geq 2\tau_{x\theta} = \frac{2TR}{J} = \frac{32T}{\pi D^3} \tag{8.11}$$

Von Mises Stress

For two-dimensional stress fields, the von Mises stress is given by

$$\sigma_{VM} = \sqrt{\sigma_1^2 + \sigma_2^2 - \sigma_1\sigma_2} \tag{8.12}$$

and the von Mises (or distortion energy) design criterion is

$$\frac{S_{YP}}{N_{FS}} \geq \sigma_{VM} \tag{8.13}$$

Substituting the principal stresses for the pure torsion case, we obtain

$$\frac{S_{YP}}{N_{FS}} \geq \sigma_{VM} = \sqrt{\sigma_1^2 + \sigma_2^2 - \sigma_1\sigma_2} = \tau_{x\theta}\sqrt{1+1+1} = \sqrt{3}\,\tau_{x\theta}$$

$$\frac{S_{YP}}{N_{FS}} \geq \frac{\sqrt{3}TR}{J} = \frac{27.71T}{\pi D^3} \tag{8.14}$$

POWER AND TORQUE

One horsepower (1 hp) is defined as a rate of work of 33,000 ft · lb/min (6600 in · lb/s). Horsepower and torque are related by the equation

$$P_{hp} = \frac{Tn}{63{,}025} \quad \text{or} \quad T = \frac{63{,}025 P_{hp}}{n} \tag{8.15}$$

where P_{hp} = power (horsepower)
T = torque (in · lb)
n = rotation speed (rpm)

Note that torque is inversely proportional to speed. For a given horsepower, if we halve the speed, the torque is doubled. The torque–power relationship should be considered when designing power transmission components.

One watt (1 W) is defined as a rate of work of one joule per second (1 J/s) where one joule is one newton · meter (1 N · m). Converting to customary machine design units, we have

$$P_{kW} = 10^{-6}T\omega \quad \text{or} \quad T = \frac{10^6 P_{kW}}{\omega} \tag{8.16}$$

where ω = angular velocity (rad/s) $= \frac{\pi n}{30}$
P_{kW} = power (kW)
T = torque (N · mm)

EXAMPLE PROBLEM Pure Torsion—Design by the Maximum Normal Stress Theory, the Maximum Shear Stress Theory, and the von Mises Theory

A shaft is to be required to transmit 40 hp at 880 rpm. Design the shaft, and find shaft weight.

Design Decisions: A solid circular shaft will be used. Steel with a yield point of 95,000 psi will be selected. A safety factor of 4 will be used.

Solution:

$$P_{hp} = 40 \qquad n = 880$$

$$S_{YP} = 105{,}000 \qquad N_{FS} = 4$$

$$T = \frac{63{,}025 \cdot P_{hp}}{n} \qquad T = 2.865 \cdot 10^3 \text{ in} \cdot \text{lb}$$

$$\gamma = 0.282 \text{ lb} \cdot \text{in}^3$$

Maximum Normal Stress Theory:

$$D = \left(\frac{16 \cdot T \cdot N_{FS}}{\pi \cdot S_{YP}}\right)^{1/3} \qquad D = 0.822$$

$$w = \frac{\pi \cdot D^2 \cdot \gamma}{4} \qquad w = 0.15 \text{ lb/in}$$

Maximum Shear Stress Theory:

$$D = \left(\frac{32 \cdot T \cdot N_{FS}}{\pi \cdot S_{YP}}\right)^{1/3} \qquad D = 1.036$$

$$w = \frac{\pi \cdot D^2 \cdot \gamma}{4} \qquad w = 0.238 \text{ lb/in}$$

Von Mises Theory:

$$D = \left(\frac{27.71 \cdot T \cdot N_{FS}}{\pi \cdot S_{YP}}\right)^{1/3} \qquad D = 0.987$$

$$w = \frac{\pi \cdot D^2 \cdot \gamma}{4} \qquad w = 0.216 \text{ lb/in}$$

Since we have selected steel, we will not use the maximum normal stress theory as a basis for design of the shaft. The diameter obtained by the maximum shear stress theory or the von Mises theory will be used.

HOLLOW SHAFTS

Hollow shafts have a more favorable torque-capacity-to-weight ratio than solid shafts. As a result, they are sometimes selected for high-torque, low-speed applications if weight is critical. Shear stress is again maximum at the outer radius R. Polar-moment-of-inertia is given by

$$J = \frac{\pi(D^4 - D_I^4)}{32} \tag{8.17}$$

where D = outside diameter

D_I = inside diameter

Hollow shafts must have adequate wall thickness to prevent failure by local bending. For high-speed operation, high precision is required. Eccentricity of the center hole may cause excessive vibration.

EXAMPLE PROBLEM Design of a Hollow Shaft

A hollow shaft is to be considered for power transmission. Bending is negligible. The maximum shear stress theory and the von Mises theory will be considered as design criteria. Design a shaft to meet the following specifications:

$$P_{kW} = 15 \qquad n = 300 \text{ rpm} \qquad N_{FS} = 3.5$$

Also find shaft weight. Compare your values with those of a solid shaft having the same capacity. (Use N, mm, kg/mm, kW, and MPa units.)

Design Decisions: Use steel with a yield point of $S_{YP} = 655$. Use a D_I/D ratio of $R_D = 0.75$.

Solution:

$$\omega = \frac{\pi \cdot n}{30} \qquad \omega = 31.4159$$

$$T = \frac{10^6 \cdot P_{kW}}{\omega} \qquad T = 4.7746 \cdot 10^5$$

$$\gamma = 7.8 \cdot 10^{-6} \text{ kg/mm}^3$$

Maximum Shear Stress Theory:

Solid Shaft

$$D = \left(\frac{32 \cdot T \cdot N_{FS}}{\pi \cdot S_{YP}}\right)^{1/3} \qquad D = 29.6203$$

$$w = \frac{\pi \cdot D^2 \cdot \gamma}{4} \qquad w = 0.0054$$

Hollow Shaft

$$D = \left[\frac{32 \cdot T \cdot N_{FS}}{\pi \cdot (1 - R_D^4) \cdot S_{YP}}\right]^{1/3} \qquad D = 33.6246$$

$$D_I = R_D \cdot D \qquad D_I = 25.2184$$

$$w_H = \frac{\pi \cdot (D^2 - D_I^2) \cdot \gamma}{4} \qquad w_H = 0.003$$

Von Mises Theory:

Solid Shaft

$$D = \left(\frac{27.71 \cdot T \cdot N_{FS}}{\pi \cdot S_{YP}}\right)^{1/3} \qquad D = 28.2326$$

$$w = \frac{\pi \cdot D^2 \cdot \gamma}{4} \qquad w = 0.0049$$

Hollow Shaft

$$D = \left[\frac{27.71 \cdot T \cdot N_{FS}}{\pi \cdot (1 - R_D{}^4) \cdot S_{YP}}\right]^{1/3} \qquad D = 32.0493$$

$$D_I = R_D \cdot D \qquad D_I = 24.037$$

$$w_H = \frac{\pi \cdot (D^2 - D_I^2) \cdot \gamma}{4} \qquad w_H = 0.0028$$

■■

ROTATING SHAFTS SUBJECT ONLY TO BENDING

A shaft carrying an idler gear or an idler pulley is torque-free if we ignore bearing friction. If the shaft rotates with the gear or pulley, then bending stress at a point on the shaft surface varies from

$$\sigma_x = -\frac{Mc}{I} \text{ to } \frac{Mc}{I} \tag{8.18}$$

We will call the amplitude of the varying stress the **range stress.** In this case, the **mean stress** is zero. For a solid circular shaft,

$$\sigma_R = \frac{Mc}{I} = \frac{32M}{\pi D^3} \quad \text{and} \quad \sigma_M = 0 \tag{8.19}$$

where σ_R = range stress (psi or MPa)

σ_M = mean stress

M = bending moment (lb · in or N · mm at a critical location)

D = shaft diameter (in or mm)

The design criterion for a rotating shaft subject only to bending (no torque) is

$$\frac{S_E}{N_{FS}} \geq \sigma_R = \frac{32M}{\pi D^3} \tag{8.20}$$

where S_E = corrected endurance limit

Corrected endurance limit is given by

$$S_E = S'_N C_F C_R C_S / K_F \tag{4.15}$$

where S'_N = endurance limit of a specially prepared test specimen

C_F = surface-finish correction factor

C_R = reliability factor

C_S = size factor

K_F = fatigue stress concentration factor (often replaced by theoretical stress concentration factor)

Calculation of these terms is described in Chapter 4. Tensile stress due to bending is the only significant stress component in this case. Thus, we can obtain the design criterion of Equation (8.20) by applying the maximum normal stress theory, the maximum shear stress theory, or the von Mises theory.

If there are no stress risers, we design the shaft on the basis of maximum bending moment. If there is a change in diameter, a keyway, or other stress concentration, then we may have to consider more than one location on the shaft. Shaft design can be governed by maximum moment or by the local moment and stress concentration factor at a different critical location.

EXAMPLE PROBLEM Design of a Rotating Shaft with an Idler Gear or an Idler Pulley

An idler gear is to be centered on a shaft, and the resultant load is 2400 lb. Design the shaft.

Design Decisions: Use steel with a yield point of 100,000 psi and an ultimate strength of 120,000 psi. Specify a ground surface, a safety factor of 1.75, and a failure rate not to exceed 1 in 100 million. Support the shaft with bearings spaced 6 in apart. Specify a shaft of constant diameter without stress risers.

Solution:

Material Properties:

$$S_U = 120{,}000 \qquad S_{YP} = 100{,}000 \qquad N_{FS} = 1.75$$

$$C_F = 0.87 \quad \text{(ground)} \qquad D = 0.5 \quad \text{(estimate)}$$

$$C_S = 0.8477 - 0.1265 \cdot \ln(D) \qquad C_S = 0.935$$

$$S_{N1} = \text{if}\,(S_U < 200{,}000,\ 0.5 \cdot S_U,\ 100{,}000) \qquad S_{N1} = 6 \cdot 10^4$$

Stress Concentration:

$$K_T = 1$$

Reliability (Samples/Failure):

$$Q = 10^8$$

Estimate of Standard Deviation:

$$X_R = 5$$

Given

$$c_{\text{norm}}(X_R) = 1 - \frac{1}{Q} \qquad SD = \text{find}(X_R) \qquad SD = 5.612$$

$$C_R(SD) = 1 - 0.08 \cdot SD \qquad C_R = C_R(SD) \qquad C_R = 0.551$$

Corrected Endurance Limit:

$$S_E = \frac{S_{N1} \cdot C_R \cdot C_F \cdot C_S}{K_T} \qquad S_E = 2.691 \cdot 10^4$$

Load:

$$F = 2400 \qquad L = 6$$

Maximum Moment:

$$M = \frac{F \cdot L}{4} \qquad M = 3.6 \cdot 10^3$$

Shaft Diameter:

$$D = \left(\frac{32 \cdot N_{FS} \cdot M}{\pi \cdot S_E}\right)^{1/3} \qquad D = 1.336$$

Corrected Size Factor and Recalculated Answer:

$$C_S = 0.8477 - 0.1265 \cdot \ln(D) \qquad C_S = 0.811$$

Corrected Endurance Limit:

$$S_E = \frac{S_{N1} \cdot C_R \cdot C_F \cdot C_S}{K_T} \qquad S_E = 2.333 \cdot 10^4$$

Shaft Diameter:

$$D = \left(\frac{32 \cdot N_{FS} \cdot M}{\pi \cdot S_E}\right)^{1/3} \qquad D = 1.401$$

Corrected Size Factor and Recalculated Answer:

$$C_S = 0.8477 - 0.1265 \cdot \ln(D) \qquad C_S = 0.805$$

Corrected Endurance Limit:

$$S_E = \frac{S_{N1} \cdot C_R \cdot C_F \cdot C_S}{K_T} \qquad S_E = 2.316 \cdot 10^4$$

Shaft Diameter:

$$D = \left(\frac{32 \cdot N_{FS} \cdot M}{\pi \cdot S_E}\right)^{1/3} \qquad D = 1.405 \text{ in}$$

The result has changed only slightly after the last correction of size factor. The final value will be used.

■■

BENDING AND TORSION LOADS ON ROTATING SHAFTS

Design criteria for combined stress were discussed in Chapter 3, "Static Stresses in Machine Members." These criteria include the maximum shear stress and von Mises theories. The

The tension ratio is typically less for flat belts. The tension ratio may be substantially higher for chain drives and toothed timing belts.

Bearing Reactions

A shaft carrying one or more pulleys is typically supported by two bearings, which may be considered simple supports. If the loading in a plane is known, then we can set the sum of the moments about one bearing (say, the left bearing) equal to zero. The resulting equation is easily solved for the reaction at the right bearing. The reaction due to the left bearing is found in a similar manner. If shaft loads do not lie in a single plane, then the vertical and horizontal components of each bearing reaction can be found separately. To find the radial load at a bearing, add the horizontal and vertical reaction vectors at that bearing; that is, the radial load on a bearing is the square root of the sum of the squares of the reaction components.

Shear and Bending Moment

Shear force and bending moment should be plotted against shaft location, using either manual or computer methods. Errors that could otherwise go undetected are often exposed by a graphical presentation. Bearing reactions and pulley loads are treated as concentrated forces, causing positive and negative jumps in shear force diagrams. We obtain bending moment by integrating the shear-force-versus-axial-location relationship. Moment and torque are combined at all critical locations according to the Soderberg–maximum shear criterion or Soderberg–von Mises criterion. If a constant-diameter shaft is to be selected, the diameter calculation is based on the most severe moment–torque combination.

SINGULARITY FUNCTION

The singularity function was introduced in Chapter 5, "Bending of Machine Members," because it lends itself well to computer solutions of beam-type problems. It is also recommended as an efficient tool for shaft design. The general form of the singularity function is $S(x,a,n)$.

Concentrated Loads

The singularity function that can be used to represent a concentrated load is

$$S(x,a,-1) = \begin{cases} 0 & \text{for } x \neq a \\ \infty & \text{for } x = a \end{cases} \tag{8.32}$$

where the function represents load density approaching infinity over a region of the shaft of length approaching zero, such that the total force is unity. For example, a pulley load F of 250 lb at $x = 3.5$ in from the left end of a shaft could be represented by

$$F = 250 \cdot S(x,3.5,-1)$$

Singularity Functions for $n \geq 0$

The singularity function is defined by

$$S(x,a,n)\ [\text{for } n \geq 0] = \begin{cases} 0 & \text{for } x < a \\ (x-a)^n & \text{for } x \geq a \end{cases} \tag{8.33}$$

Distributed Loads

Distributed loads can be represented by the singularity function with $n = 0$. For example, if the shaft itself weighs 0.5 lb/in, it contributes the following distributed load:

$$q = 0.5 \cdot S(x,0,0)$$

where q = distributed load (lb/in or N/mm), and

$$S(x,a,0) = \begin{cases} 0 & \text{for } x < a \\ 1 & \text{for } x \geq a \end{cases} \tag{8.34}$$

Shaft weight is, of course, distributed over the entire length of the shaft $0 \leq x \leq L$. If we use $S(x,a,0)$ to represent shaft weight, then $a = 0$.

Integration of Singularity Functions

Integration of singularity functions is not difficult. For example,

$$\int_0^x S(x,a,-1)\,dx = S(x,a,0) \tag{8.35}$$

and

$$\int_0^x S(x,a,0)\,dx = S(x,a,1) \tag{8.36}$$

where

$$S(x,a,1) = \begin{cases} 0 & \text{for } x < a \\ x - a & \text{for } x \geq a \end{cases} \tag{8.37}$$

In general, for $n \geq 0$,

$$\int S(x,a,n)\,dx \;[\text{for } n \geq 0] = S(x,a,n+1)/(n+1) \tag{8.38}$$

Defining the Singularity Function for Computer Use

When using the singularity function in computer programs, or with mathematics software, it is necessary to define the general form only once. Using BASIC, we can write an expression similar to the following if $n \geq 0$:

```
IF X>=A THEN S(X,A,N) = (X - A)^N ELSE S(X,A,N) = 0        (8.39)
```

A corresponding FORTRAN expression for $n \geq 0$ has the form

```
IF(X.GE.A) S(X,A,N) = (X - A)^N
ELSE S(X,A,N) = 0
END IF                                                     (8.40)
```

If *MathCAD*™ is used, the singularity function can be defined as follows for $n \geq 0$:

$$S(x,a,n) = \text{IF}[x \geq a, (x-a)^n, 0] \tag{8.41}$$

Shear Force

Shear force is given by the negative integral of the shaft loading. The equation for shear force is

$$V = -\int L\,dx \tag{8.42}$$

where V = shear force

L = loading

For example, considering only the 250 lb pulley load and the 0.5 lb/in shaft weight, the shear force contribution is

$$V_{\text{contribution}} = -\int_0^x (F + q)\,dx = -\int_0^x [250S(x,3.5,-1) + 0.5S(x,0,0)]\,dx$$
$$= -250S(x,3.5,0) - 0.5S(x,0,1)$$

Bending Moment

We obtain bending moment by integrating shear force as follows:

$$M = \int V\,dx \tag{8.43}$$

Again considering only the 250 lb pulley load and the 0.5 lb/in shaft weight, the bending moment contribution is

$$M_{\text{contribution}} = \int_0^x V\,dx = -\int_0^x [250S(x,3.5,0) + 0.5S(x,0,1)]\,dx$$
$$= -250S(x,3.5,1) - 0.5S(x,0,2)/2$$

EXAMPLE PROBLEM Loading of a Belt Driveshaft with Several Pulleys

Find torque and moment loading on a belt drive countershaft that will be designed to transmit 15 hp at 2400 rpm. The shaft will carry one input pulley and three output pulleys, specified as follows:

Pulley	*Function*	*Pitch Radius r_i (in)*	*Load Fraction R_{Ti}*	*Axial Location x (in)*
1	Input	2.8	1.00	2
2	Output	3.3	−0.55	4
3	Output	2.1	−0.30	6
4	Output	2.9	−0.15	7.4

Note that output pulleys 2, 3, and 4 transmit 55%, 30%, and 15% of the input load, respectively.

Design Decisions: The shaft will be 10 in long and will be supported by two single-row ball bearings with centers at $A = 1$ in and $B = 9$ in. The bearings will be treated as simple supports. The belts will be approximately horizontal, as in Figure 8.2. The belts on pulleys 1 and 3 will extend outward from the pulleys ($\alpha = 0$). The belts on pulleys 2 and 4 will extend inward ($\alpha = 180°$).

Solution Summary: Calculation details are given in the detailed solution which follows and are illustrated in Figure 8.3(a)–(c). Input torque T_{IN} is computed from the transmitted power and shaft speed. Torque T_i on each pulley is the product of the input torque and the load fraction for that pulley, where output torques are negative. Shaft torque is given by

$$T(x) = \sum_i T_i S(x,X_i,0)$$

where $T(x)$ = shaft torque at any location x measured from the left end of the shaft
$S(x,X_i,0)$ = step form of the singularity function, defining a unit step at each location $x = X_i$

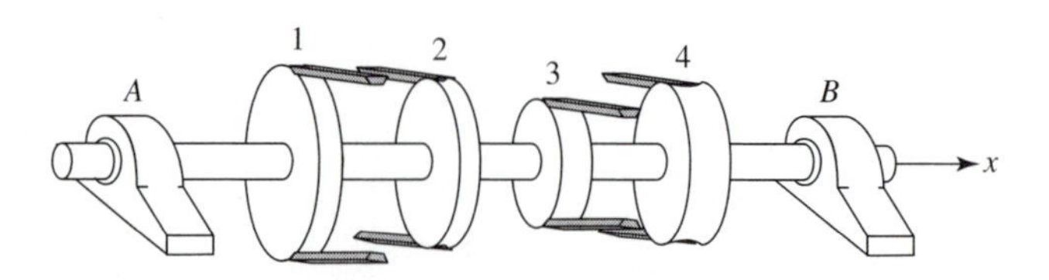

Figure 8.2 Rotating shaft with torsion and bending loads.

Figure 8.3(a) shows shaft torque $T(x)$ plotted against location x on the shaft, where x ranges from zero to 10 inches in steps of 0.05 in.

We will assume that the belts are adjusted to produce an operating tension ratio

$$F_1/F_2 = 3$$

Using the torque equation, we find total shaft load at pulley i.

$$\mathbf{F}_1 + \mathbf{F}_2 \approx F_1 + F_2 = 2T/r_i \quad \text{(approximately)}$$

We assume that the belts are approximately parallel and we use the scalar sum of the tensions. We multiply the tension sum by the cosine of the belt angle to obtain Z_i, the shaft force due to pulley i.

To find reaction B_Z at the right bearing (location B), we sum moments (due to shaft forces at the pulleys) about the left bearing (location A) and divide by the distance between bearings. The calculation of reaction A_Z at the left bearing involves summing moments about the right bearing. We now check force equilibrium. Summing the Z-direction (horizontal) forces, we obtain zero plus-or-minus a small rounding error.

Shear force is described in terms of singularity functions. The first and second terms of the shear equation represent the shear forces due to the left and right bearings. The third term describes the shear force contribution at each of the four pulleys. A plot of shear force $V_Z(x)$ versus shaft location x is shown in Figure 8.3(b).

Bending moment $M_Z(x)$ is also described by singularity functions. Note the similarity in the moment and shear equations. In the singularity function $S(x,a,n)$, n changes from 0 to 1 as we go from shear

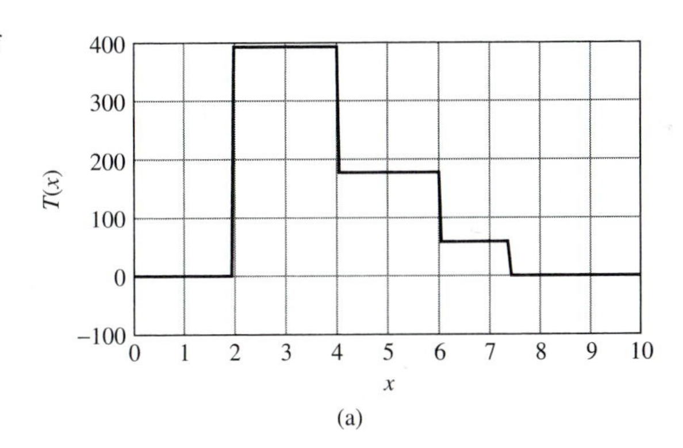

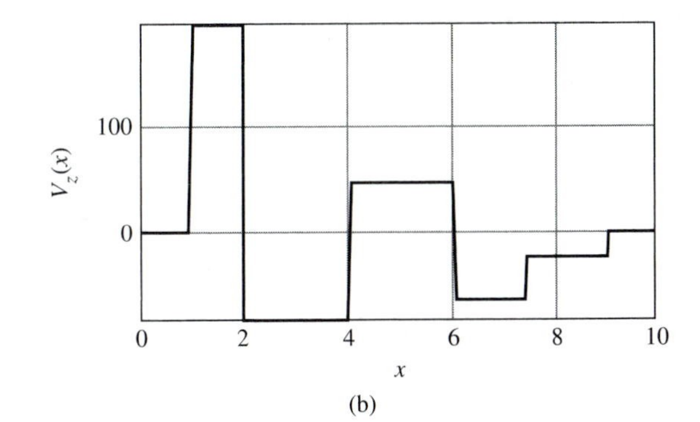

Figure 8.3 Rotating shaft. (a) Analysis of torque. (b) Shear force.

Figure 8.3 (c) Rotating shaft. Analysis of moment.

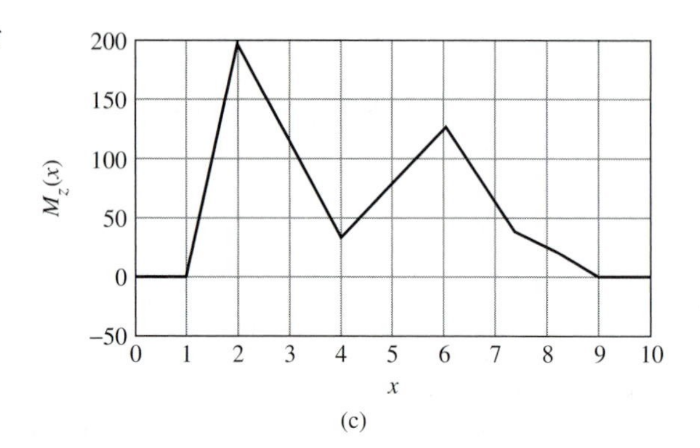

force to bending moment. The plot of $M_Z(x)$ versus x [Figure 8.3(c)] shows peaks at $x = 2$ and $x = 6$ in (pulleys 1 and 3).

The detailed solution follows.

Detailed Solution:

Shaft Design:

$$n = 4 \qquad L = 10 \qquad P_{hp} = 15 \qquad n_{rpm} = 2400$$

$$T_{IN} = 63{,}025 \cdot \frac{P_{hp}}{n_{rpm}} \qquad T_{IN} = 393.906$$

$R(T)$ is fraction of output load.

$$i = 1, \ldots, n$$

i	r_i	α_i	R_{Ti}	X_i
1	2.8	0	1	2
2	3.3	180	−0.55	4
3	2.1	0	−0.30	6
4	2.9	180	−0.15	7.4

$$T_i = T_{IN} \cdot R_{Ti}$$

$$Z_i = 2 \cdot \frac{|T_i|}{r_i} \cdot \cos(\alpha_i \cdot \deg)$$

i	T_i	Z_i
1	393.906	281.362
2	−216.648	−131.302
3	−118.172	112.545
4	−59.086	−40.749

$$x = 0, 0.05, \ldots, L$$

Singularity Function:

$$S(x,a,n) = \text{if}\,[x \geq a, (x-a)^n, 0] \quad \text{for } n \geq 0$$

$$T(x) = \sum_i T_i \cdot S(x, X_i, 0) \text{ in} \cdot \text{lb}$$

See Figure 8.3(a).

Sign Convention:

Forces and reactions: + = out
Moment: + = CW

Bearing Locations:

$$A = \frac{L}{10} \qquad B = 0.9 \cdot L$$

Shaft Loads:

$$Z = \text{horizontal} \qquad X = \text{axial location}$$

$$i = 1, \ldots, n$$

$$B_Z = \left[-\frac{1}{(B-A)}\right] \cdot \sum_i [Z_i \cdot (X_i - A)] \qquad B_Z = -23.673$$

$$A_Z = \left(\frac{1}{B-A}\right) \cdot \sum_i [Z_i \cdot (X_i - B)] \qquad A_Z = -198.182$$

Check Force Equilibrium:

$$\sum_i (Z_i) + A_Z + B_Z = -4.263 \cdot 10^{-14}$$

Shear:

$$V_Z(x) = -\left(A_Z \cdot S(x,A,0) + B_Z S(x,B,0) + \sum_i Z_i \cdot S(x,X_i,0)\right)$$

See Figure 8.3(b).

Moment:

$$M_Z(x) = -\left(A_Z \cdot S(x,A,1) + B_Z \cdot S(x,B,1) + \sum_i Z_i \cdot S(x,X_i,1)\right)$$

Combined Moment:

$$M(x) = M_Z(x)$$

$$M(2) = 198.182 \qquad M(6) = 128.068$$
$$T(2) = 393.906 \qquad T(5.99) = 177.258$$

See Figure 8.3(c).

In the preceding example, note that the computation of shear, moment, and torque involved using the singularity function 13 times, as each summation includes four pulleys. However, it was necessary to define the singularity function only once. The computer has relieved us of the repetitious calculations in this problem. Hand calculations would have been arduous. Furthermore, once the system is described in a program, we can investigate "What if . . .?" design changes with a few keystrokes. For example, "What if we change the bearing locations? Will that change reduce maximum bending moment? Should we change pulley locations?"

EXAMPLE PROBLEM Design of a Rotating Shaft with Torsion and Bending Loads

Design a shaft for the preceding application.

Design Decisions: The shaft will be made of steel with an ultimate strength of 105,000 psi and a yield point of 80,000 psi. The shaft surface will be ground. The stress concentration at the pulleys will be equivalent to a change of diameter $D/d = 1.1$ with a fillet-radius-to-diameter ratio r/d of 0.1. A safety factor of 1.5 will be used, and we will design for 1 failure in 5 million samples.

Solution Summary: The detailed calculations are given in the detailed solution which follows. The endurance limit of a polished test specimen is estimated to be one-half the ultimate strength in this case. Correction factors for size, finish, and reliability, and stress concentration factor are determined by the equations and procedures used in Chapter 4. The corrected endurance limit is 15,380 psi.

Since there are no significant forces in the vertical direction, total moment $M(x) = M_Z(x)$. At pulley 1, located 2 in from the left end of the shaft, the moment and torque are, respectively,

$$M_{\text{MAX}} = M(2) = 198.2 \text{ lb} \cdot \text{in}$$
$$T_{\text{MAX}} = T(2) = 393.9 \text{ lb} \cdot \text{in}$$

where $M(2)$ and $T(2)$ = bending moment and torque, respectively, at $x = 2$

Bending moment also peaks at pulley 3 ($x = 6$). Bending moment and torque near pulley 3 are given by

$$M(6) = 128.1 \text{ lb} \cdot \text{in}$$
$$T(5.99) = 177.3 \text{ lb} \cdot \text{in}$$

where torque is calculated just to the left of pulley 3 where torque jumps downward.

Moment and torque loading near pulley 3 are less severe than at $x = 2$. Therefore, the shaft design is based on $M(2)$ and $T(2)$. Using the Soderberg–maximum shear stress criterion, we find that the required shaft diameter is

$$D = 0.595 \text{ in}$$

Using the Soderberg–von Mises criterion, we obtain

$$D = 0.592 \text{ in}$$

In this example, the effect of bending stress fatigue dominates the design, and the two criteria produce nearly identical results.

The detailed solution follows.

Detailed Solution:

Material Properties:

$$S_U = 105{,}000 \qquad S_{\text{YP}} = 80{,}000 \qquad N_{\text{FS}} = 1.5$$
$$C_F = 0.87 \quad \text{(ground)} \qquad D = 0.625 \quad \text{(estimate)}$$
$$C_S = 0.8477 - 0.1265 \cdot \ln(D) \qquad C_S = 0.907$$
$$S_{N1} = \text{if}\,(S_U < 200000, 0.5 \cdot S_U, 100000) \qquad S_{N1} = 5.25 \cdot 10^4$$

Stress Concentration Equivalent to D/d $= 1.1$, $r/d = R_K = 0.1$

$$K_T = 0.945 \cdot R_K^{-0.229} \qquad K_T = 1.601$$

Reliability (Samples/Failure)

$$Q = 5 \cdot 10^6$$

Estimate of Standard Deviation:

$$X_R = 5$$

Given

$$c_{\text{norm}}(X_R) = 1 - \frac{1}{Q} \qquad SD = \text{find}(X_R) \qquad SD = 5.069$$

$$C_R(SD) = 1 - 0.08 \cdot SD \qquad C_R = C_R(SD) \qquad C_R = 0.594$$

Corrected Endurance Limit:

$$S_E = \frac{S_{N1} \cdot C_R \cdot C_F \cdot C_S}{K_T} \qquad S_E = 1.538 \cdot 10^4$$

Equivalent Static Moment, Soderberg–Maximum Shear:

$$M_{\text{EQ}} = \sqrt{\left(\frac{S_{\text{YP}}}{S_E} \cdot M(2)\right)^2 + T(2)^2} \qquad M_{\text{EQ}} = 1.103 \cdot 10^3 \quad \text{Critical location to the right of } X = 2$$

Shaft Diameter:

$$D = \left(\frac{32 \cdot N_{\text{FS}} \cdot M_{\text{EQ}}}{\pi \cdot S_{\text{YP}}}\right)^{1/3} \qquad D = 0.595$$

Soderberg–von Mises:

$$M_{\text{EQ VM}} = \sqrt{\left(\frac{S_{\text{YP}}}{S_E} \cdot M(2)\right)^2 + 0.75 \cdot (T(2))^2} \qquad M_{\text{EQ VM}} = 1.086 \cdot 10^3$$

$$D = \left(\frac{32 \cdot N_{\text{FS}} \cdot M_{\text{EQ VM}}}{\pi \cdot S_{\text{YP}}}\right)^{1/3} \qquad D = 0.592$$

■■

SHAFT LOADING IN TWO PLANES

Many belt drive shafts, and most shafts carrying gears, have loads in two planes. If shaft loads are not in a single plane, then horizontal shear force and bending moment should be calculated, followed by the calculation of vertical shear force and bending moment. To obtain resultant moment, take the square root of the sum of the squares of horizontal and vertical bending moment at sufficient axial locations to produce a smooth plot. Design is based on the most severe combination of resultant bending moment and torque. It may be necessary to test more than one location to find the critical location.

■■ EXAMPLE PROBLEM Shaft Design for Loads in Two Planes

A belt drive shaft carries the following loads:

Pulley i	*Function*	*Pitch Radius r_i (in)*	*Resultant Angle α_i (deg)*	*Load Fraction R_{Ti}*	*Axial Location x (in)*
1	Input	2.8	85	1.00	2
2	Output	3.3	−45	−0.30	4
3	Output	2.1	50	−0.50	6
4	Output	2.9	−60	−0.20	7.4

The shaft rotates at 1150 rpm, and the input power to pulley number 1 is 25 hp. The direction of the vector sum of the belt tensions at each pulley is identified as the resultant angle.

(a) Calculate and plot shear force and bending moment in the horizontal and vertical planes, and resultant moment and torque.

(b) Design the shaft.

Design Decisions: The shaft will be 10 in long to provide adequate space between pulleys. It will be supported by bearings at $A = 1$ in and $B = 9$ in. The shaft will be made of steel with an ultimate strength of 125,000 psi and a yield point of 100,000 psi. The surface will be ground. Stress concentration at the pulleys will be equivalent to a diameter change of $D/d = 1.1$ and a fillet-radius-to-diameter ratio of $r/d = 0.1$. We will design for a failure rate of 1 in 5 million, and we will use a safety factor of 1.25.

Solution Summary (a):

The solution follows the same general pattern as the shaft design example for loads in a single plane. However, in this case, we compute loading, shear, and moment in the horizontal plane and in the vertical plane, and then we combine the moments. Details of the solution are given in the detailed solution which follows and are illustrated in Figure 8.4(a)–(f).

Torque is tabulated and plotted against axial location x as shown in Figure 8.4(a). Resultant belt tensions are calculated based on a tension ratio of 3, as in a previous example. The horizontal and vertical components of belt tension, Z_i and Y_i, respectively, are calculated at each pulley.

Left bearing reactions are identified by A_Y (vertical component) and A_Z (horizontal component). Right bearing reactions are B_Y and B_Z. We find vertical reaction B_Y at the right bearing by setting the sum of the vertical-plane moments about the left bearing equal to zero. We find horizontal reaction B_Z by using horizontal-plane moments. To find reaction components A_Y and A_Z, we sum moments about the right bearing. The resultant reaction at the left and right bearings, R_A and R_B, are calculated for use in bearing selection. Note that the resultant reaction at the left bearing, $R_A = 967.5$ lb, is greater than R_B. Both bearings would probably be selected on the basis of load R_A. This decision could simplify fabrication and reduce the number of bearing types kept in inventory.

After checking to see that horizontal and vertical forces are in equilibrium (±rounding tolerance), we compute shear forces in the vertical and horizontal planes, V_Y and V_Z, respectively. Plots of V_Y versus x and of V_Z versus x are shown in Figure 8.4(b) and (c), respectively. Shear force equations are integrated to produce bending moments M_Y and M_Z. Note that we accomplish this step by copying the

Figure 8.4 Shaft design.

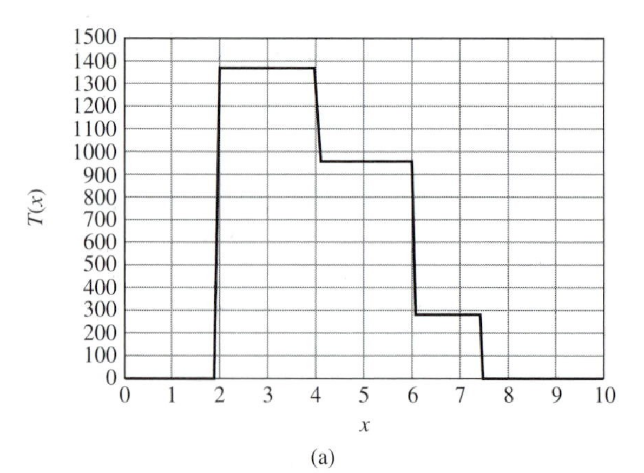

(a)

Figure 8.4 *Continued* Shaft design.

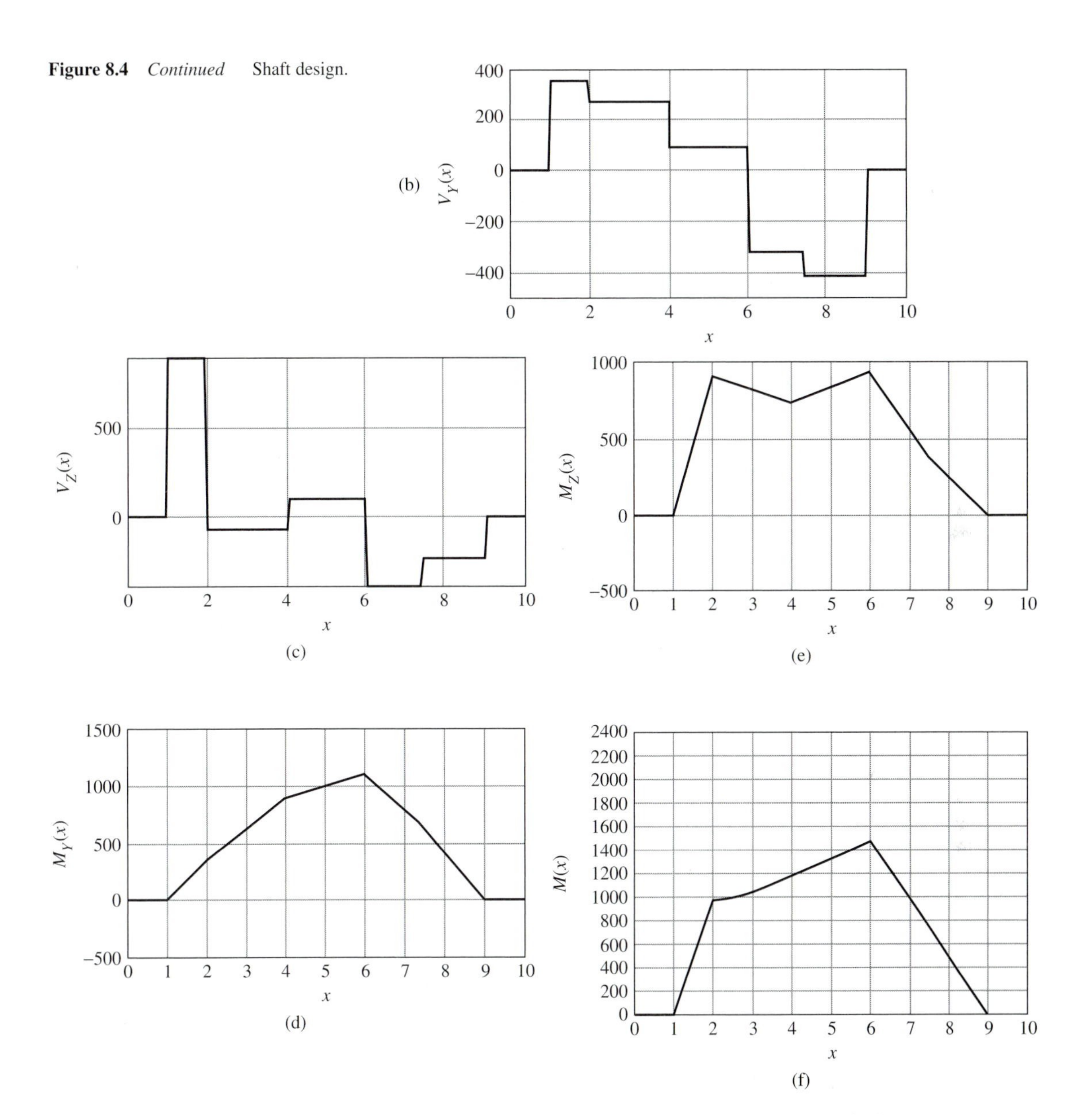

equations and making a change in the argument of the singularity function. Combined moment is then computed and plotted.

Solution Summary (b): Using the same procedure to calculate corrected endurance limit as in a previous example, we find that $S_E = 16{,}540$. Noting that torque is greatest to the left of pulley 2, while moment is greatest at pulley 3, we investigate both locations. The critical location is just to the left of pulley 3. Using the Soderberg–maximum shear theory, we determine the required shaft diameter, $D = 1.041$ in. Soderberg–von Mises theory results in a slightly smaller diameter.

The detailed solution follows.

Detailed Solution (a):

Shaft Design:

$$n = 4 \qquad L = 10 \qquad P_{hp} = 25 \qquad n_{rpm} = 1150$$

$$T_{IN} = 63{,}025 \cdot \frac{P_{hp}}{n_{rpm}} \qquad T_{IN} = 1.37 \cdot 10^3$$

Resultant force is α degrees from vertical; $R(T)$ is fraction of output load.

$$i = 1, \ldots, n$$

i	r_i	α_i	R_{Ti}	X_i
1	2.8	85	1	2
2	3.3	−45	−0.30	4
3	2.1	50	−0.50	6
4	2.9	−60	−0.20	7.4

$$T_i = T_{IN} \cdot R_{Ti} \qquad Y_i = 2 \cdot \frac{|T_i|}{r_i} \cdot \cos(\alpha_i \cdot \deg)$$

$$Z_i = 2 \cdot \frac{|T_i|}{r_i} \cdot \sin(\alpha_i \cdot \deg)$$

i	T_i	Z_i	Y_i
1	$1.37 \cdot 10^3$	974.925	85.295
2	−411.033	−176.148	176.148
3	−685.054	499.792	419.376
4	−274.022	−163.662	94.49

$$x = 0, 0.05, \ldots, L$$

Singularity Function:

$$S(x,a,n) = \text{if}[x \geq a, (x - a)^n, 0] \quad \text{for } n \geq 0$$

$$T(x) = \sum_i T_i \cdot S(x,X_i,0) \text{ in} \cdot \text{lb}$$

See Figure 8.4(a).

Sign Convention:

Forces and reactions: + = down and out
Moment: + = CW

Bearing Locations:

$$A = \frac{L}{10} \qquad B = 0.9 \cdot L$$

Shaft Loads:

Y = vertical $\quad Z$ = horizontal $\quad X$ = axial location

$$i = 0, \ldots, n$$

$$B_Y = \left[-\frac{1}{(B - A)}\right] \cdot \sum_i [Y_i \cdot (X_i - A)] \qquad B_Y = -414.419$$

$$B_Z = \left[-\frac{1}{(B-A)}\right] \cdot \sum_i [Z_i \cdot (X_i - A)] \qquad B_Z = -237.251$$

$$R_B = \sqrt{B_Y^2 + B_Z^2} \qquad R_B = 477.526$$

$$A_Y = \left(\frac{1}{B-A}\right) \cdot \sum_i [Y_i \cdot (X_i - B)] \qquad A_Y = -360.889$$

$$A_Z = \left(\frac{1}{B-A}\right) \cdot \sum_i [Z_i \cdot (X_i - B)] \qquad A_Z = -897.657$$

$$R_A = \sqrt{A_Y^2 + A_Z^2} \qquad R_A = 967.486$$

Check Force Equilibrium:

$$\sum_i (Y_i) + A_Y + B_Y = 5.684 \cdot 10^{-14} \qquad \sum_i (Z_i) + A_Z + B_Z = 2.842 \cdot 10^{-14}$$

Shear:

$$V_Y(x) = -\left(A_Y \cdot S(x,A,0) + B_Y \cdot S(x,B,0) + \sum_i Y_i \cdot S(x,X_i,0)\right)$$

See Figure 8.4(b).

$$V_Z(x) = -\left(A_Z \cdot S(x,A,0) + B_Z \cdot S(x,B,0) + \sum_i Z_i \cdot S(x,X_i,0)\right)$$

See Figure 8.4(c).

Moment:

$$M_Y(x) = -\left(A_Y \cdot S(x,A,1) + B_Y \cdot S(x,B,1) + \sum_i Y_i \cdot S(x,X_i,1)\right)$$

See Figure 8.4(d).

$$M_Z(x) = -\left(A_Z \cdot S(x,A,1) + B_Z \cdot S(x,B,1) + \sum_i Z_i \cdot S(x,X_i,1)\right)$$

See Figure 8.4(e).

Combined Moment:

$$M(x) = \sqrt{M_Y(x)^2 + M_Z(x)^2}$$

$$M(4) = 1.176 \cdot 10^3 \qquad M(6) = 1.456 \cdot 10^3$$

$$T(3.99) = 1.37 \cdot 10^3 \qquad T(5.99) = 959.076$$

See Figure 8.4(f).

Detailed Solution (b):

Material Properties:

$$S_U = 125{,}000 \quad S_{YP} = 100{,}000 \qquad N_{FS} = 1.25$$

$$C_F = 0.87 \quad \text{(ground)} \qquad D = 1.25 \quad \text{(estimate)}$$

$$C_S = 0.8477 - 0.1265 \cdot \ln(D) \qquad C_S = 0.819$$

$$S_{N1} = \text{if}(S_U < 200000, 0.5 \cdot S_U, 100000) \qquad S_{N1} = 6.25 \cdot 10^4$$

Stress Concentration Equivalent to $D/d = 1.1$, $r/d = R_K = 0.1$

$$K_T = 0.945 \cdot R_K^{-0.229} \qquad K_T = 1.601$$

Reliability (Samples/Failure):

$$Q = 5 \cdot 10^6$$

Estimate of Standard Deviation:

$$X_R = 5$$

Given

$$c_{\text{norm}}(X_R) = 1 - \frac{1}{Q} \qquad SD = \text{find}(X_R) \qquad SD = 5.069$$

$$C_R(SD) = 1 - 0.08 \cdot SD \qquad C_R = C_R(SD) \qquad C_R = 0.594$$

Corrected Endurance Limit:

$$S_E = \frac{S_{N1} \cdot C_R \cdot C_F \cdot C_S}{K_T} \qquad S_E = 1.654 \cdot 10^4$$

Equivalent Static Moment, Soderberg–Maximum Shear:

$$M_{\text{EQ}} = \sqrt{\left(\frac{S_{\text{YP}}}{S_E} \cdot M(4)\right)^2 + T(3.99)^2} \qquad M_{\text{EQ}} = 7.242 \cdot 10^3$$ Critical location to the left of $X = 6$

$$M_{\text{EQ}} = \sqrt{\left(\frac{S_{\text{YP}}}{S_E} \cdot M(6)\right)^2 + T(5.99)^2} \qquad M_{\text{EQ}} = 8.852 \cdot 10^3$$

Shaft Diameter:

$$D = \left(\frac{32 \cdot N_{\text{FS}} \cdot M_{\text{EQ}}}{\pi \cdot S_{\text{YP}}}\right)^{1/3} \qquad D = 1.041$$

Soderberg–von Mises:

$$M_{\text{EQ VM}} = \sqrt{\left(\frac{S_{\text{YP}}}{S_E} \cdot M(4)\right)^2 + 0.75 \cdot (T(3.99))^2} \qquad M_{\text{EQ VM}} = 7.21 \cdot 10^3$$

$$M_{\text{EQ VM}} = \sqrt{\left(\frac{S_{\text{YP}}}{S_E} \cdot M(6)\right)^2 + 0.75 \cdot (T(5.99))^2} \qquad M_{\text{EQ VM}} = 8.839 \cdot 10^3$$

$$D = \left(\frac{32 \cdot N_{\text{FS}} \cdot M_{\text{EQ VM}}}{\pi \cdot S_{\text{YP}}}\right)^{1/3} \qquad D = 1.04$$

■■

A FEW NOTES ON SHAFT DESIGN

It is clear that fatigue stress due to bending of rotating shafts tends to dominate the design. Failure to recognize the fatigue nature of shaft loading could result in catastrophic consequences. The following suggestions may help to reduce the probability of fatigue failure:

- Reduce bending moment by making the shaft as short as practical.
- Use three or more bearings to reduce bending moment by reducing unsupported shaft length.
- Improve fatigue strength by eliminating steps (diameter changes).

- If steps or other stress risers are necessary, reduce stress concentration by specifying generous fillets.
- Improve fatigue strength by specifying an improved shaft surface finish.
- If the shaft carries an idler gear or an idler pulley, it may be possible to specify a non-rotating shaft with a bearing at the pulley.

Design Problems

Some problems compare the results of the maximum normal stress theory, the maximum shear stress theory, and the von Mises theory. Note that the maximum normal stress theory is not a generally accepted design criterion for shafts subject to torsion. It is included for comparison purposes only.

8.1. A shaft is to be required to transmit 30 hp at 1760 rpm. Design the shaft, and find shaft weight.

Design Decisions: A solid circular shaft will be used. Steel with a yield point of 85,000 psi will be selected. A safety factor of 3 will be used. The maximum normal stress theory, the maximum shear stress theory, and the von Mises theory will be considered as design criteria.

8.2. Design a shaft to meet the following criteria:

$$P_{\text{hp}} = 10 \qquad n = 150 \qquad S_{\text{YP}} = 95{,}000 \qquad N_{\text{FS}} = 3.5$$

Also find shaft weight.

Design Decisions: A solid circular shaft will be used. The maximum normal stress theory, the maximum shear stress theory, and the von Mises theory will be considered as design criteria.

8.3. Design a shaft to meet the following criteria:

$$P_{\text{hp}} = 300 \qquad n = 4000 \qquad S_{\text{YP}} = 98{,}000 \qquad N_{\text{FS}} = 2.5$$

Also find shaft weight.

Design Decisions: A solid circular shaft will be used. The maximum normal stress theory, the maximum shear stress theory, and the von Mises theory will be considered as design criteria.

8.4. Design a hollow shaft to transmit 10 kW at 1000 rpm. Use a safety factor of 3. Use steel with a yield point of 600 MPa and a ratio of inside diameter to outside diameter of 0.7. Compare the weight with that of an equivalent solid shaft.

8.5. A hollow shaft is to be considered for power transmission. Bending is negligible. Use the maximum shear stress theory and the von Mises theory as design criteria. Design a shaft to meet the following specifications:

$$P_{\text{kW}} = 105 \qquad n = 3000 \text{ rpm} \qquad N_{\text{FS}} = 2.5$$

Also find shaft weight. Compare the values with those of a solid shaft having the same capacity. (Use N, mm, kg/mm, kW and MPa units.)

Design Decisions: Use steel with a yield point of $S_{\text{YP}} = 550$. Use a D_I/D ratio of $R_D = 0.8$.

8.6. A hollow shaft is to be considered for power transmission. Bending is negligible. Use the maximum shear stress theory and the von Mises theory as design criteria. Design a shaft to meet the following specifications:

$$P_{\text{kW}} = 85 \qquad n = 2500 \text{ rpm} \qquad N_{\text{FS}} = 2$$

Also find shaft weight. Compare the values with those of a solid shaft having the same capacity. (Use N, mm, kg/mm, kW and MPa units.)

Design Decisions: Use steel with a yield point of $S_{\text{YP}} = 585$. Use a D_I/D ratio of $R_D = 2/3$.

8.7. A pulley is to be centered on a shaft. The resultant load is 1300 lb. Design the shaft.

Design Decisions: Use steel with a yield point of 100,000 psi and an ultimate strength of 120,000 psi. Specify a ground surface, a safety factor of 2, and a failure rate not to exceed 1 in 1 million. Support the shaft with bearings spaced 7 in apart. Specify a shaft of constant diameter without stress risers.

8.8. An idler gear is to be centered on a shaft. The resultant load is 1100 lb. Design the shaft.

Design Decisions: Use steel with a yield point of 95,000 psi and an ultimate strength of 120,000 psi.

Specify a ground surface, a safety factor of 3, and a failure rate not to exceed 1 in 1 million. Support the shaft with bearings spaced 7 in apart. Specify a shaft of constant diameter without stress risers.

8.9. An idler pulley is to be centered on a shaft. The resultant load is 950 lb. Design the shaft.

Design Decisions: Use steel with a yield point of 95,000 psi and an ultimate strength of 115,000 psi. Specify a ground surface, a safety factor of 2.5, and a failure rate not to exceed 1 in 1 million. Support the shaft with bearings spaced 5.5 in apart. Specify a shaft of constant diameter without stress risers.

8.10. Find torque and moment loading on a belt drive countershaft that will be designed to transmit 29 hp at 2100 rpm. The shaft will carry one input pulley and three output pulleys specified as follows:

Pulley i	*Function*	*Pitch Radius* r_i *(in)*	*Load Fraction* R_{Ti}	*Axial Location* *x(in)*
1	Input	2.8	1.00	2
2	Output	3.3	−0.15	4
3	Output	2.1	−0.60	6
4	Output	2.9	−0.25	7.4

Design Decisions: The shaft will be 10 in long and will be supported by two single-row ball bearings with centers at $A = 1$ in and $B = 9$ in. The bearings will be treated as simple supports. The belts will be approximately horizontal as in Figure 8.2. The belts on pulleys 1 and 3 will extend outward from the pulleys ($\alpha = 0$). The belts on pulleys 2 and 4 will extend inward ($\alpha = 180°$).

8.11. Design a shaft for the application described in Problem 8.10.

Design Decisions: The shaft will be made of steel with an ultimate strength of 105,000 psi and a yield point of 80,000 psi. The shaft surface will be ground. The stress concentration at the pulleys will be equivalent to a change of diameter $D/d = 1.1$ with a fillet-radius-to-diameter ratio r/d of 0.1. A safety factor of 1.5 will be used, and we will design for 1 failure in 5 million samples.

8.12. A belt drive shaft carries the following loads:

Pulley i	*Function*	*Pitch Radius* r_i *(in)*	*Resultant Angle* α_i *(deg)*	*Axial Location* *x(in)*
1	Input	3.75	0	1.8
2	Output	2.25	45	3.2

The shaft rotates at 500 rpm, and the input power to pulley number 1 is 20 hp. The direction of the vector sum of the belt tensions at each pulley is identified as the resultant angle.

(a) Calculate and plot shear force and bending moment in the horizontal and vertical planes, and resultant moment and torque.

(b) Design the shaft.

Design Decisions: The shaft will be 5 in long and will be supported by bearings at $A = 0.75$ in and $B = 4.25$ in. The shaft will be made of steel with an ultimate strength of 120,000 psi and a yield point of 100,000 psi. The surface will be ground. Stress concentration at the pulleys will be equivalent to a diameter change D/d of 1.1 and a fillet-radius-to-diameter ratio of r/d of 0.1. We will design for a failure rate of 1 in 10^8, and we will use a safety factor of 1.25.

8.13. A belt drive shaft carries the following loads:

Pulley i	*Function*	*Pitch Radius* r_i *(in)*	*Resultant Angle* α_i *(deg)*	*Axial Location* *x(in)*
1	Input	2.75	30	2
2	Output	3.80	0	5.5

The shaft rotates at 1750 rpm, and the input power to pulley number 1 is 27.5 hp. The direction of the vector sum of the belt tensions at each pulley is identified as the resultant angle.

(a) Calculate and plot shear force and bending moment in the horizontal and vertical planes, and resultant moment and torque.

(b) Design the shaft.

Design Decisions: The shaft will be 6 in long and will be supported by bearings at $A = 0.8$ in and $B = 3.5$ in. The shaft will be made of steel with an ultimate strength of 125,000 psi and a yield point of 100,000 psi. The surface will be ground. Stress concentration at the pulleys will be equivalent to a diameter change D/d of 1.1 and a fillet-radius-to-diameter ratio r/d of 0.1. It will be assumed that there is no stress concentration at the bearing. We will design for a failure rate of 1 in 10^6, and we will use a safety factor of 1.25.

8.14. A belt drive shaft carries the following loads:

Pulley i	*Function*	*Pitch Radius* r_i *(in)*	*Resultant Angle* α_i *(deg)*	*Load Fraction* R_{Ti}	*Axial Location* *x(in)*
1	Input	3	90	1.00	3.2
2	Output	2.5	30	−0.35	4.5
3	Output	2	−30	−0.3	6
4	Output	2.2	85	−0.35	7.5

The shaft rotates at 1800 rpm, and the input power to pulley number 1 is 42 hp. The direction of the vector sum of the belt tensions at each pulley is identified as the resultant angle.

(a) Calculate and plot shear force and bending moment in the horizontal and vertical planes, and resultant moment and torque.

(b) Design the shaft.

Design Decisions: The shaft will be 10 in long and will be supported by bearings at $A = 1$ in and $B = 9$ in. The shaft will be made of steel with an ultimate strength of 125,000 psi and a yield point of 100,000 psi. The surface will be ground. Stress concentration at the pulleys will be equivalent to a diameter change of $D/d = 1.1$ and a fillet-radius-to-diameter ratio of $r/d = 0.1$. We will design for a failure rate of 1 in 5 million, and we will use a safety factor of 1.25.

8.15. A belt drive shaft carries the following loads:

Pulley i	*Function*	*Pitch Radius* r_i *(in)*	*Resultant Angle* α_i *(deg)*	*Load Fraction* R_{Ti}	*Axial Location* *x(in)*
1	Input	2.8	85	1.00	2
2	Output	3.3	180	−0.3	4
3	Output	2.1	10	−0.3	6
4	Output	2.9	−65	−0.4	7.4

The shaft rotates at 1600 rpm, and the input power to pulley number 1 is 33 hp. The direction of the vector sum of the belt tensions at each pulley is identified as the resultant angle.

(a) Calculate and plot shear force and bending moment in the horizontal and vertical planes, and resultant moment and torque.

(b) Design the shaft.

Design Decisions: The shaft will be 10 in long and will be supported by bearings at $A = 1$ in and $B = 9$ in. The shaft will be made of steel with an ultimate strength of 125,000 psi and a yield point of 100,000 psi. The surface will be ground. Stress concentration at the pulleys will be equivalent to a diameter change D/d of 1.1 and a fillet-radius-to-diameter ratio r/d of 0.1. We will design for a failure rate of 1 in 5 million, and we will use a safety factor of 1.25.

CHAPTER 9

Gears

Symbols[1]

a	addendum	N	number of teeth
b	dedendum	N_{FS}	factor of safety
c	center distance	p_c	circular pitch
C_F, C_R, C_S	corrections to endurance limit	P_d	diametral pitch
CR	contact ratio	P_{hp}	horsepower
d	gear pitch diameter, shaft diameter at shoulder	P_{kW}	power in kilowatts
		P_{num}	number of planets
D	shaft diameter	Q	samples per failure
f	coefficient of friction	Q_v	transmission accuracy level number
F	face width	r	pitch radius, fillet radius at shoulder
F_r	radial force	r_a	addendum circle radius
F_t	tangential force	r_b	base circle radius
F_x	axial (thrust) force	s	tooth space
H_B	Brinell hardness	s_{ac}	allowable contact stress number
I	geometry factor for pitting resistance (wear)	s_{at}	allowable bending stress number
		s_c	contact stress number
IP	interference point	SD	standard deviation
J	geometry factor for bending strength	S_E	endurance limit
K_T	stress concentration factor	S_{N1}	fatigue strength (polished specimen)
K_v	dynamic factor	s_t	bending stress number
ln	natural logarithm	S_U	ultimate strength
m	module	$S(x,a,n)$	singularity function
M	bending moment	S_{YP}	yield point
n	rotation speed	T	torque

[1]Some of the equations in this chapter are based on publications of the American Gear Manufacturers Association (AGMA). In those equations, AGMA notation is used except where its use could cause confusion.

v	pitch line velocity	ϕ	pressure angle
V	shear force	ω	angular velocity
w	tooth width (circular tooth thickness)	γ	pitch angle
x, X	axial location on shaft	η	efficiency
X_R	reliability	λ	lead angle
Y	vertical component of shaft load	ψ	helix angle
Z	horizontal component of shaft load		

Units

angular velocity	rad/s	module	mm
circular pitch	in or mm	moment and torque	in · lb or N · mm
diametral pitch	teeth/in	power	hp or kW
dimensions	in or mm	strength and stress	psi or MPa
forces	lb or N		

Gears are available in a seemingly unlimited variety of form and size. Gear drives utilizing spur gears or helical gears on parallel shafts are among the most common power transmission elements. Several types of gears are shown in Figure 9.1.

TYPES OF GEARS

A few common types of gears are described in the following list:

- **Spur gears** (sometimes called **straight spur gears**) have teeth cut parallel to the shaft axis. Spur gear shafts must be parallel. The smaller of a pair of gears is called a **pinion.**

Figure 9.1 A variety of gear types, including spur gears, a rack and pinion, internal gears, worms and worm gears, and bevel gears. (Courtesy of Boston Gear)

- **Helical gears** have teeth cut at angle to the shaft axis. When used to transmit power between parallel shafts, a pair of helical gears has more teeth in contact than an equivalent pair of spur gears. This results in quieter action. The helix angle of a right-hand helical gear has the same relationship to the shaft axis as the helix angle of a right-hand screw.
- **Herringbone gears** or **double helical gears** are equivalent to a right-hand and left-hand helical gear formed together. This configuration eliminates thrust loading.
- **Worms** are helical gears that resemble screws. They usually have only one or two teeth (starts). A worm drive utilizes a worm and worm gear. The worm gear is a helical gear, often cut to conform with the worm shape. The worm and worm gear shaft axes are nonintersecting and almost always perpendicular. Worm drives are often chosen when substantial speed reductions are required (typically reductions between 10-to-1 and 100-to-1 for a single worm-gear pair).
- **Bevel gears** have the form of a truncated cone. The shaft axes of a pair of bevel gears are nonparallel. Shafts that are perpendicular are common, but other shaft angles are also used.
- **Hypoid gears** are similar to bevel gears, but the teeth are curved, and the shaft axes do not intersect.
- **Internal gears** have inward-pointing teeth. They may have spur gear or helical gear form. They are commonly used in planetary gear trains.
- **Racks** are spur gears or helical gears with the teeth aligned on a straight base. A rack and pinion drive can be used to change rotational motion into rectilinear motion (or vice versa).

CHOOSING DRIVE TRAIN ELEMENTS

Gears are often the chosen drive train element when one or more of the following requirements must be met:

- exact input-to-output speed ratio
- high torque and high power transmission capacity
- high-speed operation
- high drive efficiency
- long service life

Undesirable characteristics of gear drives include the following:

- typically higher cost than some other drive train elements (e.g., belt drives)
- strict tolerance requirements on gears, shafting, and center-to-center distances (The center-to-center distance for belt and chain drive shafts is not critical.)
- transmission of shock and vibration between driver and driven elements (Shock and vibration can be absorbed by belt drives and fluid drive elements.)

SPUR GEARS

Gears with tooth elements parallel to the shaft axis are called **straight spur gears** or simply **spur gears.** Spur gears transmit power between parallel shafts. The shafts may rotate in fixed bearings, or, as in the case of planetary gears, the gear axes move in a circular path. Figure 9.2(a) shows a pair of spur gears. A pinion with a projecting hub is shown in Part (b) of the figure, and nonmetallic gears are shown in Part (c). Steel, cast iron, and brass gears are used where high reliability is required. Nonmetallic gears are corrosion-resistant, run more quietly than steel gears, and may be more cost-effective.

Spur Gear Terminology

Designers of power transmission equipment often specify standard "off-the-shelf" gears, or gears produced on standardized machinery. Such gears are identified by pitch diameter, pitch, pressure angle, and other terminology peculiar to the gear industry. Figure 9.2(d) identifies some of the dimensions of importance to gear design and specification. The basic vocabulary used in analysis and design of gear drives will be discussed below.

Pitch Diameter, Pitch Radius, and Center Distance. The **pitch diameter** (or simply the **diameter**) is the diameter of the **pitch circle**, the nominal diameter of the gear. Several other gear dimensions are related to the pitch circle. The relative motion of a pair of gears is the motion of their pitch circles rolling on one another. The **center distance** of a pair of gears is the sum of their pitch radii. If gear 1 engages gear 2, then

$$c = r_1 + r_2 = \frac{d_1 + d_2}{2} \tag{9.1}$$

where c = distance between centers of parallel shafts carrying spur or helical gears

d = pitch diameter

$r = d/2$ = pitch radius

When a pinion meshes with an internal gear, the center distance is the difference in pitch radii.

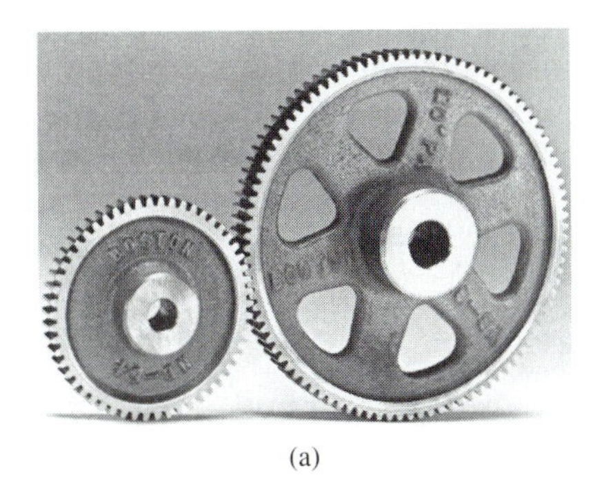

(a)

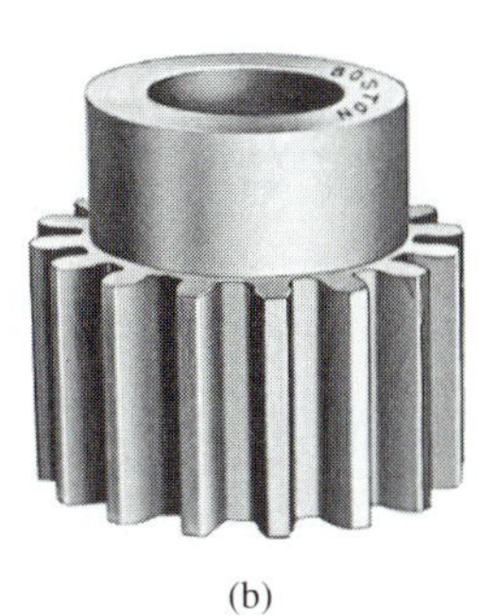

(b)

(c)

Figure 9.2 Spur gears. (a) A pair of spur gears; (b) a pinion with hub projection; (c) nonmetallic gears (Courtesy of Boston Gear)

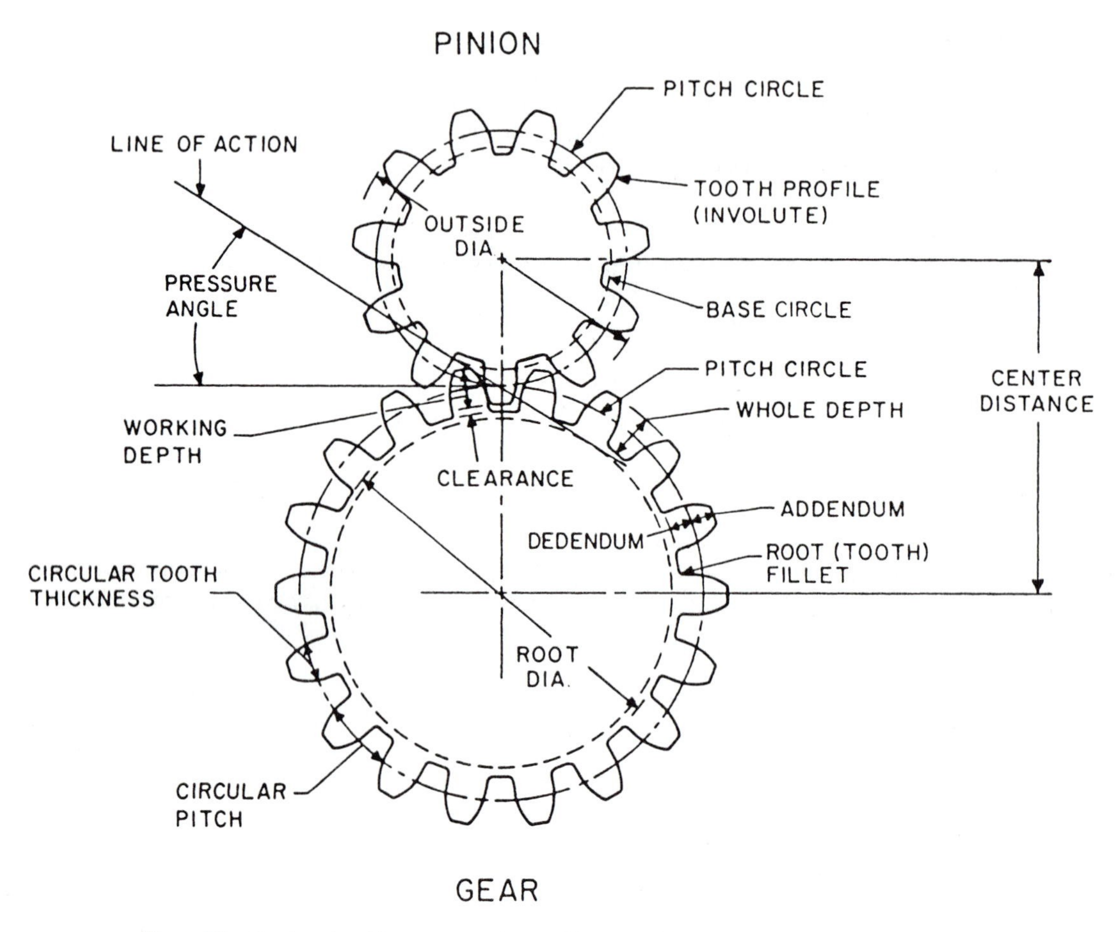

Figure 9.2 *Continued* (d) gear nomenclature. (Courtesy of Boston Gear)

Circular Pitch, Tooth Width, and Tooth Space. The **circular pitch** is the arc-distance between the centers of two adjacent teeth measured along the pitch circle. The circular pitch is divided into two nearly equal arcs: the **tooth width** and the **tooth space.**

$$p_c = \frac{\pi d}{N} \qquad w = \frac{p_c^-}{2} \qquad s = \frac{p_c^+}{2} \tag{9.2}$$

where p_c = circular pitch

w = tooth width (circular tooth thickness)

s = tooth space

N = number of teeth

When one gear drives another, both must have the same circular pitch.

Backlash. **Backlash** is the clearance or play between a pair of mating tooth surfaces of a pair of gears. It is related to the difference between tooth space and tooth width. Backlash in stock spur gears averages about 1.5% of circular pitch for circular pitch between 1/4 and 1 in. For smaller values of circular pitch, typical backlash is about 0.0025 to 0.004 in. Backlash is necessary to accommodate tolerances in machine components. If it were possible to eliminate manufacturing errors in shafts and gears, backlash would be unnecessary.

In some applications, the relationship between the input and output gear angular positions must be precise. Anti-backlash gears are designed to eliminate position error caused by backlash. An anti-backlash gear is split in the plane of the gear, and springs slightly rotate the halves relative to one another, filling the tooth space of the mating gear.

Diametral Pitch and Module. **Diametral pitch** (or simply **pitch**) is defined as the number of teeth in a gear divided by the pitch diameter. The **module** is the pitch diameter divided by the number of teeth.

$$P_d = N/d \tag{9.3}$$

$$m = d/N \tag{9.4}$$

where P_d = diametral pitch (teeth per inch of diameter, customary U.S. standard)
N = number of teeth
d = diameter
m = module (mm diameter per tooth, metric standard)

Diametral pitch and module are related to circular pitch as follows:

$$P_d = \pi/p_c \tag{9.5}$$

$$m = p_c/\pi \tag{9.6}$$

Mating spur gears dimensioned according to the customary U.S. system must have equal diametral pitches. Mating spur gears with metric dimensions must have equal modules.

Although gears could be manufactured with almost any diametral pitch, designers usually select values that are available as stock gears. For off-the-shelf steel gears, typical values of diametral pitch that are readily available include

$$P_d = 3, 4, 5, 6, 8, 10, 12, 16, 20, 24, 32, 48, \text{ and } 64 \text{ teeth/in}$$

Nylon and brass gears are available in fine pitches (e.g., P_d = 24 to 64 for brass; P_d = 24 to 48 for nylon). Cast iron gears are common in the coarser pitches (typically, P_d = 3 to 20). Off-the-shelf precision gears are available in stainless steel, aluminum, and reinforced plastics, with diametral pitch as fine as 120 teeth per in of diameter.

Metric standard gears are also available with a large range of modules. Commonly selected modules include

$$m = 1, 1.5, 2, 2.5, 3, 4, 5, 6, 8, 10, 12, 16, \text{ and } 20 \text{ mm}$$

Although some of these modules are roughly equivalent to the values of diametral pitch given above, metric and customary U.S. standard gears are not interchangeable, except at the preliminary design stage.

EXAMPLE PROBLEM Diametral Pitch and Module

A design calls for a 20 tooth gear and a 30 tooth gear with a diametral pitch of 10. Consider the possibility of replacing these gears with metric standard gears.

Design Decision: Common stock gear sizes will be considered.

Solution: The gear diameters are

$$d_1 = N_1/P_d = 20/10 = 2 \text{ in}$$

and

$$d_2 = N_2/P_d = 30/10 = 3 \text{ in}$$

The center distance is

$$c = r_1 + r_2 = \frac{d_1 + d_2}{2} = \frac{2 + 3}{2} = 2.5 \text{ in}$$

The approximate equivalent metric module is

$$m = d/N = 1/P_d = 0.1 \text{ in} = 2.54 \text{ mm} \quad \text{(approximately)}$$

If we select $m = 2.5$, then

$$d_1 = N_1 m = 20 \cdot 2.5 = 50 \text{ mm}$$

and

$$d_2 = N_2 m = 30 \cdot 2.5 = 75 \text{ mm}$$

The center distance is

$$c = r_1 + r_2 = \frac{d_1 + d_2}{2} = \frac{50 + 75}{2} = 62.5 \text{ mm} \quad \text{(about 2.461 in)}$$

It would be necessary to change the distance between the shafts and increase the gear face width to compensate for smaller teeth. These changes may be acceptable in the preliminary design stage. If the gears are intended for a gearbox that is already in production, then the suggested change should be rejected.

SPEED RATIO FOR NONPLANETARY GEARS

The speed ratio for a single pair of gears on shafts in fixed bearings is inversely proportional to the number of teeth.

$$\left|\frac{n_2}{n_1}\right| = \left|\frac{\omega_2}{\omega_1}\right| = \frac{N_1}{N_2} \tag{9.7}$$

where n = rotation speed (rpm)

ω = angular velocity (rad/s)

N = number of teeth (an integer)

A favorable power-to-weight ratio for an internal combustion engine or an electric motor is usually obtained by relatively high speed operation. Gear drives are commonly used to reduce speed to that needed for production machinery, vehicles, and other applications. Thus, the smaller of a pair of gears is usually the driver. The terms **gear** and **pinion** refer, respectively, to the larger and smaller gears in a pair of mating gears.

The speed ratio is negative for a pair of (external) spur gears; that is, the gears rotate in opposite directions. The same is true for helical gears on parallel shafts. The ratio is positive if one gear is an internal gear; that is, both gears rotate in the same direction.

If more than two gears are involved, the output-to-input speed ratio is given by

$$\left|\frac{n_{\text{OUTPUT}}}{n_{\text{INPUT}}}\right| = \left|\frac{\omega_{\text{OUTPUT}}}{\omega_{\text{INPUT}}}\right| = \frac{\text{product of driving gear teeth}}{\text{product of driven gear teeth}} \tag{9.8}$$

Note that a more complicated procedure is used to determine the speed ratio of planetary gear trains.

Rack

The teeth of a **rack** lie on a straight line. Thus, a rack is considered to have an infinite diameter. The collection of gears in Figure 9.1 includes a rack and pinion. The circular pitch and the diametral pitch or module of a rack are defined as the corresponding values for the pinion that meshes with the rack. Consider a rack driven by a pinion on a shaft supported by fixed bearings. The linear speed of the rack is equal to the tangential velocity of the pitch circle of the pinion.

INVOLUTE TOOTH FORM AND LINE-OF-ACTION

If we neglect friction, then the force transmitted by a driving gear tooth to a driven gear tooth is perpendicular to the common tangent to the teeth at the contact point. To minimize vibration, we want a constant force vector, a force that does not vary in magnitude or direction. The **involute tooth form** allows us to meet this requirement, at least theoretically. If transmitted power is constant, and if the driving gear rotates at constant speed, then the driven gear rotates at constant speed, and the force vector is constant. **Cycloidal tooth form** gears, which were popular at one time, did not produce a constant force direction. For this reason, cycloidal gears have been replaced by involute gears. Of course, we cannot totally eliminate vibration because we cannot completely eliminate elastic deflection of gear teeth and manufacturing errors.

We will discuss only gears with involute form teeth. For nonplanetary gears, the force vector lies on a line called the **line-of-action.** The line-of-action is also the locus of points of contact between driving gear teeth and driven gear teeth. Note that we are representing a gear as a two-dimensional surface to eliminate unnecessary detail in our analysis. Actual contact occurs across the face width of the tooth, which is slightly compressed as power is transmitted.

PRESSURE ANGLE AND BASE CIRCLE

The angle between the line-of-action and a common tangent to the pitch circles is called the **pressure angle.** The **base circles** of the pinion and gear are tangent to the line-of-action. Figure 9.2(d) shows the line-of-action and the base circles on a pair of gears. A large pressure angle is shown in the figure. Standard pressure angles are 14.5°, 20°, and 25°, with 14.5° and 20° most common. When a pair of gears is specified, both must have the same pressure angle.

The involute tooth form exists only outside the base circle. Base circle radii are given by

$$r_b = r\cos\phi \quad \textbf{(9.9)}$$

where r_b = base circle radius
r = pitch circle radius
ϕ = pressure angle

ADDENDUM AND DEDENDUM

The **addendum** is the radial distance a tooth projects beyond the pitch circle. The standard addendum for **full-depth teeth** is given by

$$a = 1/P_d \quad \textbf{(9.10)}$$

or

$$a = m \quad \textbf{(9.11)}$$

where a = addendum
P_d = diametral pitch
m = module

The circle representing the outside diameter of the gear is called the **addendum circle.** The addendum circle radius is

$$r_a = r + a \quad \textbf{(9.12)}$$

where r = pitch radius

The **dedendum** is the depth of the tooth space measured from the pitch circle. It must be larger than the addendum to ensure clearance. Typical dedendum values for full-depth teeth are

$$b = 1.157/P_d + 0.002 \quad \text{for } P_d \geq 20 \text{ (fine pitch)} \quad \textbf{(9.13)}$$

and

$$b = 1.20/P_d \text{ to } 1.25/P_d \quad \text{for } P_d < 20 \text{ (coarse pitch)} \quad \textbf{(9.14)}$$

where b = dedendum
Dimensions are in inches.

The flank of a gear tooth can be formed by a radial line toward the gear center, joining the dedendum circle with a fillet. As an alternative design, the base of the tooth space can be in the form of a semicircle or other clearance curve.

Gear teeth with reduced addenda are called **stub teeth.** Typically,

$$a_{\text{stub}} = 0.8/P_d \tag{9.15}$$

or

$$a_{\text{stub}} = 0.8 \cdot m \tag{9.16}$$

where a_{stub} = addendum for stub teeth (usually 20° pressure angle).

The use of gears with reduced addenda (stub teeth) is explained in the material that follows.

INTERFERENCE POINTS AND INTERFERENCE

Refer to Figure 9.3, showing a pinion rotating counterclockwise to drive a gear. Subscripts 1 and 2 refer to pinion and gear, respectively. Note base circles (BC) and addendum circles (AC). Points IP, where the line-of-action is tangent to the base circles, are called the **interference points.**

Consider a single pair of meshing teeth. The beginning of contact occurs where a point on a driver tooth contacts the tip of a driven tooth. This point is represented by point B in Figure 9.3, the intersection of the addendum circle of the driven gear with the line-of-action. Contact ends at point E where the tip of a driver tooth crosses the line-of-action.

Proper tooth action requires that contact between gear teeth occur only on the involute portion of the teeth, that is, the surface of the tooth beyond the base circles. Contact within the base circle is forbidden. This requirement is met if contact begins and ends between the interference points as in Figure 9.3. Failure to meet this condition is called **interference.** If contact begins or ends beyond line segment IP_1–IP_2, then we must redesign to eliminate interference.

Maximum Allowable Addendum Based on Interference

It is easier to express the interference criterion analytically than to use a graphical procedure. If a pinion and a gear have equal addenda, then it is necessary to check only for

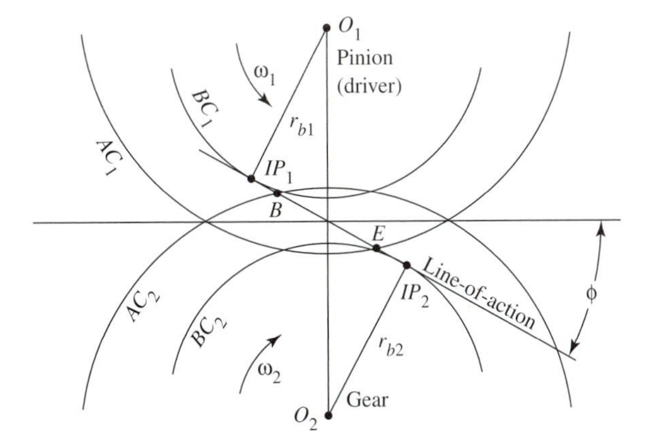

Figure 9.3 Interference points.

interference of the gear. To avoid interference, the maximum addendum circle radius is given by

$$r_{a2\text{MAX}} = \sqrt{r_{b2}^2 + (c \cdot \sin \phi)^2} \tag{9.17}$$

where $r_{a2\text{MAX}}$ = maximum addendum circle radius

$r_{b2} = r_2 \cos \phi$ = base circle radius

c = center distance

ϕ = pressure angle

Subscript 2 refers to the larger gear.

The criterion for avoiding interference is

$$r_{a2} \leq r_{a2\text{MAX}} \tag{9.18}$$

where $r_{a2} = r_2 + a$ = addendum circle radius

r_2 = pitch radius

$a = 1/P_d$ or $a = m$ for full-depth teeth

EXAMPLE PROBLEM Interference Check of a Proposed Gear Reducer

An interference check is to be done on a gear reducer with the following specifications:

Input speed:	$n_1 = 1760$ rpm
Output speed:	$n_2 = 440$ rpm
Diametral pitch:	$P_d = 4$
Pinion teeth:	$N_1 = 15$

Find the following:

(a) number of teeth in the gear
(b) gear pitch diameter
(c) center distance
(d) maximum allowable addendum circle radius for the gear
(e) standard addendum circle radius for the gear
(f) Is there interference?

Decisions: Use full-depth teeth and a pressure angle ϕ of 14.5.

Solution (a):

$$N_2 = N_1 \cdot \frac{n_1}{n_2} \qquad N_2 = 60 \text{ teeth}$$

Solution (b):

$$d_2 = \frac{N_2}{P_d} \qquad d_2 = 15$$

Solution (c):

$$d_1 = \frac{N_1}{P_d}$$

$$c = \frac{d_1 + d_2}{2} \qquad c = 9.375$$

Solution (d):

$$\cos\phi = \cos(\phi\cdot\text{deg})\qquad \sin\phi = \sin(\phi\cdot\text{deg})$$

$$r_{b2} = \frac{d_2}{2}\cdot\cos\phi \qquad r_{b2} = 7.261$$

$$r_{a2\text{MAX}} = \sqrt{r_{b2}^{\,2} + (c\cdot\sin\phi)^2} \qquad r_{a2\text{MAX}} = 7.631$$

$$a = \frac{1}{P_d}$$

Solution (e):

$$r_{a2} = \frac{d_2}{2} + a \qquad r_{a2} = 7.75$$

Solution (f):

Yes.

$$r_{a2(\text{STANDARD})} > r_{a2\text{MAX}}$$

■■

Redesigning to Eliminate Interference

We have a number of options for elimination of interference. They include the following:

- **Increasing pressure angle.** Increasing the pressure angle improves the tooth form, increasing the beam strength of the tooth. However, shaft loading is somewhat increased due to increased radial force.
- **Increasing the number of teeth.** It is assumed that the face width and the diametral pitch or module are not changed. Then increasing the number of teeth increases the gear diameter, increasing cost and weight.
- **Reducing the addendum.** This change in tooth form reduces the moment arm of bending loads, effectively increasing tooth strength. However, it results in fewer pairs of teeth in contact.
- **Undercutting.** When gears with a small number of teeth are generated with a rack cutter, some of the tooth flank may be cut away. Gears so manufactured will then mesh with other external gears without binding. However, the undercutting reduces the beam strength of the gear teeth and reduces the average number of tooth pairs in contact.

CONTACT RATIO

The average number of driving gear teeth contacting driven gear teeth is called the **contact ratio** (*CR*). For example, if $CR = 1.4$, then two pairs of teeth are in contact 40% of the time, and one pair 60% of the time. In order to ensure smooth, continuous transmission of power, *CR* must not be less than 1.1, considering worst-case tolerances. Good practice suggests that $CR \geq 1.2$, and many designers try to design for $CR = 1.4$ or higher. Other things being equal, the higher *CR* the better, because load sharing will be improved, and noise and vibration reduced.

Referring to Figure 9.3, we see that contact occurs along line-of-action segment BE. We can determine the contact ratio by dividing the length of contact by the base pitch (the arc-distance between tooth centers on the base circle). The result is

$$CR = BE/(p_c \cos \phi) \tag{9.19}$$

Equation (9.19) requires an accurate drawing. It is easier to calculate the contact ratio from the following equation:

$$CR = \frac{\sqrt{r_{a1}^2 - (r_1 \cdot \cos \phi)^2} + \sqrt{r_{a2}^2 - (r_2 \cdot \cos \phi)^2} - c \cdot \sin \phi}{p_c \cdot \cos \phi} \tag{9.20}$$

where CR = contact ratio

$r_a = r + a$ = addendum circle radius

r = pitch circle radius

a = addendum

c = center distance

ϕ = pressure angle

p_c = circular pitch

EXAMPLE PROBLEM Redesigning to Eliminate Interference and Determining Contact Ratio

The solution to the preceding example problem was not satisfactory because of interference. Suppose that we decide against undercut teeth because of strength requirements. Let us try changing the number of teeth in the pinion. After a few tries, we get the following design with 31 pinion teeth, which just avoids interference:

Input speed:	$n_1 = 1760$ rpm
Output speed:	$n_2 = 440$ rpm
Diametral pitch:	$P_d = 4$
Pinion teeth:	$N_1 = 31$

Find the following:

(a) the number of teeth in the gear
(b) gear pitch diameter
(c) center distance
(d) maximum allowable addendum circle radius for the gear
(e) standard addendum circle radius for the gear
(f) Is there interference?

Decisions: Use full-depth teeth and a pressure angle ϕ of 14.5.

Solution (a):

$$N_2 = N_1 \frac{n_1}{n_2} \qquad N_2 = 124 \text{ teeth}$$

Solution (b):

$$d_2 = \frac{N_2}{P_d} \qquad d_2 = 31$$

Solution (c):

$$d_1 = \frac{N_1}{P_d}$$

$$c = \frac{d_1 + d_2}{2} \qquad c = 19.375$$

Solution (d):

$$\cos\phi = \cos(\phi \cdot \deg) \qquad \sin\phi = \sin(\phi \cdot \deg)$$

$$r_{b2} = \frac{d_2}{2} \cdot \cos\phi \qquad r_{b2} = 15.006$$

$$r_{a2\text{MAX}} = \sqrt{r_{b2}^2 + (c \cdot \sin\phi)^2} \qquad r_{a2\text{MAX}} = 15.771$$

$$a = \frac{1}{P_d}$$

Solution (e):

$$r_{a2} = \frac{d_2}{2} + a \qquad r_{a2} = 15.75$$

Solution (f): No.

$$r_{a2(\text{STANDARD})} < r_{a2\text{MAX}}$$

$$r_2 = \frac{d_2}{2} \qquad r_1 = \frac{d_1}{2}$$

$$r_{a2} = r_2 + a \qquad r_{a1} = r_1 + a$$

$$p_c = \pi \cdot \frac{d_1}{N_1}$$

$$CR = \frac{\sqrt{r_{a1}^2 - (r_1 \cdot \cos\phi)^2} + \sqrt{r_{a2}^2 - (r_2 \cdot \cos\phi)^2} - c \cdot \sin\phi}{p_c \cdot \cos\phi} \qquad CR = 2.165$$

Thus, $CR > 1.2$, and the design is satisfactory on basis of contact ratio.

We see that the contact ratio is relatively high, which should result in quiet operation. However, the gear has 124 teeth, and the center distance is 19.375 in. The solution is satisfactory, but the redesigned speed reducer will be large, heavy, and expensive. If the solution is programmed on a computer, it is easy to try various alternatives.

Alternative Solution (1): Let us use 15 pinion teeth and 60 gear teeth but increase the pressure angle and use stub teeth. Let

$$\phi = 20° \quad \text{and} \quad a = 0.8/P_d$$

The interference problem is solved since

$$r_{a2} = 7.7 \text{ in} \qquad r_{a2\text{MAX}} = 7.743 \text{ in}$$

and the contact ratio is

$$CR = 1.343$$

which is adequate.

Alternative Solution (2): As another alternative, try 20° full-depth teeth. We find that there will be interference if the pinion has 15 teeth and the gear 60 teeth. However, with 16 pinion teeth and 64 gear teeth, we obtain the required speed ratio, and there is no interference since

$$r_{a2} = 8.25 \text{ in} \qquad r_{a2\text{MAX}} = 8.259 \text{ in}$$

The contact ratio is

$$CR = 1.647$$

an improvement over Alternative Solution (1). ■■

Interference in Rack and Pinion Drives

Figure 9.4 shows a representation of a pinion turning counterclockwise, driving a rack. Note that involute rack teeth have straight sides, sloped at an angle ϕ, the pressure angle. The figure shows the line-of-action (LA), the rack pitch line (PL), and the gear pitch circle (PC). The velocity of the rack is given by

$$v_2 = \omega_1 r_1 \tag{9.21}$$

where v_2 = rack velocity (in/s or mm/s)
ω_1 = angular velocity of pinion (rad/s)
r_1 = pitch radius of pinion (in or mm)

The rack is treated as a gear of infinite radius. As a result, Equation (9.17) for maximum addendum circle radius cannot be applied. We find the requirement for avoiding interference by examining the geometry of a rack and pinion drive. If the following condition is met, the pinion can mesh with a rack without interference and without undercutting:

$$a \le r_1 \sin^2\phi \tag{9.22}$$

where a = addendum
r_1 = pinion radius
ϕ = pressure angle

For full-depth teeth,

$$a = m \quad \text{or} \quad a = 1/P_d$$

where m = module (mm)
P_d = diametral pitch

Figure 9.4 Rack and pinion.

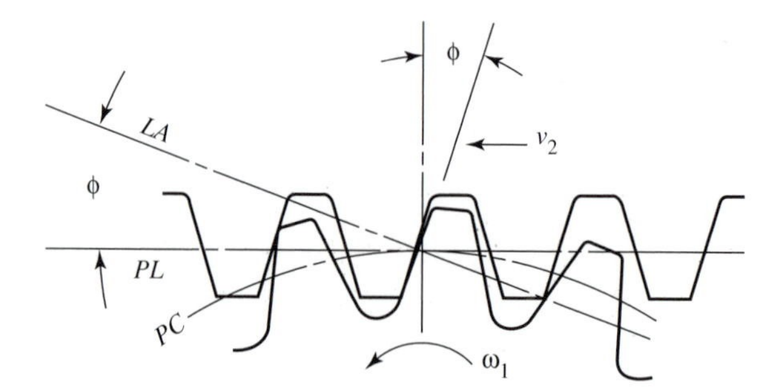

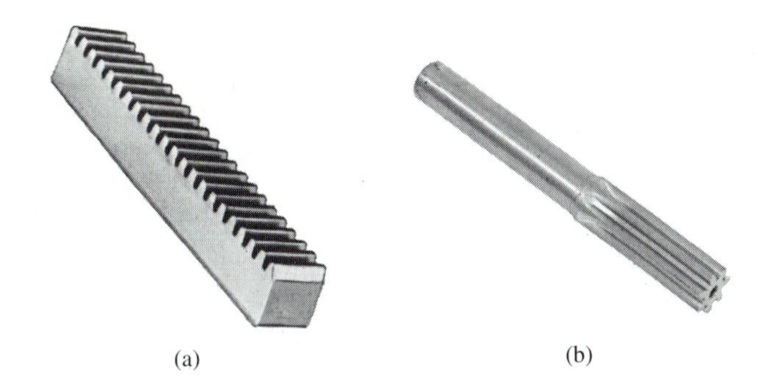

(a) (b)

Figure 9.5 (a) Rack; (b) stem pinion. (Courtesy of Boston Gear)

Making the above substitution, noting that

$$r = m \cdot N/2$$

and dividing by $m/2$, we obtain the no-interference criterion in terms of number of pinion teeth.

$$N_1 \geq 2/\sin^2\phi \qquad \textbf{(9.23)}$$

Figure 9.5(a) shows a rack. Part (b) of the figure shows a stem pinion with only seven teeth; the pinion teeth are undercut.

EXAMPLE PROBLEM Rack and Pinion Interference

Determine the smallest pinion that will mesh with a rack without interference. Consider pressure angles of 14.5°, 20°, and 25°.

Solution: Equation (9.23) applies to full-depth teeth. For stub teeth with an addendum of

$$a = 0.8 \cdot m \quad \text{or} \quad a = 0.8/P_d$$

the equivalent no-interference criterion is

$$N_{1(\text{stub})} \geq 1.6/\sin^2\phi$$

The pinion must, of course, have a whole number of teeth. Using the above equations, we obtain the following results:

Pressure Angle (degrees)	*Tooth Form*	*Minimum Pinion Teeth*
14.5	Full-depth	32
20	Full-depth	18
20	Stub	14
25	Full-depth	12

If a pinion has at least the number of teeth given in the above table, it will mesh with a rack without interference. In addition, it will mesh with any external gear without interference. When an external gear and pinion mesh, a smaller number of pinion teeth can be permitted than indicated in the table. Note that a previous example showed that a 14.5° full-depth 31 tooth pinion and 124 tooth gear will mesh without interference. It was also shown that a 20° full-depth 16 tooth pinion will mesh with a 64 tooth gear without interference.

Manufacturers' catalogs list gears with small tooth numbers that are made by rack generation or an equivalent process. Where the tooth numbers are less than those given in the above table, the gears are undercut. The undercutting process reduces strength. In addition, the average number of teeth in contact for these gears will be less than the value given by the contact ratio equation.

POWER AND TORQUE

Torque is related to transmitted power by

$$T = \frac{63{,}025 P_{\text{hp}}}{n} \tag{9.24}$$

where T = torque (in · lb)
P_{hp} = power (horsepower)
n = rotation speed (rpm)

Using metric units, the relationship is

$$T = \frac{10^6 P_{\text{kW}}}{\omega} \tag{9.25}$$

where T = torque (N · mm)
P_{kW} = power (kW)
ω = angular velocity (rad/s)

TOOTH LOADING ON SPUR GEARS

As discussed above, the line-of-action identifies the direction of force transmitted between driver and driven teeth. If we neglect friction, the transmitted force is normal to the tooth surfaces at the point of contact. Consider tooth contact occurring at the **pitch point,** the intersection of the line-of-centers and the line-of-action. The transmitted force can be broken into two components: the tangential force and the radial force, as shown in Figure 9.6. The tangential force is in the direction of the common tangent to the pitch circles.

Tangential force is given by

$$F_t = T/r \tag{9.26}$$

where F_t = tangential force
T = torque on a gear
r = pitch radius of the same gear

Tangential force on the driven gear is equal and opposite that on the driver.

Driven gear torque is given by

$$T_2 = F_t r_2 \tag{9.27}$$

Radial force is given by

$$F_r = F_t \tan \phi \tag{9.28}$$

where F_r = radial force
ϕ = pressure angle

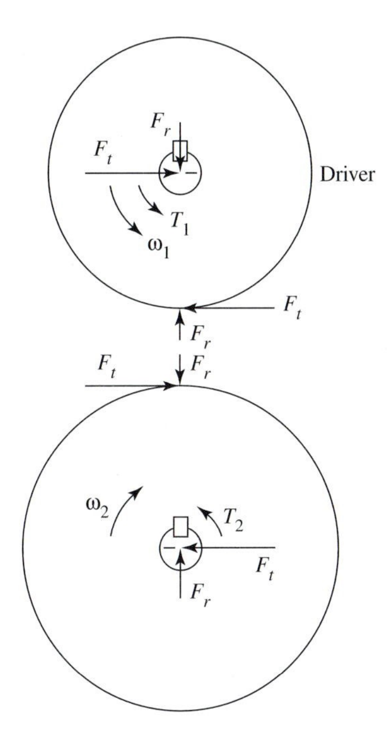

Figure 9.6 Forces and torques on spur gears (exploded view).

The radial force direction is toward the center of the gear or pinion. Radial force points away from the center of a ring gear (an internal gear). Radial force on the driven gear is equal and opposite that on the driver.

Figure 9.6 shows free-body diagrams of the driver and driven gears. The gears are separated to show forces. Note that T_1 is the input torque supplied by the driver shaft. It is in the same direction as the rotation. The shaft reactions are also shown. Torque reaction T_2 applied to the driven gear by the driven shaft (the load) opposes the rotation direction.

It is recommended that free-body diagrams be sketched in every case. If an assembly diagram is available instead of an exploded view, a separate copy can be made for each free-body diagram. A transparent-ink marker can be used to identify the subject of each free-body diagram (e.g., the gear in question).

COUNTERSHAFTS

An intermediate shaft between the driving shaft and the driven shaft in a gear drive or a belt drive is called a **countershaft.** A gear train with the input and output shafts on the same axis is called a **reverted gear train.** In Figure 9.7, countershaft C carries gears 1 and 2. This configuration is not a reverted gear train.

The symbols labeled bearings A and B in Figure 9.7 could represent ball or roller or journal bearings. Bearing friction and friction due to gear action are important when we consider lubricant viscosity and cooling of bearings and gearboxes. However, when designing spur gears and shafts, we usually neglect friction.

Figure 9.7 Gear train.

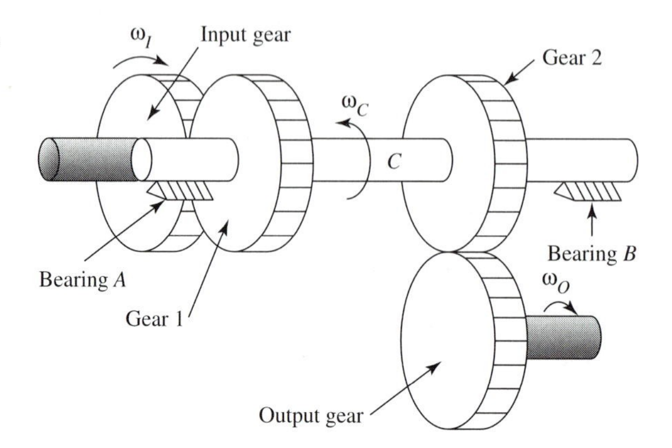

The keys to analyzing problems of this type are as follows:

1. **Action and reaction (Newton's third law).** The forces of action and reaction between bodies are equal and opposite.
2. **Static equilibrium.** For bodies at rest or moving at constant velocity, the sum of the forces in each direction is zero. The sum of the moments is also zero.

Thus, for the gear train in Figure 9.7, the following principles apply:

- The tangential forces on the input gear and gear 1 are equal and opposite.
- The radial forces on the input gear and gear 1 are equal and opposite.
- The tangential forces on the output gear and gear 2 are equal and opposite.
- The radial forces on the output gear and gear 2 are equal and opposite.
- The torque on gear 2 is equal and opposite the torque on gear 1.
- The sum of the horizontal forces on shaft C is zero.
- The sum of the vertical forces on shaft C is zero.
- The sum of the moments about any point is zero.

EXAMPLE PROBLEM Gear and Shaft Loading

We need a gear train to transmit 25 hp at 700 rpm countershaft speed. Determine the gear and shaft loading. We need this information to design the shaft and to select the gears and bearings.

Design Decisions: The gear train will be similar to the one sketched in Figure 9.7, with the input gear behind the countershaft and the output gear below the countershaft. The pitch radii of gears 1 and 2 will be $r_1 = 1.5$ and $r_2 = 2$, and the gears will be located at $X_1 = 2.5$ and $X_2 = 3.5$, where dimensions are in inches and distances are measured from the left end of the countershaft. The left and right bearings will be located, respectively, at $A = 1$ and $B = 5$, and the countershaft will be 6 in long. Spur gears with 20° pressure angle will be used.

Solution Summary: Countershaft torque T is found from Equation (9.24) relating power and speed. Tangential force F_t, equal to torque divided by pitch radius, is downward on gear 1 as shown in Figure 9.8. Radial force F_r is toward the gear center. The torque due to tangential force on gear 2 balances the torque on gear 1, thus enabling us to find the tangential and radial forces on gear 2.

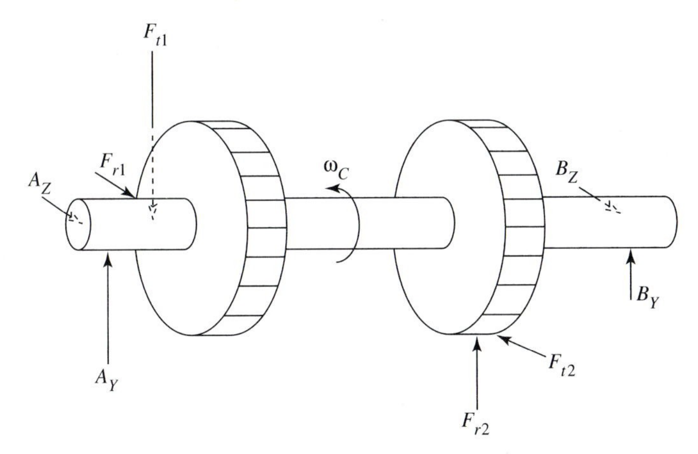

Figure 9.8 Free-body diagram of countershaft (not to scale).

We must now find bearing reactions A and B, where subscripts Y and Z refer to vertical and horizontal components, respectively. Beginning with moment equilibrium in the vertical plane, we eliminate one unknown by setting the moments of forces about the left bearing equal to zero. This yields B_Y, the vertical component of the reaction due to the right bearing. The three remaining bearing reaction components are found similarly. We find total bearing load by taking the square root of the sum of the squares of the reaction components at that bearing. As a partial check on calculations, we find that the sum of the gear and bearing loads are zero (±rounding error) in both the horizontal and vertical directions. The detailed solution follows; customary U.S. units are used.

Detailed Solution:

Gear Locations: Input gear is behind gear 1; output gear is below gear 2.

$$r_1 = 1.5 \qquad r_2 = 2 \qquad \phi = 20 \qquad \text{hp} = 25 \qquad n_{\text{rpm}} = 700$$

$$T = 63{,}025 \cdot \frac{\text{hp}}{n_{\text{rpm}}} \qquad T = 2.251 \cdot 10^3$$

$$F_{t1} = \frac{T}{r_1} \qquad Y_1 = F_{t1} \qquad Y_1 = 1.501 \cdot 10^3$$

$$Z_1 = Y_1 \cdot \tan(20 \cdot \text{deg}) \qquad Z_1 = 546.172$$

$$F_{t2} = \frac{T}{r_2} \qquad Z_2 = -F_{t2} \qquad Z_2 = -1.125 \cdot 10^3$$

$$Y_2 = Z_2 \cdot \tan(20 \cdot \text{deg}) \qquad Y_2 = -409.629$$

Sign Convention:

Forces and reactions: + = down or out
Moment: + = CW

Bearing Locations:

$$A = 1 \qquad B = 5$$

Shaft Loads:

$Y =$ vertical $\quad Z =$ horizontal $\quad X =$ axial location

$n = 2 \quad i = 0, \ldots, n - 1$

X_i	Y_i	Z_i
2.5	Y_1	Z_1
3.5	Y_2	Z_2

Bearing Reactions:

$$B_Y = \left[-\frac{1}{(B-A)}\right] \cdot \sum_i [Y_i \cdot (X_i - A)] \qquad B_Y = -306.705$$

$$B_Z = \left[-\frac{1}{(B-A)}\right] \cdot \sum_i [Z_i \cdot (X_i - A)] \qquad B_Z = 498.59$$

$$R_B = \sqrt{B_Y^2 + B_Z^2} \qquad R_B = 585.371$$

$$A_Y = \left(\frac{1}{B-A}\right) \cdot \sum_i [Y_i \cdot (X_i - B)] \qquad A_Y = -784.261$$

$$A_Z = \left(\frac{1}{B-A}\right) \cdot \sum_i [Z_i \cdot (X_i - B)] \qquad A_Z = 80.685$$

$$R_A = \sqrt{A_Y^2 + A_Z^2} \qquad R_A = 788.401$$

Check Force Equilibrium:

$$\sum_i (Y_i) + A_Y + B_Y = 1.137 \cdot 10^{-13} \qquad \sum_i (Z_i) + A_Z + B_Z = 0$$

SHAFT DESIGN

Once we know the torque and bending loads that a shaft must endure, we plot shear force and bending moment diagrams in the horizontal and vertical planes. Combined moment is then determined. Shaft diameter is governed by the location where the worst combination of moment and torque occur. Bending moment is particularly important because it results in fatigue loading.

EXAMPLE PROBLEM Design of a Countershaft with Gear Loads in Two Planes

Plot vertical and horizontal shear and moment diagrams for the shaft described in the preceding example problem. Find combined moment, and design the shaft.

Design Decisions: The bearings will be assumed to act as simple supports. We will select steel with a yield point of 120,000 psi and an ultimate strength of 140,000 psi, using a safety factor of 1.5. The shaft surface will be ground, and stress concentration at the gears will be equivalent to a diameter ratio of 1.1 and a fillet-radius-to-diameter ratio of 0.1. We will design for infinite life with a failure probability of 1 in 1 million.

Solution Summary: Using the results of the preceding example problem, we plot shear and moment diagrams as shown in Figure 9.9. Although hand calculations and plotting are possible, this boring task

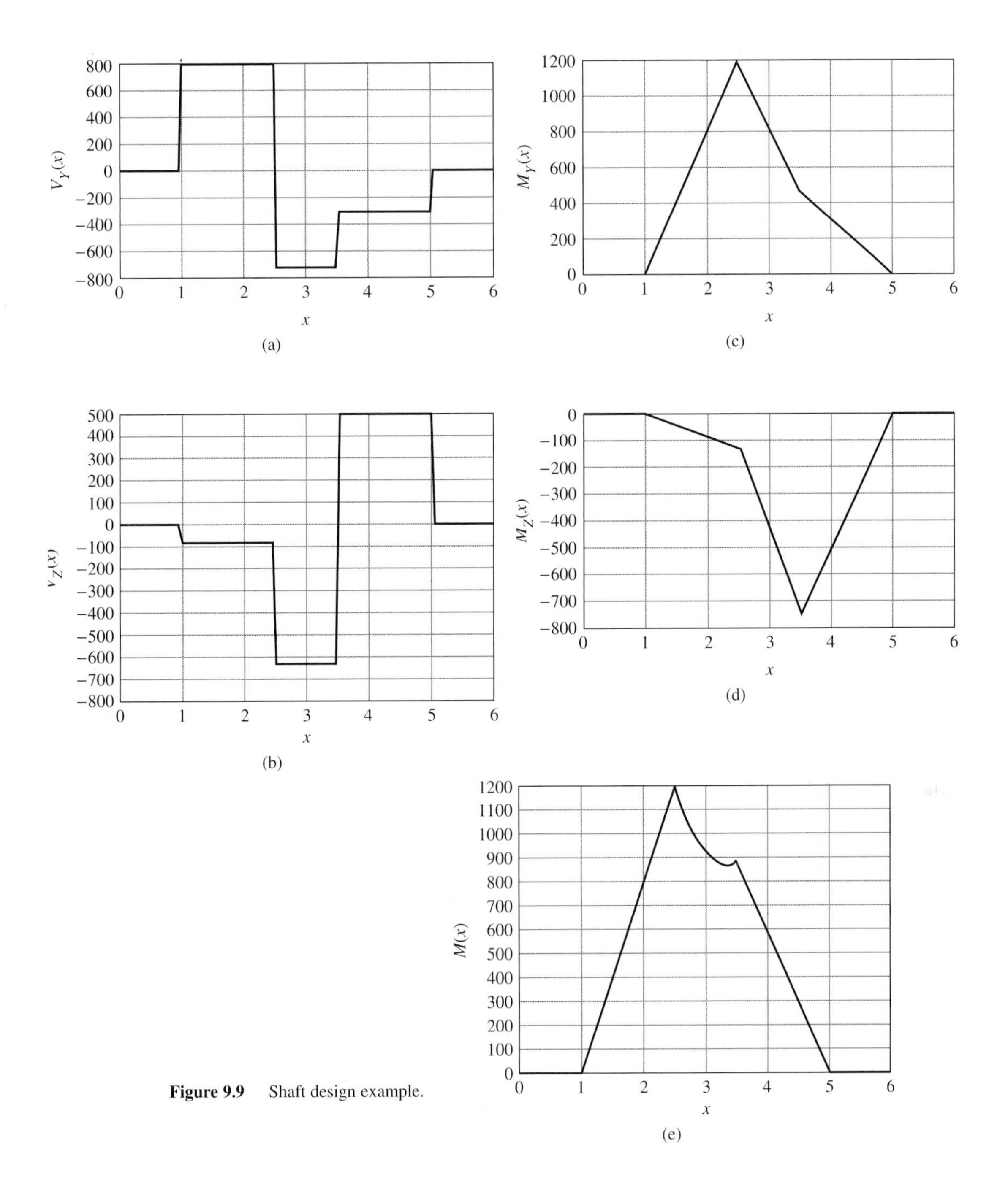

Figure 9.9 Shaft design example.

can be avoided if a computer is used. The singularity function is defined and used as in Chapter 8, "Shaft Design." Thus,

$$S(x,a,n) = \text{if } [x \geq a, (x - a)^n, 0] \quad \text{for } n \geq 0$$

where $S(x,a,n) = (x - a)^n$ if $x \geq a$ and 0 otherwise

x = axial coordinate

a = gear or bearing location

n = 0 for shear force and 1 for bending moment

The form of the logical if-statement varies according to the language or software selected.

The vertical shear force equation utilizes the singularity function four times, including the contribution of the vertical components of both bearing reactions and the forces at the two gears. To save time, this equation can be copied and slightly altered to obtain horizontal shear force. Shear force is plotted in Parts (a) and (b) of Figure 9.9. We can copy these equations and alter them again to obtain the vertical and horizontal plane moment equations, plotted in Parts (c) and (d) of the figure. For each of the 121 plotted points, the equations call for the singularity function a total of 16 times with slight changes in argument, a task that the computer can carry out quickly. If hand calculation were necessary, we would probably evaluate the functions only at the bearing and gear locations.

Combined moment is calculated from the square root of the sum of the squares of the horizontal and vertical plane moments. Figure 9.9(e) shows it to be maximum at the left gear. This location is clearly the critical location on the shaft because the gear torque applies only between the gears.

The calculations proceed as in Chapter 8; that is, first we estimate fatigue strength on the basis of ultimate strength, and then we apply corrections for surface finish, size, stress concentration, and reliability. Corrected endurance limit depends on size factor which is based on an estimate of shaft diameter. If the calculated shaft diameter is substantially different from the estimate, a second calculation should be made. The result of the first calculation is used as an improved estimate.

In this example, the bending moment dominates the design because of the fatigue effect and the high reliability requirement. The Soderberg–maximum shear criterion and the Soderberg–von Mises criterion both result in shaft diameters of about 1 in. Customary U.S. units are used in the detailed solution which follows.

Detailed Solution:

$$x = 0, 0.05, \ldots, 6$$

Singularity Function:

$$S(x,a,n) = \text{if } [x \geq a, (x - a)^n, 0] \quad \text{for } n \geq 0$$

Shear:

$$V_Y(x) = -\left(A_Y \cdot S(x,A,0) + B_Y \cdot S(x,B,0) + \sum_i Y_i \cdot S(x,X_i,0)\right)$$

See Figure 9.9(a).

$$V_Z(x) = -\left(A_Z \cdot S(x,A,0) + B_Z \cdot S(x,B,0) + \sum_i Z_i \cdot S(x,X_i,0)\right)$$

See Figure 9.9(b).

Moment:

$$M_Y(x) = -\left(A_Y \cdot S(x,A,1) + B_Y \cdot S(x,B,1) + \sum_i Y_i \cdot S(x,X_i,1)\right)$$
$$M_Y(2.5) = 1.176 \cdot 10^3$$
$$M_Z(x) = -\left(A_Z \cdot S(x,A,1) + B_Z \cdot S(x,B,1) + \sum_i Z_i \cdot S(x,X_i,1)\right)$$

Combined Moment:

$$M(x) = \sqrt{M_Y(x)^2 + M_Z(x)^2}$$
$$M(2.5) = 1.183 \cdot 10^3$$

See Figure 9.9(e).

Material Properties:

$$S_U = 140{,}000 \qquad S_{YP} = 120{,}000 \qquad N_{FS} = 1.5$$
$$C_F = 0.87 \quad \text{(ground)} \qquad D = 1 \quad \text{(estimate)}$$
$$C_S = 0.8477 - 0.1265 \cdot \ln(D) \qquad C_S = 0.848$$
$$S_{N1} = \text{if } (S_U < 200{,}000, 0.5 \cdot S_U, 100{,}000) \qquad S_{N1} = 7 \cdot 10^4$$

Stress Concentration Equivalent to D/d = *1.1,* r/d = R_K = 0.1

$K_T = 0.945 \cdot R_K^{-0.229} \qquad K_T = 1.601$

Reliability (Samples/Failure):

$$Q = 10^6$$

Estimate of Standard Deviation:

$$X_R = 5$$

Given

$$c_{\text{norm}}(X_R) = 1 - \frac{1}{10^6} \qquad SD = \text{find}(X_R) \qquad SD = 4.753$$

$$C_R(SD) = 1 - 0.08 \cdot SD \qquad C_R = C_R(SD) \qquad C_R = 0.62$$

Corrected Endurance Limit:

$$S_E = \frac{S_{N1} \cdot C_R \cdot C_F \cdot C_S}{K_T} \qquad S_E = 1.998 \cdot 10^4$$

Equivalent Static Moment, Soderberg–Maximum Shear:

$$M_{\text{EQ}} = \sqrt{\left(\frac{S_{YP}}{S_E} \cdot \text{M}(2.5)\right)^2 + T^2} \qquad M_{\text{EQ}} = 7.45 \cdot 10^3$$

Shaft Diameter:

$$D = \left(\frac{32 \cdot N_{FS} \cdot M_{\text{EQ}}}{\pi \cdot S_{YP}}\right)^{1/3} \qquad D = 0.983$$

Soderberg–von Mises:

$$M_{\text{EQ VM}} = \sqrt{\left(\frac{S_{YP}}{S_E} \cdot M(2.5)\right)^2 + 0.75 \cdot T^2} \qquad M_{\text{EQ VM}} = 7.365 \cdot 10^3$$

$$D = \left(\frac{32 \cdot N_{FS} \cdot M_{\text{EQ VM}}}{\pi \cdot S_{YP}}\right)^{1/3} \qquad D = 0.979$$

■■

GEAR TOOTH FAILURE

Gear failures result from tooth breakage, plastic flow, wear, and surface fatigue. Polishing and moderate wear may be acceptable as part of a "running in" process. Excessive wear, resulting in considerable material removal, may change the tooth profile, causing high dynamic loads and accelerating further wear. Causes of excessive wear include high loads, poor lubrication, corrosion, and the presence of abrasives. High surface stress and high-speed operation sometimes lead to overheating and lubricant breakdown. High temperature and lubricant breakdown, in turn, lead to scoring and to welding and tearing of tooth surfaces. Repeated surface stresses beyond the compressive endurance limit of material result in pitting and spalling due to surface or subsurface fatigue.

Plastic flow may occur in soft materials subject to heavy loads. Tooth breakage can result from bending fatigue failure, where tensile stresses exceed the tensile endurance limit of the material. In a typical bending fatigue failure, a crack that begins at the fillet near the base of a tooth progresses rapidly, severing the tooth.

AMERICAN GEAR MANUFACTURERS ASSOCIATION (AGMA) TECHNICAL STANDARDS

Accepted gear design procedures and rating standards are published by the American Gear Manufacturers Association (AGMA). Important standards include the following topics:

- identification and investigation of gear failure modes and a discussion of causes and failure prevention (AGMA, 1980)
- rating of the pitting resistance and the bending strength of spur and helical involute gear teeth (AGMA, 1982)
- gear materials and treatment; applications and suggested quality numbers for spur, helical, bevel, and hypoid gears, racks, and worm gearing (AGMA 1988a)
- gear classification and inspection, tolerances for spur and helical gears (AGMA 1988b)

SELECTION OF MATERIALS AND MATERIAL TREATMENT FOR GEARING

- **Safety and other considerations.** When designing gears and selecting materials for gearing, the designer must consider the consequences of possible gear failure. In some critical applications, failure could result in injury or loss of life. In addition, the designer should analyze duty cycle, ambient temperature, type of mounting and enclosure, and lubrication method.
- **Replacement gearing.** Replacement of existing gearing calls for an exercise of common sense. If the previous gearing had an inadequate service life, then the replacement gearing should have a heat treatment specification resulting in a higher hardness number. In some cases, the designer will specify a different material.
- **Uniform loading where size is not critical.** If size and weight are not important factors, pinions and gears can be made of annealed carbon steel, barstock, forgings, or castings.
- **Space and weight limitations.** The use of alloy steel pinions and gears can permit smaller face width and center distances to meet space and weight limitations.
- **High loads, impact loads, and longer life requirements.** Alloy steel pinions and gears can be used to increase service life under severe loading.
- **Speed ratio considerations.** If the gear diameter is substantially greater than that of the pinion, then the gear has a stronger tooth form, and a gear tooth has fewer bending

and wear cycles than a pinion tooth. Material choice suggestions based on the ratio of input to output speed are as follows:

Ratio	*Pinion and Gear Material*
1:1 to 2:1	Pinion and gear of the same hardness
2:1 to 8:1	Pinion hardness 40 BHN higher than gear hardness
Over 8:1	Pinion hardness more than 40 BHN higher than gear hardness

Sometimes the pinion is made of steel, and the gear of cast iron.

- **Light loads.** Stainless steel, aluminum, brass, bronze, and nonmetallic gears are often specified for cameras, control and guidance systems for aircraft and missiles, radar control systems, meters, and gages. Sometimes a stainless steel pinion is used with a gear made of a softer metal or plastic.

Quality Number

Standard tolerances on gearing are specified according to **quality number** where quality number is preceded by the letter Q on drawings and specifications. High quality numbers imply high precision (small tolerances). Consider, for example, the **index tolerance,** the allowable variation in tooth position from the theoretical or ideal position. Index tolerance is measured in ten-thousandths of an inch for a 12 in length on racks. For spur and helical racks with a normal diametral pitch of 2, index tolerance = 264 for a quality number Q3; index tolerance = 69 for Q7; and index tolerance = 7 for Q14. Standard specification nomenclature and detailed tolerance standards are given by AGMA (1988a and b).

Low quality numbers (say, Q3 through Q6) are usually selected for low-speed, noncritical applications. Where high tangential velocities are required, higher quality number gearing (say, Q7 through Q11) must be used. Gears conforming to quality numbers 12, 13, and 14 are selected only for highly specialized applications. Some of the applications and suggested quality numbers are listed in Table 9.1.

Table 9.1 Suggested quality number ranges.

Application	Quality Number*	Application	Quality Number*
Aerospace	7–13	Accelerometer	10–12
Control gearing	10–12	Antenna assembly	7–9
Loading hoists	7–11	Pressure transducer and gyroscope, computer	12–13
Small engines	12–13		
Agriculture	3–7	Engines and transmissions	8–12
Combines	5–7	Home appliances	6–10
Air compressors	10–11	Machine tool industry	8 and up
Automotive industry	10–11	Radar and missile applications	8–12
Brewing and bottling industries	6–8	Rubber and plastics industry	5–8
Cement industry	3–9	Space navigation, sextant and star tracker	13 and up
Computing machines	6–11	Steel industry, most applications	5–8
Construction equipment	3–8	Finishing rod mills	10–12
Electronic instrument control and guidance systems	7–13	High-speed rod mills	12–14

Source: Extracted from *AGMA Gear Handbook,* AGMA 390.03, with the permission of the publisher, American Gear Manufacturers Association, 1901 North Fort Myer Drive, Suite 1000, Arlington, VA 22204.

*Quality number ranges are inclusive.

DESIGN BASED ON THE BENDING STRENGTH OF GEAR TEETH

Methods of gear design discussed in the sections that follow are based largely on AGMA standards, adapted for computer use. Designers referring directly to the AGMA standards may make different decisions and may obtain different results. Gear designs based on the following methods should be considered preliminary designs. Preliminary designs should be reviewed and revised as necessary, based on

- the appropriate AGMA standard
- test results
- extrapolation of field data from similar applications

Design Factors

Basically, we will treat the gear tooth as a beam and make it strong enough so that bending stress does not exceed the strength of the material. The actual design problem is complicated by the shock and fatigue nature of tooth loading, by tooth shape and point of load application, and by other conditions affecting tooth loading and material behavior. Design factors are used in an attempt to account for these conditions.

Application Factor

Application factor K_a accounts for gear tooth loads that exceed loads based on nominal transmitted power. Designers should consider vibration, shock, and variation in system operation in selecting a value of K_a. For example, the design of gears driving a centrifugal blower, operated not more than 10 hours per day, could be based on an application factor K_a of 1 (light service). Severe service, for example, an ore crusher or a heavy-duty hoist operated more than 10 hours per day, would call for an application factor K_a of 2. In the absence of field experience, application factors can be selected from a table of service factors. If a proposed application is similar to one listed in Table 9.2, then K_a can be set equal to the corresponding service factor in that table.

Dynamic Factor

Gear teeth deflect under load. At the instant a pair of teeth is about to make contact, the tooth on the driving gear leads its theoretical position, while the tooth on the driven gear lags. The result is shock loading each time a "new" pair of teeth makes contact. Tooth errors also contribute to shock loading. Gears with lower quality numbers (corresponding to higher pitch tolerance, higher index tolerance, etc.) are subject to greater shock loading, which is accounted for by **dynamic factor** K_v.

$$K_v = \frac{50}{50 + \sqrt{v}} \quad \text{for } Q_v \le 5 \tag{9.29}$$

where Q_v = transmission accuracy level number (If transmission accuracy level number is unavailable, we will use $Q_v = Q$ = quality number.)

K_v = dynamic factor

v = pitch line velocity (ft/min) $= \pi n d/12$

n = gear rotation speed (rpm)

d = pitch diameter of the same gear (in)

Equation (9.29) can be used for pitch line velocity $v \le v_{\text{MAX}}$ where $v_{\text{MAX}} = 2500$ ft/min.

Table 9.2 Application classification for various loads.

Type of Machine to Be Driven	Service Factor — Loading: Not More than 15 Min in 2 h	Not More than 10 h per Day	More than 10 h per Day
Agitators			
Pure liquid	0.80	1.00	1.25
Semiliquids, variable-density	1.00	1.25	1.50
Blowers			
Centrifugal and vane	0.80	1.00	1.25
Lobe	1.00	1.25	1.50
Brewing and Distilling			
Bottling machinery	0.80	1.00	1.25
Brew kettles, continuous duty	—	—	1.25
Cookers, continuous duty	—	—	1.25
Mash tubs, continuous duty	—	—	1.25
Scale hopper, frequent starts	—	1.25	1.50
Can-Filling Machines	—	1.00	—
Cane Knives	—	1.50	—
Car Dumpers	—	1.75	—
Car Pullers	—	1.25	—
Clarifiers	—	1.00	1.25
Classifiers	—	1.25	1.50
Clay Working Machinery			
Brick press and briquette machine	—	1.75	2.00
Extruders and mixers	1.00	1.25	1.50
Compressors			
Centrifugal	—	1.00	1.25
Lobe, reciprocating, multicycle	—	1.25	1.50
Reciprocating, single-cycle	—	1.75	2.00
Conveyors, Uniformly Loaded and Fed			
Apron	—	1.00	1.25
Assembly belt, bucket or pan	—	1.00	1.25
Chain, flight	—	1.00	1.25
Oven, live roll, screw	—	1.00	1.25
Conveyors, Heavy-Duty, Not Uniformly Fed			
Apron		1.25	1.50
Assembly, belt, bucket or pan	—	1.25	1.50
Chain, flight	—	1.25	1.50
Live roll	—	—	—
Oven, screw	—	1.25	1.50
Reciprocating, shaker	—	1.75	2.00
Cranes and Hoists			
Main hoists			*
Bridge and trolley drive	—	1.00	1.25
Crusher			
Ore, stone	—	1.75	2.00
Sugar	—	1.50	1.50
Elevators			
Bucket, uniform load	—	1.00	1.25
Bucket, heavy load	—	1.25	1.50
Centrifugal discharge	—	1.25	1.50
Freight	—	1.25	1.50
Gravity discharge	—	1.00	1.25
Fans			
Centrifugal, light (small diam.)	—	1.00	1.25
Large industrial	—	1.25	1.50
Feeders			
Apron, belt, screw	—	1.25	1.50
Disc	—	1.00	1.25
Reciprocating	—	1.75	2.00
Food Industry			
Beef slicer	—	1.25	1.50
Cereal cooker	—	1.00	1.25
Dough mixer, meat grinder	—	1.25	1.50
Generators (not welding)	—	1.00	1.25
Hammer Mills	—	1.75	2.00
Hoists			
Heavy-duty	—	1.75	2.00
Medium-duty and skip-type	—	1.25	1.50
Laundry Tumblers	—	1.25	1.50
Line Shafts			
Uniform load	—	1.00	1.25
Heavy load	—	1.25	1.50
Machine tools			
Auxiliary drive	—	1.00	1.25
Main drive, uniform load	—	1.25	1.50
Main drive, heavy-duty	—	1.75	2.00
Metal Mills			
Draw bench carriers and main drive	—	1.25	1.50
Slitters,	—	1.25	1.50
Table Conveyors, Nonreversing			
Group drives	—	1.25	1.50
Individual drivers	—	1.75	2.00
Wire drawing, flattening, or winding	—	1.25	1.50
Mills, Rotary-Type Ball and Rod			
Spur ring gear and direct-connected	—	—	2.00
Cement kilns, pebble	—	—	1.50
Dryers and coolers	—	—	1.50
Plain and wedge bar	—	—	1.50
Tumbling barrels	—	—	2.00

Table 9.2 *continued* **Application classification for various loads.**

Type of Machine to Be Driven	Service Factor		
	Loading		
	Not More than 15 Min in 2 h	Not More than 10 h per Day	More than 10 h per Day
Mixers			
Concrete, continuous	—	1.25	1.50
Concrete, intermittent	—	1.25	1.50
Constant density	—	1.00	1.25
Semiliquid	—	1.25	1.50
Oil Industry			
Oil well pumping	—	—	*
Chillers, paraffin filter press	—	1.25	1.50
Rotary kilns	—	1.25	1.50
Paper Mills			
Agitator (mixer)	—	1.25	1.50
Agitator, pure liquids	—	1.00	1.25
Barking drums, mechanical barkers	—	1.75	2.00
Bleacher	—	1.00	1.25
Beater	—	1.25	1.50
Calender, heavy-duty	—	—	2.00
Calender, antifriction bearings	—	1.00	1.25
Cylinders	—	1.25	1.50
Chipper	—	—	2.00
Chip feeder	—	1.25	1.50
Coating rolls, couch rolls	—	1.00	1.25
Conveyors, chips, bark, chemical	—	1.00	1.25
Conveyors, log and slab	—	—	2.00
Cutter	—	—	2.00
Cylinder molds, dryers (antifriction bearing)	—	—	1.25
Felt stretcher	—	1.25	1.50
Screens, chip and rotary	—	1.25	1.50
Thickener (AC)	—	1.25	1.50
Washer (AC)	—	1.25	1.50
Winder, surface-type	—	—	1.25
Plastics Industry			
Intensive internal mixers:			
Batch-type	—	—	1.75
Continuous-type	—	—	1.50
Batch drop mill, 2 rolls	—	—	1.25
Compounding mills	—	—	1.25
Calenders	—	—	1.50
Extruder, variable-speed	—	—	1.50
Extruder, fixed-speed	—	—	1.75
Pullers			
Barge haul	—	—	2.00
Pumps			
Centrifugal	—	—	1.25
Proportioning	—	—	1.50
Reciprocating:			
Single-acting, 3 or more cycles	—	1.25	1.50
Double-acting, 2 or more cycles	—	1.25	1.50
Rotary, gear or lube	—	1.00	1.25
Rubber Industry			
Batch mixers	—	—	1.75
Continuous mixers	—	—	1.50
Calenders	—	—	1.50
Extruders, continuous	—	—	1.50
Extruders, intermittent	—	—	1.75
Tire building machines	—	—	*
Tire and tube press openers	—	—	*
Sewage Disposal Equipment			
Bar screens	—	1.00	1.25
Chemical feeders	—	1.00	1.25
Collectors	—	1.00	1.25
Dewatering screws	—	1.25	1.50
Scum breakers	—	1.25	1.50
Slow or rapid mixers	—	1.25	1.50
Thickeners	—	1.25	1.50
Vacuum filters	—	1.25	1.50
Screens			
Air washing	—	1.00	1.25
Rotary, stone or gravel	—	1.25	1.50
Traveling water intake	—	1.00	1.25
Skip Hoists	—	—	*
Slab Pushers	—	1.25	1.50
Stokers, Textile Industry	—	—	1.25
Batchers or calenders	—	1.25	1.50
Cards	—	1.25	1.50
Card machines	—	1.75	2.00
Dry cans and dryers	—	1.25	1.50
Dyeing machines	—	1.25	1.50
Looms	—	1.25	1.50
Mangles, nappers, and pads	—	1.25	1.50
Soapers, tenner frames	—	1.25	1.50
Spinners, washers, winders	—	1.25	1.50
Tumbling Barrels	1.5	1.75	2.00
Windlass	—	1.25	1.50

Source: Courtesy of Boston Gear.
Note: This list is not all-inclusive, and each application should be checked to determine if any unusual operating conditions will be encountered.
*Consult manufacturer.

Caution: Mixed units are generally to be avoided. However, expressing v in ft/min is consistent with gear design practice and standards.

Accurately cut gears have higher dynamic factors and can be operated at higher speeds. For gears with higher transmission accuracy level,

$$K_v = \left(\frac{A}{A + \sqrt{v_t}}\right)^B \quad \text{for } 6 \leq Q_v \leq 11 \tag{9.30}$$

where $A = 50 + 56(1 - B)$

$$B = \frac{(12 - Q_v)^{0.667}}{4}$$

The above equations can be used for $v \leq v_{\text{MAX}}$, where

$$v_{\text{MAX}} = [A + Q_v - 3]^2 \tag{9.31}$$

For very accurate gearing,

$$0.90 \leq K_v \leq 0.98 \quad \text{for } Q_v \geq 12 \tag{9.32}$$

Load Distribution Factor

This factor accounts for nonuniform distribution of load along the line-of-contact. Unfortunately, it depends on tooth width, and at the early design stages, we may have to estimate tooth width. We first calculate **pinion proportion factor** as follows:

$$C_{pf} = \frac{F}{10d} - 0.025 \quad \text{when } 0.5d \leq F \leq 1 \tag{9.33}$$

$$C_{pf} = \frac{F}{10d} - 0.0375 + 0.0125F \quad \text{when } 1 < F \leq 17 \tag{9.34}$$

where C_{pf} = pinion proportion factor

F = face width (in)

d = pitch diameter (in)

When $F < 0.5d$, use $F = 0.5d$ in Equation (9.33 or 9.34).

We calculate **mesh alignment factor** next. For open gearing, it is given by

$$C_{ma} = 0.247 + 0.0167F - 0.765 \cdot 10^{-4}F^2 \tag{9.35}$$

The **pinion proportion modifier** depends on pinion location relative to the center of the bearing span. It is given by

$$C_{pm} = 1 \quad \text{for } S_1/S < 0.175 \tag{9.36}$$

and

$$C_{pm} = 1.1 \quad \text{for } S_1/S \geq 0.175 \tag{9.37}$$

where S = bearing span

S_1 = pinion location measured from center of bearing span

Finally, the **load distribution factor** K_m is given by

$$K_m = 1 + C_{pf}C_{pm} + C_{ma} \quad \textbf{(9.38)}$$

Bending Strength Life Factor

Life factor adjusts allowable stress for the desired number of stress cycles. Actual service life depends on pitchline velocity and gear material ductility, fracture toughness, and residual stress. Using conservative assumptions for steel gears,

$$K_L = 1.6831N_c^{-0.0323} \quad \text{for } N_c \geq 3 \cdot 10^6 \quad \textbf{(9.39)}$$

where K_L = bending strength life factor for steel gears
N_c = number of stress cycles

The life factor for 10 million stress cycles is unity. Desired service life is often included in the application factor. When this is the case, we set $K_L = 1$.

A pinion tooth is subject to one stress cycle per revolution. For $N_c < 3 \cdot 10^6$, higher values of K_L apply, depending on gear tooth hardness and on number of cycles. However, we seldom design for short life. For example, at 2000 rpm, 3 million stress cycles correspond to only 25 hours of service.

Reliability Factor for Bending Strength

The reliability factor is based on a statistical analysis of failure data. It is given by

$$K_R = 0.5 + 0.25 \text{ lg } Q_R \quad \text{for } 100 \leq Q_R \leq 10{,}000 \quad \textbf{(9.40)}$$

where K_R = reliability factor
lg = base-ten logarithm
Q_R = number of samples per failure

For example, if $K_R = 1$, then the probability of failure is 1 in 100 (i.e., we are designing for 99% reliability). If life factor $K_L = 1$, then the design reliability applies to a life of 10 million cycles.

Temperature Factor

At gear or oil temperatures above 250°F (120°C), there may be detrimental effects on oil film or material properties. Temperatures above 350°F (175°C) may affect steel hardness and strength due to tempering. For steel gears, the suggested value of temperature factor is

$$K_T = 1 \quad \text{for } T \leq 250°\text{F} \quad \textbf{(9.41)}$$

Higher values of K_T are used for $T > 250$°F. Values depend on lubricant type and gear material and treatment.

Size Factor

Size factor is used to account for nonuniformity of material properties. It depends on tooth size; gear diameter and face width; stress pattern area; and hardenability, heat treatment, and case depth of materials. For most applications, size factor is given by

$$K_s = 1 \quad \textbf{(9.42)}$$

Higher values are used if the proportions of the gear are detrimental.

Geometry Factor for Bending Strength

If a gear has a small number of teeth, those teeth are narrow at the base, where bending moment is high. Gears with a larger number of teeth have more favorable proportions for resisting bending loads. Higher pressure angles also result in favorable proportions. These effects are taken into account in the geometry factor. The geometry factor J for bending strength of 20° full-depth spur gears is plotted in Figure 9.10(a). Part (b) shows the geometry factor for 25° full-depth spur gears.

Load Sharing

Figure 9.10(c) shows the base circles and pitch circles of a pair of gears. The line-of-action is the common tangent to the base circles. This figure applies if the pinion drives and rotates clockwise. It also applies if the gear is the driver and rotates clockwise. At this instant, contact theoretically occurs at point b, the tip of a pinion tooth, and also at point a, lower on another pinion tooth. The result is **load sharing.** Theoretically, the full tooth load is not applied

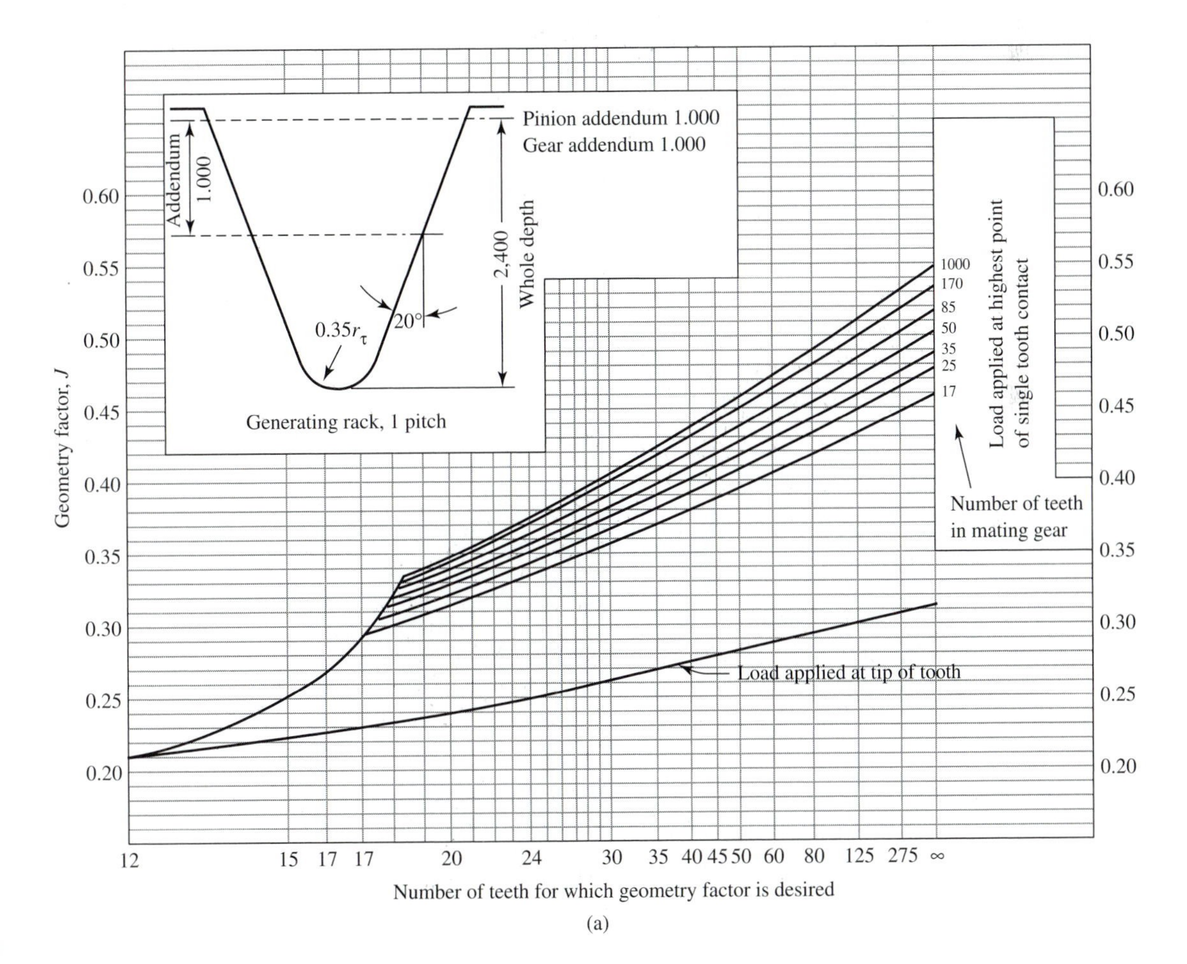

(a)

Figure 9.10 (a) Geometry factor J, 20° spur

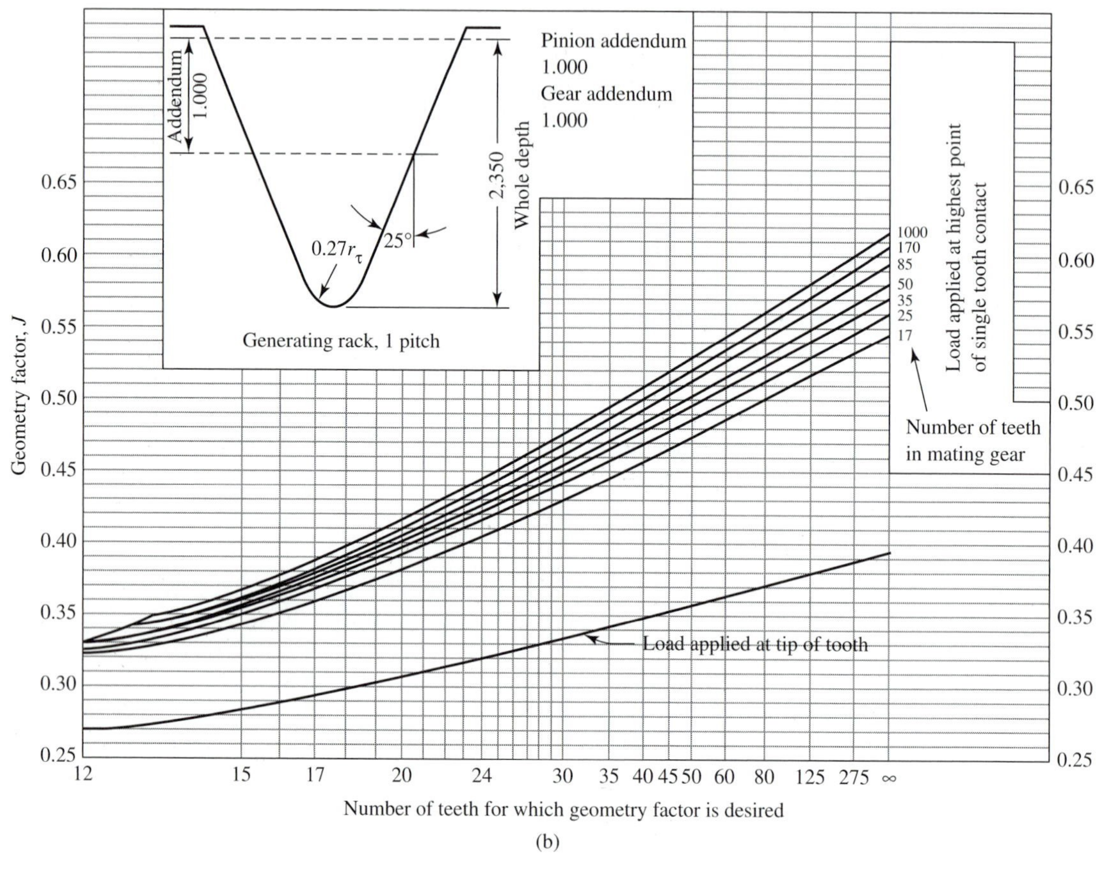

(b)

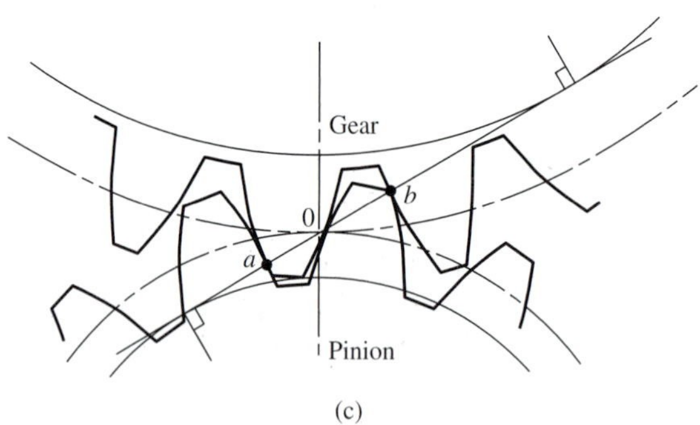

(c)

Figure 9.10 *Continued* (b) geometry factor J, 25° spur; (c) load sharing. (Extracted from *AGMA Standard for Rating the Pitting Resistance and Bending Strength of Spur and Helical Involute Gear Teeth,* AGMA 218.01, with the permission of the publisher, American Gear Manufacturers Association, 1901 North Fort Myer Drive, Suite 1000, Arlington, Virginia 22209)

to the tip of a single pinion tooth if the contact ratio exceeds 1.0. The upper curves in Parts (a) and (b) of Figure 9.10 take load sharing into account.

Real-world gears cannot be manufactured with perfect tooth form, exact pitch, zero eccentricity tolerance, and zero index tolerance. In addition, load sharing is affected by tooth deflection. Although errors based on tolerances are considered when computing dynamic factor, a conservative approach to gear design assumes that the full tooth load is applied to the tip of a single tooth. Thus, a prudent designer might select the lower curves in Parts (a) and (b) of Figure 9.10.

Analytical Approximation of Geometry Factor for Bending

The geometry factor J for bending strength can also be determined by a complicated layout procedure. An analytical expression is preferred for computer use. One approximation can be obtained by a least-squares fit to pairs of values of geometry factor and the logarithm of number of teeth. The results for 20° and 25° pressure angle full-depth spur gear teeth are

$$J = 0.0311 \ln N + 0.15 \quad (20° \text{ fd}) \qquad \textbf{(9.43)}$$

and

$$J = 0.0367 \ln N + 0.2016 \quad (25° \text{ fd}) \qquad \textbf{(9.44)}$$

where J = approximate geometry factor for bending strength (based on tip loading)

ln = natural logarithm

N = number of teeth

fd = full-depth (standard addendum) teeth

Allowable Bending Stress Number

For steel gears, allowable bending stress number is a function of hardness. The design range is bounded by

$$s_{at} \leq 6235 + 174H_B - 0.126H_B^2 \quad \text{(upper limit)} \qquad \textbf{(9.45)}$$

and

$$s_{at} \geq -274 + 167H_B - 0.152H_B^2 \quad \text{(lower limit)} \qquad \textbf{(9.46)}$$

for $180 \geq H_B \geq 400$

where s_{at} = allowable bending stress number for steel gears

H_B = Brinell hardness

Allowable bending stress number is a measure of strength. Note that a given Brinell hardness corresponds to the range of allowable bending stress numbers between the upper and lower limits. This range is based on an accumulation of gear design experience. Use of any value of allowable bending stress number within the range is considered good design practice. A value above the upper limit could result in tooth failure in bending. A value below the lower limit would result in “overdesign”; that is, the gear would be unnecessarily heavy.

AGMA standards also give a design range for allowable contact stress number, as indicated in a later section. This range represents good design practice with regard to pitting resistance or wear. In an example problem, we will arbitrarily select the conservative lower limit equations for allowable bending stress number and allowable contact stress number.

COMPUTER-AIDED DESIGN PROCEDURE BASED ON BENDING STRENGTH

The fundamental AGMA procedure involves calculating bending stress number

$$s_t = \frac{F_t K_a \cdot P_d \cdot K_s K_m}{K_v \cdot F \cdot J} \tag{9.47}$$

and comparing it with corrected allowable bending stress number, using the following criterion:

$$s_t \le \frac{s_{at} K_L}{K_T K_R} \tag{9.48}$$

If we wish to optimize the design, Equations (9.47) and (9.48) can be combined and rewritten so that the required face width is determined as a function of the other parameters. The design equation based on bending stress becomes

$$F = \frac{F_t P_d K_a K_m K_R K_s K_T}{J s_{at} K_L K_v} \tag{9.49}$$

where F = face width based on bending stress (in)
$F_t = T/r$ = tangential load (lb)
T = torque (lb · in)
r = pitch radius (in)
P_d = diametral pitch (teeth/in)
J = geometry factor for bending
s_{at} = allowable bending stress number (psi)
K_a, etc. = design factors defined on the preceding pages

If a gear and a pinion are made of different materials or have different heat treatment, then they should be designed separately, and the larger face width selected. If the materials are identical, then the geometry factor for the pinion is lower, and the pinion design governs. Then there is no need to analyze the gear.

Design of a spur-gear speed-reducer can follow these steps:

- Determine tooth numbers to satisfy speed ratio requirements. Round tooth numbers to integers.
- Estimate a reasonable diametral pitch.
- Select pressure angle and tooth form (full-depth or stub).
- Check interference and contact ratio. Change tooth numbers, pressure angle, or tooth form if necessary.

- Calculate pitch diameters.
- Calculate pitch line velocity.
- Select a quality number based on desired gear accuracy and pitch line velocity.
- Calculate dynamic factor.
- Calculate tangential load based on transmitted power and speed.
- Calculate geometry factor for bending strength. Use an analytical approximation. Assume loading at the tip of a single tooth for a conservative design. If pinion and gear material and heat treatment are identical, this calculation is necessary only for the pinion.
- Select an application factor based on anticipated use and hours-per-day service.
- Specify desired reliability and calculate reliability factor.
- Estimate face width. Calculate pinion proportion factor, mesh alignment factor, pinion proportion modifier, and load distribution factor.
- Specify desired stress cycles. Calculate bending strength life factor. Use unity if service life is considered when selecting the application factor.
- Select size and temperature factors equal to unity except for special situations.
- Specify hardness. Calculate allowable bending stress number. Use the lower limit for a conservative design.
- Calculate face width based on bending.
- Adjust the face width estimate used in calculating mesh alignment factor, and recalculate all related values if necessary.

DESIGN BASED ON PITTING RESISTANCE OF GEAR TEETH (WEAR)

The above procedure describes gear design on the basis of fracture of teeth due to bending stress. Wear is another important consideration. Corrective, nonprogressive wear is acceptable. However, high contact stresses can result in destructive pitting and progressive failure of gear teeth.

Design Factors for Pitting Resistance

Some of the wear or pitting resistance design factors are the same as the bending strength design factors. Both wear and bending failure are considered fatigue phenomena for design purposes. However, there are major differences in the two processes. Bending results in a tensile stress that is proportional to the tooth load. Wear is assumed to be proportional to (compressive) contact stresses, which are called **Hertzian stresses.** The contact stresses are, in turn, proportional to the square root of the tooth load. In this analysis, it is assumed that our design has eliminated other causes of wear, which include lubricant failures, corrosion, and the presence of abrasives.

Application Factor, Dynamic Factor, Load Distribution Factor, Reliability Factor, Temperature Factor, and Size Factor

Application factor K_a, dynamic factor K_v, load distribution factor K_m, reliability factor K_R, temperature factor K_T, and size factor K_s have the same meaning as they have in the design procedure based on bending strength. The evaluation may differ in certain cases. For example, we might select a lower value of reliability for pitting resistance. This decision could be justified if the gears were inspected frequently, so that excessive wear could be detected before failure. The sudden nature of fatigue failure due to bending would call for the higher reliability.

Pitting Resistance Life Factor

Pitting resistance life factor adjusts allowable stress for the desired number of wear cycles. Actual service life depends on pitchline velocity; lubrication method and lubricant cleanliness; and gear material ductility, fracture toughness, and residual stress. In addition, useful life depends on the definition of excess wear for a particular application and on the required smoothness of operation. For steel gears in critical applications,

$$C_L = 1.47 \quad \text{for } N_c \leq 10^4$$

and **(9.50)**

$$C_L = 2.466N_c - 0.056 \quad \text{for } N_c > 10^4$$

where C_L = pitting resistance life factor for steel gears
N_c = number of stress cycles

A pinion tooth is subject to one stress cycle per revolution. The life factor for 10 million stress cycles is unity. Gears are seldom designed for a service life of fewer than 10 million cycles. Desired service life is often included in the application factor. When this is the case, we set $C_L = 1$.

Hardness Ratio Factor

When the pinion is substantially harder than the gear, a hardness ratio factor is applied (to the gear only). It is given by

$$C_H = 1.0 + A\left(\frac{N_G}{N_P} - 1.0\right) \quad \text{for } H_{BP}/H_{BG} \leq 1.7 \tag{9.51}$$

where C_H = hardness ratio factor applied to gear
$A = 0.00898\,(H_{BP}/H_{BG}) - 0.00829$
N_P and N_G = tooth numbers
H_{BP} and H_{BG} = Brinell hardness numbers (10 mm ball @ 3000 kg load)
Subscripts P and G = pinion and gear, respectively

Elasticity Factor

Contact stress depends, in part, on the elastic coefficients of the gear and pinion. Elasticity factor is given by

$$C_E = \frac{1}{\pi\left[\left(\dfrac{1 - \nu_P^2}{E_P}\right) + \left(\dfrac{1 - \nu_G^2}{E_G}\right)\right]} \tag{9.52}$$

where C_E = elasticity factor (psi)
ν = Poisson's ratio
E = modulus of elasticity (psi)
Subscripts P and G = pinion and gear, respectively

Surface Condition Factor

Surface condition factor depends on

- surface finish as affected by the manufacturing and finishing process
- residual stress
- work hardening

If there is a detrimental surface finish effect on the teeth, a surface condition factor greater than unity is used.

Calculating the Pitting Resistance Geometry Factor

The following design factors and dimensions are used to calculate the pitting resistance geometry factor.

Ratio Factor. The pitting resistance ratio factor depends on the number of teeth in the gear and pinion. For external gears,

$$C_G = \frac{N_G}{N_G + N_P} \tag{9.53}$$

where C_G = pitting resistance ratio factor
N_G = number of gear teeth
N_P = number of pinion teeth

Curvature Factor. This design factor depends on pressure angle and ratio factor. For spur gears,

$$C_c = \frac{C_G \cos\phi \sin\phi}{2} \tag{9.54}$$

where C_c = curvature factor at the pitch line
C_G = pitting resistance ratio factor
ϕ = pressure angle

Base Circle and Addendum Circle Dimensions. The base pitch is the arc-distance between tooth centers measured along the base circle. It is given by

$$p_b = \frac{\pi \cos\phi}{P_d} \tag{9.55}$$

where p_b = base pitch
P_d = diametral pitch
ϕ = pressure angle

The base circle diameter is

$$d_b = \frac{N \cos\phi}{P_d} \tag{9.56}$$

where d_b = base circle diameter
N = number of teeth

For full-depth gears, the addendum circle diameter is

$$d_o = d + \frac{2}{P_d} \tag{9.57}$$

where d = pitch circle diameter

d_o = addendum circle diameter

Stress Point Location. These calculations give the location on the tooth profile where we determine contact stress. For spur gears, we use the lowest point of single tooth contact. The length of the pinion addendum portion of the line-of-action is

$$Z_a = 0.5\left(\sqrt{d_{PO}^2 - d_{Pb}^2} - \sqrt{d_P^2 - d_{Pb}^2}\right) \tag{9.58}$$

where subscript P = pinion

The distance along the line-of-action from the pitch point to the stress point is

$$Z_c = p_b - Z_a \tag{9.59}$$

Contact Height Factor. This factor adjusts the curvature factor for the location of the stress point. It is given by

$$C_x = \frac{R_P R_G}{R_{pP} R_{pG}} \tag{9.60}$$

where C_x = contact height factor

$R_{pP} = r_p \sin\phi$ = pinion tooth radius of curvature at pitch point

$R_{pG} = r_G \sin\phi$ = gear tooth radius of curvature at pitch point

$R_P = R_{pP} - Z_c$ = pinion tooth radius of curvature at stress point

$R_G = R_{pG} + Z_c$ = gear tooth radius of curvature at stress point for external gears

$R_G = R_{pG} - Z_c$ = gear tooth radius of curvature at stress point for internal gears

r_P = pitch radius of pinion

r_G = pitch radius of gear

Pitting Resistance Geometry Factor. This geometry factor evaluates tooth profile curvature radii for contact at the stress point. It is given by

$$I = C_c C_x \tag{9.61}$$

where I = pitting resistance geometry factor

Allowable Contact Stress Number

For steel gears, the design range for allowable contact stress number is bounded by

$$s_{ac} = 27{,}000 + 364\, H_B \quad \text{(upper limit)} \tag{9.62}$$

and

$$s_{ac} = 26{,}000 + 327\, H_B \quad \text{(lower limit)} \tag{9.63}$$

for $180 \geq H_B \geq 400$

where s_{ac} = allowable contact stress number for steel gears

H_B = Brinell hardness

COMPUTER-AIDED DESIGN PROCEDURE BASED ON PITTING RESISTANCE

The fundamental AGMA procedure involves calculating contact stress number

$$s_c = \sqrt{\frac{C_E F_t K_a C_f K_m K_s}{C_v F I d}} \tag{9.64}$$

and comparing it with corrected allowable contact stress number using the following criterion:

$$s_c \leq \frac{C_L C_H}{K_T K_R} \tag{9.65}$$

Contact stress Equations (9.64) and (9.65) can be combined and rewritten so that required face width is determined as a function of the other parameters. The design equation based on pitting resistance becomes

$$F = \frac{F_t C_E C_f K_a K_m K_R^2 K_s K_T^2}{S_{ac}^2 I C_H^2 C_L^2 K_v d_P} \tag{9.66}$$

where $F_t = T_P/r_P$ = tangential load (lb)

T_P = pinion torque (lb · in)

r_P = pinion pitch radius

d_P = pinion pitch diameter

P_d = diametral pitch (teeth/in)

I = pitting resistance geometry factor

s_{ac} = allowable contact stress number (psi)

C_E, K_a, etc. = design factors defined on the preceding pages

We now compute face width on the basis of pitting resistance, using many of the same terms used in the bending strength equations. If a gear and pinion are made of different materials or have different heat treatment, then they should be designed separately, and the larger face width selected. If the materials are identical, there is no need to analyze the gear. In that case, the pinion governs. A pinion tooth is subject to more load cycles than a gear tooth because the pinion has fewer teeth.

Evaluating the Results of the Design Based on Bending Strength and the Results of the Design Based on Pitting Resistance

The design based on bending strength and the design based on pitting resistance are compared, and the larger face width is selected. If the face width differs substantially from the value used to calculate mesh alignment factor, an adjustment and a recalculation of related values are necessary.

Gear and Pinion Weight. An approximation of the volume of a gear or pinion is given by

$$\text{Volume} \approx \frac{\pi d^2 F}{4} \tag{9.67}$$

and the approximate weight is

$$\text{Weight} \approx \frac{\pi d^2 F \gamma}{4} \tag{9.68}$$

where d = pitch diameter (in)
F = face width (in)
γ = weight density (lb/in^3)

Gear and Pinion Proportions. There are no hard-and-fast rules on gear proportions. However, the following guidelines may be helpful:

$$F \leq d_P \tag{9.69}$$

and

$$6 \leq FP_d \leq 16 \tag{9.70}$$

If face width exceeds pinion diameter, then contact may not occur over the full tooth width. The product of face width and diametral pitch falls within the indicated range for most commercially available gears. If the product is less than 6, a redesign can be considered. In that case, an increase in face width and diametral pitch could result in a more compact design.

Optimizing Speed Reducer Design. If a computer is utilized, then it is easy to optimize the design. For example, we may try to obtain minimum cost, or minimum weight, or minimum space occupied by the assembled speed reducer. We can evaluate many combinations of parameters. The effect of changes in diametral pitch, pressure angle, stub and full-depth tooth form, material, heat treatment, and so on, can be evaluated with a few keystrokes.

EXAMPLE PROBLEM Design of a Speed Reducer, Considering Bending Stress and Pitting Resistance

Design a speed reducer to meet the following requirements:

Input speed:	$n_1 = 1750$ rpm
Approximate output speed:	$n_2 \approx 440$ rpm
Power transmitted:	$P_{hp} = 3.5$

Design Decisions: We will specify some design variables and then find minimum safe tooth face width. Spur gears with 20° pressure angle full-depth teeth will be used. The gears will be made of steel with a Brinell hardness of 300. In an earlier section, we determined that an 18 tooth 20° full-depth pinion can mesh with a rack without interference. Therefore, an 18 tooth pinion is safe for this application; we will let $N_1 = 18$. Diametral pitch $P_d = 10$ will be selected. We will design for a failure probability of 1 in 5000, using an application factor K_a of 1.25 and a quality number of 5.

Solution Summary: The speed ratio equals the inverse of the tooth number ratio. This relationship gives us the number of teeth in the gear, $N_2 = 72$, using the nearest whole number. Pinion and gear diameters are 1.8 and 7.2 in, respectively. Actual output speed is $n_2 = 437.5$ rpm. We then calculate tangential load $F_t = 140$ lb and pitch line velocity $v = 825$ fpm. For this velocity, with a quality number of 5, dynamic factor $K_v = 0.635$, noting that the pitch line velocity does not exceed the limiting value.

We use an estimate of tooth width to compute load distribution factor; this factor will be recalculated if necessary. The geometry factor is approximated with an equation obtained by regression analysis. Using the conservative assumption of tip loading, we compute geometry factor for bending as $J = 0.24$. We use the more conservative equation representing the lower limit to calculate allowable bending stress number $s_{at} = 36{,}150$ psi. After determining the remaining design factors, we calculate face width. Using bending stress as the design criterion, we find that face width $F = 0.604$ in.

The next step is calculation of the design factors for pitting resistance (wear). For this potential mode of failure, we will choose the lowest point of single tooth contact as the stress point location. This choice is less conservative than the assumption of tip loading used above. However, this decision is justified if the gears are frequently inspected for excess wear. After determining the radii of the contacting teeth, we calculate pitting resistance geometry factor as $I = 0.105$. The more conservative lower limit equation for allowable contact stress number is used, resulting in the value $s_{ac} = 124{,}100$. The remaining pitting resistance design factors are calculated (some are the same as the corresponding factors based on bending). Finally, we determine face width. Using pitting resistance as a criterion, we find that the face width is $F = 1.342$ in. This value governs since it is larger than the face width based on bending. If this value had differed substantially from the estimate used when we calculated load distribution factor, we would have used the new value to revise the estimate.

The product of face width and diametral pitch is $F \cdot P_d = 13.42$, and the ratio of face width to pinion pitch diameter is $F/d_1 = 0.746$. Both of these values indicate that our design has reasonable proportions. The volume of the pinion is approximately 3.42 in^3.

The detailed solution follows.

Detailed Solution (AGMA Method):

$$\phi = 20° \text{ fd} \qquad n_1 = 1750 \qquad P_{\text{hp}} = 3.5 \qquad n_{2\text{appr}} = 440$$

$$N_1 = 18 \text{ teeth} \qquad N_2 = \text{floor}\left(0.5 + N_1 \cdot \frac{n_1}{n_{2\text{appr}}}\right) \qquad N_2 = 72$$

$$n_2 = n_1 \cdot \frac{N_1}{N_2} \qquad n_2 = 437.5 \quad \text{(ok)} \qquad \frac{n_1}{n_2} = 4$$

Select $P_d = 10$

$$d_1 = \frac{N_1}{P_d} \qquad d_1 = 1.8$$

$$r_1 = \left(\frac{d_1}{2}\right) \qquad r_1 = 0.9$$

$$d_2 = \frac{N_2}{P_d} \qquad d_2 = 7.2$$

$$c = \frac{d_1 + d_2}{2} \qquad c = 4.5$$

Torque:

$$T_1 = 63{,}025 \cdot \frac{P_{\text{hp}}}{n_1} \qquad T_1 = 126.05$$

Tangential Load:

$$F_t = \frac{T_1}{r_1} \qquad F_t = 140.056$$

Velocity:

$$v = \frac{\pi \cdot d_1 \cdot n_1}{12} \qquad v = 824.668 \text{ fpm}$$

Accuracy:

$$Q_v < 6 \quad v_{max} \le 2500$$

$$K_v = \frac{50}{50 + \sqrt{v}} \qquad K_v = 0.635$$

Geometry Factor (Bending):

For load at tip of tooth, 20° full-depth teeth,

$$J(N) = 0.0311 \cdot \ln(N) + 0.15$$
$$J(N_1) = 0.24 \qquad J(N_2) = 0.283 \qquad J = J(N_1) \qquad J = 0.24$$

Application Factor:

$$K_a = 1.25$$

Size Factor:

$$K_S = 1$$

Reliability:

$$100 \le Q \le 10{,}000 \qquad Q = 5000$$
$$K_R = 0.5 + 0.25 \cdot \log(Q) \qquad K_R = 1.425$$

Oil or Gear Temperature:

$$T_F = 210 \qquad K_T = \text{if}(T_F \le 250, 1, \infty) \qquad K_T = 1$$

Estimate:

$$F = 1.4 \qquad C_{pf} = \frac{F}{10 \cdot d_1} - 0.025 \qquad C_{pf} = 0.053$$

$$C_{pf1} = \frac{F}{10 \cdot d_1} - 0.0375 + 0.012 \cdot F \qquad C_{pf} = \text{if}(F > 1, C_{pf1}, C_{pf}) \quad C_{pf} = 0.057$$

$$C_{ma} = (0.247 + 0.0167 \cdot F) - 0.765 \cdot 10^{-4} \cdot F^2 \qquad C_{ma} = 0.27$$

$$K_m = 1 + 1.1 \cdot C_{pf} + C_{ma} \qquad K_m = 1.333$$

If K_a is used, then $K_L = 1$.

Hardness:

$$H_B = 300$$
$$s_{at} = -274 + 167 \cdot H_B - 0.152 \cdot H_B^2 \qquad s_{at} = 3.615 \cdot 10^4$$

Face Width Based on Bending:

$$F = \frac{F_t \cdot P_d \cdot K_a \cdot K_m \cdot K_R \cdot K_S \cdot K_T}{J \cdot s_{at} \cdot K_L \cdot K_v} \qquad F = 0.604$$

$$F \cdot P_d = 6.037 \qquad \frac{F}{d_1} = 0.335 \qquad \frac{\pi \cdot d_1^2 \cdot F}{4} = 1.536$$

Pitting Resistance Ratio Factor:

$$C_G = \frac{N_2}{N_1 + N_2} \qquad C_G = 0.8 \qquad \sin(\phi \cdot \text{deg}) = 0.342$$

Curvature:

$$C_c = \frac{C_G \cdot \cos(\phi \cdot \text{deg}) \cdot \sin(\phi \cdot \text{deg})}{2} \qquad C_c = 0.129$$

Base Pitch:

$$p_b - \frac{\pi \cdot \cos(\phi \cdot \text{deg})}{P_d} \qquad p_b = 0.295$$

Base Circle:

$$d_{1b} = \frac{N_1 \cdot \cos(\phi \cdot \text{deg})}{P_d} \qquad d_{1b} = 1.691$$

Full-Depth Addendum Circle:

$$d_{1o} = d_1 + \frac{2}{P_d} \qquad d_{1o} = 2$$

Stress Point:

$$Z_a = 0.5 \cdot \left(\sqrt{d_{1o}^2 - d_{1b}^2} - \sqrt{d_1^2 - d_{1b}^2}\right) \qquad Z_a = 0.226$$

$$Z_c = p_b - Z_a \qquad Z_c = 0.069$$

Curvature:

At Pitch Point *At Stress Point*

$$R_{P1} = \frac{d_1}{2} \cdot \sin(\phi \cdot \text{deg}) \qquad R_{P2} = \frac{d_2}{2} \cdot \sin(\phi \cdot \text{deg}) \qquad R_1 = R_{P1} - Z_c \qquad R_2 = R_{P2} + Z_c$$

Contact Height Factor:

$$C_x = \frac{R_1 R_2}{R_{P1} \cdot R_{P2}} \qquad C_x = 0.818$$

Pitting Resistance Geometry Factor:

$$I = C_c \cdot C_x \qquad I = 0.105$$

Elasticity Factor:

$$\nu_1 = 0.3 \qquad E_1 = 30 \cdot 10^6$$
$$\nu_2 = 0.3 \qquad E_2 = 30 \cdot 10^6$$

$$C_E = \frac{1}{\pi \cdot \left[\left(\frac{1 - \nu_1^2}{E_1}\right) + \left(\frac{1 - \nu_2^2}{E_2}\right)\right]} \qquad C_E = 5.247 \cdot 10^6$$

$$C_f = 1 \qquad C_L = 1 \qquad C_H = 1$$

Allowable Contact Stress:

$$s_{ac} = 26{,}000 + 327 \cdot H_B \qquad s_{ac} = 1.241 \cdot 10^5$$

Face Width Based on Pitting Resistance:

$$F = \frac{F_t \cdot C_E \cdot C_f \cdot K_a \cdot K_m \cdot K_R^2 \cdot K_S \cdot K_T^2}{s_{ac}^2 \cdot d_1 \cdot I \cdot C_H^2 \cdot C_L^2 \cdot K_v} \qquad F = 1.342 \quad \text{(governs)}$$

$$F \cdot P_d = 13.42 \qquad \frac{F}{d_1} = 0.746$$

$$\frac{\pi \cdot d_l^2 \cdot F}{4} = 3.415$$

HELICAL GEARS ON PARALLEL SHAFTS

We could generate a solid model of a spur gear by drawing a gear profile and extruding it in the direction of the gear shaft axis. The pitch circle of the gear profile would become the pitch cylinder. If we could twist the solid model about the shaft axis, we would form a **helical gear.**

Helical gears on parallel shafts have the following advantages. They tend to operate more quietly and smoothly than spur gears because the teeth come into contact gradually. Tooth strength and contact ratio are increased owing to the helical tooth form. Thus, the torque capacity of a pair of helical gears is greater than that of similar spur gears. However, if helical gears are selected, we must consider thrust loads when specifying bearings.

Figure 9.11(a) shows a variety of helical gears; Part(b) of the figure shows some helical gear nomenclature. The nominal diameter of a helical gear is its pitch diameter, the diameter of its pitch cylinder.

Helix Angle

Helix angle ψ is the angle that the gear teeth make with an axial element of the pitch cylinder. The gear in Figure 9.11(b) is a **left-hand** helical gear. The teeth of a left-hand helical gear lean

(a)

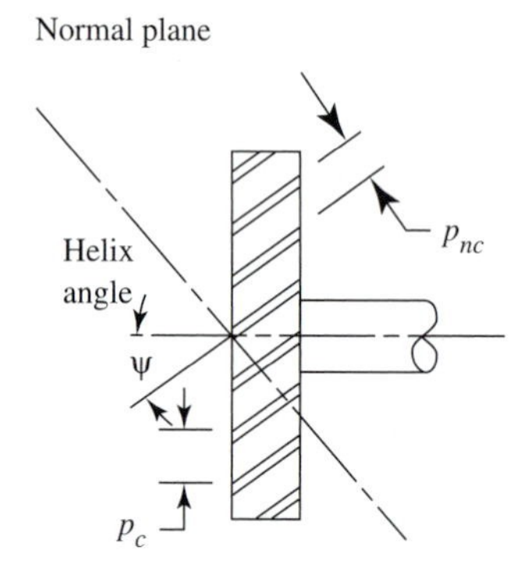

(b)

Figure 9.11 (a) Helical gears; (b) nomenclature. (Courtesy of Boston Gear)

to the left when the gear lies flat on a horizontal surface. The teeth of a **right-hand** helical gear resemble the threads of a right-hand screw. They lean to the right when the gear lies flat on a horizontal surface. Meshing helical gears on parallel shafts must have the same helix angle, but one must be right-hand and the other left-hand. Although helical gears can be manufactured in a wide range of helix angles, $\psi = 45°$ is common for standard stock helical gears.

Pitch

Circular pitch and **diametral pitch** are defined in terms of the pitch circle in the plane of rotation, as with spur gears. **Normal circular pitch** and **normal diametral pitch** are defined by

$$p_{nc} = p_c \cos \psi \tag{9.71}$$

and

$$P_{nd} = \frac{P_d}{\cos \psi} \tag{9.72}$$

where p_{nc} = normal circular pitch

$p_c = \dfrac{\pi d}{N}$ = circular pitch

d = pitch diameter

N = number of teeth

ψ = helix angle

P_{nd} = normal diametral pitch

$P_d = \dfrac{N}{d}$ = diametral pitch

Some manufacturers use the diametral pitch standard for helical gears. Such gears are typically stocked with P_d = 6, 8, 10, 12, and 24 teeth per inch of pitch diameter. Other gears are manufactured to the normal diametral pitch standard, using (typically) selected whole number values of P_{nd}.

Pressure Angle and Normal Pressure Angle

Pressure angle ϕ is defined in the plane of rotation for helical gears, just as it is with spur gears. **Normal pressure angle** is measured in the normal plane. The two are related by

$$\tan \phi_n = \tan \phi \cos \psi \tag{9.73}$$

where ϕ_n = normal pressure angle

ϕ = pressure angle

ψ = helix angle

Some manufacturers use the normal pressure angle standard for stock gears, and others standardize on pressure angle in the plane of rotation. The pressure angle in the plane of rotation is larger than the normal pressure angle. For example, if $\phi_n = 14.5°$ and $\psi = 45°$, then $\phi = 20.09°$.

Interference

The possibility of interference limits the minimum number of pinion teeth possible (without undercutting). Interference calculations are based on dimensions in the plane of rotation for helical gears, just as with spur gears. Recall that the use of larger pressure angles reduces the minimum permissible number of teeth. Note that a helical gear with a normal pressure angle of 14.5° can be made with fewer teeth than a spur gear with the same pressure angle. Thus, the designer has more flexibility and may be able to make a more compact drive.

Center Distance

Noting the relationship between diametral pitch and normal diametral pitch, we see that the pitch diameter of a helical gear is given by

$$d = \frac{N}{P_{nd}\cos\psi} \tag{9.74}$$

For a pair of helical gears on parallel shafts, the distance between shaft centers is

$$c = \frac{d_1 + d_2}{2} = \frac{N_1 + N_2}{2P_d} = \frac{N_1 + N_2}{2P_{nd}\cos\psi} \tag{9.75}$$

A wide range of helix angles is possible if a rack-type hob is used to generate helical gears. Such gears would be made to the normal diametral pitch standard. Then, as indicated by Equation (9.75), center distance is related to helix angle.

Contact Ratio

Contact ratio is the average number of pairs of teeth in contact in a gear pair. Axial overlap contributes to contact ratio in helical gears. It is given by

$$CR_x = \frac{F}{p_x} = \frac{F\tan\psi}{p_c} = \frac{FP_d\tan\psi}{\pi} \tag{9.76}$$

where CR_x = axial overlap

F = face width (gear blank thickness)

p_x = axial pitch = $p_c/\tan\psi$

p_c = circular pitch (in plane of rotation)

P_d = diametral pitch (in plane of rotation)

Contact ratio for helical gears on parallel shafts is given by

$$CR = CR_x + CR_{\text{spur}} \tag{9.77}$$

where CR_{spur} = contact ratio as determined for a pair of spur gears with the same dimensions in the plane of rotation

Speed Ratio

As with spur gears, the speed ratio of a pair of helical gears on parallel shafts is

$$\frac{n_2}{n_1} = -\frac{N_1}{N_2} = -\frac{d_1}{d_2} \tag{9.78}$$

where n = speed

N = number of teeth

d = diameter

The sign indicates that the shafts turn in opposite directions unless one of the gears is an internal gear.

In general, the output-to-input speed ratio for all nonplanetary gear trains is

$$\left|\frac{n_{\text{out}}}{n_{\text{in}}}\right| = \frac{\text{product of driving gear teeth}}{\text{product of driven gear teeth}} \tag{9.79}$$

Equivalent Number of Teeth

Helical gear teeth are stronger than the teeth of a similar spur gear. We account for the difference by basing the strength of helical gear teeth on an equivalent number of teeth. The equivalent or virtual number of teeth is given by

$$N_e = \frac{N}{\cos^3\psi} \tag{9.80}$$

where N_e = equivalent number of teeth
N = actual number of teeth

Consider a 20 tooth helical gear with a 45° helix angle. The equivalent number of teeth is

$$N_e = \frac{20}{\cos^3 45°} = 56.6$$

The gear has a tooth form equivalent to that of a spur gear with 56 or 57 teeth. The equivalent number of teeth has no effect on speed ratio.

Tangential, Radial, and Thrust Loads

The resultant force whereby one helical gear transmits power to another is separated into three perpendicular components: the tangential, the radial, and the thrust force components. Figure 9.12, an exploded view of a pair of helical gears, shows how these force components act on each gear. Shaft reactions are not shown on the sketch. The equations for tangential and radial load on spur gear teeth apply to helical gears as well. The helical tooth orientation

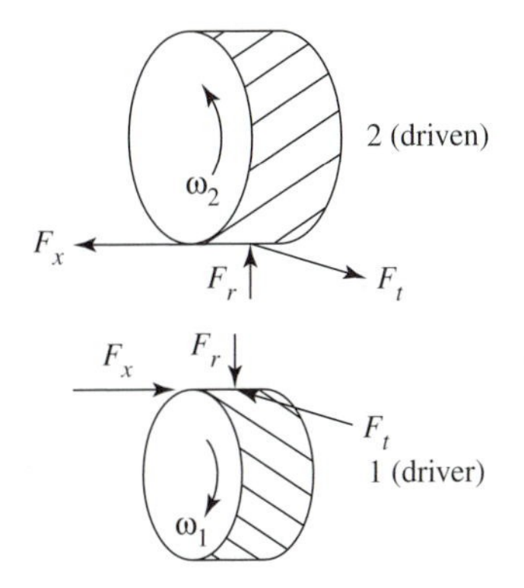

Figure 9.12 Tooth loading on helical gears.

produces an additional force, a thrust proportional to the tangent of the helix angle. The force components on the gear teeth are given by

$$F_t = \frac{T}{r} \qquad F_r = F_t \tan \phi \qquad F_x = F_t \tan \psi \tag{9.81}$$

where F_t = tangential force
T = torque on one of the gears
r = pitch radius (of the same gear)
F_r = radial force
ϕ = pressure angle (in plane of rotation)
F_x = thrust force
ψ = helix angle

Tooth Load Directions for Helical Gears on Parallel Shafts. Tangential force is tangent to the pitch circle and lies in the plane of rotation. It opposes the motion of the driving gear. For the gear pair shown in Figure 9.12, the driver turns clockwise, and the tangential force direction is inward. Tangential force on the driven gear is equal and opposite. Radial force direction is toward the center of a gear except for internal gears. Radial forces on the driver and driven gear are equal and opposite.

Thrust direction is given by treating the driving gear as a screw. A right-hand screw turned clockwise will advance. The (right-hand) driving gear in Figure 9.12 turns clockwise. The resistance of the driven gear causes a thrust force to the right on the driver. The thrust on the driven gear is equal and opposite.

Balancing Thrust Forces. Figure 9.13 is an exploded view of a **reverted gear train** speed reducer. Gear 1 drives, and gear 4 is on the output shaft. When assembled, the input and out-

Figure 9.13 Thrust balance.

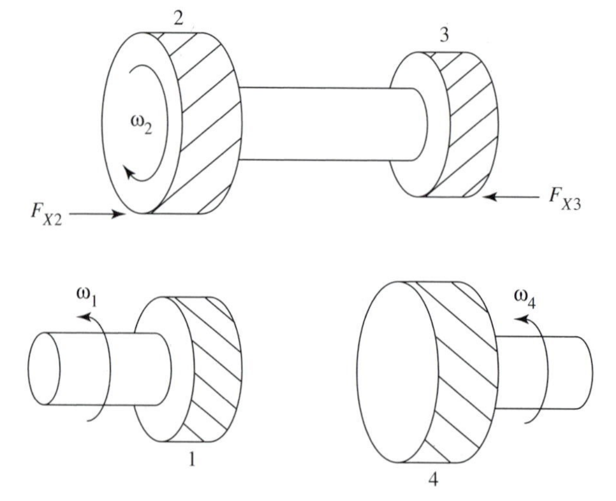

put shafts are collinear. Gears 1 and 2 must have equal helix angles of opposite hand. The tangential force on gear 1 is

$$F_{t1} = T_1/r_1 \quad \text{(outward)}$$

where T_1 = input torque
r_1 = pitch radius

Number subscripts refer to gear numbers.

Tangential force F_{t2} is inward with the same magnitude as F_{t1}. Countershaft torque is

$$T_2 = F_{t2}r_2$$

The same torque is applied to gear 3, and the tangential force on gear 3 is

$$F_{t3} = T_2/r_3 \quad \text{(outward)}$$

The tangential force on gear 4 is equal that on gear 3, but inward. Radial forces, which are directed toward the center of each gear, can be found from

$$F_r = F_t \tan \phi$$

Output torque is given by

$$T_4 = F_{t4}r_4$$

The thrust forces on the countershaft gears are

$$F_{x2} = F_{t2} \tan \psi_2$$

and

$$F_{x3} = F_{t3} \tan \psi_3$$

Note that tangential forces on gears 2 and 3 on the countershaft are inversely proportional to pitch radius. If we wish to balance countershaft thrust forces, the helix angles should be related by

$$\frac{\tan \psi_3}{\tan \psi_2} = \frac{r_3}{r_2} \tag{9.82}$$

where ψ = helix angle
r = pitch radius

Gears 2 and 3 are on the same shaft and have the same hand.

We must consider unbalanced thrust loads on the input and output shafts when selecting bearings.

If helical gears are shifted axially to produce different speed ratios, a helical spline is used. The hand of the spline is the same as the hand of the gear. Equation (9.82) can be used to relate the spline helix angle and the gear helix angle. If the equation is satisfied, then the thrust force at the gear teeth is balanced by thrust forces at the spline teeth. There is no tendency for the gear to slide along the shaft due to tooth loads.

EXAMPLE PROBLEM Design of a Reverted Helical Gear Reducer with Countershaft Thrust Balance

Speed is to be reduced from 2400 rpm to approximately 260 rpm; 3 hp will be transmitted. Input and output shafts should be collinear. Design the speed reducer.

Design Decisions: A reverted helical gear train will be used (similar to that in Figure 9.13). The diametral pitch of gears 1 and 2 will be $P_{d1} = 8$, and the diametral pitch of gears 3 and 4 will be $P_{d3} = 6$ (in the plane of rotation). The normal pressure angle will be $\phi_n = 14.5°$ for all gears. We will select a helix angle $\psi_1 = 45°$ for gears 1 and 2. We will try gear tooth numbers $N_1 = N_3 = 18$ and a range of values of N_2 and N_4. Countershaft thrust forces will be balanced.

Restraints: We are limited to whole numbers of gear teeth. In addition, the center distance between the input shaft and the countershaft must equal the center distance between the countershaft and the output shaft.

Solution Summary: One possible design procedure is described below. Other approaches can also produce a satisfactory design. Considering the center distance restraint, we have

$$c = \frac{N_1 + N_2}{2P_{d1}} = \frac{N_3 + N_4}{2P_{d3}}$$

from which

$$N_4 = (N_1 + N_2) \cdot \frac{P_{d3}}{P_{d1}} - N_3$$

Certain values of N_2 must be selected so that N_4 is a whole number. We will try $N_2 = 26, 30, \ldots, 78$. Since N_2 is varied, output values that depend on it are so labeled [e.g., $N_4(N_2)$, etc.]. We then calculate the pitch radii of the gears and check to see that the center distance requirement is met.

Shaft speeds, torques, and tooth load components are calculated. Note the following:

- Tooth load components on gear 2 are equal and opposite those on gear 1.
- Tooth load components on gear 4 are equal and opposite those on gear 3.
- Torque is balanced on the countershaft.
- Thrust can be balanced on the countershaft if gears 2 and 3 have the same hand and if the tangents of the helix angles for gears 2 and 3 are proportional to the gear radii.

The results for the selected values of tooth numbers are given in Table 9.3. We see that 66 and 45 teeth on gears 2 and 4, respectively, result in output speed n_4 of 261.8 rpm, which is acceptable. The center distance requirement is met, and the thrust force on gear 2 is balanced by the thrust force on gear 3. Gear face width depends on application, duty cycle, and material selection (AGMA, 1982).

The detailed solution follows.

Detailed Solution:

Transmitted Power:

$$P_{\text{hp}} = 3$$

Input Speed:

$$n_1 = 2400$$

Table 9.3 **Results for selected values of tooth numbers.**

N_2	$N_4(N_2)$	$r_2(N_2)$	$r_4(N_2)$	$n_2(N_2)$	$n_4(N_2)$	$c(N_2)$	$C(N_2)$	$F_{t3}(N_2)$	$T_4(N_2)$	$\psi_{3deg}(N_2)$	$F_{x3}(N_2)$	$\phi_{3deg}(N_2)$	$F_{r3}(N_2)$	
26	15	1.625	1.25	1661.538	1993.846	2.75	2.75	75.863	94.829	42.709	70.028	19.39	26.7	
30	18	1.875	1.5	1440	1440	3	3	87.535	131.302	38.66	70.028	18.324	28.991	
34	21	2.125	1.75	1270.588	1089.076	3.25	3.25	99.206	173.611	35.218	70.028	17.566	31.404	
38	24	2.375	2	1136.842	852.632	3.5	3.5	110.877	221.755	32.276	70.028	17.008	33.915	
42	27	2.625	2.25	1028.571	685.714	3.75	3.75	122.549	275.734	29.745	70.028	16.587	36.503	
46	30	2.875	2.5	939.13	563.478	4	4	134.22	335.55	27.553	70.028	16.262	39.152	
50	33	3.125	2.75	864	471.273	4.25	4.25	145.891	401.201	25.641	70.028	16.006	41.851	
54	36	3.375	3	800	400	4.5	4.5	157.562	472.687	23.962	70.028	15.802	44.592	
58	39	3.625	3.25	744.828	343.767	4.75	4.75	169.234	550.01	22.479	70.028	15.636	47.366	
62	42	3.875	3.5	696.774	298.618	5	5	180.905	633.168	21.161	70.028	15.5	50.168	
66	45	4.125	3.75	654.545	261.818	5.25	5.25	192.576	722.161	19.983	70.028	15.386	52.994	**Final Design**
70	48	4.375	4	617.143	231.429	5.5	5.5	204.248	816.991	18.925	70.028	15.291	55.84	
74	51	4.625	4.25	583.784	206.041	5.75	5.75	215.919	917.656	17.969	70.028	15.21	58.704	
78	54	4.875	4.5	553.846	184.615	6	6	227.59	1024.156	17.103	70.028	15.141	61.582	

Diametral Pitch:

Gears 1 and 2

$$P_{d1} = 8$$

Gears 3 and 4

$$P_{d3} = 6$$

Helix Angle (Gears 1 and 2):

$$\psi_1 = 45°$$

Gear 1: Right-hand   Gear 2: Opposite hand

Normal Pressure Angle, All Gears:

$$\phi_n = 14.5$$

Pressure Angle, Gears 1 and 2:

$$\phi_1 = \operatorname{atan}\left(\frac{\tan(\phi_n \cdot \mathrm{deg})}{\cos(\psi_1 \cdot \mathrm{deg})}\right) \qquad \phi_1 = 20.09 \cdot \mathrm{deg}$$

Tooth Numbers:

$$N_1 = 18 \qquad N_2 = 26, 30, \ldots, 78 \qquad N_3 = 18$$

Center Distance Requirement:

$$N_4(N_2) = \frac{P_{d3}}{P_{d1}} \cdot (N_1 + N_2) - N_3$$

Pitch Radii:

$$r_1 = \frac{N_1}{2 \cdot P_{d1}} \qquad r_1 = 1.125 \qquad r_2(N_2) = \frac{N_2}{2 \cdot P_{d1}}$$

$$r_3 = \frac{N_3}{2 \cdot P_{d3}} \qquad r_3 = 1.5$$

$$r_4(N_2) = \frac{N_4(N_2)}{2 \cdot P_{d3}}$$

Center Distance:

Input–Countershaft	*Countershaft–Output*
$c(N_2) = r_1 + r_2(N_2)$	$C(N_2) = r_3 + r_4(N_2)$

Countershaft rpm:

$$n_2(N_2) = n_1 \cdot \frac{N_1}{N_2}$$

Output rpm:

$$n_4(N_2) = \frac{n_1 \cdot N_1 \cdot N_3}{N_2 \cdot N_4(N_2)}$$

Input Torque (lb · in):

$$T_1 = \frac{63{,}025 \cdot P_{hp}}{n_1} \qquad T_1 = 78.781$$

Tangential Force, Gears 1 and 2 (lb):

$$F_{t1} = \frac{T_1}{r_1} \qquad F_{t1} = 70.028$$

Radial Force, Gears 1 and 2 (lb):

$$F_{r1} = F_{t1} \cdot \tan(\phi_1) \qquad F_{r1} = 25.612$$

Thrust, Gears 1 and 2 (lb):

$$F_{x1} = F_{t1} \cdot \tan(\psi_1 \cdot \mathrm{deg}) \qquad F_{x1} = 70.028$$

Countershaft Torque:

$$T_2(N_2) = F_{t1} \cdot r_2(N_2)$$

Tangential Force, Gears 3 and 4:

$$F_{t3}(N_2) = \frac{T_2(N_2)}{r_3}$$

Output Torque:

$$T_4(N_2) = F_{t3}(N_2) \cdot r_4(N_2)$$

Helix Angle, Gears 3 and 4:
For thrust balance,

Gear 3: same hand as 2 Gear 4: opposite hand

$$\psi_3(N_2) = \mathrm{atan}\left(\tan(\psi_1 \cdot \mathrm{deg}) \cdot \frac{r_3}{r_2(N_2)}\right)$$

Gear 3 Thrust (Opposes Gear 2):

$$F_{x3}(N_2) = F_{t3}(N_2) \cdot \tan(\psi_3(N_2))$$

Pressure Angle, Gears 3 and 4:

$$\phi_3(N_2) = \mathrm{atan}\left(\frac{\tan(\phi_n \cdot \mathrm{deg})}{\cos(\psi_3(N_2))}\right)$$

Radial Force, Gears 3 and 4:

$$F_{r3}(N_2) = F_{t3}(N_2) \cdot \tan(\phi_3(N_2))$$

Angles (Degrees):

$$\psi_{3\mathrm{deg}}(N_2) = \psi_3(N_2) \cdot \frac{180}{\pi} \qquad \phi_{3\mathrm{deg}}(N_2) = \phi_3(N_2) \cdot \frac{180}{\pi}$$

■■

HELICAL GEARS ON CROSSED SHAFTS

When spur gears or helical gears on parallel shafts are transmitting power, the stress pattern is equivalent to two parallel cylinders in contact. We call this **line contact.** Because of elastic deflection, the actual line-of-contact has a width dimension; it is not a *mathematical line.*

Helical gears can be operated on parallel or nonparallel shafts. Figure 9.14 shows several types of stock gears, some for operation on parallel shafts and some for operation on nonparallel shafts. A helical gear pair near the center of the figure has its axes at right angles. Both gears are stock left-hand gears with 45° helix angles, oriented as **crossed helical gears.**

The stress pattern of crossed helical gears is equivalent to two nonparallel contacting cylinders. We call this **point contact.** Unlike the *mathematical point,* the contact point of crossed helical gears has nonzero area owing to elastic deflection. Point contact causes higher stresses than line contact for the same transmitted load. As a result, crossed helical gears have lower capacity than similar helical gears on parallel shafts. In many cases, careful attention to lubrication is required because of high sliding velocities of crossed helical gear teeth.

Shaft Angle

Crossed helical gears can be used to transmit power between nonintersecting shafts at any angle. Usually, both gears in a crossed helical gear drive have the same hand, but the helix angles need not be equal. Shaft angle and helix angles are related by

$$\psi_{\text{shaft}} = \psi_1 + \psi_2 \tag{9.83}$$

Figure 9.14 Gears for operation on parallel and nonparallel shafts. (Courtesy of Boston Gear)

where ψ_{shaft} = shaft angle

ψ_1 and ψ_2 = helix angles (if both have the same hand)

If the shaft angle is small, the gears may have opposite hands. In that case, one of the helix angles in Equation (9.83) is given a negative sign.

Center Distance

The distance between shaft centers is given by

$$c = r_1 + r_2 = \frac{\dfrac{N_1}{\cos \psi_1} + \dfrac{N_2}{\cos \psi_2}}{2P_{nd}} \qquad \textbf{(9.84)}$$

where c = center distance

r = pitch radii

N = tooth numbers

ψ = helix angle

P_{nd} = normal diametral pitch

The normal diametral pitch must be the same for both gears. If the helix angles are unequal, the diametral pitch in the plane of rotation for one gear will not equal that of the other.

Speed Ratio

The speed ratio is given by

$$\frac{n_2}{n_1} = \frac{N_1}{N_2} = \frac{d_1 \cos \psi_1}{d_2 \cos \psi_2} \qquad \textbf{(9.85)}$$

where n = rotation speeds

N = tooth numbers

d = diameters

ψ = helix angles

WORM DRIVES ON PERPENDICULAR SHAFTS

Ordinary helical gears are seldom used to transmit power between crossed shafts, mainly because of the limited load capacity of crossed helical gears. However, when a large speed reduction is required, a worm drive is often selected. Worm drives are a special case of helical gears on crossed shafts.

Several worm drive elements are shown in Figure 9.15. One gear, the **worm,** resembles a screw and usually has one to four teeth. The number of teeth (or threads) is the number of starts, which we can count by looking at one end of the worm. The other gear in a worm drive is called a **worm gear** or **worm wheel.** A single-tooth worm driving a large worm gear can produce low output speeds and high output torque.

Figure 9.15 Worm drive elements. (Courtesy of Boston Gear)

Worm drives transmit power between nonintersecting crossed shafts. Although worm drives can be designed for shaft angles other than 90°, such designs are rare. We will consider only 90° shaft angles in the material that follows. Figure 9.16(a) shows a right-hand worm; Part(b) of the figure shows a worm gear. Note that the worm-gear teeth are curved to increase the contact surface with the worm. A close-up of tooth mesh is shown in Part(c) of the figure. If the worm addendum forms a cylindrical surface, and if the worm-gear teeth are curved, then the drive is called **single-enveloping.** If the worm addendum has an hourglass form, and if the worm-gear teeth are also curved, then the drive is **double-enveloping.**

Lead Angle

The helix angle of a worm gear is measured from the axial direction. Instead of identifying the helix angle of a worm, we identify the **lead angle,** which is the orientation of the worm teeth with respect to the plane of rotation. For a 90° shaft angle,

$$\lambda = \psi_G \tag{9.86}$$

where λ = lead angle of the worm

ψ_G = helix angle of the gear

The worm and gear have the same hand.

Speed Ratio

The speed ratio is

$$\frac{n_G}{n_W} = \frac{N_W}{N_G} = \frac{d_w \sin \lambda}{d_G \cos \psi_G} = \frac{d_w}{d_G} \cdot \tan \lambda \tag{9.87}$$

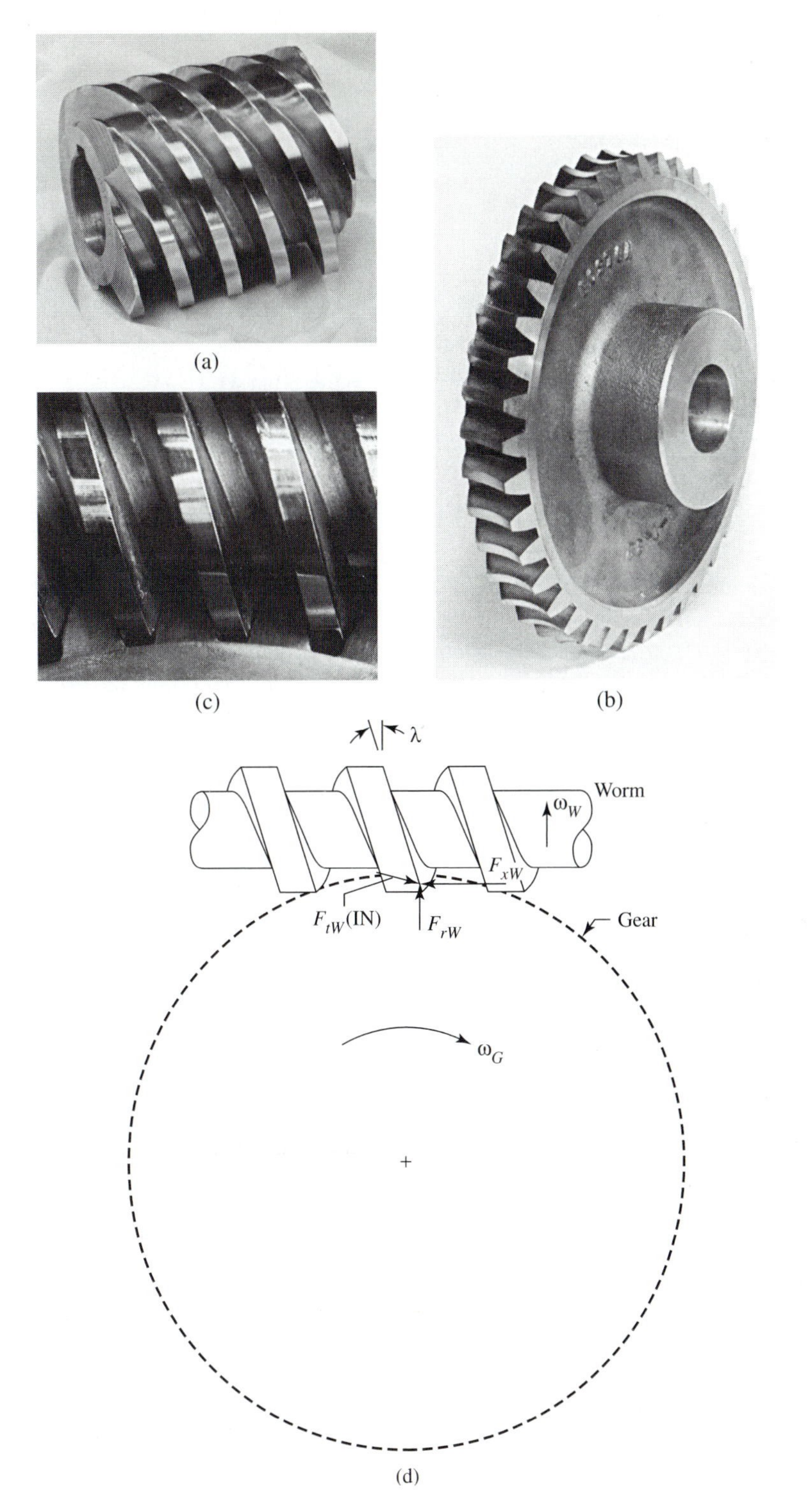

Figure 9.16 (a) Right-hand worm; (b) worm gear; (c) close-up of tooth mesh; (d) forces on a worm tooth. (Photographs courtesy of Boston Gear)

where n = speed

N = tooth number

d = pitch diameter

Worm Drive Efficiency

High sliding velocities are typical of worm drives. As a result, drive efficiency is sometimes a concern. Efficiency can be estimated from

$$\eta = \frac{\tan\lambda(1 - f\tan\lambda)}{f + \tan\lambda} \quad \textbf{(9.88)}$$

where η = efficiency of the drive

λ = lead angle of the worm

f = coefficient of friction

The coefficient of friction, which depends on gear materials, sliding velocity, and lubrication, typically ranges from 0.03 to 0.12. For a hardened steel worm driving a bronze worm gear, $f = 0.03$ to 0.05 is reasonable if lubrication is adequate. If a significant amount of power is lost to friction, it may be necessary to design a system for cooling the drive.

Self-Locking and Overrunning Drives

Almost all worm drives are designed to reduce speed; the worm drives the gear. The drive is called **self-locking** if the worm gear *cannot* drive the worm. If

$$f > \tan\lambda$$

then the drive is self-locking; friction force exceeds the driving force when we try to make the worm gear the driver. If there is a loss of power to the worm in a self-locking drive, then the drive acts as a brake.

If the worm gear *can* act as driver, then we have an **overrunning** or **backdriving** gearset. The theoretical criterion for an overrunning worm drive is

$$f < \tan\lambda$$

The practical situation is more complicated than the above criterion because coefficient of friction may vary. Vibration may have the effect of reducing coefficient of friction, while friction in other parts of the drive increases the tendency toward self-locking. Worm drives can usually be expected to be self-locking if $\lambda < 5°$. The experience of one manufacturer indicates that for a hardened worm and bronze gear properly manufactured, mounted, and lubricated, the drive will usually be overrunning if $\lambda > 11°$ (*Boston Gears,* 1992).

Tangential, Radial, and Thrust Loads for Worm Drives on Perpendicular Shafts

Worm drives, of course, follow the basic laws of physics, including Newton's third law: *When one body exerts a force on another, the second exerts an equal and opposite force on the first.* However, the tangential direction on a worm corresponds to axial (thrust) direction on the worm gear, and vice versa.

Figure 9.16(d) shows a right-hand worm driving a right-hand helical gear (dashed). The shafts are perpendicular, and the worm and gear rotate as shown. The gear applies the following reaction forces to a worm tooth:

Tangential force:	F_{tW}	inward
Radial force:	F_{rW}	upward
Thrust force:	F_{xW}	to left

The forces that the worm applies to a gear tooth are not shown in the figure. They are as follows:

Tangential force:	F_{tG}	to right
Radial force:	F_{rG}	downward
Thrust force:	F_{xG}	outward

The worm and gear forces can be calculated from the following equations:

$$F_{tW} = \frac{T_W}{r_W} \qquad F_{rW} = F_{tW} \tan \phi_W \qquad F_{xW} = \frac{F_{tW}}{\tan \lambda}$$

$$F_{tG} = -F_{xW} \qquad F_{rG} = -F_{rW} \qquad F_{xG} = -F_{tW} \tag{9.89}$$

where F_t = tangential force
$T_W = 63{,}025 P_{hp}/n_W$ = worm torque
P_{hp} = horsepower input to worm
n_W = rpm of worm
r = pitch radius
F_r = radial force
ϕ = pressure angle (in plane of rotation)
F_x = thrust force
λ = lead angle of the worm
Subscripts W and G = worm and worm gear, respectively

BEVEL GEARS

Many applications require power transmission between shafts with intersecting centerlines. An outboard motor, for example, may have a vertical driveshaft and a horizontal propeller shaft. Bevel gears are often selected for such applications. Bevel gears are also commonly used in automotive differentials.

Figure 9.17(a) shows a pair of straight bevel gears; Part(b) of the figure shows commonly used bevel gear nomenclature. The **pitch cone** of a bevel gear is analogous to the pitch circle or pitch cylinder of a spur gear. The nominal diameter is the pitch diameter measured on the pitch cone at the outer edge of a tooth. The **pitch angle** is measured from the shaft centerline to the pitch cone. The angle at which the gear shafts intersect is given by

$$\sum = \gamma_P + \gamma_G \tag{9.90}$$

where $\sum$ = shaft angle
γ_P and γ_G = pitch angles of the pinion and gear, respectively

Figure 9.17 (a) Bevel gears; (b) bevel gear nomenclature. (Courtesy of Boston Gear)

(a)

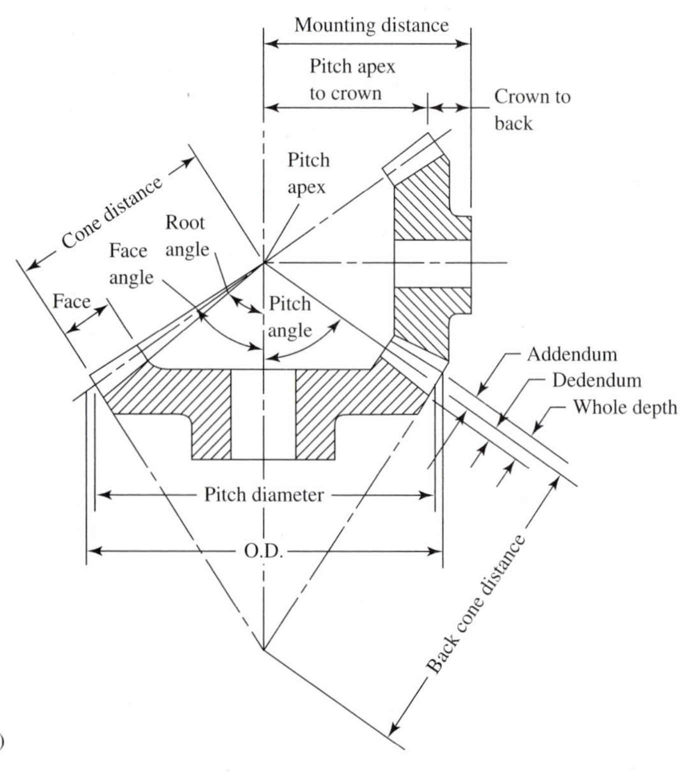

(b)

Miter gears have 45° pitch angles and are designed to operate in pairs, with a shaft angle of 90°.

Speed Ratio

The speed ratio for bevel gears is given by

$$\frac{n_G}{n_P} = \frac{N_P}{N_G} = \frac{d_P}{d_G} = \frac{\sin \gamma_P}{\sin \gamma_G} \tag{9.91}$$

where n = speed

N = tooth number

d = pitch diameter

γ = pitch angle

The speed ratio of a miter gear pair is 1:1, and the tooth numbers are equal. These speed ratio relationships do not apply to bevel gear differentials, which must be treated as planetary trains.

Most bevel gear pairs operate on perpendicular shafts. In that case, the pitch angles can be determined from tooth numbers or speed ratio as follows:

$$\gamma_P = \arctan\left(\frac{N_P}{N_G}\right) = \arctan\left(\frac{n_G}{n_P}\right)$$

and **(9.92)**

$$\gamma_G = \arctan\left(\frac{N_G}{N_P}\right) = \arctan\left(\frac{n_P}{n_G}\right)$$

where $\sum = 90° =$ shaft angle

The **mounting distance** is critical since we must attempt to have the pitch cone apexes touch at the intersection of the shaft centerlines. Accurately mounted bevel gears will operate smoothly with line contact. Substantial error in mounting distance can cause binding and excessive wear or tooth breakage. If the mounting distance is significantly greater than the ideal dimension, then the contact area will be reduced, and backlash will be excessive. Excessive backlash causes shock loading when starting and reversing, and it may cause tooth damage.

Tooth Loading of Bevel Gears on Perpendicular Shafts

Power and Torque. As with other gears, torque and horsepower are related by

$$T_P = \frac{63{,}025 P_{\text{hp}}}{n_P} \tag{9.93}$$

where T_P = pinion torque (lb · in)

P_{hp} = transmitted power (hp)

n_P = pinion speed (rpm)

Tangential Force. The resultant tangential force at the approximate tooth center on the pitch cone is

$$F_t = \frac{T_P}{r_{mP}} \tag{9.94}$$

where F_t = tangential force

r_{mP} = mean pinion radius = $(d_P - F \sin \gamma_P)/2$

F = face width

The tangential force on the driving gear opposes its motion. The tangential force on the driven gear is equal and opposite. Torque on the gear is given by

$$T_G = F_t r_{mG} \tag{9.95}$$

where r_{mG} = mean gear radius = $(d_G - F \sin \gamma_G)/2$

Radial Force and Thrust Force. Radial and thrust (axial) forces depend on pressure angle and pitch angle. For the pinion,

$$F_{rP} = F_t \tan \phi \cos \gamma_P \tag{9.96}$$

and

$$F_{xP} = F_t \tan \phi \sin \gamma_P \tag{9.97}$$

where F_{rP} = radial force on the pinion

F_{xP} = thrust force on the pinion

For bevel gears on perpendicular shafts, the radial force on the gear is equal and opposite the thrust force on the pinion, and vice versa; that is,

$$F_{rG} = -F_{xP} \tag{9.98}$$

and

$$F_{xG} = -F_{rP} \tag{9.99}$$

The radial forces are toward the center of the gear and toward the center of the pinion. Thrust forces are directed away from the shaft intersection point.

PLANETARY GEAR TRAINS

Planetary or epicyclic gear trains are available in a wide range of configurations. A common design, which we will call a **simple planetary train,** is sketched in Figure 9.18. It includes a sun gear (*S*), three or four planet gears (*P*), a ring or internal gear (*R*), and a planet carrier (*C*). The planets are mounted on shafts in the carrier before assembly. Either spur gears or helical gears can be used in this type of train.

From the geometry of the simple planetary train, we see that

$$r_C = r_S + r_P \tag{9.100}$$

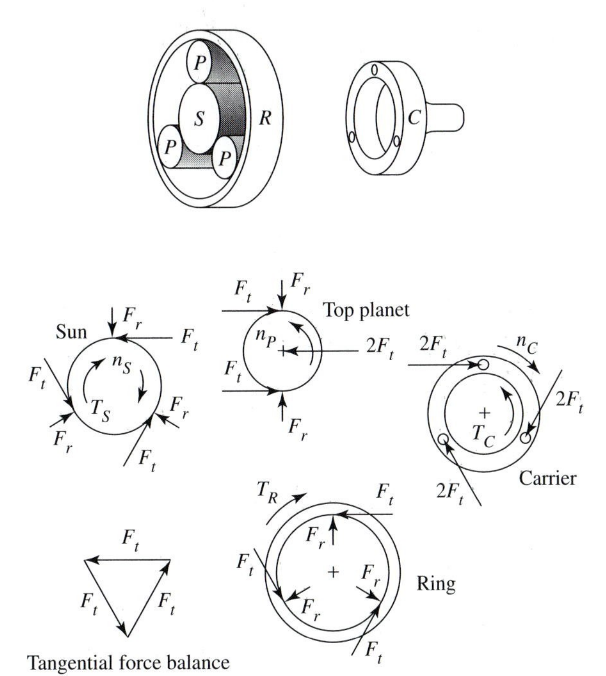

Figure 9.18 Planetary train.

and

$$r_R = r_S + 2r_P \qquad \textbf{(9.101)}$$

where r_C = radius of the planet shaft circle in the planet carrier
The other terms refer to gear pitch radii.

These equations do not depend on the number of planets.

The same diametral pitch or module must be used for all gears in a simple planetary train. This condition and the Equation (9.101) lead to

$$N_R = N_S + 2N_P \qquad \textbf{(9.102)}$$

where N = number of teeth

Planetary gear trains have certain advantages over nonplanetary trains. Planetary train features include

- the ability to balance lateral forces, reducing shaft loads
- a sharing of tooth load
- shifting schemes using band brakes and internal clutches, making automatic speed changing practical

Balanced Planetary Trains

Fatigue loading of shafting must be considered in the design of nonplanetary gear drives. In balanced planetary trains, there is no resultant lateral load on the sun and ring gears and the planet carrier. A train is balanced if the planets are equally spaced within the planet carrier (e.g., 3 planets at 120° intervals or 4 at 90°). Balanced spacing is possible only if

$$\frac{N_S + N_R}{P_{\text{num}}} = I \tag{9.103}$$

where P_{num} = number of planets

I = an integer

Speed Ratios

Speeds in planetary gear trains can be determined from the relationship

$$\frac{n_Y - n_C}{n_X - n_C} = \frac{n_{YC}}{n_{XC}} \tag{9.104}$$

where two gears which are not planets are identified as X and Y

n_X and n_Y = actual speeds of gears X and Y, respectively

n_{XC} and n_{YC} = speeds of gears X and Y with respect to the carrier

n_{XC}/n_{YC} = speed ratio when the carrier is stationary

n_C = actual planet carrier speed

Figure 9.19 shows a precision gearhead incorporating a planetary gear train and a reverted spur gear train. Gearheads of this type are often used with stepper motors or servo motors in control systems. They are available with speed reductions of 3:1 to 100:1.

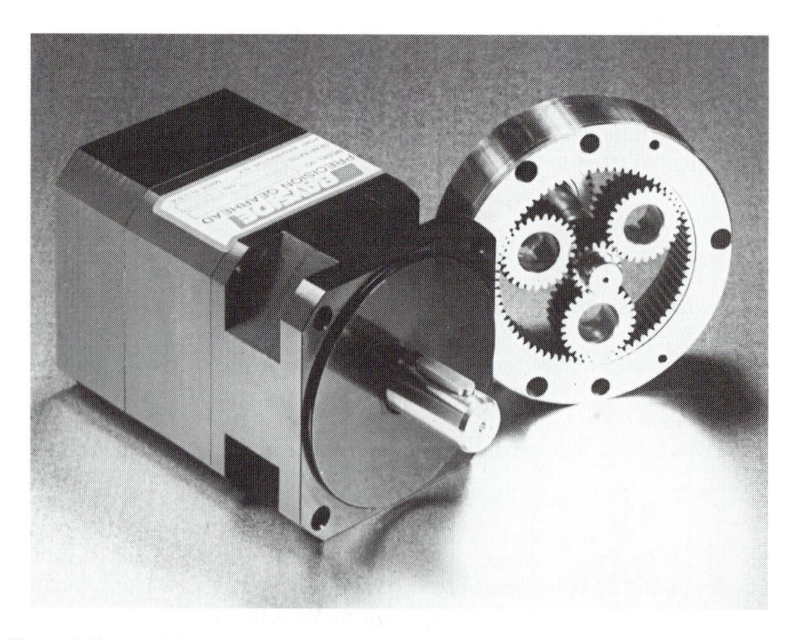

Figure 9.19 Precision gearhead incorporating a planetary train. (Courtesy of Bayside Controls, Inc.)

The Simple Planetary Train as a Speed Changer

Planetary trains provide a fairly compact design that is adaptable to speed-ratio changing. The sun gear can be used as an input gear and the carrier as output in a simple planetary train. Using an internal clutch, the sun gear is locked to the carrier to produce a 1:1 speed ratio. In this mode, the ring rotates with the other gears. In order to produce a speed reduction, the internal clutch is released and a band brake locks the ring gear.

Torque and Tooth Loading in a Planetary Train

The power–torque relationship is the same as for other gear trains, but tooth loading depends on the number of planets. Consider a simple planetary spur gear train in which the sun is the input gear, the ring gear is held stationary, and the carrier drives the output shaft. The tangential force is

$$F_t = \frac{T_S}{r_S P_{\text{num}}} \quad \textbf{(9.105)}$$

where F_t = tangential force at one tooth mesh

T_S = torque on sun gear

r_S = sun gear pitch radius

P_{num} = number of planets

The reduction in tangential force due to load sharing makes it possible to design planetary trains with smaller teeth than would be required in similar, nonplanetary drives.

Radial force is given by

$$F_r = F_t \tan \phi \quad \textbf{(9.106)}$$

where F_r = radial force at each tooth mesh

ϕ = pressure angle

Since the sun is the driver, tangential loads on the sun oppose its rotation. Radial loads are toward the center of the sun and toward the center of each planet. Radial loads are directed away from the center of the ring. Figure 9.18 shows radial and tangential loads for a train with three planets.

The loads on a planet where it contacts the sun are equal and opposite those on the sun. Tangential forces on the planets cause opposing torques, since there is no net torque on the planet shafts. The carrier reaction on the planet balances these forces with a force of $2F_t$. A reaction torque on the ring balances the effect of the tangential forces from the planets. The ring could be bolted in place or held stationary with a band brake. Note that the same magnitude tangential load is applied at each tooth mesh on the sun, the planets, and the ring.

EXAMPLE PROBLEM Two-Speed Balanced Planetary Gear Train

Make a preliminary design of a gear train to produce output speeds of about 460 rpm and 1800 rpm with an input speed of 1800 rpm. Find tooth loads and torques for 2.5 hp transmitted power.

Design Decisions: A simple planetary train similar to the one in Figure 9.18 will be used. The sun gear will be the input gear, and the planet carrier will be the output. We will obtain the 1800 rpm output speed by using a clutch to lock the train, with the ring free to turn. We will obtain the lower output speed by releasing the clutch and holding the ring stationary with a band brake. We will select spur gears with a diametral pitch of 10.

Solution: We will first attempt to determine tooth numbers that will give us an output speed of about 460 rpm. There are many possible solutions. Using the speed ratio equation with Y the ring gear and X the sun gear, we obtain

$$\frac{n_R - n_C}{n_S - n_C} = \frac{n_{RC}}{n_{SC}}$$

We set actual ring gear speed n_R equal to 0. The right side of the equation is the ring-to-planet speed ratio with the carrier fixed. We treat the planet as both a driving and a driven gear, and we note that the planet and ring turn in the same direction if the carrier is fixed. The result is

$$\frac{n_R - n_C}{n_S - n_C} = \frac{0 - n_C}{n_S - n_C} = \frac{n_{RC}}{n_{SC}} = -\frac{N_S N_P}{N_P N_R} = \frac{-N_S}{N_R}$$

We manipulate this equation to obtain the output-to-input speed ratio for a simple planetary train with the ring fixed.

$$\frac{n_C}{n_S} = \frac{1}{1 + \dfrac{N_R}{N_S}}$$

Rewriting the above equation in terms of the ratio of gear teeth, and inserting the required speeds, we obtain

$$\frac{N_R}{N_S} = \frac{n_S}{n_C} - 1 = \frac{1800}{460} - 1 = 2.913 \quad \text{(approximately)}$$

We can try various pairs of numbers on a calculator. One solution is $N_R = 58$ and $N_S = 20$, where $N_R/N_S = 2.900$, which is close enough. In order to fit the train together, we must satisfy the equation

$$N_R = N_S + 2N_P$$

from which

$$N_P = (N_R - N_S)/2 = 19$$

The balanced train equation is satisfied by using $P_{\text{num}} = 3$ planets which will, of course, be equally spaced. Actual output speed is

$$n_C = \frac{n_S}{1 + \dfrac{N_R}{N_S}} = \frac{1800}{1 + \dfrac{58}{20}} = 461.5 \text{ rpm} \quad \text{clockwise}$$

Using the definition of diametral pitch, we find pitch radii of the gears

$$r = N/(2P_d) = N/(2 \cdot 10) = 1, 0.95, \text{ and } 2.90 \text{ in}$$

for the sun, planets, and ring, respectively. Carrier radius is

$$r_C = r_S + r_P = 1 + 0.95 = 1.95 \text{ in}$$

We will apply a clockwise torque to the sun gear, causing it to rotate clockwise. The sun gear torque is

$$T_S = 63{,}025 P_{\text{hp}}/n_S = 63{,}025 \cdot 2.5/1800 = 87.54 \text{ in} \cdot \text{lb}$$

The tangential forces are

$$F_t = \frac{T_S}{r_S P_{\text{num}}} = 87.54/(1 \cdot 3) = 29.18 \text{ lb}$$

and the radial forces are

$$F_r = F_t \tan \phi = 29.18 \tan 20° = 10.62 \text{ lb}$$

These forces, applied at each tooth mesh between the sun and planets and between the planets and ring, are shown in Figure 9.18. Note that there is no torque on the planets, but each planet applies a force $2F_t$ on the carrier. Carrier torque is

$$T_C = 2F_t r_C P_{\text{num}} = 2 \cdot 29.18 \cdot 1.95 \cdot 3 = 341.385 \text{ in} \cdot \text{lb}$$

The output shaft (load) applies a counterclockwise torque to the carrier; the carrier applies a clockwise torque to the output shaft. Output power is

$$P_{\text{hp}} = \frac{T_C n_C}{63{,}025} = \frac{341.385 \cdot 461.54}{63{,}025} = 2.5$$

The computed output power equals the input power because we have neglected friction losses in the gear train.

The band brake must be designed to apply a reaction torque to the ring. The required torque is

$$T_R = F_t r_R P_{\text{num}} = 29.18 \cdot 2.9 \cdot 3 = 253.85 \text{ in} \cdot \text{lb}$$

The band applies a clockwise torque to the ring, adding to the input torque. The torques on the sun, the ring, and the carrier balance. Note that the tangential forces on the sun are equal at each of the three gear mesh locations. The force balance is represented by an equilateral triangle as shown in Figure 9.18. The same applies to the ring gear and carrier. Radial forces on the gears also balance. Thus, there is no bending load applied to the input and output shafts.

■■

The preceding example problem is easily solved with a calculator if we make a lucky guess of tooth numbers to satisfy the speed ratio, assembly, and balanced train requirements. A computer solution is more convenient if we consider a redesign or wish to explore different requirements. In the example that follows, hand calculation would be oppressively repetitious.

■■ EXAMPLE PROBLEM Preliminary Design Investigation of a Series of Planetary Trains

A manufacturer wishes to offer a series of gear trains to produce output speeds ranging from about 500 to 700 rpm with an input speed of 2500 rpm. A 1:1 speed ratio will also be available. The trains will be designed to transmit 11 hp. Suggest a set of designs, and determine tooth forces and shaft torques for these gear trains.

Design Decisions: The design will call for simple balanced planetary trains using three or four planets. The gear trains will operate as described in the preceding example; the sun gear will be the input gear, and the carrier will be the output. We will use a diametral pitch of 10 and a 20° pressure angle, and the sun will have 21 teeth.

Solution Summary: We will vary the number of teeth in the planet gears as we investigate a set of possible gear trains. Thus, gear sizes, forces, torques, and so on, will be expressed as functions of number of planet teeth. The detailed solution that follows describes gear trains with the ring gear held stationary. We decide the number of planets by testing to see if four planets produce a balanced train. If so, four planets are used. If not, we try three planets. If neither three nor four planets produce a balanced train, we reject the solution, indicating a zero in the P_{num} column. (This calculation is further discussed after the example in the section "Conditional Functions.") Planets having 17 to 30 teeth provide 10 possible gear trains with the required output speeds. Four of our tries are rejected.

The results for the selected values of tooth numbers are given in Table 9.4. The detailed solution follows.

Table 9.4 Results for selected values of tooth numbers.

N_P	$N_R(N_P)$	$r_P(N_P)$	$r_C(N_P)$	$r_R(N_P)$	$P_{num}(N_P)$	$F_t(N_P)$	$F_r(N_P)$
17	55	0.85	1.9	2.75	4	66.026	24.032
18	57	0.9	1.95	2.85	3	88.035	32.042
19	59	0.95	2	2.95	4	66.026	24.032
20	61	1	2.05	3.05	0	0	0
21	63	1.05	2.1	3.15	4	66.026	24.032
22	65	1.1	2.15	3.25	0	0	0
23	67	1.15	2.2	3.35	4	66.026	24.032
24	69	1.2	2.25	3.45	3	88.035	32.042
25	71	1.25	2.3	3.55	4	66.026	24.032
26	73	1.3	2.35	3.65	0	0	0
27	75	1.35	2.4	3.75	4	66.026	24.032
28	77	1.4	2.45	3.85	0	0	0
29	79	1.45	2.5	3.95	4	66.026	24.032
30	81	1.5	2.55	4.05	3	88.035	32.042

N_P	$T_C(N_P)$	$T_R(N_P)$	$T_{sum}(N_P)$	$n_C(N_P)$	$P_C(N_P)$	
17	−1003.598	726.288	0	690.789	−11	
18	−1030.009	752.699	0	673.077	−11	
19	−1056.419	779.109	0	656.25	−11	
20	0	0	0	640.244	0	**Reject**
21	−1109.24	831.93	0	625	−11	
22	0	0	0	610.465	0	**Reject**
23	−1162.061	884.751	0	596.591	−11	
24	−1188.471	911.161	0	583.333	−11	
25	−1214.882	937.572	0	570.652	−11	
26	0	0	0	558.511	0	**Reject**
27	−1267.703	990.393	0	546.875	−11	
28	0	0	0	535.714	0	**Reject**
29	−1320.524	1043.214	0	525	−11	
30	−1346.934	1069.624	0	514.706	−11	

Detailed Solution:

Power, rpm, and Torque:

$$P_{hp} = 11 \qquad n_S = 2500$$

$$T_S = \frac{63{,}025 \cdot P_{hp}}{n_S} \qquad T_S = 277.31$$

Pressure Angle:

$$\phi = 20°$$

Diametral Pitch, Number of Teeth, and Pitch Radius:

$$P_d = 10 \qquad N_S = 21$$

$$r_S = \frac{N_S}{2 \cdot P_d} \qquad r_S = 1.05$$

$$N_P = 17, 18, \ldots, 30 \qquad N_R(N_P) = N_S + 2 \cdot N_P \qquad r_P(N_P) = \frac{N_P}{2 \cdot P_d}$$

$$r_C(N_P) = r_S + r_P(N_P) \qquad r_R(N_P) = r_S + 2 \cdot r_P(N_P)$$

Number of Planets for a Balanced Train:

$$Z(N_P) = \frac{N_S + N_R(N_P)}{4} \qquad Y(N_P) = Z(N_P) - \text{floor}(Z(N_P))$$

$$z(N_P) = \frac{N_S + N_R(N_P)}{3} \qquad y(N_P) = z(N_P) - \text{floor}(z(N_P))$$

$$P_{num}(N_P) = \text{if}[(Y(N_P) \neq 0) \cdot (y(N_P) \neq 0), 0, 1]$$

$$P_{num}(N_P) = \text{if}(Y(N_P) = 0,\ 4 \cdot P_{num}(N_P),\ 3 \cdot P_{num}(N_P))$$

Tangential and Radial Force:

$$F_t(N_P) = \text{if}\left(P_{num}(N_P) \neq 0, \frac{T_S}{P_{num}(N_P) \cdot r_S}, 0\right)$$

$$F_r(N_P) = F_t(N_P) \cdot \tan(\phi \cdot \text{deg})$$

Torque:

$$T_C(N_P) = -2 \cdot F_t(N_P) \cdot r_C(N_P) \cdot P_{num}(N_P)$$

$$T_R(N_P) = F_t(N_P) \cdot r_R(N_P) \cdot P_{num}(N_P)$$

$$T_S(N_P) = \text{if}(P_{num}(N_P) \neq 0,\ T_S,\ 0)$$

Torque Balance:

$$T_{sum}(N_P) = T_S(N_P) + T_C(N_P) + T_R(N_P)$$

Carrier Speed:

$$n_C(N_P) = \frac{n_S}{1 + \dfrac{N_R(N_P)}{N_S}}$$

Output Power:

$$P_C(N_P) = \frac{T_C(N_P) \cdot n_C(N_P)}{63{,}025}$$

CONDITIONAL FUNCTIONS

Conditional functions are applicable to gear design and to many other machine design problems. When designing and problem solving with a pencil and paper, we unconsciously make decisions. We know, for example, that a gear cannot have 20.3 teeth and that a planetary train cannot have $3^1/_2$ planets. Computers use conditional functions to make decisions. By introducing conditional functions into our programs, we make it unnecessary to stop the computation to make decisions.

An IF operator was used in the preceding example problem. In that problem, we could have tried three or four planets in the balanced train equation,

$$\frac{N_S + N_R}{P_{\text{num}}} = Z \quad \textbf{(9.107)}$$

We could have then tabulated the values of Z and rejected noninteger values simply by scanning. This method is often the quickest way of handling such problems. Conditional functions or logical functions are an alternative method.

Conditional functions usually involve **Boolean operators** such as

`IF, AND, OR, NOT, EXCEPT, and THEN`

with the mathematical operators

$$=, \quad \neq, \quad <, \quad >, \quad \leq, \quad \geq, \quad \text{and so on}$$

For example, using MathCAD™, the conditional statement

```
A:= IF(x > y, 1,0)
```

returns **A** = 1 if true; that is, if $x > y$
A = 0 otherwise

Two conditions can be indicated with an **AND gate,** using multiplication as follows:

```
B: = IF((w > x)*(y > z), 1,0)
```

returns B = 1 if both are true; that is, if $w > x$ ***and*** $y > z$
B = 0 otherwise

An **OR gate** uses a plus sign. If we want to know if ***either*** condition is true, we key

```
C: = IF((w > x) + (y > z), 1,0)
```

which returns C = 1 if ***either*** is true; that is, if $w > x$ ***or*** $y > z$
C = 0 otherwise

In the above example, we let

$$Y = \text{the remainder} = [Z - \text{the largest integer that is} \leq Z]$$

where Z is calculated using Equation (9.107), for four planets

If $Y \neq 0$, the train cannot be balanced with four planets. The corresponding remainder for three planets is y.

If $Y \neq 0$ and $y \neq 0$, we reject the configuration, setting $P_{\text{num}} = 0$; otherwise, we set $P_{\text{num}} = 1$ for now. Then if $Y = 0$, we multiply the original value of P_{num} by 4; otherwise by 3. In some cases, either three or four planets could be used, but we have favored four planets.

FORTRAN and other computer languages contain explicit logical operators such as **AND, OR,** and **NOT.** These can be used in **IF–THEN–ELSE** blocks along with **GOTO** commands for efficient decision making.

References and Bibliography

AGMA Gear Classification and Inspection Handbook. ANSI/AGMA 2000–A88. Arlington, VA: American Gear Manufacturers Association, June 1988b.

AGMA Standard Nomenclature of Gear Tooth Failure Modes. ANSI/AGMA 110.04. Arlington, VA: American Gear Manufacturers Association, August 1980.

AGMA Standard for Rating the Pitting Resistance and Bending Strength of Spur and Helical Involute Gear Teeth. AGMA 218.01. Arlington, VA: American Gear Manufacturers Association, December 1982.

American Gear Manufacturers Association Gear Handbook. AGMA 390.03. Arlington, VA: American Gear Manufacturers Association, June 1988a.

Boston Gears. Quincy, MA: Imo Industries Incorporated, Boston Gear Division, 1992.

Deutschman, A. D., W. J. Michels, and C. E. Wilson. *Machine Design: Theory and Practice.* New York: Macmillan, 1975.

Stock Drive Products, Stirling Instrument. *Handbook of Gears.* New Hyde Park, NY: DSG Designtronics, Inc., 1990.

Wilson, C. E., and J. P. Sadler. *Kinematics and Dynamics of Machinery.* 2d ed. New York, HarperCollins, 1993.

Design Problems

9.1. You are asked to design a gear train to reduce speed from 1500 to 500 rpm.

Tentative Design Decisions:

Module = 3 mm   Pressure angle = 14.5°
Number of pinion teeth = 12

Check to see if the proposed design is satisfactory. Find the following:

(a) number of teeth in the gear
(b) gear pitch diameter
(c) center distance
(d) maximum allowable addendum circle radius for the gear
(e) standard addendum circle radius for the gear
(f) Is there interference?
(g) Find contact ratio
(h) Is *CR* satisfactory?

9.2. You are asked to design a gear reducer. The given data are as follows:

Input speed:	$n_1 = 1500$
Output speed:	$n_2 = 1000$
Module:	$m = 4$
Pinion teeth:	$N_1 = 20$

Use full-depth teeth and a pressure angle ϕ of 20°.

Find the following:

(a) number of teeth in the gear
(b) gear pitch diameter
(c) center distance
(d) maximum allowable addendum circle radius for the gear
(e) standard addendum circle radius for the gear
(f) Is there interference?

9.3. Refer to the gear reducer of Problem 9.2.

(a) Find contact ratio.
(b) Is *CR* satisfactory?

9.4. We require a machine component to convert rotational motion into translation, and vice versa.

Design Decisions: After considering slider-crank linkages and other alternatives, we select a rack and pinion drive. Based on loading and other details, we decide on 20° pressure angle, full-depth teeth, with 10 mm metric module and 25 mm face width.

Construct a solid model of the rack.

9.5. You are asked to design a gear reducer. The given data are as follows:

Input speed:	$n_1 = 1760$
Output speed:	$n_2 = 440$
Diametral pitch:	$P_d = 4$
Pinion teeth:	$N_1 = 32$

Use full-depth teeth and a pressure angle ϕ of 14.5°. Find the following:

(a) number of teeth in the gear
(b) gear pitch diameter
(c) center distance
(d) maximum allowable addendum circle radius for the gear
(e) standard addendum circle radius for the gear
(f) Is there interference?
(g) contact ratio
(h) Is *CR* satisfactory?

9.6. Suppose that we try 20° stub teeth in the gear reducer design of Problem 9.1. Will there still be interference? Is contact ratio a problem?

9.7. Speed is to be reduced from 1800 to 600 rpm using full-depth spur gears with a module of 5. The pinion will have 14 teeth. Find drive dimensions, and compare 14.5° and 25° pressure angle gears for this application.

9.8. We must design the countershaft of a reverted gear train to transmit 40 hp at 800 rpm countershaft speed.

(a) Determine gear and shaft loading.
(b) Find the required countershaft diameter using appropriate fatigue failure theories.

Design Decisions: Refer to Figure 9.7, except that the input gear will be above gear 1 and the output gear will be above gear 2. The following data apply:

	Pitch Radius (in)	*Location (in)*	*Pressure Angle (degrees)*
Gears	2.75	2	20
	3.5	3	20
Bearings		0.5	
		4.5	

Shaft

Ultimate strength = 120,000 psi
Yield point = 100,000 psi
Safety factor = 1.75
Surface: ground
Stress concentration equivalent to $D/d = 1.1$ and $r/d = 0.1$
Samples/failure = 10^8
Length = 5 in

9.9. Design a 7 in long countershaft for a reverted gear train that transmits 40 hp at 1800 rpm countershaft speed.

(a) Determine gear and shaft loading.
(b) Find the required countershaft diameter using appropriate fatigue failure theories.

Refer to Figure 9.7.

Design Decisions: The gear train will be similar to that in Figure 9.7, except that the input gear will be above gear 1 and the output gear will be above gear 2. The following data applies:

	Pitch Radius (in)	*Location (in)*	*Pressure Angle (degrees)*
Gears	2	2.5	20
	3.5	6.5	20
Bearings		0.5	
		5	

Shaft

Ultimate strength = 125,000 psi
Yield point = 110,000 psi
Safety factor = 1.5
Surface: ground
Stress concentration equivalent to $D/d = 1.1$ and $r/d = 0.1$
Samples/failure = 10^7

9.10. Design the countershaft of a gear train that transmits 30 hp. The countershaft speed is 1800 rpm, and the countershaft length is 7 in.

(a) Determine gear and shaft loading.
(b) Find the required countershaft diameter using appropriate fatigue failure theories.

Refer to Figure 9.7 for gear orientation.

Design Decisions: The following data apply:

	Pitch Radius (in)	*Location (in)*	*Pressure Angle (degrees)*
Gears	2.5	2	20
	3.5	6.5	20
Bearings		1	
		5	

Shaft

Ultimate strength = 140,000 psi
Yield point = 120,000 psi
Safety factor = 2
Surface: ground
Stress concentration equivalent to $D/d = 1.1$ and $r/d = 0.1$
Samples/failure = 10^7

9.11. Speed must be reduced from 4000 to about 2400 rpm; transmitted power is 2.5 hp. Design the speed reducer. Consider bending stress and wear (pitting resistance).

Design Decisions: We will use the AGMA method, adapted for computer use. We will use 20° pressure angle, full-depth, 16 diametral pitch steel spur gears of 300 Brinell hardness. A 20 tooth pinion will be used. The failure probability will be 1 in 10,000; the application factor is 1.25; and the quality number is 5.

9.12. Speed must be reduced from 3000 to about 1250 rpm; transmitted power is 40 hp. Design the speed reducer. Consider bending stress and wear (pitting resistance).

Design Decisions: We will use the AGMA method, adapted for computer use. We will use 20° pressure angle, full-depth, 6 diametral pitch steel spur gears of 300 Brinell hardness. An 18 tooth pinion will be used. The failure probability will be 1 in 1000. The design calls for an application factor of 1.5 and a quality number of 5.

9.13. Speed must be reduced from 3200 to about 950 rpm; transmitted power is 10 hp. Design the speed reducer. Consider bending stress and wear (pitting resistance).

Design Decisions: We will use the AGMA method, adapted for computer use. We will use 20° pressure angle, full-depth, 8 diametral pitch steel spur gears of 300 Brinell hardness. An 18 tooth pinion will be used. The failure probability will be 1 in 1000; the application factor is 1.5; and the quality number is 5.

9.14. Design a 2 hp capacity speed reducer for an input speed of 1800 rpm and an output speed of about 650 rpm. Try to obtain a face-width-to-pinion-diameter ratio between 0.75 and 1.0. Select an appropriate diametral pitch.

Design Decisions: We will use the AGMA method, adapted for computer use. We will use 20° pressure angle, full-depth, steel spur gears of 300 Brinell hardness. An 18 tooth pinion will be used. The failure probability will be 1 in 10,000; the application factor is 1.5; and the quality number is 5.

9.15. Refer to Figure 9.10(a) and (b), the AGMA bending strength geometry factor curves that apply to 20° and 25° full-depth teeth with loading at the tip. Select representative points on these curves. Obtain regression equations to represent the data. Plot the regression equations and the selected curve points. Calculate the correlation coefficients.
Note: Regression analysis is discussed in Chapter 4. Regression procedures are available in mathematical software and preprogrammed on scientific calculators. There are many possible solutions. The results are likely to differ from those given in the section on geometry factor for bending.

9.16. Design a reverted gear train speed reducer to produce an output speed of approximately 135 rpm. The input speed is 1450 rpm, and 2 hp is transmitted.

Design Decisions: Helical gears will be used in a reverted gear train similar to that of Figure 9.13 (exploded view). The diametral pitch in the plane of rotation will be 8, and the normal pressure angle will be 14.5° for all gears. Gear 1 will be a 20 tooth right-hand gear with 45° helix angle. Gear 3 will have 18 teeth. Thrust forces will be balanced on the countershaft.

Suggested Procedure: Try a range of tooth numbers for gear 2. Find the corresponding number of teeth in gear 4, noting that the input and output shafts must be collinear. Calculate pitch radii and center distance. Check collinearity by comparing center distance of gears 1 and 2 with that of 3 and 4. Calculate torques, tooth loads, and helix angles. Check the thrust balance. Select a train with output speed closest to the desired value.

9.17. Design a reverted gear train speed reducer to produce an output speed of approximately 200 rpm. The input speed is 1760 rpm, and 1.5 hp is transmitted.

Design Decisions: Helical gears will be used in a reverted gear train similar to that of Figure 9.13 (exploded view). The diametral pitch in the plane of rotation will be 10 for gears 1 and 2, and 8 for gears 3 and 4. The normal pressure angle will be 14.5° for all gears. Gear 1 will be a 20 tooth right-hand gear with 45° helix angle. Gear 3 will have 18 teeth. Thrust forces will be balanced on the countershaft.

Suggested Procedure: Try a range of tooth numbers for gear 2. Find the corresponding number of teeth in gear 4, noting that the input and output shafts must be collinear.

Calculate pitch radii and center distance. Check collinearity by comparing center distance of gears 1 and 2 with that of 3 and 4. Calculate torques, tooth loads, and helix angles. Check the thrust balance. Select a train with output speed closest to the desired value.

9.18. Design a reverted gear train speed reducer to produce an output speed of approximately 270 rpm. The input speed is 4000 rpm, and 1.3 hp is transmitted.

Design Decisions: Helical gears will be used in a reverted gear train similar to that of Figure 9.13 (exploded view). The diametral pitch in the plane of rotation will be 10, and the normal pressure angle will be 14.5° for all gears. Gear 1 will be a 19 tooth right-hand gear with 45° helix angle. Gear 3 will have 19 teeth. Thrust forces will be balanced on the countershaft.

Suggested Procedure: Try a range of tooth numbers for gear 2. Find the corresponding number of teeth in gear 4, noting that the input and output shafts must be collinear. Calculate pitch radii and center distance. Check collinearity by comparing center distance of gears 1 and 2 with that of 3 and 4. Calculate torques, tooth loads, and helix angles. Check the thrust balance. Select a train with output speed closest to the desired value.

9.19. We need a two-speed drive to produce output speeds of 3000 and about 535 rpm with 3000 rpm input speed; 10 hp is to be transmitted. Perform preliminary design calculations.

Design Decisions: The design will be based on a simple balanced planetary spur gear train with three or four planets, as shown in Figure 9.18. The sun gear will be the input gear, and the carrier will be the output. An internal clutch will lock the train to produce the 1:1 speed ratio. A band brake will lock the ring to produce the speed reduction. We will use a diametral pitch of 10, a 20° pressure angle, and a 20 tooth sun gear. We will specify gear tooth numbers and radii and number of planets. We will determine torques and will check torque and power balance.

9.20. Design a two-speed gear train to produce a 1:1 speed ratio and a low-speed output torque of about 620 lb · in; 4 hp is to be transmitted, and input speed is 2000 rpm. Perform preliminary design calculations.

Design Decisions: The design will be based on a simple balanced planetary spur gear train with three or four planets. The sun gear will be the input gear, and the carrier will be the output, as shown in Figure 9.18. An internal clutch will lock the train to produce the 1:1 speed ratio. A band brake will lock the ring to produce the speed reduction. We will use a diametral pitch of 10, a 20° pressure angle, and a 17 tooth sun gear. We will specify gear tooth numbers and radii and number of planets. We will determine torques and will check torque and power balance.

9.21. Do a preliminary design investigation for a series of two-speed drives to produce 1:1 speed ratios as well as output speeds ranging from about 500 to 675 rpm; 12 hp is to be transmitted, and input speed is 2400 rpm. Perform preliminary design calculations.

Design Decisions: The designs will be based on a simple balanced planetary spur gear trains with three or four planets. The sun gear will be the input gear, and the carrier will be the output for each train, as shown in Figure 9.18. An internal clutch will lock the train to produce the 1:1 speed ratio. A band brake will lock the ring to produce the speed reduction. We will use a diametral pitch of 8, a 20° pressure angle, and a 22 tooth sun gear. We will specify gear tooth numbers and radii and number of planets. We will determine torques and will check torque and power balance.

9.22. Do a preliminary design investigation for a series of two-speed drives to produce 1:1 speed ratios as well as low-speed output torques ranging from about 250 to 300 in · lb; 1.5 hp is to be transmitted, and input speed is 1500 rpm. Perform preliminary design calculations.

Design Decisions: The designs will be based on a simple balanced planetary spur gear trains with three or four planets. The sun gear will be the input gear, and the carrier will be the output for each train, as shown in Figure 9.18. An internal clutch will lock the train to produce the 1:1 speed ratio. A band brake will lock the ring to produce the speed reduction. We will use a diametral pitch of 12, a 20° pressure angle, and an 18 tooth sun gear. We will specify gear tooth numbers and radii and number of planets. We will determine torques and will check torque and power balance.

CHAPTER 10

Belt Drives

Symbols

A, B	regression constants
c	center distance
E	elastic modulus
f	coefficient of friction
F_1, F_2	belt tension
F_c	inertia effect
F_{limit}	maximum allowable tension
g	acceleration of gravity
i	summation counter
K	service factor
ln	natural logarithm
L	length of contact
m'	mass per unit length
n	speed, number of pairs of data points
P	power
R	radius
T	torque
v	velocity
w'	weight per unit length
β	half the included angle of sheave groove
σ	stress
$\sum$	summation
θ_1	angle of contact
ω	angular velocity

Units

acceleration of gravity	in/s^2
angles	rad
angular velocity	rad/s
dimensions	in or m
elastic modulus	psi or Pa
mass per unit length	$lb \cdot s^2/in^2$ or kg/m
power	hp or kW
speed	rpm
tension and inertia effect	lb or N
torque	in · lb or N · m
weight per unit length	lb/in
velocity	in/s or m/s

Internal combustion engines and electric motors are ordinarily designed for a high power-to-weight ratio. This ratio is accomplished through high operating speeds; that is, speeds usually higher than required by the driven device. Belt drives are a low-cost means of power transmission and speed reduction. Other belt drive advantages include isolation of shock and vibration, quiet operation, and the ability to accommodate a wide range of distances between driver and driven shafts. Disadvantages include a lower service life than that typical of gears and, for belt drives relying on friction, a speed ratio that cannot be controlled precisely. V-belts riding in grooved sheaves (pulleys) are the most common type of belt drive. Figure 10.1(a) shows a drive using two V-belts; a three-belt drive is shown in Figure 10.1(b). A cut section of a typical single V-belt is shown in Figure 10.1(c); Part (d) of the figure shows a cut section of a V-band belt, and Part (e) shows a V-cog belt. A V-belt sheave with split tapered bushings is shown in Figure 10.1(f).

SPEED RATIO

The speed ratio of a belt drive is given by

$$n_2/n_1 = \omega_2/\omega_1 = R_1/R_2 \qquad \textbf{(10.1)}$$

where n = rotation speed (rpm)

ω = angular velocity (rad/s)

R = sheave pitch radius

Subscripts 1 and 2 refer, respectively, to sheaves 1 and 2.

The pitch radius differs, in general, from the outside radius of a sheave.

For the configuration shown in Figure 10.2, both sheaves turn in the same direction. Belt tension varies from minimum (slack side) to maximum (tight side) while the belt is in contact with a sheave. Belt strain varies as well, causing some slippage in belts transmitting power by friction. Thus, the ratio given in Equation (10.1) is approximate. Applications requiring precise speed ratios utilize gears, chain drives, or toothed belts and sheaves called **timing** or **synchronous** belt drives.

BELT DRIVE GEOMETRY: CONTACT ANGLE AND BELT LENGTH

Referring to Figure 10.2(a), we see that for a 1:1 speed ratio, the sheaves are equal ($R_1 = R_2 = R$). Neglecting belt deflection, the belt contacts each sheave over an angle π radians, and belt pitch length is given by $2c + 2\pi R$ where c = center distance. However, sheaves are usually unequal, and the smaller one ordinarily drives.

Drive capacity depends on contact angle and is governed by the smaller contact angle, usually the contact angle on the drive sheave. For unequal sheaves, shown in Figure 10.2(b), the angle between the drive centerline and a straight section of belt is given by

$$\alpha = \arcsin[(R_2 - R_1)/c]$$

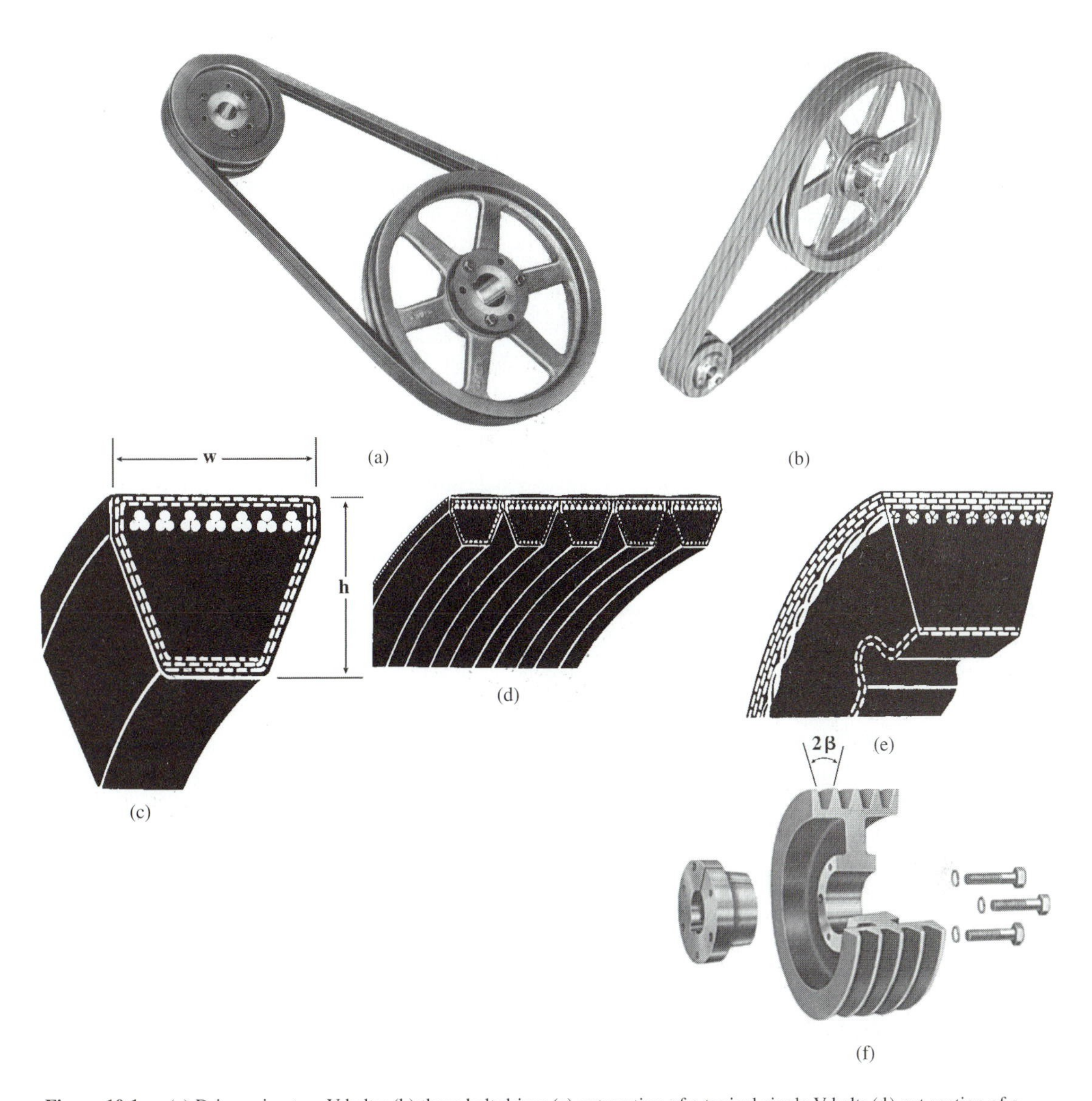

Figure 10.1 (a) Drive using two V-belts; (b) three-belt drive; (c) cut section of a typical single V-belt; (d) cut section of a V-band belt; (e) V-cog belt; (f) V-belt sheave with split tapered bushings. (Courtesy of T. B. Wood's Sons Company)

and the contact angle on the smaller (critical) sheave is

$$\theta_1 = \pi - 2\alpha \text{ rad} \tag{10.2}$$

The length of a straight section of belt is

$$L_s = [c^2 - (R_2 - R_1)^2]^{1/2}$$

and total pitch length is given by

$$L = 2L_s + R_1(\pi - 2\alpha) + R_2(\pi + 2\alpha) \tag{10.3}$$

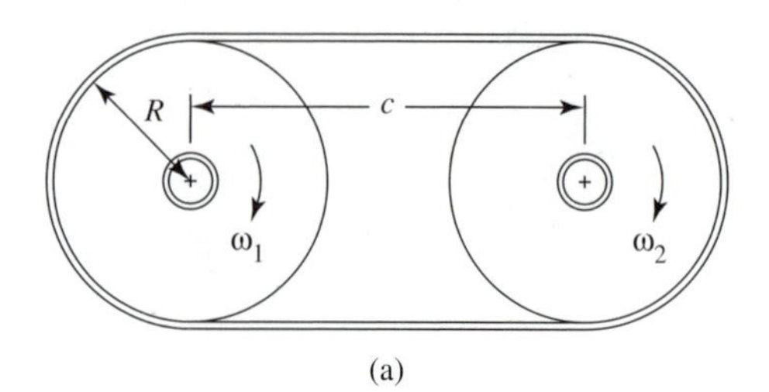

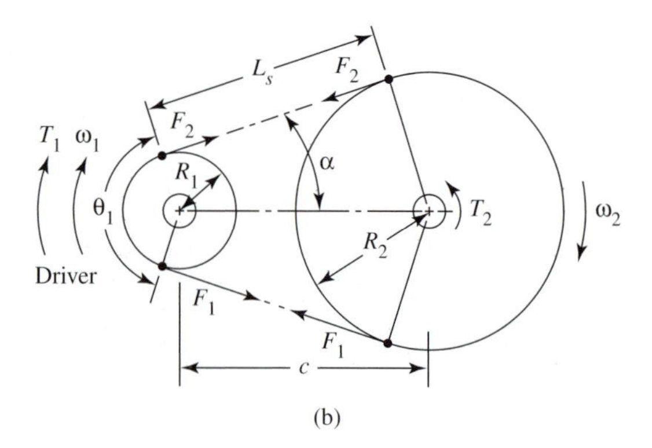

Figure 10.2 Belt drive geometry. (a) 1:1 speed ratio; (b) speed reduction.

Ordinarily, standard belts will be selected from stock, and the center distance adjusted accordingly. Since belts stretch under loading, there should be provision for adjusting center distance, or an idler pulley must be provided to maintain optimum tension.

POWER, ROTATIONAL SPEED, AND TORQUE

Power, speed, and torque are related as follows:

$$P_{\text{hp}} = Tn/63{,}025 \tag{10.4}$$

where P_{hp} = power (hp)

n = shaft speed (rpm)

T = torque *on the same shaft* (in · lb)

Alternatively, using metric units,

$$P_{\text{kW}} = 10^{-3}T\omega \tag{10.5}$$

where P_{kW} = power (kW)

ω = shaft angular velocity (rad/s)

T = torque *on the same shaft* (N · m)

These equations are applicable to all drives.

BELT TENSIONS

Refer again to Figure 10.2 which shows a typical belt drive configuration. Torques are related to belt tension as follows:

$$T_1 = (F_1 - F_2)R_1 \tag{10.6}$$

$$T_2 = (F_1 - F_2)R_2 \tag{10.7}$$

where T_1 and T_2 = torque on shafts 1 and 2, respectively (in · lb or N · m)

R_1 and R_2 = pitch radius of sheaves 1 and 2, respectively (in or m)

F_1 = maximum, i.e., tight-side, belt tension (lb or N)

F_2 = minimum, i.e., slack-side, belt tension (lb or N)

Shaft torque on the driver sheave is in the same direction as angular velocity. A free-body diagram of the driven sheave will show a shaft torque (due to the load) which opposes angular velocity. Shaft force reactions are not shown in Figure 10.2.

INERTIA EFFECTS

As a belt rides on a sheave, there is a radial acceleration toward the sheave center, resulting in an inertia effect which tends to cause the belt to pull away from the sheave. The inertia effect is equivalent to

$$F_c = m'\omega^2R^2 = m'v^2 \tag{10.8}$$

where F_c = equivalent inertia-force (lb or N)

ω = angular velocity of the sheave (rad/s)

R = radius of the same sheave (in or m)

v = belt velocity (in/s or m/s)

$m' = w'/g$ = mass per unit length of belt (lb · s²/in² or kg/m)

w' = weight per unit length of belt (lb/in)

g = acceleration of gravity = 386 in/s²

Caution: When metric units are used, the relationship between mass and force requires that m' be in kg/m and that R be in meters. Avoid millimeter units here.

Note that F_c, the equivalent inertia-force, is proportional to the square of the rotation speed. The inertia effect influences our design of high-speed belt drives; it is negligible in low-speed drives.

V-BELTS

V-belts are ordinarily made of reinforced rubber and may be fabric-wrapped to reduce abrasion. Nylon, polyamide, steel, or other reinforcing material carries most of the tensile load.

The sheaves are grooved so that the tapered sides of the belt make contact with adequate friction force, but not so steep that the belt jams. A sheave is identified by its outside diameters (OD), but speed ratios and other calculations depend on pitch diameter (PD). Commercially available V-belts include A, B, C, D, E, 3V, 5V, and 8V cross sections. Type 3V, 5V, and 8V V-belts are designed to transmit higher loads in less space than the classical (A through E) types. Types 3VX, 5VX, and 8VX are cog V-belts which are interchangeable with types 3V, 5V, and 8V. Table 10.1 gives V-belt dimensions and approximate unit weights.

V-Belt Drive Capacity

Transmitted power depends on sheave speed and torque; but torque is limited by maximum belt tension, friction, and angle of contact, while inertia effects limit speed. For V-belts, the tension relationship at the verge of slippage is

$$\ln[(F_1 - F_c)/(F_2 - F_c)] = f \cdot \theta_1/\sin\beta \tag{10.9}$$

or

$$(F_1 - F_c)/(F_2 - F_c) = e^{f \cdot \theta_1/\sin\beta} \tag{10.10}$$

where F_1 = maximum, i.e., tight-side, belt tension based on drive capacity (lb or N)

F_2 = minimum, i.e., slack-side, belt tension (lb or N)

F_c = equivalent inertia-force (lb or N)

ln = natural (Naperian) logarithm

f = coefficient of friction of the belt on the sheave

θ_1 = minimum angle of contact ordinarily on the smaller sheave (rad)

β = half the included angle of the sheave groove

The tension relationship given in Equations (10.9) and (10.10) occurs when the drive is operating at maximum torque capacity. When the drive is operating below capacity, the ratio $(F_1 - F_c)/(F_2 - F_c)$ is lower.

Service Factor and Design Power

When a belt drive is subject to severe loading or shock loading, or when it is operated continuously, a service factor should be used. The service factor is 1.0 for a normal-torque electric motor driving a steady load such as a light-duty blower for a few hours per day. A crusher or hoist in continuous service may require a service factor of 1.8. Typical service factors for various applications are given in Table 10.2. We multiply the drive unit rating by the service factor to determine the design power.

Drive Sheaves and Idlers

V-belt drive sheaves are available with a single groove or with multiple grooves. If separate belts are used on a multigroove sheave, they should be purchased in matched sets. When necessary, all should be replaced at the same time because belts undergo permanent stretching

Table 10.1 V-belts and sheaves.

Belt Type	Dimension* w	h	Weight (lb/in)	Minimum OD of Sheave	Outside Diameter of Sheave Minus Pitch Diameter (OD − PD)
3V	3/8	5/16	0.0044	2.2	0.05
5V	5/8	17/32	0.012	7.1	0.1
8V	1	7/8	0.035	12.5	0.2
A	1/2	11/32	0.0072	3	0
B	21/32	7/16	0.011	3.4	−0.01
C	7/8	17/32	0.018	7	0
D	1 1/4	3/4	0.037	12	0

Typical Sheave Groove Included Angles

Belt Type	Sheave Size OD (in)	Included Angle 2β (degrees)*
3V	<3.5	36
3V	3.5 to 6	38
3V	6.01 to 12	40
3V	>12	42
5V	<10	38
5V	10 to 16	40
5V	>16	42
8V	<16	38
8V	16 to 22.4	40
8V	>22.4	42

*Refer to Figure 10.1.

during use. Band-belts consisting of two or more V-belts joined with a single back are also available.

Light-duty fractional horsepower drives often employ sheaves made of aluminum, stamped steel, plastic, or die-cast zinc. Cast iron sheaves are used in industrial service for belt speeds up to about 1200 in/s or 30 m/s for classical belts (types A, B, C, and D) and up to about 1300 in/s or 33 m/s for narrow belts (types 3V, 5V, and 8V). Ductile iron sheaves can be used at higher speeds.

We obtain optimum belt tensions by adjusting the center distance between the driver and driven sheaves or by using an idler. Idlers are used to reduce belt whip for long-center-distance drives, to tension drives with fixed center distance, and to clear obstructions that would otherwise be in the belt path. The slack portion of the belt is preferable to the tight portion for idler location. An idler on the outside of a belt (as in Figure 10.3) should be a flat sheave; the recommended idler diameter is about 4/3 the diameter of the smaller driving sheave. This location improves the angle of contact on the driving sheaves, but it causes reversed bending which reduces belt life. A small idler could result in severe bending.

V-belt sheaves can be used as idlers on the inside of a belt. Idler diameter should equal the smaller driving sheave diameter in this case. Although the inside location reduces angle of contact on the loaded sheaves, it is preferred because reverse bending is avoided.

Table 10.2 Service Factors.

Driven Machine	Driver					
	ac Normal-Torque Electric Motor (NEMA Design A–B)[2]			**ac High-Torque Electric Motor (NEMA Design C–D)[3]**		
	Intermittent Service[4]	**Normal Service[5]**	**Continuous Service[6]**	**Intermittent Service[4]**	**Normal Service[5]**	**Continuous Service[6]**
Agitators for liquids Blowers and exhausters Centrifugal pumps and compressors Conveyors (light duty) Fans (up to 10 hp)	1.0	1.1	1.2	1.1	1.2	1.3
Belt conveyors for sand, grain, etc. Fans (over 10 hp) Generators Laundry machinery Line shafts Machine tools Mixers (dough) Positive-displacement rotary pumps Printing machinery Punches, presses, shears[1] Revolving and vibrating screens	1.1	1.2	1.3	1.2	1.3	1.4
Blowers (positive-displacement) Brick machinery Compressors (piston)[1] Conveyors (drag, pan, screw) Elevators (bucket) Exciters Hammer mills Paper mill beaters Pulverizers Pumps (piston) Sawmill and woodworking machinery Textile machinery	1.2	1.3	1.4	1.4	1.5	1.6

Table 10.2 *continued*

Driven Machine	Driver: ac Normal-Torque Electric Motor (NEMA Design A–B)[2]			Driver: ac High-Torque Electric Motor (NEMA Design C–D)[3]		
	Intermittent Service[4]	Normal Service[5]	Continuous Service[6]	Intermittent Service[4]	Normal Service[5]	Continuous Service[6]
Crushers (giratory, jaw, roll)[1] Mills (ball, rod, tube)[1] Hoists[1] Rubber calenders, extruders, mills[1]	1.3	1.4	1.5	1.5	1.6	1.8

Source: Courtesy of T. B. Wood's Sons Company.

[1]The driven machines listed are representative samples only. When one of the sheaves of the drive is used as a flywheel to reduce speed fluctuations and equalize the energy exerted at the shaft, or when it is used for applications involving impact or jam loads, specially constructed sheaves may be required. Consult the manufacturer.

[2]Included under this heading are the following electric motors: synchronous and squirrel-cage ac normal-torque, ac split-phase, dc shunt-wound, and internal combustion engines.

[3]Included under this heading are the following electric motors: ac High-Torque, ac High-Slip, ac Repulsion, Induction, ac single-phase series-wound, ac slip-ring, and dc compound-wound.

[4]Intermittent service refers to 3–5 hours of daily or seasonal operation.

[5]Normal service indicates 8–10 hours of daily operation.

[6]Continuous Service refers to 16–24 hours of daily operation.

[7]If idlers are used, add the following to the service factor:

Idler on slack side (inside)	None
Idler on slack side (outside)	0.1
Idler on tight side (inside)	0.1
Idler on tight side (outside)	0.2

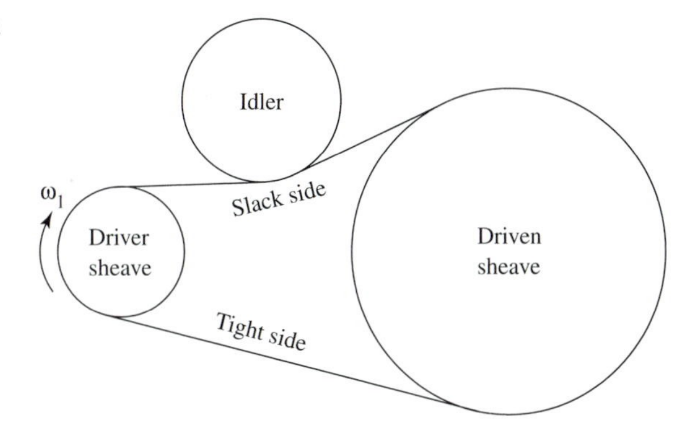

Figure 10.3 Belt drive with idler.

V-BELT DRIVE DESIGN AND SELECTION

One possible V-belt drive design procedure is as follows:

1. Typical V-belt drives are 90% to 98% efficient. Taking this into account, we select a suitable drive unit. We then multiply the service factor for the particular application (Table 10.2) by the power rating of the drive unit to obtain design power.
2. We tentatively select a belt type; Figure 10.4 can be used for selection of 3V, 5V, and 8V belt cross sections. We select driver and driven sheaves to produce the required speed ratio, observing recommended minimum diameters (Table 10.1). For electric motor drives, Table 10.3 gives minimum sheave diameter recommendations based on motor horsepower and speed. If possible, both sheaves should be selected from standard stock sizes. If the required speed ratio cannot be obtained with standard sheaves, it is usually most economical to use a custom-made small sheave and select a standard size for the larger sheave.
3. If the center distance between sheaves is not governed by other considerations, then the recommended center distance can be obtained from

$$c = \text{the larger of} \begin{cases} (3D_1 + D_2)/2 \\ D_2 \end{cases} \tag{10.11}$$

where D_1 and D_2 = small and large sheave diameters, respectively

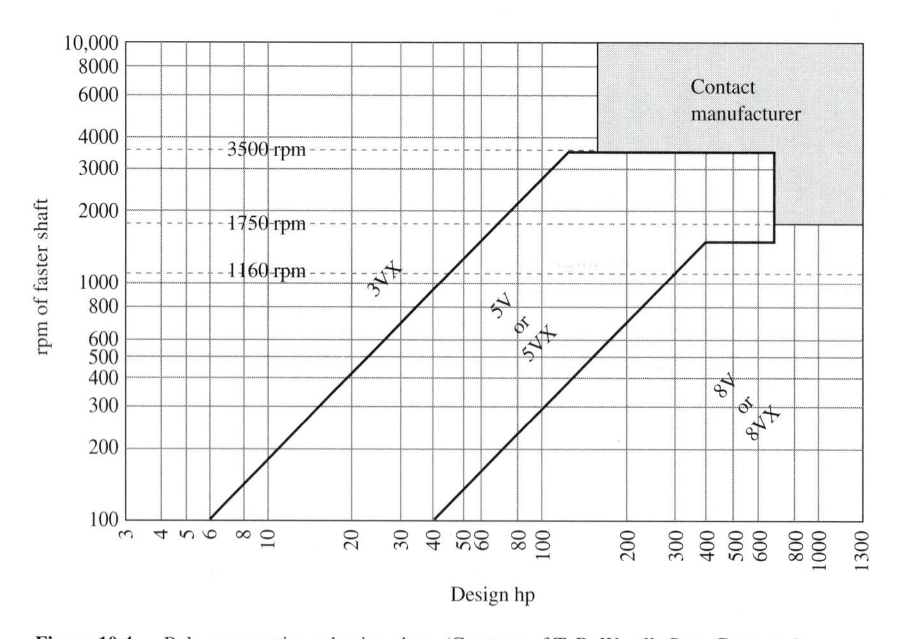

Figure 10.4 Belt cross section selection chart. (Courtesy of T. B. Wood's Sons Company)

Table 10.3 Minimum recommended sheave diameters in inches for electric motors*

	Motor rpm			
Motor hp	**870**	**1160**	**1750**	**3500**
½	2.2	—	—	—
¾	2.4	2.2	—	—
1	2.4	2.4	2.2	—
1½	2.4	2.4	2.4	2.2
2	3.0	2.4	2.4	2.4
3	3.0	3.0	2.4	2.4
5	3.8	3.0	3.0	2.4
7½	4.4	3.8	3.0	3.0
10	4.4	4.4	3.8	3.0
15	5.2	4.4	4.4	3.8
20	6.0	5.2	4.4	4.4
25	6.8	6.0	4.4	4.4
30	6.8	6.8	5.2	—
40	8.2	6.8	6.0	—
50	8.4	8.2	6.8	—
60	10.0	8.2	7.4	—
75	10.0	10.0	8.6	—
100	12.0	10.0	8.6	—
125	—	12.0	10.5	—
150	—	—	10.5	—
200	—	—	13.2	—
250	—	—	—	—
300	—	—	—	—

Source: Courtesy of T. B. Wood's Sons Company.
*NEMA standards.

This value can be adjusted to use a standard belt length. If a definite center distance is specified, then it may be necessary to use an idler.

4. We next calculate the contact angle for the smaller sheave and the inertia effect. If an idler is used, it may affect contact angle.
5. We calculate small sheave torque from design power and rotation speed. When a V-belt drive is operating at capacity (on the verge of slipping), then we find maximum tension by combining Equations (10.6) and (10.10). The result is

$$F_1 = F_c + [T_1/R_1][e^{f\cdot\theta_1/\sin\beta}/(e^{f\cdot\theta_1/\sin\beta} - 1)] \qquad \textbf{(10.12)}$$

where F_1 = maximum belt tension based on drive capacity and design power

6. Belt stress includes the effects of direct tension and bending at the sheaves, where small sheaves cause the greater bending stress. The bending effect can be assumed to be inversely proportional to sheave radius. In the absence of reliable test data, the

designer can use one of the following equations obtained from regression analysis of belt manufacturers' data:

Belt Selection	*Maximum Allowable Direct Tension*
3VX	$F_{\text{limit}} = 161.5 - 104.4/R_1$
5V	$F_{\text{limit}} = 369.7 - 467.5/R_1$
5VX	$F_{\text{limit}} = 421.5 - 487.7/R_1$
8V and 8VX	$F_{\text{limit}} = 613.3 - 1677.9/R_1$

(10.13)

where F_{limit} = maximum allowable direct tension based on belt strength (lb)

R_1 = pitch radius of smaller sheave; that is, approximate radius of curvature of the belt on the smaller sheave (in)

Recall that 1 lb = 4.448 N and 1 in = 0.0254 m = 25.4 mm.

7. We now compare the maximum allowable direct tension based on belt strength with the maximum tension based on design power (i.e., drive unit capacity adjusted with a service factor). If

$$F_{\text{limit}} \geq F_1$$

then our belt selection is satisfactory. If not, we may consider a larger cross-section belt or use more than one belt in the drive. Figure 10.5 shows a drive using eight matched V-belts.

Alternative Design Procedure for Multiple V-Belt Drives

Multiple V-belt drives are used to satisfy high power transmission requirements. The suggested design procedure involves finding the torque capacity of a single belt. We begin by rearranging Equation (10.10) to obtain

$$F_2 = F_c + (F_1 - F_c)e^{-f \cdot \theta_1/\sin\beta} \tag{10.14}$$

where F_2 = minimum (slack-side) belt tension for a single belt (lb or N)

F_1 = maximum (tight-side) belt tension for a single belt (lb or N)

F_c = equivalent inertia-force (a function of belt speed and mass) for a single belt (lb or N)

F_1 and F_2 apply when the belt is on the verge of slipping (i.e., at capacity).

f = coefficient of friction

θ_1 = angle of contact, ordinarily on smaller sheave, which depends on drive geometry (rad)

β = half of included angle of sheave groove (rad)

The design procedure follows:

1. Determine the drive requirements and perform Steps (1)–(4) as in the preceding V-belt drive design and selection procedure.
2. Calculate F_2 from Equation (10.14), where $F_1 = F_{\text{limit}}$ for the selected belt.

Figure 10.5 Multiple V-belt drive. (Courtesy of T. B. Wood's Sons Company)

3. Calculate small sheave torque capacity per belt as follows:

$$T_{1B} = (F_1 - F_2)R_1$$

where R_1 = sheave 1 pitch radius (in or m)

4. Calculate design torque from

$$T_{D1} = \frac{63{,}025P_{\text{hp}}}{n_1} \quad \text{or} \quad T_{D1m} = \frac{10^3 P_{\text{kW}}}{\omega_1}$$

where T_{D1} and T_{D1m} = torque on sheave 1 (lb · in or N · m)
P_{hp} and P_{kW} = design power with service factor applied (hp or kW)
n_1 = speed of sheave 1 (rpm)
ω_1 = angular velocity of sheave 1 (rad/s)

5. The required number of belts is given by

$$N_{\text{belts}} \geq \frac{T_{D1}}{T_{1B}} \quad \text{or} \quad N_{\text{belts}} \geq \frac{T_{D1m}}{T_{1B}}$$

where torque units must be consistent.

Note that actual belt tensions depend on the power transmitted at a given instant. Our calculations of belt tensions F_1 and F_2 are based on design power, which depends on the drive unit (usually an electric motor or internal combustion engine) and the service factor. These values are also limited by the coefficient of friction, the sheave groove angle, and the angle of contact. In most belt drives, the smaller sheave has the smaller angle of contact, and it governs the design. In some situations, we might have to consider both driver and driven sheaves when calculating tensions. These include designs incorporating an idler or sheaves made of different materials.

EXAMPLE PROBLEM Belt Drive Design

Design a belt drive to transmit 20 hp. Driver sheave (pulley) speed is 450 rpm, and driven sheave speed is to be approximately 350 rpm. Power will be supplied by an ac high-torque electric motor; the driven unit is a crusher, which will be operated about 16 hours per day.

Table 10.4 V-belt spreadsheet.

	A	B	C	D	E	F	G	H	I	J	K	L
1	Belt drive design. in-lb-s-rad, rpm, hp units											
2	Smaller OD's available depending on number of grooves.											
3	Larger diameters required for higher power applications.											
4												
5	Power hp	Speed n1	Speed n2	Torque1	Frict f	Serv K	Omega1	Grav g				
6	20	450	350	2801.111	.25	1.8	47.12389	386				
7												
8	Belt sec	OD	OD-PD	Weight w'	R1	R2	OD2	c dist	Alpha	L1 str	Length L	Theta1
9												
10	3VX	6	.05	.0044	2.975	3.825	7.7	12.85	.0661962	12.82186	47.11908	3.0092
11	3VX	6.9	.05	.0044	3.425	4.403571	8.857143	14.77857	.066264	14.74614	54.21615	3.009065
12	3VX	8	.05	.0044	3.975	5.110714	10.27143	17.13571	.0663262	17.09804	62.89034	3.00894
13	3VX	10.6	.05	.0044	5.275	6.782143	13.61429	22.70714	.0664219	22.65707	83.39299	3.008749
14	3VX	14	.05	.0044	6.975	8.967857	17.98571	29.99286	.0664934	29.92658	110.2041	3.008606
15												
16												
17	5V	7.1	.1	.012	3.5	4.5	9.1	15.2	.065837	15.16707	55.59855	3.009919
18	5V	12.5	.1	.012	6.2	7.971429	16.04286	26.77143	.066217	26.71276	98.18097	3.009159
19	5V	16	.1	.012	7.95	10.22143	20.54286	34.27143	.0663262	34.19607	125.7807	3.00894
20	5V	21.2	.1	.012	10.55	13.56429	27.22857	45.41429	.0664219	45.31414	166.786	3.008749
21												
22												
23	5VX	7.1	.1	.012	3.5	4.5	9.1	15.2	.065837	15.16707	55.59855	3.009919
24	5VX	12.5	.1	.012	6.2	7.971429	16.04286	26.77143	.066217	26.71276	98.18097	3.009159
25	5VX	16	.1	.012	7.95	10.22143	20.54286	34.27143	.0663262	34.19607	125.7807	3.00894
26	5VX	21.2	.1	.012	10.55	13.56429	27.22857	45.41429	.0664219	45.31414	166.786	3.008749

```
A10: ' 3VX
B10: 6
C10: .05
D10: .0044
E10: (B10-C10)/2
F10: +E10*$B$6/$C$6
G10: 2*F10+C10
H10: @IF((G10+3*B10)/2>G10,(G10+3*B10)/2,G10)
I10: @ASIN((F10-E10)/H10)
J10: (H10^2-(F10-E10)^2)^.5
K10: 2*J10+E10*(@PI-2*I10)+F10*(@PI+2*I10)
L10: @PI-2*I10
M10: 38
N10: @PI*M10/360
O10: +D10*$G$6^2*E10^2/$H$6
P10: @EXP($E$6*L10/@SIN(N10))
Q10: +O10+($D$6/E10)*(P10/(P10-1))
R10: +Q10-$D$6/E10
S10: ((Q10+R10*@COS(2*I10))^2+(R10*@SIN(2*I10))^2)^.5
T10: (Q10-R10)*E10*$B$6/63025
U10: (Q10-R10)*F10
V10: +Q10*$F$6
W10: 161.5-104.4/(B10/2)
X10: @INT(V10/W10+.99)
```

Solution: In order to evaluate different possibilities, we will use a spreadsheet in the solution. We will estimate the coefficient of friction f to be 0.25 for this application. According to Table 10.2, a service factor of $K = 1.8$ applies. According to the recommendations of Table 10.3, the smaller sheave diameter should not be less than 6 in. Various sheave diameters using 3VX, 5V, and 5VX belts will be considered, because the power and speed requirements are close to the line dividing 3VX from 5V and 5VX belts in Figure 10.4.

Table 10.4, a spreadsheet for comparison of various V-belt drive combinations, is based on the preceding equations in this chapter. The spreadsheet formulas represent the equations directly, except that each constant or variable is identified by its cell, labeled according to the intersection of a column and

M	N	O	P	Q	R	S	T	U	V	W	X
2Beta-deg	Beta	Fc	Gamma	Tens F1	Tens F2	Res Load	HP chk	Torque2	K F1	F(nom)	Belts req (approx)
38	.3316126	.2240379	10.08178	1045.449	103.8986	1148.52	20	3601.429	1881.807	126.7	15
40	.3490659	.2969401	9.020313	920.111	102.2684	1021.572	20	3601.429	1656.2	131.2391	13
40	.3490659	.3999649	9.019493	792.9532	88.27111	880.5262	20	3601.429	1427.316	135.4	11
40	.3490659	.7043566	9.018232	597.9468	66.93047	664.3467	20	3601.429	1076.304	141.8019	8
42	.3665191	1.231505	8.156577	458.9397	57.34674	515.8362	20	3601.429	826.0915	146.5857	6
38	.3316126	.8456921	10.08734	889.2326	88.91516	977.4478	20	3601.429	1600.619	238.0099	7
40	.3490659	2.653747	9.020934	510.7725	58.98037	569.2897	20	3601.429	919.3905	294.9	4
40	.3490659	4.363254	9.019493	400.6398	48.29883	448.5598	20	3601.429	721.1517	311.2625	3
42	.3665191	7.68389	8.157391	310.2877	44.77955	354.7223	20	3601.429	558.5179	325.5962	2
38	.3316126	.8456921	10.08734	889.2326	88.91516	977.4478	20	3601.429	1600.619	284.1197	6
40	.3490659	2.653747	9.020934	510.7725	58.98037	569.2897	20	3601.429	919.3905	343.468	3
40	.3490659	4.363254	9.019493	400.6398	48.29883	448.5598	20	3601.429	721.1517	360.5375	2
42	.3665191	7.68389	8.157391	310.2877	44.77955	354.7223	20	3601.429	558.5179	375.4906	2

row. Only one row of spreadsheet formulas is shown. Most of these formulas were copied (replicated) in the other rows by a single command. The equation for driven sheave radius

$$R_2 = R_1 n_1 / n_2$$

is represented in cell F10 by the spreadsheet formula

```
+ E10 * $B$6/$C$6
```

When this formula is copied (replicated) in cell F11, it becomes

```
+ E11 * $B$6/$C$6
```

The absolute cell references, `$B$6` and `$C$6`, remain the same, while the relative cell reference (no `$` signs) changes to the value of R_1 in the next row.

The center distance is not specified for this drive. The formula for center distance in cell H10 is equivalent to the statement

$$\text{If } (3D_1 + D_2)/2 > D_2, \text{ then use center distance} = (3D_1 + D_2)/2;$$

$$\text{if not, use center distance} = D_2$$

Belt tension F_1 is calculated in two steps, formulas P10 and Q10.

Belt tension in column Q is based on the specified power (20 hp). We multiply it by the service factor to obtain the belt tension values in column V. These values exceed the allowable tensions given in column W. A single belt will be inadequate for any of the drive combinations we considered. The approximate number of belts required for each configuration is given in column X. Note that the equivalent inertia-force calculation was based on a single belt. However, inertia-force is not an important factor at the low speeds considered in the problem.

After checking stock belt lengths and sheave diameters, we might select a drive made up of one 5V12.5–4 sheave, one 5V16–4 sheave, and four 5V-belts about 100 in long. Sheave designation begins with belt type, followed by outside diameter and number of grooves. Center distance is then adjusted to accommodate the standard belt length. There are many other possible solutions to this design problem.

Check: To check our belt selection in this design problem, we will find the capacity of a single 5V-belt by setting maximum tension F_1 equal to maximum allowable direct tension F_{limit}. Equation (10.10) can be rewritten in the form

$$F_2 = F_c + (F_1 - F_c)/e^{f \cdot \theta_1 / \sin \beta}$$

where F_c is given by Equation (10.8) and

$$e^{f \cdot \theta_1 / \sin \beta} = e^{0.25 \cdot 3.009 / \sin(0.349)}$$

from which minimum tension is given by

$$F_2 = 2.654 + (294.9 - 2.654)/9.021 = 35.05 \text{ lb}$$

Torque capacity of a single belt on sheave 1 is given by

$$T_{D1} = (F_1 - F_2)R_1 = (294.9 - 35.05)6.2 = 1611 \text{ in} \cdot \text{lb}$$

Using the horsepower–torque relationship, we find required torque on sheave 1.

$$T_{1(\text{REQUIRED})} = 2801 \text{ in} \cdot \text{lb}$$

The required torque multiplied by the service factor is greater than three times the torque capacity of the single belt. As a result, we specify four 5V-belts, verifying the above solution. ■■

Actual service conditions are difficult to predict. Coefficient of friction, for example, is dependent on environmental conditions. We assumed a controlled environment in the preceding example problem and used $f = 0.25$. Some manufacturers are less optimistic and use a lower value to account for the possibility of moisture or oil in the sheaves. One manufacturer recommends about 6 hp/belt for the belt and sheave sizes selected and the speeds specified. We find the design power by multiplying the 20 hp nominal power by a service factor of 1.8 to obtain 36 hp. Based on this recommendation, we would use 36/6 = 6 belts for the drive instead of four.

REGRESSION ANALYSIS: USER-WRITTEN PROGRAMS

We often obtain sets of data by experimental or other means. These data can be simply tabulated and "looked up" when needed, or they can be entered into a computer database system. For computer use, it is sometimes convenient to "fit" an equation to the data. This process, called **regression analysis,** was used to obtain the equations for allowable direct tension in a belt. For hints on using preprogrammed regression formulas, see Chapter 4.

In **linear regression,** given a series of data points, $x_1, y_1; x_2, y_2$; and so on, we can try to find a linear equation

$$y = A + Bx \tag{10.15}$$

that will reproduce the data points. Of course, two points determine a line; if there are more points, the equation will, in general, be approximate. The constants are given by

$$B = [n \Sigma xy - \Sigma x \Sigma y]/[n \Sigma x^2 - (\Sigma x)^2] \tag{10.16}$$

$$A = [\Sigma y - B \Sigma x]/n \tag{10.17}$$

where n = number of pairs of data points

$\Sigma x = \Sigma_{i=1}^{n} x_i = x_1 + x_2 + \cdots + x_n$, and so on.

These equations are coded on some software, and some scientific calculators "know" regression equations.

With a bit of ingenuity, the linear regression equations can be used to fit other curves. For example, if we pair the natural logarithm $\ln x$ with y, the resulting equation is a **logarithmic regression**

$$y = A + B \ln x \tag{10.18}$$

We obtain an **exponential regression** by pairing x with $\ln y$, resulting in the form

$$\ln y = \ln A + Bx \quad \text{or} \quad y = Ae^{Bx} \tag{10.19}$$

We obtain a **power regression** by pairing $\ln x$ and $\ln y$, producing the regression equations

$$\ln y = \ln A + B \ln x \quad \text{or} \quad y = Ax^B \tag{10.20}$$

The form of the equation can be based on the result of a number of trials of different regression curves, or it can be based on some knowledge of the physical problem. In the discussion of V-belts, it was observed that bending stress was approximately proportional to inverse sheave radius, leading to an equation in the form

$$F_{\text{limit}} = A + B/R_1 \tag{10.21}$$

where it is expected that constant B will be negative.

EXAMPLE PROBLEM Regression Analysis to Determine Maximum Allowable Direct Tension in a Belt

Tabulated speed and horsepower ratings given by a belt manufacturer were used along with other data to calculate the following relationship:

R_1	F_{limit}
1.1	64.54
1.5	93.21
2.25	117.52
3	128.46
5.3	138.46

(a) Write an equation to approximate this relationship.
(b) How good is the result?

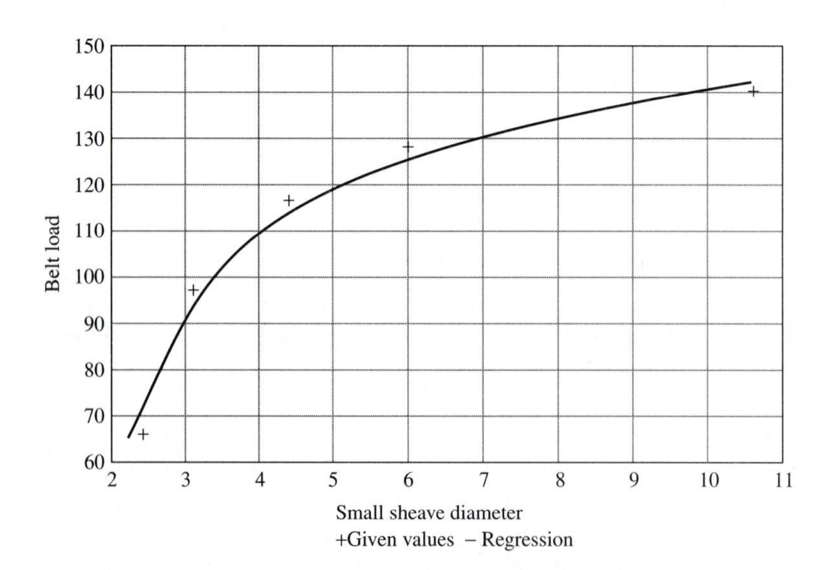

Figure 10.6 Regression analysis.

Solution (a): Let $1/R_1 = x$ and $F_{limit} = y$. Then

$$B = [n \Sigma xy - \Sigma x \Sigma y]/[n \Sigma x^2 - (\Sigma x)^2]$$
$$= [5 \times 241.99 - 2.542 \times 542.19]/5 \times 1.615 - (2.542)^2]$$
$$= -104.4$$
$$A = [\Sigma y - B \Sigma x]/n$$
$$= [542.19 + 104.4 \times 2.542]/5 = 161.5$$

resulting in the regression equation

$$y = A + Bx = 161.5 - 104.4x$$

or

$$F_{limit} = 161.5 - 104.4/R_1$$

This part of the problem was solved using a scientific calculator, and the solution was repeated using a spreadsheet. A user-written computer program would have worked as well.

Solution (b): This appears to be a fair approximation when the calculated values (using the regression equation) are compared with the given values. Thus, we have the following:

R_1	F_{limit} *(Given)*	F_{limit} *(Calculated)*
1.1	64.54	66.6
1.5	93.21	91.9
2.25	117.52	115.1
3	128.46	126.7
5.3	138.46	141.8

Figure 10.6 is a plot of the regression equation (F_{limit} versus D_1) with the given values shown as +'s.

FLAT BELTS

The equations used earlier in this chapter for speed ratio, belt drive geometry, power, torque, and inertia effects apply to flat belts as well as V-belts. The pitch radius of a flat belt is given by the sheave radius plus one-half of the belt thickness.

Flat Belt Drive Capacity

While V-belts contact the sheave sides, flat belts contact the outer surface of the sheaves. Both types depend on friction. V-belts have the advantage of wedging action that produces greater normal forces. Thus, flat belt drives require higher belt tensions for the same power capacity. Flat belts are far less common in power transmission applications than they once were. They are, however, used for conveyors. Conveyor loads include friction along the length of the belt and the work of moving materials against gravity when the conveyor is inclined.

Using the V-belt drive equations with an included sheave angle of 180°, the result is the following tension relationship for flat belts at the verge of slipping:

$$\ln[(F_1 - F_c)/(F_2 - F_c)] = f \cdot \theta_1 \quad (10.22)$$

or

$$(F_1 - F_c)/(F_2 - F_c) = e^{f \cdot \theta_1} \quad (10.23)$$

These equations apply when a flat belt is operating at maximum torque capacity.

EXAMPLE PROBLEM Design of a Flat Belt Drive

Design a drive to transmit 3.5 kW of mechanical power with a speed reduction from 400 to 180 rad/s. The drive is to power a generator in continuous service. Consider the feasibility of a flat steel belt drive.

Decisions: We will examine the feasibility of a 0.2 mm thick by 40 mm wide steel belt. Drive sheave diameters ranging from 50 to 850 mm will be considered. A service factor of $K = 1.4$ will be used for this application. Assume that we have considered the belt material and the service environment and have determined that a coefficient of friction f of 0.1 will be reasonable in this study.

Solution Summary: The spreadsheet shown in Table 10.5 is used in the analysis. Most of the formulas are similar to those used in the example problem "Belt Drive Design." Resultant load, column S, is the vector sum of belt tensions F_1 and F_2. This would be necessary for design of the shaft and selection of bearings. Column W gives direct stress

Table 10.5 Flat belt spreadsheet.

A	B	C	D	E	F	G	H	I	J	K	L
Flat belt drive design. mks mm rad/s kw units											
Thickness .2 Width = 40 Dens = 7750 kg/m^3											
Power	Speed 1	Speed 2	Torque1	Frict f	Serv K		Grav g				
3.5	400	180	8750	.1	1.4		9.8				
	OD	OD - PD	m' kg/m	R1	R2	OD2	c dist	Alpha	L1 str	Length L	Theta1
	50	-.2	.062	25.1	55.77778	111.3556	130.6778	.2369706	127.0258	522.6761	2.667651
	100	-.2	.062	50.1	111.3333	222.4667	261.2333	.2366023	253.9554	1044.044	2.668388
	150	-.2	.062	75.1	166.8889	333.5778	391.7889	.2364794	380.8849	1565.413	2.668634
	200	-.2	.062	100.1	222.4444	444.6889	522.3444	.236418	507.8145	2086.781	2.668757
	250	-.2	.062	125.1	278	555.8	652.9	.2363811	634.744	2608.149	2.66883
	300	-.2	.062	150.1	333.5556	666.9111	783.4556	.2363566	761.6736	3129.518	2.66888
	350	-.2	.062	175.1	389.1111	778.0222	914.0111	.236339	888.6031	3650.886	2.668915
	400	-.2	.062	200.1	444.6667	889.1333	1044.567	.2363258	1015.533	4172.254	2.668941
	450	-.2	.062	225.1	500.2222	1000.244	1175.122	.2363156	1142.462	4693.623	2.668961
	500	-.2	.062	250.1	555.7778	1111.356	1305.678	.2363074	1269.392	5214.991	2.668978
	550	-.2	.062	275.1	611.3333	1222.467	1436.233	.2363007	1396.321	5736.36	2.668991
	600	-.2	.062	300.1	666.8889	1333.578	1566.789	.2362951	1523.251	6257.728	2.669002
	650	-.2	.062	325.1	722.4444	1444.689	1697.344	.2362904	1650.18	6779.096	2.669012
	700	-.2	.062	350.1	778	1555.8	1827.9	.2362863	1777.11	7300.465	2.66902
	750	-.2	.062	375.1	833.5556	1666.911	1958.456	.2362828	1904.04	7821.833	2.669027
	800	-.2	.062	400.1	889.1111	1778.022	2089.011	.2362797	2030.969	8343.201	2.669033
	850	-.2	.062	425.1	944.6667	1889.133	2219.567	.236277	2157.899	8864.57	2.669039

$$\sigma_{\text{direct}} = F_1/(tw) \tag{10.24}$$

where t = belt thickness

w = width

Maximum bending stress, column X, is given by

$$\sigma_{\text{bending}} = Ec/R_1 \tag{10.25}$$

where E = modulus of elasticity

$c = t/2$

Maximum tensile stress in the belt, given by

$$\sigma_{\text{max}} = \sigma_{\text{direct}} + \sigma_{\text{bending}} \tag{10.26}$$

occurs at the outer surface of the belt as it is bent around the smaller (driver) sheave. Minimum stress on that surface is given by

$$\sigma_{\text{MIN}} = \frac{F_2}{tw}$$

when it is straight (between sheaves) on the slack side. Thus, the belt is subject to fatigue loading. The results are tabulated in the spreadsheet of Table 10.5 and are plotted in Figure 10.7.

If the driver sheave is small (say, 50 mm diameter), then the values of F_1 and F_2 must be large to obtain the required torque, and the bending stress is high as well. Bending stress decreases monotonically

M	N	O	P	Q	R	S	T	U	V	W	X	Y
2Beta-deg	Beta	Fc	Gamma	Tens F1	Tens F2	Res Load	kw chk	Torque2	K F1	Dir Str	Bending	Max Str
180	1.570796	6.249699	1.305734	1495	1146	2569	3.50	19444	2093	262	825	1086
180	1.570796	24.8993	1.30583	771	596	1329	3.50	19444	1079	135	413	548
180	1.570796	55.9489	1.305862	553	437	963	3.50	19444	775	97	276	372
180	1.570796	99.3985	1.305878	473	385	834	3.50	19444	662	83	207	289
180	1.570796	155.2481	1.305888	454	384	815	3.50	19444	635	79	165	245
180	1.570796	223.4977	1.305894	472	414	862	3.50	19444	661	83	138	221
180	1.570796	304.1473	1.305899	517	468	958	3.50	19444	724	91	118	209
180	1.570796	397.1969	1.305902	584	540	1093	3.50	19444	817	102	103	206
180	1.570796	502.6465	1.305905	669	630	1262	3.50	19444	936	117	92	209
180	1.570796	620.4961	1.305907	770	735	1463	3.50	19444	1078	135	83	217
180	1.570796	750.7457	1.305909	887	855	1693	3.50	19444	1241	155	75	230
180	1.570796	893.3953	1.30591	1018	989	1951	3.50	19444	1425	178	69	247
180	1.570796	1048.445	1.305911	1163	1136	2236	3.50	19444	1629	204	64	267
180	1.570796	1215.894	1.305912	1323	1298	2547	3.50	19444	1852	231	59	291
180	1.570796	1395.744	1.305913	1495	1472	2885	3.50	19444	2093	262	55	317
180	1.570796	1587.994	1.305914	1681	1659	3248	3.50	19444	2354	294	52	346
180	1.570796	1792.643	1.305915	1881	1860	3637	3.50	19444	2633	329	49	378

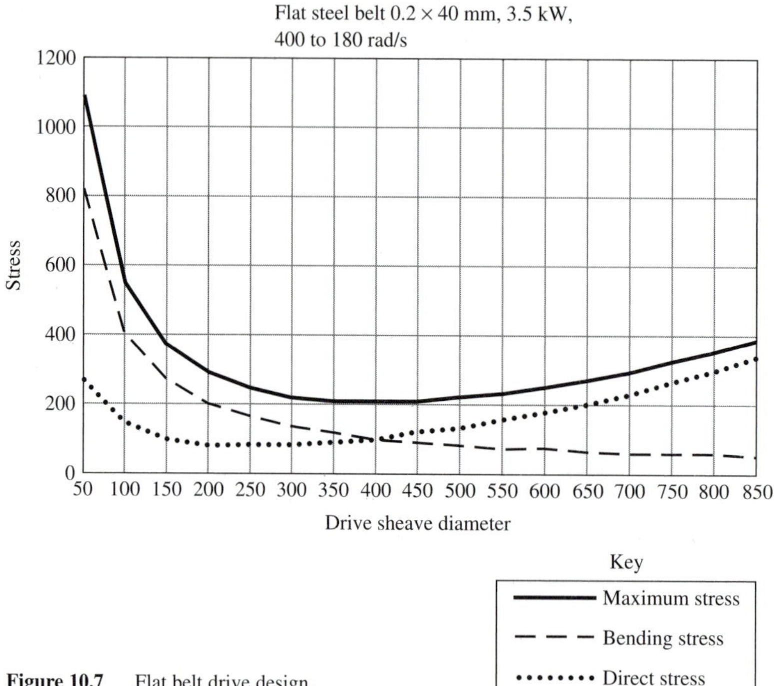

Figure 10.7 Flat belt drive design.

as driver sheave radius is increased. Direct stress decreases to a minimum of about 79 MPa at driver sheave diameter near 250 mm; then it increases because the effect of inertia on larger sheaves dominates. The smallest value of σ_{MAX}, about 206 MPa, occurs with a driver sheave diameter of about 400 mm. Thus, considering only σ_{MAX}, the optimum design for the drive may be a 400 mm diameter driver sheave and a driven sheave diameter of 889 mm.

Note that V-belt and gear drives are usually used for applications of this type. This feasibility study only touched on limited aspects of the proposed design. Extensive analysis is required, including safety and performance studies.

SYNCHRONOUS BELT (TIMING BELT) DRIVES

V-belt and flat belt drives which rely on friction cannot provide a precise speed ratio, because slippage occurs at the sheaves. Certain applications require an exact output-to-input speed ratio. One such example is the timing system of a four-stroke-cycle engine, where the camshaft must turn at precisely one-half of the crankshaft speed. Synchronous belts have teeth that mesh with grooves in the sheaves. Most synchronous belts are reinforced with glass fiber, steel, or aramid. Output-to-input speed ratio is given by

$$n_2/n_1 = \omega_2/\omega_1 = N_1/N_2 \tag{10.27}$$

where N_1 and N_2 = tooth numbers for the driver and driven sheaves, respectively

Figure 10.8 shows a belt drive that will provide a precise speed ratio.

Figure 10.8 Belt drive for precise speed ratio. (Courtesy of T. B. Wood's Sons Company)

VARIABLE-SPEED DRIVES

The speed ratio of a V-belt drive can be varied if one or both sheaves have a variable pitch diameter. Variable-speed sheaves are shown in Figure 10.9. If the flanges are moved closer together, the pitch diameter is effectively increased. Some sheaves incorporate a rack and pinion drive to move the flanges in opposite directions, causing the belt to travel in a fixed centerline. The opposite pulley may incorporate spring-loaded flanges so that belt tension is maintained. Flange movement of spring-loaded sheaves can also be accomplished by changing the distance between the driver and driven shafts. Variable-speed drives can be pneumatically or electrically controlled, allowing programming of a predetermined cycle of speed ratio variation.

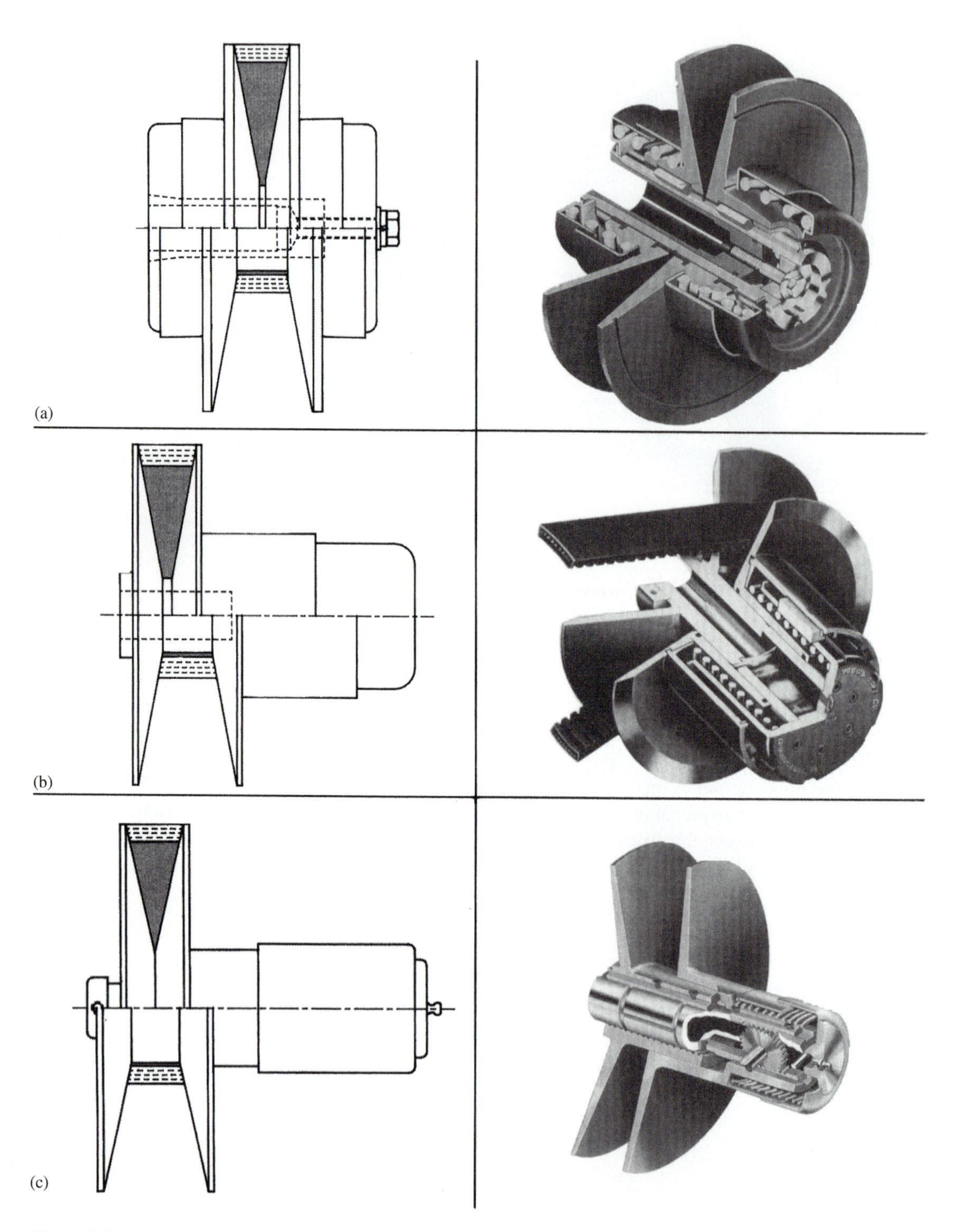

Figure 10.9 Variable-speed sheaves. (a) two-flange movable sheave; (b) one-flange movable sheave; (c) two-flange movable sheave with rack-and-pinion drive. (Courtesy of T. B. Wood's Sons Company)

Design Problems

10.1. Design a V-belt drive for the following requirements:

10 hp transmitted

Input speed = 1160 rpm

Output speed = 800 rpm

$f = 0.25$

Service factor = 1.5

Consider various sheave diameters.

10.2. Design a V-belt drive for the following requirements:

45 hp transmitted

Input speed = 1750 rpm

Output speed = 900 rpm

$f = 0.2$

Service factor = 1.6

Consider various sheave diameters.

10.3. Suppose that allowable direct tension for a particular V-belt section is given for various sheave diameters as follows:

Diameter D_l (in)	*Tension $F_{1(\text{limit})}$ (lb)*
7.1	232.84
9	272.87
16	315.40
18.7	313.48

Approximate this relationship with a regression equation.

Suggestion: Use the form $F_{1(\text{limit})} = A + B/R_1$.

10.4. Suppose that allowable direct tension for a particular V-belt section is given for various sheave diameters as follows:

Diameter D_l (in)	*Tension $F_{1(\text{limit})}$ (lb)*
4.4	194.28
8	310.60
12.5	349.93
18.7	357.33

Approximate this relationship with a regression equation.

Suggestion: Use the form $F_{1(\text{limit})} = A + B/R_1$.

10.5. Suppose that allowable direct tension for a particular V-belt section is given for various sheave diameters as follows:

Diameter D_l (in)	*Tension $F_{1(\text{limit})}$ (lb)*
12.5	339.84
15	393.31
18	433.21
24.8	472.75

Approximate this relationship with a regression equation.

Suggestion: Use the form $F_{1(\text{limit})} = A + B/R_1$.

10.6. Design a flat belt to transmit 3 hp. The input speed is 500 rpm, and the output speed 250 rpm. Consider 0.005 in thick by 2 in wide steel (density = 0.28 lb/in^3). Assume a coefficient of friction of 0.05 and a service factor of 1.8.

10.7. Design a flat belt to transmit 3 hp. The input speed is 2500 rpm, and the output speed 1200 rpm. Consider 0.008 in thick by 2 in wide steel (density = 0.28 lb/in^3). Assume a coefficient of friction of 0.1 and a service factor of 1.4.

CHAPTER 11

Chain Drives

Symbols

c center distance
d pitch diameter
K_C chordal action ratio
K_S service factor
n speed
N number of teeth in sprocket or chainwheel
P pitch
P_B horsepower rating based on roller bushing failure
P_L horsepower rating based on linkplate failure

Units

pitch and distances	in
power	hp
speed	rpm
velocity	fpm

Chain drives were probably developed more than 2000 years ago for moving and lifting materials. They have been used in bicycles and automobiles for about 100 years. Chains drove the propellers on the first powered flight and have been used in recent years in the drive train of a human-powered aircraft.

Commercially available drive chain types include the following:

- roller chain, an assembly of linkplates, pins, rollers, and bushings
- silent chain, an assembly of pins, bushings, and links with inward-pointing teeth
- bead chain, an assembly of hollow beads and wire links
- ladder chains, an assembly of interlocking wire links

DRIVE SELECTION

The most common machine drives are based on gears, belts, or chains, or on combinations of these. Most chain drives for power transmission and conveyors are of the roller chain type or silent chain (inverted-tooth) type. Figure 11.1 shows a silent chain drive. Power transmission capacity of chain drives ranges from a fraction of a horsepower to 2000 hp. Each drive type has advantages and drawbacks, some of which are described in the following lists.

Considerations Favoring Selection of Chain Drives

- **Timing and phase relationships.** Chain drives, like gear drives, can produce exact input-to-output speed ratios. V-belts and flat belts that rely on friction cannot be used where precise phase relationships must be maintained.
- **Tolerances and relative motion between shafts.** Chain drives and belt drives have the advantage of being less sensitive than gears to shaft placement.
- **Large spans.** Chains can span relatively long distances between driver and driven shafts.

Figure 11.1 Chain drive. (Courtesy of Ramsey Products Corporation)

- **Capacity and service life.** If we consider size constraints, chain drives tend to have higher capacity and longer service lives than belts, particularly at elevated temperatures.
- **Cost.** Chain drives are usually less expensive than comparable gear drives.
- **Efficiency.** Chain drives are usually more efficient than belt drives.
- **Space limitations.** Chain drives often require less space than belt drives.
- **Shaft loading.** Shaft loading on chain drives is less than that on equivalent belt drives because the slack-side tension is less on chain drives.

Considerations When Choosing between Types of Chain Drives

- **Cost.** For a given capacity, roller chain drives are usually lower in cost than silent chain drives.
- **Weight.** For a given capacity, roller chain is narrower and lighter than silent chain.
- **Tolerances and misalignment.** Roller chain is more flexible and can accept more misalignment than silent chain.
- **Speed.** Silent chain can operate at higher speed than roller chain (up to 7000 feet per minute in some applications).
- **Conveying applications.** Silent chain has a smooth flat surface which can be used for conveying.
- **Special low-power applications.** Bead chain is made with hollow beads and dumbbell-shaped wire links. Ladder chain is made of interlocking links of wire. Both are very flexible, allowing almost unlimited sprocket orientation. Proper operation of roller chain and silent chain drives requires that a pair of pulleys operate in the same plane, or very nearly so. One bead chain or ladder chain sprocket plane can be 90° relative to the other sprocket plane.

Considerations Favoring Selection of Other Drive Types

- **Service life.** Typical gear drives have longer service lives than chain drives.
- **Space limitations.** For a given capacity, gear drives are usually more compact than chain or belt drives.
- **Cost.** Belt drives usually cost less than equivalent chain drives.
- **Lubrication.** Chain and gear drives may require lubrication systems.
- **Speed ratio.** Single-step input-to-output speed ratios are usually limited to about 8:1. By comparison, 50:1 or 100:1 ratios are common with worm drives.
- **Speed changing.** Although variable-speed chain and belt drives are available, gear drives are usually selected for high-capacity speed changers.
- **Chordal action.** The effective diameter of a sprocket varies due to the finite length of chain links, resulting in some speed variation and vibration. Chordal action is most significant if the sprocket has a small number of teeth.

ROLLER CHAIN

Power transmission roller chain is an assembly of steel linkplates, pins, rollers, and bushings. Figure 11.2 shows a roller chain link assembly. Linkplates, which are the primary tension members, can be made of thru-hardened, medium carbon, fine-grain steel. Linkplates and

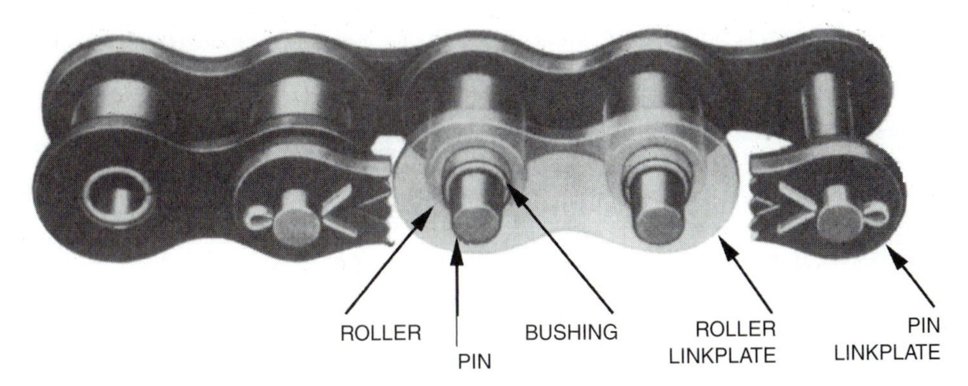

Figure 11.2 Roller chain link assembly. (Courtesy of Diamond Chain Company)

rollers may be shot-peened to produce residual compressive stresses on the surface. Roller chain sprockets are toothed wheels shaped to mesh with the chain.

Pitch and Pitch Diameter

Chain **pitch** is the distance between pin centers. The **pitch diameter** of a sprocket is the diameter of a circle that is the locus of chain pin centers. Pitch diameter is given by

$$d = \frac{P}{\sin\left[\dfrac{\pi}{N}\right]} \tag{11.1}$$

where d = sprocket pitch diameter
P = pitch
N = number of teeth

Note that π/N, the argument of the sine function, is in radians, not degrees. If using a calculator without a radian setting, replace π with 180°.

Speed Ratio and Chain Velocity

Speed ratio depends on the numbers of teeth in the driver and the driven sprockets; that is,

$$\frac{n_2}{n_1} = \frac{N_1}{N_2} \tag{11.2}$$

and chain velocity is given by

$$V = PN_1n_1/12 \tag{11.3}$$

where n = rotation speed (rpm)
N = number of teeth
P = pitch (in)
V = chain velocity (fpm)
Subscripts 1 and 2 refer to driver and driven sprockets, respectively.

Caution: Industry practice is to express chain velocity in feet per minute (fpm). As a result, Equation (11.3) has mixed units.

Power transmission roller chain sprockets are generally available with a minimum of 6 teeth. The maximum number of teeth in off-the-shelf sprockets usually ranges up to about 120 or 150 teeth.

The driver and driven sprockets in power transmission chain drives ordinarily rotate in the same direction. One exception is a chain drive with one input sprocket and two or more output sprockets. If one of the sprockets engages the chain on the outside of the loop, then its rotation direction will be opposite that of the other sprockets. **Idler sprockets** are sometimes used to take up chain slack. Unlike idler gears, idler sprockets do not affect rotation direction of the driven shaft.

Figure 11.3 shows several preferred chain drive arrangements. The arrows indicate driver sheave rotation direction. If one sheave is directly above another, an idler sprocket may be necessary. Idler sprockets are located on the slack span of the chain. Special-purpose chains, including bead chains and ladder chains, can operate on sprockets at almost any angle, even running crossed in a figure-eight configuration. Out-of-plane configurations may require

Figure 11.3 Suggested chain drive arrangements (preferred rotation direction of driver shown). (Courtesy of Diamond Chain Company)

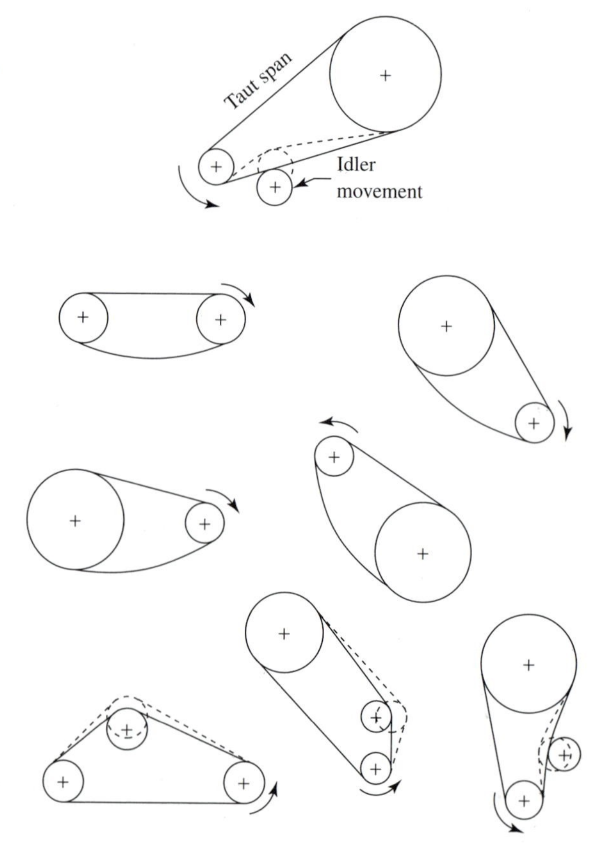

idlers or guides to prevent the chain from touching itself. Ladder and bead chains are used in recorders, tuners, and other low-torque devices.

Chordal Action. Consider the pitch circle of a sprocket, the locus of chain pin centers. The links can be thought of as chords of that circle. Let a straight, horizontal section of chain approach the top of the sprocket. The section of chain is at its highest position when a pin is at the top of the sprocket. It falls to a lower position when the center of a link is at the top. This effect, called **chordal action,** is most severe when a sprocket has a small number of teeth. The total radius change divided by the sprocket pitch radius is given by

$$K_C = 1 - \cos(\pi/N) \tag{11.4}$$

where $K_C = \delta/r$ = chordal action ratio

δ = total motion of chain axis

r = sprocket pitch radius

N = number of sprocket teeth

The argument of the cosine is in radians.

For smooth operation, the recommended minimum numbers of teeth in the small sprocket are as follows:

12 teeth for slow-speed operation

17 teeth for medium-speed operation

25 teeth for high-speed operation

ROLLER CHAIN POWER RATING

Low-Speed Failure

At relatively low speeds, linkplate failure will govern chain drive capacity. Horsepower rating is proportional to the 0.9 power of rotation speed in the low-speed range. It is given by

$$P_L = 0.004 \cdot n_1^{0.9} \cdot N_1^{1.08} \cdot P^{3-0.07 \cdot P} \tag{11.5}$$

where P_L = horsepower rating based on linkplate failure

n_1 = speed of small sprocket (rpm)

N_1 = number of teeth in small sprocket

P = pitch (in)

High-Speed Failure

At high speeds, roller bushing failure will govern. Horsepower rating is proportional to the -1.5 power of rotation speed in the high-speed range. It is given by

$$P_B = \frac{17{,}000 \cdot N_1^{1.5} \cdot P^{0.8}}{n_1^{1.5}} \tag{11.6}$$

where P_B = horsepower rating based on roller bushing fatigue

Power Ratings

Power rating is based on the lower of the two results based on linkplate failure and roller bushing failure. Peak power is obtained at the speed where the two curves cross.

EXAMPLE PROBLEM Roller Chain Velocity, Chordal Action, and hp Rating

A 1 in pitch roller chain drive will utilize a 1000 rpm, 25 tooth drive sprocket and a 50 tooth driven sprocket.

Find chain velocity.
Evaluate chordal action.
Determine the hp rating.

Solution:

Small Sprocket Teeth:

$$N_1 = 25$$

Pitch:

$$P = 1$$

Speed (rpm):

$$n_1 = 1000$$

Velocity (fpm):

$$V = \frac{P \cdot N_1 \cdot n_1}{12} \qquad V = 2.083 \cdot 10^3$$

Chordal Action:

$$K_C = 1 - \cos\left(\frac{\pi}{N_1}\right) \qquad K_C = 0.008$$

Capacity Based on Linkplate Fatigue:

$$P_L = 0.004 \cdot n_1^{0.9} \cdot N_1^{1.08} \cdot P^{3-0.07 \cdot P} \qquad P_L = 64.839$$

Capacity Based on Roller Bushing Fatigue:

$$P_B = \frac{17{,}000 \cdot N_1^{1.5} \cdot P^{0.8}}{n_1^{1.5}} \qquad P_B = 67.198$$

$$P_{LB} = \begin{pmatrix} P_L \\ P_B \end{pmatrix}$$

Capacity (hp):

$$P_{\text{hp}} = \min(P_{LB}) \qquad P_{\text{hp}} = 64.839$$

The chordal action ratio is less than 1%, indicating smooth operation at medium or high speed. The drive rating is 64.8 hp, governed by linkplate fatigue. Further calculations would show that we are

Table 11.1 Pitch, strength, and power rating of single-strand roller chain.

ANSI Number	Pitch (in)	Average Strength (lb)	Maximum Rating 17 Tooth Sprocket (hp)	At Speed* (rpm)
40	1/2	4000	8.96	1800
50	5/8	6600	14.4	1400
60	3/4	8500	21.6	1200
80	1	14,500	38.9	900
100	1 1/4	24,000	63	800
120	1 1/2	34,000	89	700
140	1 3/4	46,000	150	600
160	2	58,000	166	500
180	2 1/4	76,000	188	400
200	2 1/2	95,000	163	250

*Approximate maximum ratings given for 1/2 to 2 in chain pitch, using a 17 tooth small sprocket. Ratings will be lower at higher or lower sprocket speeds. Decreasing the number of teeth in the small sprocket will decrease the rating; increasing the number of teeth will increase the rating. The ratings for 2 1/4 and 2 1/2 pitch are limited by maximum recommended speed. Manual or drip lubrication can be used for chain operating at low linear velocity. Oil bath or slinger lubrication is recommended for intermediate velocities. Chain drives operating at high velocity should incorporate an oil pump.

using this chain and sprocket near its optimum speed (i.e., the maximum rating of this chain and sprocket corresponds to about 1000 rpm). In most designs, the average power transmitted would be lower to account for variations in the power source and load.

ANSI Designation, Pitch, Strength, and Peak Power Rating

Table 11.1 lists some American National Standards Institute (ANSI) designation numbers and their corresponding chain pitch. The table also shows average tensile strength, approximate maximum power rating, and corresponding speed based on a 17 tooth small sprocket. The sprocket speed to produce maximum power is not the maximum permissible speed.

EXAMPLE PROBLEM Effect of Rotation Speed on hp Rating

A roller chain drive consists of 5/8 in pitch roller chain and 15 and 25 tooth sprockets. Check the chordal action ratio. Investigate the effect of sprocket speed on hp rating.

Solution Summary: A detailed solution with a plot of hp rating (see Figure 11.4) follows. The chordal action ratio is 2.2%. As a result, smooth operation can be expected only at low speeds. The horsepower rating is calculated as a function of small-sprocket speed in the range from 100 to 4000 rpm. Linkplate fatigue governs up to about 1400 rpm for this drive. At higher speeds, roller bushing fatigue governs.

Detailed Solution:

Small-Sprocket Teeth:

$$N_1 = 15$$

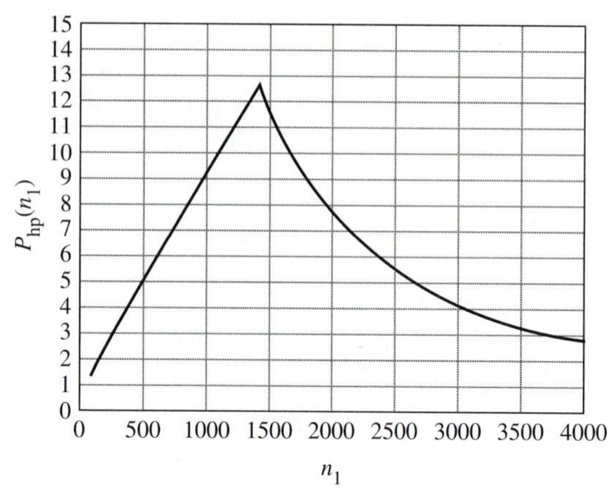

Figure 11.4 Effect of rotation speed on hp rating.

Pitch:

$$P = \frac{5}{8}$$

Speed (rpm):

$$n_1 = 100, 200, \ldots, 4000$$

Chordal Action:

$$K_C = 1 - \cos\left(\frac{\pi}{N_1}\right) \qquad K_C = 0.022$$

Capacity Based on Linkplate Fatigue:

$$P_L(n_1) = 0.004 \cdot n_1^{0.9} \cdot N_1^{1.08} \cdot P^{3 - 0.07 \cdot P}$$

Capacity Based on Roller Bushing Fatigue:

$$P_B(n_1) = \frac{17{,}000 \cdot N_1^{1.5} \cdot P^{0.8}}{n_1^{1.5}} \qquad P_{LB}(n_1) = \begin{pmatrix} P_L(n_1) \\ P_B(n_1) \end{pmatrix}$$

Capacity (hp):

$$P_{hp}(n_1) = \min(P_{LB}(n_1))$$

See Figure 11.4.

Service Factors and Design Power

Horsepower ratings are based on

- very smooth driving conditions (a service factor of 1)
- a chain length of 100 pitches
- use of recommended lubrication
- a two-sprocket drive
- sprockets aligned in the same plane, mounted on parallel horizontal shafts

Table 11.2 Service factors for roller chain.

Type of Driven Equipment	Power Source Type* A	B	C
Agitators for Liquid	1.0	1.0	1.2
Beaters	1.2	1.3	1.4
Blowers and Fans, Centrifugal	1.0	1.0	1.2
Boat Propellers	1.2	1.3	1.4
Compressors			
Centrifugal and lobe	1.2	1.3	1.4
Reciprocating, 3+ cylinders	1.2	1.3	1.4
Reciprocating, 1- and 2-cylinder	1.4	1.5	1.7
Conveyors			
Belt or chain, smoothly loaded	1.0	1.0	1.2
Heavy-duty, not uniformly loaded	1.2	1.3	1.4
Clay-working Machinery			
Pug mills	1.2	1.3	1.4
Brick presses, briquetting machinery	1.4	1.5	1.7
Crushers	1.4	1.5	1.7
Dredges			
Cable, reel, and conveyor drives	1.2	1.3	1.4
Cutter head, jig, and screen drives	1.4	1.5	1.7
Elevators, Bucket			
Smoothly loaded or fed	1.0	1.0	1.2
Not uniformly loaded or fed	1.2	1.3	1.4
Feeders			
Rotary table	1.0	1.0	1.2
Apron, screw, rotary vane	1.2	1.3	1.4
Reciprocating	1.4	1.5	1.7
Food Processing			
Slicers, dough mixers, grinders	1.2	1.3	1.4
Kilns and Dryers	1.2	1.3	1.4
Machine Tools			
Drills, grinders, lathes	1.0	1.0	1.2
Boring mills, milling machines	1.2	1.3	1.4
Punch presses, shears	1.4	1.5	1.7
Machinery, General			
Uniform load, nonreversing	1.0	1.0	1.2
Moderate shock load, non-reversing	1.2	1.3	1.4
Severe shock load, reversing	1.4	1.5	1.7
Mills			
Ball, pebble, tube	1.2	1.3	1.4
Hammer, rolling	1.4	1.5	1.7
Pumps			
Centrifugal	1.0	1.0	1.2
Reciprocating, 3+ cylinders	1.2	1.3	1.4
Reciprocating, 1- and 2-cylinder	1.4	1.5	1.7
Paper Industry			
Pulp grinders	1.2	1.3	1.4
Calenders, mixers, sheeters	1.4	1.5	1.7
Printing Presses, Magazine and Newspaper	1.4	1.5	1.7
Textile Industry			
Calenders, mangles, nappers	1.2	1.3	1.4
Carding machinery	1.4	1.5	1.7
Woodworking Machinery	1.2	1.3	1.4

Source: Courtesy of Diamond Chain Company.
*Power source types: A—Internal combustion engine with hydraulic drive
B—Electric motor or turbine.
C—Internal combustion engine with mechanical drive

- operation in a nonabrasive environment
- a service life of about 15,000 hours

Service factors are used to adjust for actual drive conditions. Roller chain service factors for various types of equipment and power sources are shown in Table 11.2. We obtain **design power** by multiplying average power by the service factor.

DESIGN AND SELECTION OF ROLLER CHAIN

A designer can choose single-strand chain with a chain pitch just adequate to meet the drive requirements. In some cases, a designer will select multistrand chain in order to reduce the required sprocket size.

Design for Single-Strand Chain

When average power requirement, speed, and type of service are given, we can determine the required chain pitch. First, we select the number of teeth in the smaller sprocket, considering chordal action. The capacity equations based on linkplate fatigue and roller bushing fatigue are set equal to the design power. It would be difficult to rearrange the first of these to solve for chain pitch. An iterative solution is easier, particularly if mathematical software is available. After solving both equations for chain pitch, we select a standard-size pitch equal to or greater than the larger of the two solutions.

EXAMPLE PROBLEM Design of a Drive Using Single-Strand Chain

A two-cylinder reciprocating compressor is to be powered by an electric motor. Average power is 10 hp, the motor speed is 1000 rpm, and the compressor speed will be 500 rpm. Design a drive for this application.

Design Decisions: Single-strand roller chain will be used to drive the compressor. A 17 tooth drive sprocket will be selected for smooth operation.

Solution Summary: The driven sprocket will have 34 teeth. The service factor for the compressor will be 1.5, resulting in a design horsepower of 15. A numerical solution yields a pitch of 0.701 in based on linkplate fatigue and 0.316 in based on roller bushing fatigue. A standard chain pitch of 0.750 is selected. We check the result by inserting the pitch in the capacity equations, resulting in a capacity of 18.3 hp.

The detailed solution follows.

Detailed Solution: Find the pitch for single-strand chain.

Power Requirement:

$$P_{avg} = 10$$

Service Factor:

$$K_S = 1.5$$

Small-Sprocket Teeth:

$$N_1 = 17$$

Speed (rpm):

$$n_1 = 1000$$

Begin iteration at $P = 1$.

Design Horsepower:

$$P_{des} = P_{avg} \cdot K_S \qquad P_{des} = 15$$

Capacity Based on Linkplate Fatigue: Given

$$0.004 \cdot n_1^{0.9} \cdot N_1^{1.08} \cdot P^{3 - 0.07 \cdot P} = P_{des}$$

$$P_1 = \text{find}(P) \qquad P_1 = 0.701$$

Begin iteration at $P = 1$.

Capacity Based on Roller Bushing Fatigue: Given

$$\frac{17{,}000 \cdot N_1^{1.5} \cdot P^{0.8}}{n_1^{1.5}} = P_{des}$$

$$P_2 = \text{find}(P) \qquad P_2 = 0.316$$

Select standard size $P = 3/4$.

Velocity (fpm):

$$V = \frac{P \cdot N_1 \cdot n_1}{12} \qquad V = 1.063 \cdot 10^3$$

Chordal Action:

$$K_C = 1 - \cos\left(\frac{\pi}{N_1}\right) \qquad K_C = 0.017$$

Capacity Based on Linkplate Fatigue:

$$P_L = 0.004 \cdot n_1^{0.9} \cdot N_1^{1.08} \cdot P^{3 - 0.07 \cdot P} \qquad P_L = 18.31$$

Capacity Based on Roller Bushing Fatigue:

$$P_B = \frac{17{,}000 \cdot N_1^{1.5} \cdot P^{0.8}}{n_1^{1.5}} \qquad P_B = 29.934$$

$$P_{LB} = \begin{pmatrix} P_L \\ P_B \end{pmatrix}$$

Capacity (hp):

$$P_{hp} = \min(P_{LB}) \quad P_{hp} = 18.31$$

Design of Multiple-Strand Chain Drives

Double-, triple-, and quadruple-strand roller chain and sprockets are stocked in most standard sizes. In some sizes, up to 10-strand chain width is available. ANSI designations apply, with the number of strands indicated. For example, 4-strand 1 in pitch chain is given ANSI Number 80-4. Four-strand chain has four times the average tensile strength of single-strand chain of the same pitch. Horsepower ratings, however, do not increase proportionately. The hp ratings for single-strand chain can be multiplied by the rating factors in Table 11.3.

Table 11.3 Horsepower rating factors for multiple-strand roller chain.

Number of Strands	Rating Factor
1	1
2	1.7
3	2.5
4	3.3

EXAMPLE PROBLEM Selection of Multiple-Strand Chain for Severe Service

A piece of machinery that is subject to severe shock and reversing is powered by an internal combustion engine with a mechanical drive. The average power requirements is 73 hp. The engine speed is 600 rpm, and the machinery speed is 200 rpm. Space limitations are a consideration. Select a drive for this application.

Design Decisions: We will try a 2 in pitch roller chain with 10 teeth in the drive sprocket and 30 in the driven sprocket.

Solution Summary: The applicable capacity equations are the same as those used in the preceding problems. The capacity of a single-strand chain is found to be 63.7 hp. However, the drive conditions correspond to a service factor of 1.7, resulting in a design power of 124.1 hp. The design horsepower is almost twice the single-strand rating. A 3-strand chain, which has a rating factor of 2.5, is selected. Although the 10 tooth drive sprocket will cause substantial chordal action, larger sprockets could cause the drive to exceed space limitations.

The detailed solution follows.

Detailed Solution:

Power Requirement (hp):

$$P_{avg} = 73$$

Service Factor:

$$K_S = 1.7$$

Small-Sprocket Teeth:

$$N_1 = 10$$

Pitch:

$$P = 2$$

Speed (rpm):

$$n_1 = 600$$

Velocity (fpm):

$$V = \frac{P \cdot N_1 \cdot n_1}{12} \qquad V = 1 \cdot 10^3$$

Chordal Action:

$$K_C = 1 - \cos\left(\frac{\pi}{N_1}\right) \qquad K_C = 0.049$$

Capacity Based on Linkplate Fatigue:

$$P_L = 0.004 \cdot n_1^{0.9} \cdot N_1^{1.08} \cdot P^{3 - 0.07 \cdot P} \qquad P_L = 110.495$$

Capacity Based on Roller Bushing Fatigue:

$$P_B = \frac{17{,}000 \cdot N_1^{1.5} \cdot P^{0.8}}{n_1^{1.5}} \qquad P_B = 63.686$$

$$P_{LB} = \begin{pmatrix} P_L \\ P_B \end{pmatrix}$$

Capacity (hp):

$$P_{hp} = \min(P_{LB}) \qquad P_{hp} = 63.686$$

Design Horsepower:

$$P_{des} = P_{avg} \cdot K_S \qquad P_{des} = 124.1$$

$$\frac{P_{des}}{P_{hp}} = 1.949$$

Use three strands. ■■

SILENT CHAIN

Silent or inverted-tooth chain has inward-pointing teeth that engage toothed chain wheels, as shown in Figures 11.1 and 11.5. Silent chain drives are often selected for high-power, high-speed service. The most common silent chain form has central guide links that fit central grooves in the chain wheels, preventing the chain from riding off the wheels. Double-guide chain has a double row of guide links. Side-flange silent chain, with outside guide links, is available in narrow chain widths (usually less than 3/4 in). Nonflange silent chain is available for use with flanged chain wheels. Duplex silent chain has teeth on both sides. Duplex chain can be used in serpentine drives, with driven chain wheels both inside and outside the chain loop.

Factors for Choosing Single-Pin or Two-Pin Silent Chain

Single-pin chain joints usually consist of a single oval or round pin. Two-pin chain joints consist of a pair of convex surfaces that roll on one another. The rolling action causes an effective change in chain pitch, producing smoother motion as the chain engages the sprocket. As a result, two-pin joints tend to reduce chordal action.

Two-pin silent chain is available in a wide range of pitches and widths. It is favored for the following conditions and applications:

- high-speed operation
- requirements for smooth operation

However, two-pin chain requires lubrication and is sensitive to environmental conditions, including high temperature and dirty conditions.

Single-pin silent chain is more tolerant of environmental conditions. It is favored in the following situations:

- high temperatures

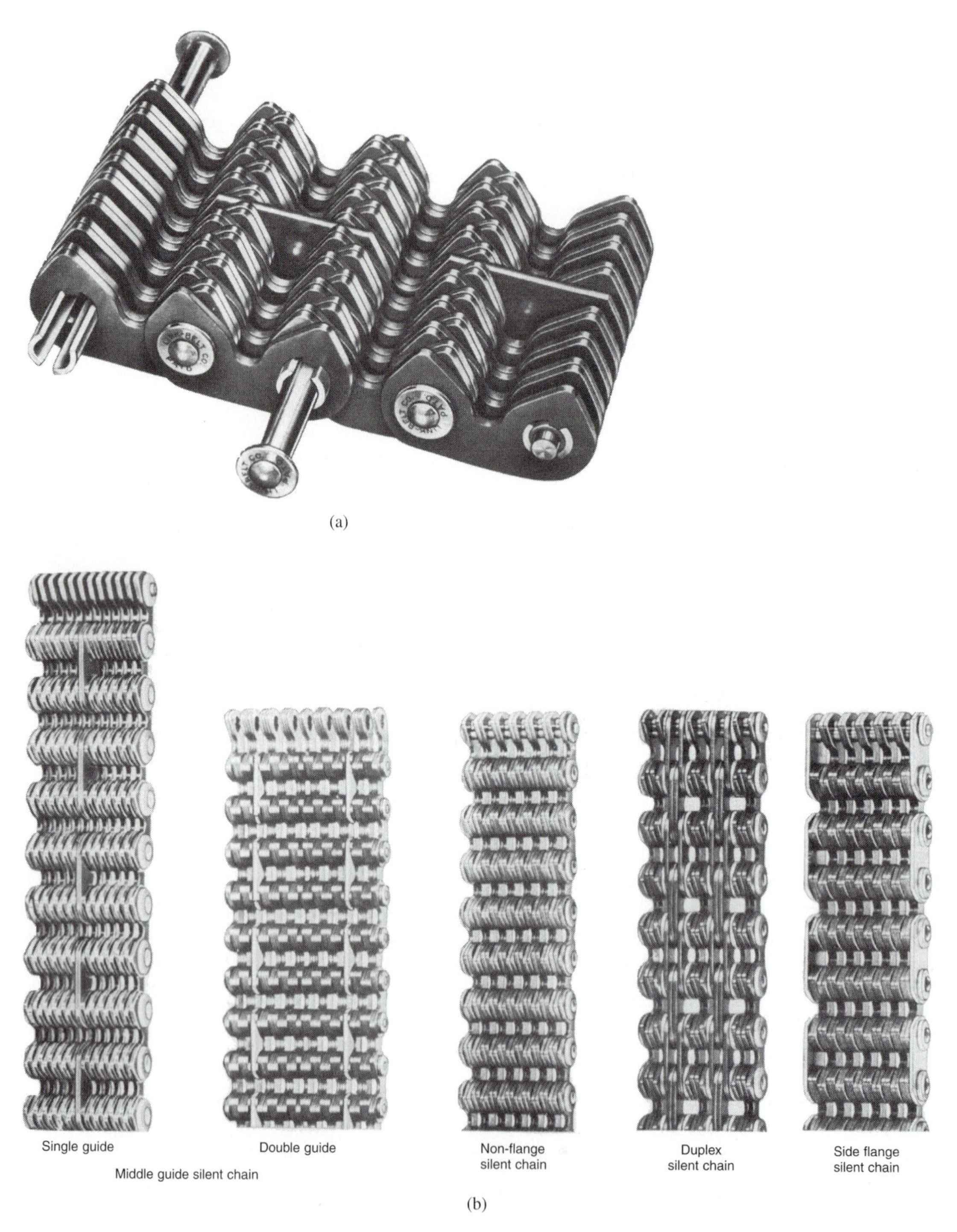

Figure 11.5 Silent chain. (a) Assembly; (b) silent chain configurations. (Courtesy of Rexnord Corporation)

- dirty environments
- applications where lubrication is likely to be infrequent

However, single-pin chain is limited to low-speed applications.

Power, Torque, Tension, and Speed

The relationship between power, torque, and speed is the same as for other drives.

$$P_{\mathrm{hp}} = \frac{Tn}{63{,}025} \tag{11.7}$$

or

$$P_{\mathrm{kW}} = 10^{-6}T\omega \tag{11.8}$$

where P_{hp} and P_{kW} = power (hp or kW)
T = torque on one chainwheel (lb · in or N · mm)
n and ω = speed of the same chainwheel (rpm or rad/s, respectively)

The tension on the slack side of a chain drive is usually quite low. Power can also be expressed in terms of tight-side tension and chain speed as follows:

$$P_{\mathrm{hp}} = \frac{FV}{33{,}000} \tag{11.9}$$

or

$$P_{\mathrm{kW}} = 10^{-6}F_N v \tag{11.10}$$

where F and F_N = tight-side chain tension (lb or N, respectively)
V and v = chain velocity (fpm or mm/s, respectively)

Speed ratio is given by the inverse ratio of the number of teeth. That is,

$$\frac{n_2}{n_1} = \frac{N_1}{N_2} \tag{11.11}$$

where N_1 and N_2 = number of teeth in the pinion (the small chainwheel) and large chainwheel, respectively

Speed ratios of 1:12 are possible. However, if the output speed is to be less than 1/8 of the input speed, a two-step reduction is preferred.

Note from the power equations that torque is increased with a speed reducer. For example, if the large chainwheel has three times the number of teeth as the pinion, the torque on the large chainwheel will be about three times the torque on the pinion. Output power will equal input power less a few percent due to friction.

Chain Length and Center Distance

Chain length and center distance are usually measured in pitches. The center distance is the distance between chainwheel centers. It should be large enough to prevent the sprockets from touching one another, and it should result in at least 120° chain contact with the pinion. If practical, the center distance should not exceed 60 pitches.

Chain length must be an integer number of pitches, preferably an even number. If the number of pitches is odd, then an offset section is required. In that case, the chain width should be increased 25% to account for the reduced strength of the offset link.

After determining an acceptable chain length, we can approximate the center distance from the following formula:

$$c = \frac{L - \dfrac{N_1 + N_2}{2} + \sqrt{\left(L - \dfrac{N_2 + N_1}{2}\right)^2 - \dfrac{2(N_2 - N_1)^2}{\pi^2}}}{4} \qquad \textbf{(11.12)}$$

where c = center distance in pitches

L = chain length in pitches

N = number of teeth

EXAMPLE PROBLEM Speed Ratio, Chain Length, and Center Distance

Speed is to be reduced from 1600 to 400 rpm. Select a drive.

Design Decisions: A 3/4 in pitch silent chain will be used. The pinion will have 21 teeth. Chain length will be 122 pitches.

Solution: Using the speed ratio equation, we select an 84 tooth driven chainwheel. Center distance is given by

$$c = \frac{122 - \dfrac{21 + 84}{2} + \sqrt{\left(122 - \dfrac{21 + 84}{2}\right)^2 - \dfrac{2(84 - 21)^2}{\pi^2}}}{4} \qquad \textbf{(11.13)}$$

$$c = 33.24 \text{ pitches} \quad \text{or} \quad 33.24 \cdot 0.75 = 24.93 \text{ in}$$

Power Ratings for Silent Chain Drives

Silent chain drives are usually rated on the basis of horsepower per inch of chain width. Chain pitch is the length of one link, the distance between joint centers. Small-pitch chains can be operated at higher pinion (small chainwheel) speeds than large-pitch chains. Silent chain is available in a wide range of pitches. Typical power ratings for 1/2 in pitch, 3/4 in pitch, and 1 in pitch are given in Table 11.4.

Normal lubrication systems include the following:

- Manual and drip-cup lubrication for low-power drives with chain speeds below 1500 fpm. There should be no exposure to abrasive dust if manual lubrication is specified.
- Splash lubrication and oil-disc lubrication for low- and moderate-speed drives.
- Forced lubrication for large horsepower drives, heavily loaded drives, and high-speed drives.

The chain manufacturer should be consulted for drives that exceed the speed and power requirements for normal lubrication.

Table 11.4 Power ratings for silent chain (hp per in of chain width).

$\frac{1}{2}$ in Pitch Chain (Typically Available in $\frac{3}{4}$, 1, $1\frac{1}{2}$, 2, $2\frac{1}{2}$, 3, $3\frac{1}{2}$, 4, 5 in Widths)

Number of Teeth in Pinion	Maximum Bore (in)	hp/in of Chain Width* rpm of Pinion 100	500	700	1000	1200	1800	2000	2500	3000	3500	4000
17†	$1\frac{3}{8}$	0.83	3.8	5.0	6.3	7.5	10	11	11	11	11	
19†	$1\frac{5}{8}$	0.93	3.8	5.0	7.5	8.8	11	13	14	14	14	
21	$1\frac{7}{8}$	1.0	5.0	6.3	8.8	10	14	14	15	16	16	
23	$2\frac{1}{8}$	1.1	5.0	7.5	10	11	15	16	18	19	19	18
25	$2\frac{3}{8}$	1.2	5.0	7.5	10	13	16	18	20	21	21	20
27	$2\frac{5}{8}$	1.3	6.3	8.8	11	13	18	19	21	24	24	23
29	$2\frac{13}{16}$	1.4	6.3	8.8	13	14	19	21	24	25	25	25
31	$3\frac{1}{16}$	1.5	7.5	10	13	15	21	23	25	28	28	28
33	$3\frac{5}{16}$	1.6	7.5	10	14	16	23	24	28	29	30	29
35	—	1.8	7.5	11	15	18	24	25	29	31	31	30
37	—	1.9	8.8	11	16	19	25	26	30	33	33	
40	—	2.0	8.8	13	18	20	28	29	33	35	35	
45	—	2.5	10	14	19	23	30	30	36	39		
50	—	2.5	11	15	21	25	34	36	40			
Lubrication				△					□			

$\frac{3}{4}$ in Pitch Chain (Typically Available in $1\frac{1}{2}$, 2, $2\frac{1}{2}$, 3, 4, 5, 6, 7, 8 in Widths)

Number of Teeth in Pinion	Maximum Bore (in)	hp/in of Chain Width* rpm of Pinion 100	500	700	1000	1200	1500	1800	2000	2500
17†	2	1.9	8.1	11	14	15	16	18	18	
19†	$2\frac{3}{8}$	2.0	9.3	13	15	18	20	21	21	
21	$2\frac{3}{4}$	2.3	10	14	18	20	23	24	25	24
23	$3\frac{1}{4}$	2.5	11	15	20	23	25	28	28	28
25	$3\frac{5}{8}$	2.8	13	16	21	25	29	31	31	30
27	$3\frac{7}{8}$	2.9	14	18	24	28	31	34	35	35
29	$4\frac{3}{8}$	3.1	15	20	26	30	34	36	38	38
31	$4\frac{5}{8}$	3.4	15	21	28	31	36	40	41	41
33	$5\frac{1}{8}$	3.6	16	23	30	34	39	43	44	44
35	—	3.8	18	24	31	36	41	45	46	46
37	—	4.0	19	25	34	39	44	48	49	49
40	—	4.4	20	28	36	41	48	51	53	53
45	—	4.9	23	30	40	46	53	56	58	
50	—	5.4	25	34	45	51	58	61		
Lubrication			△					□		

1 in Pitch Chain (Typically Available in 2, 3, 4, 5, 6, 7, 8, 9, 10, 12 in Widths)

Number of Teeth in Pinion	Maximum Bore (in)	hp/in of Chain Width* rpm of Pinion 100	200	300	400	500	700	1000	1200	1500	1800	2000
17†	$2\frac{3}{4}$	3.8	6.3	8.8	11	14	18	21	23			
19†	$3\frac{1}{4}$	3.8	7.5	10	13	15	20	25	26	28		
21	$3\frac{3}{4}$	3.8	7.5	11	15	18	23	29	31	33	33	
23	$4\frac{3}{8}$	3.8	8.8	13	16	19	25	31	35	38	38	
25	$4\frac{3}{4}$	5.0	8.8	14	18	21	28	35	39	41	41	41
27	$5\frac{3}{8}$	5.0	10	15	19	24	30	39	43	46	46	45
29	$5\frac{3}{4}$	5.0	11	16	20	25	33	41	46	50	51	50
31	$6\frac{3}{8}$	6.3	11	16	23	28	35	45	50	54	55	54
33	7	6.3	13	18	24	29	38	49	54	59	59	58
35	—	6.3	13	19	25	30	40	51	56	61	63	61
37	—	6.8	14	20	26	33	43	54	60	65	66	
40	—	7.5	15	23	29	35	45	59	65	70		
45	—	8.8	16	25	31	39	51	65	71	76		
50	—	10	19	28	35	43	56	71	78			
Lubrication				△					□			

Source: Courtesy of Rexnord Corporation.

†For best results, the pinion should have at least 21 teeth.

*Ratings are based on a service factor of 1. For listing of service factors, refer to Table 11.5.

△ Normal.

□ Consult manufacturer for approved application and type of lubrication.

Service Factors and Design Power for Silent Chain Drives

Power ratings for chain drives are based on ideal conditions without shock loading transmitted by the driving or driven equipment. Such conditions are typical of an electric-motor-driven lathe or drill press operated 8 to 10 hours per day. Service factors are used to account for more severe service. For example, a service factor of 2.5 is applied to the average power of a concrete mixer operated more than 10 hours per day and driven by a four-cylinder engine. In selecting a drive, we multiply the service factor by the average power to obtain the design power. Silent chain drive service factors are given in Table 11.5.

Table 11.5 Service factors for silent chain drives.

	Source of Power							
			Internal Combustion Engine					
	Electric Motor		With Hydraulic Coupling		6 Cylinders or More		4 Cylinders or Less	
	Service Requirement, h/Day							
Driven Machine	**8–10**	**10–24**	**8–10**	**10–24**	**8–10**	**10–24**	**8–10**	**10–24**
Agitators, Paddle or Propeller								
Liquid, semiliquid	1.0	1.3	1.2	1.5	1.4	1.7	1.6	1.9
Bakery Machinery								
Dough mixer	1.2	1.5	—	—	—	—	—	—
Brewing and Distilling Equipment								
Bottling machinery	1.0	1.3	—	—	—	—	—	—
Kettles, cookers	1.0	1.3	—	—	—	—	—	—
Mash tubs, scale hoppers	1.0	1.3	—	—	—	—	—	—
Brick and Clay Machinery								
Auger machines, cutting tables	1.3	1.6	1.4	1.7	1.5	1.8	1.7	2.0
De-airing machines	1.3	1.6	1.4	1.7	1.5	1.8	1.7	2.0
Brick machines, dry presses	1.4	1.7	1.5	1.8	1.6	1.9	1.8	2.2
Granulators, mixers	1.4	1.7	1.5	1.8	1.6	1.9	1.8	2.2
Pug mills, rolls	1.4	1.7	1.5	1.8	1.6	1.9	1.8	2.2
Cement Plants								
Kilns	1.4	1.7	1.5	1.8	1.6	1.9	1.8	2.2
Kominuters	1.5	1.8	1.6	1.9	1.7	2.0	2.0	2.5
Centrifuges	1.4	1.7	1.5	1.8	1.6	1.9	1.8	2.2
Compressors								
Centrifugal, rotary	1.0	1.3	1.2	1.5	1.4	1.7	1.6	1.9
Reciprocating								
3 or more cylinders	1.3	1.6	1.4	1.7	1.5	1.8	1.7	2.0
1 or 2 cylinders	1.6	1.9	1.7	2.0	1.8	2.2	2.0	2.5
Conveyors								
Apron	1.4	1.7	1.5	1.8	1.6	1.9	1.8	2.2
Pan, bucket	1.4	1.7	1.5	1.8	1.6	1.9	1.7	2.0
Belt (ore, coal, sand)	1.2	1.5	1.3	1.6	1.4	1.7	1.6	1.9
Belt (light-package)	1.0	1.3	1.1	1.4	1.2	1.5	1.4	1.7
Oven	1.0	1.3	1.2	1.5	1.3	1.6	1.4	1.7
Screw, flight	1.6	1.9	1.7	2.0	1.8	2.2	2.0	2.5

	Source of Power							
			Internal Combustion Engine					
	Electric Motor		With Hydraulic Coupling		6 Cylinders or More		4 Cylinders or Less	
	Service Requirement, h/Day							
Driven Machine	**8–10**	**10–24**	**8–10**	**10–24**	**8–10**	**10–24**	**8–10**	**10–24**
Cotton Oil Plants								
Linters, cookers	1.4	1.7	1.5	1.8	1.6	1.9	1.8	2.2
Cranes and Hoists								
Main hoist								
Medium-duty	1.0	1.3	1.2	1.5	1.4	1.7	1.8	2.2
Heavy-duty	1.3	1.6	1.4	1.7	1.5	1.8	2.0	2.5
Skip hoist	1.3	1.6	1.4	1.7	1.5	1.8	2.0	2.5
Crushing Machinery								
Coal breakers and pulverizers	1.4	1.7	1.5	1.8	1.6	1.9	1.8	2.2
Linseed crushers	1.4	1.7	1.5	1.8	1.6	1.9	1.8	2.2
Ball Mills, Hardinge mills	1.5	1.8	1.6	1.9	1.7	2.0	2.0	2.5
Rod mills, tube mills	1.5	1.8	1.6	1.9	1.7	2.0	2.0	2.5
Cone and gyratory crushers	1.5	1.8	1.6	1.9	1.7	2.0	2.0	2.5
Jaw crushers, crushing rolls	1.5	1.8	1.6	1.9	1.7	2.0	2.0	2.5
Dredges								
Conveyors, pumps, stackers	1.4	1.7	1.5	1.8	1.7	2.0	1.8	2.2
Jigs, screens	1.6	1.9	1.7	2.0	1.8	2.2	2.0	2.5
Elevators								
Bucket, uniformly fed	1.1	1.4	1.2	1.5	1.3	1.6	1.4	1.7
Bucket, heavy-duty	1.3	1.6	1.5	1.8	1.7	2.0	2.0	2.5
Fans and Blowers								
Propeller, induced draft	1.1	1.4	1.3	1.6	1.6	1.9	1.8	2.2
Centrifugal	1.2	1.5	1.4	1.7	1.6	1.9	1.8	2.2
Exhausters	1.2	1.5	1.4	1.7	1.6	1.9	1.8	2.2
Mine fans	1.3	1.6	1.5	1.8	1.7	2.0	2.0	2.5
Positive blowers	1.5	1.8	1.6	1.9	1.7	2.0	2.0	2.5
Flour, Feed, Cereal Mill Machinery								
Lofter legs	1.0	1.3	1.2	1.5	1.3	1.6	1.4	1.7
Bolters, sifters	1.1	1.4	1.2	1.5	1.3	1.6	1.5	1.8
Purifiers, reels	1.1	1.4	1.2	1.5	1.3	1.6	1.5	1.8
Separators	1.1	1.4	—	—	—	—	—	—
Grinders, hammer mills	1.2	1.5	1.3	1.6	1.4	1.7	1.6	1.9
Roller mills	1.3	1.6	1.4	1.7	1.5	1.8	1.7	2.0
Mainline shafts	1.4	1.7	1.5	1.8	1.6	1.9	1.7	2.0
Generators and Exciters	1.2	1.5	1.3	1.6	1.4	1.7	1.6	1.9
Hoists (See Cranes and Hoists)								
Ice Machines	1.5	1.8	1.6	1.9	1.7	2.0	2.0	2.5
Laundry Machinery								
Dampeners, washers	1.1	1.4	1.2	1.5	1.4	1.7	1.6	1.9
Extractors, flat-work ironers	1.1	1.4	1.2	1.5	1.4	1.7	1.6	1.9
Tumblers	1.2	1.5	1.3	1.6	1.4	1.7	1.6	1.9
Line Shafts								
Grain elevators	1.0	1.3	1.1	1.4	1.2	1.5	1.4	1.7

Table 11.5 ***Continued.***

Driven Machine	Source of Power							
	Electric Motor		Internal Combustion Engine					
			With Hydraulic Coupling		6 Cylinders or More		4 Cylinders or Less	
	Service Requirement, h/Day							
	8–10	10–24	8–10	10–24	8–10	10–24	8–10	10–24
Line Shafts–cont'd								
Cotton gins, cotton oil plant	1.1	1.4	1.2	1.5	1.3	1.6	1.5	1.8
Coal-handling plants	1.2	1.5	1.3	1.6	1.4	1.7	1.5	1.8
Paper mills	1.3	1.6	1.4	1.7	1.5	1.8	1.7	2.0
Rubber plants, steel mills	1.4	1.7	1.5	1.8	1.6	1.9	1.8	2.2
Brick plants	1.5	1.8	1.6	1.9	1.7	2.0	2.0	2.5
Miscellaneous	1.2	1.5	1.3	1.6	1.4	1.7	1.6	1.9
Machine Tools								
Boring mills, cam cutters	1.0	1.3	—	—	—	—	—	—
Drill presses, lathes	1.0	1.3	—	—	—	—	—	—
Grinders, milling machines	1.1	1.4	—	—	—	—	—	—
Drop hammers	1.1	1.4	—	—	—	—	—	—
Punch presses, shears	1.2	1.5	—	—	—	—	—	—
Mills								
Ball, rod, tube, Hardinge, etc.	1.6	1.9	1.7	2.0	1.8	2.2	—	—
Tumbling barrels	1.6	1.9	1.7	2.0	1.8	2.2	—	—
Mixers								
Liquid, semiliquid	1.0	1.3	1.2	1.5	1.4	1.7	1.6	1.9
Concrete	1.4	1.7	—	—	1.8	2.2	2.0	2.5
Oil Field Machinery								
Compounding units	1.0	1.3	1.2	1.5	1.3	1.6	1.4	1.7
Pipeline pumps	1.2	1.5	1.3	1.6	1.4	1.7	1.5	1.8
Slush pumps	1.3	1.6	1.4	1.7	1.5	1.8	1.6	1.9
Draw works	1.4	1.7	1.6	1.9	1.8	2.2	2.0	2.5
Oil Refinery Equipment								
Chillers, rotary kilns	1.4	1.7	1.6	1.9	1.8	2.2	2.0	2.5
Paraffin-filter presses	1.4	1.7	1.6	1.9	1.8	2.2	2.0	2.5
Paper Machinery								
Agitators	1.0	1.3	1.2	1.5	1.3	1.6	1.5	1.8
Paper machines	1.2	1.5	1.3	1.6	1.5	1.8	1.6	1.9
Calenders, dryers	1.2	1.5	1.4	1.7	1.5	1.8	1.6	1.9
Jordan engines	1.2	1.5	1.4	1.7	1.5	1.8	1.6	1.9
Yankee dryers	1.3	1.6	1.4	1.7	1.5	1.8	—	—
Beaters	1.3	1.6	1.5	1.8	1.6	1.9	1.7	2.0
Washers	1.4	1.7	—	—	—	—	—	—
Nash pumps	1.4	1.7	1.6	1.9	1.7	2.0	1.8	2.2
Winder drums	1.5	1.8	1.6	1.9	1.7	2.0	—	—
Chippers	1.5	1.8	1.6	1.9	1.7	2.0	1.8	2.2
Printing Machinery								
Paper cutters, rotary presses	1.1	1.4	1.2	1.5	1.3	1.6	1.4	1.7
Linotype machines	1.1	1.4	1.2	1.5	1.3	1.6	1.5	1.8
Embossing and flat bed presses	1.2	1.5	1.3	1.6	1.5	1.8	1.6	1.9

	Source of Power							
			Internal Combustion Engine					
	Electric Motor		With Hydraulic Coupling		6 Cylinders or More		4 Cylinders or Less	
	Service Requirement, h/Day							
Driven Machine	**8–10**	**10–24**	**8–10**	**10–24**	**8–10**	**10–24**	**8–10**	**10–24**
Printing Machinery—cont'd								
Folders	1.2	1.5	1.3	1.6	1.5	1.8	1.6	1.9
Magazine and newspaper presses	1.5	1.8	—	—	—	—	—	—
Pumps								
Centrifugal, rotary	1.1	1.4	1.3	1.6	1.5	1.8	1.6	1.9
Gear	1.2	1.5	1.3	1.6	1.5	1.8	1.6	1.9
Reciprocating								
3 or more cylinders	1.3	1.6	1.4	1.7	1.5	1.8	1.7	2.0
1 or 2 cylinders	1.6	1.9	1.7	2.0	1.8	2.2	2.0	2.5
Pipeline	1.4	1.7	1.5	1.8	1.6	1.9	1.8	2.2
Dredge	1.5	1.8	1.6	1.9	1.7	2.0	2.0	2.5
Miscellaneous	1.5	1.8	1.6	1.9	1.7	2.0	2.0	2.5
Rubber Mill Equipment								
Calenders	1.5	1.8	1.6	1.9	1.7	2.0	1.8	2.2
Tubers, tire-building	1.5	1.8	1.6	1.9	1.7	2.0	1.8	2.2
Mixers, sheeters, mills	1.6	1.9	1.7	2.0	1.8	2.2	2.0	2.5
Rubber Plant Machinery								
Banbury mills, mixers	1.5	1.8	1.6	1.9	1.7	2.0	1.8	2.2
Calenders, rolls	1.5	1.8	1.6	1.9	1.7	2.0	1.8	2.2
Screens								
Air-washing, traveling	1.0	1.3	1.2	1.5	1.3	1.6	1.4	1.7
Conical, revolving	1.2	1.5	1.3	1.6	1.4	1.7	1.5	1.8
Rotary, gravel, stone	1.3	1.6	1.5	1.8	1.6	1.9	1.7	2.0
Vibrating	1.4	1.7	1.6	1.9	1.7	2.0	1.8	2.2
Steel Plants								
Wire benches	1.2	1.5	1.3	1.6	1.5	1.8	1.6	1.9
Rolling mills	1.3	1.6	1.4	1.7	1.5	1.8	1.7	2.0
Stokers	1.0	1.3	—	—	—	—	—	—
Textile Machinery								
Twisters, warpers	1.0	1.3	—	—	—	—	—	—
Spinning frames, reels	1.0	1.3	—	—	—	—	—	—
Looms	1.1	1.4	—	—	—	—	—	—
Batchers, calenders	1.1	1.4	—	—	—	—	—	—

Source: Courtesy of Rexnord Corporation.

EXAMPLE PROBLEM Design of a Silent Chain Drive

A centrifugal fan is operated by an electric motor 12 hours per day. Motor speed is 1600 rpm, and fan speed is to be about 450 rpm. Average power is 20 hp. Design the drive.

Design Decisions: We will use silent chain, selecting a 21 tooth pinion. It is assumed that drive noise could be objectionable in this application. The silent chain should be reasonably quiet, and a 21 tooth pinion is less likely to cause excessive vibration than a smaller pinion.

Solution Summary: We use the speed ratio equation to obtain the number of teeth in the fan chainwheel. We round the number to the nearest integer by adding 1/2 and dropping the remainder. Using 75 teeth in the fan chainwheel, the fan speed is 448 rpm, close enough to the desired speed. Trying 3/4 in pitch chain and interpolating in Table 11.4, we obtain a rating of 23.33 hp/in. Table 11.5 indicates a service factor of 1.5 for this application. Multiplying by average power, we find that the design horsepower is 30. Dividing design horsepower by the hp/in rating, we obtain a chain width of 1.286. Standard 1.5 in wide chain is selected, resulting in a drive rating of 35 hp, somewhat higher than required. The detailed solution follows. Solutions using different chain pitch are possible.

Detailed Solution:

Transmitted Horsepower:

$$P_{avg} = 20$$

Speed (rpm):

$$n_1 = 1600 \qquad n_{2approx} = 450$$

Service Factor:

$$K_S = 1.5$$

Design Horsepower:

$$P_{des} = P_{avg} \cdot K_S \qquad P_{des} = 30$$

Pinion Teeth:

$$N_1 = 21$$

Large Chainwheel:

$$N_2 = \text{floor}\left(0.5 + N_1 \cdot \frac{n_1}{n_{2approx}}\right) \qquad N_2 = 75$$

Output Speed:

$$n_2 = n_1 \cdot \frac{N_1}{N_2} \qquad n_2 = 448$$

Select 3/4 in pitch.

Rating/in:

$$P_1 = 23 + \frac{100}{300} \qquad P_1 = 23.333$$

Chain Width:

$$w_{approx} = \frac{P_{des}}{P_1} \qquad w_{approx} = 1.286$$

Use standard size $w = 1.5$.

Drive Rating:

$$P_{hp} = P_1 \cdot w \qquad P_{hp} = 35$$

■■

References and Bibliography

American Chain Association. *Chains for Power Transmission and Material Handling.* New York: Marcel Dekker, Inc., 1982.

Buschbaum, F., F. Freudenstein, and P. Thornton, eds. *Design and Application of Small Standardized Components.* Vol. 2. New Hyde Park, NY: Stock Drive Products, 1983.

Catalog 490. Charlotte, NC: Ramsey Products Corporation, 1994.

Link-Belt. *Silent Chain Drives.* Indianapolis: Rexnord Corporation, 1989.

Product Guide 1092. Indianapolis: Diamond Chain Company, 1992.

Design Problems

11.1. A 1.5 in pitch roller chain drive will utilize a 700 rpm, 20 tooth drive sprocket and a 35 tooth driven sprocket. Find chain velocity. Evaluate chordal action. Determine the hp rating.

11.2. A 1/2 in pitch roller chain drive will utilize a 4000 rpm, 17 tooth drive sprocket and a 2000 rpm driven sprocket. Find chain velocity. Evaluate chordal action. Determine the hp rating. Which failure mode governs?

11.3. A roller chain drive consists of 2 in pitch roller chain and 12 and 40 tooth sprockets. Check the chordal action ratio. Investigate the effect of sprocket speed on hp rating. Plot the results.

11.4. A roller chain drive consists of 1 in pitch roller chain and 20 and 45 tooth sprockets. Check the chordal action ratio. Investigate the effect of sprocket speed on hp rating.

11.5. A roller chain drive consists of 3/4 in pitch roller chain and 17 and 42 tooth sprockets. Check the chordal action ratio. Investigate the effect of sprocket speed on hp rating. Plot the results for speeds of 100 to 2400 rpm.

11.6. A four-cylinder reciprocating compressor is to be powered by an internal combustion engine with a mechanical drive. Average power is 2 hp, the motor speed is 3000 rpm, and the compressor speed will be 1500 rpm. Design the drive. Check the results.

Design Decisions: A 12 tooth drive sprocket with single-strand roller chain will be used to drive the compressor.

11.7. The power source of a milling machine drive is an electric motor, and the loading is nonuniform. Average power is 1.8 hp, the motor speed is 4000 rpm, and the miller speed will be 1600 rpm. Design the milling machine drive. Check the results.

Design Decisions: A 16 tooth drive sprocket with single-strand roller chain will be used.

11.8. A reciprocating four-cylinder pump is to be powered by an electric motor. Average power is 17 hp, the motor speed is 800 rpm, and the compressor speed will be 400 rpm. Design the drive. Check the results.

Design Decisions: A 14 tooth drive sprocket with single-strand roller chain will be used to drive the pump.

11.9. A printing press is to be powered by an electric motor. The average power requirement is 26 hp, the motor speed is 1200 rpm, and the press speed is 300 rpm. Select a drive for this application. Check the chordal action.

Design Decisions: We will try a 3/4 in pitch roller chain with 21 teeth in the drive sprocket for smooth operation.

11.10. A reciprocating feeder is to be powered by an electric motor. The average power requirement is 70 hp, the motor speed is 800 rpm, and the feeder driveshaft speed is to be 400 rpm. Select a drive for this application. Check the chordal action.

Design Decisions: We will try a 1.25 in pitch roller chain with a 19 tooth drive sprocket.

11.11. A four-cylinder internal combustion engine drives a vibrating screen. Average power is 22 hp for 13 hours per day, input speed is 800 rpm, and output speed is 150 rpm. Design the drive.

Design Decisions: Use 1 in pitch silent chain with a 21 tooth pinion.

11.12. A boring mill is operated 12 hours per day. Average power is 25 hp, supplied by an electric motor. Input speed is 1800 rpm, and output speed is 525 rpm. Design the drive.

Design Decisions: Use 1/2 in pitch silent chain with a 23 tooth pinion.

11.13. A centrifuge is to be driven by an electric motor. Average power is 31 hp for 6 hours per day, input speed is 3600 rpm, and output speed is 2700 rpm. Design the drive.

Design Decisions: Use 1/2 in pitch silent chain with a 23 tooth pinion.

CHAPTER 12

Clutches

Symbols

ceil(x)	smallest integer not less than x
f	coefficient of friction
F	actuating force
INT(x)	integer part of x
K_{load}	loading factor
n	speed
N_A	number of teeth in gear A, etc.
N_{pr}	number of pairs of friction surfaces
p	pressure
P_{hp}	horsepower
P_{kW}	power kilowatts
r	radius
T	torque

Units

dimensions	in, m, or mm
force	lb or N
pressure	psi, Pa, or MPa
torque	in · lb, N · m, or N · mm

Clutches allow an input shaft to be connected to and disengaged from an output shaft as required. Positive clutches have jaws or toothed surfaces. In most cases, smooth engagement is required, and friction surfaces are employed. A time interval of a few milliseconds, up to a few seconds, may be required to bring the driven shaft to the drive shaft speed. Other clutch types include hysteresis clutches, magnetic-particle clutches, eddy current clutches, and fluid couplings.

Clutches can be actuated mechanically, electrically, hydraulically, or pneumatically. Centrifugal clutches are designed to begin engagement at a predetermined speed and to disengage if speed is reduced below that value. Clutches that engage and disengage according to drive conditions include torque-limiting clutches, overrunning clutches, and reverse-locking clutches.

DISC CLUTCHES

A disc clutch can consist of a single drive clutch plate brought into contact with a driven plate. Disc clutches designed for high torque capacity have several pairs of contacting surfaces, with plates alternately connected to the drive and driven shafts. Figure 12.1 shows exploded views of typical mechanically operated clutches. When the clutch is assembled, the protruding driving lugs of the outer discs fit in a slot in the driving cup or ring. The inner discs are separated by springs and are keyed or splined to the shaft.

If a clutch consists of 7 inner discs alternated with 6 outer discs, there are 12 friction pairs, that is, 12 pairs of contacting surfaces that transmit power between the driver and driven shafts. When the clutch is engaged, contact is made on both sides of each of the 6 discs driving the ring and on both sides of 5 of the discs splined to the shaft. Each of the 2 end discs splined to the shaft makes contact on only one side.

Typically, mechanical clutches are operated by moving a sleeve to compress the discs. Levers are sometimes incorporated for mechanical advantage. The shaft is then engaged to the cup or ring. The ring can be connected to a gear, sheave, sprocket, or other shaft. Double-throw clutches employ two sets of discs, each set assembled in a separate ring. Moving the sleeve in one direction engages one ring, and moving in the other direction engages a second ring. A partially assembled double-throw clutch is included in Figure 12.1. A shifting fork (not shown) can be used to move the sleeve.

APPLICATIONS OF SINGLE- AND DOUBLE-THROW CLUTCHES

A double-throw clutch can be used in a gear train designed to change speed ratio or to reverse direction. Double-throw clutches are sometimes used to brake the output shaft. Figure 12.2 shows schematic diagrams of single- and double-throw clutch applications. For the single-throw clutch, shown in Part (a) of the figure, the V-belt sheave is engaged to the shaft by compressing the clutch plates.

Figure 12.2(b) shows a gear train utilizing a double-throw clutch. This configuration, with input and output shafts collinear, is called a **reverted gear train.** Compression of the left disc assembly engages the output shaft to the input shaft, producing a 1:1 speed ratio. In this

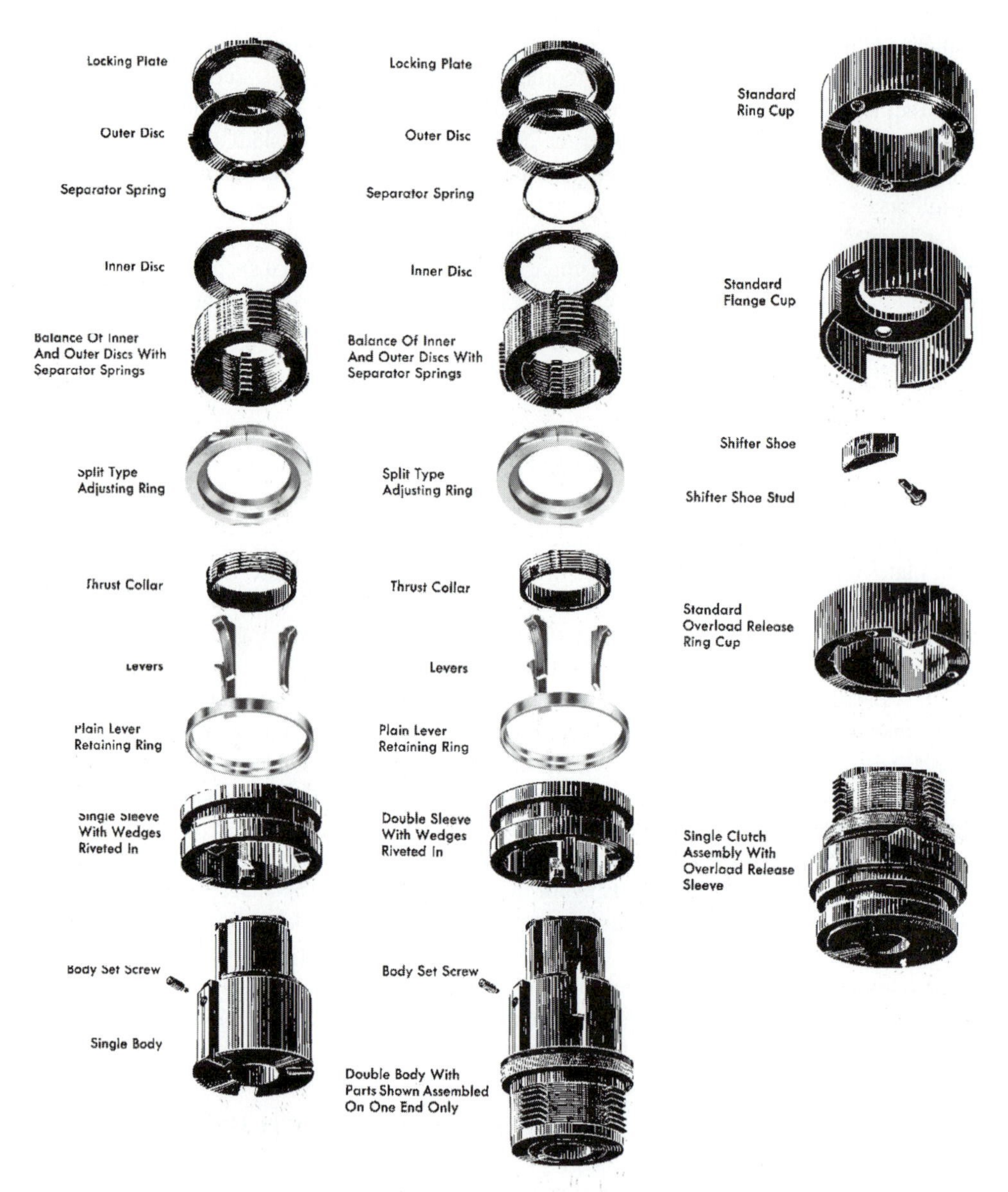

Figure 12.1 Exploded view of single- and double-throw clutches. (Courtesy of Carlyle Johnson Machine Company)

mode, gears A, B, C, and D turn freely without transmitting power since the right disc assembly does not engage the output shaft to gear D. Compression of the right disc assembly (with the left assembly disengaged) engages gear D to the output shaft, producing a speed reduction of

$$n_{out}/n_{in} = N_A N_C/(N_B N_D) \tag{12.1}$$

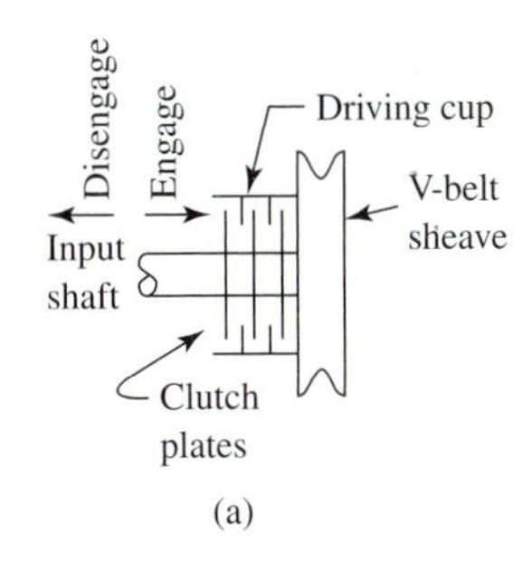

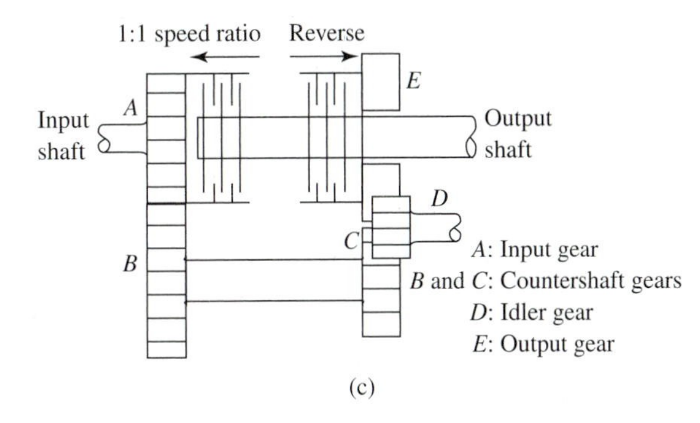

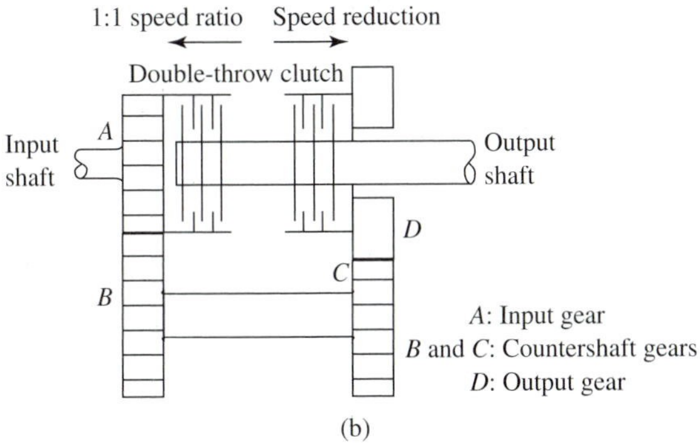

Figure 12.2 Schematics of single- and double-throw clutch applications. (a) Single-throw clutch; (b) double-throw clutch, two-speed application; (c) double-throw clutch, 1:1 speed ratio and reverse.

where n = shaft speed (rad/s or rpm)

N_A = number of teeth in gear A, etc.

A modification of the above configuration, shown in Figure 12.2(c), utilizes an idler gear to reverse output shaft direction. Compression of the right clutch assembly engages gear E to the output shaft, producing a speed ratio of

$$n_{\text{out}}/n_{\text{in}} = -N_A N_C/(N_B N_E) \tag{12.2}$$

where the minus sign indicates that the input and output shafts turn in opposite directions.

The reader may have already considered additional applications of single- and double-throw clutches. The following are suggested as exercises: the design of drives to produce two different speed reduction ratios with a minimum number of gears; two different speed reduction ratios in a reverted gear train; and three speed ratios, including 1:1. Note that belt drives or chain drives can be substituted for gear drives in some applications.

POWER AND TORQUE

Power, torque, and speed of any drive system are related as follows:

$$T = 63{,}025 P_{\text{hp}}/n \tag{12.3}$$

where T = torque (in · lb)

P_{hp} = power (hp)

n = speed in (rpm)

The equivalent expression in SI units is

$$T = 1000P_{kW}/\omega \tag{12.4}$$

or

$$T = 9549P_{kW}/n \tag{12.5}$$

where T = torque (N · m)

P_{kW} = power (kW)

ω = speed (rad/s)

1 rpm = $\pi/30$ rad/s

TORQUE CAPACITY OF A CLUTCH

Clutch design is commonly based on one of the following assumptions:

1. pressure is assumed to be uniform over the contact surfaces; or
2. wear is assumed to be uniform over the contact surfaces and proportional to contact pressure times relative (sliding) velocity.

The **uniform wear model,** which results in a more conservative clutch design, will be used. The resulting contact pressure is inversely proportional to radius, reaching a maximum at the inner radius. Torque capacity for a single pair of friction surfaces is given by

$$T_{pr} = \pi f p_{MAX} r_i (r_o^2 - r_i^2) \tag{12.6}$$

and, for a multiple-disc clutch,

$$T = \pi f N_{pr} p_{MAX}\, r_i (r_o^2 - r_i^2) \tag{12.7}$$

where T_{pr} = torque capacity per pair of friction surfaces (in · lb or N · m)

T = total torque capacity (in · lb or N · m)

f = coefficient of friction

p_{MAX} = maximum contact pressure (psi or Pa)

r_i and r_o = inner and outer radius, respectively, of the contact surfaces on the discs (in or m)

N_{pr} = number of pairs of friction surfaces

Clutch capacity may exceed actual torque transmitted by the drive train.

ACTUATING FORCE

The actuating force is the axial force required to maintain the pressure distribution over the clutch faces. Actuating force can be expressed in terms of pressure and geometry as follows:

$$F = 2\pi p_{MAX} r_i (r_o - r_i) \quad \textbf{(12.8)}$$

where F = actuating force (lb or N)

Writing the torque equation in terms of actuating force instead of pressure, we have

$$T = fFN_{pr}(r_o + r_i)/2 \quad \textbf{(12.9)}$$

The torque capacity contribution of each friction pair is equivalent to applying the actuating force at the midradius of the disc. For a given total torque capacity, the use of more discs allows us to reduce actuating force.

Some applications require a normally engaged clutch, with the actuating force applied by a spring. Some require a normally disengaged clutch. Clutch assemblies may include separator springs between clutch discs to reduce friction when a clutch is disengaged. When designing a clutch operating system, we add the separator spring force to the required actuating force to obtain the **operating force,** the actual force required to engage the clutch.

EXAMPLE PROBLEM Specifying the Number of Discs in a Multiple-Disc Clutch—The Effect of Speed of Rotation

A clutch is to transmit 5 hp (peak or design power) at 150 rpm. An internal disc radius of 1.0 in, an external radius of 3.0 in, and a 550 lb actuating force are specified. Assume a coefficient of friction of 0.17.

(a) Determine the required number of pairs of friction surfaces.

(b) Suppose that similar clutches are required for 5 hp (design power) at 50 rpm, at 100 rpm, and so on. How does rotation speed affect the design?

Solution Summary (a): The necessary torque capacity is given by

$$T = 63{,}025 P_{hp}/n = 63{,}025 \cdot 5/150 = 2100.8 \text{ in} \cdot \text{lb}$$

Equation (12.9) can be rewritten in the form

$$N_{pr} = 2T/[fF(r_o + r_i)]$$

Substituting the known values gives the following result:

$$N_{pr} = 2 \times 2100.8/[0.17 \times 550\,(3 + 1)] = 11.2$$

We will use 12 pairs of friction surfaces.

Solution Summary (b): We see from the first equation used in Part (a) that torque is inversely proportional to speed. A higher capacity clutch is needed to handle the same power at lower speed. If clutch location is optional in a power train that includes a speed reducer, then the clutch will be located in the high-speed part of the system.

We will use mathematics software to aid in specifying clutches for speeds from 50 to 1000 rpm in 50 rpm steps, as shown in the detailed solution which follows. We tell the software what speeds to use by entering the first, second, and final values (rpm). Torque per pair and required number of pairs are functions of speed n. We identify them as such for computer calculations: $T(n)$ and $N_{pr}(n)$.

Torque capacity per pair of friction surfaces is given by

$$T_{pr} = f \cdot F(r_o + r_i)/2 = 187 \text{ in} \cdot \text{lb/pair}$$

Then the required number of pairs is the required torque capacity divided by torque per pair.

$$N_{pr} = \text{ceil}[T/T_{pr}]$$

where the function ceil[x] yields the smallest integer that is not less than x; that is, $x \leq \text{ceil}[x] < x + 1$

If this function is not available, an alternative expression can be programmed.

$$N_{pr} = \text{INT}[T/T_{pr} + 0.99999]$$

where the integer function INT[x] truncates the fractional part of x.

The detailed solution includes the tabulated results. Although a clutch can be designed with an odd number of friction pairs, it is more common to use an even number (i.e., $N_{pr} = 3$ would result in 4 friction pairs, etc.). Figure 11.3 shows the plot of N_{pr} versus operating speed n where $N_{pr} = T/T_{pr}$ is used to obtain a continuous plot.

Detailed Solution:

Capacity (hp):

$$P_{\text{hp}} = 5$$

Speed (rpm):

$$n = 50, 100, \ldots, 1000$$

Dimensions (in):

$$r_i = 1 \qquad r_o = 3$$

Friction Coefficient:

$$f = 0.17$$

Actuating Force (lb):

$$F = 550$$

Torque (in · lb):

$$T(n) = 63{,}025 \cdot \frac{P_{\text{hp}}}{n}$$

Torque per Pair:

$$T_{pr} = f \cdot F \cdot \frac{r_o + r_i}{2} \qquad T_{pr} = 187$$

Number of Pairs:

$$N_{pr}(n) = \text{ceil}\left[\frac{T(n)}{T_{pr}}\right]$$

n	$N_{pr}(n)$
50	34
100	17
150	12
200	9
250	7
300	6
350	5
400	5
450	4
500	4
550	4
600	3
650	3
700	3
750	3
800	3
850	2
900	2
950	2
1000	2

Maximum Pressure (psi):

$$p_{\text{MAX}} = \frac{F}{2 \cdot \pi \cdot r_i \cdot [r_o - r_i]} \qquad p_{\text{MAX}} = 43.768$$

Number of Pairs:

$$N_{pr}(n) = \frac{T(n)}{T_{pr}} \quad \text{(for continuous plot)}$$

See Figure 12.3.

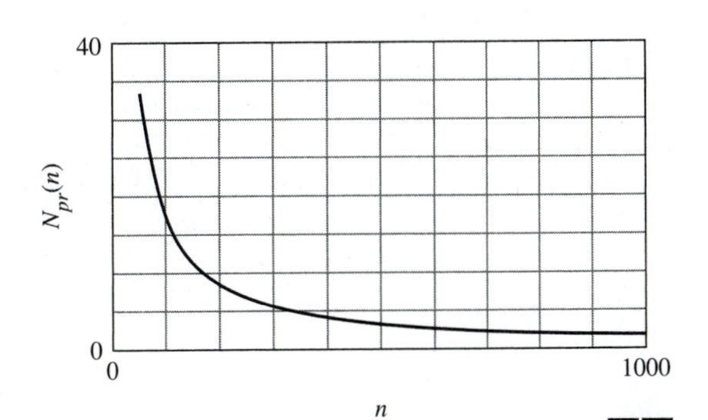

Figure 12.3 Number of pairs of friction surfaces versus operating speed.

LOADING FACTOR

Power trains may be subject to shock loading when starting or during the duty cycle. Thus, to obtain design power, we multiply the nominal (steady-state) power by a loading factor

$$P = P_{nom} K_{load} \tag{12.10}$$

where $P = P_{hp}$ or P_{kW}, the design power used to design or specify the clutch
P_{nom} = nominal or steady-state power transmitted
K_{load} = loading factor

In the absence of test data to determine the value of the loading factor, the range $1.5 \leq K_{load} \leq 3$ is suggested. A machine tool that is brought up to speed before a working load is applied is an example of a light-duty application, for which a load factor K_{load} of 1.5 can be used. Light-duty applications include lathes and milling machines. A heavy-duty application, where maximum torque is required during acceleration to bring a machine up to speed, suggests a load factor K_{load} of 3. A concrete mixer is an example of a heavy-duty application. For other applications, the designer must weigh all available information to assign a load factor. Large masses, large mass-moments-of-inertia, and abrupt (shock) loading suggest high load factors. Smooth operation and flexibility in a drive train (e.g., belt drives) suggest a lower load factor.

DESIGN AND SPECIFICATION OF DISC CLUTCHES

If speed and power requirements for a clutch are given, then a designer must choose clutch materials and determine the required number and dimensions of the discs. If inside radius is selected, then we can compute the optimum outside radius corresponding to a given number of friction pairs. Rewriting the torque equation, we obtain

$$r_o = [r_i^2 + T/(N_{pr} \pi f r_i p_{MAX})]^{1/2} \tag{12.11}$$

Material properties can be determined from tests or data supplied by manufacturers. If other data are lacking, the ranges in Table 12.1 can be used for preliminary design.

The metal surfaces should be smooth. If operating conditions are uncertain, design calculations can be based on a "conservative" coefficient of friction based on the lower limits of

Table 12.1 Material properties.

Material Steel or Cast Iron versus:	Maximum Pressure		Coefficient of Friction	
	psi	MPa	Dry	Wet
Steel or cast iron	100–250	0.7–1.7	0.15–0.42	0.029–0.12
Sintered metal	100–300	0.7–2.1	0.1–0.45	0.05–0.1
Phosphor bronze	100–250	0.7–1.7	0.3–0.34	0.15–0.17
Woven metal and fiber	50–100	0.3–0.7	0.25–0.5	0.08–0.2

the values given in Table 12.1 or obtained by dividing the average values by a safety factor of 1.3 to 2.0. For longer clutch life, maximum pressure can be limited to values lower than those given in Table 12.1

"Wet" coefficient of friction refers to operation in an oil bath or spray. Wet multiple-disc clutches are often employed in automotive and industrial automatic transmissions. Although we see from the above equations that torque capacity is proportional to coefficient of friction, other factors can lead to design of a clutch with a relatively low coefficient of friction. A few such considerations include the following:

- **Heat dissipation capabilities.** Operation in an oil bath may improve heat dissipation.
- **Temperature limitations.** Rapid cycling may result in high temperatures. Sintered metal versus cast iron or steel can be allowed to reach temperatures of 500°F to 1000°F (260°C to 538°C). Woven materials are ordinarily limited to lower temperatures.
- **Stability and consistency.** Dynamic coefficients of friction are given above. Static coefficients of friction are somewhat higher. If there is a wide difference, chatter or slip-stick operation may result, with the clutch plates alternately fully engaging and slipping. Also, operation should be relatively insensitive to environmental changes such as temperature, humidity, and dust.
- **Environmental and health considerations.** Although woven and molded asbestos surfaces on clutch plates have a relatively high coefficient of friction, their use results in environmental pollution and creates a health hazard to manufacturing personnel and mechanics.

EXAMPLE PROBLEM Design of a Multiple-Disc Clutch

Design a clutch to transmit $P_{\text{nom}} = 50$ hp (nominal) at $n = 1000$ rpm with moderate shock loading. The inner radius of the discs is to be $r_i = 1.5$ in.

Design Decisions: The clutch will be designed to be operated dry. Phosphor bronze and hardened steel discs will be used. Design will be based on coefficient of friction of $f = 0.3$ and a maximum pressure of $p_{\text{MAX}} = 90$ psi to ensure reasonably long life. A loading factor of $K_{\text{load}} = 2.5$ will be used for moderate shock. Potential clutch designs will have N_{pr} of 2 to 20 pairs of friction surfaces. Ordinarily, contact is made on both sides of the discs that engage the cup or ring, resulting in an even number of pairs.

Solution Summary: The design power is given by

$$P = P_{\text{nom}} K_{\text{load}} = 50 \times 2.5 = 125 \text{ hp}$$

The design torque is

$$T = 63{,}025P/n = 63{,}025 \times 125/1000 = 7878 \text{ in} \cdot \text{lb}$$

The detailed solution which follows shows the calculation of required disc outside radius, using mathematics software. Since the number of pairs is varied in this example, outside radius and other variables are identified as functions of N_{pr}. Actuating force is also calculated, and the results are tabulated. Figure 12.4 shows a plot of outside disc radius r_o (in) and actuating force F_{kip} (kilopounds).

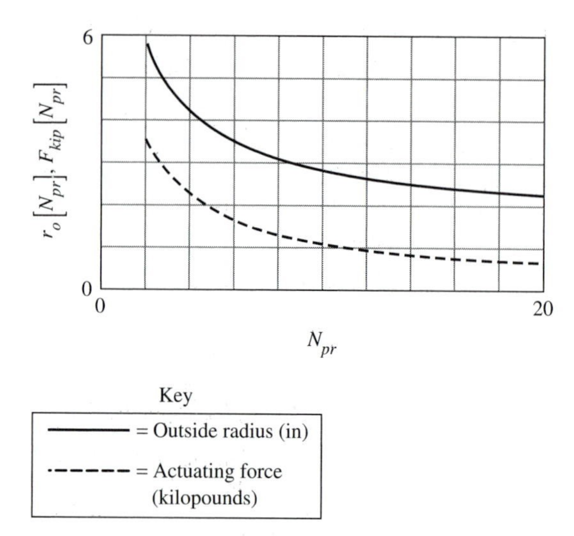

Figure 12.4 Required outer radius and actuating force versus number of pairs.

Torque capacity is then recalculated from actuating force, friction coefficient, and clutch geometry. The result checks with the required torque capacity based on power and speed. This check is a partial one; it could fail to detect some types of error, including errors in the assumptions.

The results show the advantage of using a large number of discs.

1. The clutch will fit in a smaller diameter radial space.
2. A smaller actuating force is required for the smaller diameter discs.

In selecting a final design, the designer would consider actuating force, space limitations, weight, manufacturing costs, and other factors. These considerations might, for example, lead to a clutch with seven phosphor bronze discs engaged to the cup or ring, and eight steel discs engaged to a splined shaft. Each of the seven phosphor bronze discs engaged to the cup or ring is contacted on both sides by a steel disc engaged to the splined shaft, resulting in $N_{pr} = 14$ friction pairs. As indicated in the table for this selection, outside radius $r_o = 2.583$ in, and actuating force $F = 919$ lb. A hydraulic system might be employed to provide the operating force. If spring separators are used, the operating force must be somewhat larger than the actuating force calculated above.

The detailed solution follows.

Detailed Solution:

Maximum Pressure (psi):

$$p_{\text{MAX}} = 90$$

Inner Radius (in):

$$r_i = 1.5$$

Speed (rpm):

$$n = 1000$$

Speed (rad/s):

$$\omega = \pi \cdot \frac{n}{30} \qquad \omega = 104.72$$

Capacity:

$$P = 125 \text{ hp}$$

Number of Pairs:

$$N_{pr} = 2, 4, \ldots, 20$$

Coefficient of Friction:

$$f = 0.3$$

Torque (in · lb):

$$T = 63{,}025 \cdot \frac{P}{n} \qquad T = 7.878 \cdot 10^{3}$$

Outer Radius (in):

$$r_o[N_{pr}] = \left[r_i^2 + \frac{T}{N_{pr} \cdot \pi \cdot f \cdot r_i \cdot p_{\text{MAX}}} \right]^{0.5}$$

Actuating Force:

Pounds

$$F[N_{pr}] = 2 \cdot \pi \cdot p_{\text{MAX}} \cdot r_i \cdot [r_o[N_{pr}] - r_i]$$

Kilopounds

$$F_{kip}[N_{pr}] = \frac{F[N_{pr}]}{1000}$$

Check:

$$T_{\text{check}}[N_{pr}] = f \cdot F[N_{pr}] \cdot N_{pr} \cdot \frac{r_o[N_{pr}] + r_i}{2}$$

N_{pr}	$r_o[N_{pr}]$	$F[N_{pr}]$	$T_{check}[N_{pr}]$
2	5.763	3616	7878
4	4.211	2299	7878
6	3.545	1735	7878
8	3.161	1409	7878
10	2.905	1192	7878
12	2.722	1037	7878
14	2.583	918.77	7878
16	2.474	826.04	7878
18	2.385	750.981	7878
20	2.312	688.866	7878

EXAMPLE PROBLEM Design of a Series of Clutches for Different Power Capacities

A series of clutches is to be manufactured to transmit from 0.25 to 2.5 kW mechanical power (nominal) at 175 rpm. Loading conditions suggest a service factor of 2. Each clutch assembly will consist of 10 friction pairs resulting from 5 discs driven by a splined shaft and 6 discs driving the cup or ring. A 20 mm inner radius will be specified. The coefficient of friction will be not less than 0.25, and maximum pressure will be 0.5 MPa.

(a) Determine the required disc dimensions.

(b) Evaluate the design from a manufacturing standpoint.

Solution Summary (a): Power capacity (design power) for the clutch series will be

$$P = P_{\text{nom}} K_{\text{load}} = 0.5, 1, \ldots, 5 \text{ kW}$$

The detailed solution, which utilizes mathematics software, follows. Torque T (N · m) is given by Equation (12.4) or (12.5); disc outside radius r_o (m) is given by Equation (12.11); and actuating force F (N) is given by Equation (12.8). The results are tabulated along with torque capacity T_{check} calculated by Equation (12.9). The result of this calculation of torque capacity (the last column of the table) is compared with required torque capacity (the second column) as a partial check. Disc outside radius ranges from 27 to 62 mm, and actuating force from 461 to 2654 N. Outside radius, torque, and actuating force are plotted against design power in Figure 12.5. Values are scaled so that the top of the grid represents an outside radius of 70 mm, a torque of 700 N · m, and an actuating force of 7000 N.

Solution Summary (b)—A Comment on the Design: Manufacturers and suppliers find it economical to use common parts wherever possible. Solution Summary (a) for the 10 design power ratings (0.5 to 5 kW) results in 10 different disc sizes. The possibility of standardizing on one or two different disc sizes for all power ratings should be considered. This would require removing the specification of 10 friction pairs. Then different power ratings would call for different numbers of discs. Note that the above calculations were not entirely wasted. Reviewing the table of results, we might decide to standardize on discs with 55 or 60 mm outside radius. This change could reduce tooling, manufacturing, and inventory costs.

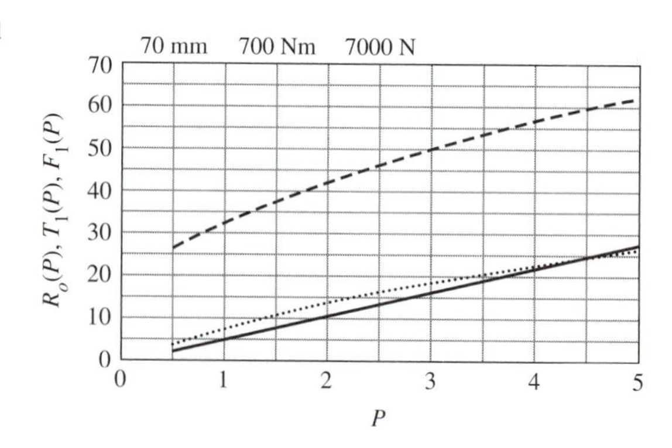

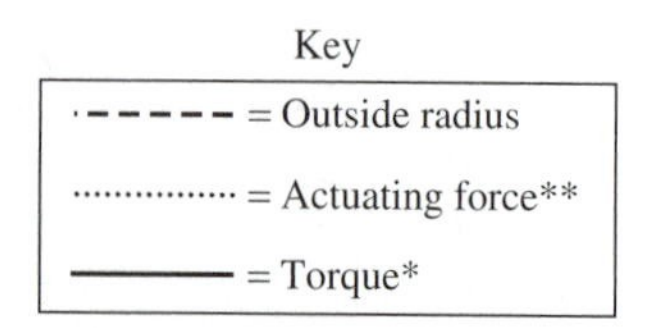

Note: *Torque is to be multiplied by 10
**Actuating force is to be multiplied by 100

Figure 12.5 Outside radius, torque, and actuating force versus design power.

The detailed solution follows.

Detailed Solution:

Maximum Pressure (pa):

$$p_{\mathrm{MAX}} = 0.5 \cdot 10^6$$

Inner Radius (m):

$$r_i = 0.02$$

Speed (rpm):

$$n = 175 \text{ rpm}$$

Speed (rad/s):

$$\omega = \pi \cdot \frac{n}{30} \qquad \omega = 18.326$$

Capacity:

$$P = 0.5, 1, \ldots, 5 \text{ kW}$$

Number of Pairs:

$$N_{pr} = 10$$

Coefficient of Friction:

$$f = 0.25$$

Torque (N · m):

$$T(P) = 1000 \cdot \frac{P}{\omega}$$

Outer Radius (m):

$$r_o(P) = \left[r_i^2 + \frac{T(P)}{N_{pr} \cdot \pi \cdot f \cdot r_i \cdot P_{\mathrm{MAX}}} \right]^{0.5}$$

Actuating Force (N):

$$F(P) = 2 \cdot \pi \cdot p_{\mathrm{MAX}} \cdot r_i \cdot [r_o(P) - r_i]$$

Check:

$$T_{\mathrm{check}}(P) = f \cdot F(P) \cdot N_{pr} \cdot \frac{r_o(P) + r_i}{2}$$

p	$T(P)$	$r_o(P)$	$F(P)$	$T_{check}(P)$
0.5	27.284	0.027	461.084	27.284
1	54.567	0.033	822.303	54.567
1.5	81.851	0.038	1129	81.851
2	109.135	0.042	1401	109.135
2.5	136.419	0.046	1648	136.419
3	163.702	0.05	1875	163.702
3.5	190.986	0.053	2087	190.986
4	218.27	0.056	2286	218.27
4.5	245.553	0.059	2475	245.553
5	272.837	0.062	2654	272.837

Scale Values for Plotting:

$$T_1(P) = \frac{T(P)}{10} \qquad R_o(P) = 1000 \cdot r_o(P) \qquad F_1(P) = \frac{F(P)}{100}$$

Clutches for various application are shown in Figure 12.6. These include clutches designed for air or hydraulic operation, electric operation, and mechanical operation. Also shown are automatic overload and torque limiter clutches and a clutch disc assembly. Figure 12.7 shows a high-performance multidisc clutch designed for a racing car with a manual transmission. Most automobiles with manual transmissions utilize single-disc clutches. Contact is made on both sides of the disc; that is, there are two friction pairs in the single-disc clutch.

ENGINE PERFORMANCE

Clutch design and selection can be based on engine performance. Referring to torque-versus-speed Equations (12.3) to (12.5), we see that torque is inversely proportional to speed, if power is held constant. However, the power output of an engine depends on operating speed. We cannot produce unlimited torque just by slowing down an engine. Figure 12.8 shows torque-versus-speed and power-versus-speed curves for several six-, eight-, and ten-cylinder gasoline and diesel engines. The scale at the left is torque in ft · lb; the scale at the right is hp.

Figure 12.6 Clutches for various applications. (Courtesy of Carlyle Johnson Machine Company)

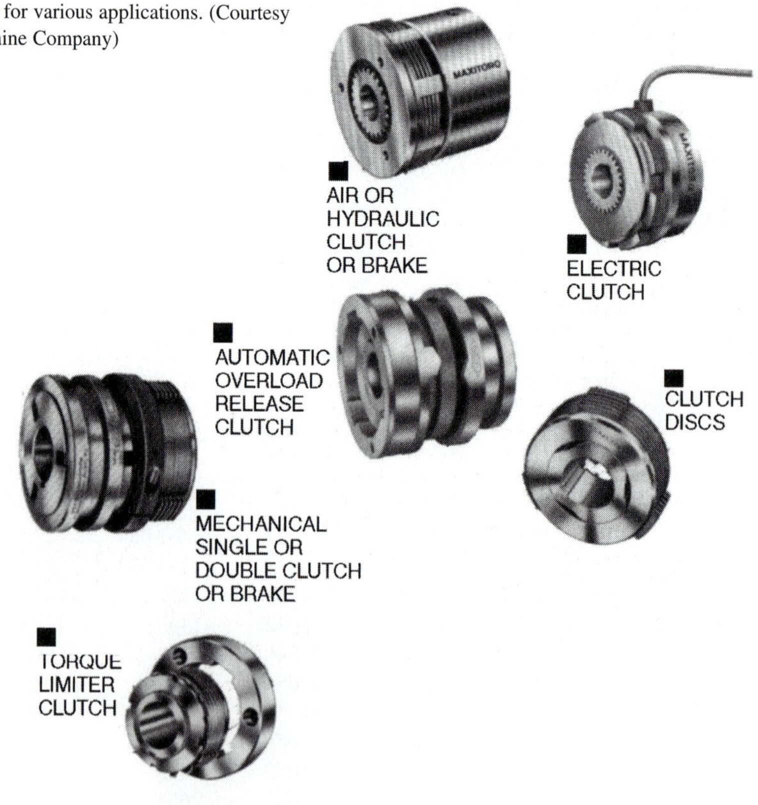

AIR OR HYDRAULIC CLUTCH OR BRAKE

ELECTRIC CLUTCH

AUTOMATIC OVERLOAD RELEASE CLUTCH

CLUTCH DISCS

MECHANICAL SINGLE OR DOUBLE CLUTCH OR BRAKE

TORQUE LIMITER CLUTCH

Figure 12.7 High-performance automotive clutch. (Courtesy of Chrysler Corporation)

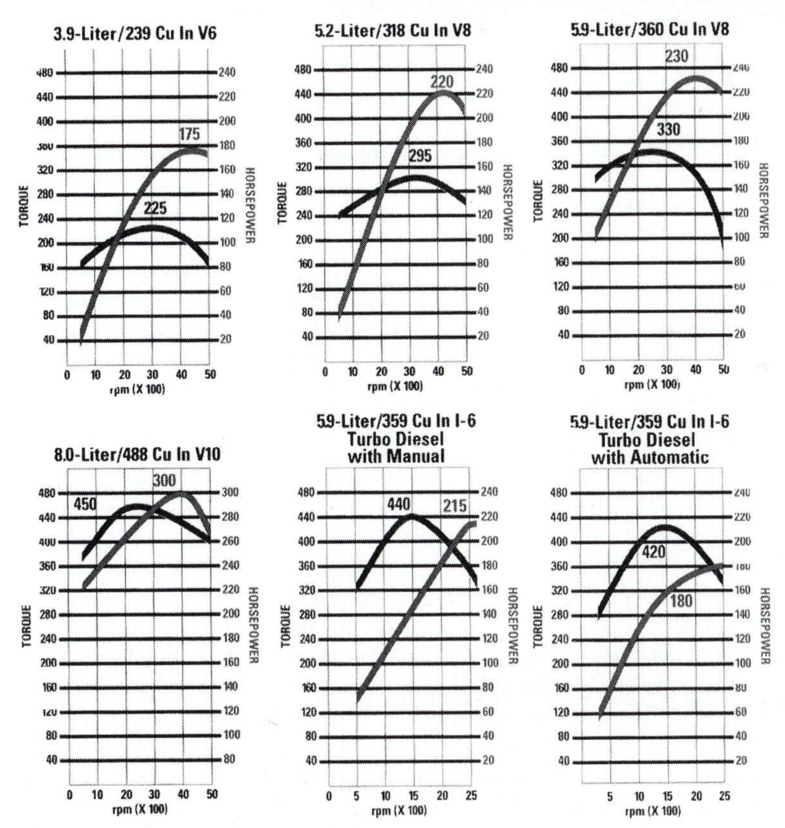

Figure 12.8 Power and torque curves for selected gasoline and diesel engines. (Courtesy of Chrysler Corporation)

Key: Torque

Horsepower

Design Problems

12.1. Sketch a gear drive that will produce two different speed ratios (excluding 1:1). Use a double-throw clutch and a minimum number of gears.

12.2. Sketch a belt drive that will produce two different speed ratios. Use a double-throw clutch and a minimum number of sheaves.

12.3. Sketch a reverted gear drive that will produce two different speed ratios (excluding 1:1).

12.4. Sketch a gear drive that will produce three different speed ratios, including 1:1.

12.5. A series of clutches is to be designed to transmit 2 hp (peak or design power) at operating speeds ranging from 50 to 1000 rpm in 50 rpm increments. A 400 lb actuating force is specified. The friction surfaces are to have a 0.75 in internal radius and 2 in outer radius. The coefficient of friction will be 0.2.

(a) Determine required torque and maximum contact pressure. Specify the number of friction pairs for each clutch in the series.

(b) Plot the required number of friction pairs against operating speed.

12.6. A series of clutches is to be designed to transmit 3 hp (peak or design power) at operating speeds ranging from 50 to 1000 rpm in 50 rpm increments. A 575 lb actuating force is specified. The friction surfaces are to have a 0.75 in internal radius and 2.5 in outer radius. The coefficient of friction will be 0.2.

(a) Determine required torque and maximum contact pressure. Specify the number of friction pairs for each clutch in the series.

(b) Plot the required number of friction pairs against operating speed.

12.7. A series of clutches is to be designed to transmit 2.5 kW (peak or design power) at operating speeds ranging from 50 to 1000 rpm in 50 rpm increments. A 2500 N actuating force is specified. The friction surfaces are to have a 20 mm internal radius and 60 mm outer radius. The coefficient of friction will be 0.2.

(a) Determine required torque and maximum contact pressure. Specify the number of friction pairs for each clutch in the series.

(b) Plot the required number of friction pairs against operating speed.

12.8. Peak or design power for a series of clutches is to range from 0.5 to 5 hp in 0.5 hp increments at 110 rpm. Each clutch in the series will have an inner radius of 1.125 in and 12 friction pairs. Assume a coefficient of friction of 0.22, and allow a maximum pressure of 100 psi.

(a) Find required torque capacity, disc outer radius, and actuating force for each clutch in the series. Check by recalculating torque capacity in terms of actuating force.

(b) Plot outer radius, torque capacity, and actuating force against power capacity.

(c) Evaluate the designs in terms of manufacturability.

12.9. Peak or design power for a series of clutches is to range from 1 to 10 hp at 225 rpm in 1 hp increments. Each clutch in the series will have an inner radius of 1.1 in and 12 friction pairs. Assume a coefficient of friction of 0.25, and allow a maximum pressure of 115 psi.

(a) Find required torque capacity, disc outer radius, and actuating force for each clutch in the series. Check by recalculating torque capacity in terms of actuating force.

(b) Plot outer radius, torque capacity, and actuating force against power capacity.

(c) Evaluate the designs in terms of manufacturability.

12.10. Peak or design power for a series of clutches is to range from 1 to 20 hp at 440 rpm in 1 hp increments. Each clutch in the series will have an inner radius of 1.2 in and 12 friction pairs. Assume a coefficient of friction of 0.26, and allow a maximum pressure of 112 psi.

(a) Find required torque capacity, disc outer radius, and actuating force for each clutch in the series. Check by recalculating torque capacity in terms of actuating force.

(b) Plot outer radius, torque capacity, and actuating force against power capacity.

(c) Evaluate the designs in terms of manufacturability.

12.11. Peak or design power for a series of clutches is to range from 0.5 to 5 kW at 150 rpm in 0.5 kW increments. Each clutch in the series will have an inner radius of 25 mm and 8 friction pairs. Assume a coefficient of friction of 0.3, and allow a maximum pressure of 0.6 MPa.

(a) Find required torque capacity, disc outer radius, and actuating force for each clutch in the series. Check by recalculating torque capacity in terms of actuating force.

(b) Plot outer radius, torque capacity, and actuating force against power capacity.

(c) Evaluate the designs in terms of manufacturability.

12.12. A clutch is to be designed with a capacity of 25 hp at 240 rpm. The inner radius of the discs is 1 in, and the maximum pressure is 125 psi.

(a) Calculate required torque. Determine required disc outside radius for different numbers of pairs. Find actuating force. Check the torque value.

(b) Plot various designs, showing outside radius and actuating force plotted against number of friction pairs. Consider only even numbers of friction pairs.

12.13. A clutch is to be designed with a capacity of 295 hp at 560 rpm. The inner radius of the discs is 1.625 in, and the maximum pressure is 95 psi.

(a) Calculate required torque. Determine required disc outside radius for different numbers of pairs. Find actuating force. Check the torque value.

(b) Plot various designs, showing outside radius and actuating force plotted against number of friction pairs. Consider only even numbers of friction pairs.

CHAPTER 13

Brakes

Symbols

A	acceleration
E_K	kinetic energy
E_P	potential energy
f	coefficient of friction
F_A	actuating force
F_1, F_2	band tension
g	acceleration of gravity
J	mass-moment-of-inertia
ln	natural logarithm
M	mass moving in translation
M_N, M_f	normal and friction moment on brake shoe
N_c	number of caliper brake shoes
N_{pr}	number of pairs of friction surfaces
p	pressure
P	power
r, R	radius
R_G	radius of gyration
T	torque
t	time
v	velocity
W	weight or force
X_B	braking distance
z	elevation change
α	angular acceleration
γ	mass density
γ_W	weight density
δ_t	temperature rise
θ	rotation of brake drum, angular location on drum brake, angle of contact of band brake
ϕ	roadway grade
ω	angular velocity

Units

acceleration	in/s^2 or m/s^2
angles	rad
dimensions and distances	in or m
energy	in · lb or N · m (J)
force and tension	lb or N
mass	lb · s^2/in or kg
mass density	lb · s^2/in^4 or kg/m^3
moment and torque	in · lb or N · m
pressure	psi or Pa
temperature rise	F°, C°, or K

Brakes are used to stop or control the speed of machinery and vehicles. Major design concerns include

- providing adequate braking torque or force
- absorbing and dissipating the heat produced by the kinetic energy of the vehicle or machine.

TYPES OF BRAKES

Common **mechanical (friction) brake types** include caliper disc brakes, plate-type disc brakes, internal-shoe drum brakes, band brakes, and expanding- and constricting-type brakes. These rely on mechanical friction and can be actuated by hydraulic, mechanical, pneumatic, or electric systems.

Actuating methods are sometimes combined. Automotive braking systems may include "power assist" or "hydraulic boost" which utilize intake-manifold vacuum or power-steering pump pressure to provide actuating force. Some brakes are designed with metal-to-metal contact. More commonly, a friction lining on one surface contacts a smooth metal surface. Friction linings may consist of sintered metal (pressure-molded iron and copper powder), cermets (pressure-molded ceramic and metal powders), and resin-impregnated cotton. Asbestos-based brake linings have been used extensively. However, many manufacturers are replacing asbestos with other materials in order to reduce health risks during manufacture and health risks to the general public.

Nonmechanical brakes are used for special applications and are virtually wear-free. They include hysteresis brakes which provide a braking torque proportional to control current; eddy-current brakes which are used to control speed, but not for stopping; and magnetic-particle brakes. These three brake types do not depend on mechanical friction, but they do convert kinetic or potential energy into heat. They may require cooling systems if use is frequent or continuous, particularly if they are used as tension brakes (drag load machinery speed-control braking).

Typical **fail-safe brakes** are engaged by a spring or a permanent magnet. They are disengaged by applying a voltage to a coil to produce a magnetic force strong enough to counteract the permanent magnet or spring force. Some are controlled by air pressure. When electric power or the compressed air system fails, the brake engages. When power to an electric hoist is interrupted, a fail-safe brake can be used to stop the load from falling. Obviously, fail-safe brakes are inappropriate for stopping vehicles.

KINETIC ENERGY AND POTENTIAL ENERGY

A brake must absorb kinetic energy to control or stop a moving mass. The kinetic energy contribution due to rectilinear motion has the form $Mv^2/2$, and the contribution due to rotation has the form $J\omega^2/2$. If the mass descends during the braking, we must consider potential energy Wz. The sum of kinetic and potential energy in a system is given by

$$E_K + E_P = Mv^2/2 + J\omega^2/2 + Wz \qquad \textbf{(13.1)}$$

where E_K = kinetic energy

E_P = potential energy (Energy is measured in in · lb or N · m, where 1 N · m = 1 J.)

$M = W/g$ = mass moving in translation (lb · s²/in or kg)

v = velocity of mass moving in translation (in/s or m/s)

$W = Mg$ = weight or force (lb or N)

g = acceleration of gravity = 386 in/s² = 9.8 m/s²

J = mass-moment-of-inertia (lb · s² · in or kg · m²)

ω = angular velocity (rad/s)

z = elevation change of mass (in or m)

ENERGY DUE TO BRAKING

Some brake systems are designed to stop rectilinear motion directly. For example, a brake shoe can directly contact a track. However, most brakes include a rotating disc or drum and nonrotating bands, pads, or other friction members. At any instant, the power dissipated in the brake is

$$P_B = T_B\omega \qquad \textbf{(13.2)}$$

where P_B = power dissipated in braking (in · lb/s or W)

T_B = brake torque (in · lb or N · m)

For a friction brake that applies constant torque, the energy converted to heat is

$$E_B = T_B\theta \qquad \textbf{(13.3)}$$

where E_B = heat energy due to brake friction (in · lb or N · m where 1 N · m = 1 J)

θ = rotation of the brake (rad)

TOTAL ENERGY

If we account for the kinetic, potential, and heat energy in the machinery or vehicle, including the possibility of masses rotating at different speeds, the result is

$$E_T = E_K + E_P + E_B = Mv^2/2 + \sum_{i=1}^{n} J_i\omega_i^2/2 + Wz + T_B\theta \qquad \textbf{(13.4)}$$

where E_T = total energy in the system

n (in the summation) = number of rotating masses

Since we have included energy converted to heat, then total energy E_T should be constant.

CALCULATION OF MASS-MOMENT-OF-INERTIA

The torque required to decelerate a rotating body is proportional to the mass-moment-of-inertia, the second moment of mass about the rotation axis. Mass-moment-of-inertia is defined by

$$J = \int_m r^2 \, dm \tag{13.5}$$

where dm = an element of mass

r = radius from the rotation axis

For a solid disc of radius R and thickness b,

$$dm = \gamma \cdot b \cdot 2\pi r \, dr$$

$$J = 2\pi\gamma b \int_0^R r^3 \, dr = \pi\gamma b R^4/2 \tag{13.6}$$

where $\gamma = \gamma_W/g$ = mass density ($\text{lb} \cdot \text{s}^2 \cdot \text{in}^{-1}/\text{in}^3 = \text{lb} \cdot \text{s}^2/\text{in}^4$ or kg/m^3)

γ_W = weight density (lb/in^3)

g = acceleration of gravity = 386 in/s^2

Radius of Gyration

Mass-moment-of-inertia is also expressed in the form

$$J = MR_G^2 \tag{13.7}$$

where M = total mass ($\text{lb} \cdot \text{s}^2/\text{in}$ or kg)

R_G = radius of gyration (in or m)

For the solid disc, total mass is given by

$$M = \gamma \cdot b \cdot \pi R^2 \tag{13.8}$$

Thus,

$$R_G^2 = J/M = \pi\gamma b R^4/(2\gamma b \pi R^2) = R^2/2 \tag{13.9}$$

In the example problem that follows, we will examine a disc consisting of a heavy rim with a lighter inner section.

EXAMPLE PROBLEM Mass-Moment-of-Inertia

Find the weight, mass-moment-of-inertia, and radius of gyration of an 8 in diameter steel disc. The thickness is

$$b_1 = 0.75 \text{ in} \quad \text{for } 0 \le r < 2.5$$

$$b_2 = 2.25 \text{ in} \quad \text{for } 2.5 < r \le 4 \text{ in}$$

Solution: The approximate weight density of steel is 0.284 lb/in³. The volume of the inner part is $\pi R_1^2 b_1$, and the volume of the outer part is $\pi(R_2^2 - R_1^2)b_2$. Adding and multiplying by the weight density, we obtain the weight.

$$W = \gamma_W \pi[R_1^2 b_1 + (R_2^2 - R_1^2)b_2]$$
$$= 0.284\pi[2.5^2 \cdot 0.75 + (4^2 - 2.5^2)2.25] = 23.76 \text{ lb}$$

Dividing by the acceleration of gravity, we find mass.

$$M = W/g = 23.76/386 = 0.62 \text{ lb} \cdot \text{s}^2/\text{in}$$

Mass-moment-of-inertia is given by

$$J = \pi[R_1^4 b_1 + (R_2^4 - R_1^4)b_2]\gamma_W/(2g) = 0.598 \text{ lb} \cdot \text{s}^2 \cdot \text{in}$$

Radius of gyration is

$$R_G = (J/M)^{1/2} = (0.598/0.62)^{1/2} = 3.117 \text{ in}$$

■■

In the preceding example problem, the disc configuration concentrates the mass near the outer radius, as indicated by the radius of gyration. This produces a large value of mass-moment-of-inertia, which is desirable in some cases. For example, flywheels are designed with a high mass-moment-of-inertia in order to store energy and limit speed variations.

HOISTS AND OTHER MATERIALS-HANDLING MACHINERY

Hoist systems often include self-locking worm drives. If power to the drive motor is interrupted, then the friction of a self-locking drive prevents the load from driving the motor. Brakes may still be required because the worm or gear teeth may fail in bending or wear. One such failure, due to lack of lubricant, occurred in a scaffold drive. A fatal accident resulted because the system lacked an independent brake. Overrunning hoist drives are not self-locking; they almost always require a brake for speed control and emergency stopping.

Consider a mass that is to be raised and lowered by winding a cable or chain on a drum. Total energy as given by Equation (13.4) is constant. Thus, the sum of potential and kinetic energy when the brake is first applied equals the heat energy when the system is finally stopped. If the system rotates through an angle θ during brake application, required brake torque is given by

$$T_B = (J\omega^2 + Mv^2)/(2\theta) + WR_{\text{pitch}} \tag{13.10}$$

where T_B = brake torque (constant, in · lb or N · m)

R_{pitch} = drum radius measured to cable center (in or m)

θ = drum rotation during stopping (rad)

■■ EXAMPLE PROBLEM Hoist Brake

A 140 kg mass is to be raised and lowered by a cable on a 195 mm radius drum. The drum is made of steel and aluminum with thickness as follows:

$b_1 = 25$ mm for $0 \le r < 185$ mm (steel part)

$b_2 = 250$ mm for $185 < r \le 195$ mm (aluminum part)

Maximum drum speed is 440 rpm, and $R_{pitch} = 200$ mm. A brake is to be designed to stop the mass within a distance of $z = 2.5$ m. Find required brake torque.

Solution: The approximate mass densities of aluminum and steel are 2770 and 7860 kg/m^3, respectively. Mass-moment-of-inertia of the drum is given by

$$\begin{aligned} J &= \pi[\gamma_1 b_1 R_1^4 + \gamma_2 b_2 (R_2^4 - R_1^4)]/2 \\ &= \pi[2770 \cdot 0.025 \cdot 0.185^4 + 7860 \cdot 0.250(0.195^4 - 0.185^4)]/2 \\ &= 0.975 \text{ kg} \cdot \text{m}^2 \end{aligned}$$

Total drum mass is 30.9 kg, and the radius of gyration is 0.178 m. The force due to the translating mass is

$$W = Mg = 140 \cdot 9.8$$

The 140 kg mass is initially moving at a speed of

$$v = \omega R_{pitch}$$

The drum turns through an angle

$$\theta = z/R_{pitch}$$

The required brake torque is

$$T_B = (J\omega^2 + Mv^2)/(2\theta) + WR_{pitch} = 832.8 \text{ N} \cdot \text{m}$$

■■

It is important to use consistent units. Note that rotational speed (rpm) must be converted to angular velocity (rad/s).

VEHICLE BRAKES

Moving vehicles possess kinetic energy owing to translating vehicle mass and rotating masses. In this case, the rotating masses translate as well. If the vehicle is descending a hill, potential energy must also be considered. The potential energy change is given by Wz where W = weight (lb) in the customary U.S. system and force Mg (N) in the metric system, and z = elevation change (in or m). Be sure to use consistent units throughout. Convert speeds in mph or km/h to in/s or m/s.

When a vehicle is brought to a stop, the kinetic energy and potential energy are converted into heat. This process is so rapid that little of the heat energy is dissipated, causing a rapid rise in brake temperature. The designer must consider the temperature rise and must plan for heat dissipation so that the brake cools before the next use.

Consider the inertia effect at the center-of-gravity of a forward-moving automobile as the brakes are applied. The moment of the inertia-force causes the auto to pitch, putting more force on the front wheels and less on the rear wheels. As a result, engineers design the front brakes to provide a greater braking torque than the rear brakes.

When a brake stops the rotation of a wheel, the wheel is said to be "locked." However, friction between the tires and the road surface usually governs the stopping distance for a vehicle. One or more locked wheels may result in loss of vehicle control as the tire(s) skid on the road surface. Antilock brake systems are designed to sense a loss of traction and adjust braking torque in an attempt to minimize stopping distance while maintaining vehicle control.

Road-tire friction conditions vary widely owing to road surface, tire condition, precipitation, and other factors. If we determine a limiting deceleration for a controlled stop, then braking time is given by

$$t_B = v/A_B \tag{13.11}$$

where t_B = time the brake is applied to bring the vehicle to a stop (s)
v = initial vehicle velocity (in/s or m/s)
A_B = deceleration (in/s^2 or m/s^2)

Reaction and Response Time

In a laboratory setting, a subject can respond to a stimulus in a fraction of a second. In field conditions where the stimulus is not expected, the response time is greater. McCormick and Sanders (1982) reported on a study of automobile drivers applying the brake pedal following an auditory signal. When the signal was anticipated within the next 10 km, the mean response time was 0.54 s, and when the signal was a surprise, mean response time was 0.73 s. Under surprise conditions, the median time was about 0.90 s, with some subjects taking more than 2 s to respond. Considering both braking time and reaction time,

$$t = t_B + t_R \tag{13.12}$$

where t = total stopping time
t_R = reaction time

If vehicle velocity decreases uniformly during braking, then braking distance is given by

$$X_B = v^2/(2A_B) \tag{13.13}$$

If vehicle velocity is constant during the reaction time, total stopping distance is

$$X = v^2/(2A_B) + vt_R \tag{13.14}$$

where X_B = braking distance (in or m)
X = total stopping distance (in or m)

Kinetic energy at the start of braking is

$$E_K = (Mv^2 + J\omega^2)/2 \tag{13.15}$$

where M = total vehicle mass including rotating masses (kg, or W/g lb · s^2/in)
J = mass-moment-of-inertia of rotating masses (lb · s^2 · in or kg · m^2)

At the start of braking, potential energy is

$$E_P = WX_B \sin \phi \tag{13.16}$$

where W = total vehicle weight or force (lb, or Mg N)
ϕ = roadway downgrade angle
E_K and E_P = kinetic and potential energy, respectively (in · lb or N · m = J)

If total brake torque is constant in time, then the sum of the potential and kinetic energy absorbed by the brake is equal to brake torque times angular rotation in radians. This leads to

$$T_B = (E_K + E_P)R/X_B \tag{13.17}$$

where T_B = sum of the torques on all wheels
R = tire radius

Heat generated in stopping the vehicle is given by the brake torque times the angle of rotation, or simply by the sum of the kinetic and potential energy at the start of braking.

$$E_B = T_B X_B/R = E_K + E_P \tag{13.18}$$

where E_B = heat energy due to stopping the vehicle (in · lb or N · m)

We can convert inch · pound units to British thermal units (Btu) by dividing by (12 · 778.2).

EXAMPLE PROBLEM Stopping Distance and Time; Brake Torque; and Heat Generated in a Vehicle Brake System

A 3000 lb vehicle is to be stopped from various speeds on a 2° downgrade. The wheels, tires, and associated masses are equivalent to 250 lb of rotating mass with a radius of gyration of 9.5 in. Tire radius is 14.5 in. Determine stopping time and distance; braking torque; and heat generated.

Assumptions: Calculations will be based on a 1 second driver reaction time. It will be assumed that a controlled stop can be made at 0.7g, that is, at a deceleration of 0.7 the acceleration of gravity.

Solution Summary: To avoid repetitious calculations, mathematics software was used to plot and tabulate results which are given in the detailed solution that follows. We set initial vehicle speed

$$s = 0, 5, \ldots, 65$$

instructing the program to calculate values for speeds from zero to 65 miles per hour (mph) with 5 mph increments. For computer calculations, we identify variables such as stopping distance, stopping time, and heat generated as functions of initial vehicle speed s. We convert speeds in mph to in/s by multiplying by [5280 · 12/3600]. We add braking time

$$t_B = v/A_B$$

to reaction time to obtain total stopping time.

Stopping time (plotted in hundredths of a second in Figure 13.1) varies linearly with speed. Braking distance and stopping distance are given by Equations (13.13) and (13.14). Note that braking distance is proportional to the square of vehicle velocity. Stopping distance was converted into feet before tabulating and plotting. We find angular velocity of the wheels by dividing vehicle velocity by wheel radius. Equations (13.15) and (13.16) are used to determine kinetic and potential energy, and Equation (13.18) is used to determine heat generated in stopping the vehicle. Heat generated in braking is plotted and tabulated in Btu. The sum of the torque on all wheels is

$$T_B = 33{,}060 \text{ in} \cdot \text{lb}$$

as given by Equation (13.17). Note that T_B does not depend on vehicle speed; the numerator and the denominator of the equation are both proportional to v^2.

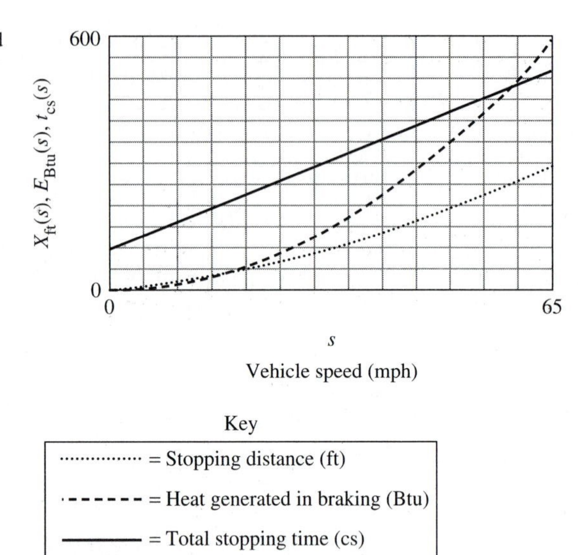

Figure 13.1 Stopping time and distance and heat generated in braking versus vehicle speed.

The detailed solution follows.

Detailed Solution: Stopping distance, heat generated in braking, and time to stop are plotted against vehicle speed in Figure 13.1. These quantities are also tabulated below.

s *(mph)*	*X(s)* *ft*	*E(s)* *Btu*	*Time to Stop (s)* *t(s)*
0	0	0	1
5	8.528	3.498	1.326
10	19.443	13.994	1.651
15	32.748	31.486	1.977
20	48.44	55.975	2.303
25	66.521	87.462	2.628
30	86.99	125.945	2.954
35	109.848	171.425	3.28
40	135.094	223.902	3.605
45	162.728	283.376	3.931
50	192.751	349.847	4.257
55	225.162	423.314	4.583
60	259.962	503.779	4.908
65	297.149	591.241	5.234

CALIPER DISC BRAKES

Caliper disc brakes are used in vehicles and conveyors, as tension brakes and flywheel brakes, and in other high-torque and high-kinetic-energy stopping applications. When actuated by air or fluid pressure, calipers force friction shoes against each side of the brake disc. Two or four pistons are used in **fixed-caliper-type brakes** to ensure equal force on either side of the disc. Figure 13.2(a) shows a caliper disc brake designed for a racing car. Part (b) of the figure is a view of the calipers. An exploded view of a partial caliper disc brake assembly is shown in Figure 13.2(c). When brake lining wear is excessive, the wear sensor rubs against the disc, emitting an audible warning.

A single piston can be used in **floating-caliper-type** and **sliding-caliper-type brakes.** The piston forces one pad (shoe) into contact with the disc; then the reaction force pulls the other pad into contact on the opposite side of the disc. The floating caliper or sliding caliper moves slightly along the shaft or axle axis in the process.

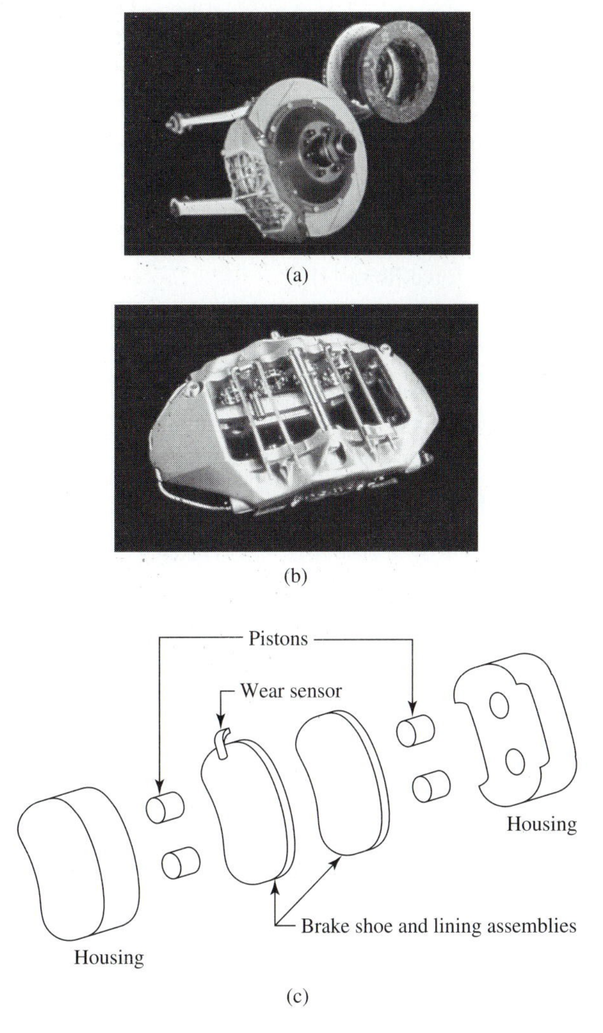

Figure 13.2 (a) Caliper disc brake; (b) a view of the calipers; (c) sketch of some caliper components. (Photos courtesy of Chrysler Corporation)

Industrial caliper brakes utilize either air or hydraulic actuation systems. Air actuation systems are used in heavy trucks, while automobile systems are hydraulic. When the hydraulic or pneumatic pressure is released, springs disengage the brake shoes from the disc. The spring force, which opposes actuating force, should be considered in the design.

Automotive braking is a complicated process that depends on the driver, the road surface, and other factors. During "panic" stops, one or more wheels may lock, causing skidding and loss of control. Antilock brakes modulate brake pressure in an attempt to prevent wheel lockup.

CALIPER BRAKE DESIGN

Caliper brakes usually have two brake shoes, but any even number of shoes can be used. Brake torque is given by

$$T_B = fFN_c r_M \tag{13.19}$$

where T_B = brake torque (in · lb or N · m)

f = coefficient of friction

F = effective actuating force (lb or N)

r_M = mean radius of shoe (shoe center to disc center, in or m)

N_c = number of brake shoes

Actuating force is limited by allowable pressure on the friction material.

EXAMPLE PROBLEM Caliper Brake

A mass rotating at $n = 5200$ rpm is to be stopped in $t_B = 2.5$ s (excluding reaction time). The mass (including estimated brake disc mass) is $M = 45$ kg with a radius of gyration of $R_G = 100$ mm. Plan a tentative brake design, and determine temperature rise in the disc.

Design Decisions: A pair of brake pads will be located at a mean radius of $R_M = 95$ mm. The brake disc will be steel with a mass of $M_D = 3$ kg. Pressure on the pads will be limited to $p = 500{,}000$ Pa, and the coefficient of friction is estimated to be 0.2.

Solution: Initial angular velocity is given by

$$\omega = n\pi/30 = 5200\pi/30 = 544.5 \text{ rad/s}$$

During brake application, there will be a constant angular deceleration, given by

$$\alpha = \omega/t_B = 544.5/2.5 = 217.8 \text{ rad/s}^2$$

The mass-moment-of-inertia is

$$J = MR_G^2 = 45 \cdot 0.100^2 = 0.45 \text{ kg} \cdot \text{m}^2$$

and initial kinetic energy is

$$E_K = J\omega^2/2 = 0.45 \cdot 544.5^2/2 = 66{,}720 \text{ N} \cdot \text{m (J)}$$

All of the kinetic energy results from rotation. Thus, required brake torque is given simply by

$$T_B = J\alpha = 0.45 \cdot 217.8 = 98 \text{ N} \cdot \text{m}$$

Rearranging Equation (13.19), we find effective actuating force as follows:

$$F_A = T_B/(fN_cR_M) = 98/(0.2 \cdot 2 \cdot 0.095) = 2579 \text{ N}$$

The required area of each brake pad is

$$A = F/p = 2579/500{,}000 = 5.159 \cdot 10^{-3} \text{ m}^2 = 5159 \text{ mm}^2$$

The specific heat of steel is about 0.107 cal/(g · K). Multiplying by 4197, we obtain specific heat in the appropriate units: $H_s = 448$ J/(kg · K). Temperature rise in the disc is given by

$$\delta_t = E_K/(H_sM_D) = 66{,}720/(448 \cdot 3) = 49.6 \text{ K}$$

The temperature rise is 49.6 K (49.6 C° or 49.6 · 1.8 = 89 F°).

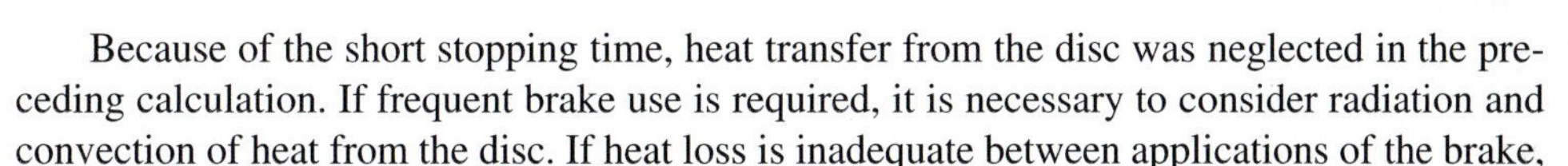

Because of the short stopping time, heat transfer from the disc was neglected in the preceding calculation. If frequent brake use is required, it is necessary to consider radiation and convection of heat from the disc. If heat loss is inadequate between applications of the brake, then a redesign is required to prevent brake overheating.

DRUM BRAKES

The term **drum brake** usually refers to a design incorporating internal shoes that contact a rotating drum. Figure 13.3 is a simplified sketch of an automotive-type drum brake. When the service brake is applied, a small diameter master cylinder provides hydraulic pressure to each wheel cylinder. The wheel cylinders are larger in diameter than the master cylinder, thus magnifying the force applied at the master cylinder. The wheel cylinder pistons force the brake shoes against the drum, which is an integral part of the wheel. Contact occurs at a pad or lining of friction material that is bonded or riveted to the shoe. When the force is released on the master cylinder, the retracting springs return the brake shoes so that they no longer contact the drum.

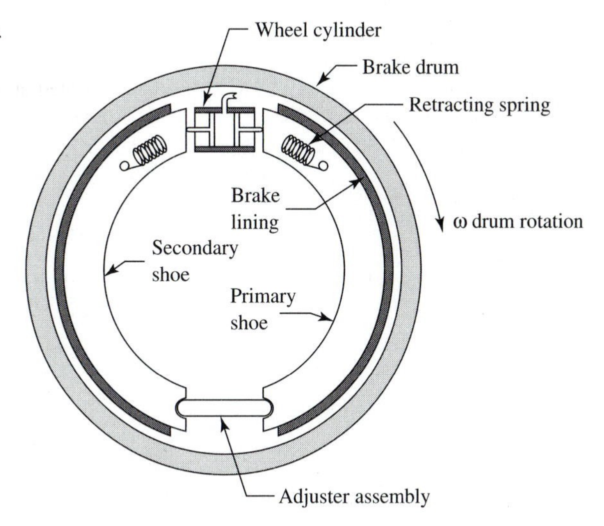

Figure 13.3 Drum brake (simplified sketch).

The brake shoe assembly must be anchored to prevent it from rotating. A cable system can be used to operate the rear wheel brakes as parking brakes. However, some brake systems have been designed with independent parking brakes. Parking brakes were called emergency brakes before the general availability of dual-master-cylinder service brake systems.

Dual master cylinders separate an automotive hydraulic braking system into two separate systems. Usually one system serves the front wheel brakes, and the other the rear wheel brakes. If fluid is lost because of rupture of a brake line in one system, the other system will continue to operate. Obviously, the fault must be corrected because stopping distance will be increased and continued vehicle operation would be unsafe. The driver is alerted to an unsafe condition by a signal controlled by a transducer which senses a difference in pressure between the two systems.

DRUM BRAKE ANALYSIS

Forces and pressures in a drum brake are difficult to determine because of the manner in which the shoe contacts the drum. Figure 13.4, which treats the pivot point on each shoe as a fixed bearing, can be used for analysis. However, the configuration in the figure is not intended to portray the actual parts in a typical drum brake.

Primary Shoe (Self-Actuating Shoe)

With the wheel cylinder located near the top of the drum, the forward brake shoe in each wheel is called the **primary shoe.** We will begin by examining the primary shoe. The brake lining extends from θ_1 to θ_2 on each shoe in Figure 13.4. Contact pressure is assumed to vary along the shoe according to

$$p = P \sin \theta / \sin \theta_P \quad \textbf{(13.20)}$$

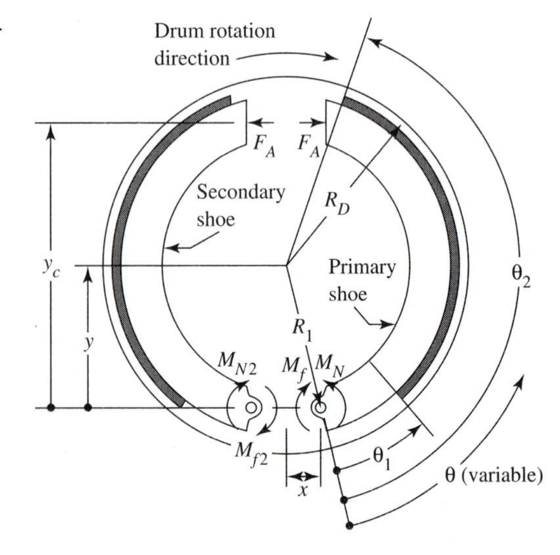

Figure 13.4 Drum brake analysis.

where p = contact pressure at any location θ

P = maximum allowable contact pressure, which depends on lining material (psi or Pa)

$$\theta_P = \begin{cases} \theta_2 & \text{for } \theta_2 < \pi/2 \text{ rad} \\ \pi/2 & \text{for } \theta_2 \geq \pi/2 \text{ rad} \end{cases}$$

At any location θ, the pressure of the drum on the lining causes a normal force contribution of

$$pbR_D\, d\theta$$

with a moment arm about the pivot of

$$r_N = R_1 \sin \theta$$

The moment of the normal forces on the brake shoe is given by the integral of the product of the normal force contributions and their moment arm.

$$M_N = \int_{\theta_1}^{\theta_2} pbR_D r_N\, d\theta \qquad \textbf{(13.21)}$$

where M_N = moment of normal forces (lb · in or N · m)

b = width of lining (in or m)

R_D = drum radius (in or m)

r_N = moment arm of normal force contribution (in or m)

R_1 = distance from pivot point to drum center (in or m)

We obtain the friction force contributions by multiplying the normal force contributions by the coefficient of friction. The moment arm of friction force contributions is

$$r_f = R_c - R_1 \cos \theta$$

and the moment due to friction is

$$M_f = \int_{\theta_1}^{\theta_2} pbR_D f r_f\, d\theta \qquad \textbf{(13.22)}$$

where M_f = moment of friction forces (lb · in or N · m)

f = coefficient of friction between brake lining and drum

The wheel cylinder provides the actuating force. The moment of actuating force about the shoe pivot is $F_A y_c$. Consider the moments on the primary shoe. The friction moment aids the moment of actuating force, while the normal moment opposes it. Thus, actuating force is given by

$$F_A = (M_N - M_f)/y_c \qquad \textbf{(13.23)}$$

where F_A = actuating force on primary shoe (lb or N)

y_c = distance from line of action of wheel cylinder force to pivot point (in or m)

Since friction moment aids the actuating force, the primary shoe can be called the **self-actuating shoe.**

Brake torque due to the primary shoe is the friction moment about the wheel center, given as

$$T_1 = fR_D^2 b \int_{\theta_1}^{\theta_2} p\, d\theta \tag{13.24}$$

where T_1 = braking torque on one wheel cylinder due to the primary shoe

Secondary Shoe

Equilibrium of moments about the pivot yields

$$F_A y_c - M_{N2} - M_{f2} = 0$$

where M_{N2} and M_{f2} = normal force moment and friction moment on the secondary shoe, respectively

This equation shows that the secondary shoe is not self-actuating; friction moment and normal force moment both oppose actuating force moment about the secondary shoe pivot point. If the wheel cylinder diameter is the same on both sides, then the actuating force will be the same on both shoes. The secondary shoe will be less effective than the primary shoe. Maximum pressure on the secondary shoe is given by

$$P_2 = PF_A y_c/(M_N + M_f) \tag{13.25}$$

where P_2 = maximum contact pressure on the secondary shoe

M_N and M_f = normal force moment and friction moment on the primary shoe, respectively

The local pressure is

$$p_2 = P_2 \sin\theta/\sin\theta_P \tag{13.26}$$

where p_2 = contact pressure at any location on the secondary shoe

Secondary shoe braking torque is given by

$$T_2 = fR_D^2 b \int_{\theta_1}^{\theta_2} p_2\, d\theta \tag{13.27}$$

where the values of θ_1 and θ_2 can be the same as for the primary shoe.

Total braking torque on one brake drum is

$$T = T_1 + T_2 \tag{13.28}$$

As noted above, the friction moment on the primary shoe helps to hold that shoe against the drum, adding to its effectiveness. Friction moment can be increased by increasing coefficient of friction and changing the pivot point of the shoe. Examining Equation (13.23) for actuating force, note that the required actuating force becomes zero if friction moment equals normal moment. For $M_f \geq M_N$, the brake shoe becomes **self-locking.** A self-locking brake can be used for emergency stopping of industrial machinery. Automotive service brakes are designed to ensure that self-locking cannot occur, since steering control can be lost if the brakes lock.

Large actuating forces are required to produce high braking torque. Force applied to an automotive brake pedal can be magnified by the use of a small diameter master cylinder and large diameter wheel cylinders and by mechanical leverage. However, force magnification increases the required pedal travel. The self-actuating feature of drum brakes accounts for their earlier popularity. After power-assisted brake systems became generally available, caliper disc brakes, which are not self-actuating, became standard for automotive use, particularly for front-wheel brakes.

EXAMPLE PROBLEM Drum Brake

Design a brake to produce 400 N · m torque on a drum of radius $R_D = 135$ mm. Base the design on a coefficient of friction of $f = 0.25$, and allow a maximum contact pressure of $P = 0.8$ MPa.

Design Decisions: Refer to Figure 13.4, "Drum Brake Analysis." Let the lining extend on the shoe from 15° to 120°. Use the following dimensions (mm):

$$x = 40 \qquad y = 100 \qquad y_c = 200$$

Solution (a)—First Approximation: We will try a lining width of 30 mm. The location where contact pressure is maximum is obviously

$$\theta_P = \pi/2 \text{ rad}$$

for this brake. The logical IF statement,

$$\theta_P = \text{IF}(\theta_2 < \pi/2, \theta_2, \pi/2)$$

produces the same result. Local pressure is given by Equation (13.20).

$$p = P \sin \theta / \sin(\pi/2) = 0.8 \cdot 10^6 \sin \theta$$

Shoe pivot location is given by

$$R_1 = (0.040^2 + 0.100^2)^{1/2} = 0.108 \text{ m}$$

Normal force moment arm is

$$r_N = R_1 \sin \theta$$

and friction force moment arm is

$$r_f = R_D - R_1 \cos \theta$$

If a user-written or mathematics software integration routine is available, we can let the computer or calculator compute the moments. On the primary shoe, the moment of the normal forces is

$$M_N = \int_{\theta_1}^{\theta_2} p b R_D r_N \, d\theta = 438.92 \text{ N} \cdot \text{m}$$

and the moment due to friction is

$$M_f = \int_{\theta_1}^{\theta_2} pbR_D f r_f \, d\theta = 130.51 \text{ N} \cdot \text{m}$$

Note that the primary shoe is self-actuating, but since $M_N > M_f$, it is not self-locking.

Actuating force for the primary shoe is

$$F_A = (M_N - M_f)/y_c = 1542 \text{ N}$$

Brake torque due to the primary shoe is

$$T_1 = fR_D^2 b \int_{\theta_1}^{\theta_2} p \, d\theta = 160.30 \text{ N} \cdot \text{m}$$

The secondary shoe will be designed as a mirror image of the primary shoe. Since the secondary shoe is not self-actuating, maximum pressure on it is given by

$$P_2 = PF_A y_c/(M_N + M_f) = 0.433 \cdot 10^6 \text{ Pa}$$

and local pressure is

$$p_2 = P_2 \sin\theta/\sin\theta_P = 0.433 \cdot 10^6 \sin\theta$$

Secondary shoe braking torque is given by

$$T_2 = fR_D^2 b \int_{\theta_1}^{\theta_2} p_2 \, d\theta = 86.82 \text{ N} \cdot \text{m}$$

Total braking torque on one brake drum is

$$T = T_1 + T_2 = 247.12 \text{ N} \cdot \text{m}$$

for a 30 mm shoe width.

Solution (b)—Corrected Design: Since braking torque is inadequate, a design change is necessary. Because a change in maximum pressure might require a material change, and because some geometry changes such as an increase in drum diameter could be unacceptable, we will increase shoe width. To produce the required torque $T_R = 400$ N · m, the corrected shoe width is

$$b_c = bT_R/T = 0.030 \cdot 400/247.12 = 48.56 \quad \text{(say, 49 mm)}$$

Rerunning the program with $b = 0.049$ m and the other dimensions unchanged, we obtain

$$M_f = 213.2 \qquad M_N = 716.9 \text{ N} \cdot \text{m} \qquad F_A = 2519 \text{ N}$$

$$T_1 = 261.8 \qquad T_2 = 141.8 \qquad T = 403.6 \text{ N} \cdot \text{m}$$

If an integration routine is not available, then a table of integrals will be helpful for putting the moment and torque equations in usable form.

DRUM BRAKE DESIGN OPTIONS

Refer to Figures 13.3 and 13.4, assuming that drum rotation is always, or almost always, as shown. We see from the preceding example problem that maximum pressure on the secondary shoe is substantially lower than maximum pressure on the primary shoe. This condition leads us to consider various design options such as the following:

1. Increase actuating force on the secondary shoe. This change increases braking torque but requires a more complicated wheel cylinder design.
2. Reduce the width of the secondary shoe. This change reduces weight without affecting braking torque.
3. Invert the secondary shoe. This change results in two self-acting shoes, substantially increasing brake torque. Two separate wheel cylinders are required.
4. After evaluating the above options, we may choose to go back to the original configuration as sketched in Figure 13.3. Option (1) requires an asymmetric wheel cylinder design; option (2) results in brake shoes of two different sizes; and option (3) requires two separate wheel cylinders per brake rather than one cylinder with two pistons. Designers must consider inventory costs and the cost of manufacture and assembly of the final product. The advantages of options (1), (2), or (3) may not be worth the additional cost of producing and assembling asymmetric parts or the additional inventory costs due to increasing the number of different parts. Guidelines for decisions of this type are found in Chapter 19, "Design for Manufacture," and in the references cited in that chapter.

CLUTCH-TYPE DISC BRAKES

Clutches are designed to connect and disconnect an input and output shaft as required. If one set of discs is fixed, a multiple-disc clutch acts as a brake. Braking torque is given by

$$T = \pi f N_{pr} P_{\text{MAX}} r_I (r_O^2 - r_I^2) \tag{13.29}$$

where T = braking torque (in · lb or N · m)

f = coefficient of friction

P_{MAX} = maximum contact pressure (psi or Pa)

N_{pr} = number of pairs of friction surfaces

r_O and r_I = outer and inner radius, respectively, of the contact surfaces on the discs (in or m)

A double-throw clutch can be used as a brake if one set of discs is fixed. It can then be used to engage an output shaft to an input shaft, allow both shafts to rotate freely, or stop the output shaft.

BAND BRAKES

A typical band brake consists of a metal band contacting the exterior of a rotating drum, as shown schematically in Figure 13.5. Applications of band brakes include changing of planetary gear train ratios in automatic transmissions.

Band brake analysis is similar to analysis of belt drives. Braking torque is given by

$$T = (F_1 - F_2)R \tag{13.30}$$

Figure 13.5 Band brake.

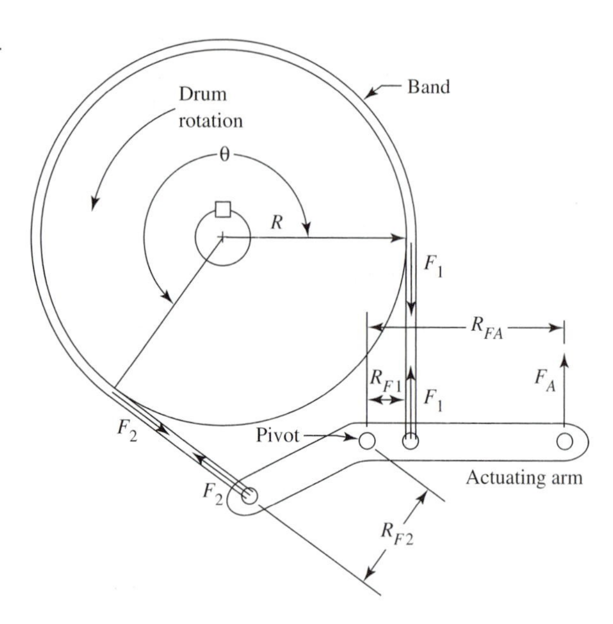

where T = brake torque (in · lb or N · m)

R = drum radius (in or m)

F_1 and F_2 = maximum and minimum band tension, respectively (lb or N)

Note that the tight (maximum tension) side of the brake depends on drum rotation direction.

At maximum braking capacity, the tension relationship is

$$\ln(F_1/F_2) = f\theta \tag{13.31}$$

or

$$F_1/F_2 = e^{f\theta} \tag{13.32}$$

where f = coefficient of friction

θ = angle of contact of band on drum (rad)

e = base of natural logarithms

ln = natural logarithm

Note: $\ln(x) = \lg(x)/\lg(e) = \lg(x)/0.43429$ where $\lg = \log_{10}$ = common logarithm.

Equation (13.32) gives the maximum possible tension ratio F_1/F_2. If the brake band is partially tightened on the drum, the actual tension ratio F_1/F_2 will be less. Maximum tension F_1 is also limited by the strength of the band.

The band can be tightened by a differential actuating arm as in Figure 13.5. Using equilibrium of moments, we find the required actuating force as follows:

$$F_A = (F_2R_{F2} - F_1R_{F1})/R_{FA} \tag{13.33}$$

where moment arms R_{F1}, R_{F2}, and R_{FA} are the perpendicular distances from their respective forces to the actuating arm pivot.

This arrangement results in the maximum tension aiding the actuating force. The brake is self-actuating, somewhat like the primary shoe in a drum brake.

If $F_2R_{F2} \leq F_1R_{F1}$, then the brake is self-locking. An engineer must be cautious in the design of brakes with differential actuating arms. If the brake locks, the band or lever may rupture, with serious consequences. Note that a reversal in drum rotation changes the force relationship and that wear and brake band deformation can change moment arms.

EXAMPLE PROBLEM Band Brake

A braking torque, T of 700 in · lb is to be applied to a drum with a radius of $R = 3.5$ in. Design the brake.

Design Decisions: Use a band brake, incorporating a steel band with thickness $t = 0.02$ in and working strength $S_W = 12{,}000$ psi. Let the band contact three-fourths of the drum, and estimate a coefficient of friction f of 0.15.

Solution: The angle of contact is

$$\theta = (3/4)2\pi = 1.5\pi \text{ rad}$$

When the brake is operating at design capacity, the tension ratio will be

$$F_1/F_2 = e^{f\theta} = 2.028$$

Using this ratio and rearranging Equation (13.30), we find maximum (tight-side) tension

$$F_1 = T/[R(1 - F_2/F_1)] = 394.6 \text{ lb}$$

and minimum (slack-side) tension

$$F_2 = 194.6 \text{ lb}$$

If stress in the brake band is relieved after bending, then the design can be based on direct tension in the band. The required width of the band is

$$b = F_1/(S_W t) = 1.65 \text{ in}$$

The required actuating force depends on the geometry of the actuating arm.

BELAYS

We can use the band brake equations to approximate the effect of a belay or snub, where a rope of cable is wound on a pin, cleat, or other projection.

EXAMPLE PROBLEM Belay

A person is to belay a 2000 lb load by winding a cable around a large pin. How many turns would this belay require?

Assumptions: Estimate a coefficient of friction of 0.1, and assume that a person could exert a force of 50 lb.

Solution: Maximum tension $F_1 = 2000$, and minimum tension $F_2 = 50$ lb. Rearranging Equation (13.31), we have

$$\theta = \ln(F_1/F_2)/f = 36.9 \text{ rad}$$

and

$$\theta/(2\pi) = 5.9 \quad \text{(about 6 turns of cable)}$$

Reference

McCormick, E. J., and M. S. Sanders. *Human Factors in Engineering Design.* 5th ed. New York: McGraw-Hill, 1982, pp. 197–200.

Design Problems

13.1. A 5.5 in radius steel disc is 0.5 in thick for $0 \leq r < 5$ in and 3 in thick for $5 < r \leq 5.5$ in. Find the weight, mass-moment-of-inertia, and radius of gyration.

13.2. An 8 in radius aluminum disc is 0.875 in thick for $0 \leq r < 7$ in and 4.5 in thick for $7 < r \leq 8$ in. Find the weight, mass-moment-of-inertia, and radius of gyration.

13.3 A 200 mm radius steel-and-aluminum disc is 22 mm thick for $0 \leq r < 175$ mm and 115 mm thick for $175 < r \leq 200$ mm. The inner section is aluminum, and the rim steel. Find the mass, mass-moment-of-inertia, and radius of gyration.

13.4. A 195 mm radius steel-and-aluminum disc is 18 mm thick for $0 \leq r < 150$ mm and 100 mm thick for $150 < r \leq 195$ mm. The inner section is aluminum, and the rim steel. Find the mass, mass-moment-of-inertia, and radius of gyration.

13.5. A 100 kg mass is to be raised and lowered by a cable on a 205 mm radius drum. The drum is made of steel with thickness as follows:

$$b_1 = 20 \text{ mm} \quad \text{for } 0 \leq r < 185 \text{ mm}$$
$$b_2 = 250 \text{ mm} \quad \text{for } 185 < r \leq 205 \text{ mm}$$

Maximum drum speed is 440 rpm, and $R_{pitch} = 210$ mm. A brake is to be designed to stop the mass within a distance z of 1.2 m. Find required brake torque.

13.6. A 100 kg mass is to be raised and lowered by a cable on a 205 mm radius drum. The drum is made of steel with thickness as follows:

$$b_1 = 20 \text{ mm} \quad \text{for } 0 \leq r < 185 \text{ mm}$$
$$b_2 = 250 \text{ mm} \quad \text{for } 185 < r \leq 205 \text{ mm}$$

Maximum drum speed is 250 rpm, and $R_{pitch} = 210$ mm. A brake is to be designed to stop the mass in 3 s (plus reaction time). Find required brake torque and stopping distance.

13.7. A 140 kg mass is to be raised and lowered by a cable on a 195 mm radius aluminum drum. The drum thickness is as follows:

$$b_1 = 25 \text{ mm} \quad \text{for } 0 \leq r < 185 \text{ mm}$$
$$b_2 = 250 \text{ mm} \quad \text{for } 185 < r \leq 195 \text{ mm}$$

Maximum drum speed is 440 rpm, and $R_{pitch} = 200$ mm. A brake is to be designed to stop the mass within a distance z of 2.5 m. Find required brake torque.

13.8. A 140 kg mass is to be raised and lowered by a cable on a 195 mm radius steel drum. The drum thickness is as follows:

$$b_1 = 25 \text{ mm} \quad \text{for } 0 \leq r < 185 \text{ mm}$$
$$b_2 = 250 \text{ mm} \quad \text{for } 185 < r \leq 195 \text{ mm}$$

Maximum drum speed is 440 rpm, and $R_{pitch} = 200$ mm. A brake is to be designed to stop the mass within a distance z of 2.5 m. Find required brake torque.

13.9. A 6000 lb vehicle is to be stopped from various speeds on an 8° downgrade. Vehicle weight includes 200 lb of rotating masses with a radius of gyration of 9 in. Tire radius is 13.5 in.

(a) Find required braking torque.

(b) Plot and tabulate stopping distance, stopping time, and heat generated versus vehicle speed.

Assumptions: Results are to be based on a 1 s driver reaction time and a deceleration of $0.75g$.

13.10. A 1500 lb vehicle is to be stopped from various speeds on a 10° downgrade. Vehicle weight includes 100 lb of rotating masses with a radius of gyration of 8 in. Tire radius is 14 in.

(a) Find required braking torque.

(b) Plot and tabulate stopping distance, stopping time, and heat generated versus vehicle speed.

Assumptions: Results are to be based on a 1 s driver reaction time and a deceleration of 0.5*g*.

13.11. A 2000 lb vehicle is to be stopped from various speeds on a 5° downgrade. Vehicle weight includes 100 lb of rotating masses with a radius of gyration of 8 in. Tire radius is 14 in.

(a) Find required braking torque.

(b) Plot and tabulate stopping distance, stopping time, and heat generated versus vehicle speed.

Assumptions: Results are to be based on a 1 s driver reaction time and a deceleration of 0.6*g*.

13.12. A mass rotating at $n = 2200$ rpm is to be stopped in $t_B = 1.5$ s (excluding reaction time). The mass (including estimated brake disc mass) is $M = 45$ kg with a radius of gyration R_G of 100 mm. Plan a tentative brake design, and determine temperature rise in the disc.

Design Decisions: A pair of brake pads will be located at a mean radius of $R_M = 120$ mm. The brake disc will be steel with a mass of $M_D = 3$ kg. Pressure on the pads will be limited to $p = 450{,}000$ Pa, and the coefficient of friction is estimated to be 0.25.

13.13. A mass rotating at $n = 4800$ rpm is to be stopped in $t_B = 2.5$ s (excluding reaction time). The mass (including estimated brake disc mass) is $M = 75$ kg with a radius of gyration R_G of 100 mm. Plan a tentative brake design, and determine temperature rise in the disc.

Design Decisions: A pair of brake pads will be located at a mean radius R_M of 95 mm. The brake disc will be steel with a mass M_D of 1.75 kg. Pressure on the pads will be limited to $p = 500{,}000$ Pa, and the coefficient of friction is estimated to be 0.2.

13.14. A mass rotating at $n = 5200$ rpm is to be stopped in $t_B = 2.5$ s (excluding reaction time). The mass (including estimated brake disc mass) is $M = 65$ kg with a radius of gyration R_G of 100 mm. Plan a tentative brake design, and determine temperature rise in the disc.

Design Decisions: A pair of brake pads will be located at a mean radius R_M of 95 mm. The brake disc will be steel with a mass M_D of 1.5 kg. Pressure on the pads will be limited to $p = 500{,}000$ Pa, and the coefficient of friction is estimated to be 0.2.

13.15. A mass rotating at $n = 1200$ rpm is to be stopped in $t_B = 3.5$ s (excluding reaction time). The mass (including estimated brake disc mass) is $M = 45$ kg with a radius of gyration R_G of 100 mm. Plan a tentative brake design.

Design Decisions: A pair of brake pads will be located at a mean radius R_M of 95 mm. The brake disc will be steel. Pressure on the pads will be limited to $p = 450{,}000$ Pa, and the coefficient of friction is estimated to be 0.25.

13.16. Design a brake to produce 325 N · m torque on a drum of radius $R_p = 140$ mm. Base the design on a coefficient of friction f of 0.3, and allow a maximum contact pressure p of 0.9 MPa. Find required shoe width and actuating force.

Design Decisions: Refer to Figure 13.4, "Drum Brake Analysis." Let the lining extend on the shoe from 20° to 130°. Use the following dimensions (mm):

$$x = 45 \qquad y = 110 \qquad y_c = 210$$

13.17. Design a brake to produce 350 N · m torque on a drum of radius $R_D = 130$ mm. Base the design on a coefficient of friction f of 0.28, and allow a maximum contact pressure p of 0.85 MPa. Find required shoe width and actuating force.

Design Decisions: Refer to Figure 13.4, "Drum Brake Analysis." Let the lining extend on the shoe from 10° to 125°. Use the following dimensions (mm):

$$x = 45 \qquad y = 115 \qquad y_c = 200$$

13.18. Design a brake to produce 750 N · m torque on a drum of radius $R_D = 145$ mm. Base the design on a coefficient of friction f of 0.3, and allow a maximum contact pressure p of 1.05 MPa. Find required shoe width and actuating force.

Design Decisions: Refer to Figure 13.4, "Drum Brake Analysis." Let the lining extend on the shoe from 10° to 130°. Use the following dimensions (mm):

$$x = 45 \qquad y = 110 \qquad y_c = 215$$

13.19. A braking torque T of 80 in · lb is to be applied to a drum with a radius R of 2.5 in. Design the brake.

Design Decisions: Use a band brake, incorporating a steel band with thickness $t = 0.01$ in and working strength $S_W = 10{,}000$ psi. Let the band contact three-fourths of the drum, and estimate a coefficient of friction f of 0.1.

13.20. A braking torque T of 100 in · lb is to be applied to a drum with a radius R of 2 in. Design the brake.

Design Decisions: Use a band brake, incorporating a steel band with thickness $t = 0.02$ in and working strength $S_W = 10{,}000$ psi. Let the band contact three-fourths of the drum, and estimate a coefficient of friction f of 0.1.

13.21. A braking torque T of 250 in · lb is to be applied to a drum with a radius R of 3 in. Design the brake.

Design Decisions: Use a band brake, incorporating a steel band with thickness $t = 0.015$ in and working strength $S_W = 10{,}000$ psi. Let the band contact three-fourths of the drum, and estimate a coefficient of friction f of 0.1.

13.22. A person is to belay a 1000 lb load by winding a cable around a large pin. How many turns would this belay require?

Assumptions: Estimate a coefficient of friction of 0.15, and assume that a person could exert a force of 80 lb.

13.23. A person is to belay a 1500 lb load by winding a cable around a large pin. How many turns would this belay require?

Assumptions: Estimate a coefficient of friction of 0.12, and assume that a person could exert a force of 60 lb.

13.24. Five turns of cable are wound around a large pin in order to belay a 2500 lb load. What is the required minimum force on the other end of the cable? Assume a coefficient of friction of 0.15.

13.25. Three turns of cable are wound around a large pin in order to belay a 3000 lb load. What is the required minimum force on the other end of the cable? Assume a coefficient of friction of 0.2.

CHAPTER 14

Power Screws and Linear Actuators

Symbols

d_i	screw diameter at base of threads
d_O	outside diameter of screw
E	elastic modulus
f	coefficient of friction
F	load or thrust force
I	moment-of-inertia
L	lead ($N_t p$)
n_w	worm speed
N_g	number of gear teeth
N_t	number of threads
N_w	number of worm teeth
p	pitch
P	power
r_m	mean thread radius $(d_O + d_i)/4$
T	torque
T_{pi}	turns per inch
v	linear velocity of the screw or traveling nut
λ	lead angle
ϕ	thread angle
η	efficiency

Units

angles	degrees or radians
elastic modulus	psi or MPa
lead, pitch	in or mm
load	lb or N
moment-of-inertia	in^4 or mm^4
power	hp or kW
radius, diameter	in or mm
torque	lb · in or N · mm
velocity	in/s or mm/s

Power screws and linear actuators are screw-and-nut systems used to change rotational motion into linear motion. They are used for positioning, leveling, lifting, and other drive and actuation applications. The term **linear actuator** is usually applied when linear motion is required for actuating other components or systems. Linear actuators are used in medical devices, automobiles, and many other applications. The term **power screw** is usually applied when substantial force is to be transmitted; power screws can be machine-powered or manually operated. Force capacities in commercially available power screws range up to hundreds of tons. When power screws are used for positioning, they may be called **lead screws.** The lead screw of a lathe is used for precision positioning of the cutting tool.

Power screws are often called **screw jacks** when used for lifting and leveling. In the usual configuration, the nut rotates in place, and the screw moves axially. Figure 14.1 shows a worm-gear screw jack, where the gear teeth are cut on the outside of the nut. Screw rotation can be prevented by keying the screw to the base. In nonkeyed jacks, the load pad at the top of the screw is restrained to prevent screw rotation. In an alternative design, the screw rotates in place, and the nut moves axially. The nut must then be restrained to prevent it from rotating.

Most power screws are **self-locking.** An axial force will not cause relative motion in a self-locking power screw. This is often a desirable feature. In positioning applications, self-locking screws prevent inadvertent motion. Self-locking screw jacks hold the load in place without an additional brake. Some power screws are **overrunning.** Axial force on an over-

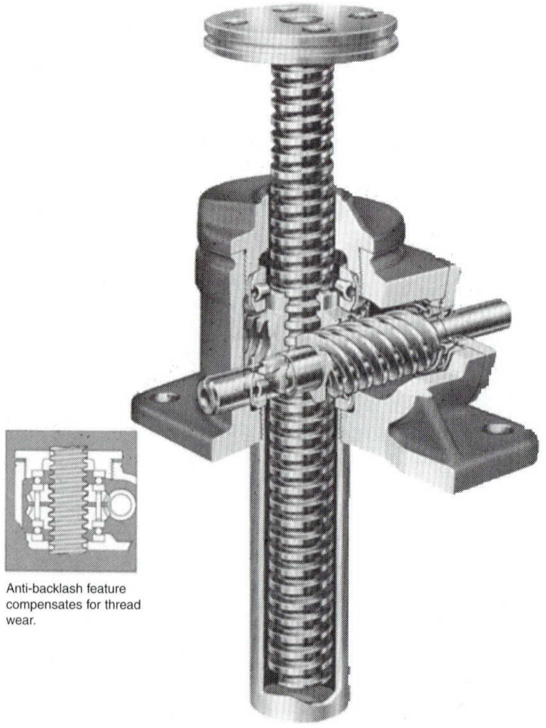

Figure 14.1 Worm-gear screw jack. (Courtesy of Joyce/Dayton Corporation)

Anti-backlash feature compensates for thread wear.

running power screw can cause relative rotation. In this case, linear motion is converted to rotation. Overrunning power screws are analogous to rack and pinion drives. In an automotive rack and pinion steering system, the pinion drives the rack to initiate a turn. However, if the pinion is free to rotate when the turn is completed, wheel caster will force the rack to drive the pinion back to its straight-ahead position.

The nut and screw threads slide directly on one another in ordinary power screws. These power screws are almost always self-locking. They are rugged and low in cost. However, sliding friction in ordinary power screws results in low efficiencies.

Backlash is the axial tolerance between the screw threads and the nut threads. As threads wear, backlash increases. Precision applications require anti-backlash features. For example, certain operations require a precise relationship between cutting tool position and lead screw rotation in a lathe. However, with backlash, the relationship depends on the rotation direction of the lead screw. A split nut can be used to compensate for wear and eliminate backlash. The lead screw nut may be split in an axial plane or a plane perpendicular to the screw axis. In either case, the two nut halves are squeezed together to eliminate backlash.

BALL-SCREWS

Ball-screws have a train of balls that ride between the screw and the nut in a recirculating track. Figure 14.2 shows a cutaway illustration of a ball-screw and two precision ball-screw assemblies. As the balls roll out of one end of the nut, they go through a ball-return tube and resupply the other end of the nut. Ball-screws have higher efficiencies than ordinary sliding friction power screws. Overrunning ball-screws require an auxiliary brake for some applications. Some automotive steering systems utilize ball-screws.

LINEAR MOTION SYSTEMS

Ball-screws are also used in precision linear motion systems. Figure 14.3 shows three linear motion systems. Two of the systems include ball-screws, and one of these includes a motor and controller. Industries utilizing linear motion systems include aerospace, automotive, machine tool, packaging, medical, automated assembly, materials handling, and robotic. A few examples are computer-aided tomography (CAT scan) devices, aircraft wing flaps, and satellite dishes.

SPECIAL-APPLICATION POWER SCREWS AND LINEAR ACTUATORS

Special-application power screws and linear actuators include Ball Reversers™, linear clutches, and recirculating roller screws. **Ball Reversers™** are ball-screw actuators that provide automatic reciprocating action (see Figure 14.4). The screw, which has both right-hand and left-hand helical ball tracks, rotates in only one direction. The nut moves axially along the screw, reversing when the balls move from a right-hand to a left-hand helical track. Figure 14.4(b) is a cutaway drawing of a Ball Reverser™. Part (c) of the figure shows the reversing sequence. The two slot balls move relative to the fixed ball as they follow the

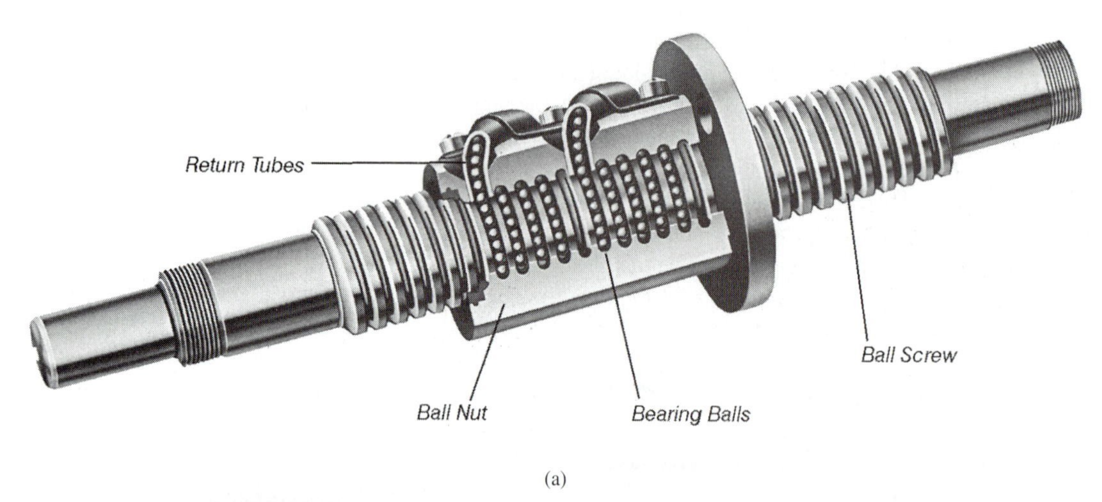

(a)

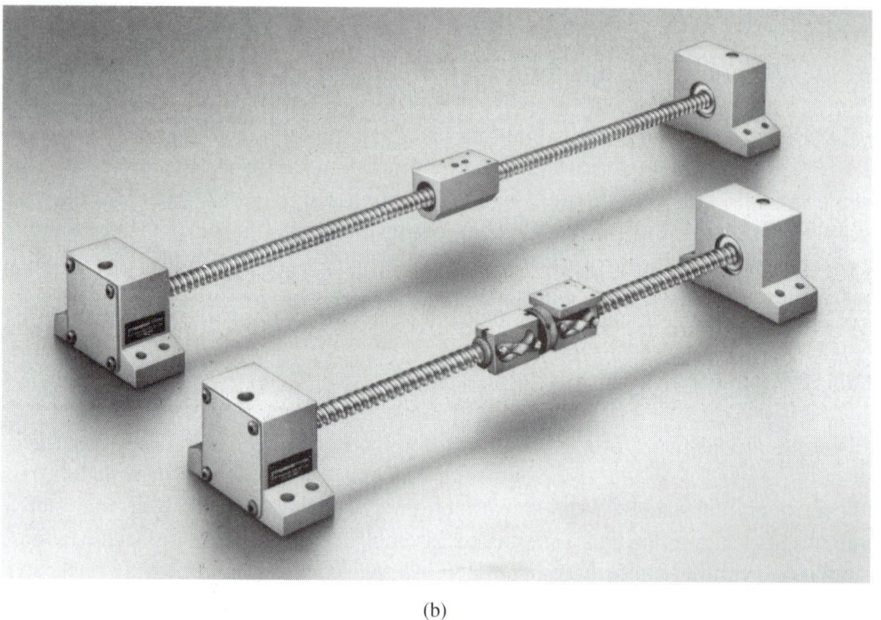

(b)

Figure 14.2 (a) Cutaway of a ball-screw; (b) ball-screw assemblies. (Courtesy of Thomson Industries, Inc.)

turnaround curve. The standard turnaround curve is designed for optimum acceleration of the mass. An instant turnaround curve and a turnaround curve with a dwell are also available, as shown in part (d) of the figure. Cable winding is one application of Ball Reversers™. The cable guide is driven by the ball reducer so that cable is wound evenly over the drum. Agitators are another Ball Reverser™ application; they are used to move stirrer paddles back and forth to prevent settling or hardening of mixtures. Ball Reversers™ are also used to bend automotive radiator fins on a production line and to fold bolts of cloth for packaging.

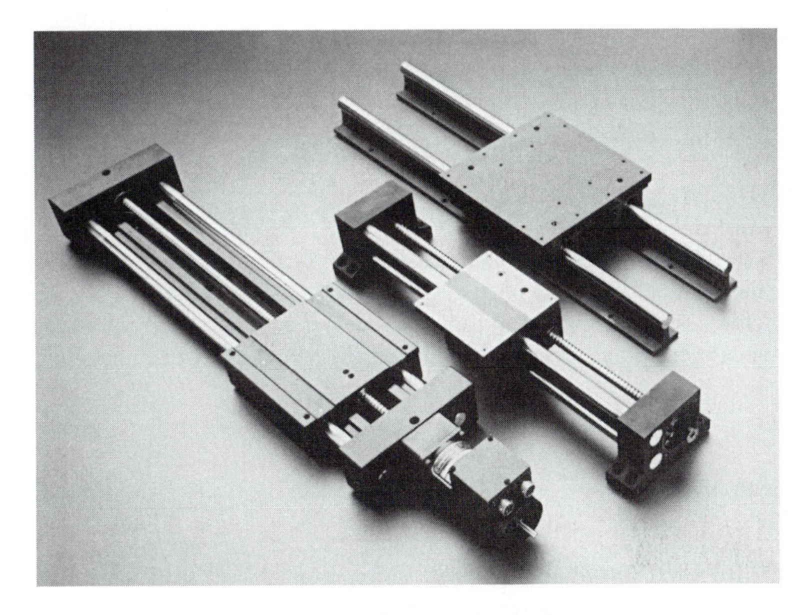

Figure 14.3 Linear motion systems. (Courtesy of Thomson Industries, Inc.)

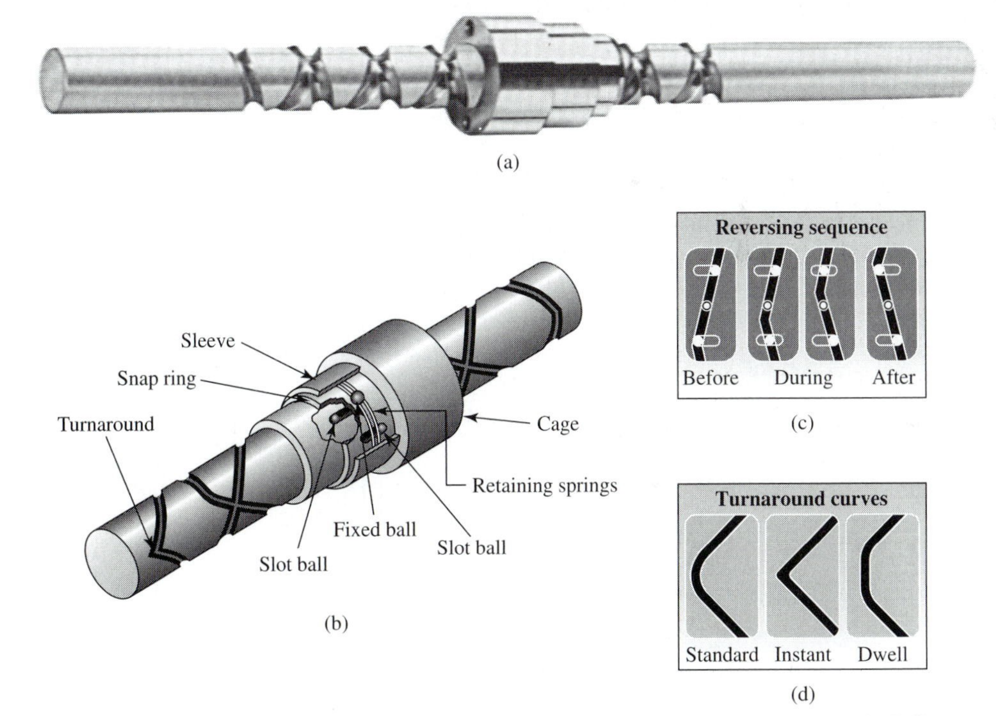

Figure 14.4 (a) Ball Reverser™ reciprocating actuator; (b) cutaway drawing showing elements of a Ball Reverser™; (c) reversing sequence; (d) turnaround curves. (Courtesy of Flennor, Inc.)

Recirculating roller screws utilize rollers instead of balls. They are used for special precision applications. **Linear clutches** in a form similar to ball-screw nuts operate on smooth shafts. They convert rotational motion to linear motion, but they slip when overloaded. The overload setting is adjustable.

PITCH, LEAD, AND LEAD ANGLE

Screw **pitch** is the axial distance between thread centers. **Lead** is the relative axial motion for one screw or nut rotation. Lead and pitch are equal for single-thread screws. In general, the lead is the pitch times the number of threads (the number of starts); lead equals twice pitch for a double-thread screw; and so on. **Lead angle** is the helix angle of the thread.

$$\tan \lambda = \frac{L}{2\pi r_m} \quad \textbf{(14.1)}$$

where λ = lead angle

$L = N_t p$ = lead

$r_m = (D_O + d_i)/4$ = mean thread radius

d_O = outside diameter of screw

d_i = screw diameter at base of threads

p = pitch

N_t = number of threads or starts (A single thread is most common.)

THREAD FORM

Screw thread forms were standardized in the 1800s in order to encourage uniformity and interchangeability in machine tools. Some of these forms are still in use. The most common power screw thread forms are Acme, stub Acme, and modified square.

Figure 14.5(a) shows the general-purpose Acme thread with thread angle $\phi = 14.5°$. A modified square thread with thread angle $\phi = 5°$ is shown in Part (b). The width of either

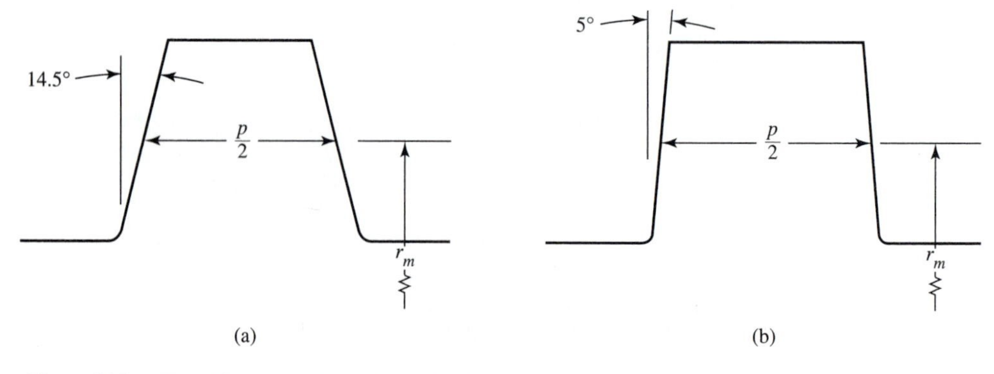

Figure 14.5 Thread forms. (a) Acme; (b) modified square.

thread at mean radius is about one-half the pitch, and the height is also about one-half the pitch. The height of a stub thread is about 0.4 times the pitch. Thread forms are often identified by the included angle 2ϕ. Thus, the general-purpose Acme thread is called a 29° Acme thread, and the modified square thread a 10° modified square thread. If we used a square thread with zero thread angle, then we would not be able to accommodate for wear by tightening the halves of a split nut against the thread.

Right-hand threads are most common, but left-hand threads are also available. Both the worm and the screw are right-hand in the worm-gear screw jack shown in Figure 14.1.

POWER SCREW EFFICIENCY

Efficiency is defined as the power output divided by the power input. An alternative definition for power screws is the ratio of zero friction input torque to actual input torque. Power loss due to sliding friction is substantial and must be considered in our designs. In some cases, the duty cycle must be limited to permit cooling.

If we consider the power screw only, efficiency is given by

$$\eta = \frac{\cos\phi - f\tan\lambda}{\cos\phi + f/\tan\lambda} \qquad \textbf{(14.2)}$$

where η = efficiency of power screw
ϕ = thread angle
λ = lead angle
f = coefficient of friction

Efficiency is very sensitive to lead angle and coefficient of friction, and less sensitive to thread form.

■■ EXAMPLE PROBLEM Power Screw Efficiency—The Effect of Lead Angle, Coefficient of Friction, and Thread Form

Compare the efficiency of power screws for various values of lead angle and coefficient of friction. Compare Acme and modified square thread forms.

Decisions: We will use computer-generated plots of Equation (14.2). The plots require little effort and are easy to interpret; they are shown in Figure 14.6.

Solution: Refer to Figure 14.6. Part (a) shows efficiency versus lead angle for Acme screws with various values of coefficient of friction. Typical sliding friction power screws have coefficients of friction between 0.10 and 0.15. Friction in ball-screws is lower. As expected, high coefficients of friction produce low efficiencies. Small lead angles also result in low efficiencies because there is more sliding. Efficiency drops off for very large lead angles. Efficiency cannot actually be negative. Negative values on the plot indicate that friction would prevent conversion of rotational motion into linear motion.

Power screws are commonly used to produce large forces and for precise positioning. Both of these applications lead to designs with small lead angles. The curves are replotted for small lead angles in Figure 14.6(b). Part (c) shows the effect of thread angle on efficiency for 0.10 coefficient of friction. The upper curve describes the efficiency of a 5° thread angle modified square thread, and the lower curve a

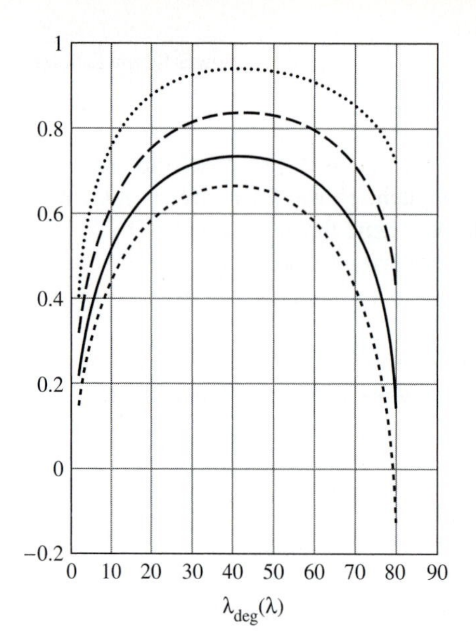

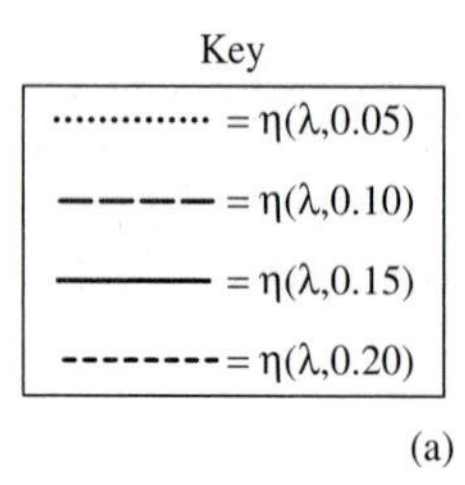

(a)

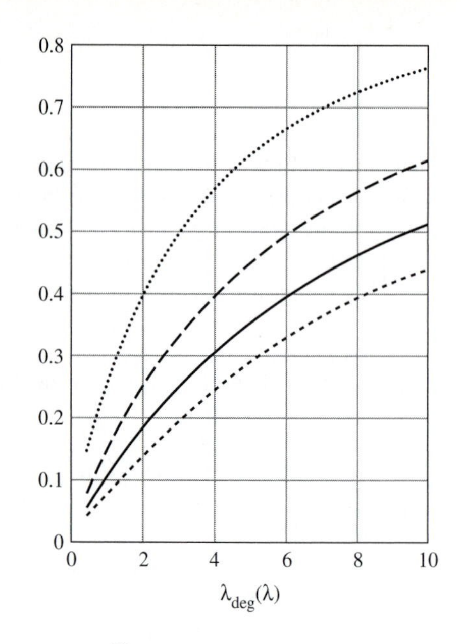

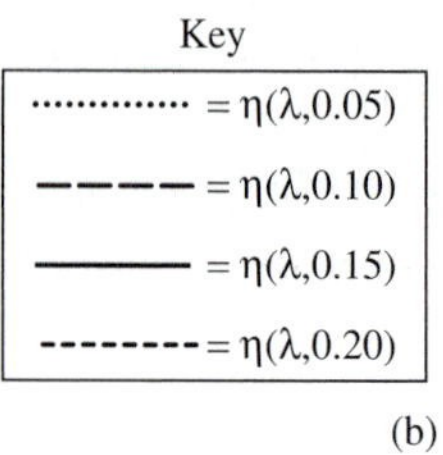

(b)

Figure 14.6 (a) Power screw efficiency; (b) power screw efficiency for small lead angles; (c) effect of thread angle on efficiency.

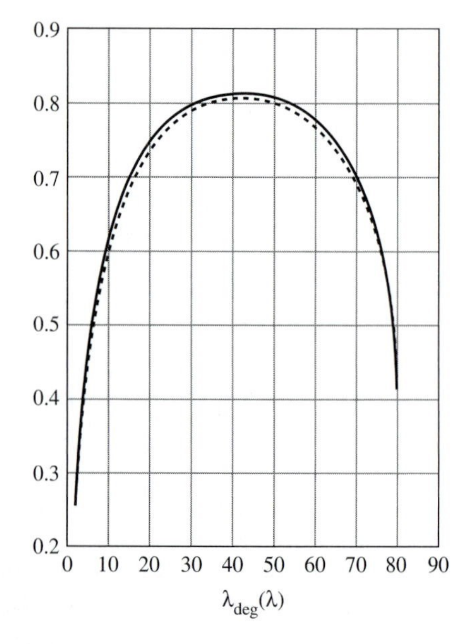

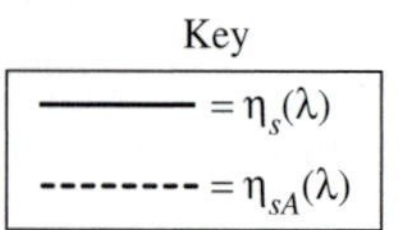

(c)

14.5° Acme thread. Comparing the curves in Part (c), note that the difference in efficiency due to thread angle is small. Power screw systems incorporate bearings, gears, and other components. Thus, actual system efficiency will be lower than indicated by these curves.

BEVEL AND WORM DRIVES FOR POWER SCREWS

Power screws can be operated by electric motors, internal combustion engines, air gear motors, or hydraulic motors. Manually operated screw jacks can utilize handwheels or ratchets.

Bevel gears are used for high-speed, high-efficiency power screw drives. A small bevel gear drives a larger bevel gear which may be an integral part of the nut. Screw rotation is prevented by a key or by restraining the load against rotation. In an alternative design, the traveling nut configuration, the larger bevel gear is fixed to the screw. The screw rotates in place while the nut moves axially. The traveling nut must be restrained against rotation.

Worm drives are used for power screw applications requiring high thrust and precise positioning. In the usual configuration, the worm drives the gear which is an integral part of the nut. The screw moves axially, but screw rotation is prevented by a key or by restraining the load. Traveling nut configurations are also available. Large ratios of input to output speed result in significant friction losses. Thus, input power requirements are sensitive to worm drive efficiency as well as power screw efficiency. Worm drive efficiency can be estimated from

$$\eta_w = \frac{\cos \phi_n - f \tan \lambda_w}{\cos \phi_n + f/\tan \lambda_w} \qquad \textbf{(14.3)}$$

where η_w = efficiency of worm drive

ϕ_n = normal pressure angle

$\lambda_w = \arctan[(N_w d_g)/(N_g d_w)]$ = lead angle of worm

f = coefficient of friction

The efficiency of bearings and collars must also be considered. The efficiency of a worm drive power screw is given by

$$\eta = \eta_s \eta_b \eta_w \qquad \textbf{(14.4)}$$

where η = overall drive efficiency

η_s = power screw efficiency

η_b = bearing efficiency

η_w = worm drive efficiency

POSITIONING PRECISION AND LINEAR VELOCITY

If a power screw is to be used for leveling or other precision positioning applications, a large number of input rotations per unit screw travel is desirable. A large number of input rotations per unit screw travel will also produce large thrust forces. For a worm drive power screw, the number of worm revolutions per inch of screw travel is given by

$$T_{pi} = \frac{N_g}{N_w L} \qquad \textbf{(14.5)}$$

where T_{pi} = turns per inch

N_g = number of gear teeth

N_w = number of worm teeth

L = screw lead (in)

The linear velocity of the screw is

$$v = \frac{n_w N_w L}{60 N_g} \tag{14.6}$$

where v = linear velocity of the screw or traveling nut (in/s)

n_w = worm speed (rpm)

If the lead is given in millimeters, T_{pi} becomes turns per millimeter, and the velocity units are millimeters per second.

OUTPUT AND INPUT POWER

Output power is given by

$$P_{\text{out}} = \frac{Fv}{6600} \tag{14.7}$$

where P_{out} = output power (hp)

F = load or thrust force (lb)

v = velocity (in/s)

In metric units, we have

$$P_{\text{out(kW)}} = \frac{Fv}{10^6} \tag{14.7, metric}$$

where $P_{\text{out(kW)}}$ = output power (kW)

F = load or thrust force (N)

v = velocity (mm/s)

The input power requirement is

$$P_{\text{in}} = \frac{P_{\text{out}}}{\eta} \tag{14.8}$$

INPUT TORQUE REQUIREMENT

Torque applied to the input gear is given by

$$T_{\text{in}} = \frac{63{,}025 P_{\text{in}}}{n_{\text{in}}} \tag{14.9}$$

where T_{in} = input torque (lb · in)

P_{in} = input horsepower

n_{in} = speed of input gear (rpm)

In metric units, the torque is

$$T_{in} = \frac{10^6 P_{in}}{\omega_{in}} \quad \textbf{(14.9, metric)}$$

where T_{in} = input torque (N · mm)

P_{in} = input power (kW)

ω_{in} = angular velocity of input gear (rad/s)

SCREW REACTION TORQUE

A worm drive power screw substantially magnifies the input torque. If the power screw nut rotates in place, then the screw must be keyed to prevent it from rotating. As an alternative, it can be fastened to the load, and the load must prevent screw rotation. In the traveling nut design, the nut must be restrained against rotation. The screw (or nut) reaction torque that must be considered in the design of a worm drive power screw is

$$T_s = \frac{T_{in} N_g \eta_w}{N_w} \quad \textbf{(14.10)}$$

where T_s = reaction torque

A similar equation can be used to calculate reaction torque in a bevel gear drive power screw.

EXAMPLE PROBLEM Power Screw Design

Design a system to provide 8000 lb thrust at 0.060 to 0.075 in/s for an input speed of 200 rpm.

Design Decisions: Many satisfactory designs are possible. We will try a power screw with 2.25 in outside diameter, a 0.5 in pitch, 14.5° thread angle, single, full-depth Acme thread. We will use a worm drive to provide the necessary speed reduction.

Solution Summary: The calculations are given in the detailed solution which follows. For a full-depth thread, the approximate diameter at the base of the threads is

$$d_i = d_O - p$$

and the mean screw diameter is

$$r_m = (d_O + d_i)/4 = (2d_O - p)/4 = (2 \cdot 2.25 - 0.5)/4 = 1 \text{ in}$$

For a single-thread screw, the lead equals the pitch. We calculate a screw lead angle of about 0.079 rad and a screw efficiency of 0.387 (38.7%), assuming a coefficient of friction of 0.12.

We rewrite the screw velocity equation to obtain the required gear ratio, which is

$$\frac{N_g}{N_w} = \frac{n_w L}{} = \frac{200 \cdot 0.5}{60[0.060 \text{ to } 0.075]} = 22.2 \text{ to } 27.8$$

We will use a 24 tooth gear and a single tooth worm with 20°normal pressure angle. Assuming a ratio of gear diameter to worm diameter of approximately 2, the worm lead angle is 0.083 rad. Assuming the same coefficient of friction for the screw and worm drive, we obtain a worm drive efficiency of 0.391. It is assumed that friction at all bearings will result in a bearing efficiency of 0.85. Multiplying the screw, worm, and bearing efficiencies together, we obtain a system efficiency of 0.129.

Output power is based on the selected gear ratio. Dividing by system efficiency, we obtain the input power requirement, which is 0.655 hp. We must provide an input torque to the worm shaft of 206.33 in · lb. The key or other restraint to prevent screw rotation must be able to withstand 1935 in · lb torque.

The detailed solution follows.

Detailed Solution:

Outside Diameter:

$$d_O = 2.25$$

Pitch:

$$p = 0.5$$

Mean Radius:

$$r_m = \frac{2 \cdot d_O - p}{4} \qquad r_m = 1$$

Thread Angle, Acme:

$$\phi = 14.5 \cdot \text{deg} \qquad \phi = 0.253$$

Threads:

$$N_t = 1$$

Lead:

$$L = p \cdot N_t$$

Lead Angle:

$$\lambda = \text{atan}\left(\frac{L}{2 \cdot \pi \cdot r_m}\right) \qquad \lambda = 0.079$$

Friction:

$$f = 0.12$$

Screw Efficiency:

$$\eta_s = \frac{\cos(\phi) - f \cdot \tan(\lambda)}{\cos(\phi) + \dfrac{f}{\tan(\lambda)}} \qquad \eta_s = 0.387$$

Worm Drive:

$$N_w = 1 \qquad N_g = 24 \qquad \phi_n = 20 \cdot \text{deg}$$

Lead Angle for dg/dw = 2:

$$\lambda_w = \text{atan}\left(2 \cdot \frac{N_w}{N_g}\right) \qquad \lambda_w = 0.083$$

Worm Turns/in of Travel:

$$T_{pi} = \frac{N_g}{N_w \cdot p} \qquad T_{pi} = 48$$

Worm Drive Efficiency:

$$\eta_w = \frac{\cos(\phi_n) - f \cdot \tan(\lambda_w)}{\cos(\phi_n) + \dfrac{f}{\tan(\lambda_w)}} \qquad \eta_w = 0.391$$

Bearings Efficiency:

$$\eta_b = 0.85$$

System Efficiency:

$$\eta = \eta_s \cdot \eta_b \cdot \eta_w \qquad \eta = 0.129$$

Load:

$$F = 8000$$

Input Speed (rpm):

$$n_w = 200$$

Velocity (in/s):

$$v = \frac{n_w \cdot N_w}{60 \cdot N_g} \cdot L \qquad v = 0.069$$

Horsepower:

$$P_{\text{out}} = \frac{F \cdot v}{6600} \qquad P_{\text{out}} = 0.084 \qquad P_{\text{in}} = \frac{P_{\text{out}}}{\eta} \qquad P_{\text{in}} = 0.655$$

Torque (lb · in):

$$T_{\text{in}} = 63{,}025 \cdot \frac{P_{\text{in}}}{n_w} \qquad T_{\text{in}} = 206.33$$

Screw Reaction:

$$T_s = T_{\text{in}} \cdot \frac{N_g}{N_w} \cdot \eta_w \qquad T_s = 1.935 \cdot 10^3$$

■■

ELASTIC STABILITY (BUCKLING) OF POWER SCREWS

Elastic stability failures are called **column buckling** or simply **buckling.** When power screws and screw jacks carry high compressive loads, the designer must guard against the possibility of buckling. Buckling can occur in power screws in any orientation; a vertical column is not necessary for buckling failure to occur. The reader may wish to refer to Chapter 7, "Elastic Stability."

The Euler Column Design Criterion

If a power screw is guided at both ends, as with the lead screw of a lathe, then the Euler column design criterion can be used. It is

$$P_{\text{COMPRESSIVE}} \leq \frac{\pi^2 EI}{L^2 N_{\text{FS}}} \tag{14.11}$$

For design purposes, we can rearrange Equation (14.11) to find the required moment-of-inertia

$$I \geq \frac{L^2 N_{\text{FS}} P_{\text{COMPRESSIVE}}}{\pi^2 E} \tag{14.12}$$

or to find maximum allowable unsupported length

$$L \leq \sqrt{\frac{\pi^2 EI}{N_{\text{FS}} P_{\text{COMPRESSIVE}}}} \tag{14.13}$$

where $P_{\text{COMPRESSIVE}}$ = maximum axial compressive load on power screw
E = elastic modulus
L = unsupported length of power screw
N_{FS} = factor of safety
$I = \pi d_i^4/64$ = moment-of-inertia
d_i = screw diameter at the base of threads

Other Boundary Conditions

The restraints at the ends of a power screw are critical. Careful design and analysis are required because inadequate restraints can lead to unsafe conditions.

In the case of a screw jack mounted on a well-supported base, clamped–free support conditions may apply. The result is

$$P_{\text{COMPRESSIVE}} \leq \frac{\pi^2 EI}{4L^2 N_{\text{FS}}} \tag{14.14}$$

We can rearrange Equation (14.14) to find the required moment-of-inertia for the clamped–free case

$$I \geq \frac{4L^2 N_{\text{FS}} P_{\text{COMPRESSIVE}}}{\pi^2 E} \tag{14.15}$$

or to find maximum allowable unsupported length for the clamped–free case

$$L \leq \frac{\sqrt{\dfrac{\pi^2 EI}{N_{FS} P_{COMPRESSIVE}}}}{2} \tag{14.16}$$

MITER GEAR BOXES AND MULTIPLE LIFTING SYSTEMS

Screw jacks and actuators are used in various configurations for lifting, positioning, holding, and clamping. A pair of worm-gear screw jacks can be synchronized by coupling both to the same drive shaft, as shown in Figure 14.7. Sometimes it is necessary to use three or more

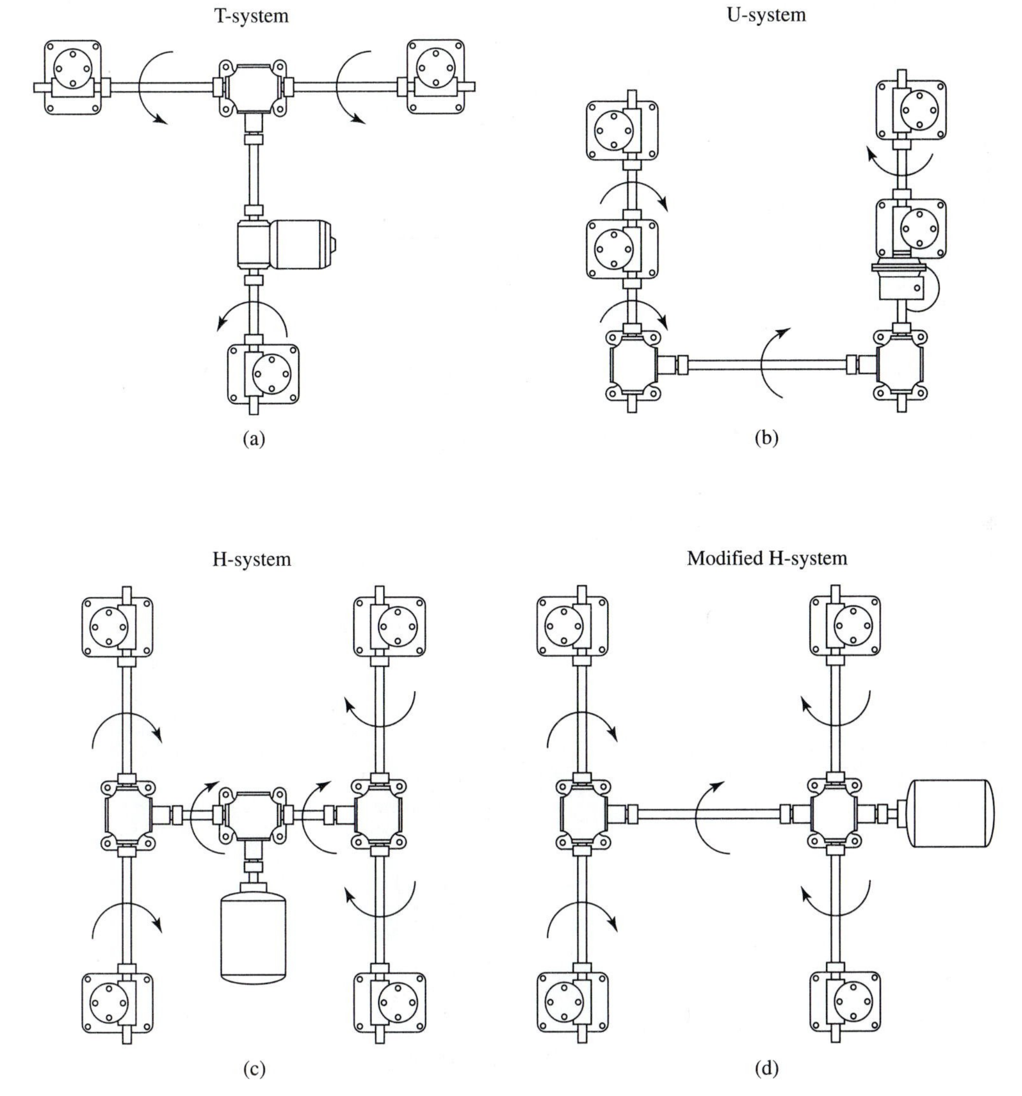

Figure 14.7 System arrangements for synchronizing worm gear screw jacks. (Courtesy of Joyce/Dayton Corporation)

synchronized jacks to raise an unevenly distributed load. T-, U-, and H-systems can be used as shown in the figure, where the arrows indicate the rotation direction to raise the load. Note the orientation of the worm-gear screw jacks in each system.

The T-system of Figure 14.7(a) shows three worm-gear screw jacks, one at each shaft end. A miter gear box is located at the shaft intersection. A motor and a gear reducer drive are in the leg of the T between the miter gear box and the jack.

The U-system, shown in Figure 14.7(b), utilizes two miter gear boxes, one at each shaft intersection. The right-hand leg of the U includes a single unit combining a worm-gear screw jack, a motor, and a speed reducer. The other three jacks are synchronized with this unit.

Figure 14.7(c) and (d) shows H- and modified H-systems. Four worm-gear screw jacks are located at the ends of the legs of the H. Miter gear boxes are located at the shaft intersections. If the system is driven from the center of the crossbar of the H, an additional miter gear box is required. The modified H-system utilizes only two miter gear boxes.

A cutaway and a section view of a miter gear box are shown in Figure 14.8. Miter gear boxes are the key to synchronization of the worm-gear screw jacks. The input and output

Figure 14.8 Miter gear box, section view and cutaway. (Courtesy of Joyce/Dayton Corporation)

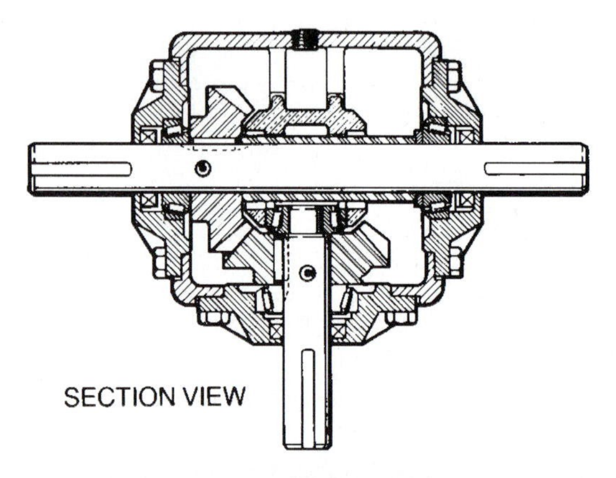

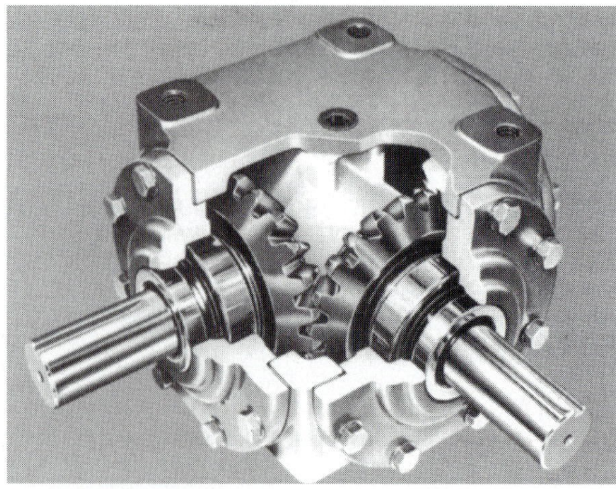

bevel gears have the same number of teeth. The modified H-system requires one special miter gear box (different from that shown in the figure). The reader is encouraged to sketch a cross-sectional view of the required miter gear box and to determine motor shaft rotation direction.

Design Problems

14.1. A full-depth, 14.5° thread angle, single-thread Acme power screw has an outside diameter of 1.5 in and a pitch of 0.375 in. Based on torque measurements, the efficiency was found to be 44%. Assume that bearing losses are negligible. Find the coefficient of friction.

14.2. A full-depth, 14.5° thread angle, single-thread Acme power screw has an outside diameter of 1.5 in and a pitch of 0.375 in. Based on torque measurements, the efficiency was found to be 40%. Assume that bearing losses are negligible. Find the coefficient of friction.

14.3. A modified square thread power screw has a 5° thread angle. Estimate a 0.12 coefficient of friction. Plot screw efficiency for lead angles of 2° to 80°.

14.4. A modified square thread power screw has a 5° thread angle. Plot screw efficiency for lead angles of 0.5° to 10°. Make separate plots for coefficients of friction equal to 0.04, 0.08, 0.12, and 0.16.

14.5. An Acme thread form power screw has a 14.5° thread angle. Plot screw efficiency for lead angles of 0.5° to 10°. Make separate plots for coefficients of friction equal to 0.04, 0.08, 0.12, and 0.16.

14.6. Compare the efficiencies of 5° thread angle modified square and 14.5° Acme power screw threads. Plot the results for a coefficient of friction of 0.15 and lead angles from 2° to 80°.

14.7. Compare the efficiencies of 5° thread angle modified square and 14.5° Acme power screw threads. Plot the results for a coefficient of friction of 0.12 and lead angles from 2° to 80°.

14.8. Plot ball screw efficiency versus lead angle for lead angles from 2° to 84°. Assume a coefficient of friction of 0.05 based on zero thread angle.

14.9. A single-thread power screw incorporates a worm drive. The screw has an outside diameter of 1.5 in, 0.375 in pitch, and 14.5° Acme thread form. The worm drive consists of a single-tooth worm driving a 24 tooth gear. Gear pitch diameter is about twice worm pitch diameter, and the normal pressure angle is 20°. Assume a coefficient of friction of 0.12 for the screw and the worm drive, and estimate 95% efficiency for the bearings. Find power screw efficiency, worm drive efficiency, and overall system efficiency.

14.10. A single-thread power screw incorporates a worm drive. The screw has an outside diameter of 1.5 in, 0.375 in pitch, and 14.5° Acme thread form. The worm drive consists of a single-tooth worm driving a 24 tooth gear. Gear pitch diameter is about twice worm pitch diameter, and the normal pressure angle is 20°. Assume a coefficient of friction of 0.13 for the screw and the worm drive, and estimate 95% efficiency for the bearings. Find power screw efficiency, worm drive efficiency, and overall system efficiency.

14.11. A 7 in outside diameter, single-thread power screw incorporates a worm drive. The screw has 1 in pitch and 5° modified square thread form. The worm drive consists of a single-tooth worm driving a 36 tooth gear. Gear pitch diameter is about twice worm pitch diameter, and the normal pressure angle is 20°. Assume a coefficient of friction of 0.13 for the screw and the worm drive, and estimate 85% efficiency for the bearings. Find power screw efficiency, worm drive efficiency, and overall system efficiency.

14.12. The worm drive power screw described in Problem 14.11 is to raise a load of 150 tons with a worm speed of 60 rpm. Find worm turns per inch of screw travel, linear screw velocity, input and output power, input torque, and screw reaction torque.

14.13. We must position a load of 8 tons at a linear speed of about 0.3 in/s. Design the system for an input speed of 300 rpm.

Design Decisions: We will use a 2 in outside diameter, single-thread power screw incorporating a worm drive. The screw will have 1/2 in pitch and 14.5° Acme thread form. The worm drive will use a 24 tooth gear. Gear pitch diameter will be about twice worm pitch diameter, and the normal pressure angle 20°. Assume a coefficient of friction of 0.12 for the screw and the worm drive, and estimate 85% efficiency for the bearings. Find required number of worm teeth, power screw efficiency, worm drive efficiency, and overall system efficiency. Find worm turns per inch of screw travel, actual linear screw velocity, input and output power, input torque, and screw reaction torque.

14.14. An 8000 lb load is to be positioned at a linear speed of about 0.3 in/s. Design the system for an input speed of 300 rpm.

Design Decisions: We will use a 1.5 in outside diameter, single-thread power screw incorporating a worm drive. The screw will have 3/8 in pitch and 14.5° Acme thread form. The worm drive will use a 24 tooth gear. Gear pitch diameter will be about twice worm pitch diameter, and the normal pressure angle 20°. Assume a coefficient of friction of 0.12 for the screw and the worm drive, and estimate 85% efficiency for the bearings. Find required number of worm teeth, power screw efficiency, worm drive efficiency, and overall system efficiency. Find worm turns per inch of screw travel, actual linear screw velocity, input and output power, input torque, and screw reaction torque.

CHAPTER 15

Fasteners

Symbols

A	area	P	strength
d	diameter	P_i, X_i, Y_i	load on ith fastener
e	eccentricity	R	ratio of part stiffness to bolt stiffness
F	force	R_S	ratio of yield point strength to endurance limit
g_x, g_y	center-of-gravity coordinates	S	strength
I	moment-of-inertia (based on unit area)	t	time, thickness
J	polar-moment-of-inertia (based on unit area)	T	torque
K	stiffness	w	width
M	moment	x, y	coordinates, coordinate location
n	number of rivets	η	efficiency
N_{FS}	factor of safety	σ	tensile stress
p	pitch	τ	shear stress

Units

area	in^2 or mm^2
coordinate locations, dimensions	in or mm
force	lb or N
moment and torque	lb · in or N · mm
moment-of-inertia (based on unit area)	in^2 or mm^2
strength and stress	psi or MPa

Fasteners are available in seemingly endless variety. Common fastener types used in machine design include screws, bolts and nuts, rivets, pin-type fasteners, eyelets, spring retaining rings, quick-release fasteners, and special plastics fasteners. Welds and adhesive joints will be considered in Chapter 16.

JOINT FAILURES

We will design joints and select fasteners on the basis of several failure modes. Bolts statically loaded in tension can fail because of thread pullout or tension failure at a critical cross section. Fatigue is a common failure mode with fluctuating loads. Bolts and rivets subject to shear loading can fail in shear or bearing. Failure can also occur in tension or bearing in the plates joined by fasteners. If the load does not go through the center of the fastener group, then we must consider loading eccentricity. Some fastener failures have severe consequences. If a joint failure could be life-threatening, then we must choose a safety factor accordingly.

FASTENER MATERIALS

Most bolts and other mechanical fasteners are made of carbon steel, alloy steel, stainless steel, brass, aluminum, or titanium. Typical ultimate strengths of steel bolt materials range around 55,000 psi for plain carbon steel bolts to about 300,000 psi for high-strength alloy steels. Corrosion is an important concern for many fastener applications, particularly if the fastener and base material are different. Coatings and platings are used to provide corrosion resistance. Recommended surface treatment, temperature limits, and ultimate strength of selected fastener materials are given in Table 15.1.

Table 15.1 Fastener materials.

Material	Surface Treatment	Temperature Limit (°F)	Ultimate Strength at Room Temperature (ksi)
Carbon steel	Zinc plate	−65 to 250	55 and up
Alloy steels	Cadmium, nickel, zinc, or chromium plate	−65 to plating limit	Up to 300
A-286 stainless	Passivated	−423 to 1200	Up to 220
17-4PH stainless	None	−300 to 600	Up to 220
17-7PH stainless	Passivated	−200 to 600	Up to 220
300 series stainless	Furnace-oxidized	−423 to 800	70 to 140
410, 416, and 430 stainless	Passivated	−250 to 1200	Up to 180
U-212 stainless	Passivated	1200	185
Titanium	None	−350 to 500	Up to 160

Source: Adapted from Barrett (1990).

THREADED FASTENERS

Threaded fastener geometry is usually based on American National Standards Institute (ANSI) standards. A 30° thread angle (60° included angle) is common. Our discussion of threaded fasteners will refer to the common type which has a single start (a single thread). With single-start fasteners, the **number of threads** refers to the number of full turns of thread. The **pitch** is the axial distance between thread centers. Thus, the pitch of a common bolt is the relative axial motion resulting from one rotation of a nut.

The major diameter of a bolt is the largest diameter of the threaded portion. The root diameter is measured at the base of the threads. For Unified National threads types UN, UNR, and UNK, the approximate root diameter is

$$d_r = d - 1.3p \tag{15.1}$$

where d_r = root diameter

d = major diameter

p = pitch

Bolts of 1/4 in and larger diameter are usually designated by nominal (major) diameter and number of threads per inch. For example, a coarse-thread series 3/4–10 bolt has 3/4 in nominal diameter and a pitch of 1/10 inch. This diameter bolt is also available with fine threads (3/4–16) and extra-fine threads (3/4–20). Bolts smaller than 1/4 in are designated by number. The major diameter of small bolts is given by

$$d = 0.060 + 0.013N \tag{15.2}$$

where d = major diameter

N = number designation

For example, a 12–24 bolt has a 0.216 in major diameter and a pitch of 1/24 inch.

BOLTED JOINTS IN TENSION

Because of elastic deformation under load, bolt threads are not uniformly loaded. If we assume that one turn takes all of the load, shear stress in a thread at the mean bolt diameter is approximated by

$$\tau_t = \frac{2F}{\pi d_m p} \tag{15.3}$$

where τ_t = shear stress

F = axial load

$d_m = (d + d_r)/2$ = mean diameter

It is assumed that at least one turn of thread is engaged in a nut or tapped hole.

A less conservative approach assumes uniform thread loading. For static loading, yielding at locations of highest stress causes a favorable stress redistribution. This approach is equivalent to dividing the stress calculated in Equation (15.3) by the actual number of threads engaged in a nut or tapped hole. Or we can divide by engaged length instead of pitch. In addition, we will change the constant to conform with the pullout strength given by Barrett (1990) to obtain

$$\tau_t = \frac{3F}{\pi d_m L} \tag{15.4}$$

where L = axial length of engaged threads (nut thickness or length of threaded portion of a bolt in a tapped hole)

For bolts subject to tensile loads, a nut thickness about equal to bolt major diameter is suggested.

Tensile stress at the root area of a bolt is approximated by

$$\sigma = \frac{F}{A_r} \tag{15.5}$$

where σ = approximate tensile stress at minimum bolt area

$A_r = \pi d_r^2/4$ = approximate minimum cross section of bolt

Unless the shank of the bolt has been reduced to the root diameter, the actual minimum cross-sectional area will be larger than this value. To account for this, A_r in Equation (15.5) can be replaced by the **stress area,** the minimum cross-sectional area of the threaded part of the bolt. The stress area of standard bolts is tabulated in some handbooks.

JOINT DESIGN FOR STATIC TENSILE LOADS

In most static tensile loading cases, joint design can be based on the pullout strength (shear strength) of the threads and the tensile strength of the bolts at the weakest point. We obtain the design equations by replacing the stress term in the preceding equations with the applicable material strength divided by the safety factor. After rearranging, we find pullout strength for a single bolt as follows:

$$F_p = \frac{\pi \cdot d_m \cdot S_{SYP} \cdot L}{3 \cdot N_{FS}} \tag{15.6}$$

We find tensile strength of a single bolt as follows:

$$F_t = \frac{\pi \cdot d_r^2 \cdot S_{YP}}{4 \cdot N_{FS}} \tag{15.7}$$

where F_p = pullout strength of a bolt (lb or N)

S_{SYP} = shear yield point strength of the internal or external threads (psi or MPa)

F_t = tensile strength of the bolt at its weakest cross section (lb or N)

S_{YP} = yield point strength of the bolt (psi or MPa)

N_{FS} = factor of safety

The shear yield point can be taken as one-half the yield point of the weaker material (the bolt, nut, or tapped threads).

The smaller value (either pullout strength or tensile strength) governs the design. Most joints utilize several fasteners to distribute the load. The following procedure can be used for **joints uniformly loaded in tension:**

- Tentatively select a fastener.
- Determine governing fastener strength.
- Divide total joint load by governing strength to find required number of fasteners.
- If the number of fasteners is not reasonable, try a different fastener, and repeat the above steps.

EXAMPLE PROBLEM Joint Design for Uniform Static Tensile Load

The head of a pressure vessel is to carry a total static load of 6000 lb. Select appropriate fasteners, and determine the required number.

Design Decisions: Customary U.S. units will be used. We will try 3/8–16 coarse-thread bolts with 3/8 in thread engagement in tapped holes. Figure 15.1 shows a cross section of the head and vessel. A 30,000 psi yield point strength will apply to the bolts and the base material, and a safety factor of 2 will be used. The bolts will be equally spaced around the head to ensure equal loads on all bolts. In most applications, the bolts would receive an initial tightening. In this solution, we ignore additional bolt tension due to initial tightening. Preloading is discussed in the following section.

Solution Summary: As shown in the detailed solution which follows, seven bolts are required on the basis of pullout strength.

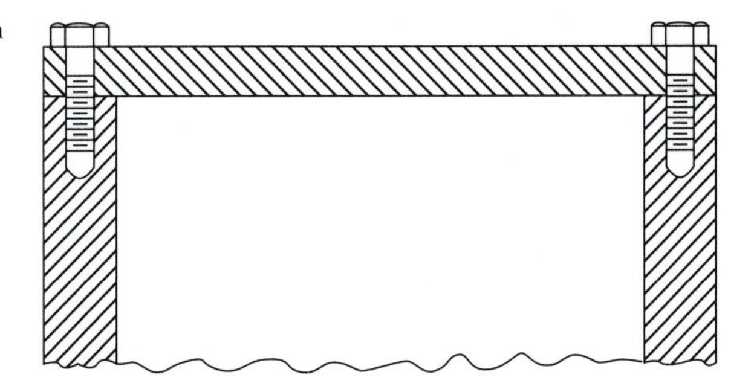

Figure 15.1 Joint design for uniform static load (not to scale).

Detailed Solution:

Total Load:

$$F = 6000$$

Major Diameter:

$$d = \frac{3}{8}$$

Pitch:

$$p = \frac{1}{16}$$

Thread Engagement:

$$L = \frac{3}{8}$$

Yield Strength:

In Tension	*In Shear*
$S_{YP} = 30{,}000$	$S_{SYP} = \frac{S_{YP}}{2}$

Safety Factor:

$$N_{FS} = 2$$

Root Diameter:

$$d_r = d - 1.3 \cdot p \qquad d_r = 0.294$$

Root Area:

$$\pi \cdot \frac{d_r^2}{4} = 0.068$$

Mean Diameter:

$$d_m = \frac{d + d_r}{2} \qquad d_m = 0.334$$

Thread Pullout Strength:

$$F_p = \frac{\pi \cdot d_m \cdot S_{SYP} \cdot L}{3 \cdot N_{FS}} \qquad F_p = 984.816$$

$$N_b = \text{ceil}\left(\frac{F}{F_p}\right) \qquad N_b = 7 \quad \text{(governs)}$$

Tensile Strength Based on Stress at Root:

$$F_t = \frac{\pi \cdot d_r^2 \cdot S_{YP}}{4 \cdot N_{FS}} \qquad F_t = 1.017 \cdot 10^3$$

$$N_b = \text{ceil}\left(\frac{F}{F_t}\right) \qquad N_b = 6$$

■■

PRELOADING BOLTED JOINTS FOR FATIGUE LOAD SHARING

There are a number of reasons for tightening bolts, including the following:

1. maintaining friction between a bolt and nut or tapped hole to prevent loosening due to vibration (with or without a lockwasher)
2. retaining pressure in a vessel (usually with a gasket or an O-ring)
3. sharing fluctuating load with a part

When our calculated results and common sense are at odds, it usually indicates a calculation error. However, common sense notwithstanding, severe bolt preload can actually reduce the probability of bolt failure. Consider, for example, the bolts that hold a connecting rod together. If the bolts are only "hand tight," they must resist the entire fluctuating connecting rod load. If we preload the bolts, the part is compressed. As the joint is loaded, part compression is decreased as bolt tension increases. The part is sharing the fatigue component of the load.

Bolt preloading is most important when the external load has a large fluctuating component. Preloading is most effective when the part is much stiffer than the bolt. Part stiffness is difficult to measure. In the absence of other information, we can estimate that the effective cross-sectional area of the part that is compressed by each bolt is about five times one bolt cross-sectional area. If bolt and part have the same elastic modulus, the above estimate results in a stiffness ratio of $R = 5$, where R is the ratio of part stiffness to bolt stiffness. The part and bolt share the load as follows, provided that the preload is adequate to keep the part in compression:

$$F_b(t) = \frac{F_e(t)}{1 + R} + F_i$$

$$F_p(t) = \frac{F_e(t) \cdot R}{1 + R} - F_i \tag{15.8}$$

where $F_b(t)$ = time-varying bolt load

$F_e(t)$ = time-varying external load

R = ratio of part stiffness to bolt stiffness (Stiffness and the effect of gaskets are discussed in a later paragraph.)

F_i = preload

$F_p(t)$ = time-varying part load

The effect of preload is best illustrated by an example.

EXAMPLE PROBLEM Preloading of Bolts

Each bolt in a joint is subject to a load that varies from 100 to 1500 lb at a rate of 600 cycles per minute. The bolts are not preloaded. Redesign the joint.

Design Assumptions: The loading is assumed to be sinusoidal, and the stiffness ratio is assumed to be 5.

Design Decisions: The only design change will be a preload. We will try a 1700 lb preload on each bolt.

Solution Summary: The calculations are given in the detailed solution which follows. Without preload, each bolt is subject to a load varying from 100 to 1500 lb, as shown in Figure 15.2 (the middle curve). The mean load is 800 lb, and the range load is 700 lb. Preloading increases mean bolt load but reduces the varying component, as shown in the upper curve. The lower curve shows that the part is always in compression with the preload that we selected.

Detailed Solution:

Stiffness Ratio:

$$R = 5$$

External Load:

Maximum *Minimum*

$F_{eMAX} = 1500$ $F_{eMIN} = 100$

Mean

$$F_{em} = \frac{F_{eMAX} + F_{eMIN}}{2} \qquad F_{em} = 800$$

Range

$$F_{er} = \frac{F_{eMAX} - F_{eMIN}}{2} \qquad F_{er} = 700$$

$$\omega = 62.8 \qquad F_e(t) = F_{em} + F_{er} \cdot \sin(\omega \cdot t)$$

Figure 15.2 Fatigue loading of bolts.

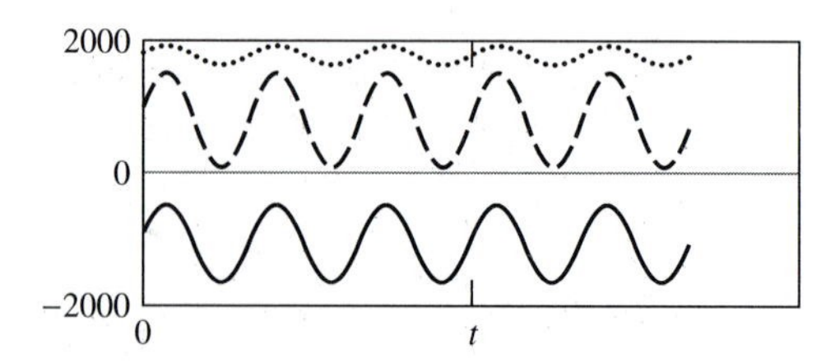

Key

$\cdots\cdots = F_b(t)$

$- - - - = F_e(t)$

$—— = F_p(t)$

Preload:

$$F_i = 1700$$

Bolt Load:

$$F_b(t) = \frac{F_e(t)}{1 + R} + F_i \quad \text{if and only if part is always in compression}$$

Part Load:

$$F_p(t) = \frac{F_e(t) \cdot R}{1 + R} - F_i \quad \text{if and only if part is always in compression}$$

$$t = 0, 0.005, \ldots, 0.5$$

See Figure 15.2. ■■

DETERMINATION OF OPTIMUM PRELOAD AND EVALUATION OF THE DECISION

As noted above, the load sharing equations apply only if the part is in compression. The minimum preload that will keep the part in compression is

$$F_{i0} = \frac{F_{e\text{MAX}} R}{1 + R} \tag{15.9}$$

where F_{i0} = minimum preload for part compression

$F_{e\text{MAX}}$ = maximum external load

R = ratio of part stiffness to bolt stiffness

Considering possible errors in our loading and stiffness estimates, we could use a 10% margin of safety. In that case, we have

$$F_i = 1.10 F_{i0} \tag{15.10}$$

where F_i = selected preload

The range load on the bolt is

$$F_{br} = \frac{F_{er}}{1 + R} \tag{15.11}$$

where F_{br} = range load on bolt

$F_{er} = (F_{e\text{MAX}} - F_{e\text{MIN}})/2$ = external range load

The mean bolt load is

$$F_{bm} = \frac{F_{em}}{1 + R} + F_i \tag{15.12}$$

where F_{bm} = mean bolt load

$F_{em} = (F_{e\text{MAX}} + F_{e\text{MIN}})/2$ = external mean load

Stiffness and the Effect of Gaskets

Stiffness values for bolts and parts depend on length, cross-sectional area, and elastic modulus. The stiffness or spring rate is approximated by

$$K = \frac{EA}{L} \tag{15.13}$$

where K = stiffness (lb/in or N/mm)

E = elastic modulus (psi or MPa)

A = effective cross-sectional area (in^2 or mm^2)

L = effective length (in or mm)

Effective bolt length refers to the length of the bolt that is in tension. The cross-sectional area of a part is an estimate of the area that is compressed by the bolt head. Effective part length refers to the thickness of the part where it is compressed by the bolt.

Sometimes a gasket is used to improve the seal of two mating surfaces. In that case, the stiffness of the part (e.g., head with gasket) is given by

$$K_p = \frac{1}{\dfrac{1}{K_1} + \dfrac{1}{K_2}} \tag{15.14}$$

where K_p = stiffness of the head with a gasket

K_1 = stiffness of head only

K_2 = stiffness of gasket only

Note that a gasket decreases the part stiffness. A very soft gasket material will dominate, resulting in a low part-to-bolt stiffness ratio.

Evaluation of Preloaded Joints

The significance of the preload depends on loading, on stiffness ratio, and on the mechanical properties of the bolts. Fatigue loading models were considered in Chapter 4, "Dynamic Loading of Machine Members." If we adapt the Soderberg criterion to this problem, we can obtain an equivalent load as follows:

$$F_{eq} = F_m + R_s F_r \tag{15.15}$$

where F_{eq} = equivalent load (i.e., static equivalent to the varying load)

F_m = mean load

F_r = range load

$R_s = S_{YP}/S_E$

S_{YP} = yield point strength

S_E = endurance limit corrected for stress concentration, reliability, etc.

EXAMPLE PROBLEM Comparing Different Values of Preload

External load varies from 1000 to 4000 lb. The part-to-bolt stiffness ratio is 5, and the ratio of yield point to corrected endurance limit is 4.

(a) Compare the effect of different values of preload.

(b) A soft gasket is used, reducing the stiffness ratio to only 0.75. Investigate the effect.

Solution Summary (a): The mean and range external load components are, respectively, 2500 and 1500 lb. Without preload, the bolts are subject to the external load, resulting in an equivalent bolt load of

$$F_{beq0} = F_{em} + R_s F_{er} = 8500 \text{ lb}$$

represented by the top line on the graph of Figure 15.3(a). The second line on the graph shows equivalent bolt load for various values of preload. The minimum preload to ensure part compression is

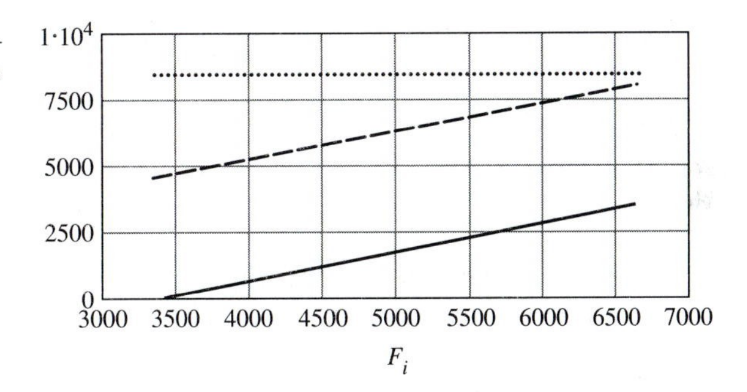

(a)

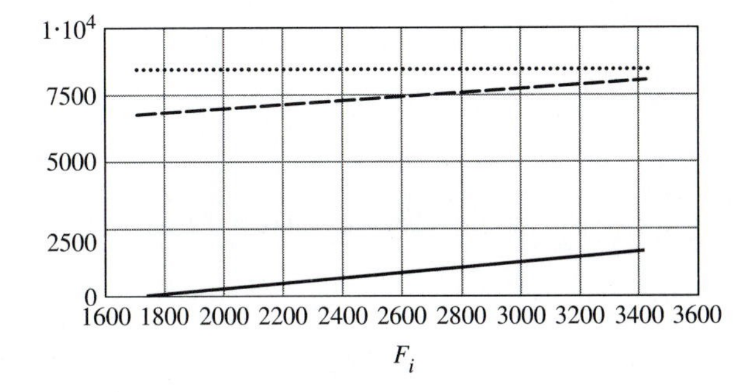

(b)

Figure 15.3 (a) Fatigue loading of bolts—comparing different values of preload; (b) the effect of a soft gasket.

$F_{i0} = 3333$ lb. The lowest line on the graph is the part load F_{pMAX} (shown as compression) for maximum external load. Note that $F_{pMAX} = 0$ for 3333 lb preload.

Equivalent bolt load F_{beq} increases with increasing preload. We see that very high preloads are not desirable. If we use 3667 lb preload (110% of minimum value), part load is 333 lb compression at maximum external load, and equivalent bolt load is 5083 lb. The bolts are far safer than they would be without preload.

Solution Summary (b): The detailed solution which follows illustrates that preloading is less effective owing to the low stiffness ratio. The equivalent bolt load at 110% of minimum preload is 6743 lb.

Detailed Solution (a):

Stiffness Ratio:

$$R = 5$$

External Load:

Maximum *Minimum*

$F_{eMAX} = 4000$ $F_{eMIN} = 1000$

Mean

$$F_{em} = \frac{F_{eMAX} + F_{eMIN}}{2} \qquad F_{em} = 2.5 \cdot 10^3$$

Range

$$F_{er} = \frac{F_{eMAX} - F_{eMIN}}{2} \qquad F_{er} = 1.5 \cdot 10^3$$

Preload for $F_{pMAX} = 0$:

$$F_{i0} = \frac{F_{eMAX} \cdot R}{1 + R} \qquad F_{i0} = 3.333 \cdot 10^3$$

Preload:

$$F_i = F_{i0}, 1.05 \cdot F_{i0}, \ldots, 2 \cdot F_{i0} \qquad 1.1 \cdot F_{i0} = 3.667 \cdot 10^3$$

Bolt Load:

Range

$$F_{br} = \frac{F_{er}}{1 + R} \qquad F_{br} = 250 \quad \text{if and only if } F_{pMAX} \leq 0$$

Mean

$$F_{bm}(F_i) = \frac{F_{em}}{1 + R} + F_i \quad \text{if and only if } F_{pMAX} \leq 0$$

Part Load at F_{eMAX}:

$$F_{pMAX}(F_i) = \frac{F_{eMAX} \cdot R}{1 + R} - F_i \qquad F_{PMAX}(1.1 \cdot F_{i0}) = -333.333$$

Ratio of Yield Point to Endurance Limit:

$$R_S = 4$$

Equivalent Bolt Load:

With Preload

$F_{beq}(F_i) = F_{bm}(F_i) + R_s \cdot F_{br}$ $\quad F_{beq}(1.1 \cdot F_{i0}) = 5.083 \cdot 10^3$

Without Preload

$F_{beq0} = F_{em} + R_s \cdot F_{er}$ $\quad F_{beq0} = 8.5 \cdot 10^3$

See Figure 15.3(a).

Detailed Solution (b):

Stiffness Ratio:

$$R = 0.75$$

External Load:

Maximum *Minimum*

$F_{eMAX} = 4000$ $\quad F_{eMIN} = 1000$

Mean

$$F_{em} = \frac{F_{eMAX} + F_{eMIN}}{2} \qquad F_{em} = 2.5 \cdot 10^3$$

Range

$$F_{er} = \frac{F_{eMAX} - F_{eMIN}}{2} \qquad F_{er} = 1.5 \cdot 10^3$$

Preload for $F_{pMAX} = 0$:

$$F_{i0} = \frac{F_{eMAX} \cdot R}{1 + R} \qquad F_{i0} = 1.714 \cdot 10^3$$

Preload:

$$F_i = F_{i0}, 1.05 \cdot F_{i0}, \ldots, 2 \cdot F_{i0} \qquad 1.1 \cdot F_{i0} = 1.886 \cdot 10^3$$

Bolt load:

Range

$$F_{br} = \frac{F_{er}}{1 + R} \qquad F_{br} = 857.143 \quad \text{if and only if } F_{pMAX} \le 0$$

Mean

$$F_{bm}(F_i) = \frac{F_{em}}{1 + R} + F_i \quad \text{if and only if } F_{pMAX} \le 0$$

Part Load at $F_{e\text{MAX}}$:

$$F_{p\text{MAX}}(F_i) = \frac{F_{e\text{MAX}} \cdot R}{1 + R} - F_i \qquad F_{p\text{MAX}}(1.1 \cdot F_{i0}) = -171.429$$

Ratio of Yield Point to Endurance Limit:

$$R_s = 4$$

Equivalent Bolt Load:

With Preload

$F_{beq}(F_i) = F_{bm}(F_i) + R_s \cdot F_{br} \qquad F_{beq}(1.1 \cdot F_{i0}) = 6.743 \cdot 10^3$

Without Preload

$F_{beq0} = F_{em} + R_s \cdot F_{er} \qquad F_{beq0} = 8.5 \cdot 10^3$

See Figure 15.3(b). ■■

SHEAR LOADING OF FASTENERS

Figure 15.4(a) shows a cross section of a riveted joint, with the load indicated by the arrows. A rivet joining plates *A* and *B* has failed in shear at the common plate surface. This failure mode is called **single shear.** The strength of one rivet in single shear is given by

$$P_{ss} = \frac{\pi d^2 S_{SW}}{4} \tag{15.16}$$

where P_{ss} = strength of a rivet in single shear (lb or N)
d = rivet diameter (in or mm)
S_{SW} = working strength in shear (psi or MPa)

If a working strength in shear is not specified, we can use

$$S_{SW} = \frac{S_{\text{YP}}}{2N_{\text{FS}}} \tag{15.17}$$

where S_{YP} = yield point strength of the rivet
N_{FS} = factor of safety

Double Shear

Figure 15.4(b) shows a cross section of a butt joint, with the arrows indicating the load direction. A rivet has failed in double shear at the surfaces of plate *D* and the upper and lower butt plates. The strength of a rivet in double shear is given by

$$P_{ds} = \frac{\pi d^2 S_{SW}}{2} \tag{15.18}$$

where P_{ds} = the strength of one rivet in double shear

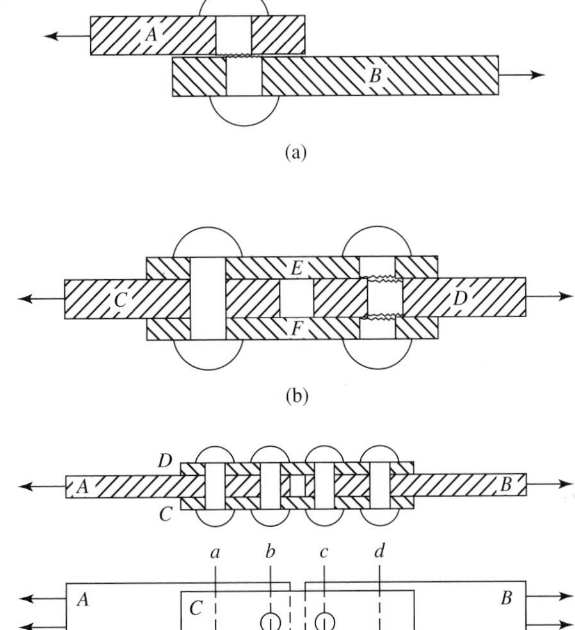

Figure 15.4 Rivet failure. (a) Single shear; (b) double shear; (c) butt joint analysis.

Note that a moment is induced in the single shear case but not in the double shear case. Some designers take this into account and use a higher working strength in double shear. On that basis, a rivet in double shear can carry more than twice the load of a rivet in single shear.

Bearing Failure

A part or a fastener can also fail in bearing (compression). The strength in bearing is based on the projected area as follows:

$$P_b = S_{CW}td \tag{15.19}$$

where P_b = bearing strength of a rivet or of the plate at one rivet

S_{CW} = working strength in bearing

d = rivet diameter

t = plate thickness

For the single shear case shown in Figure 15.4(a), t is equal to the thickness of plate A or B, whichever is smaller. For the double shear case shown in Part (b) of the figure, t is equal to the thickness of the main plate (C or D) or the sum of the thicknesses of the butt plates (E and F).

If a working strength in bearing is not specified, we can use

$$S_{CW} = \frac{S_{\mathrm{YP}}}{N_{\mathrm{FS}}} \tag{15.20}$$

where S_{YP} = yield point strength of the rivet or plate, whichever is smaller

JOINT DESIGN FOR SYMMETRIC LOAD

Figure 15.4(c) shows a butt joint, with plates A and B joined by rivets that pass through butt plates C and D. Possible modes of failure, include

- tensile failure of net plate A through row a
- tensile failure of both butt plates through row b
- double shear failure of all the rivets in both rows a and b
- bearing failure of all the rivets in rows a and b (against plate A)
- bearing failure of plate A where the rivets in rows a and b bear against it
- bearing failure where the rivets of rows a and b contact both butt plates
- tensile failure of net plate A through row b and double shear failure of all the rivets in row a

Other failure modes are possible as well.

Joint Efficiency

The efficiency of a riveted joint is defined by

$$\eta = \frac{P_{\mathrm{MIN}}}{P_{\text{gross plate}}} \tag{15.21}$$

where P_{MIN} = lowest value of strength, considering all possible failure modes

$P_{\text{gross plate}}$ = strength of the main plate without holes

The strength of the gross plate is given by

$$P_{\text{gross plate}} = twS_W \tag{15.22}$$

where t = plate thickness

w = plate width

$S_W = S_{\mathrm{YP}}/N_{\mathrm{FS}}$ = working strength of the plate in tension

S_{YP} = yield point

N_{FS} = factor of safety

Note that the efficiency of riveted joints is always less than unity owing to reduction in cross section at rivet holes.

Optimum Joint Design

Optimum joint design involves selection of a rivet pattern and rivet size to produce high joint efficiency at low cost. The strength of the net plate through a given row of rivets is

$$P_{\text{net}} = t(w - n_{\text{row}}d)S_W \quad \textbf{(15.23)}$$

where P_{net} = strength of the net plate (reduced by the rivet holes)

n_{row} = number of rivets in a given row

d = hole diameter (the rivet diameter, or a somewhat larger diameter if the holes are punched)

We can identify the **limiting joint strength** P_{limit} as the strength of the net plate at a critical cross section. Row a of Figure 15.4(c) is a critical cross section in the joint because failure through row a represents total joint failure.

We usually assume that all of the rivets are equally loaded in the case of symmetric loading. This assumption is reasonable for static loading if the rivets and plate are ductile, allowing for yielding to equalize loading. The optimum number of rivets can be determined by

$$n_{\text{total}} \geq \frac{P_{\text{limit}}}{P_{\text{rivet}}} \quad \textbf{(15.24)}$$

where n_{total} = number of rivets in one-half of the joint [e.g., rows a and b in Figure 15.4(c)]

P_{rivet} = minimum strength at one rivet based on shear failure of a rivet or bearing failure of a rivet or the plate

Various rivet patterns can be tried, and all possible failure modes considered, to obtain an optimum design.

SHEAR LOADING OF BOLTED JOINTS

The design of symmetrically loaded bolted joints is almost identical to the design of riveted joints as described above. Where possible, we limit the length of the threaded portion, so that the full cross section of a bolt resists the shear load. When bolts are used in tapped holes, the shear load is resisted by a cross section reduced by the threads.

IN-PLANE ECCENTRIC LOADS

Sometimes the load does not go through the center of a fastener group. In that case, we must consider a torque or twisting effect as well as a direct load.

Center-of-Gravity of a Fastener Group

We assume that the twisting effect occurs about the center-of-gravity of the fastener group. The center-of-gravity is defined as follows:

$$g_x = \frac{1}{n} \cdot \sum_{i=1}^{n} x_i$$

$$g_y = \frac{1}{n} \cdot \sum_{i=1}^{n} y_i \qquad \textbf{(15.25)}$$

where g_x and g_y = center-of-gravity coordinates

n = number of fasteners

x_i and y_i = fastener coordinates

The xy-coordinate origin location is arbitrary.

More often than not, the calculation of Equation (15.25) is unnecessary. If the fasteners are uniformly spaced, as in Figure 15.5, the center-of-gravity can usually be located by inspection.

Torque

A torque is produced if the load vector does not go through the center-of-gravity. The torque is the product of the load and its (perpendicular) distance from the center-of-gravity. If the load is not parallel to a coordinate axis, it can be broken into components before torque is calculated. The result is

$$T = F_y \cdot (x_F - g_x) - F_x \cdot (y_F - g_y) \qquad \textbf{(15.26)}$$

where T = torque (lb · in or N · mm)

F_x and F_y = load components (lb or N)

x_F and y_F = coordinates of the point of load application (in or mm)

Figure 15.5 In-plane eccentric loads.

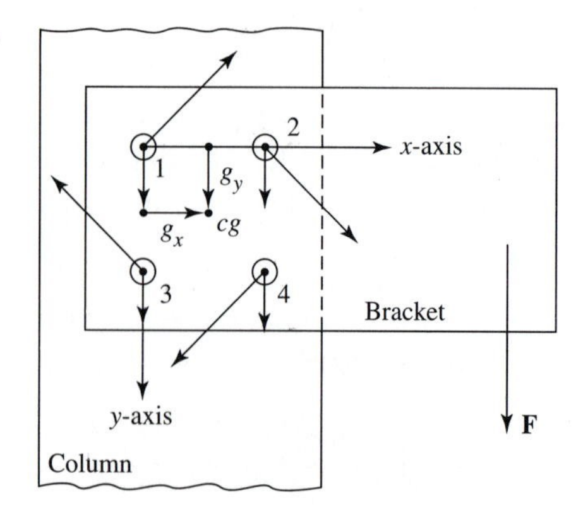

Moment-of-Inertia

The ability of a fastener group to withstand twisting is approximately proportional to its polar-moment-of-inertia. However, if we are designing the joint, the fastener size may be unknown. Therefore, we calculate moment-of-inertia for this application as if each fastener had a cross-sectional area of unity. The following equations can be used:

$$I_x = \sum_{i=1}^{n} (y_i - g_y)^2$$

$$I_y = \sum_{i=1}^{n} (x_i - g_x)^2 \quad \textbf{(15.27)}$$

$$J = I_x + I_y \quad \textbf{(15.28)}$$

where I_x and I_y = moments-of-inertia about axes through the center-of-gravity (in^2 or mm^2 based on unit fastener cross section)

J = polar-moment-of-inertia (in^2 or mm^2 based on unit fastener cross section)

Fastener Load

The direct load and the twisting load both contribute to the force on the fasteners. It is assumed that the direct load is shared equally by the fasteners and that the effect of the torque is proportional to the distance (of an individual fastener) from the center-of-gravity.

The four fasteners in Figure 15.5 share vertical load **F** as shown by the vertical force vectors on each fastener. The twisting effect due to load eccentricity is represented by force vectors on each fastener perpendicular to a line between center-of-gravity *cg* and the individual fastener center. We could add the direct load vectors and the twisting effect vectors. The resultants would show that fasteners 2 and 4 in this figure are more severely loaded than the other two fasteners.

However, if we wish to "work smarter," we can reduce the number of calculations by calculating fastener force components in the x- and y-directions. The following equations apply:

$$X_i = \frac{F_x}{n} - T \cdot \frac{y_i - g_y}{J} \qquad Y_i = \frac{F_y}{n} + T \cdot \frac{x_i - g_x}{J} \qquad P_i = \sqrt{(X_i)^2 + (Y_i)^2} \quad \textbf{(15.29)}$$

where X_i and Y_i = components of shear force on the ith fastener (lb or N)

F_x and F_y = components of total applied load (lb or N)

n = number of fasteners

T = torque (lb · in or N · mm)

x_i and y_i = coordinates of ith fastener (in or mm)

g_x and g_y = coordinates of center-of-gravity of fastener group (in or mm)

J = polar-moment-of-inertia of fastener group (in^2 or mm^2)

P_i = resultant shear force on the ith fastener (lb or N)

Note that the y-distance $y_i - g_y$ contributes to the $-x$-component of fastener force. The x-distance $x_i - g_x$ contributes to the $+y$-component.

Fastener Size

All of the fasteners in a pattern will be identical, even though they may not be equally loaded. To specify otherwise would result in extra manufacturing and inventory expense. Fastener selection is based on the most severely loaded fastener. The required rivet or bolt diameter can be found with the following equation:

$$d = \sqrt{\frac{4 \cdot P}{\pi \cdot S_{SW}}} \tag{15.30}$$

where d = diameter at the section subject to shear[1]

P = shear load on the most severely loaded fastener

S_{SW} = working strength in shear

If the working strength value is not specified, we can use

$$S_{SW} = \frac{S_{YP}}{2N_{FS}}$$

Bearing on the Plate

The bearing stress on the plate or fastener is assumed to equal the load divided by projected area, from which

$$t = \frac{P}{S_{bW} \cdot d} \qquad \textbf{(15.31)}^{2}$$

where t = minimum allowable plate thickness (in or mm)

S_{bW} = working strength in bearing for the plate or fastener, whichever is less (psi or MPa).

If working strength in bearing is not specified, we can use

$$S_{bW} = \frac{S_{YP}}{N_{FS}}$$

EXAMPLE PROBLEM Eccentrically Loaded Bracket

A bracket is required to support a 3000 lb load, oriented at 77° as shown in Figure 15.6. The x- and y-coordinates of the point of load application are 9 and 3.5 in, respectively. Design a fastener system to join the bracket to a column.

Design Decisions: Eight fasteners will be used. Fastener coordinates will be 0.0; 2,0; 4,0; 0,2; 2,2; 4,2; 0,4; and 4,4. Note that the coordinate origin is at the center of fastener number 1.

---

[1]This is the nominal diameter of a rivet or bolt if the threaded portion does not extend into the shear section. If bolts are used in tapped holes, then the calculated diameter refers to the minimum diameter at the base of the threads. In any case, the value is rounded upward so that standard size fasteners can be used.

[2]Other design aspects will usually govern plate thickness.

Figure 15.6 Eccentrically loaded bracket.

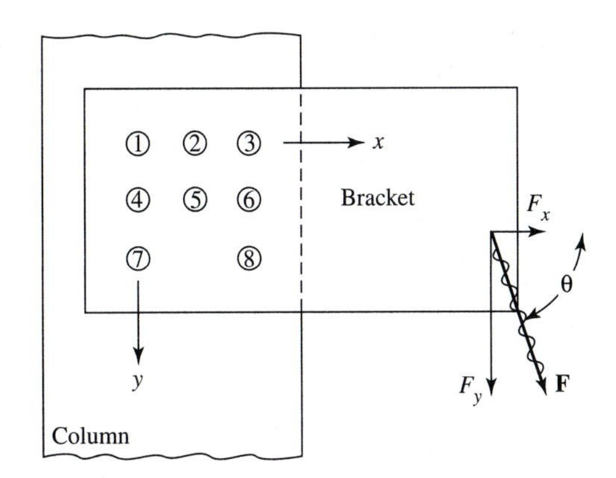

Solution: It is obvious that the center-of-gravity of the fastener group falls in the middle column of fasteners. Our calculations show it to be 1.75 in below fastener 2. The y-component of force produces a clockwise torque contribution, while the x-component produces a counterclockwise contribution. The result is a clockwise torque of 19,281 lb · in on the fastener group. We calculate the moments-of-inertia about horizontal and vertical axes through the center-of-gravity, adding to obtain the polar-moment-of-inertia, 43.5 in^2.

Each fastener is subject to a shear force of 1/8 of the 3000 lb applied load and a torque-induced shear force depending on the distance from the fastener to the center-of-gravity. These shear forces combine as vectors. We expect fasteners 3 and 8 on the "load side" of the fastener group to be heavily loaded because they are far from the center-of-gravity and because the two vector force contributions are nearly in the same direction. Our expectations are borne out by the calculations (given in the detailed solution which follows) which show fastener 8 as most heavily loaded, with a total shear force of about 1550 lb. All of the fastener loads are calculated because it requires little effort when using a computer. If we had to rely on hand calculations, we might have checked only those expected to have high loads. Note that we avoided extra calculation steps by separating X_i and Y_i, the load components on the ith fastener.

Finally, we must determine fastener size and plate thickness. We will use rivets or bolts and nuts, ensuring that bolt threads extend only as far as necessary. Suppose that we decide to use working strengths of 10,000 psi in shear and 18,000 psi in bearing. We then obtain a fastener diameter of 0.444 in and a minimum thickness of 0.194 in for the bracket and column. In both cases, we would use standard sizes not smaller than these values.

The detailed solution follows.

Detailed Solution:

Load Orientation: Clockwise from x-axis.

$$\theta = 77 \cdot \text{deg}$$

Load (lb):

$$F = 3000$$

Load Components:

$$F_x = F \cdot \cos(\theta) \qquad F_x = 674.853$$
$$F_y = F \cdot \sin(\theta) \qquad F_y = 2923.11$$

Location:

$$x_F = 9 \qquad y_F = 3.5$$

Number of Fasteners:

$$n = 8$$
$$i = 1, \ldots, n$$

Fastener Locations:

i	x_i	y_i
1	0	0
2	2	0
3	4	0
4	0	2
5	2	2
6	4	2
7	0	4
8	4	4

Location of Center-of-Gravity:

$$g_x = \frac{1}{n} \cdot \sum_{i=1}^{n} x_i \qquad g_x = 2$$

$$g_y = \frac{1}{n} \cdot \sum_{i=1}^{n} y_i \qquad g_y = 1.75$$

Torque about cg*:*

$$T = F_y \cdot (x_F - g_x) - F_x \cdot (y_F - g_y) \qquad T = 19{,}280.778$$

Moment-of-Inertia (Based on Unit Area):

$$I_x = \sum_{i=1}^{n} (y_i - g_y)^2 \qquad I_x = 19.5$$

$$I_y = \sum_{i=1}^{n} (x_i - g_x)^2 \qquad I_y = 24$$

$$J = I_x + I_y \qquad J = 43.5$$

Fastener Load:

$$X_i = \frac{F_x}{n} - T \cdot \frac{y_i - g_y}{J} \qquad Y_i = \frac{F_y}{n} + T \cdot \frac{x_i - g_x}{J} \qquad P_i = \sqrt{(X_i)^2 + (Y_i)^2}$$

i	X_i	Y_i	P_i
1	860.02	−521.084	1005.566
2	860.02	365.389	934.422
3	860.02	1251.861	1518.813
4	−26.452	−521.084	521.755
5	−26.452	365.389	366.345
6	−26.452	1251.861	1252.141
7	−912.925	−521.084	1051.171
8	−912.925	1251.861	1549.383

Working Strength:

In Shear

$S_{SW} = 10{,}000$

In Bearing

$S_{bW} = 18{,}000$

Fastener Diameter:
(Round upward to standard size.)

$$d = \frac{4 \cdot P_8}{\pi \cdot S_{SW}} \qquad d = 0.444$$

Minimum Plate Thickness:
(Other considerations may govern.)

$$t = \frac{P_8}{S_{bW} \cdot d} \qquad t = 0.194 \text{ in}$$

DESIGN FOR SYMMETRY

When we are designing a bracket for a hoist or other overhung load, it may be impractical to avoid eccentric loads. When possible, however, a symmetric design is often best. Consider the bracket shown in Figure 15.7, but without fastener 7. The load line goes through the center-of-gravity, which is obviously the geometric center of the six fasteners. If we add another fastener without regard to symmetry, it may have adverse effects.

EXAMPLE PROBLEM Symmetric and Asymmetric Design

A horizontal load of 6000 lb is to be supported by a bracket similar to the one in Figure 15.7.
(a) Design the bracket, using a symmetric fastener pattern.
(b) Consider the effect of adding an additional fastener.

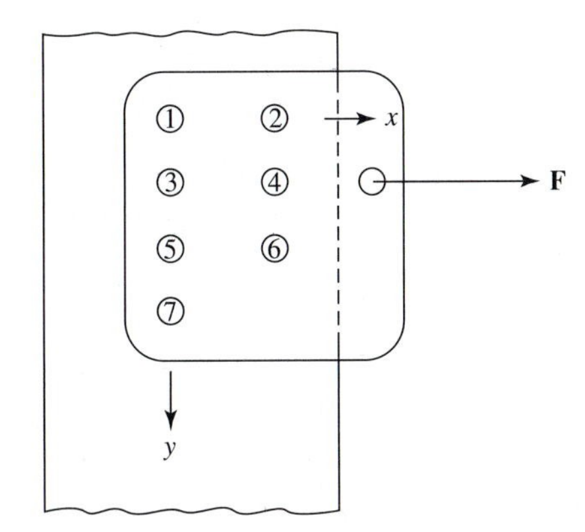

Figure 15.7 Symmetric and asymmetric design.

Solution Summary (a):

Design Decisions: Six fasteners will be used. They will be located at *x,y* coordinates 0,0; 5,0; 0,3; 5,3; 0,6; and 5,6 as in Figure 15.7 (without fastener 7). The load line goes through the middle row of fasteners.

The center-of-gravity lies at the geometric center of the six fasteners. Thus, the load line goes through the center-of-gravity, and there is no torsional effect. The shear load is the same on each fastener.

$$P = \frac{F}{n} = \frac{6000}{6} = 1000 \text{ lb/fastener}$$

Fastener and plate size can be determined as in the preceding example problem.

Solution Summary (b):

Design Decision: A seventh fastener will be added at coordinates 0,9.

The center-of-gravity now lies below the load line, producing a torque load on all fasteners. Shear forces P_i are tabulated for all of the fasteners in the detailed solution which follows. We see that the shear force now exceeds 1000 lb on both fasteners 1 and 2. The additional fastener has weakened the joint rather than strengthening it. The design decision was unwise because the center-of-gravity no longer lies on the load line.

Detailed Solution (b):

Load Orientation: Clockwise from *x*-axis.

$$\theta = 0 \cdot \text{deg}$$

Load:

$$F = 6000$$

Load Components:

$$F_x = F \cdot \cos(\theta) \qquad F_x = 6000$$
$$F_y = F \cdot \sin(\theta) \qquad F_y = 0$$

Location:

$$x_F = 0 \qquad y_F = 3$$

Number of Fasteners:

$$n = 7$$
$$i = 1, \ldots, n$$

Fastener Locations:

i	x_i	y_i
1	0	0
2	5	0
3	0	3
4	5	3
5	0	6
6	5	6
7	0	9

Location of Center-of-Gravity:

$$g_x = \frac{1}{n} \cdot \sum_{i=1}^{n} x_i \qquad g_x = 2.143$$

$$g_y = \frac{1}{n} \cdot \sum_{i=1}^{n} y_i \qquad g_y = 3.857$$

Torque about cg*:*

$$T = F_y \cdot (x_F - g_x) - F_x \cdot (y_F - g_y) \qquad T = 5142.857$$

Moment-of-Inertia (Based on Unit Area):

$$I_x = \sum_{i=1}^{n} (y_i - g_y)^2 \qquad I_x = 66.857$$

$$I_y = \sum_{i=1}^{n} (x_i - g_x)^2 \qquad I_y = 42.857$$

$$J = I_x + I_y \qquad J = 109.714$$

Fastener Load:

$$X_i = \frac{F_x}{n} - T \cdot \frac{y_i - g_y}{J} \qquad Y_i = \frac{F_y}{n} + T \cdot \frac{x_i - g_x}{J} \qquad P_i = \sqrt{(X_i)^2 + (Y_i)^2}$$

i	X_i	Y_i	P_i
1	1037.946	−100.446	1042.795
2	1037.946	133.929	1046.551
3	897.321	−100.446	902.926
4	897.321	133.929	907.261
5	756.696	−100.446	763.334
6	756.696	133.929	768.457
7	616.071	−100.446	624.206

■■

FASTENERS WITH OUT-OF-PLANE LOADS

Out-of-plane loads can cause both shear and tensile loading in fasteners. The tensile loads depend on the rigidity of the connected parts and on how well they fit together.

Consider, for example, the angle bracket bolted to an I-section column shown in Figure 15.8. The fastener group is subject to a moment

$$M = Fe \tag{15.32}$$

where M = moment (lb · in or N · mm)

F = applied force (lb or N)

e = eccentricity (in or mm)

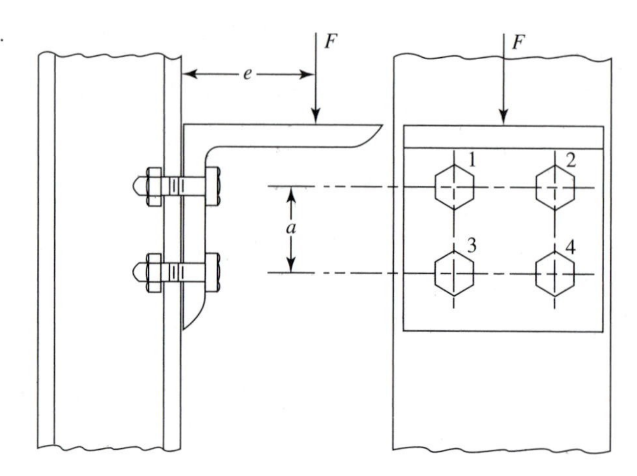

Figure 15.8 Out-of-plane loads.

Moment-of-Inertia

We must now try to model the behavior of the angle and I-section when loaded. The moment-of-inertia of the fastener group is based on unit fastener area as follows:

$$I = \sum_{i=1}^{n} y_i^2 \tag{15.33}$$

where I = moment-of-inertia about selected axis (in^2 or mm^2)
y_i = distance from axis to the ith fastener (in or mm)

We could select a horizontal line through the center of the fastener group. Instead, we will select an axis through the lowest row of fasteners. For the four-bolt group shown in Figure 15.8, the moment-of-inertia about an axis through bolts 3 and 4 is

$$I = a^2 + a^2 + 0 + 0 = 2a^2$$

Tensile Load

Tensile load is given by

$$P_{ti} = \frac{My_i}{I} + P_{pre} \tag{15.34}$$

where P_{ti} = tensile load on the ith fastener (lb or N)
P_{pre} = preload

We should select a preload adequate to keep the part in compression at all times. Referring to the angle bracket in Figure 15.8, note that the tensile load on fastener 1 or 2 is given by

$$P_{t1} = \frac{My_1}{I} + P_{pre} = \frac{Fea}{2a^2} + P_{pre} = \frac{Fe}{2a} + P_{pre}$$

Shear Load

We assume that the shear load is shared equally by the fasteners, resulting in

$$P_S = \frac{F}{n} \tag{15.35}$$

where P_S = shear force per fastener (lb or N)

The shear force per fastener in the angle bracket shown is

$$P_S = \frac{F}{4}$$

Actually, friction should somewhat reduce the rivet shear load, but we cannot count on friction owing to the possibility of load variation and vibration.

Fastener Selection

At the surface common to the bracket and the column, the tensile and shear loads combine. The required cross-sectional area is

$$A \geq \frac{2N_{FS}}{S_{YP}} \sqrt{\left(\frac{P_{ti}}{2}\right)^2 + P_S^2} \tag{15.36}$$

where A = cross-sectional area at common surface (in^2 or mm^2)

N_{FS} = safety factor

S_{YP} = yield point strength (psi or MPa)

If bolts are used, then failure can occur at minimum cross section at the base of the threads. If the threads do not extend to the common surface, only the tension load applies. Minimum cross section is given by

$$A_{root} \geq \frac{N_{FS} P_{ti}}{S_{YP}}$$

where A_{root} = root area of the threads

FASTENER TYPES

An optimum choice of fasteners and fastening methods should result in a safe product that is economical to manufacture. A poor choice of fasteners can result in premature product failure or a product that is noncompetitive because of assembly costs. The designer should be familiar with many fastener options. A few are noted below. Detailed information can be found in machine design and manufacturing periodicals and product catalogs.

Steel Bolts with Nuts or Tapped Holes

Ordinary steel bolts used with nuts or tapped holes are used in countless applications. Reasons for their selection include

- relatively high strength
- reliability
- low fastener cost
- ease of disassembly and reassembly
- ease of customer/consumer assembly with common tools
- control of preloading

Disadvantages of bolts used with nuts or tapped holes include

- unintended loosening due to vibration
- relatively high assembly cost
- difficulty in automating assembly

Self-Drilling Screws

There is a self-contained drill bit on the end of each self-drilling screw. The bit is followed by hardened self-tapping threads. Advantages of self-drilling screws include the following:

- They eliminate separate drilling operation.
- They eliminate separate tapping operation.
- They eliminate the problem of tolerances on hole location.
- They reduce assembly cost.

Disadvantages of and limitations on self-drilling screws include the following:

- The total thickness of the material being fastened must not exceed the length of the drill-bit portion.
- The total thickness of the material being fastened must be less than the length of the thread-forming portion of the screw.
- The drill-bit portion of the screw extends through the assembly.

Rivets

Rivets are used for permanent fastening. Factors favoring the selection of rivets include

- low fastener cost
- low assembly cost
- the ability to automate assembly
- an almost flush surface with some rivets
- high shear strength
- vibration resistance

Disadvantages of rivets include

- difficult disassembly of riveted joints
- tensile strength usually lower than that of comparable bolts
- access to both sides of an assembly required for installation (except for blind rivets)

Blind Rivets

The typical blind rivet configuration is a hollow rivet containing a mandrel. When the mandrel is pulled through the rivet, the mandrel head upsets the rivet on the far side of the joint. Factors favoring selection of blind rivets include the following:

- Blind rivets can be used where joints are accessible from only one side.
- Low-cost, portable, blind riveting tools are available for field work.

Disadvantages of blind rivets include the following:

- Hollow shank blind rivets have low shear strength. (However, some blind rivets retain all or part of the mandrel, improving shear strength.)
- They have low tensile strength.

Threaded Inserts and Caged Nuts

Internally threaded inserts and caged nuts are used for special applications, including blind fastening locations and joining of thin metal or relatively soft material. If a material is too soft or too thin to form strong threads, or if one side of a joint will be inaccessible during assembly, threaded inserts and caged nuts may be considered. A threaded insert takes the place of a nut. A description of available types follows.

Wire-thread inserts are screwed into tapped holes to provide an internal thread. They are used to provide a strong thread in aluminum and other soft metals and to repair tapped holes.

Thread-cutting or thread-forming inserts are used in materials that are machinable but of insufficient strength to sustain loads in threads tapped directly in these materials. They are used in soft, weak metals, plastics, and wood. They are also used where excessive thread wear occurs because of frequent assembly and disassembly. The internal threads of the inserts accommodate standard machine screws. Inserts are designed to withstand vibration and are used in airdrop equipment, chain saws, outboard motors, and helicopter engines. The insert shown in Figure 15.9(a) taps its own threads with the edges of the insert holes. It is designed

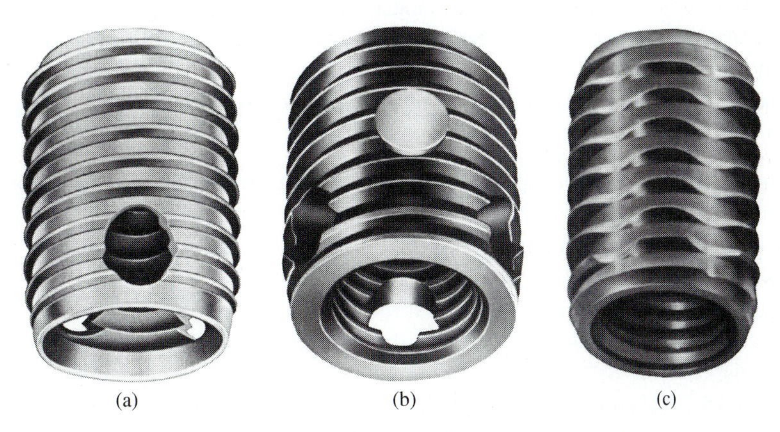

Figure 15.9 Thread-cutting and thread-forming inserts. (a) Thread-cutting insert; (b) thread-cutting insert with nylon locking pellet; (c) thread-forming insert. (Courtesy of Groov-Pin Corporation)

primarily for difficult-to-tap materials, including wrought aluminum, magnesium, and tough aluminum alloy castings. This type of insert is suitable for high production rates. The installation rate using a semiautomatic installer is about 800 inserts per hour in drilled or cored holes. Figure 15.9(b) shows an insert with a nylon pellet that locks the mating fastener against loosening. It is suited for use with adjustment screws. Part (c) shows a bushing with external wave-form threads. The wave-form threads roll the base material without cutting. This insert type can be used where the formation of chips could cause a problem.

Molded-in inserts can be used in thermosetting plastics and thermoplastics and in rubber and ceramics.

Knurled press-in inserts are pressed into a prepared hole, causing a cold flow of metal into a locking groove.

Ultrasonic inserts are used in thermoplastics. They incorporate annular and longitudinal grooves to prevent pullout and rotation after insertion. The ultrasonic installation operation melts the thermoplastic into the grooves.

Caged nuts are useful in sheet-metal and blind fastening applications. A cage retainer prevents rotation or loss of the nut during assembly.

Self-piercing nuts have external undercuts. They are installed into sheet-metal panels by automatic machinery. During installation, the nuts are fed continuously under a plunger, driving the nut into the panel and clinching the panel into the undercuts. Some pierce nuts are wired together in a continuous strip. This simplifies automatic assembly, since the nuts are fed into position with the required orientation.

Floating nuts and inserts are used where bolt hole misalignment is expected during assembly. Various nut and insert types are available as floating fasteners.

Self-Locking Screws and Nuts and Other Locking Methods

Vibration, inertia effects, shock, temperature variation, and other conditions can cause fasteners to loosen, sometimes with catastrophic consequences. **Lockwashers** are used to provide a spring force in an attempt to ensure a steady friction force opposing loosening of fasteners. **Locknuts** are jammed against the primary nut to increase friction torque to prevent relative motion. **Cotter pins** through castellated nuts and drilled bolts are a positive locking device, preventing relative rotation. **Safety wire** through a group of drilled bolts is another positive locking method.

A few of the many configurations of **self-locking screws and nuts** are described in the following list:

- Ramp-shaped radial serrations on the underside of the screw head in a proprietary design result in a higher off-torque than on-torque. This configuration prevents loss of preload in vibration situations.
- Bolt threads containing a polymer locking element.
- Nuts with a plastic locking element at the top.
- Nuts with a flexible locking section at the top which causes inward pressure on the bolt threads.
- Figure 15.10(a) shows a proprietary thread form with a 30° wedge ramp at the root of the nut thread. When tightened, the crests of the male fastener draw against the wedge, con-

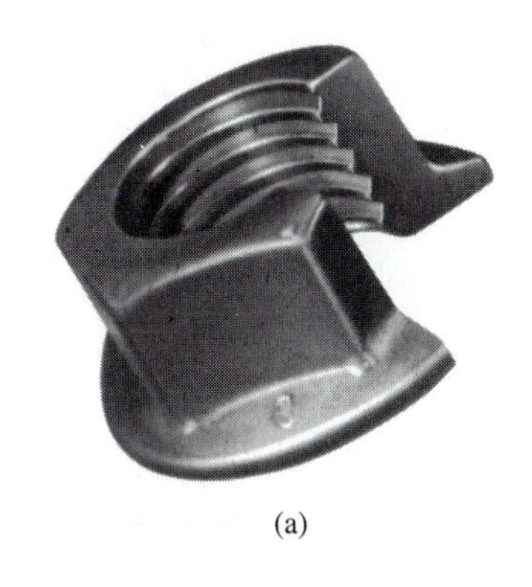

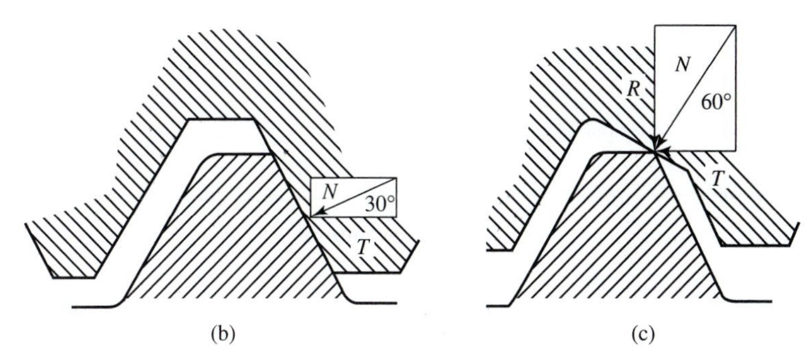

Figure 15.10 Self-locking thread form. (a) Nut cut to show thread form; (b) standard thread; (c) self-locking thread. (Courtesy of Detroit Tool Industries).

tacting along the length of the engaged threads. This contact force prevents vibration-induced loosening. Parts (b) and (c) of the figure compare the loading of a standard thread pair with that of a standard male thread contacting a female thread with the proprietary thread form.

When standard threads are engaged, a major portion of the load is carried by the first thread of the bolt, as shown in Figure 15.11(a). Part (b) of the figure shows improved load distribution with a standard male thread meshing with the proprietary female thread. The load distribution is based on photoelastic studies, illustrated in Parts (c) and (d) of the figure.

Applications of the special thread form include space shuttle engines, seismic guns for offshore oil exploration, motorcycle engine/drive compensator assemblies, stethoscopes, and prosthetics. Figure 15.12 shows an artificial knee joint secured by titanium screws into Spiralock™ threaded holes. The human knee is subject to about 1 million loading cycles per year. Since knee implant surgery is intended to last a lifetime, the possibility of fastener loosening is a critical consideration.

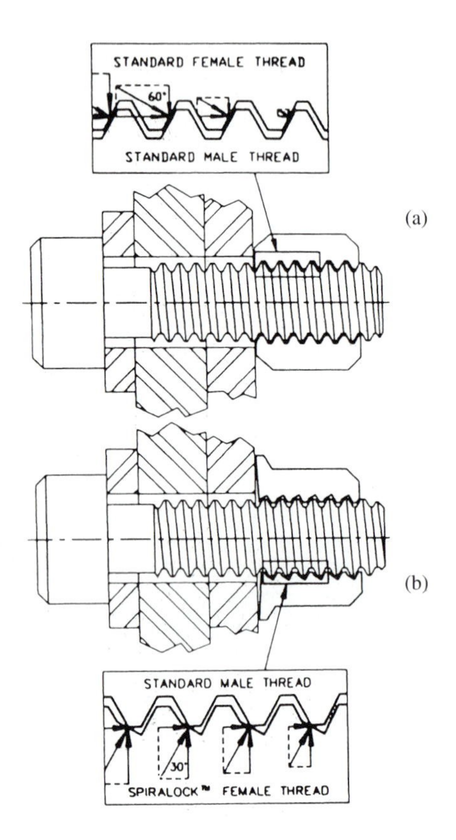

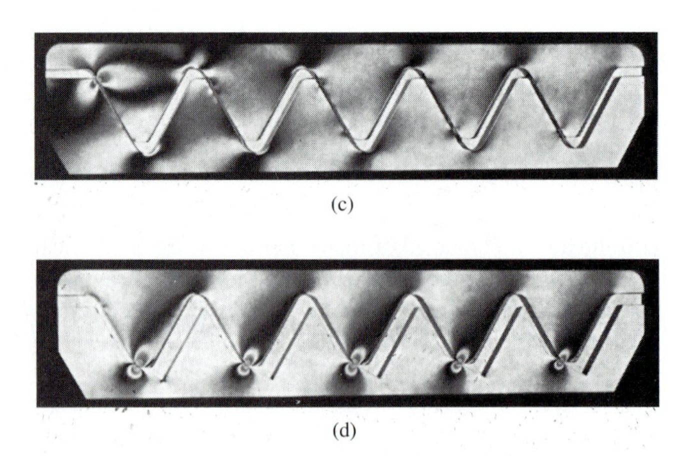

Figure 15.11 Thread load sharing. (a) Standard thread forms; (b) standard male thread with self-locking female thread; (c) photoelastic study of standard thread forms; (d) photoelastic study of standard male thread with self-locking female thread. (Courtesy of Detroit Tool Industries).

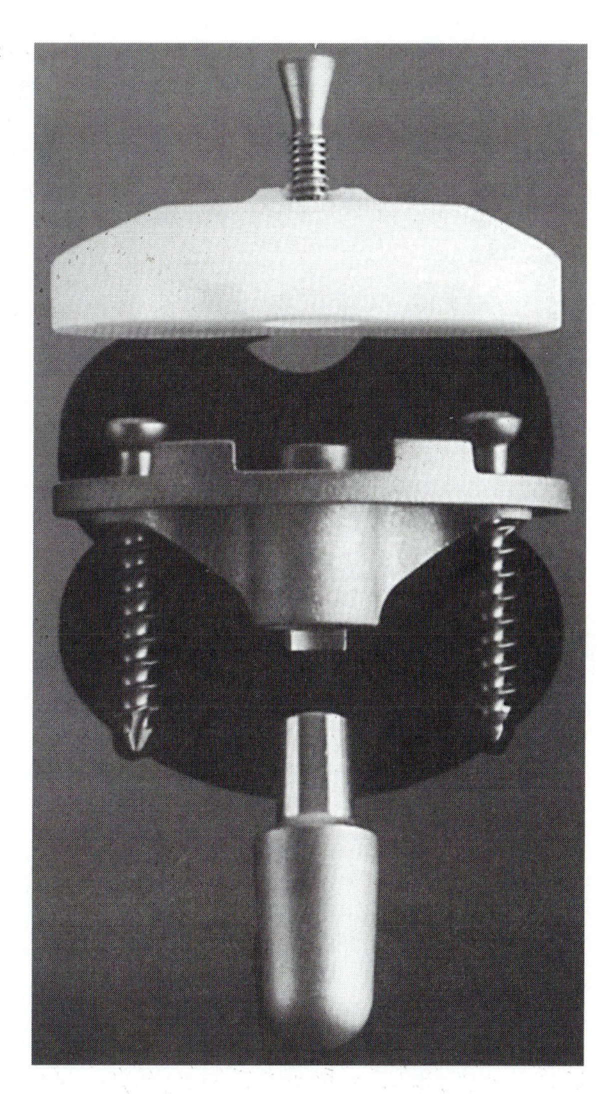

Figure 15.12 Artificial knee joint. (Courtesy of Detroit Tool Industries)

Snap-in Design

In some cases it is practical to reduce the number of fasteners or eliminate traditional fasteners entirely. By incorporating grooves and tabs into a design, designers ensure that parts can be fit together with a reduced number of fasteners. When designed with flexible plastics, some assemblies can be snapped together. Snap-in design eliminates the cost of fasteners and reduces assembly costs.

References and Bibliography

Barrett, R. T. *Fastener Design Manual.* NASA Reference Publication 1228. Cleveland, OH: Lewis Research Center, March 1990.

Chaddock, D. H. *Introduction to Fastening Systems.* Oxford: Oxford University Press, 1974.

Keeley, F. M. *Miscellaneous Fasteners.* Oxford: Oxford University Press, 1974.

Design Problems

15.1. Design a joint for a uniform static tensile load of 4000 lb.

Design Decisions: Try 1/4–20 bolts with 1/4 in thread engagement in tapped holes. A 30,000 psi yield point strength will apply to the bolts and the base material, and a safety factor of 1.5 will be used. The joint design will ensure equal loads on all bolts.

15.2. Design a joint for a uniform static tensile load of 45,000 lb.

Design Decisions: Try 1.25–7 bolts with 1.25 in thread engagement in tapped holes. A 30,000 psi yield point strength will apply to the bolts and the base material, and a safety factor of 2.5 will be used. The joint design will ensure equal loads on all bolts.

15.3. A head is subject to a total load of 35,000 lb. The load should be shared by 8 to 12 bolts. Design the joint.

Design Decisions: Try 1/2–13 bolts with 1/2 in thread engagement in tapped holes. A 70,000 psi yield point strength will apply to the bolts and the base material, and a safety factor of 2 will be used. The joint design will ensure equal loads on all bolts.

15.4. A bolt is subject to a load that varies from 500 to 2000 lb at a rate of 6.37 cycles per second. Plot the bolt load without preload. Consider the effect of preloading the bolt. Plot the bolt and part load with preload.

Design Decisions: Try 2000 lb preload and assume a stiffness ratio of 5.

15.5. Joint load varies from 200 to 1000 N per bolt at a rate of 7.96 cycles per second. Plot the bolt load without preload. Consider the effect of preloading the bolt. Plot the bolt and part load with preload.

Design Decisions: Try 900 N preload and assume a stiffness ratio of 5.

15.6. A bolt is subject to a load that varies from zero to 900 lb at a rate of 7.96 cycles per second. Plot the bolt load without preload. Consider the effect of preloading the bolt. Plot the bolt and part load with preload.

Design Decisions: Try 1000 lb preload and assume a stiffness ratio of 5.

15.7. The external tensile load on a bolted joint varies from zero to 1000 lb per bolt.

(a) The part-to-bolt stiffness ratio is estimated to be 5.0, and the ratio of yield point to corrected endurance limit is 2.5. Find minimum effective preload. Compare the effect of various values of preload. Plot the results.

(b) A gasket reduces the effective stiffness ratio to 1.0. Find minimum effective preload. Compare the effect of various values of preload. Plot the results.

15.8. The external tensile load on a bolted joint varies from 1500 lb per bolt to 10,000 lb per bolt. The part-to-bolt stiffness ratio is estimated to be 5.0, and the ratio of yield point to corrected endurance limit is 3.5. Find minimum effective preload. Compare the effect of various values of preload. Plot the results.

15.9. The external tensile load on a bolted joint varies from zero to 1000 lb per bolt. The part-to-bolt stiffness ratio is estimated to be 4.0, and the ratio of yield point to corrected endurance limit is 2.0. Find minimum effective preload. Compare the effect of various values of preload. Plot the results.

15.10. A flat plate must support an overhung vertical load of 12,000 lb located at $x = 10$ inches in the plane of the plate. Design a fastener system to join the plate to a column. Locate the center-of-gravity of the fastener group. Find torque and moment-of-inertia. Find shear load on each fastener, and determine minimum fastener and plate size.

Design Decisions: Six fasteners will be used, located at 0,0; 5,0; 0,4; 5,4; 0,6; and 5,6. Working strengths of 9000 psi in shear and 16,000 psi in bearing will be used.

15.11. A flat plate bracket is needed to support an 8 ton load. The load lies in the plane of the plate, 60° from the horizontal, at coordinates 10,0. Design a fastener system to join the plate to a column. Locate the center-of-gravity of the fastener group. Find torque and moment-of-inertia. Find shear load on each fastener, and determine minimum fastener and plate size.

Design Decisions: Eight fasteners will be used, located at 0,0; 2,0; 4,0; 0,2; 2,2; 4,2; 0,4; and 2,4. Working strengths of 10,000 psi in shear and 20,000 psi in bearing will be used.

15.12. A flat plate must support an overhung vertical load of 7000 lb located at x = 10 inches in the plane of the plate. Design a fastener system to join the plate to a column. Locate the center-of-gravity of the fastener group. Find torque and moment-of-inertia. Find shear load on each fastener, and determine minimum fastener and plate size.

Design Decisions: Eight fasteners will be used, located at 0,0; 2,0; 4,0; 0,2; 2,2; 4,2; 0,5; and 4,5. Working strengths of 10,000 psi in shear and 20,000 psi in bearing will be used.

CHAPTER 16

Welds and Adhesive Joints

Symbols

Symbol	Meaning
F	force
g_x, g_y	coordinates of weld group center-of-gravity
h	weld leg dimension (the nominal weld size)
I	moment-of-inertia based on a unit throat dimension
J	polar-moment-of-inertia based on a unit throat dimension
L	total weld length
M	moment
N_{FS}	factor of safety
S_{SW}	working strength in shear
t	weld throat dimension
T	torque
X, Y	dimensions of weld group
τ, τ_x, τ_y	shear stress based on a unit throat dimension

Units

Quantity	Units
coordinate locations, dimensions	in or mm
force	lb or N
moment and torque	lb · in or N · mm
moment-of-inertia based on a unit throat dimension	in^3 or mm^3
strength	psi or MPa
stress based on a unit throat dimension	lb/in or N/mm

Welding involves joining of materials (with or without filler material) by heating the common surfaces to a molten or near-molten state. Welding can also be accomplished with high pressure alone. Welding dates back to about 4000 years ago when forge welding was used to join copper and bronze. Arc welding was performed as early as 1881, while gas welding was developed in 1903.

The designer should compare the cost of welding with the cost of fabrication using fasteners or adhesives. Sometimes, weldments are found to be more economical than complex castings or forgings. About 70 different welding, brazing, and soldering processes are in current use. A few of them are listed below:

- shielded metal arc welding (SMAW)
- submerged arc welding (SAW)
- gas metal arc welding (GMAW)
- gas tungsten arc welding (GTAW)
- resistance welding (RW)
- laser beam welding (LBW)
- oxyfuel gas welding (OFW)
- electron beam welding (EBW)
- friction welding (FRW)
- ultrasonic welding (USW)
- vibration welding
- brazing (B)

Most of these methods are used for joining metal parts, particularly steel. Vibration welding and ultrasonic welding are commonly used to join thermoplastics. The welding process can be manual, semiautomated, or fully automated.

SHIELDED METAL ARC WELDING (SMAW)

Manual shielded metal arc welding, also called **stick welding,** is a common method for welding steel. The welder uses a flux-coated metal rod electrode in a holder. The holder and the workpiece are attached to a direct current or alternating current power supply. The welder strikes an arc between the workpiece and the electrode, completing the electrical circuit. The arc melts the end of the electrode and the surface of the workpiece to form the welded joint. The flux cleanses the weld area and vaporizes to form a gas shield, protecting the weld area from atmospheric gases and stabilizing the arc. The availability of portable equipment for manual shielded metal arc welding makes it an ideal process for fabrication and repair of large equipment in the field.

SUBMERGED ARC WELDING (SAW)

Another common method for joining steel, **submerged arc welding,** uses a bare-wire consumable electrode fed into the weld. Powdered welding flux is carried in a flux tube and is deposited ahead of the weld. The flux covers the arc and prevents splatter, while stabilizing the arc and shielding the weld from atmospheric gases. The process is usually fully machine-automated but can be semiautomatic, with the operator guiding the electrode and flux tube. Automated submerged arc welding is faster than manual welding and is likely to produce a better-quality weld.

GAS METAL ARC WELDING (GMAW)

GMAW involves joining metals by heating them with an electric arc, using a wire electrode. The electrode, of the same material that is being welded, is fed continuously from a reel. Shielding gas is carried in a nozzle surrounding the electrode. Because of the continuous electrode used in GMAW, the process is faster than manual shielded metal arc welding. Carbon dioxide (sometimes mixed with argon) can be used as the shielding gas when carbon steel is being welded. Argon or helium is used for welding aluminum, titanium, and nickel alloys. GMAW is sometimes called **metal inert gas (MIG)** welding.

The continuous electrode feed also makes GMAW suitable for automation. Figure 16.1 shows GMAW (MIG) applications using a robot controller (a dedicated computer) to control welding operations. Part (a) of the figure shows welding of forklift truck components. Part (b) shows welding of a front end loader to be used for underground mining. The operator is holding a teach-pendant which can be used to program the robot or change running parameters such as welding voltage, amperage, and wire feed speed.

(a)

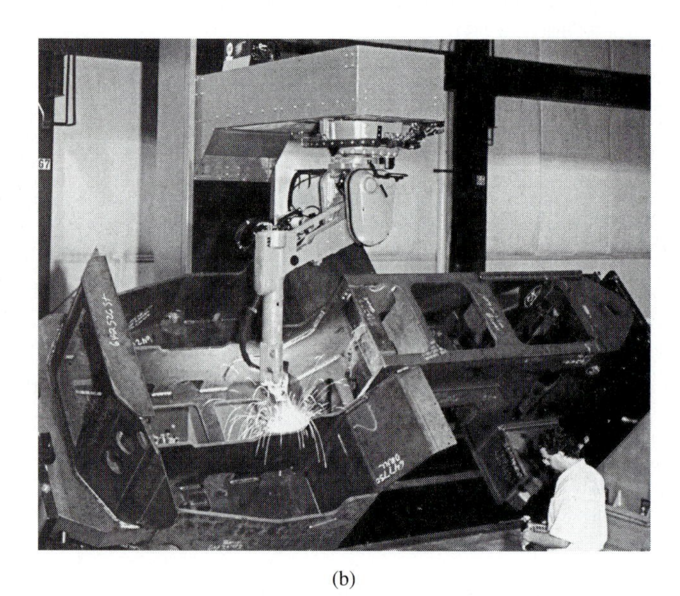

(b)

Figure 16.1 Applications of automated welding systems. (a) GMAW (MIG) welding of forklift truck components; (b) GMAW (MIG) welding of a front end loader for underground mining. (Courtesy of ABB Flexible Automation, Inc.)

GAS TUNGSTEN ARC WELDING (GTAW)

GTAW is the process of joining metals using a shielding gas and an arc formed between the workpiece and a tungsten electrode. The tungsten electrode is essentially nonconsumable; it is not intended to form a part of the weld. Filler metal can be used. GTAW is sometimes called **tungsten inert gas (TIG)** welding or **heliarc** welding. GTAW is commonly selected for welding of stainless steel or aluminum.

RESISTANCE WELDING (RW)

Spot welding is a form of **resistance welding** usually used to join thin metals. One metal sheet is stacked upon another. Pressure is applied above and below the sheets by a pair of electrodes, and a current is passed between the electrodes. The sheets are melted at the common surface between the electrodes, forming a spot weld. Spot welding is successful if both the current and the resistance are high at the common surface. **Seam welders** use wheel electrodes that roll on the metals to be joined. The principle is the same as with spot welding, but seam welding forms a continuous weld.

Resistance welding is practical for fabrication of metal cabinets and other sheet-metal items. Spot and seam welds can be used instead of fasteners if there is no need to disassemble the joint.

LASER BEAM WELDING (LBW)

LBW utilizes a high-energy-density beam of light. The light energy is released to produce heat energy for the weld. Some LBW systems operate in a pulsed mode, and others operate continuously. Lasers can be computer-controlled along several axes. As a result, a number of manufacturers offer computer numerically controlled (CNC) laser welding systems.

Advantages of LBW systems include good weld quality, good controllability, the ability to weld dissimilar materials, ease of automation, and the production of a narrow heat-affected zone (HAZ). By comparison, gas tungsten arc welding generates more heat, resulting in a wide HAZ and the possibility of more shrinkage and distortion. Disadvantages of LBW include higher initial cost than with equivalent arc welding, gas welding, or resistance welding systems.

ELECTRON BEAM WELDING (EBW)

EBW involves an electron gun that focuses and directs electrons at the weld joint. The kinetic energy of the electrons is converted into heat to form the weld. Most commercial applications of EBW require a vacuum chamber enclosure for both the electron beam gun and the object to be welded. The vacuum chamber limits EBW to small parts. Advantages of EBW include high weld quality and high welding speed. Disadvantages of EBW include a high initial cost for specialized equipment.

Lewis and Dimino (1975) describe a proposal for a small, portable vacuum chamber designed for electron beam skate welding. Elastomer drive tracks and end seals provide adequate vacuum for continuous electron beam welding of large parts. A self-contained EBW gun has been designed for experiments in joining parts under zero gravity (Elder, Lowry, and Miller, 1975). More conventional uses are also proposed for this welder, including use in research institutions and the automotive and aerospace industries.

FRICTION WELDING (FRW)

In **friction welding,** parts are welded by the heat generated by friction. For example, the end of a metal tube can be rotated and pressed against the end of a stationary tube. When the welding temperature is reached, the rotation is stopped, and the parts are held together with a high force. Advantages of friction welding include the ability to produce welds rapidly and the ability to weld dissimilar metals. A major disadvantage is that most weld configurations are not suitable for friction welding.

ULTRASONIC WELDING (USW)

Humans can hear airborne sound within the frequency range of approximately 20 to 20,000 hertz (Hz), where 1 Hz = 1 cycle/second. **Ultrasonic welding** utilizes high-frequency, solid-borne sound waves (vibrations).

A typical USW unit involves a power supply to convert 50 or 60 Hz electrical current to higher frequency, a converter to produce mechanical vibrations, a booster to modify vibration amplitude, and a horn to transmit the vibration to the parts to be joined. The horn vibrates longitudinally and maintains the required pressure on the parts to form a weld. Typical welding frequencies are 20,000 and 40,000 Hz, and horn length is one-half wavelength at the operating frequency. A schematic of an ultrasonic assembly system is shown in Figure 16.2(a). Part (b) of the figure illustrates ultrasonic welding and related ultrasonic processes.

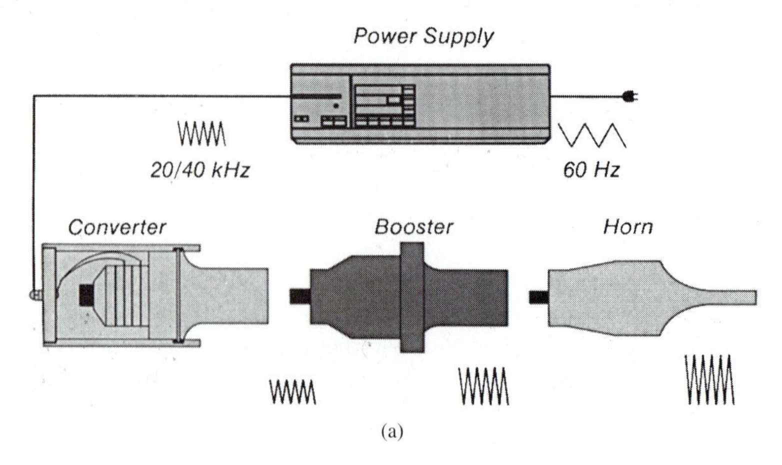

(a)

Figure 16.2 (a) Ultrasonic assembly system schematic;

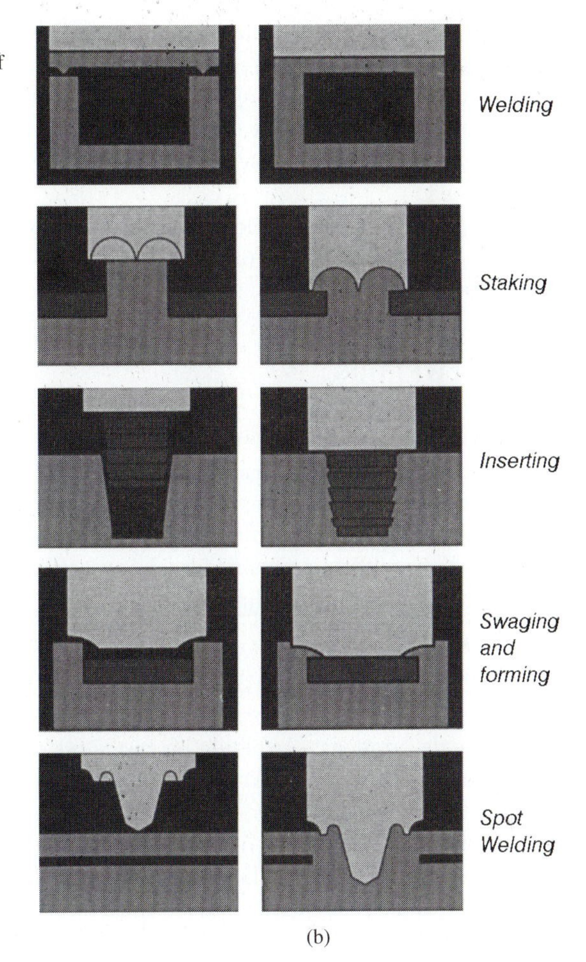

Figure 16.2 ***continued*** (b) ultrasonic welding and related processes. (Courtesy of Branson Ultrasonics Corporation)

USW works well with thermoplastics, particularly amorphous resins which have a broad glass transition temperature (T_g). These materials soften gradually, and they melt and flow without hardening prematurely before the welding is complete. Welding of dissimilar resins is practical if they have like molecular structure and similar melt temperatures. More detailed information on ultrasonic plastic assembly and compatibility of thermoplastics can be found in product literature (e.g., Branson Ultrasonics Corporation, 1991, 1992, 1993).

Other Ultrasonic Joining and Forming Techniques

Ultrasonic energy can also be used for staking, inserting, spot welding, swaging, and forming.

Staking involves melting and re-forming a thermoplastic stud. This process can be used to lock another (dissimilar) part in place. It is often used to join a metal part to a thermoplastic part.

Inserting is the embedding of a metal component into a hole in a thermoplastic part. Threaded metal inserts are commonly used in plastic parts.

Ultrasonic spot welding is similar to resistance spot welding in that parts can be joined at localized points. It is used to join sheets of extruded or cast thermoplastic and to weld thermoplastic parts with complicated geometry and hard-to-reach joining surfaces.

Swaging and **forming** involve ultrasonic melting and re-forming a ridge in a plastic part in order to lock in another part. Swaging of plastic tubing involves a diameter change at the area to be joined.

Advantages of ultrasonic welding include the avoidance of distortion due to high temperature and its compatibility with automated assembly line production. Disadvantages include limitations on weld size.

VIBRATION WELDING

Vibration welding typically involves frequencies of 120 or 240 Hz, much lower than ultrasonic welding frequencies. One part is vibrated against a stationary part while pressure is applied. The relative motion is in the plane of the surfaces to be joined. Friction-generated heat causes the parts to melt at the common surface. The weld joining the parts is then allowed to solidify. Vibration welding is suitable for joining large parts and for joining resins that are difficult to join with ultrasonic welding.

SELECTION OF WELDING METHODS

The preceding descriptions of welding processes are too brief to enable a designer to select the best fabrication process. Books on welding technology give more detail on the processes and the practice of welding (e.g., Bowditch and Bowditch, 1990; Koellhoffer, Manz, and Hornberger, 1988). Although the designer is unlikely to do the actual welding, a knowledge of welding techniques found in "how-to" books is still useful. This knowledge helps the designer select practical fabrication methods and avoid designs that cannot be fabricated.

BUTT JOINTS

Figure 16.3(a) shows a cross section of a butt weld joining two metal plates. The heat-affected area is shown by dashed lines. The weld material buildup at the top and bottom is called **reinforcement**; however, the word *reinforcement* is a misnomer in that it does not increase weld strength but may cause stress concentration. For welds subject to severe fatigue loads, the surface is often ground flush to prevent stress concentration.

Before arc welding plates that are between 1/8 and 1/4 in thick (about 3 to 6 mm thick), the welder usually leaves a gap of about one-half the plate thickness. If the plates thicker than 1/8 in meet without a gap, the weld may not fuse in the center.

Thick plates may be V-grooved on both sides of the joint. If only one side of the joint is accessible for welding, then the plates are V-grooved on that side, as shown in Figure 16.3(b). When the weld material is applied from only one side of the joint, there is a greater tendency toward distortion owing to uneven cooling. Note that several arc welding passes may be

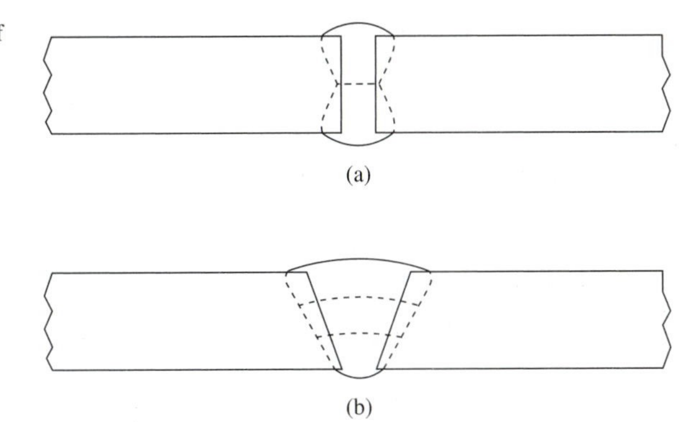

Figure 16.3 Butt welds. (a) Both sides of joint accessible; (b) one side accessible.

needed to get full weld penetration. Electron beam processes are capable of producing welds in relatively thick material in a single pass. Thus, these processes can be selected to reduce weld distortion.

Where possible, the designer specifies a weld filler material with a tensile strength no less than that of the plates to be welded. In that case, a full-penetration butt joint is approximately as strong as the plates that are joined. Otherwise, we must consider the tensile strength of the filler material and the joint dimensions.

FILLET WELDS

Figure 16.4(a) shows a fillet weld joining two members. For design purposes, the weld material cross section is assumed to be an equilateral triangle with legs h and throat t, as shown in Part (b). Shear failure is assumed to occur through the throat dimension, which is given by

$$t = \frac{h}{\sqrt{2}} \tag{16.1}$$

where t = weld throat dimension

h = weld leg dimension (the nominal weld size)

If the weld cross section is concave, the throat dimension is as defined by Part (c) of the figure.

Figure 16.5 shows a fillet weld subject to shear loading. The weld lies along near-side edges, AB, BC, and CD, and along far-side edge DA. Weld shear stress is approximated by

$$\tau = \frac{F}{Lt} \tag{16.2}$$

where τ = shear stress as the weld throat

F = force

L = total weld length $= AB + BC + CD + DA$ in this case

Figure 16.4 (a) Fillet weld; (b) cross section showing throat; (c) concave weld cross section.

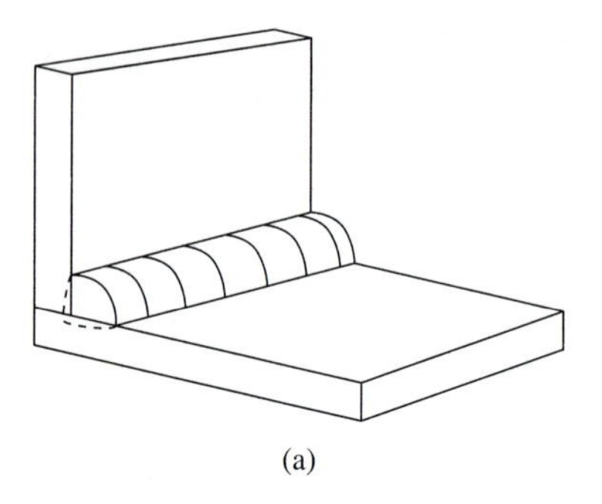

(a)

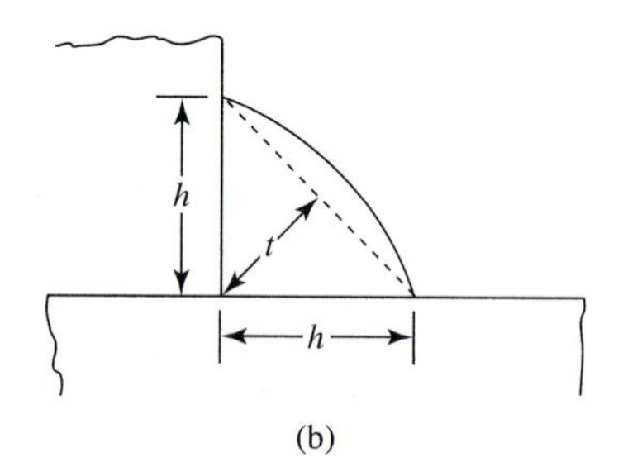

(b)

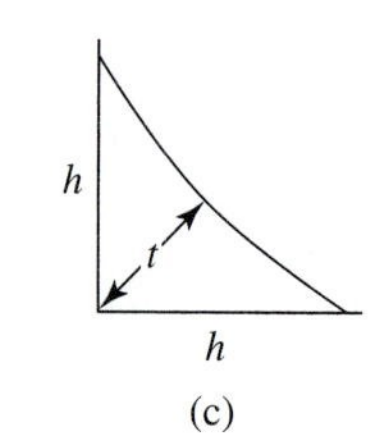

(c)

Figure 16.5 Shear loading of a fillet weld.

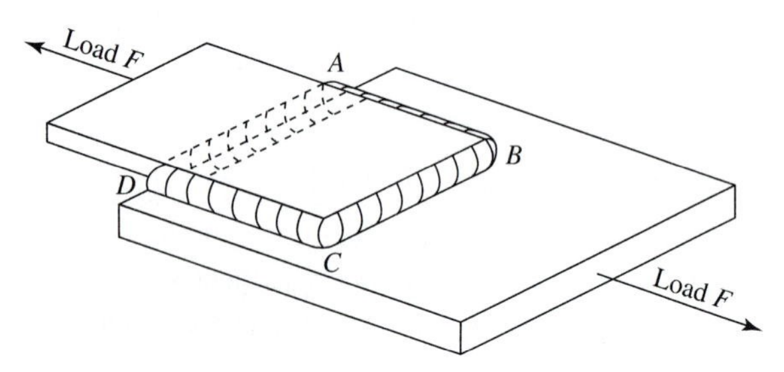

The design criterion for symmetrically loaded fillet welds becomes

$$h = t\sqrt{2} = \frac{F\sqrt{2}}{LS_{SW}} \tag{16.3}$$

where S_{SW} = working strength in shear = S_{SYP}/N_{FS}

S_{SYP} = shear yield point of weld material

N_{FS} = factor of safety

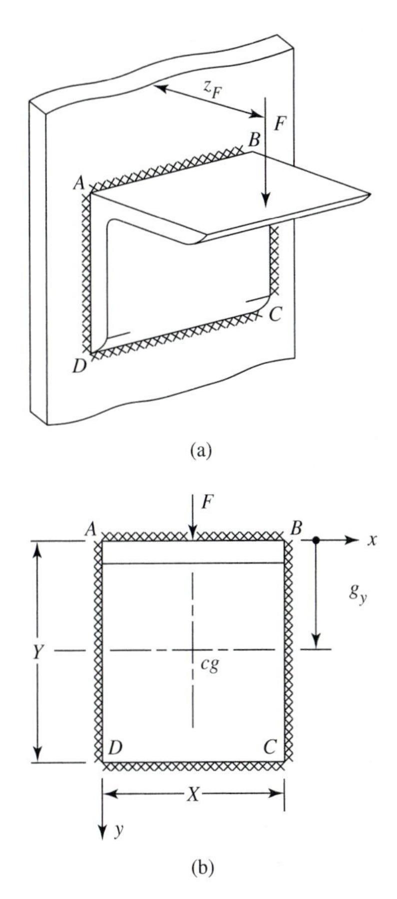

Figure 16.6 Out-of-plane loading.

Note that there are tensile stresses as well as shear stresses in the welds along edges *BC* and *DA* in Figure 16.5. Nevertheless, it is customary to design the entire weld pattern as if it were in shear.

DESIGN OF WELDS WITH OUT-OF-PLANE ECCENTRIC LOADS

When the load does not go through the center-of-gravity of a weld group, we must modify the design procedure. Figure 16.6 shows a weld group with load F at a distance z_F from the plane of the welds. Several simplifying assumptions will be used in the design. They include the following:

1. We will base the weld design on the assumption that the load produces only shear stress in the throat section of each weld.
2. The welds will be represented as lines (in this case, lines lying along edges *AB*, *BC*, *CD*, and *DA*).
3. Stresses in the welds are produced by rotation about an axis through the weld group center-of-gravity (*cg*). (This is a safe assumption. For a thick, rigid angle-section, we

could assume rotation about *CD*, the bottom of the angle. The alternative assumption is less conservative).

4. We will initially assume that the throat dimension is unity (i.e., 1 in or 1 mm). We make this initial assumption so that we can analyze the design and determine optimum weld size. This assumption is only a device to simplify design calculations. Calculation results determine actual size of the fillet weld to be produced by a welder or an automatic welding system.

Direct Shear

The direct shear in the weld group is given by

$$\tau_y = \frac{F}{L} \qquad \textbf{(16.4)}$$

where τ_y = direct shear stress based on unit throat

F = load

L = total weld length

Moment-Induced Shear

Moment-induced shear is given by

$$\tau_x = \frac{Mc}{I_x} \qquad \textbf{(16.5)}$$

where $M = Fz_F$ = moment due to offset load

z_F = distance from load to plane of weld group

c = maximum weld distance from axis through center-of-gravity of weld group

I_x = moment-of-inertia of weld group about central axis (based on unit throat)

Moment-of-Inertia

For out-of-plane loading, we treat the weld group like the cross section of a beam in bending. If the loading is vertical, we are interested in the moment-of-inertia about a horizontal axis through the center-of-gravity. The moment-of-inertia of a weld about a given axis depends on the weld moment-of-inertia about its own central axis and the distance from the weld center to the given axis. Let the weld group moment-of-inertia about a horizontal axis through the weld center-of-gravity be identified as I_x. Then the contribution of weld *AB* to I_x is given by

$$I_{xAB} = X\left(\frac{Y}{2}\right)^2 \qquad \textbf{(16.6)}$$

where I_{xAB} = contribution of weld *AB* to I_x

X and Y = weld group dimensions

The moment-of-inertia of weld *AB* about its own central axis is neglected.

The contribution of weld AD is

$$I_{xAD} = \frac{Y^3}{12} \tag{16.7}$$

where I_{xAD} = contribution of weld AD to I_x

The center of weld AD lies on a horizontal axis through the weld group center-of-gravity. If it did not lie on that axis, another term would have been required. Moment-of-inertia I_x is given by the sum of the contributions from all welds.

Determination of Required Weld Size

Direct shear and moment-induced shear can be combined using

$$\tau_{MAX} = \sqrt{\tau_x^2 + \tau_y^2} \tag{16.8}$$

where τ_{MAX} = shear on the most heavily loaded weld, based on unit throat

Finally, we compare shear based on unit throat with the shear working strength of the weld, multiplying by $2^{1/2}$ to obtain the required nominal size of the weld.

$$h = \frac{\tau_{MAX}\sqrt{2}}{S_{SW}} \tag{16.9}$$

where h = nominal size of weld-leg dimension (in or mm)

S_{SW} = working strength of weld in shear (psi or MPa)

Units Used in Weld Design for Eccentric and Out-of-Plane Loads

As noted above, we initially assumed a unit weld throat dimension in order to (finally) obtain the most economical weld size. Shear stress based on a unit throat dimension appears to have the units of lb/in or N/mm rather than the usual units (psi or MPa). Moments-of-inertia appear to have units of in^3 or mm^3 rather than the expected units (in^4 or mm^4). Equation (16.9) for weld size results in a nominal size given in inches or millimeters.

EXAMPLE PROBLEM Weld Design for Out-of-Plane Loading

Design a weld to carry a 30,000 N load with a 220 mm offset.

Design Decisions: The weld design will be similar to that shown in Figure 16.6, with

$$X = AB = CD = 90 \text{ mm} \quad \text{and} \quad Y = BC = DA = 110 \text{ mm}$$

The weld will have a working strength of 155 MPa in shear. In order to design for the optimum weld size, we will leave it as an unknown, assuming a unit throat dimension for now.

Solution: The weld group center-of-gravity is obviously in the center of the group. The weld is located so that a plane through the center-of-gravity and the force is perpendicular to the weld group plane. This placement eliminates twisting in the weld group plane. Total weld length is given by

$$L = AB + BC + CD + DA = 400 \text{ mm}$$

We need to calculate I_x, the moment-of-inertia about a horizontal axis through the center-of-gravity, because the load is vertical. The moment-of-inertia contribution of weld AB is

$$I_{xAB} = X\left(\frac{Y}{2}\right)^2 = 90\left(\frac{110}{2}\right)^2 = 272{,}250$$

and the contribution of weld AD is

$$I_{xAD} = \frac{Y^3}{12} = \frac{110^3}{12} = 110{,}917$$

The contribution of CD is the same as that of AB, and the contribution of BC is the same as that of AD. Adding the four contributions, we obtain

$$I_x = 766{,}334$$

The direct shear is

$$\tau_y = F/L = 30{,}000/400 = 75$$

Moment is given by

$$M = Fz_F = 30{,}000 \cdot 220 = 6.6 \cdot 10^6$$

and the moment-induced shear at a distance of $c = 55$ mm from the center-of-gravity is

$$\tau_x = Mc/I_x = 6.6 \cdot 10^6 \cdot 55/766{,}334 = 473.7$$

Combining direct and moment-induced shear, we have

$$\tau_{\text{MAX}} = \sqrt{\tau_x^2 + \tau_y^2} = \sqrt{473.7^2 + 75^2} = 479.6$$

where moment-of-inertia and stresses are based on a unit weld throat dimension.

The required nominal size (fillet weld leg dimension) is obtained from

$$h = \frac{\tau_{\text{MAX}}\sqrt{2}}{S_{SW}} = \frac{479.6\sqrt{2}}{155} = 4.4 \text{ mm}$$

We would probably specify 5 mm (nominal size) fillet welds. ■■

OUT-OF-PLANE LOADS ON ASYMMETRIC WELD GROUPS

The weld group designed in the preceding example was symmetric about a horizontal axis and a vertical axis. If a weld group lacks symmetry, then the center-of-gravity of the weld group cannot be found by inspection. If weld elements are parallel to the x- and y-axes, the coordinates of the center-of-gravity are given by the following equations:

$$g_x = \frac{\sum_{i=1}^{n} (X_i + Y_i)x_i}{L} \quad \textbf{(16.10)}$$

$$g_y = \frac{\sum_{i=1}^{n} (X_i + Y_i)y_i}{L} \quad \textbf{(16.11)}$$

where we have arbitrarily selected a coordinate origin (e.g., the top left corner of the weld group)

g_x and g_y = x- and y-coordinates of the weld group center-of-gravity

$\sum_{i=1}^{n}$ = tells us (or the computer) to sum all of the weld element contributions from $i = 1$ to $i = n$

n = number of weld elements

X_i and Y_i = weld lengths in the x- and y- directions

x_i and y_i = coordinates of the center-of-gravity of individual weld elements

Moment-of-inertia calculation is more difficult when the weld group lacks symmetry. For vertical loading, the neutral axis will be horizontal. If all the welds are parallel to either the x- or the y-axis, then the moment-of-inertia about the neutral axis is given by

$$I_x = \left[\sum_{i=1}^{n} \left[(X_i + Y_i) \cdot (y_i - g_y)^2 + \frac{(Y_i)^3}{12} \right] \right] \qquad \textbf{(16.12)}$$

where I_x = moment-of-inertia about the neutral axis, based on a unit throat dimension

The neutral axis is perpendicular to the loading direction, and it goes through the center-of-gravity of the weld group. Note that $(y_i - g_y)$ in Equation (16.12) is the distance from the center of an individual weld to the neutral axis. Direct shear and moment-induced shear are combined as in the preceding example problem. Finally, we obtain the required weld leg dimension (nominal size) from

$$h = \frac{\tau_{\mathrm{MAX}} \sqrt{2}}{S_{SW}}$$

EXAMPLE PROBLEM Design of an Asymmetric Weld

Design a weld to carry a 30,000 N load with a 220 mm offset as in Figure 16.6. The lower edge of the bracket is inaccessible for welding.

Design Decisions: The weld design will be similar to that shown in Figure 16.6, except that there will be no weld along edge CD. Dimensions will be $AB = 90$ mm and $BC = DA = 110$ mm.

Solution Summary: The detailed solution follows. The equations and table are set up to include a weld along edge CD in case we want to consider another problem where that edge is accessible. For the present problem, we simply set that weld length equal to zero. The total weld length is 310 mm, and the center-of-gravity of the weld group is about 39 mm from the top of the bracket. The moment-of-inertia about the neutral axis is $I_x = 415{,}000$ based on a unit throat. This value is substantially less than the value of I_x when all four edges are welded. Moment-induced shear and direct shear are combined to obtain maximum shear on the weld group. Finally, we find that the weld leg dimension must be at least 10.33 mm. All three weld segments (AB, BC, and DA) will have the same nominal size, about 11 mm.

Detailed Solution:

Load (N):

$$F = 30{,}000$$

Direction: Vertical.

Location:

$$z_F = 220$$

Working Strength (MPa):

$$S_{SW} = 155$$

$$i = 1, \ldots, 4$$

		Dimensions (mm)		Center location (mm)	
Weld	*i*	X_i	Y_i	x_i	y_i
AB	1	90	0	45	0
BC	2	0	110	90	55
CD	3	0	0	45	110
DA	4	0	110	0	55

Total Weld Length (mm):

$$L = \sum_{i=1}^{4} X_i + Y_i \qquad L = 310$$

Weld Group Center-of-Gravity:

$$g_x = \frac{\sum_{i=1}^{4} (X_i + Y_i) \cdot x_i}{L} \qquad g_x = 45$$

$$g_y = \frac{\sum_{i=1}^{4} (X_i + Y_i) \cdot y_i}{L} \qquad g_y = 39.032$$

Weld Group Moment-of-Inertia about Horizontal Axis through Center-of-Gravity (mm)[1]

$$I_x = \left[\sum_{i=1}^{4} \left[(X_i + Y_i) \cdot (y_i - g_y)^2 + \frac{(Y_i)^3}{12}\right]\right] \qquad I_x = 4.15 \cdot 10^5$$

Moment (N · mm):

$$M = F \cdot z_F \qquad M = 6.6 \cdot 10^6$$

Maximum Weld Distance from Central Axis:

$$c = 110 - g_y \qquad c = 70.968$$

Moment-Induced Shear:[1]

$$\tau_x = \frac{M \cdot c}{I_x} \qquad \tau_x = 1.129 \cdot 10^3$$

Direct Shear:

$$\tau_y = \frac{F}{L} \qquad \tau_y = 96.774$$

Maximum Shear:[1]

$$\tau_{\text{MAX}} = \sqrt{\tau_x^2 + \tau_y^2} \qquad \tau_{\text{MAX}} = 1.133 \cdot 10^3$$

Required Nominal Size: The weld leg dimension should not be less than this value.

$$h = \frac{\tau_{\text{MAX}} \cdot \sqrt{2}}{S_{SW}} \qquad h = 10.334 \text{ mm}$$

WELDS WITH IN-PLANE ECCENTRIC LOADS

Whenever practical, we try to design welds so that the line of force goes through the center of the weld group. When that condition is met, the weld is usually most economical and easiest to design. Figure 16.7 illustrates a situation where we were unable to meet that condition. There is an eccentric load in the plane of the weld group. As a result, there is a twisting moment (torque) about the center of the weld group.

Center-of-Gravity and Moments-of-Inertia

We begin with a trial weld group configuration. As in previous cases, we have not yet determined the optimum weld size at this stage of the design. If the center-of-gravity of the weld group cannot be located by inspection, we use the same equations as in the previous example problem. Since there is torsional loading, we must determine the polar-moment-of-inertia of the weld group. We first find the moments-of-inertia about horizontal and vertical axes, I_x and I_y. The first is given in Equation (16.12). The second is given by

$$I_y = \left[\sum_{i=1}^{n} \left[(X_i + Y_i) \cdot (x_i - g_x)^2 + \frac{(X_i)^3}{12} \right] \right] \tag{16.13}$$

where I_y = moment-of-inertia about a vertical axis through the center-of-gravity

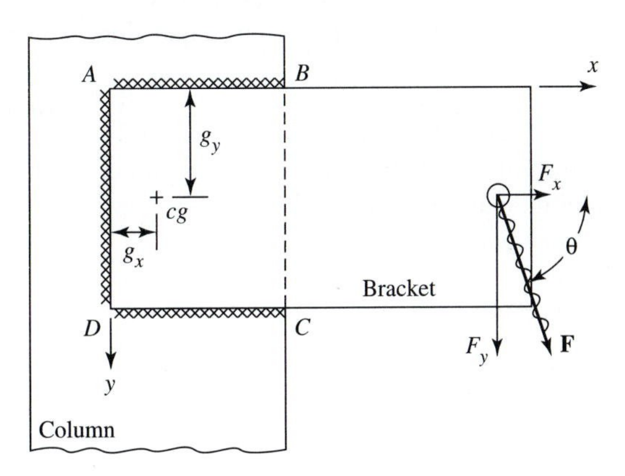

Figure 16.7 Weld with in-plane eccentric load.

[1]Based on unit throat.

Equations (16.12) and (16.13) apply if the weld elements are parallel to the *x*- and *y*-axes. The polar-moment-of-inertia is given by

$$J = I_x + I_y \tag{16.14}$$

where J = polar-moment-of-inertia

Like I_x and I_y, it is based on a unit throat dimension.

Torque

Torque on the weld group is the product of the load and the (perpendicular) distance from the line of force to the center-of-gravity. The perpendicular distance is easy to find graphically. When we are using a computer, it is more convenient to break up the force into horizontal and vertical components. We will use a procedure similar to the one that we used for fasteners with eccentric loads. The equation is

$$T = -F_x(y_F - g_y) + F_y(x_F - g_x) \tag{16.15}$$

where T = torque

F_x and F_y = horizontal and vertical force components

x_F, y_F = location of a point of application of the force

g_x, g_y = location of the center-of-gravity

Weld Stress

The weld size is still undetermined, so we will base stress calculations on a unit weld throat dimension. The load divided by total weld length produces a direct shear throughout the weld group.

$$\tau_{\text{DIRECT}} = F/L$$

where τ_{DIRECT} = direct shear in the direction of the load

At any point in the weld group, there is a torque-induced shear

$$\tau_{\text{TORQUE}} = Tr/J$$

where τ_{TORQUE} = torque-induced shear which is perpendicular to a line from the center-of-gravity to the point in question

r = distance from the center-of-gravity to the point in question

Resultant shear is the vector sum of direct and torsion-induced shear.

Instead of calculating and combining direct and torsion-induced shear, we will calculate and combine horizontal and vertical shear components, saving calculation steps. The recommended procedure follows.

Shear at location *x,y* is given by

$$\tau_x = T\frac{g_y - y}{J} + \frac{F_x}{L} \tag{16.16}$$

$$\tau_y = -T\frac{g_x - x}{J} + \frac{F_y}{L} \tag{16.17}$$

$$\tau = \sqrt{\tau_x^2 + \tau_y^2} \tag{16.18}$$

where τ_x, τ_y, and τ = horizontal, vertical, and resultant shear, respectively

J = polar-moment-of-inertia (Shear and moment-of-inertia are based on unit throat.)

T = torque

x,y = location of a point on the weld

g_x,g_y = location of the center-of-gravity

L = total weld length

F_x and F_y = force components

Required Weld Size

Maximum shear will occur closest to the load line and/or farthest from the center-of-gravity. We check the values at various points in the weld pattern to find the critical location, and we base our entire design on it. The required weld size is

$$h = \frac{\tau_{\text{MAX}}\sqrt{2}}{S_{SW}}$$

where h = leg dimension for all segments of the weld (in or mm)

τ_{MAX} = greatest shear stress at any point on weld (psi or MPa)

S_{SW} = working strength of weld (psi or MPa)

EXAMPLE PROBLEM Design of a Weld with an Eccentric Load

Refer to Figure 16.7. The coordinate origin will be taken as point A. Design a weld to support a load of 11,000 lb at an angle of 75° from the horizontal. The load is applied at location $x_F = 10$ in, $y_F = 3$ in.

Design Decisions: We will weld along near-side bracket edges AB, CD, and DA, since the far side of the bracket is not easily accessible. Dimensions are $AB = CD = 5$ and $DA = 6$ in. The working strength of the weld will be 29,000 psi.

Solution Summary: The center-of-gravity of the weld group is calculated to fall at $g_x = 1.563$, $g_y = 3$. We did not need to calculate g_y; it is obvious because of the symmetry. The moments-of-inertia about horizontal and vertical axes through the center-of-gravity are calculated and added to obtain the polar-moment-of-inertia $J = 152.27$ based on a unit throat dimension. The load produces a torque $T = 89{,}650$ lb · in about the center-of-gravity.

We then calculate horizontal, vertical, and resultant shear for various locations on the weld group. The selected locations are x,y = 0,0; 5,0; 5,6; and 0,6. The most severe stress is

$$\tau_{\text{MAX}} = \tau(5,0) = 3317$$

based on a unit throat dimension. Finally, we calculate required nominal size, a weld throat dimension of $h = 0.162$ in. We could specify a size of 3/16 or 1/4 in.

In some cases, our initial design decisions might be faulty. For example, a poor selection of bracket size could lead to a very large weld size. The calculated weld leg dimension might even exceed bracket thickness. We can correct this error by making the bracket thicker or by increasing weld group dimensions AB and CD and/or increasing DA.

The detailed solution follows.

Detailed Solution:

Load (lb):

$$F = 11{,}000$$

Direction:

$$\theta = 75 \cdot \text{deg} \qquad \theta = 1.309 \text{ rad}$$
$$F_x = F \cdot \cos(\theta) \qquad F_x = 2.847 \cdot 10^3$$
$$F_y = F \cdot \sin(\theta) \qquad F_y = 1.063 \cdot 10^4$$

Location (in):

$$x_F = 10 \qquad y_F = 3$$

Working Strength:

$$S_{SW} = 29{,}000$$
$$i = 1, \ldots, 3$$

	Dimensions (in)		*Center location (in)*	
Weld	X_i	Y_i	x_i	y_i
AB	5	0	2.5	0
CD	5	0	2.5	6
DA	0	6	0	3

Weld Group Center-of-Gravity:

$$g_x = \frac{\sum_{i=1}^{3} (X_i + Y_i) \cdot x_i}{\left[\sum_{i=1}^{3} X_i + Y_i\right]} \qquad g_x = 1.563$$

$$g_y = \frac{\sum_{i=1}^{3} (X_i + Y_i) \cdot y_i}{\left[\sum_{i=1}^{3} X_i + Y_i\right]} \qquad g_y = 3$$

Weld Group Moment-of-Inertia about Center-of-Gravity:[1]

$$I_x = \left[\sum_{i=1}^{3} \left[(X_i + Y_i) \cdot (y_i - g_y)^2 + \frac{(Y_i)^3}{12}\right]\right] \qquad I_x = 108$$

$$I_y = \left[\sum_{i=1}^{3} \left[(X_i + Y_i) \cdot (x_i - g_x)^2 + \frac{(X_i)^3}{12}\right]\right] \qquad I_y = 44.271$$

$$J = I_x + I_y \qquad J = 152.271$$

Torque about Center-of-Gravity (lb · in):

$$T = -F_x \cdot (y_F - g_y) + F_y \cdot (x_F - g_x) \qquad T = 8.965 \cdot 10^4$$

Total Weld Length:

$$L = \left[\sum_{i=1}^{3} X_i + Y_i\right] \qquad L = 16$$

Shear:[1]

$$\tau_x(xx,yy) = T \cdot \frac{g_y - yy}{J} + \frac{F_x}{L}$$

$$\tau_y(xx,yy) = T \cdot \frac{xx - g_x}{J} + \frac{F_y}{L}$$

$$\tau(xx,yy) = \sqrt{\tau_x(xx,yy)^2 + \tau_y(xx,yy)^2}$$

Location	*Shear*	*Maximum Shear*	
A	$\tau(0,0) = 1.961 \cdot 10^3$	$\tau_{MAX} = \tau(5,0)$	$\tau_{MAX} = 3.317 \cdot 10^3$
B	$\tau(5,0) = 3.317 \cdot 10^3$		
C	$\tau(5,6) = 3.122 \cdot 10^3$		
D	$\tau(0,6) = 1.609 \cdot 10^3$		

Required Nominal Size: The weld leg dimension should not be less than this value.

$$h = \frac{\tau_{MAX} \cdot \sqrt{2}}{S_{SW}} \qquad h = 0.162 \text{ in}$$

WELDING SYMBOLS

The American Welding Society (AWS) publishes standards for welding and specifies standard arrow symbols. The AWS symbols are used on drawings to indicate the required size and type of weld, material preparation, and weld location. A few commonly used welding symbols are shown in Figure 16.8.

ADHESIVES

Industrial and structural adhesives are used to bond a wide range of materials, including metal, plastic, glass, ceramics, and wood. They are used in the manufacture of automobiles, aircraft, electronic devices, fixed structures, industrial tools, and machinery.

Adhesives Terminology

A few terms related to design with adhesives and adhesives testing are defined below.

- **Adherend.** Substance bonded to another by an adhesive.
- **Adhesion.** Forces holding two surfaces together.

---

[1]Based on unit throat.

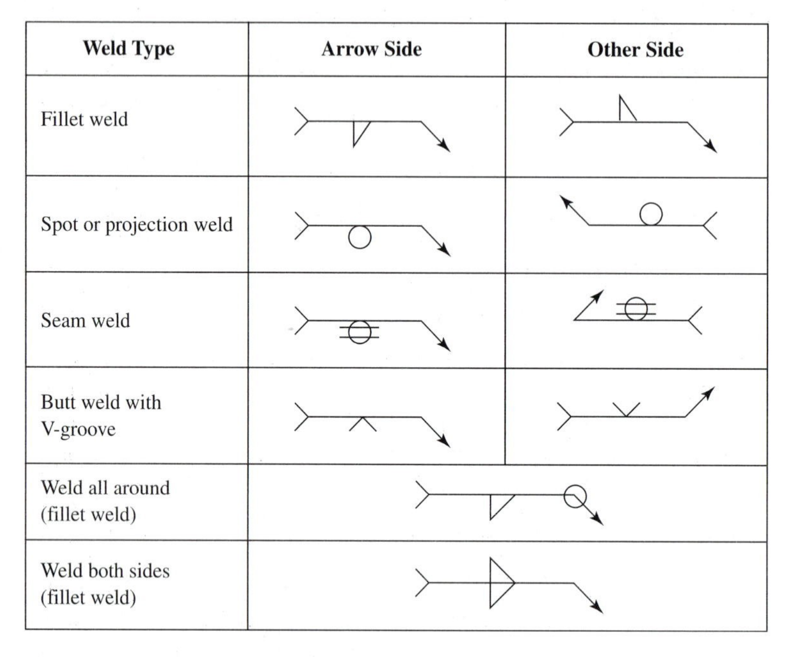

Figure 16.8 Common welding symbols.

- **Adhesive.** Substance used to hold two materials together.
- **Adhesive failure.** Separation at the bond line between adhesive and adherend.
- **Catalyst.** Substance to speed curing.
- **Cohesive failure.** Failure within an adhesive.
- **Cure time.** Interval from start of reaction until specified adhesive strength is realized.
- **Peel strength.** Load per unit width of bond line at failure. Applies to test of flexible adherends.
- **Primer.** Surface coating applied before applying the adhesive (to improve adhesive performance).
- **Tensile lap shear strength.** Load per unit area at shear failure due to tensile loading of test specimens.

Advantages of Adhesives

Advantages of adhesives over other fastening methods include

- more uniform transfer of stress
- cost-effectiveness
- sealing ability
- surface smoothness and appearance

Special Considerations for Designs Incorporating Adhesives

Effective use of adhesives requires attention to a number of environmental conditions and design considerations, including

- high- or low-temperature exposure
- exposure to moisture or immersion in water
- chemical exposure
- material compatibility
- surface preparation
- part design

ADHESIVE SELECTION

Adhesives are selected on the basis of strength requirements, environmental considerations, and processing requirements. Features of selected adhesive types are summarized in *Adhesives* (1990). Table 16.1 is based on that reference.

ADHESIVE TESTS

Whenever possible, adhesives should be evaluated under actual conditions of use. Standard test results help us to make preliminary adhesive selections to meet expected loading conditions. Many such tests have been developed by industry, governmental agencies, and professional societies. The American Society for Testing Materials (ASTM) publishes testing procedures for adhesives. These testing standards specify sample shape and size, preparation,

Table 16.1 Characteristics of selected adhesives.

	Adhesive Type			
	Epoxy	**Urethane**	**Silicone**	**UV-Cure Acrylate**
Process				
Room-temp. or elev.-temp. cure	X	X	X	
Environmental				
Moisture resistance	X		X	X
Low-temp. performance		X	X	
High-temp. performance	X		X	
Chemical resistance	X	X		X
Strength				
Low thermal stress		X	X	
High strength	X	X		X
Flexibility		X	X	X
Peel strength	X	X	X	X
Lap shear strength	X	X		X
Surface				
Fills gaps	X	X	X	X

Source: Based on *Adhesives* (1990).

testing techniques, and reporting details. A few of these tests are noted below. Extensive descriptions and analyses of these tests are given by Anderson, Bennett, and DeVries (1977).

Tensile Tests

Figure 16.9(a) shows a standard tensile test configuration that is often used with metal adherends. The adherend cross-sectional area at the adhesive is usually 1 in^2.

Some materials cannot be easily machined into the configuration shown in Figure 16.9(a). In that case, the test specimen can be sandwiched between two adhesive layers bonded to standard test device halves, as in Part (b) of the figure. This test can be used to test adhesive bond strengths with glass specimens or polymer films. Of course, the bond strength at the test device halves must be greater than the bond strength at the specimen under test.

Figure 16.9(c) shows another test configuration, the cross-lap tensile test. This test has been used for glass, wood, and honeycomb and sandwich materials. The results of this test may not be comparable to the results of other tests, owing to bending of the adherends.

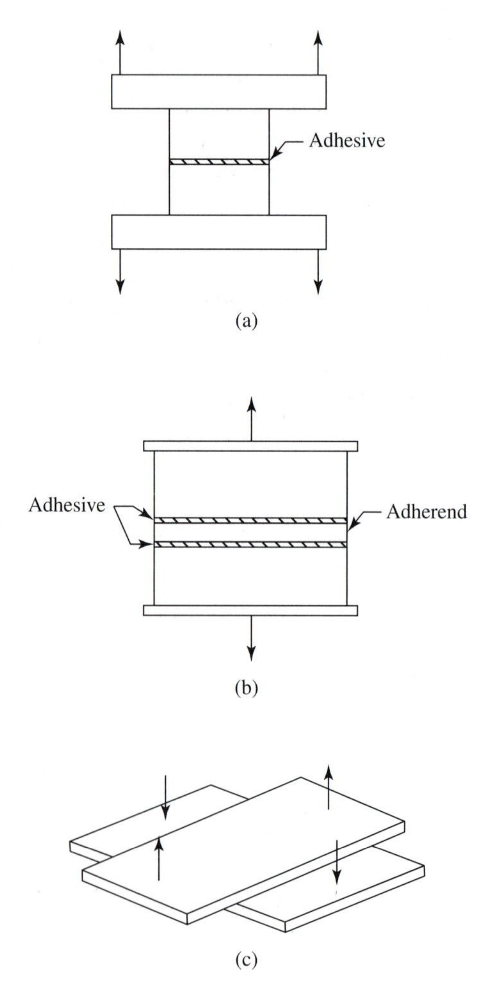

Figure 16.9 Tensile tests for adhesives. (a) Metal-to-metal test; (b) sandwich test; (c) cross-lap test.

Inspection of the joint after the test will indicate adhesive or cohesive failure. Strength can sometimes be improved by surface preparation of the adherend (e.g., physical and/or chemical treatment).

Joint Design

The significance of a test relates to how closely it can predict performance of an adhesive under typical loading. Adhesive bonds have substantially lower tensile strengths than steel and other metals. Thus, when designing with adhesives, we try to avoid direct tensile stresses in the adhesive. Instead, joints are usually designed with a large adhesive area and are oriented so that they resist shear.

Lap Shear Tests

Lap shear tests determine the shear properties of adhesion by tensile loading of the adherend parts. Figure 16.10(a) shows a single lap shear test, the most common. The nominal size of each adherend strip is 4 in (106.4 mm) long by 1 in (25.4 mm) wide by 0.064 in (1.62 mm) thick. The recommended overlap is 0.5 in (12.7 mm). The strips are gripped at the ends, producing a tensile force in the adherends and a shear stress in the adhesive joint. This test is used to evaluate adhesives for metal-to-metal bonding and also for plastics bonding. Test results are reported as *tensile lap shear strength* or simply *shear strength.*

Sometimes adhesives do not behave the same in practical applications as they do in common tests. Differences may be the result of mixing of large adhesive quantities, exothermic chemical reactions, lack of control under field conditions, or other variables. One way to evaluate factory-applied or field-applied adhesives is shown in Figure 16.10(b). A piece is cut from the assembled part, and sawcuts are made to reduce the test area. The part is loaded in tension to produce shear failure in the region between the sawcuts.

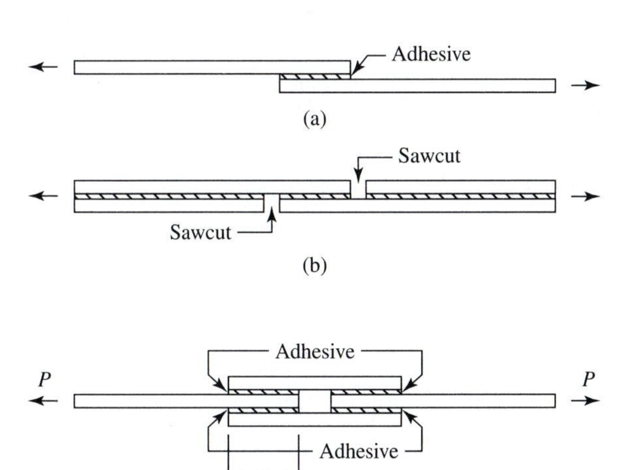

Figure 16.10 Shear tests. (a) Single lap shear test; (b) laminated assembly test; (c) double lap shear test.

Loading in the single lap shear test will tend to deform the adherends to line up with the forces at the ends. This problem is eliminated in the double lap shear test sketched in Figure 16.10(c). Nominal shear strength is given by

$$S_S = \frac{P}{2ab} \tag{16.19}$$

where S_S = nominal shear strength of the bond in double shear (psi or MPa)

P = load on parts (lb or N)

a = overlap on one side of the joint (in or mm); see Figure 16.10(c)

b = part width (in or mm)

Peel Tests

Peel tests are used to evaluate bonds on flexible materials. Test specimens are bonded over part of their length. The unbonded ends are bent at 90°, and these ends are gripped in a tensile testing machine. Peel strength is reported in lb/in of width of bond line or N/mm of width of bond line.

References and Bibliography

Adhesives. Woburn, MA: Emerson and Cuming, Inc., 1990.

Anderson, G. P., S. J. Bennett, and K. L. DeVries. *Analysis and Testing of Adhesive Bonds.* New York: Academic Press, 1977.

Blodgett, O. W. *Design of Weldments.* Cleveland, OH: James F. Lincoln Arc Welding Foundation, 1963.

Bowditch, W. A., and K. E. Bowditch. *Welding Technology Fundamentals.* South Holland, IL: Goodheart-Willcox Company, 1991.

Characteristics and Compatibility of Plastics for Ultrasonic Assembly. Danbury, CT: Branson Ultrasonics Corporation, 1992.

Elder, F. A., J. F. Lowry, and R. A. Miller. "Self-Contained Electron Beam Welding Gun." *Welding and Joining,* p. 1. NASA SP-5978(03). Springfield, VA: National Aeronautics and Space Administration, NTIS, 1975.

Hicks, J. G. *Welded Joint Design.* New York: Halsted Press/John Wiley, 1979.

Houldcroft, P. T. *Welding Processes.* Oxford: Oxford University Press, 1975.

Kennedy, G. A. *Welding Technology.* 2d ed. Indianapolis, IN: Bobbs-Merrill Educational Publishing, 1982.

Koellhoffer, L., A. F. Manz, and E. G. Hornberger. *Welding Processes and Practices.* New York: John Wiley, 1988.

Lewis, J. R., and J. M. Dimino. "Portable Electron Beam Weld Chamber." *Welding and Joining,* p. 4. NASA SP-5978(03). Springfield VA: National Aeronautics and Space Administration, NTIS, 1975.

Ultrasonic Insertion. Danbury, CT: Branson Ultrasonics Corporation, 1993.

Ultrasonic Plastics Assembly. Danbury, CT: Branson Ultrasonics Corporation, 1991.

Design Problems

16.1. Design a weld to carry a 45,000 N load with a 400 mm offset.

Design Decisions: The weld design will be similar to that of Figure 16.6 with

$$X = AB = CD = 140 \text{ mm} \quad \text{and} \quad Y = BC = DA = 125 \text{ mm}$$

The working strength of the weld material will be 175 MPa in shear.

16.2. A bracket is to support a 9000 lb load with an overhang of 9 in. Design the bracket.

Design Decisions: The weld design will be similar to that of Figure 16.6, with

$$X = AB = CD = 4 \text{ in} \quad \text{and} \quad Y = BC = DA = 4.6 \text{ in}$$

The working strength of the weld in shear will be 25,000 psi.

16.3. A bracket is to support a 35,000 N load with an overhang of 300 mm. Design the bracket.

Design Decisions: The weld design will be similar to that of Figure 16.6, with

$$X = AB = CD = 100 \text{ mm} \quad \text{and} \quad Y = BC = DA = 125 \text{ mm}$$

Use a working strength of 185 MPa in shear.

16.4. Refer to Figure 16.6, showing out-of-plane loading. Design a welded support for a 30,000 N load with a 220 mm overhang.

Design Decisions: We will use a heavy angle-section for a bracket and will locate it as in the figure. The bracket dimensions will be

$$AB = CD = 90 \text{ mm} \quad \text{and} \quad BC = DA = 110 \text{ mm}$$

but only the horizontal edges will be welded. The working strength of the weld material will be 155 MPa.

16.5. Refer to Figure 16.6, showing out-of-plane loading. Design a welded support for a 9000 lb load with a 9 in overhang.

Design Decisions: We will use a heavy angle-section for a bracket and will locate it as in the figure. The bracket dimensions will be

$$AB = CD = 4 \text{ in} \quad \text{and} \quad BC = DA = 4.6 \text{ in}$$

but the bottom edge of the bracket will not be welded. The working strength of the weld material will be 25,000 psi.

16.6. Refer to Figure 16.6, showing out-of-plane loading. Design a welded support for a 45,000 N load with a 400 mm overhang.

Design Decisions: We will use a heavy angle-section for a bracket and will locate it as in the figure. The bracket dimensions will be

$$AB = CD = 140 \text{ mm} \quad \text{and} \quad BC = DA = 125 \text{ mm}$$

but the bottom edge of the bracket will not be welded. The working strength of the weld material will be 175 MPa.

16.7. A plate is to be welded to a column to support an overhung load. Refer to Figure 16.7, showing a weld with an in-plane eccentric load. The coordinate origin will be taken as point *A*. Design a weld to support a load of 10,000 lb at an angle of 80° from the horizontal. The load is applied at location $x_F = 8$ in, $y_F = 2.2$ in.

Design Decisions: We will weld along near-side bracket edges *AB*, *CD*, and *DA*, since the far side of the bracket is not easily accessible. Dimensions are

$$AB = CD = 3 \quad \text{and} \quad DA = 4 \text{ in}$$

The working strength of the weld will be 25,000 psi.

16.8. A plate is to be welded to a column to support an overhung load. Refer to Figure 16.7, showing a weld with an in-plane eccentric load. The coordinate origin will be taken as point *A*. Design a weld to support a vertical load of 7000 lb. The load is applied at location $x_F = 5$ in, $y_F = 3.5$ in.

Design Decisions: We will weld along near-side bracket edges AB, CD, and DA, since the far side of the bracket is not easily accessible. Dimensions are

$$AB = CD = 2.5 \quad \text{and} \quad DA = 3.5 \text{ in}$$

The working strength of the weld will be 23,000 psi.

16.9. A plate is to be welded to a column to support a 45,000 N vertical overhung load. Refer to Figure 16.7 showing a weld with an in-plane eccentric load. The coordinate origin will be taken as point A. Design a weld to support the load that is applied at $x_F = 320$ mm.

Design Decisions: We will weld along near-side bracket edges AB, CD, and DA, since the far side of the bracket is not easily accessible. Dimensions are

$$AB = CD = 100 \text{ mm} \quad \text{and} \quad DA = 125 \text{ mm}$$

The working strength of the weld will be 190 MPa.

16.10. A plate is to be welded to a column to support an overhung load at an angle of 85° from the horizontal. Refer to Figure 16.7, showing a weld with an in-plane eccentric load. The coordinate origin will be taken as point A. Design the weld to support a load of 25,000 N. The load is applied at location $x_F = 250$ mm, $y_F = 35$ mm.

Design Decisions: We will weld along near-side bracket edges AB, CD, and DA, since the far side of the bracket is not easily accessible. Dimensions are

$$AB = CD = 100 \text{ mm} \quad \text{and} \quad DA = 75 \text{ mm}$$

The working strength of the weld will be 190 psi.

CHAPTER 17

Springs and Torsion Bars

Symbols

A_c	clash allowance
C	spring index, damping coefficient
d	spring wire diameter, torsion bar diameter
f	frequency of forcing function
f_n	natural frequency
g	acceleration of gravity
h_f	free height
h_s	solid height
K	spring rate
M	mass
N_a	number of active coils
N_{FS}	factor of safety
N_t	total number of coils
p	pitch
P	load
R	mean radius of spring, moment arm of torsion bar load
S_f	fatigue stress number
S_{SE1}	one-way shear endurance limit
S_{SW}	working strength in shear
S_{YP}	yield point strength (in tension)
TR	transmissibility
δ	displacement of load
ϕ	relative rotation of ends of torsion bar
τ_{MAX}	maximum shear stress

Units

acceleration of gravity	in/s^2
dimensions and displacement	in or mm
frequency	Hz
load	lb or N
mass	lb · s^2/in or kg
spring rate	lb/in or N/mm (N/m when based on kg mass)
strength and stress	psi or MPa

Rigidity is desirable in most machine elements and structures. For example, excessive deflection of an automotive camshaft would prevent optimum engine performance. Springs and torsion bars are important exceptions. They are intentionally designed to deflect to store energy, to isolate vibration and shock loads, and to provide forces and torques. Figure 17.1 shows a variety of compression springs, tension springs, and torsion springs.

Torsion Bars

In some cases, a torsion bar fits the available space better than any other type of spring. Torsion bars are sometimes used in automobile and bus suspension systems as the main suspension springs or as sway bars. Torsion bars are also used to provide the torque necessary to open the lid of automobile trunks. Figure 17.2 is a sketch showing the loading and support of a torsion bar. When designing a torsion bar, we can determine the required bar diameter by considering strength. Then we can determine the bar length necessary to provide the required deflection.

Figure 17.1 Springs. (Courtesy of Rockford Spring Company)

Figure 17.2 Torsion bar.

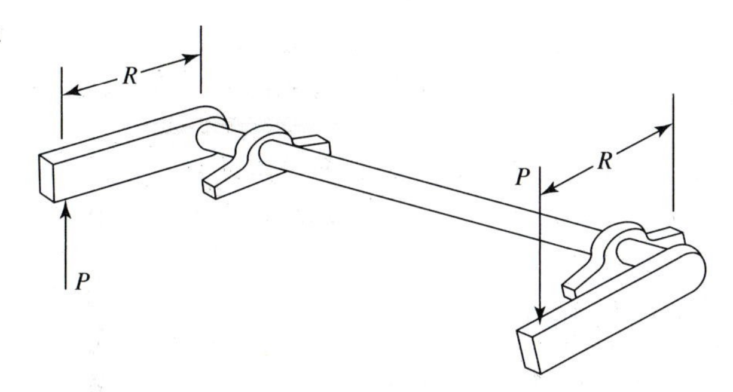

Bar Diameter for Static Loading

The following procedure can be used for design of torsion bars for static loading. Note that the unsupported ends of the bar are subject to both torsion and direct shear. Adding the torsional shear stress to the direct shear stress, we find that the maximum shear stress is

$$\tau_{\mathrm{MAX}} = \frac{16PR}{\pi d^3}\left(1 + \frac{0.3075d}{R}\right) \tag{17.1}$$

and the design criterion is

$$S_{SW} \geq \frac{16PR}{\pi d^3}\left(1 + \frac{0.3075d}{R}\right) \tag{17.2}$$

where τ_{MAX} = maximum shear stress
P = load
R = moment arm of load
d = bar diameter
S_{SW} = working strength in shear

For dimensions in inches and load in pounds, stress and strength are given in $\mathrm{lb/in^2}$.

For dimensions in mm and load in N, stress and strength are given in MPa.

Working Strength

If the working strength in shear is not specified, it can be determined from

$$S_{SW} = \frac{S_{\mathrm{YP}}}{2N_{\mathrm{FS}}} \tag{17.3}$$

where S_{YP} = yield point (in tension)
N_{FS} = factor of safety

Bar Length

The relative rotation of the ends of a torsion bar is proportional to the torque and the length. Relative rotation is inversely proportional to shear modulus and polar-moment-of-inertia. For a solid circular bar, the polar-moment-of-inertia is

$$J = \frac{\pi d^4}{32} \tag{17.4}$$

where J = polar-moment-of-inertia (in^4 or mm^4)

For a given relative rotation, the required torsion bar length is

$$L = \frac{GJ\phi}{PR} \tag{17.5}$$

where L = torsion bar length (in or mm)
G = shear modulus (psi or MPa)
ϕ = relative rotation of ends of bar (radians)

It is often economical to select standard material sizes. Note, however, that required torsion bar length is proportional to the fourth power of bar diameter. If we increase diameter by 10%, for example, we must increase the length by about 46% to obtain the same stiffness.

EXAMPLE PROBLEM Design of a Torsion Bar for Given Rotation under Static Load

Design a torsion bar for a rotation of 0.3 rad under a static load of 1000 lb with a torque arm of 4 in.

Design Decisions: A steel bar with a yield point strength of 120,000 psi will be used. The shear modulus of steel is $11.5 \cdot 10^6$ psi. A safety factor of 4 will be used owing to uncertainty about the load.

Solution Summary: The equation relating working strength to bar diameter is rewritten in the form

$$S_{SW} - \frac{16PR}{\pi d^3}\left(1 + \frac{0.3075d}{R}\right) = 0$$

We estimate a root of $d = 1$ for this equation and solve using mathematics software to obtain $d = 1.139$ in; the calculations are given in the detailed solution which follows. The corresponding length is 142.4 in. Note the effect of rounding the diameter up to 1.25 in.

For hand calculation, the above equation can be rewritten in standard cubic equation form. The solutions to cubic equations are found in a number of handbooks. Or we can estimate the term in parentheses and solve, using that result to update our estimate, until the solution converges.

Detailed Solution:

Static Load:

$$P = 1000$$

Torque Arm:

$$R = 4$$

Shear Modulus:

$$G = 11.5 \cdot 10^6$$

Yield Point:

$$S_{YP} = 120{,}000$$

Safety Factor:

$$N_{FS} = 4$$

Rotation under Static Load (rad):

$$\phi = 0.3$$

Shear Working Strength:

$$S_{SW} = \frac{S_{YP}}{2 \cdot N_{FS}} \qquad S_{SW} = 15{,}000$$

Estimate $d = 1$.

Design Equation for Diameter:

$$d_1 = \mathrm{root}\left[S_{SW} - \frac{16 \cdot P \cdot R}{\pi \cdot d^3} \cdot \left(1 + 0.3075 \cdot \frac{d}{R}\right), d\right]$$

Find optimum diameter: $d_1 = 1.139$.

Polar-Moment-of-Inertia:

$$J = \frac{\pi \cdot d_1^{\,4}}{32}$$

Length of Torsion Bar:

$$L = \frac{G \cdot J \cdot \phi}{P \cdot R} \qquad L = 142.432$$

If the diameter is rounded to the next higher standard size, L must be recalculated. For diameter $d_1 = 1.25$,

$$J = \frac{\pi \cdot d_1^{\,4}}{32}$$

$$L = \frac{G \cdot J \cdot \phi}{P \cdot R} \qquad L = 206.728$$

■■

A similar procedure can be used if the static deflection of a torsion bar is specified. For small rotations, the displacement of the load is approximately equal to the arc distance ϕR. Thus, we use

$$\phi = \frac{\delta}{R} \tag{17.6}$$

where ϕ = rotation (rad)
δ = load displacement
R = moment arm of load

EXAMPLE PROBLEM Design of a Torsion Bar for Given Static Deflection

A spring is to deflect 55 mm under a load of 400 N. Select spring type and design the spring if space beneath the load is limited.

Design Decisions: Owing to the space constraint, a torsion bar will be selected for this application. A torque arm of 150 mm will be used to ensure that the rotation angle is small. Steel with a yield point of 760 MPa will be used, with a safety factor of 3. The shear modulus of steel is approximately 79,300 MPa.

Solution: The rotation angle is found to be

$$\phi = \frac{\delta}{R} = \frac{55}{150} = 0.3667 \text{ rad}$$

Torsion bar dimensions are found as in the preceding example. The calculated diameter is 13.54 mm, and the corresponding length is 1.6 m. If a 15 mm diameter bar is used, the length is increased to 2.41 m. A later example, "Design of a Helical Compression Spring for Static Loading," considers a helical spring for this application.

HELICAL COMPRESSION SPRINGS: DESIGN FOR STATIC LOADING

Figure 17.3 shows a helical compression spring with mean coil radius R and axial load P. Stress and deflection are caused principally by torsion, as in a torsion bar.

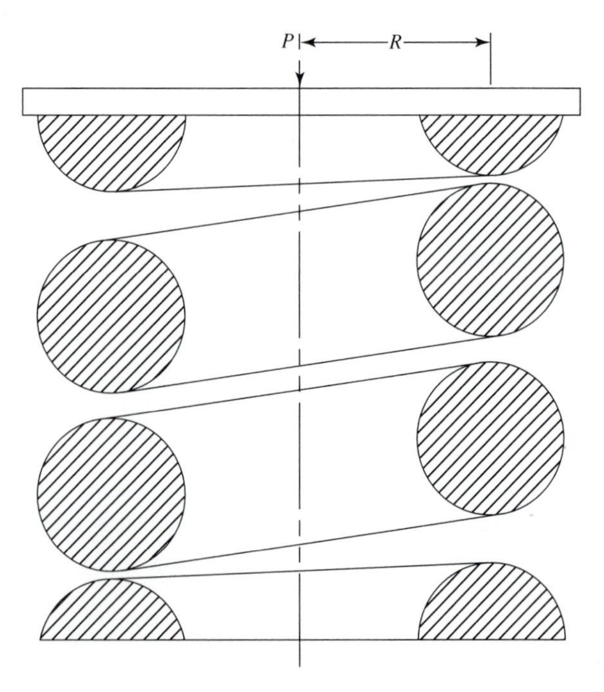

Figure 17.3 Helical compression spring.

Shear Stress

The sum of the torsional shear stress and the direct shear stress is greatest at points on the spring wire nearest to the center of the coil. For static loading, we want to be sure that shear stress does not exceed working strength in shear. The following equations apply:

$$S_{SW} \geq \tau_{\text{MAX}} = \frac{16PR}{\pi d^3}\left[\frac{4C-1}{4C-4} + \frac{0.615}{C}\right] \qquad \textbf{(17.7)}$$

or the equivalent expression

$$S_{SW} \geq \tau_{\text{MAX}} = \frac{8PC}{\pi d^2}\left[\frac{4C-1}{4C-4} + \frac{0.615}{C}\right] \qquad \textbf{(17.8)}$$

where $S_{SW} = S_{\text{YP}}/(2N_{\text{FS}})$ = working strength in shear

P = static load

R = mean coil radius

d = wire diameter

$C = 2R/d$ = spring index

These equations are essentially the same as the torsion bar equations except for the first term in the brackets. This term, called the **curvature factor** or **Wahl factor,** adjusts for the change in stress distribution due to coiling the wire into a helical spring.

Determination of Wire Diameter Based on Stress

If there are no precise restraints on spring coil outside diameter or inside diameter, then it is convenient to select a value of spring index. Values in the range

$$6 \leq C \leq 12$$

are often used, but values outside this range can also be used. After selecting spring index, we can easily solve Equation (17.8) for wire diameter by writing it in the form

$$d = \sqrt{\frac{8PC}{\pi S_{SW}}\left[\frac{4C-1}{4C-4} + \frac{0.615}{C}\right]} \qquad \textbf{(17.9)}$$

Spring Rate

Spring rate (or spring constant) is the load per unit deflection of the spring. It is given by

$$K = \frac{P}{\delta} \qquad \textbf{(17.10)}$$

where K = spring rate (lb/in or N/mm)

P = load (lb or N)

δ = deflection at that load (in or mm)

Spring rate is one of the terms used to describe a spring for use as a vibration isolator or an energy storage device.

Active and Inactive Coils

The spring in Figure 17.3 has **squared** and **ground ends.** The top and bottom coils have a reduced helix angle, and they are ground to lie on a flat surface. Thus, they cannot contribute to the spring deflection, and we call them **inactive coils.** For squared and ground ends, there are approximately two inactive coils. It is often more cost-effective to square the ends, but not to grind them flat. A stamping can be designed to conform to the spring ends, keeping the spring axis vertical (or in the desired orientation). Again, we would estimate that there are two inactive coils. The **active coils** participate in spring deflection. Thus, deflection is proportional to the number of active coils, and spring rate is inversely proportional to the number of active coils. For squared and ground ends (and springs set in a conforming stamping), the total number of coils is

$$N_t = N_a + 2 \qquad \textbf{(17.11)}$$

where N_t = total number of coils (turns)

N_a = active coils

Specification of Active Coils Based on Spring Rate and Deflection

Like a torsion bar, a helical compression spring deflects by twisting of the wire. Each coil represents a wire length of about $2\pi R$. The number of active coils needed for a specified deflection or a specified spring rate is given by

$$N_a = \frac{G\delta d^4}{64PR^3} = \frac{Gd}{8C^3K} \qquad \textbf{(17.12)}$$

Free Height, Solid Height, and Clash Allowance

Free height is the height of a spring when it is not loaded. **Solid height** is the total height at maximum deflection, when the coils are compressed together. Solid height is approximated by

$$h_s = N_t d \qquad \textbf{(17.13)}$$

When a spring is compressed to its solid height, it is protected from additional overload because it cannot deflect further. Of course, extreme overloads would cause a stability failure or compressive yielding.

Solid deflection is the difference between free height and solid height. **Working deflection** is the deflection at maximum expected load. It is given by

$$\delta = \frac{P}{K} \qquad \textbf{(17.14)}$$

Clash allowance is the additional percent or fraction of deflection between the working deflection and the solid deflection. It is given by

$$A_c = \frac{\delta_s}{\delta} - 1 \qquad \textbf{(17.15)}$$

Commercially available compression springs often have a clash allowance of 0.40 to 0.60 (40% to 60%). If we can estimate maximum load accurately, a clash allowance of 0.20 to 0.30 can be used. If, for example, we select a clash allowance of 0.25, then the spring is protected against overloads of more than 25%. However, there is a trade-off. Springs are often used for vibration isolation or for protection of delicate equipment against shock. If the percentage of overload exceeds the clash allowance, then the spring fails to perform its intended function.

When a clash allowance is selected, we can calculate solid deflection, free height, and pitch. They are

$$\delta_s = \delta(1 + A_c) \qquad \textbf{(17.16)}$$

$$h_f = h_s + \delta_s \qquad \textbf{(17.17)}$$

$$p = d + \frac{\delta_s}{N_a} \qquad \textbf{(17.18)}$$

where δ_s = solid deflection

δ = deflection at maximum expected load

A_c = clash allowance

h_f = free height

h_s = solid height

p = pitch = axial center-to-center distance between coils when the spring is unloaded

N_a = number of active coils

EXAMPLE PROBLEM Design of a Helical Compression Spring for Static Loading

It is decided to use a helical coil spring to replace the torsion bar designed in the previous example "Design of a Torsion Bar for Given Static Deflection." The spring is to deflect 55 mm under a load of 400 N. Design the spring.

Design Decisions: We will again select steel with a yield point of 760 MPa and use a safety factor of 3. We will use a spring index of 22, so that mean radius of the spring will be roughly the same as the torque arm of the torsion bar. These design decisions will be reconsidered in the example that follows.

Solution Summary: As shown in the following detailed solution, the calculated wire diameter is 13.72 mm, and we will select a diameter of 15 mm, as with the torsion bar. Thus, the mean coil radius is 165 mm. We will need 1.92 active coils to obtain the required static deflection which corresponds to a spring rate of 7.27 N/mm. Using squared and ground ends, the total number of coils will be 3.92. The solid height is about 58.8 mm, and the free height about 127.6 mm, if we select a clash allowance of 0.25.

Detailed Solution:

Static Load:

$$P = 400$$

Shear Modulus:

$$G = 79{,}300$$

Spring Index:

$$C = 2R/d \qquad C = 22$$

where R = mean coil radius

Yield Point:

$$S_{YP} = 760$$

Safety Factor:

$$N_{FS} = 3$$

Static Deflection:

$$\delta = 55$$

Spring Rate:

$$K = \frac{P}{\delta} \qquad K = 7.273$$

Shear Working Strength:

$$S_{SW} = \frac{S_{YP}}{2 \cdot N_{FS}} \qquad S_{SW} = 126.667$$

Design Equation for Wire Diameter:

$$d = \left[\frac{8 \cdot P \cdot C}{\pi \cdot S_{SW}} \cdot \left(\frac{4 \cdot C - 1}{4 \cdot C - 4} + \frac{0.615}{C}\right)\right]^{1/2} \qquad d = 13.718$$

The diameter can be rounded to the next higher standard size, $d = 15$.

Mean Coil Radius:

$$R = C \cdot \frac{d}{2} \qquad R = 165$$

Polar-Moment-of-Inertia:

$$J = \frac{\pi \cdot d^4}{32}$$

Number of Active Coils:

$$N_a = \frac{G \cdot d}{8 \cdot C^3 \cdot K} \qquad N_a = 1.92$$

Total Number of Coils for Squared and Ground Ends:

$$N_t = N_a + 2 \qquad N_t = 3.92$$

Approximate Length of Wire:

$$L = 2 \cdot \pi \cdot R \cdot N_t \qquad L = 4.064 \cdot 10^3$$

Approximate Solid Height:

$$h_s = N_t \cdot d \qquad h_s = 58.801$$

Clash Allowance:

$$A_c = 0.25$$

Solid Deflection:

$$\delta_s = \delta \cdot (1 + A_c) \qquad \delta_s = 68.75$$

Approximate Free Height:

$$h_f = h_s + \delta_s \qquad h_f = 127.551$$

Approximate Spring Pitch:

$$p = d + \frac{\delta_s}{N_a}$$

EXAMPLE PROBLEM Evaluating and Improving a Spring Design

Evaluate the design in the preceding example for possible changes.

Design Decisions: A lighter, more compact spring design would be desirable. To this end, we will select a spring index of 6. A clash allowance of 0.25 will be used, as in the preceding design. Since the spring is protected in case of overloads in excess of 25%, we will reduce the safety factor to 1.25. Wire with a yield point strength of 1000 MPa will be specified. We expect that our design changes will result in a smaller wire diameter, which can be obtained with a higher unit strength.

Solution: The equations are the same as used in the previous example. The calculated wire diameter is 4.375 mm. We will use a 5 mm wire diameter and a mean coil radius of 15 mm. To obtain the required deflection and spring rate, we need 31.6 active coils and 33.6 total coils, resulting in a free height of 236.5 mm. The resulting design is clearly lighter and more compact than the previous design.

The preceding examples illustrate three possible solutions to the same problem. In some cases, a torsion bar uses space most efficiently, and in other cases, a helical spring fits the available space best. Compression springs with a large number of coils require internal or external restraints to prevent buckling. Long torsion bars must also be supported against buckling.

Commercially available springs are commonly specified according to outside diameter, wire diameter, approximate free height, working load, working height, solid height, and spring rate. Outside diameter is given by

$$D_O = 2R + d \tag{17.19}$$

HELICAL COMPRESSION SPRINGS: DESIGN FOR FATIGUE LOADING

In many instances, springs are subject to repeated loading. The Soderberg criterion (which was introduced in Chapter 4) can be adapated to aid in the design of these springs.

Testing and Service Loading of Spring Materials

A common test for helical spring materials loads the material in one-way torsion. This loading produces a shear stress that varies in time from zero to a maximum value as in Figure 17.4(a). On the basis of such tests, using corrections for surface finish, reliability, and so on, we estimate the one-way shear endurance limit S_{SE1}, the limiting value of fluctuating one-way shear stress for infinite life.

Figure 17.4(b) shows typical service loading which varies in time between two nonzero values. **Mean** and **range loads** are defined by

$$P_m = \frac{P_{\text{MAX}} + P_{\text{MIN}}}{2} \tag{17.20}$$

$$P_r = \frac{P_{\text{MAX}} - P_{\text{MIN}}}{2} \tag{17.21}$$

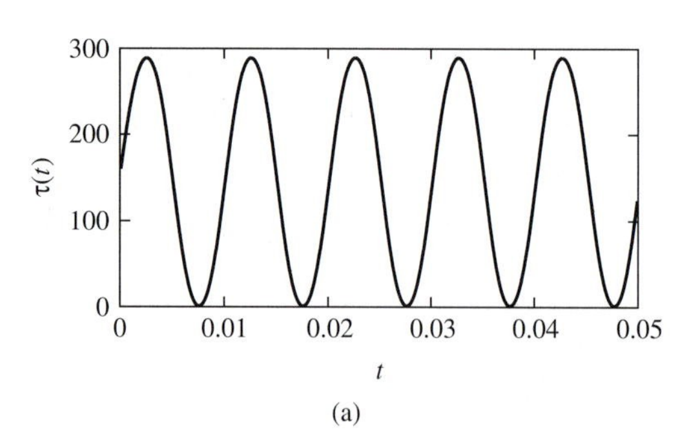

(a)

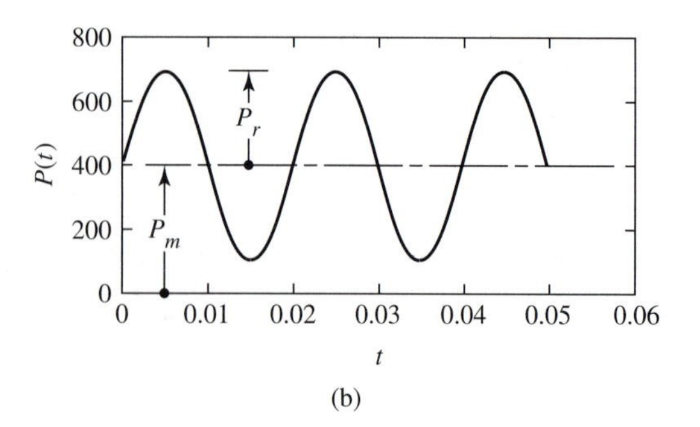

(b)

Figure 17.4 Fatigue loading of helical springs. (a) Testing of helical spring materials; (b) loading in service.

where P_m = mean load

P_r = range load

P_{MAX} = maximum load

P_{MIN} = minimum load

Allowable Mean Shear Stress

In order to obtain a safe design for fatigue loading, we will locate safe stress points on a plot of range shear stress versus mean shear stress, as shown in Figure 17.5. For zero range stress, the first safe stress point is the shear working strength, where

$$S_{SW} = \frac{S_{SYP}}{N_{FS}} \tag{17.22}$$

or

$$S_{SW} = \frac{S_{YP}}{2N_{FS}} \tag{17.23}$$

where S_{SW} = shear working strength

S_{SYP} = shear yield point strength

S_{YP} = yield point strength (in tension)

N_{FS} = factor of safety

If the shear yield point strength is known, we use Equation (17.22). If the shear yield point strength is unknown, we can use Equation (17.23).

Test values of range and mean stress both correspond to one-half the one-way shear endurance limit. Incorporating the safety factor, we define a fatigue stress number as

$$S_f = \frac{S_{SE1}}{2N_{FS}} \tag{17.24}$$

The second safe stress point is S_f,S_f on the plot of range shear stress versus mean shear stress in Figure 17.5. We now join the two points with a line. The line is the safe stress limit.

Figure 17.5 Allowable mean shear stress.

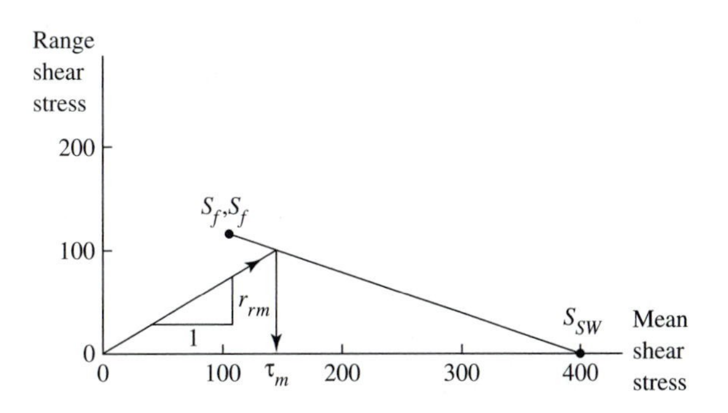

Since we have yet to design the spring, we do not know the stresses. However, the ratio of range to mean stress should be the same as the ratio of range to mean load. Thus, we define

$$r_{rm} = \frac{P_r}{P_m} \tag{17.25}$$

and draw a line with slope r_{rm} from the coordinate origin to the safe stress limit. A vertical line from the intersection point gives us the allowable mean shear stress τ_m as shown in Figure 17.5. This is the optimum value; a higher value would be unsafe, and a lower value would result in overdesigning the spring.

Calculating Optimum Wire Diameter for Fatigue Loading

The design equation for fatigue loading is similar to Equation (17.9) for static loading except that we use mean load instead of static load, and allowable mean shear stress instead of shear working strength. The result is

$$d = \sqrt{\frac{8P_mC}{\pi\tau_m}\left(\frac{4C-1}{4C-4} + \frac{0.615}{C}\right)} \tag{17.26}$$

where d = wire diameter
P_m = mean load (mean value of time-varying load)
$C = 2R/d$ = spring index
τ_m = allowable mean shear stress

Determining Spring Height

The number of active coils is based on spring rate or load and deflection, as with static loading. Maximum deflection corresponds to maximum load as follows:

$$\delta_{MAX} = \frac{P_{MAX}}{K} \tag{17.27}$$

where δ_{MAX} = maximum expected deflection (in or mm)
$P_{MAX} = P_m + P_r$ = maximum expected load (lb or N)
K = spring rate (lb/in or N/mm)

Solid deflection is based on maximum expected deflection and clash allowance.

$$\delta_s = \delta_{MAX}(1 + A_c) \tag{17.28}$$

Free height and pitch are as defined for static loads.

■■ EXAMPLE PROBLEM Graphical Determination of Allowable Mean Shear Stress for Fatigue Loading

A spring mounting system supports a weight of 400 N per spring and an additional varying load (range load) of 300 N per spring amplitude. Find allowable mean shear stress.

Decisions: We will select a material with 1000 MPa yield point in tension and 290 MPa endurance limit in one-way shear. A safety factor of 1.25 will be used.

Solution: The working strength in shear is

$$S_{SW} = S_{YP}/(2N_{FS}) = 1000/(2 \cdot 1.25) = 400 \text{ MPa}$$

The fatigue stress number is

$$S_f = S_{SE1}/(2N_{FS}) = 290/(2 \cdot 1.25) = 116 \text{ MPa}$$

We draw the safe stress limit as a line between points 400,0 and 116,116 on the coordinates of range shear stress versus mean shear stress, as in Figure 17.5.

The range-stress-to-mean-stress ratio equals the load ratio

$$r_{rm} = P_r/P_m = 300/400 = 0.75$$

We draw a line of 0.75 slope from the origin until it intersects the safe stress limit line. A careful plot will show an allowable mean shear stress of about 141 MPa. ■■

Designing the Spring Analytically

In the previous example, we determined allowable mean shear stress by plotting two lines and finding the intersection. An analytical solution is preferred because it allows us to investigate possible changes in the design with only a keystroke. For computer solutions, we write the equations of the lines and solve the equations simultaneously for allowable mean shear stress. After some rearrangement, we find

$$\tau_m = \frac{S_{SW}}{1 + r_{rm}\left(\dfrac{S_{SW}}{S_f} - 1\right)} \tag{17.29}$$

where τ_m = allowable mean shear stress

S_{SW} = working strength in shear

$r_{rm} = P_r/P_m$ = range-stress-to-mean-stress ratio

$S_f = S_{SE1}/(2N_{FS})$ = fatigue stress number

After calculating τ_m, we find wire diameter and the other spring dimensions.

■■ **EXAMPLE PROBLEM Designing a Spring for Fatigue Loads**

A spring mounting system must support a static load of 400 N per spring and an additional varying load (range load) of 300 N per spring amplitude. Spring deflection is to be 55 mm under the 400 N load. Design the springs.

Design Decisions: A spring index of 6 and a clash allowance of 0.25 will be used. We will select a material with 1000 MPa yield point in tension and 290 MPa endurance limit in one-way shear. A safety factor of 1.25 will be used.

Solution Summary: The equation for allowable mean shear stress gives us the same value that we obtained graphically. Using that value and the supported weight (mean load), we obtain the required wire diameter, which is rounded upward to 7.5 mm. For a spring index of 6, the result is a mean coil radius of 22.5 mm. Using the mean load and the static deflection under that load, we find that the spring rate is 7.273 N/mm. To obtain this spring rate, we need 47.3 active coils and a total of 49.3 coils for squared and ground ends. The solid height is about 370 mm. The deflection at maximum load is about 96 mm. Clash allowance is applied to deflection at maximum load, not static deflection. Adding a 25% clash allowance to deflection at maximum load and adding the result to the solid deflection, we obtain a free height of about 490 mm. When the spring is unloaded, the spring pitch is about 10 mm. It may be necessary to restrain the spring internally or externally to prevent buckling.

The detailed solution follows.

Detailed Solution:

Supported Load (N/spring):

$$P_m = 400$$

Range Load:

$$P_r = 300$$

Shear Modulus:

$$G = 79{,}300$$

Spring Index:

$$C = 2R/d \qquad C = 6$$

where R = mean coil radius

Yield Point:

$$S_{\mathrm{YP}} = 1000$$

Safety Factor:

$$N_{\mathrm{FS}} = 1.25$$

Endurance Limit in One-Way Shear:

$$S_{SE1} = 290$$

Static Deflection (at Mean Load):

$$\delta = 55$$

Maximum Load:

$$P_{\mathrm{MAX}} = P_m + P_r \qquad P_{\mathrm{MAX}} = 700$$

Spring Rate:

$$K = \frac{P_m}{\delta} \qquad K = 7.273$$

Deflection at Maximum Load:

$$\delta_{\mathrm{MAX}} = \frac{P_{\mathrm{MAX}}}{K} \qquad \delta_{\mathrm{MAX}} = 96.25$$

Shear Working Strength:

$$S_{SW} = \frac{S_{YP}}{2 \cdot N_{FS}} \qquad S_{SW} = 400$$

Range-stress-to-mean-stress ratio:

$$r_{rm} = \frac{P_r}{P_m} \qquad r_{rm} = 0.75$$

Fatigue Stress Number:

$$S_f = \frac{S_{SE1}}{2 \cdot N_{FS}} \qquad S_f = 116$$

Soderberg Criterion Modified for Springs:

$$\frac{S_{SW} - \tau_m}{r_{rm} \cdot \tau_m} = \frac{S_{SW} - S_f}{S_f}$$

Allowable Mean Shear Stress:

$$\tau_m = -S_f \cdot \frac{S_{SW}}{(-S_f - r_{rm} \cdot S_{SW} + r_{rm} \cdot S_f)}$$

Allowable Mean Shear Stress Equation Rearranged:

$$\tau_m = \frac{S_{SW}}{1 + r_{rm} \cdot \left(\dfrac{S_{SW}}{S_f} - 1\right)} \qquad \tau_m = 141.033$$

Design Equation for Wire Diameter:

$$d = \left[\frac{8 \cdot P_m \cdot C}{\pi \cdot \tau_m} \cdot \left(\frac{4 \cdot C - 1}{4 \cdot C - 4} + \frac{0.615}{C}\right)\right]^{1/2} \qquad d = 7.367$$

Select the diameter: $d = 7.5$.

Mean Coil Radius:

$$R = C \cdot \frac{d}{2} \qquad R = 22.5$$

Polar-Moment-of-Inertia:

$$J = \frac{\pi \cdot d^4}{32}$$

Number of Active Coils:

$$N_a = \frac{G \cdot d}{8 \cdot C^3 \cdot K} \qquad N_a = 47.325$$

Total Number of Coils for Squared and Ground Ends:

$$N_t = N_a + 2 \qquad N_t = 49.325$$

Approximate Length of Wire:

$$L = 2 \cdot \pi \cdot R \cdot N_t \qquad L = 6.973 \cdot 10^3$$

Approximate Solid Height:

$$h_s = N_t \cdot d \qquad h_s = 369.94$$

Clash Allowance:

$$A_c = 0.25$$

Solid Deflection:

$$\delta_s = \delta_{\text{MAX}} \cdot (1 + A_c) \qquad \delta_s = 120.313$$

Approximate Free Height:

$$h_f = h_s + \delta_s \qquad h_f = 490.252$$

Approximate Spring Pitch:

$$p = d + \frac{\delta_s}{N_a} \qquad p = 10.042$$

■■

HELICAL EXTENSION SPRINGS

Helical extension springs normally have tightly wound coils, with the coil at each end bent to serve as a hook. A number of extension springs are shown in Figure 17.1.

Determining Wire Diameter

The bend in the end coils results in stress concentration. As a result, we must modify design Equation (17.9) as follows:

$$d = \sqrt{\frac{8CK_fP}{\pi S_{SW}}\left[\frac{4C-1}{4C-4} + \frac{0.615}{C}\right]} \qquad \textbf{(17.30)}$$

where K_f = stress concentration factor

$S_{SW} = S_{YP}/(2N_{FS})$ = working strength in shear

P = static load

R = mean coil radius

d = wire diameter

$C = 2R/d$ = spring index

N_{FS} = safety factor

A stress concentration factor of about 1.35 can be used if the bend in the end turns is not too sharp. When a helical compression spring is overloaded, the coils will ordinarily compress to their solid height without fracturing. Extension springs lack that built-in safety

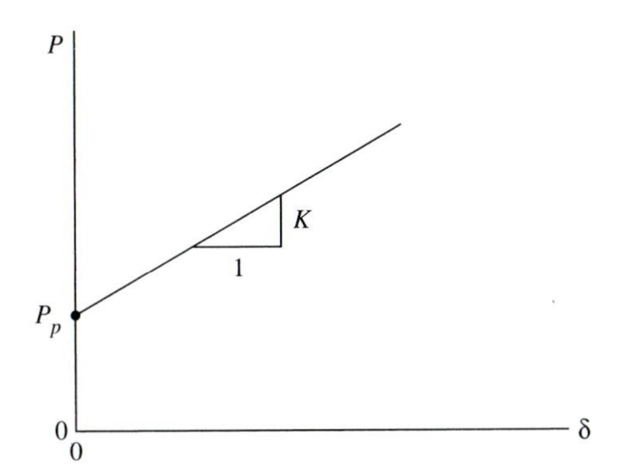

Figure 17.6 Loading of extension springs.

feature in that overload may cause catastrophic failure. Thus, we would select a higher safety factor for helical tension springs.

Spring Rate and Preload

The coils in a typical helical extension spring are tightly wound, resulting in a preload or an initial tension. For commercially available extension springs, preload is usually about 8% to 10% of the working load. The typical helical extension spring load–deflection relationship is shown in Figure 17.6. Deflection is zero for loads up to the preload P_p. Spring rate is defined by the slope of the load–deflection plot for loads exceeding the preload. For loads greater than the preload,

$$K = \frac{P - P_p}{\delta} \tag{17.31}$$

where K = spring rate
P = load
δ = corresponding deflection

Active Coils

The number of active coils required to produce a given deflection or spring rate is

$$N_a = \frac{G\delta d^4}{64(P - P_p)R^3} = \frac{Gd}{8C^3K} \tag{17.32}$$

Active coils exclude the end turns which act as hooks. Spring rates are approximate, because the end turns bend when loaded, adding to the deflection.

LEAF SPRINGS

Leaf springs are used to provide tension in switches and other assemblies. Some automobile suspension systems incorporate multileaf springs.

Cantilever Spring

Consider a beam of constant cross section, which is clamped at one end and free at the other, as in Figure 17.7(a). A load at the free end will produce a spring rate of

$$K = \frac{P}{\delta} = \frac{bh^3E}{4L^3} \tag{17.33}$$

and maximum bending stress of

$$\sigma_{\text{MAX}} = \frac{6K_fLP}{bh^2} \tag{17.34}$$

where K = spring rate (lb/in or N/mm)

P = load at free end (lb or N)

δ = deflection at that load (deflection and dimensions: in or mm)

b = beam width

h = beam thickness

E = elastic modulus (modulus and stress: psi or MPa)

σ_{MAX} = maximum tensile stress

K_f = stress concentration factor at the clamped end

Simply Supported Spring

If a beam of constant cross section is simply supported at the ends and loaded at the center as in Figure 17.7(b), then the spring rate is

$$K = \frac{P}{\delta} = \frac{4bh^3E}{L^3} \tag{17.35}$$

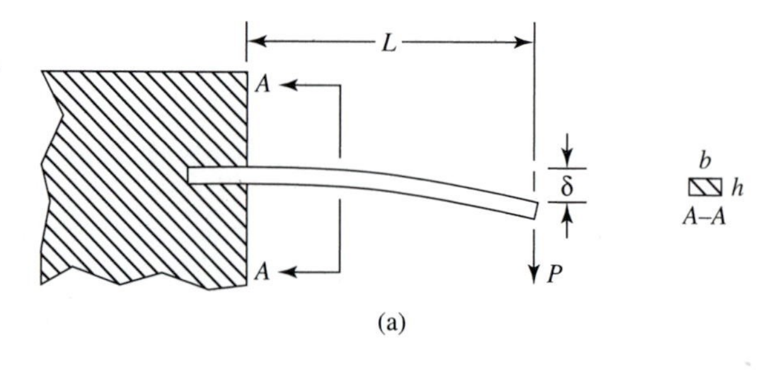

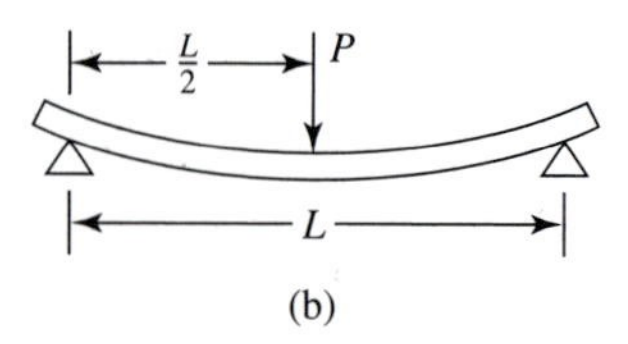

Figure 17.7 Leaf springs. (a) Clamped–free; (b) simply supported.

and maximum bending stress is

$$\sigma_{\text{MAX}} = \frac{3LP}{2bh^2} \qquad \textbf{(17.36)}$$

where P = load at center

δ = deflection at center

Multileaf Springs

The constant-cross-section cantilever and simply supported springs do not use material efficiently, because stress on the surface varies along the length of the springs. Springs of varying thickness have been tried, but this innovation results in manufacturing difficulties. Springs of varying width present a space problem.

Multileaf springs, illustrated in Figure 17.8, are another attempt at efficient material use. One-half the load is supported by each end of the spring when it is used in automobile suspension systems. The leaves of typical vehicle springs are thinner than the proportions in the sketch, and they are usually curved when unloaded. The leaves have graduated length so that the number of leaves at any location on the spring is roughly proportional to the bending moment at that location. The leaves are lubricated, allowing them to slide on one another. The spring rate is

$$K = \frac{P}{\delta} = \frac{8n_L bh^3 E}{3L^3} \qquad \textbf{(17.37)}$$

and maximum bending stress is

$$\sigma_{\text{MAX}} = \frac{3LP}{2n_L bh^2} \qquad \textbf{(17.38)}$$

where n_L = number of leaves

b = width of a single leaf

h = thickness of a single leaf

Note that the spring rate is lower than that of a single simply supported spring of width $n_L b$. Maximum stress is the same. Thus, the spring is softer and lighter than a comparable simply supported spring of width $n_L b$. Both spring rate and maximum stress values are approximate.

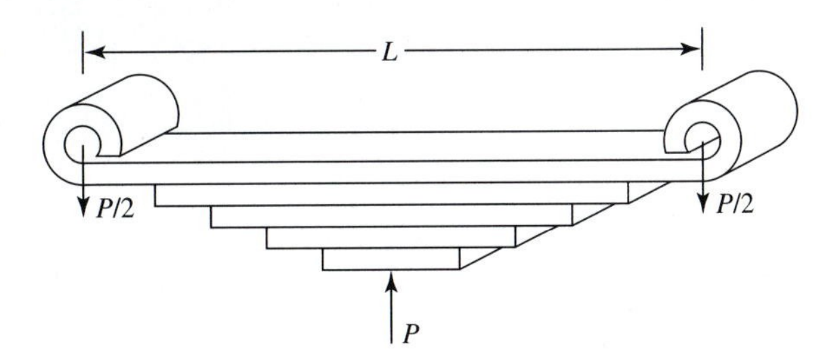

Figure 17.8 Multileaf spring (not to scale).

AIR SPRINGS

A typical air spring is a rubber and fabric bellows. The load is supported by compressed air. A variety of air spring configurations are shown in Figure 17.9.

Figure 17.9 Air springs. (Courtesy of Firestone Industrial Products Company)

Air springs are used in truck and trailer suspensions and in passenger rail cars. Air helper springs are used in conjunction with steel leaf springs to provide a more controlled ride. They allow side-to-side and front-to-back leveling of vehicles by varying the air pressure in the springs. Air helper springs are used in pickup trucks, vans, recreational vehicles, tow trucks, ambulances, and small buses. In industrial applications, air springs can be used as actuators and lifting devices for materials handling. Industrial and research uses of air springs include vibration and shock isolation of lasers, spectrometers, anechoic chambers, holographs, forging hammers, vibratory conveyors, and other industrial machinery.

The natural frequency of air spring isolators is only slightly sensitive to variations in pressure and load. However, the spring rate is not constant but depends on the change in effective area, volume, and pressure. Spring rate can be determined from calculations based on effective area as published in manufacturers' catalogs and design guides (e.g., Firestone, 1992).

DISC SPRINGS

Disc springs, also called Belleville springs, are conical, washerlike springs. They are used for high loads when space is limited. Disc springs are available in high carbon spring steel, alloy steel, stainless and heat-resisting steel, and nonferrous materials.

Disc springs of various sizes are shown in Figure 17.10. They are often stacked to produce the effect of a harder spring (higher spring rate) or a softer spring (lower spring rate). When stacking, it is preferable to guide the springs with a rod through the inside diameter. As the outside diameter changes under loading, a sleeve over the outside diameter could bind. The ends to the stack should contact a hardened thrust washer at the spring outer diameter, as shown in Figure 17.11. It is particularly important that the spring at the moving end of the stack contact a hardened thrust washer at its outer diameter.

Figure 17.10 Disc springs. (Courtesy of Rolex Company, National Disc Spring® Division)

Figure 17.11 Guiding disc springs. (Courtesy of Rolex Company, National Disc Spring® Division)

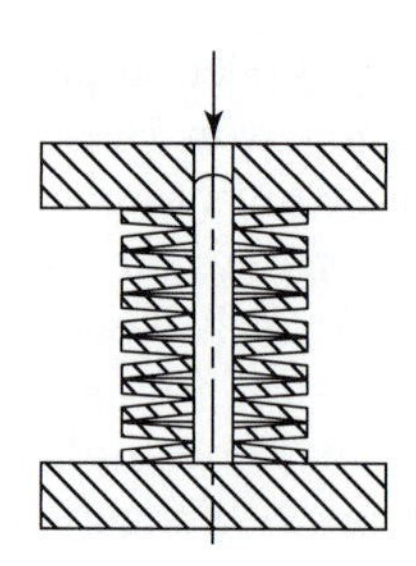

Load and Deflection

The load-versus-deflection relationship is complicated. The following equations are given by Almen and Lazzlo (1936) and Rolex (1991):

$$P = \frac{E \cdot \delta}{(1 - \nu^2) \cdot M \cdot R^2} \cdot \left[\left(h - \frac{\delta}{2}\right) \cdot (h - \delta) \cdot t + t^3\right] \quad \textbf{(17.39)}$$

where $M = \dfrac{6}{\pi \cdot \ln \alpha} \cdot \dfrac{(\alpha - 1)^2}{\alpha^2}$

In the flattened condition, the deflection δ is equal to the **conical height** h, and the equation becomes

$$P_f = \frac{E \cdot h \cdot t^3}{(1 - \nu^2) \cdot M \cdot R^2} \quad \textbf{(17.40)}$$

where the terms in the equations refer to Figure 17.12 and are as follows:

O.D. = maximum outside diameter (upper surface)
I.D. = minimum inside diameter (bottom surface)
h = conical disc height (cone height)
O.H. = overall height = $Y + h$
t = actual thickness of disc
β = cone angle of disc
R = radius from centerline to load bearing circle (bottom surface)
M = ratio factor
ν = Poisson's ratio (0.3 for steel)
E = Young's modulus (30,000,000 for steel)
δ = deflection of disc

Figure 17.12 Disc spring dimensions and nomenclature. (Courtesy of Rolex Company, National Disc Spring® Division)

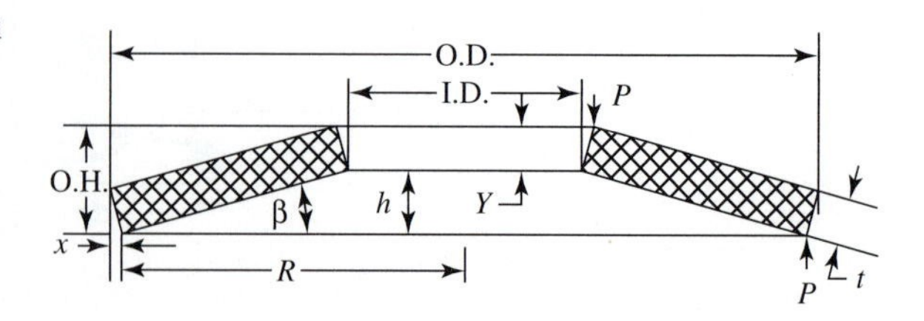

α = ratio of diameters (O.D./I.D.)

P = load at a given deflection (lb)

P_f = load at flat (lb)

$X = \sin \beta \cdot t$

$Y = \cos \beta \cdot t$

Stacking Disc Springs in Parallel and Series

When the disc springs in a stack all face in the same direction, as in Figure 17.13(a), we call this **parallel stacking.** Spring rates are additive for parallel stacking. That is, a parallel stack of six identical disc springs will produce a spring rate of $6K$ where K is the spring rate of a single disc spring.

In series stacking, the disc springs face in alternating directions, with the small ends contacting one another and the large ends contacting one another, as in Figure 17.13(b). The

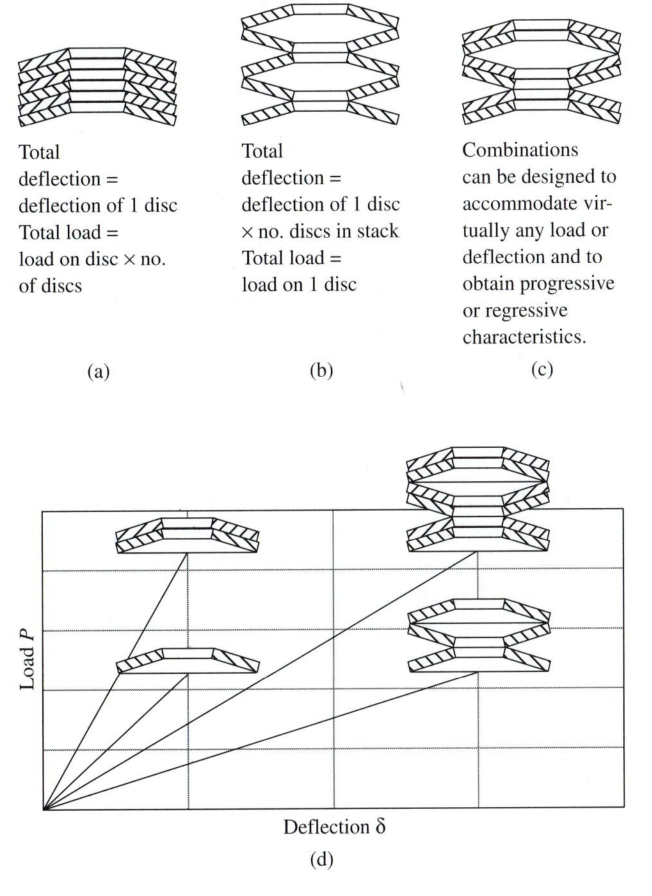

Figure 17.13 Response of disc springs (a) in parallel; (b) in series; (c) in parallel-series. (d) Plot of load versus deflection. (Courtesy of Rolex Company, National Disc Spring® Division)

spring rate of the stack is the reciprocal of the sum of the reciprocals of the individual spring rates.

$$K_{\text{series}} = \frac{1}{\dfrac{1}{K_1} + \dfrac{1}{K_2} + \dfrac{1}{K_3} + \cdots} \tag{17.41}$$

For example, for five identical disc springs stacked in series as in Figure 17.13(b), the spring rate is

$$K_{\text{series}} = K/5$$

where K = spring rate of one disc spring

EXAMPLE PROBLEM Parallel-Series Stacking for a Given Spring Rate

Disc springs are available with spring rate K. Obtain a spring rate of $2/3K$.

Design Decision: Instead of ordering thinner springs, we will stack the available springs.

Solution: A pair of parallel springs yields a spring rate of $2K$. Three such pairs in series, stacked as in Figure 17.13(c), yield a spring rate of

$$K_{\text{combined}} = 2K/3$$

See Figure 17.13(d), which shows the effect of various combinations on spring rate.

Spring Linearity

Springs with a linear load-versus-deflection relationship are called **linear springs.** That is, linear springs have a constant spring rate for all loads up to the design load. The linearity of disc springs depends on the height-to-thickness ratio h/t. For $h/t \leq 0.4$, disc springs are nearly linear. Disc springs with a height-to-thickness ratio of $h/t \approx 1$ are softening springs. That is, the spring rate decreases with increasing load. For $h/t > 1.4$, the disc spring may snap through.

Shock mounts are often designed as hardening springs. The spring rate of a hardening spring increases with increasing load. This property is beneficial if a wide range of loads will be encountered. Disc springs of different thicknesses can be stacked in series, as in Figure 17.14(a). The thinner discs will compress flat with increasing load, reducing the effective number of discs in the stack. The spring rate of the stack then jumps to a higher value as one after another disc spring is compressed flat. Part (b) of the figure shows another option, using disc springs of equal thickness.

A hardening helical compression spring can be designed by varying the pitch along the height of the spring. As the load is increased, the closest coils clash, decreasing the number of active coils and increasing the spring rate.

VIBRATION TRANSMISSIBILITY AND ISOLATION

Shock and vibration loading can damage delicate equipment or interfere with the operation of sensitive instruments and equipment. In addition, vibration causes human discomfort and may result in audible noise.

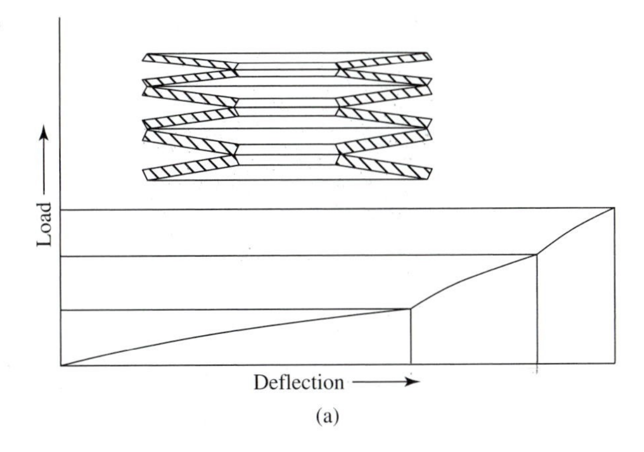

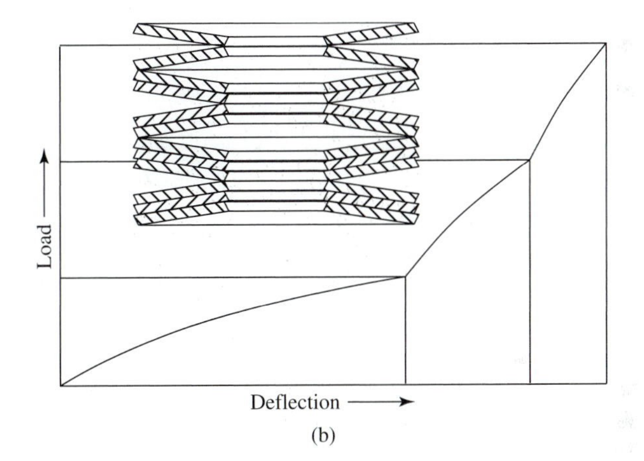

Figure 17.14 Combining disc springs to obtain a hardening spring. (Courtesy of Rolex Company, National Disc Spring® Division)

Natural Frequency

Figure 17.15(a) is a schematic of a single-degree-of-freedom system with negligible damping. If the mass is displaced, it will tend to vibrate harmonically at circular frequency

$$\omega_n = \sqrt{\frac{K}{M}} \tag{17.42}$$

where ω_n = undamped natural circular frequency (rad/s)

$f_n = \dfrac{\omega_n}{2\pi}$ = undamped natural frequency (Hz, where 1 Hz = 1 cycle/s)

$M = w/g$ = supported mass (lb · s²/in or kg)

w = weight (lb)

$g = 386\ \text{in/s}^2$ = acceleration of gravity

K = spring rate (lb/in or N/m)

Figure 17.15 One-degree-of-freedom vibrating systems. (a) Free vibration; (b) base-excited; (c) mass-excited.

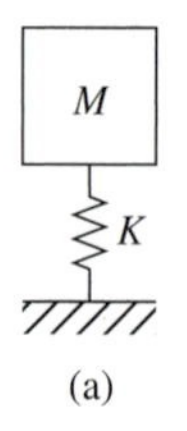

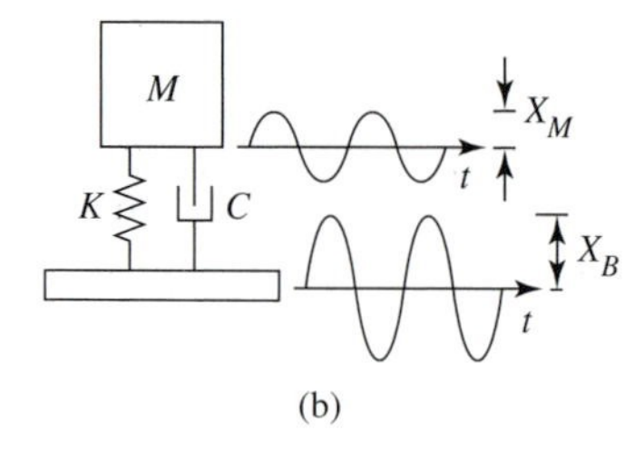

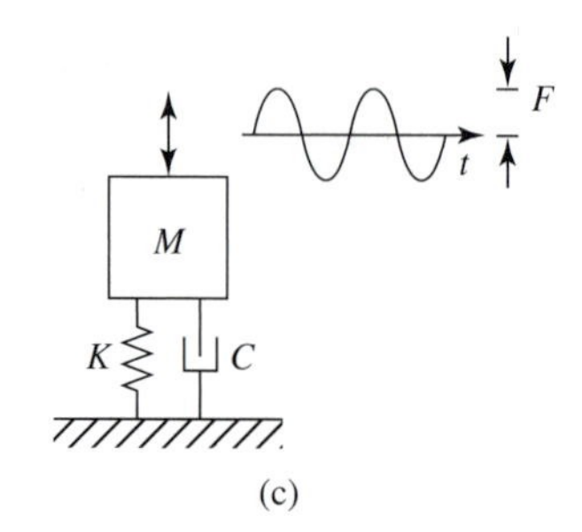

Vibration Damping

Energy loss, called **damping,** is inherent in all vibrating systems. Loss mechanisms include internal hysteresis in springs, air friction as a mass moves, and solid-to-solid friction as between disc springs. The actual resistance to motion can be a constant force, or a force proportional to velocity or velocity squared, or a combination of these. In each case, the force direction opposes the motion. A viscous damping force proportional to velocity is commonly assumed because it leads to the most convenient mathematical representation.

Vibration and Shock Isolation

Isolation involves using mountings to reduce the transmission of vibration and shock from one point to another. Springs are used in shock mounts to protect against sudden impact loads. Typical vibration isolation problems involve base-excited and mass-excited systems.

Base-excited Systems. A base-excited system includes a mass that we wish to isolate from vibrations in its supporting structure, for example, the floor. Spring mounts can be used to protect base-excited systems. Examples of precise equipment and delicate instrumentation requiring isolation from floor or structural vibrations include navigation systems, laser cutting apparatus, electron microscopes, precision measuring equipment, and precision machinery.

Figure 17.15(b) shows a schematic of a base-excited, mass spring damper system, where

C = damping coefficient (lb · s/in or N · s/m)
X_B = forced vibration amplitude of the base
X_M = vibration amplitude response of the mass

The motion of the base is given by

$$x_B = X_B \cos(\omega t) \tag{17.43}$$

where x_B = base position at any time
$\omega = 2\pi f$ = circular frequency of the forcing function (rad/s)
f = frequency of forcing function (Hz)
t = time (s)

Motion Transmissibility

The ratio of steady-state vibration amplitude of the mass to the vibration amplitude of the base is the transmissibility. It is given by

$$TR = \sqrt{\frac{1 + (2r_d r_f)^2}{(1 - r_f)^2 + (2r_d r_f)^2}} \tag{17.44}$$

where $TR = X_M/X_B$ = transmissibility
$r_d = C/C_c$ = damping ratio
$C_c = 2(KM)^{1/2}$ = critical damping coefficient
$r_f = \omega/\omega_n = f/f_n$ = frequency ratio

Force Transmissibility

Another common vibration problem is the transmission of vibratory forces to a floor or other supporting structure. Heavy machinery can transmit unacceptable forces to a factory floor, interfering with the operation of other machinery. Unbalanced forces in consumer products present a similar problem. A refrigerator, for example, cycles on and off, day and night. The forces transmitted to the floor of a residence must be reduced to a level that does not interfere with sleep. In this case, we define transmissibility as the ratio of transmitted force to the forcing function applied to the mass.

$$TR = \frac{F_{tr}}{F} \tag{17.45}$$

where TR = transmissibility as given in the motion transmissibility Equation (17.44)
F_{tr} = force amplitude transmitted to the supporting structure (lb or N)
F = forcing function amplitude (lb or N)

Other terms are defined as in Equation (17.44).

A schematic of a mass-excited system is shown in Figure 17.15(c).

EXAMPLE PROBLEM Motion Transmissibility

It is necessary to isolate a precision instrument from vibration of the factory floor. Make general recommendations for design or selection of vibration mounts.

Decisions: We will examine various frequency ratios and damping ratios.

Solution: As shown in Figure 17.16, we plot transmissibility against frequency ratio for damping ratios of 0.1, 0.2, 0.4, and 0.8. Note that there is amplification in the region $0 < r_f < 2^{1/2}$. Occasionally, we wish to amplify vibrations. For example, if we wanted to shake railroad cars to empty them, we would operate near resonance, that is, near $r_f = 1$.

In this case, we want isolation. The isolation region is $r_f > 2^{1/2}$, with higher values of r_f providing better isolation. That is, we will specify soft mounts for the precision instrument, so that its natural frequency is well below the forcing frequency of the factory floor.

Figure 17.16 Transmissibility.

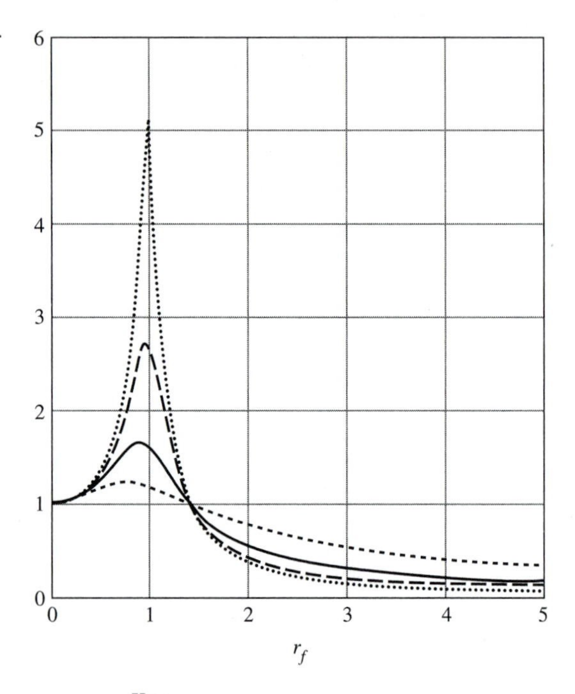

Specification of damping is a trade-off. A low damping ratio results in the lowest transmissibility, provided that we are in the isolation region. However, it is assumed that the vibrations of the factory floor result from vibration of a nearby machine. During machine runup, our delicate instrument will be subject to resonance conditions. Lightly damped systems have a high amplification factor at resonance. Steel compression springs alone have very low damping. If we selected steel springs, we would probably use snubbers to add damping and limit vibration amplitudes during runup of nearby machinery.

EXAMPLE PROBLEM Force Transmissibility

An unbalanced piece of machinery is rigidly mounted to the floor, causing unacceptable structural vibration. We must limit transmitted force to 15% of the forcing function. Find the required frequency ratio.

Design Decision: We will specify a damping ratio of 0.1.

Solution Summary: The transmissibility is set equal to 0.15, and the equation is solved for frequency ratio. A numerical solution yields a frequency ratio of $r_f = 2.95$. That is, we must select a mounting system so that the natural frequency of the mass is not greater than $f/2.95$ where f = forcing function frequency. The detailed solution follows.

Detailed Solution:

Transmissibility:

$$TR(r_f, r_d) = \sqrt{\frac{1 + (2 \cdot r_d \cdot r_f)^2}{(1 - r_f^2)^2 + (2 \cdot r_d \cdot r_f)^2}}$$

where TR = ratio of transmitted force amplitude to forcing function amplitude

Damping Ratio:

$$r_d = 0.1$$

Find the required frequency ratio for a transmissibility of 0.15. Estimate $r_f = 2$. Then

$$r_{f1} = \text{root}\left[\sqrt{\frac{1 + (2 \cdot r_d \cdot r_f)^2}{(1 - r_f^2)^2 + (2 \cdot r_d \cdot r_f)^2}} - 0.15, r_f\right]$$

$$r_{f1} = 2.953$$

The transmissibility equation was solved numerically in the preceding example. This method is the easiest one, if we have mathematical software. If we prefer an analytical solution, we can work on the equation to obtain a fourth-order equation in r_f (with odd-order terms missing). That is, we have a quadratic equation with r_f^2 as a variable. Finally, this equation can be solved with the quadratic formula.

Selection of Spring Mounts to Limit Motion or Force Transmissibility

In most cases, we use several springs on a system for isolation. A machine or an instrument is commonly supported at each corner by spring mounts. In the simplest vibration case, the springs act in parallel; that is, they deflect equally during vibration. Then we simply divide the total mass by the number of springs and solve the problem as if each spring supported a fraction of the load. The natural frequency equation is

$$f_n = \frac{\omega_n}{2\pi} = \frac{1}{2\pi}\sqrt{\frac{K}{M}} \tag{17.46}$$

from which we find spring rate

$$K = 4\pi^2 f_n^2 M \tag{17.47}$$

where f_n = system natural frequency (Hz)

K = spring rate of a single spring (lb/in or N/m)

M = mass supported by a single spring (lb · s²/in or kg)

The illustrations above refer to simple single-degree-of-freedom systems, an important special case. Many vibration problems involve structures with significant elasticity and motion that cannot be described by assuming that springs act in parallel. These more general problems require detailed analysis. References that may be of interest include James et al. (1989); Steidel (1989); Timoshenko, Young, and Weaver (1974); and Wilson (1994).

EXAMPLE PROBLEM Spring Selection to Limit Transmissibility

A 190 lb machine operating at 960 rpm transmits unacceptable forces to the building in which it operates. Solve the problem.

Design Decisions: We will specify a mounting system that will transmit only 1/16 of the unbalanced machine force to the building. We will use four vibration mounts and will specify a damping ratio of 0.11.

Solution Summary: A numerical solution of the transmissibility equation (given in the detailed solution which follows) yields a frequency ratio of $f/f_n = 4.935$. Dividing machine speed by 60, we find the forcing function frequency of $f = 16$ Hz, from which the required natural frequency is 3.24 Hz. Dividing the total supported weight by the acceleration of gravity and the number of mounts, we find that each mount supports a mass of 0.123 lb · s²/in. Using Equation (17.47) for spring rate based on natural frequency, we find a required spring rate of 51.07 lb/in for each mount and a static deflection of 0.93 in. In order to realize the assumed damping ratio, we must specify a damping coefficient of each mount of 0.55 lb · s/in.

Detailed Solution:

Transmissibility Equation:

$$TR(r_f, r_d) = \sqrt{\frac{1 + (2 \cdot r_d \cdot r_f)^2}{(1 - r_f^2)^2 + (2 \cdot r_d \cdot r_f)^2}}$$

where TR = ratio of transmittted force amplitude to forcing function amplitude

Total Supported Weight:

$$w_t = 190$$

Gravity:

$$g = 386$$

Number of Parallel Springs:

$$N_S = 4$$

Supported Mass per Spring:

$$M = \frac{w_t}{g \cdot N_S} \qquad M = 0.123$$

Damping Ratio:

$$r_d = 0.11$$

Forcing Function Frequency:

$$f = 16$$

Find the required frequency ratio for a transmissibility of $TR = 1/16$. Estimate $r_f = 2$. Then

$$r_{f1} = \text{root}\left[\sqrt{\frac{1 + (2 \cdot r_d \cdot r_f)^2}{(1 - r_f^2)^2 + (2 \cdot r_d \cdot r_f)^2}} - TR, r_f\right]$$

$$r_{f1} = 4.935$$

Maximum Natural Frequency:

$$f_n = \frac{f}{r_{f1}} \qquad f_n = 3.242$$

Maximum Spring Rate (One Spring):

$$K = 4 \cdot \pi^2 \cdot f_n^2 \cdot M \qquad K = 51.07$$

Static Deflection:

$$\delta = \frac{w_t}{K \cdot N_S} \qquad \delta = 0.93$$

Check of Natural Frequency:

$$f_n = \frac{1}{2 \cdot \pi} \cdot \sqrt{\frac{g}{\delta}} \qquad f_n = 3.242$$

Critical Damping Coefficient:

$$C_c = 2 \cdot \sqrt{K \cdot M} \qquad C_c = 5.014$$

Damping Coefficient:

$$C = C_c \cdot r_d \qquad C = 0.552$$

■■

References and Bibliography

Airstroke Actuators/Airmount Isolators Engineering Manual and Design Guide, EMDG492. Carmel, IN: Firestone Industrial Products Company, 1992.

Almen, J., and A. Lazzlo. *Transactions.* New York: American Society of Mechanical Engineers, May 1936.

James, M. L., G. M. Smith, J. C. Wolford, and P. W. Whaley. *Vibration of Mechanical and Structural Systems.* New York: Harper and Row, 1989.

National Disc Springs. Hillside, NJ: Rolex Company, 1991.

Steidel, R. F. *Introduction to Mechanical Vibrations.* 3d ed. New York: John Wiley, 1989.

Timoshenko, S. P., D. H. Young, W. Weaver. *Vibration Problems in Engineering.* 4th ed. New York: John Wiley, 1974.

Wilson, C. E. *Noise Control—Measurement, Analysis and Control of Sound and Vibration.* Malabar, FL: Krieger Publishing, 1994.

Design Problems

17.1. Design a torsion bar for a deflection of 2 in under a static load of 80 lb.

Design Decisions: Use a torque arm of 5 in. Select steel with a yield point of 105,000 psi, and use a safety factor of 2.5.

17.2. Design a torsion bar for a rotation of 0.6 rad under a static load of 1000 lb at a torque arm of 4 in.

Design Decisions: Select steel with a yield point of 120,000 psi, and use a safety factor of 4.

17.3. Design a torsion bar for a deflection of 40 mm under a static load of 2500 N.

Design Decisions: Use a torque arm of 90 mm. Select steel with a yield point of 800 MPa, and use a safety factor of 1.75.

17.4. Design a torsion bar for a rotation of 40° under a static load of 3300 N at a torque arm of 200 mm.

Design Decisions: Select steel with a yield point of 900 MPa, and use a safety factor of 2.25. Consider the effect of specifying rod diameters larger than the calculated value.

17.5. Design a spring for a deflection of 1.5 in under a static load of 800 lb.

Design Decisions: We will design a helical compression spring with a spring index of 6 and a clash allowance of 0.25. We will select wire with a yield point of 120,000 psi and a safety factor of 1.25.

17.6. Design a compression spring for a deflection of 2 in under a static load of 80 lb.

Design Decisions: We will design a helical compression spring with a spring index of 21.6 and a clash allowance of 0.25. We will select wire with a yield point of 105,000 psi and a safety factor of 2.5.

17.7. Design a spring for a deflection of 1.5 in under a static load of 960 lb.

Design Decisions: We will design a helical compression spring with a spring index of 6 and a clash allowance of 0.25. We will select wire with a yield point of 120,000 psi and a safety factor of 1.5.

17.8. A spring mounting system must support a static load of 800 lb per spring and an additional varying load (range load) of 300 lb per spring amplitude. Spring deflection is to be 1.5 in under the 800 lb load. Design the springs.

Decisions: A spring index of 6 and a clash allowance of 0.25 will be used. We will select a material with 120,000 psi yield point in tension and 40,000 psi endurance limit in one-way shear. A safety factor of 1.25 will be used.

17.9. A spring mounting system must support a static load of 80 lb per spring and an additional varying load (range load) of 50 lb per spring amplitude. Spring deflection is to be 2 in under the 80 lb load. Design the springs.

Decisions: A spring index of 10 and a clash allowance of 0.25 will be used. We will select a material with 120,000 psi yield point in tension and 40,000 psi endurance limit in one-way shear. A safety factor of 1.25 will be used.

17.10. A spring mounting system must support a static load of 960 lb per spring and an additional varying load (range load) of 480 lb per spring amplitude. Spring deflection is to be 2 in under the 960 lb load. Design the springs.

Decisions: A spring index of 6 and a clash allowance of 0.25 will be used. We will select a material with 120,000 psi yield point in tension and 40,000 psi endurance limit in one-way shear. A safety factor of 1.25 will be used.

17.11. A spring mounting system must support a static load of 550 N per spring and an additional varying load (range load) of 200 N per spring amplitude. Spring deflection is to be 25 mm under the 550 N load. Design the springs.

Decisions: A spring index of 8 and a clash allowance of 0.25 will be used. We will select a material with 1000 MPa yield point in tension and 290 MPa endurance limit in one-way shear. A safety factor of 1.25 will be used.

17.12. It is necessary to isolate a precision instrument from vibration of the factory floor. We wish to investigate possible solutions. Plot transmissibility for frequency ratios up to 4 and damping ratios of 0.05, 0.10, 0.15, and 0.20.

17.13. A rigidly mounted appliance transmits unacceptable vibratory force to the floor. We wish to obtain general information on transmissibility in preparation for solving the problem. Plot transmissibility for frequency ratios $0 \leq r_f \leq 4$ and damping ratio $r_d = 0.25, 0.50, 0.75$, and 1.00.

17.14. An unbalanced piece of machinery is rigidly mounted to the floor, causing unacceptable structural vibration. We must limit transmitted force to 20% of the forcing function. Find the required frequency ratio.

Design Decision: We will use mounts that will yield an approximate damping ratio of 0.2.

17.15. Floor vibrations result in unacceptable motion of a precision machine. We must limit motion of the machine to 20% of the floor vibration. Find the required frequency ratio.

Design Decision: We will use mounts that will yield an approximate damping ratio of 0.1.

17.16. An unbalanced piece of machinery is rigidly mounted to the floor, causing unacceptable structural vibration. We must limit transmitted force to 25% of the forcing function. Find the required frequency ratio.

Design Decision: We will use mounts that will yield an approximate damping ratio of 0.05.

17.17. A 350 lb machine transmits unacceptable forces to the building in which it operates. The forced vibration frequency is 18 Hz. Solve the problem. Find required spring rate and damping coefficient for each spring. Find static deflection.

Design Decisions: We will specify a mounting system that will transmit only 1/12 of the unbalanced machine force to the building. We will use six springs and will specify a damping ratio of 0.1.

17.18. A 50 lb machine transmits unacceptable forces to the building in which it operates. The forced vibration frequency is 25 Hz. Solve the problem. Find required spring rate and damping coefficient for each spring. Find static deflection.

Design Decisions: We will specify a mounting system that will transmit only 1/15 of the unbalanced machine force to the building. We will use four springs and will specify a damping ratio of 0.08.

17.19. A 500 lb machine transmits unacceptable forces to the building in which it operates. The forced vibration frequency is 10 Hz. Solve the problem. Find required spring rate and damping coefficient for each spring. Find static deflection.

Design Decisions: We will specify a mounting system that will transmit only 5% of the unbalanced machine force to the building. We will use four springs and will specify a damping ratio of 0.12.

17.20. A factory floor is vibrating at a frequency of 15 Hz. A 250 lb testing device must be isolated from the floor. Specify a suitable mounting system.

Design Decisions: We will allow a motion transmissibility of 15% and will specify a damping ratio of 0.05. Four mounts will be used.

17.21. A factory floor is vibrating at 1800 cycles/min. A 440 lb precision machine must be isolated from the floor. Select a suitable mounting system. Specify spring rate and damping coefficient.

Design Decisions: We will allow a motion transmissibility of 25% and will specify a damping ratio of 0.1. Four mounts will be used.

CHAPTER 18

Bearings and Lubrication

Symbols

a	ball bearing life adjustment factor	l	bearing length
c	radial clearance	L_{10r}	predicted bearing life (revolutions) with 10% failure probability
C	basic load rating	n	rotation speed (rpm)
d_I	inner race diameter	p_n	nominal bearing pressure
d_O	outer race diameter	P	radial load on a bearing
D	ball diameter	P_{hpf}	power loss due to friction (hp)
e	eccentricity	P_{kWf}	power loss due to friction (kW)
e_r	eccentricity ratio	P_1	pressure number
E_K	kinetic energy	r	bearing radius
f	coefficient of friction	r_d	diameter ratio for ball bearings
F_f	friction force	R	reliability
g	ball bearing geometry factor	S	Sommerfeld number
h_{MIN}	minimum film thickness	T	friction torque
h_r	reciprocal film thickness ratio	T_1	friction torque number
H_f	heat generated due to friction	z	absolute viscosity (cP)
H_{10}	predicted bearing life (hours) with 10% failure probability	μ	absolute viscosity (reyn)

Units[1]

absolute viscosity	reyn ($lb \cdot s/in^2$) or cP where 1 cP = 10^{-3} $N \cdot s/m^2$
dimensions, clearance, and eccentricity	in or m
force	lb or N
heat generated	Btu/h
kinetic energy	in · lb or N · m
predicted life	h or rev
torque	in · lb or N · m

[1]Many of the terms are dimensionless ratios.

The invention of the wheel must have been accompanied by the invention of some type of bearing. Lubrication of bearings and sliding members has been practiced since ancient times. Dorinson and Ludima (1985) note that a chariot found in a tomb of circa 1400 B.C. still had some of the original lubricant on its axle. The most common bearing types are ball bearings, roller bearings, and journal bearings or sleeve bearings. Bearings assembled in housings are called **pillow blocks.** Some examples are shown in Figure 18.1.

FRICTION

Friction is the force that resists sliding motion or the tendency toward motion of one body relative to another. Friction enables us to walk and is necessary for the operation of vehicles and machinery. However, bearing friction consumes energy and generates heat which must be dissipated.

Figure 18.1 An assortment of bearings. (Courtesy of Rexnord Corporation, Link-Belt Bearing Division)

Coefficient of Friction

The coefficient of friction is the ratio of the friction force to the normal force between two sliding bodies or two bodies at impending relative motion. It is given by

$$f = \frac{F_f}{F_n} \tag{18.1}$$

where f = coefficient of friction

F_f = friction force tangent to direction of relative motion

F_n = force normal to the common surface of the two bodies

If friction force is measured as the bodies are sliding, f is called the **coefficient of kinetic friction** or the **coefficient of sliding friction.** If there is no relative motion, and friction force is measured at impending motion, then f is called the **coefficient of static friction.** For solid-to-solid contact, the coefficient of static friction is usually higher than the coefficient of dynamic friction. A friction force exists only if a tangential force is applied to the bodies. If we multiply the normal force by the coefficient of friction, we obtain the friction force at impending sliding, that is, the available resistance to motion.

Friction and Dissipation of Energy

Kinetic energy is given by

$$E_K = \frac{Mv^2}{2} + \frac{J\omega^2}{2} \tag{18.2}$$

where E_K = kinetic energy (lb · in or N · m)

M = mass of body in rectilinear motion (lb · s^2/in or kg)

v = velocity of body in rectilinear motion (in/s or m/s)

J = mass-moment-of-inertia of rotating parts (lb · s^2 · in or kg · m^2)

ω = angular velocity of rotating parts (rad/s)

Brakes for stopping vehicles and machinery are examples of beneficial effects of friction. Kinetic energy is converted to heat by braking friction. When a vehicle is skidding, or if skidding is impending, then the friction force equals the vehicle weight times the appropriate tire-to-roadway coefficient of friction. Work due to friction equals friction force times distance traveled while braking. When we are stopping a vehicle, friction within the brakes and tire-to-roadway friction both work in our favor. When a brake is stopping rotating machinery, braking capacity may be limited by geometry and coefficient of friction of the brake and by the brake's capacity to dissipate heat. Vehicle stopping distances are usually controlled by the coefficient of friction of the tires on the roadway. Time to stop a vehicle, stopping distance, required braking torque, and heat generated in braking were considered in Chapter 13.

BALL BEARINGS

Ball bearings are selected for many applications where a reduction of friction is important. In bicycles, for example, ball bearings are used because of the importance of minimizing power loss.

Figure 18.2 identifies some components of typical ball bearing units. The balls in a ball bearing roll on the inner and outer races. The inner race appears concave in a cross section through the bearing axis and convex in a cross section through the plane of the bearing. The outer race appears concave in both major cross sections. Thus, contact between a ball and the inner race is critical to ball bearing stresses and ball bearing life.

Load Distribution

The theoretical load distribution in a ball bearing is shown in Figure 18.3. Based on the early work of Stribeck and Hertz (*New Departure Handbook,* 1951), the total compression of a ball and its races varies as the 2/3 power of the normal load. Permissible static loading on a ball is proportional to the square of the ball diameter. As the balls are not loaded in tension, no more than half of the balls can be loaded at one time, and these balls do not share the load equally.

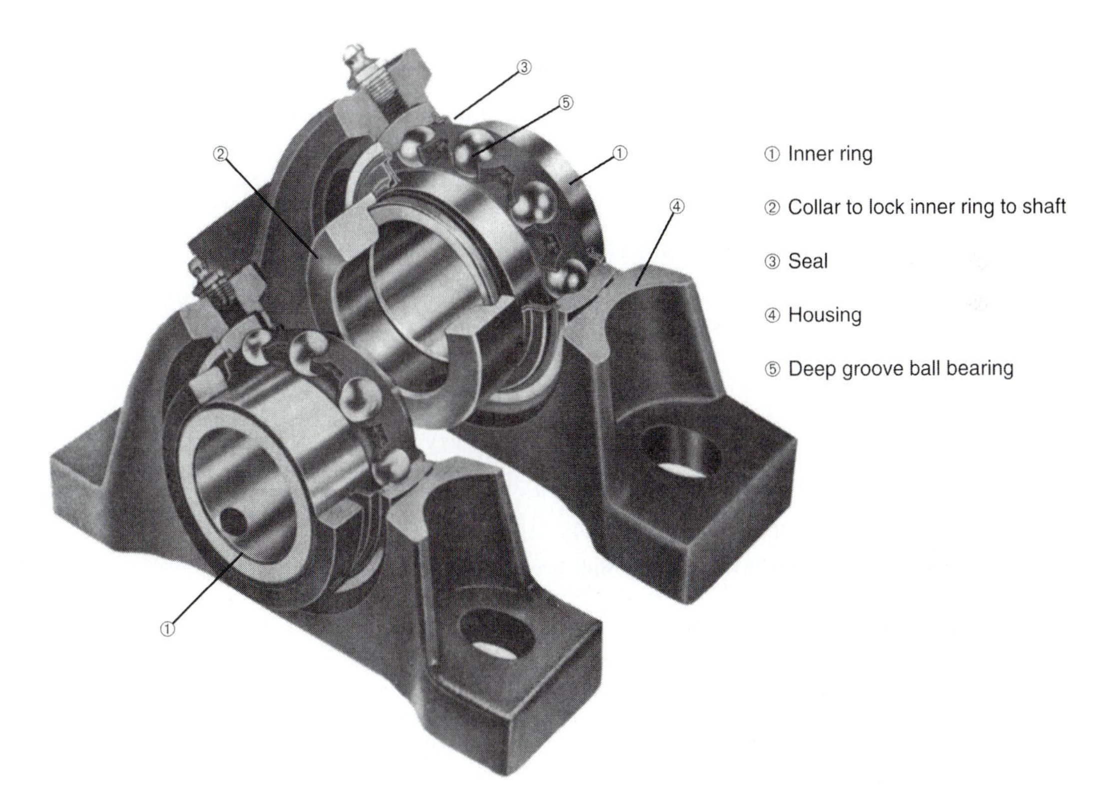

Figure 18.2 Some ball bearing components. (Courtesy of Rexnord Corporation, Link-Belt Bearing Division)

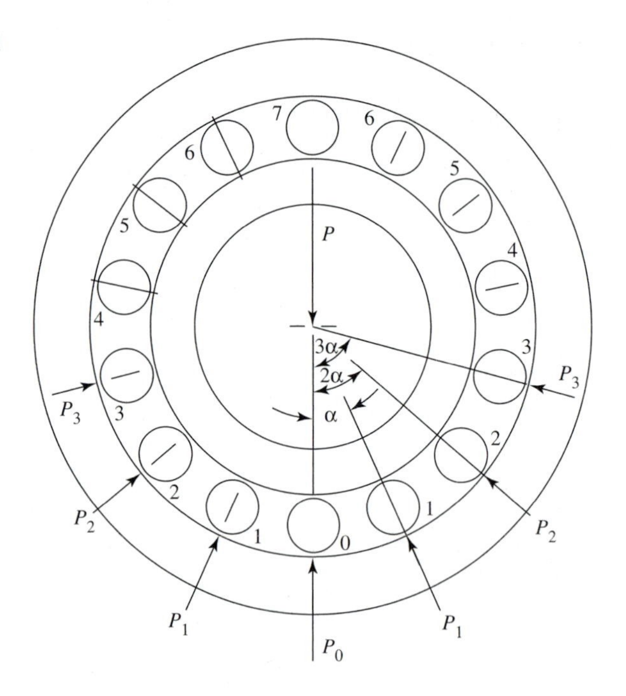

Figure 18.3 Load distribution in a ball bearing.

Using the work cited above, we obtain the following load distribution equation:

$$\frac{P_0}{P} = \frac{1}{1 + 2\cos^{5/2}\alpha + 2\cos^{5/2}(2\alpha) + \cdots + 2\cos^{5/2}(x\alpha)} \qquad \textbf{(18.3)}$$

where P_0 = load on one ball when it is most heavily loaded

P = radial load on the bearing

Z = number of balls

x = greatest integer $\leq Z/4$

The bearing in Figure 18.3 has 14 balls, and the shaft load is downward. The ball at the lowest position (ball 0 in the figure) has the greatest instantaneous load. The pairs of balls numbered 1, 2, and 3 have lesser loads, and the remaining balls (4 through 7) are not loaded at this instant.

EXAMPLE PROBLEM Ball Bearing Load Distribution

Examine the load sharing of various ball bearings.

Decisions: We will determine the fraction of the bearing load taken by one ball when it is most heavily loaded. Bearings with 7 to 16 balls will be considered.

Solution Summary: Load sharing Equation (18.3) is used, where

$Q = P_0/P$
S = the denominator of the equation

The results indicate that one ball, when most heavily loaded, carries from 27% to 62% of the bearing load. The detailed solution follows.

Detailed Solution:

$$Q = P_0/P = \text{fraction of load taken by most severely loaded ball}$$

Number of Balls:

$$Z = 7, \ldots, 16$$

$$\alpha(Z) = \frac{2 \cdot \pi}{Z} \qquad x(Z) = \text{floor}\left(\frac{Z}{4}\right)$$

$$S(Z) = 1 + \sum_{k=1}^{x(Z)} 2 \cdot (\cos(k \cdot \alpha(Z)))^{\frac{5}{2}} \qquad Q(Z) = \frac{1}{S(Z)}$$

Z	$\alpha(Z)$	$x(Z)$	$Q(Z)$
7	0.898	1	0.62
8	0.785	2	0.543
9	0.698	2	0.487
10	0.628	2	0.438
11	0.571	2	0.397
12	0.524	3	0.364
13	0.483	3	0.336
14	0.449	3	0.312
15	0.419	3	0.291
16	0.393	4	0.273

■■

Ball Bearing Load Carrying Capacity

The load carrying capacity of a ball bearing depends on many factors. They include number of balls, ball diameter, mean race diameter, required life, required reliability, bearing materials, environmental conditions, manufacturing precision, and surface finish.

In order to make it easier to select ball bearings for a given application, manufacturers list load ratings in their catalogs. These ratings are based on extensive testing and on records of service life of bearings in various applications. The basis of these ratings varies from manufacturer to manufacturer. Thus, it is important to note the life expectancy and reliability values used to determine the ratings. We can then make corrections to account for the required life expectancy and reliability for our application.

Basic Load Rating of Ball Bearings

Basic load ratings can be calculated if we know the bearing geometry and the number and diameter of the balls. Several steps are involved.

1. We begin by calculating a diameter ratio

$$r_d = \frac{D \cos \alpha_c}{d_m} \tag{18.4}$$

where r_d = diameter ratio

D = ball diameter

$d_m = 0.5(d_o + d_I)$ = mean race diameter

d_O = outer race diameter (at ball contact)

d_I = inner race diameter (at ball contact)

α_c = normal contact angle (measured from the load line to the plane of the bearing)

2. A geometry factor is tabulated against diameter ratio in publications and standards, including Anti-Friction Bearing Manufacturers Association (1978) and Buchsbaum, Freudenstein, and Thornton (1983). For single-row radial contact bearings, single-row and double-row angular contact bearings, and groove-type bearings, the geometry factor is approximated by

$$g = 83{,}940r_d^3 - 85{,}850r_d^2 + 23{,}710r_d + 2597 \quad \textbf{(18.5)}$$

where g = geometry factor to be used with customary U.S. units (lb and in)

r_d = diameter ratio where $0.05 \leq r_d \leq 0.40$

If metric units are to be used, the geometry factor is given by

$$g_m = 0.013166g \quad \textbf{(18.6)}$$

where g_m = metric geometry factor to be used with mm and N units

3. Basic load rating is given by

$$C = (n_r \cos\,\alpha_c)^{0.7}Z^{2/3}D^{1.8}g \quad \textbf{(18.7)}$$

where C = basic load rating (lb)

g = geometry factor

n_r = number of rows of balls

α_c = normal contact angle

Z = number of balls per row

D = ball diameter (in)

The equation is valid for $D \leq 1$ in.

Metric basic load rating is given by

$$C_m = (n_r \cos\,\alpha_c)^{0.7}Z^{2/3}D^{1.8}g_m \quad \textbf{(18.8)}$$

where C_m = metric basic load rating (N)

g_m = metric geometry factor

n_r = number of rows of balls

α_c = normal contact angle

Z = number of balls per row

D = ball diameter (mm, where $D \leq 25.4$ mm)

The basic load rating is the radial load on a bearing that corresponds to a predicted life of 1 million revolutions with a 10% failure probability. A greater life expectancy is required for most applications. Thus, in most cases, we select bearings with a basic load rating that is higher than the actual service load.

EXAMPLE PROBLEM Geometry Factor for Ball Bearings

We wish to design and select ball bearings without the need to look up numbers in a table. Selected values of geometry factor f_c versus diameter ratio r_d reported in the publications Anti-Friction Bearing Manufacturers Association (1978) and Buchsbaum et al. (1983) are tabulated below in the detailed solution which follows. Write an equation to fit these values.

Decisions: There are many possible solutions. A simple linear regression does not look promising because the relationship is not monotonic. We will try to fit the data to a third-order equation in r_d, using mathematical software.

Alternative Decisions: Other equation forms can be used to approximate the data if the available software is limited. Linear regression procedures are available on scientific calculators and can even be done manually if time is no object. Possibilities include expressing geometry factor in terms of the absolute value $|r_d - 0.20|$ or in terms of $(r_d - 0.20)^2$. These forms are suggested in Problem 18.5 at the end of the chapter. Another possibility is a piece-wise fit, which can be attempted with IF-statements.

Solution Summary: We instruct our mathematical software to fit the data to a third-order equation. The software returns a solution with the coefficients of r_d^3, r_d^2, r_d, and a constant term, all listed as the vector **S**. The resulting regression equation was already used to approximate the geometry factor in the preceding section. In Figure 18.4 the tabulated values (identified as f_c versus r_d) are plotted as squares, and the regression equation (identified as g versus r) is plotted as a line. Note that the third-order equation is a fairly good fit to the published data.

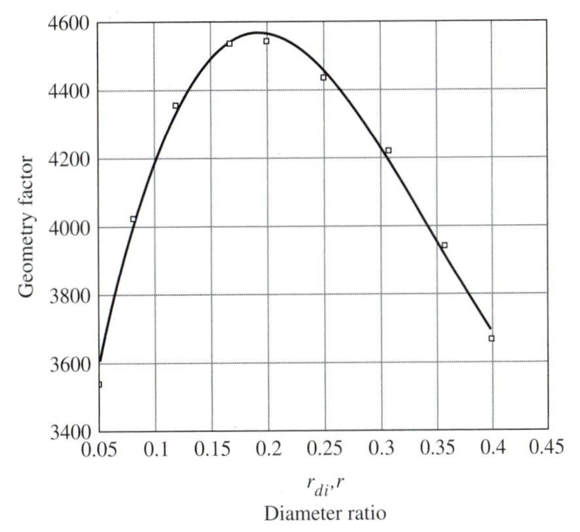

Figure 18.4 Geometry factor for basic load rating.

Detailed Solution:

Diameter Ratio $r_d = D \cos \alpha / d_m$	*Geometry Factor*
r_{d_i}	f_{c_i}
0.05	3550
0.08	4020
0.12	4370
0.16	4530
0.20	4550
0.24	4480
0.30	4250
0.36	3930
0.40	3670

$$i = 0, \ldots, 8$$

$$F(x) = \begin{bmatrix} x^3 \\ x^2 \\ x \\ 1 \end{bmatrix}$$

$$S = \text{linfit}(r_d, f_c, F) \qquad S = \begin{bmatrix} 8.349 \cdot 10^4 \\ -8.585 \cdot 10^4 \\ 2.371 \cdot 10^4 \\ 2.597 \cdot 10^3 \end{bmatrix}$$

$$g(t) = F(t) \cdot S$$

$$r = 0.05, 0.06, \ldots, 0.4$$

See Figure 18.4

■■

Ball Bearing Life Expectancy

Extensive experience has indicated the approximate relationship between ball bearing life and load: Life is inversely proportional to the cube of load. Thus, we can predict bearing life from basic load rating by using the following equations:

$$L_{10r} = 10^6 \left(\frac{C}{P} \right)^3 \tag{18.9}$$

or

$$H_{10} = \frac{10^6}{60n} \left(\frac{C}{P} \right)^3 \tag{18.10}$$

where L_{10r} = predicted bearing life (revolutions) with 10% failure probability

C = basic load rating, based on 1 million revolutions predicted life with 10% failure probability (lb or N)

P = actual radial load (lb or N)

H_{10} = predicted bearing life (h) with 10% failure probability

n = rotation speed (rpm)

Reliability

A 10% failure probability corresponds to a reliability of 0.90 which is not good enough for many applications. Consider a machine designed with several ball bearings, where failure of a single bearing would cause a breakdown in the machine. Then the reliability of the machine based on bearing failure alone would be

$$R_{\text{combined}} = R_1^{N_B} \tag{18.11}$$

where R_{combined} = reliability of the system based on bearing failure alone

R_1 = reliability of a single ball bearing

N_B = number of bearings

For example, if a design utilizes four bearings with 10% failure probability for each bearing, then the reliability of the machine (based on bearing failure alone) is

$$R_{\text{combined}} = R_1^{N_B} = 0.90^4 = 0.656$$

That is, there is a bearing failure probability of about 34% for the design life. Other failure modes of the machine would further reduce reliability.

Life Adjustment Factor for Reliability. A life adjustment factor is used when we want to select bearings with a reliability different from 10%. Operating time based on mean life (50% failure probability) is about five times operating life at 10% failure probability. However, we are more likely to require a high reliability. Reliability factor versus life adjustment factor is tabulated in various references, including PT Components (1982). Table 18.1 and Figure 18.5 show percent failure probability, life adjustment factor, and reliability. A linear regression of the values yields

$$a = 0.0974x + 0.118 \tag{18.12}$$

where $R = 1 - \dfrac{x}{100}$ = reliability

a = life adjustment factor for reliability

x = percent failure probability

Table 18.1 Percent failure probability, life adjustment factor, and reliability.

Failure x_i (%)	Life Adjustment Factor a_i	Reliability R_i $R_i = 1 - \dfrac{x_i}{100}$
1	0.21	0.99
2	0.33	0.98
3	0.44	0.97
4	0.53	0.96
5	0.62	0.95
10	1	0.9
50	5	0.5

Figure 18.5 Life adjustment factor versus percent failure.

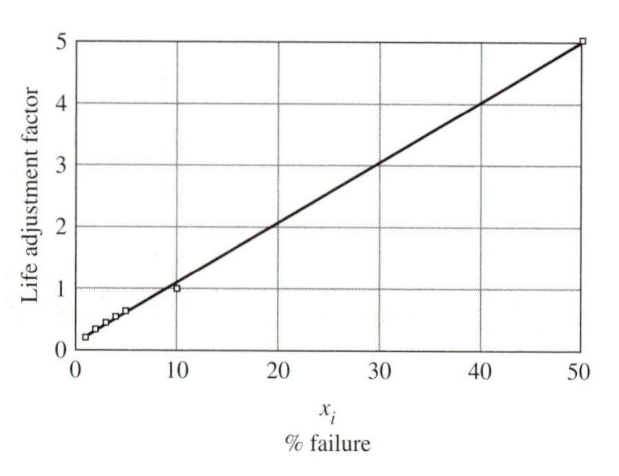

EXAMPLE PROBLEM Ball Bearing Load, Life, and Reliability

A proposed machine includes a shaft that rotates at 880 rpm and causes a 45 lb radial load on the bearings at each end of the shaft. The machine will operate 40 hours per week, and we hope to obtain a life expectancy of a few years with a reliability of 0.98 (for a single bearing). Specify the bearings.

Design Decisions: We will try a single-row ball bearing with eleven 1/8 in diameter balls and a 3/4 in diameter mean ball path.

Solution Summary: This bearing will have a basic load rating of about 533 lb. The life expectancy of an individual bearing is about 246 work weeks with a 2% failure probability. As shown in the detailed solution that follows, the tentative bearing selection appears to be satisfactory.

Detailed Solution:

Regression Equation:

$$g(r) = 83{,}490 \cdot r^3 - 85{,}850 \cdot r^2 + 23{,}710 \cdot r + 2597$$

Rows:

$$n_r = 1$$

Contact Angle:

$$\alpha_c = 0 \cdot \text{deg}$$

Balls/Row:

$$Z = 11$$

Ball Diameter:

$$D = \frac{1}{8}$$

Mean Ball Path Diameter:

$$d_m = 0.75$$

Diameter Ratio:

$$r_d = \frac{D \cdot \cos(\alpha_c)}{d_m} \qquad r_d = 0.167$$

Geometry Factor:

$$g(r_d) = 4.55 \cdot 10^3$$

Basic Load Rating (lb):

$$C = g(r_d) \cdot (n_r \cdot \cos(\alpha))^{0.7} \cdot Z^{2/3} \cdot D^{1.8} \qquad C = 533.036$$

Speed (rpm):

$$n = 880$$

Load (lb):

$$P = 45$$

Life (rev) for 10% Failure:

$$L_{10r} = 10^6 \cdot \left(\frac{C}{P}\right)^3 \qquad L_{10r} = 1.662 \cdot 10^9$$

Life (h) for 10% Failure:

$$H_{10} = \frac{10^6}{60 \cdot n} \cdot \left(\frac{C}{P}\right)^3 \qquad H_{10} = 3.148 \cdot 10^4$$

Reliability:

$$R = 0.98$$

Percent Failure:

$$x = 100 \cdot (1 - R) \qquad x = 2$$

Reliability Life Adjustment Factor:

$$a = 0.09739 \cdot x + 0.118 \qquad a = 0.313$$

Life (h) for Reliability R*:*

$$H = a \cdot H_{10} \qquad H = 9.845 \cdot 10^3$$

Life (40 h Work Weeks) for Reliability R*:*

$$W = \frac{H}{40} \qquad W = 246.137$$

■■

EXAMPLE PROBLEM Specifying a Bearing Using Metric Units

We need to specify shaft bearings that will carry a 1200 N radial load. The shaft will operate 40 h/week at 450 rpm. A service life of at least 1 year is required, with 95% reliability of individual bearings.

Design Decisions: We will try a single-row ball bearing with nine 6 mm diameter balls and a 30 mm diameter mean ball path.

Solution Summary: The basic load rating is determined to be 6554 N. The life expectancy of an individual bearing is about 91 work weeks with a 5% failure probability. As shown in the detailed solution that follows, the tentative bearing selection appears to be satisfactory. The geometry factor is adjusted for metric units. A plot of geometry factor versus diameter ratio is shown in Figure 18.6.

Detailed Solution:

Diameter Ratio:

$$r_d = D\cos(\alpha/d_m)$$

Regression Equation:

$$g(r) = (83{,}490 \cdot r^3 - 85{,}850 \cdot r^2 + 23710 \cdot r + 2597) \cdot 0.013166$$

Rows:

$$n_r = 1$$

Contact Angle:

$$\alpha_c = 0 \cdot \mathrm{deg}$$

Balls/Row:

$$Z = 9$$

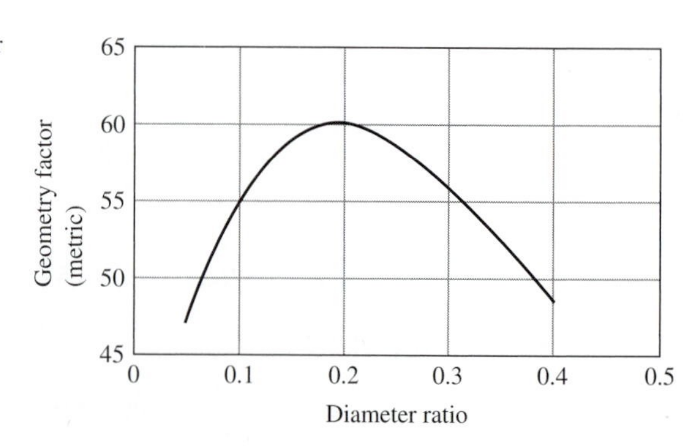

Figure 18.6 Metric geometry factor for basic load rating.

Ball Diameter:

$$D = 6$$

Mean Ball Path Diameter:

$$d_m = 30$$

Diameter Ratio:

$$r_d = \frac{D \cdot \cos(\alpha_c)}{d_m} \qquad r_d = 0.2$$

Geometry Factor:

$$g(r_d) = 60.207$$

Basic Load Rating (N):

$$C = g(r_d) \cdot (n_r \cdot \cos(\alpha))^{0.7} \cdot Z^{2/3} \cdot D^{1.8} \qquad C = 6.554 \cdot 10^3$$

Speed (rpm):

$$n = 450$$

Load (N):

$$P = 1200$$

Life (rev) for 10% Failure:

$$L_{10r} = 10^6 \cdot \left(\frac{C}{P}\right)^3 \qquad L_{10r} = 1.629 \cdot 10^8$$

Life (h) for 10% Failure:

$$H_{10} = \frac{10^6}{60 \cdot n} \cdot \left(\frac{C}{P}\right)^3 \qquad H_{10} = 6.033 \cdot 10^3$$

Reliability:

$$R = 0.95$$

Percent Failure:

$$x = 100 \cdot (1 - R) \qquad x = 5$$

Reliability Life Adjustment Factor:

$$a = 0.09739 \cdot x + 0.118 \qquad a = 0.605$$

Life (h) for Reliability R:

$$H = a \cdot H_{10} \qquad H = 3.65 \cdot 10^3$$

Life (40 h Work Weeks) for Reliability R:

$$W = \frac{H}{40} \qquad W = 91.242$$

■■

Ball Bearing Selection from Manufacturers' Catalogs

Most users find it cost-effective to specify off-the-shelf ball bearings from a catalog. In this case, it is necessary only to decide on an acceptable life and reliability and then compute the required basic load rating. The load rating can be computed as follows:

$$C = P\left(\frac{L_r}{10^6 a}\right)^{1/3} = 0.01P\left(\frac{L_r}{a}\right)^{1/3} \qquad \textbf{(18.13)}$$

where C = basic load rating to be used for ball bearing selection

P = expected radial load on the bearing

L_r = desired life expectancy (rev)

a = life adjustment factor for reliability other than 0.90

P and C can be both in lb or both in N.

EXAMPLE PROBLEM Ball Bearing Selection

A ball bearing rotating at 300 rpm will support a 110 lb load. Find the required basic load rating so that a bearing can be selected from a catalog.

Design Decisions: We will plan on a 5000 h life expectancy with 0.98 reliability (2% failure rate).

Solution Summary: We calculate a life adjustment factor of 0.313 for reliability. A bearing with a basic load rating of at least 726 lb will be selected, as shown in the following detailed solution.

Detailed Solution:

Speed (rpm):

$$n = 300$$

Load:

$$P = 110$$

Required Life Expectancy (h):

$$H = 5000$$

Required Life Expectancy (rev):

$$L_r = 60 \cdot H \cdot n \qquad L_r = 9 \cdot 10^7$$

Reliability:

$$R = 0.98$$

Percent Failure:

$$x = 100 \cdot (1 - R) \qquad x = 2$$

Reliability Life Adjustment Factor:

$$a = 0.09739 \cdot x + 0.118 \qquad a = 0.313$$

Basic load rating should not be less than

$$C = 0.01 \cdot P \cdot \left(\frac{L_r}{a}\right)^{1/3} \qquad C = 726$$

■■

Special Considerations for Ball Bearing Selection

Outer Race Rotation. The above procedures apply to bearings with rotating inner races and stationary outer races. If the design calls for a rotating outer race, the same equations can be used, except that the value of P is increased to 1.20 times actual radial load.

Thrust Loads. It is assumed that the loads on the bearings are principally radial, that is, in the plane of the bearing. Spur gear drives and belt drives produce loads that are essentially radial. Worms and other helical gears, bevel gears, hypoid gears, and power screws produce thrust loads, that is, loads in the direction of the shaft axis. If the thrust loads are large, they must be accounted for by use of an equivalent load (see PT Components, 1982, or Deutschman, Michels, and Wilson, 1975). In some cases, thrust bearings should be incorporated in the design.

Rating Criteria. The basic load rating is based on a life expectancy of 1 million revolutions with 10% failure rate (0.90 reliability). Some manufacturers base their ratings on other criteria, for example, 3800 h average life (50% reliability) at various speeds. When using their catalogs, designers must adjust the selection equations accordingly.

Environmental Conditions, Materials, and Loading. The load rating equations assume that the bearings are made of good quality steel and are adequately lubricated, protected from foreign matter and extreme temperatures, and properly supported and aligned. Failure to meet these conditions may have a significant effect on bearing performance.

Static Load Rating of Ball Bearings

In most applications, bearings are expected to survive for many millions, or even billions, of revolutions. Thus, fatigue life governs ball bearing selection. Basic load rating C as determined above is based on fatigue failure experience.

Static load rating is the maximum allowable static force. In a limited number of applications, particularly those involving intermittent service, static loads may cause bearing failure. The static load rating may be as low as one-fourth of the basic load rating for small bearings. If the required service life of a bearing is low (less than 50 million revolutions in some cases), static load rating could govern bearing selection. Catalogs should be consulted for static load ratings when high shock loads are possible and when bearing selection is based on limited service life.

ROLLER BEARINGS

Cylindrical roller bearings are shown in Figure 18.7. The inner and outer rings of high-capacity roller bearings are made of high-quality bearing steel, with microfinished raceways. The rollers can be separated with a retainer made of formed steel, or segmented steel, or glass-reinforced polymer.

Procedures for selecting roller bearings are similar to those for selecting ball bearings. However, many years of experience and testing have indicated that roller bearing life is inversely proportional to the 10/3 power of load. Thus, required load rating is proportional to the 3/10 power of life expectancy. The basic roller bearing selection equation is

$$C = P\left(\frac{L_r}{10^6 a}\right)^{3/10} \tag{18.14}$$

where C = basic load rating to be used for roller bearing selection

P = expected radial load

L_r = desired life expectancy (rev)

a = life adjustment factor for reliability other than 0.90 [Use Equation (18.12) for ball bearings.]

P and C can be both in lb or both in N.

Figure 18.7 Cylindrical roller bearings. (Courtesy of Rexnord Corporation, Link-Belt Bearing Division)

Basic load rating is based on an expected life of 1 million revolutions with a reliability of 0.90.

EXAMPLE PROBLEM Roller Bearing Selection

Select bearings for 600 lb load and 2000 rpm shaft speed.

Design Decisions: Roller bearings will be used. We will select bearings for a 3800 h life expectancy with 1 failure in 100.

Solution Summary: The calculated reliability factor is 0.215. A roller bearing with a basic load rating of at least 5969 lb will be selected, as the following detailed solution illustrates.

Detailed Solution:

Speed (rpm):

$$n = 2000$$

Load:

$$P = 600$$

Required Life Expectancy (h):

$$H = 3800$$

Required Life Expectancy (rev):

$$L_r = 60 \cdot H \cdot n \qquad L_r = 4.56 \cdot 10^8$$

Reliability:

$$R = 0.99$$

Percent Failure:

$$x = 100 \cdot (1 - R) \qquad x = 1$$

Reliability Life Adjustment Factor:

$$a = 0.09739 \cdot x + 0.118 \qquad a = 0.215$$

Basic load rating should not be less than

$$C = P \cdot \left(\frac{L_r}{10^6 \cdot a} \right)^{3/10} \qquad C = 5969$$

Spherical roller bearings are often selected for heavy-duty applications, including earth-moving equipment, steel mill and paper mill equipment, and torque converters. Figure 18.8 shows a spherical roller bearing. The bearing is self-aligning owing to outer ring geometry.

Figure 18.8 Spherical roller bearing. (Courtesy of Rexnord Corporation, Link-Belt Bearing Division)

BEARING SEALS

Seals are designed to retain bearing lubricant and exclude metal chips, abrasives, and other foreign matter. Rubber and/or metal seals can be fastened to the outer race of a ball bearing and ride on the inner race. Some external bearing seals produce a labyrinth-like sealing surface on the inner and outer race. They are used in contaminated environments and where rubber seals would not perform properly.

LINEAR MOTION SYSTEMS

Linear ball bearings and roller-type linear systems are used when applications call for precise linear motion. Some lightweight units employ an aluminum extruded rail and carrier. Figure 18.9(a) shows linear ball bearing systems, and Part (b) of the figure shows roller-type linear systems.

LUBRICATION

Lubricants are used to reduce friction and wear in machine parts. Although oil and grease are the most commonly used lubricants, there are other options. Special applications call for air bearings and water-lubricated bearings. Lubricant-impregnated porous metal bearings, and

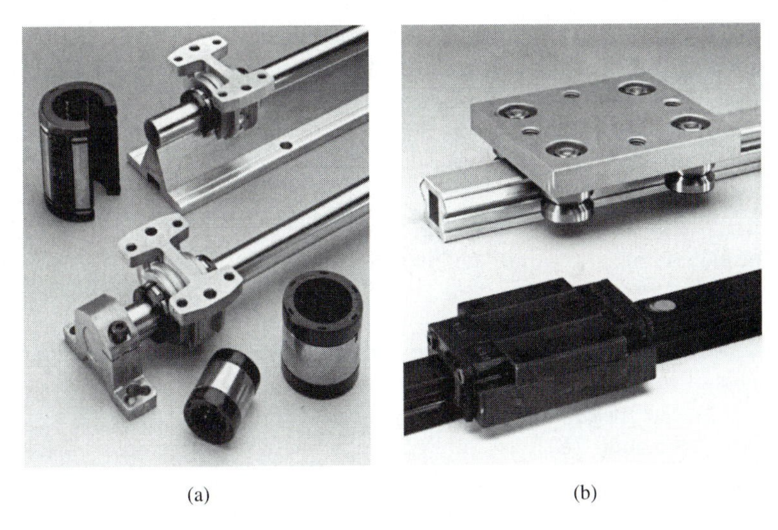

(a) (b)

Figure 18.9 (a) Linear ball bearing systems; (b) roller-type linear systems. (Courtesy of SKF Specialty Products Company)

unlubricated bushings made of nylon, carbon, or poly-tetra-fluoroethylene (PTFE), are also used.

Lubrication Regimes

Several different lubrication regimes are of interest to machine designers, including

- hydrodynamic lubrication
- elastohydrodynamic lubrication
- boundary lubrication
- mixed lubrication
- hydrostatic lubrication

Hamrock (1991) discusses these lubrication regimes in detail. Some characteristics of each are discussed in the following paragraphs.

Hydrodynamic Lubrication. Hydrodynamic lubrication (HDL) usually requires conformal surfaces, typically a rotating shaft surrounded by fluid in a journal bearing. The rotation causes a positive pressure in the lubricant, separating the shaft and bearing. That is, the lubricant supports the shaft load. Typical values of coefficient of friction range from $6 \cdot 10^{-4}$ to $2 \cdot 10^{-3}$. With HDL, the minimum lubricant film thickness is usually greater than 1 micrometer (1 micron). The symbol for micrometer is μm, where 1 μm $= 10^{-6}$ m $= 39.37$ μin (about 40 microinches).

Elastohydrodynamic Lubrication. The deformation of the lubricated parts is an important factor with elastohydrodynamic lubrication (EHL). Typical values of coefficient of friction range from $2 \cdot 10^{-3}$ to $60 \cdot 10^{-3}$.

When the bearing materials have a high elastic modulus (metals, for example), this lubrication condition is called **hard EHL.** This lubrication type usually occurs with nonconformal surfaces, such as ball and roller bearings. Minimum film thickness may be as low as 4 μin (0.1 μm). In heavily loaded situations, maximum pressure may be between about 70,000 and 400,000 psi. This range corresponds to about 1/2 to 3 GPa where 1 GPa = 10^9 Pa and 1 psi = 6895 Pa.

Soft EHL occurs with materials having a low elastic modulus. Examples where soft EHL is present include rubber seals, tires, and human joints.

Boundary Lubrication. It is impossible to finish a surface to perfect smoothness. The surface roughness of finished bearing surfaces usually falls between 0.4 and 400 μin (between 0.01 and 10 μm). The roughness high points are called **asperities.**

Boundary lubrication (BL) is the regime involving considerable asperity contact. The contact can be metal-to-metal, or the asperities can be separated by a thin lubricant film. High loads can lead to substantial wear rates under BL conditions. Such conditions can occur briefly in starting in the case of bearings designed for hydrodynamic lubrication. Otherwise, bearings designed for boundary lubrication are ordinarily restricted to infrequent operation. Surface films in BL can range in thickness from about 0.04 to 0.4 μin. This range corresponds to about 1 to 10 nm, where 1 nm (1 nanometer) = 10^{-9} m = 0.03937 μin. Typical values of coefficient of friction with BL range from 0.06 to 0.15.

Mixed Lubrication. Mixed lubrication (ML) or partial lubrication refers to a combination of boundary and fluid film effects. Consider a journal bearing operating under conditions of hydrodynamic lubrication. An increase in load and/or a reduction in speed may cause the journal bearing to operate in the mixed lubrication regime. In the case of a roller bearing or a ball bearing, a similar change may change the lubrication regime from elastohydrodynamic to mixed. Typical film thicknesses in the ML regime range from 0.4 to 40 μin (0.01 to 1 μm).

Hydrostatic Lubrication. Hydrostatic lubrication (HSL) refers to a system in which externally pressurized lubricant is supplied between the moving parts. Thus, HSL can separate the moving parts even at starting speeds. HSL is particularly effective in thrust bearings and other situations where it is difficult to establish conditions for HDL.

Bearing Selection

Bearing selection for a particular application depends on shaft speed, required life and reliability, cost and space limitations, expected maintenance, and environmental factors. A few characteristics typical of specific bearing types are listed below.

Ball and Roller Bearings

- low starting friction
- low operating friction
- support radial and thrust loads
- usually prelubricated
- large radial space requirement (except for needle bearings)
- finite life

Unlubricated Sleeve Bearings

- low cost
- small space requirement
- no maintenance required
- limited to low loads
- suitable for intermittent use

Journal Bearings with Hydrodynamic Lubrication

Hydrodynamically lubricated journal bearings are sleevelike bearings separated from shafts by a lubricant film.

- reliable (in the presence of adequate lubrication)
- very long life expectancy
- high load capacity
- regular inspection required to ensure presence of adequate lubrication
- high starting friction
- lubricant film that may break down when starting, stopping, or changing direction

Hydrostatically Lubricated Bearings

- low starting friction
- high load capacity, even at low speed
- external system required to pressurize lubricant
- reliability that depends on performance of external pressure source

Lubricant Viscosity

Lubricants are selected principally for their ability to reduce friction and wear. Other functions include cooling moving parts, providing corrosion resistance, and removing wear particles.

In a hydrodynamically lubricated bearing, the most important descriptor of the lubricant is its viscosity. **Viscosity** is a resistance to flow, the shear stress in a lubricant divided by the velocity gradient. Viscosity is temperature-dependent. An increase in temperature causes a decrease in viscosity in most lubrication situations.

Absolute Viscosity. Absolute viscosity is expressed in reyn or centipoise. The two units are related by

$$\mu = \frac{z}{6.895 \cdot 10^6} \tag{18.15}$$

where μ = absolute viscosity (reyn)

1 reyn = 1 lb · s/in²

z = absolute viscosity in centipoise (cP)

$1 \text{ cP} = 10^{-2} \text{ dyne} \cdot \text{s/cm}^2 = 10^{-3} \text{ N} \cdot \text{s/m}^2 = 10^{-3} \text{ Pa} \cdot \text{s}$

Kinematic Viscosity. Experimental measurements often return the kinematic viscosity of a lubricant. Kinematic viscosity is related to absolute viscosity by

$$z_k = \frac{z}{\rho} \tag{18.16}$$

where z_k = kinematic viscosity in centistokes (cS)

ρ = mass density = specific gravity (g/cm^3)

1 cS = 10^{-2} stokes = 10^{-6} m^2/s

Typical engine oils have a specific gravity that decreases with temperature from about 0.90 at 0°C to about 0.875 at 40°C and 0.80 at 170°C.

Viscosity Grade. The International Organization for Standards/American Society for Testing and Materials (ISO/ASTM) grading system classifies lubricating oils by kinematic viscosity. The ISO/ASTM viscosity grade is the viscosity in centistokes at a temperature of 40°C (104°F). A few Society of Automotive Engineers (SAE) crankcase oil grades and their approximately equivalent American Gear Manufacturers Association (AGMA) grades and ISO/ASTM grades are listed in Table 18.2.

Viscosity–Temperature Dependence. Oil and other liquids, greases and other solids, and air and other gases can be used as lubricants. Gas viscosity is low compared with the viscosity of other lubricant types, and it is not very sensitive to temperature. Gas viscosities tend to increase with increasing temperature.

The viscosity of lubricating oil decreases sharply with increasing temperature. This property complicates lubricant selection for automotive applications and other uses where temperatures vary widely. High-viscosity-grade, "thick" oil makes startup difficult at low temperatures. However, if low-viscosity-grade, "thin" oil is used to reduce starting load, then high-temperature running conditions may reduce the viscosity to a point where the lubricant film cannot support the bearing loads.

Figure 18.10 shows typical viscosity-versus-temperature relationships for a few SAE single-grade oils. Lubricant properties vary widely, even within a given grade. Thus, these curves can be used to indicate only a general trend. Multigrade oils follow a similar trend, but the viscosity of a multigrade lubricant is less sensitive to changes in temperature. Therefore, multigrade oils are often selected for cold-starting, hot-running applications.

Table 18.2 Approximate viscosity grade equivalence.

ISO/ASTM Viscosity	AGMA Grade	SAE Grade
220	5	50
150	4	40
100	3	30
68	2	20W
32	—	10W

Figure 18.10 Typical viscosity-versus-temperature relationships for a few SAE single-grade oils. (Based on data from NASA Reference Publication 1255; see Hamrock, 1991)

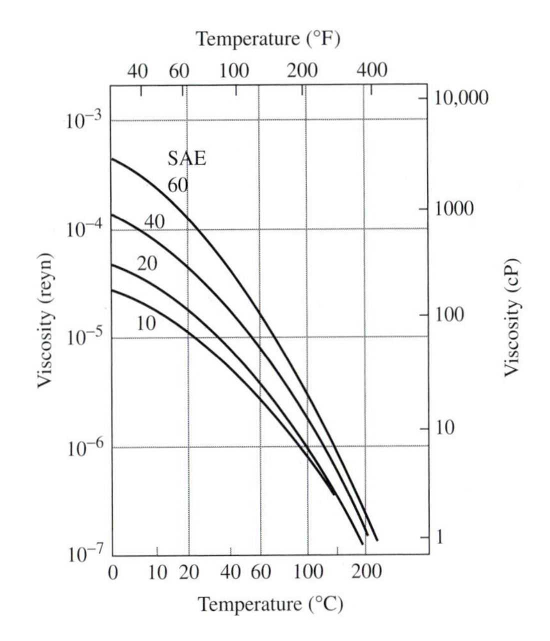

EXAMPLE PROBLEM Viscosity

Find the absolute viscosity of SAE 10W crankcase oil at 40°C.

Solution Summary: We estimate a specific gravity of 0.875. Using Table 18.2, we find a kinematic viscosity of about 32 cS. The viscosity equations yield an absolute viscosity of 28 cP or 4.06 μreyn, as illustrated in the following detailed solution.

Detailed Solution:

Kinematic Viscosity:

cS *m/s²*

$z_k = 32$ $Z_k = 10^{-6} \cdot z_k$ $Z_k = 3.2 \cdot 10^{-5}$

Specific Gravity:

$$\rho = 0.875$$

Absolute Viscosity:

cP *(N · s/m²)*

$z = \rho \cdot z_k$ $z = 28$ $Z = 10^{-3} \cdot z$ $Z = 0.028$

reyn

$$\mu = \frac{Z}{6.895 \cdot 10^6} \qquad \mu = 4.061 \cdot 10^{-6}$$

Hydrodynamic Lubrication: Design for Light Loads

Newton–Petrov Theory. Newton's law of lubrication states that the shear stress in a lubricant film is proportional to the relative velocity divided by the film thickness. Petrov extended this theory to determine friction torque (Newton, 1668; Petrov, 1883; and Fuller, 1984).

Petrov assumed that the shaft rotated exactly in the center of the bearing. As a result, the Newton–Petrov theory is fairly accurate if

- the load is light;
- the projected area of the bearing is large;
- the rotation speed is high;
- the lubricant viscosity is high.

Friction Torque for Lightly Loaded Bearings. If the above conditions are reasonably met, the friction torque on a shaft rotating in a bearing is given by the Newton–Petrov theory in the following equation:

$$T = \frac{\pi^2 \mu l n r^3}{15c} \tag{18.17}$$

where T = friction torque (in · lb or N · m)
l = bearing length (in or m)
n = shaft speed (rpm)
r = shaft radius (in or m)
c = radial clearance (in or m)
μ = absolute viscosity (lb · s/in^2 or N · s/m^2)

Note that 1 lb · s/in^2 = 1 reyn and 1 N · s/m^2 = 1000 cP.

Power Loss Due to Friction. The power lost to friction can be determined from the friction torque by

$$P_{hpf} = \frac{nT}{63{,}025} \tag{18.18}$$

or

$$P_{kWf} = 10^{-6} T\omega \tag{18.19}$$

where P_{hpf} = power loss (hp)
P_{kWf} = power loss (kW)
ω = angular velocity (rad/s)

Coefficient of Friction. The actual friction torque results from friction forces distributed about the shaft as it rotates in the bearing. Thus, there is no single force vector. However, we can define an effective friction force and a coefficient of friction as follows:

$$F_f = \frac{T}{r} \tag{18.20}$$

and

$$f = \frac{F_f}{P} \tag{18.21}$$

where F_f = effective friction force (lb or N)
f = coefficient of friction
P = radial bearing load (lb or N)

Heat Generated within the Bearing. The rate of heat generation within the bearing is related to the power loss by

$$H_f = 2544P_{\text{hp}f} \tag{18.22}$$

where H_f = heat generated (Btu/h)
The rate of heat generation in SI units is given directly by $P_{\text{kW}f}$ where 1 kW = 1000 J/s = 3412 Btu/h.

For stable conditions, the heat generated within a bearing must be dissipated at the same rate. In some cases, the rate of heat loss from a bearing by convection and radiation is adequate to maintain a reasonable lubricant temperature. In other cases, it may be necessary to circulate oil through the bearing, cool the oil in a heat exchanger or sump, and then return the cooled oil to the bearing.

Nominal Bearing Pressure and Sommerfeld Number as Design Criteria. The Newton–Petrov torque equation assumes light loading; the actual radial load is not explicitly present in that equation. The nominal bearing pressure is the radial load divided by the projected area of the bearing.

$$p_n = \frac{P}{ld} \tag{18.23}$$

where p_n = nominal bearing pressure (psi or Pa)
l = bearing length (in or m)
d = shaft diameter (in or m)

The Sommerfeld number (Deutschman et al., 1985) is a dimensionless descriptor, combining terms that distinguish lightly loaded bearings from heavily loaded bearings. It can be defined by

$$S = \left(\frac{r}{c}\right)^2 \cdot \frac{n\mu}{60p_n} \tag{18.24}$$

where S = Sommerfeld number
r = shaft radius (in or m)
c = radial bearing clearance (in or m)
n = shaft speed (rpm)
μ = absolute viscosity (reyn or $\text{N} \cdot \text{s/m}^2$)
p_n = nominal bearing pressure (psi or Pa)

When the Sommerfeld number is equal to or greater than unity, then the shaft rides near the center of the bearing and the Newton–Petrov theory is fairly accurate. For heavily loaded bearings, where $S < 1$, the more complicated theory discussed in the section following the example problem should be considered.

EXAMPLE PROBLEM Analysis of a Lightly Loaded Bearing

Bearings are needed to support a 3 in diameter shaft. The bearing load will be 400 lb at each bearing, and shaft speed can range up to 4000 rpm. Propose a bearing. Determine friction torque and power loss. Evaluate your results.

Design Decisions: We will select a 4 in long journal bearing with 0.003 in radial clearance. Lubricating oil with a viscosity of 24 cP at the operating temperature will be used. We will assume that the Newton–Petrov theory is valid for this design.

Solution Summary: Based on the Newton–Petrov theory, friction torque increases linearly with shaft speed. Friction torque reaches about 41 lb · in at 4000 rpm as shown in the plot of Figure 18.11. Power loss is proportional to shaft speed squared, reaching about 2.6 hp at 4000 rpm. Sommerfeld number is

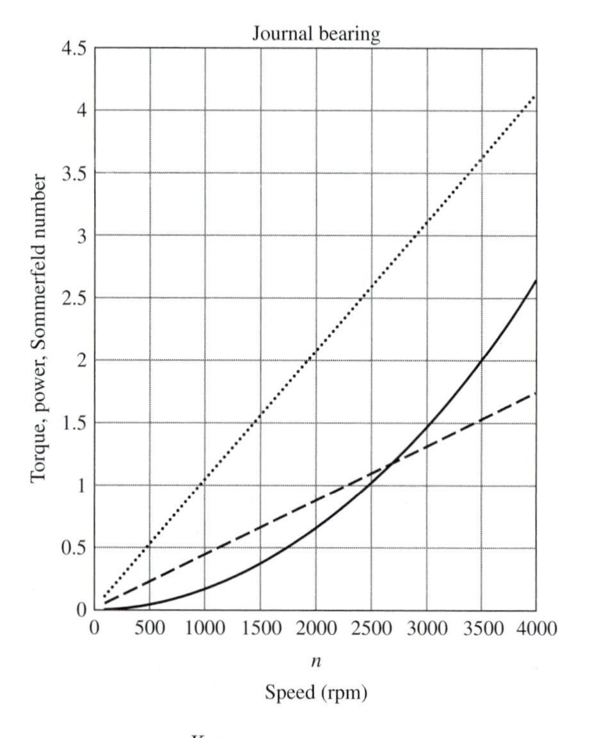

Figure 18.11 Lightly loaded bearing example: friction torque, friction horsepower, and Sommerfeld number plotted against speed.

also plotted against shaft speed in Figure 18.11. If we assume the theory valid for $S \geq 1$, then the solution can be used for the speed range 2300 rpm $\leq n \leq$ 4000 rpm. Below 2300 rpm, we can consider the method described in the following section, "Hydrodynamic Lubrication: Design for Heavy Loads." The detailed solution follows.

Detailed Solution:

Diameter (in):

$$d = 3$$

Radial Clearance:

$$c = 0.003$$

Radius (in):

$$r = \frac{d}{2} \qquad r = 1.5$$

Clearance Modulus:

$$m_c = \frac{c}{r} \qquad m_c = 0.002$$

Length (in):

$$l = 4$$

Speed (rpm):

$$n = 100, 110, \ldots, 4000$$

Angular Velocity (rad/s):

$$\omega(n) = \frac{\pi \cdot n}{30}$$

Viscosity:

cP *reyn*

$$z = 24 \qquad \mu = \frac{z}{6.9 \cdot 10^6} \qquad \mu = 3.47826 \cdot 10^{-6}$$

Bearing Load (lb):

$$P = 400$$

Nominal Pressure:

$$p_n = \frac{P}{l \cdot d} \qquad p_n = 33.33333$$

Sommerfeld Number:

$$S(n) = \left(\frac{r}{c}\right)^2 \cdot \frac{n \cdot \mu}{60 \cdot p_n}$$

Friction Torque (lb · in):

$$T(n) = \frac{\pi^2 \cdot \mu \cdot l \cdot n \cdot r^3}{15 \cdot c}$$

Friction Horsepower:

$$P_{\text{hp}f}(n) = \frac{n \cdot T(n)}{63{,}025}$$

Effective Friction Force (lb):

$$F_f(n) = \frac{T(n)}{r}$$

Coefficient of Friction:

$$f(n) = \frac{F_f(n)}{P}$$

Heat Generated (Btu/h):

$$H_f(n) = 2544 \cdot P_{\text{hp}f}(n)$$

See Figure 18.11. ■■

Hydrodynamic Lubrication: Design for Heavy Loads

The above discussion involving the Newton–Petrov theory assumed that the shaft remained in the center of the bearing, surrounded by a lubricant film of uniform thickness. If we have heavy loads, low operating speed, and low lubricant viscosity, then we must consider the shaft eccentricity in the bearing. Unfortunately, the analysis requires several steps.

Shaft eccentricity influences results when the Sommerfeld number $S < 1$. Figure 18.12 shows a vertically loaded shaft rotating clockwise in a bearing. The shaft may initially climb to the right. Finally, with hydrodynamic lubrication conditions, the shaft will reach equilibrium in a position similar to that shown in the figure. Clearance c, the difference between shaft radius and bearing radius, is exaggerated. Minimum film thickness of the lubricant is given by

$$h_{\text{MIN}} = c - e \tag{18.25}$$

where h_{MIN} = minimum film thickness

e = shaft eccentricity, the location of the shaft center relative to the exact bearing center

Instead of using eccentricity (in or m), we often use eccentricity ratio. It is given by

$$e_r = \frac{e}{c} \tag{18.26}$$

where e_r = eccentricity ratio

c = radial clearance in the bearing

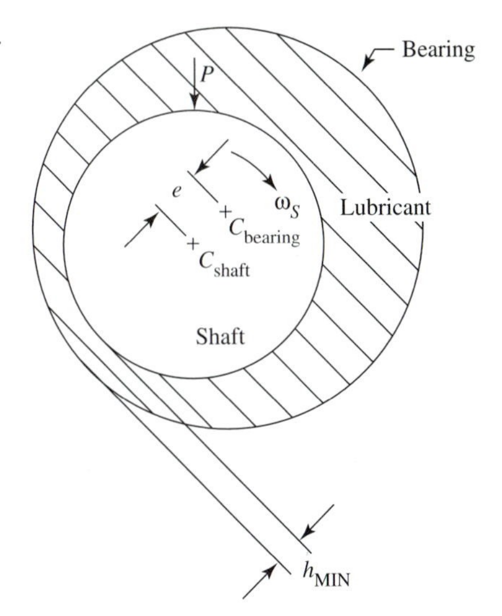

Figure 18.12 Eccentricity.

Pressure Number. The procedure that follows is based on analysis and tables given by Dennison (1936) and Fuller (1984). We begin by defining a pressure number that combines nominal pressure, geometry, speed, and viscosity parameters as follows:

$$P_1 = \frac{m_c^2 p_n}{\mu n} \tag{18.27}$$

where P_1 = pressure number
$m_c = c/r$ = clearance modulus
$p_n = P/(ld)$ = nominal pressure (psi or Pa)
μ = absolute viscosity (lb · s/in^2 or N · s/m^2)
n = speed (rpm)
c = radial clearance (in or m)
r = shaft radius (in or m)
l = bearing length (in or m)

Reciprocal Film Thickness Ratio. Reciprocal film thickness ratio is the radial clearance divided by the minimum film thickness. It is given by

$$h_r = \frac{c}{h_{\text{MIN}}} = \frac{1}{1 - e_r} \tag{18.28}$$

Calculation of Friction Torque. The references cited above relate pressure number to film thickness, and film thickness to torque, using tables and design charts. For computer calculations, we usually prefer equations. Regression equations that can be used to find friction torque are developed in the following example problem.

EXAMPLE PROBLEM Regression Equations for Bearing Analysis

Values relating to lubricant film thickness, friction torque, and lubricant pressure are tabulated in the detailed solution. Develop regression equations from these tables.

Decisions: Many equation forms could give acceptable results. We will try third-order polynomials and then plot the values and equations to verify the results.

Solution Summary: Mathematics software was used to develop regression equations relating the variables, as shown in the detailed solution which follows. One equation expresses reciprocal film thickness ratio as a function of pressure number. The other expresses friction torque number as a function of reciprocal film thickness ratio. Figure 18.13 shows friction torque number and pressure number plotted against reciprocal film thickness ratio. Also, eccentricity ratio is tabulated against reciprocal film thickness ratio in the detailed solution. The regression equations are written in standard form and applied to bearing analysis in the section "Eccentricity and Torque Calculation Using Regression Equations" following this example problem.

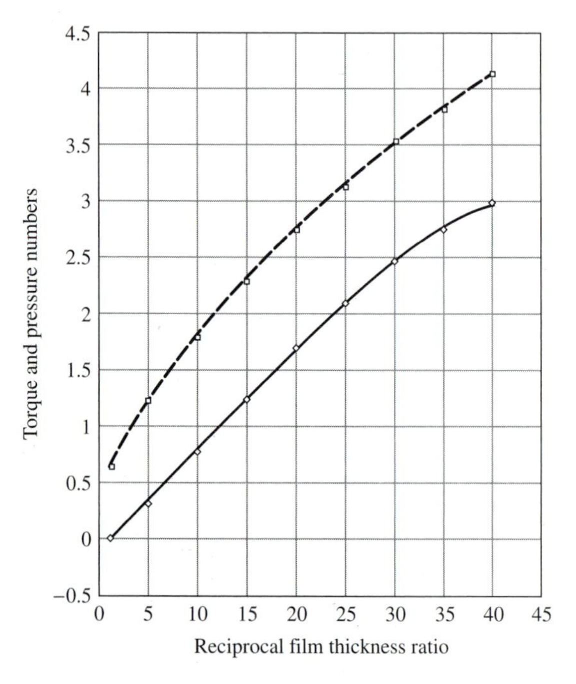

Figure 18.13 Torque number and pressure number plotted against reciprocal film thickness number.

Detailed Solution:

Length/diameter = 1

Data Pairs:

$$i = 0, \ldots, 8$$

Reciprocal Film Thickness Ratio h_{ri}	*Friction Torque Number* T_{1i}	*Pressure Number* P_{1i}
1	0.658	0
5	1.252	0.293
10	1.858	0.778
15	2.36	1.27
20	2.80	1.76
25	3.18	2.14
30	3.53	2.48
35	3.85	2.73
40	4.14	2.96

Eccentricity Ratio:

$$e_{ri} = 1 - \frac{1}{h_{ri}}$$

h_{ri}	e_{ri}
1	0
5	0.8
10	0.9
15	0.933
20	0.95
25	0.96
30	0.967
35	0.971
40	0.975

Regression Equation Form:

$$F(h_r) = \begin{bmatrix} h_r^3 \\ h_r^2 \\ h_r \\ 1 \end{bmatrix} \qquad S = \text{linfit}(h_r, T_1, F) \qquad S_1 = \text{linfit}(h_r, P_1, F)$$

Regression Equation Constants:

$$S = \begin{bmatrix} 2.39856 \cdot 10^{-5} \\ -0.00263 \\ 0.15771 \\ 0.51276 \end{bmatrix} \qquad S_1 = \begin{bmatrix} -2.85107 \cdot 10^{-5} \\ 9.44479 \cdot 10^{-4} \\ 0.08408 \\ -0.11181 \end{bmatrix}$$

$$T_2(h) = F(h) \cdot S \qquad P_2(h) = F(h) \cdot S_1 \qquad h = 1, 1.1, \ldots, 40$$

See Figure 18.13.

■■

Eccentricity and Torque Calculation Using Regression Equations

When the torque and the eccentricity are expressed in equation form, the design equations can be programmed on a computer. Then only a few keystrokes are required to investigate proposed design changes. This helps us in our attempts to optimize the design, that is, to find the best bearing for a given situation.

A regression equation for reciprocal film thickness ratio in terms of pressure number was obtained from the tabulated data of the preceding example problem. The torque number equation obtained in this example was written in standard form. Both equations are as follows:

$$h_r = 1.328P_1^3 - 4.297P_1^2 + 14.241P_1 + 1.043 \quad \textbf{(18.29)}$$

$$T_1 = 23.99 \cdot 10^{-6}h_r^3 - 0.00263h_r^2 + 0.1577h_r + 0.5128 \quad \textbf{(18.30)}$$

These equations are based on a bearing-length-to-shaft-diameter (l/d) ratio of 1. They will give reasonable results for

$$0.90 < l/d < 1.10$$

Outside that range, it is suggested that the designer consult one of the following references: Dennison (1936) or Fuller (1984).

Eccentricity and Minimum Film Thickness. Eccentricity ratio and eccentricity can be found from reciprocal film thickness ratio by the following equations:

$$e_r = 1 - \frac{1}{h_r} \quad \textbf{(18.31)}$$

$$e = ce_r \quad \textbf{(18.32)}$$

Minimum film thickness is given by

$$h_{\text{MIN}} = \frac{c}{h_r} \quad \textbf{(18.33)}$$

Limitations of Theory. As noted above, hydrodynamic lubrication theory for light loads (HDL–LL) based on the Newton–Petrov theory is valid only if the eccentricity is small. This restriction corresponds to requiring a minimum value of Sommerfeld number, say, $S \geq 1$.

Hydrodynamic lubrication theory for heavy loads (HDL–HL) considered in this section is valid in a wider region that includes the Newton–Petrov validity region. However, HDL–HL requires a minimum lubricant film thickness of at least 40 μin (about 1 μm). In practice, HDL–HL is more restricted. We must consider surface roughness effects and account for the fact that the shaft axis cannot be exactly parallel to the bearing axis. To be on the safe side, we can require that

$$h_{\text{MIN}} \geq 120\ \mu\text{in} \quad \text{or} \quad h_{\text{MIN}} \geq 3\ \mu\text{m}$$

If the calculated value of h_{MIN} is less than 80 or 120 μin (2 or 3 μm), the bearing may fall in the partial or mixed lubrication regime (ML). In this regime, viscosity may not be the controlling parameter, and HDL–HL results will not be valid.

Friction Torque. After calculating torque number, we find friction torque from

$$T = \frac{\mu l n r^2 T_1}{m_c} \tag{18.34}$$

where T = friction torque (in · lb or N · m)
μ = absolute viscosity (reyn or N · s/m²)
l = bearing length (in or m)
n = shaft speed (rpm)
r = shaft radius (in or m)
T_1 = torque number
$m_c = c/r$ = clearance modulus
c = radial clearance

Power Loss and Heat Generation. Calculation of power loss due to friction, effective friction force, coefficient of friction, and heat generated in the bearing follows the same procedure as used with Newton–Petrov bearing analysis. Special considerations for heat dissipation are often required for heavily loaded bearings operated at high speed.

EXAMPLE PROBLEM Bearing Selection for Heavy Loading

Bearings are required to support a shaft load of 1050 lb/bearing. The shaft is 2 inches in diameter and rotates at 2000 rpm. Select bearings for this application, and determine eccentricity, friction loss, heat generated, and other relevant values.

Design Decisions: We will tentatively select a 2 in long journal bearing with 0.002 in radial clearance. We will use a lubricant with 22 cP viscosity at the expected operating temperature. Hydrodynamic lubrication theory for heavy loads (HDL–HL) will be used.

Solution Summary: Calculations (given in the detailed solution which follows) show an eccentricity ratio of almost 70%. As a result, the Newton–Petrov (HDL–LL) theory would not have produced valid results. Minimum film thickness ($h_{MIN} = c - e$) is well above the value for the mixed lubrication regime. Thus, the assumption of HDL–HL is valid. The friction torque is 6.4 lb · in, and the power loss is about 1/5 hp. The required rate of heat dissipation is 516 Btu/h.

Detailed Solution:

Length/diameter = 1

Radial Clearance:

$$c = 0.002$$

Diameter (in):

$$d = 2$$

Radius (in):

$$r = \frac{d}{2} \qquad r = 1$$

Length (in):

$$l = 2$$

Clearance Modulus:

$$m_c = \frac{c}{r} \qquad m_c = 0.002$$

Speed (rpm):

$$n = 2000$$

Angular Velocity (rad/s):

$$\omega = \frac{\pi \cdot n}{30} \qquad \omega = 209.43951$$

Viscosity:

cP *reyn*

$z = 22$ $\quad \mu = \frac{z}{6.9 \cdot 10^6} \qquad \mu = 3.18841 \cdot 10^{-6}$

Bearing Load (lb):

$$P = 1050$$

Nominal Pressure:

$$p_n = \frac{P}{l \cdot d} \qquad p_n = 262.5$$

Pressure Number:

$$P_1 = \frac{m_c^2 \cdot p_n}{\mu \cdot n} \qquad P_1 = 0.16466$$

Reciprocal Film Thickness Ratio:

$$h_r = 1.328 \cdot P_1^3 - 4.297 \cdot P_1^2 + 14.241 \cdot P_1 + 1.043 \qquad h_r = 3.27734$$

Eccentricity Ratio:

$$e_r = 1 - \frac{1}{h_r} \qquad e_r = 0.69487$$

Eccentricity (in):

$$e = c \cdot e_r \qquad e = 0.00139$$

Torque Number:

$$T_1 = 23.99 \cdot 10^{-6} \cdot h_r^3 - 0.00263 \cdot h_r^2 + 0.1577 \cdot h_r + 0.5128 \qquad T_1 = 1.00223$$

Friction Torque (lb · in):

$$T = \frac{\mu \cdot l \cdot n \cdot t^2 \cdot T_1}{m_c} \qquad T = 6.39104$$

Torque (N · mm):

$$T_m = 112.98 \cdot T \qquad T_m = 722.1$$

Friction Horsepower:

$$P_{\text{hp}f} = \frac{n \cdot T}{63{,}025} \qquad P_{\text{hp}f} = 0.20281$$

Friction Power (kW):

$$P_{\text{kW}f} = 10^{-6} \cdot \omega \cdot T_m \qquad P_{\text{kW}f} = 0.15123$$

Effective Friction Force (lb):

$$F_f = \frac{T}{r} \qquad F_f = 6.39104$$

Coefficient of Friction:

$$f = \frac{F_f}{P} \qquad f = 0.00609$$

Heat Generated (Btu/h):

$$H_f = 2544 \cdot P_{\text{hp}f} \qquad H_f = 515.94804$$

■■

Bearing Characteristics versus Rotation Speed. Consider a shaft rotating in a bearing in the hydrodynamic lubrication–heavy load (HDL–HL) regime. At low speeds, there is a high eccentricity ratio (the minimum film thickness is small). With increasing speed, shaft rotation has the effect of pumping more oil into the minimum clearance space, decreasing eccentricity. Friction torque and friction power loss increase as speed increases. At high speeds, the lubrication regime may approach the hydrodynamic lubrication–light load (HDL–LL) region. In the HDL–LL region, friction torque is approximately proportional to speed, and power loss is approximately proportional to speed squared.

■■ EXAMPLE PROBLEM Performance versus Speed for a Heavily Loaded Bearing

Select bearings to support a 3.5 in diameter shaft. Estimate a load of 4000 lb per bearing. Determine eccentricity ratio, friction torque, and power loss as a function of speed.

Decisions: We will select a 3.5 in long bearing, with a radial clearance of 0.0025 in. We will specify a lubricant with a viscosity of 3 μreyn at operating temperature.

Solution Summary: The HDL–HL equations are used as in the previous example problem (see the detailed solution which follows). In Figure 18.14, results are plotted for speeds ranging from 100 to 6000 rpm. As discussed in a previous section, HDL–HL theory is not valid for very small values of minimum

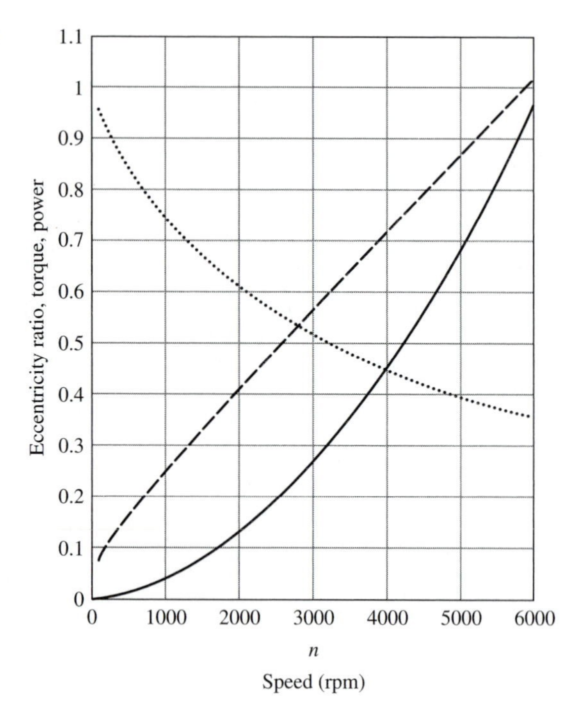

Figure 18.14 Performance versus shaft speed for a heavily loaded bearing.

lubricant thickness. For this bearing, the results may not be valid for an eccentricity ratio greater than 0.95. Thus, the results do not apply to speeds of 100 rpm or lower.

Detailed Solution:

Length/diameter = 1

Diameter (in):

$$d = 3.5$$

Radial Clearance:

$$c = 0.0025$$

Radius (in):

$$r = \frac{d}{2} \qquad r = 1.75$$

Length (in):

$$l = 3.5$$

Clearance Modulus:

$$m_c = \frac{c}{r} \qquad m_c = 0.00143$$

Speed (rpm):

$$n = 100, 120, \ldots, 6000$$

Angular Velocity (rad/s):

$$\omega(n) = \frac{\pi \cdot n}{30}$$

Viscosity:

cP *reyn*

$$z = 20.7 \qquad \mu = \frac{z}{6.9 \cdot 10^6} \qquad \mu = 3 \cdot 10^{-6}$$

Bearing Load (lb):

$$P = 4000$$

Nominal Pressure:

$$p_n = \frac{P}{l \cdot d} \qquad p_n = 326.53061$$

Pressure Number:

$$P_1(n) = \frac{m_c^2 \cdot p_n}{\mu \cdot n}$$

Reciprocal Film Thickness Ratio:

$$h_r(n) = 1.328 \cdot P_1(n)^3 - 4.297 \cdot P_1(n)^2 + 14.241 \cdot P_1(n) + 1.043$$

Eccentricity Ratio:

$$e_r(n) = 1 - \frac{1}{h_r(n)}$$

Eccentricity (in):

$$e(n) = c \cdot e_r(n)$$

Torque Number:

$$T_1(n) = 23.99 \cdot 10^{-6} \cdot h_r(n)^3 - 0.00263 \cdot h_r(n)^2 + 0.1577 \cdot h_r(n) + 0.5128$$

Friction Torque (lb · in):

$$T(n) = \frac{\mu \cdot l \cdot n \cdot r^2 \cdot T_1(n)}{m_c}$$

Friction Horsepower:

$$P_{hpf}(n) = \frac{n \cdot T(n)}{63{,}025}$$

Effective Friction Force (lb):

$$F_f(n) = \frac{T(n)}{r}$$

Coefficient of Friction:

$$f(n) = \frac{F_f(n)}{P}$$

Heat Generated (Btu/h):

$$H_f(n) = 2544 \cdot P_{hpf}(n)$$

See Figure 18.14. ■■

Speed and Heat Dissipation Considerations. Continuous operation of heavily loaded bearings at low speed is not advisable if we intend to rely on hydrodynamic lubrication. It may be necessary to design a system to provide hydrostatic lubrication under such conditions. At high speeds, HDL theory is usually valid, but note that power loss increases sharply with speed. Thus, heat dissipation must be considered. It is often necessary to design a system to circulate and cool the lubricant.

References and Bibliography

Anti-Friction Bearing Manufacturers Association. ANSI-AFBMA Standard 9-1978. New York: American National Standards Institute, 1978.

Buchsbaum, F., F. Freudenstein, and P. J. Thornton, eds. *Design and Application of Small Standardized Components.* Data Book 757, vol. 2. New Hyde Park, NY: Stock Drive Products, 1983.

Dennison, E. S. "Film Lubrication Theory and Engine Bearing Design." *Transactions of the ASME* 58(1936):25–36.

Deutschman, A. D., W. J. Michels, and C. E. Wilson. *Machine Design—Theory and Practice.* New York: Macmillan, 1975.

Dorinson, A., and K. C. Ludema. *Mechanics and Chemistry in Lubrication.* New York: Elsevier, 1985.

Fuller, D. D. *Theory and Practice of Lubrication for Engineers.* 2d ed. New York: Wiley-Interscience, 1984.

Hamrock, B. J. *Lubrication of Machine Elements.* NASA Reference Publication 1126. Washington, DC: National Aeronautics and Space Administration, 1984.

Hamrock, B. J. *Fundamentals of Fluid Film Lubrication.* NASA Reference Publication 1255. Washington, DC: National Aeronautics and Space Administration, 1991.

New Departure Handbook. 7th ed., vol II. Bristol, CT: New Departure, 1951.

Newton, I. *Mathematical Principles.* London, 1668, *See also* Fuller, 1984.

Petrov, N. "Friction in Machines and the Effect of the Lubricant." *Engineering Journal* (in Russian). St. Petersburg, 1883. *See also* Fuller, 1984.

PT Components. *Bearing Technical Journal.* 4th ed. Indianapolis, IN: PTC, Link-Belt Bearing Division, 1982.

Design Problems

18.1. Find the stopping distance for a vehicle on a level surface if the coefficient of friction is 0.6.

Decisions and Assumptions: We will assume a reaction/response time of 2 s and will consider speeds from 0 to 70 mph. We will assume that the brakes are in good working order and that stopping distance is governed by friction between the tires and the roadway.

Suggestion: Refer to Chapter 13 and to the section "Friction and Dissipation of Energy" in this chapter.

18.2. Find the stopping distance for an alert person driving a vehicle on a wet level surface.

Decisions and Assumptions: We will assume a reaction/response time of 1 s and will consider speeds from 0 to 70 mph. Using a coefficient of friction of 0.5, we will assume that the brakes are in good working order and that stopping distance is governed by friction between the tires and the roadway. (See the suggestion following Problem 18.1.)

18.3. Find the stopping distance for a vehicle on a level concrete surface if the coefficient of friction is 0.8.

Decisions and Assumptions: We will assume a reaction/response time of 1.5 s and will consider speeds from 0 to 80 mph. We will assume that the brakes are in good working order and that stopping distance is governed by friction between the tires and the roadway. (See the suggestion following Problem 18.1.)

18.4. Find the fraction of the bearing load taken by a single ball when it is most heavily loaded. Consider single-row ball bearings with 4 to 20 balls. Note that the results may indicate that it is impractical to specify a small number of balls.

18.5. Consider the data given in the example problem "Geometry Factor for Ball Bearings." Use linear regression equations to express the tabulated data in equation form as follows. Express geometry factor **(a)** in terms of the absolute value $|r_d - 0.20|$ and **(b)** in terms of $(r_d - 0.20)^2$.

18.6. A single-row ball bearing has nine 1/16 in diameter balls, 0.33 in inner race diameter, and 0.452 in outer race diameter at ball contact.

(a) Find basic load rating.

(b) Find the life expectancy with 0.99 reliability if the radial load is 25 lb and the speed is 1800 rpm.

18.7. A proposed machine includes a shaft that rotates at 400 rpm and causes a 300 lb radial load on the bearings at each end of the shaft. The machine will operate 40 h/week, and we hope to obtain a life expectancy of at least 1 year with a reliability of 0.975 (for a single bearing). Specify the bearings.

Design Decisions: We will try a single-row ball bearing with ten 1/4 in diameter balls and a 1 in diameter mean ball path. Will this choice meet the requirements?

18.8. A proposed machine includes a shaft that rotates at 2000 rpm and causes a 55 lb radial load on the bearings at each end of the shaft. The machine will operate 40 h/week. Specify the bearings and determine predicted life based on a reliability of 0.95 (for a single bearing).

Design Decisions: We will try a single-row ball bearing with eight 0.1 in diameter balls and a 0.6 in diameter mean ball path.

18.9. We need to specify shaft bearings that will carry a 200 N radial load. The shaft will operate 40 h/week at 220 rpm. Specify the bearings, and determine service life, with 96% reliability of individual bearings.

Design Decisions: We will try a single-row ball bearing with ten 1.5 mm diameter balls and a 10 mm diameter mean ball path.

18.10. A shaft is to rotate at 3500 rpm and carry a 200 N radial load. The shaft will operate 40 h/week. Specify the bearings. Determine service life based on a 2% failure probability for an individual bearing.

Design Decisions: We will try a single-row ball bearing with eleven 3 mm diameter balls and a 25 mm diameter mean ball path.

18.11. A ball bearing rotating at 1760 rpm will support a 240 lb load. Find the required basic load rating so that a bearing can be selected from a catalog.

Design Decisions: We will plan on a 600 h life expectancy with 0.95 reliability (5% failure rate).

18.12. A ball bearing rotating at 720 rpm will support a 90 lb load. Find the required basic load rating so that a bearing can be selected from a catalog.

Design Decisions: We will plan on a 3800 h life expectancy with a 2.5% failure rate.

18.13. A ball bearing rotating at 400 rpm will support a 95 lb load. Find the required basic load rating so that a bearing can be selected from a catalog.

Design Decisions: We will plan on a 2000 h life expectancy with a predicted failure rate of 4 bearings in 100.

18.14. Select bearings for 440 lb load and 300 rpm shaft speed.

Design Decisions: Roller bearings will be used. We will select bearings for a 4800 h life expectancy with 2% failure probability.

18.15. Select roller bearings for 2300 N load and 300 rpm shaft speed.

Design Decisions: We will select bearings for a 7000 h life expectancy with 2.5% failure probability. Find basic load rating in N.

18.16. Select bearings for 1900 lb load and 5500 rpm shaft speed.

Design Decisions: Roller bearings will be used. We will select bearings for a 6500 h life expectancy with 0.98 reliability.

18.17. Find the absolute viscosity of AGMA grade 4 oil at 104°F (40°C). Give values in cP and reyn.

18.18. Find the absolute viscosity of a lubricating oil with an ISO/ASTM viscosity of 68. Give values in cP and reyn.

18.19. Find the absolute viscosity of a typical SAE 30 crankcase oil at 104°F (40°C). Give values in cP and reyn.

18.20. Bearings are needed to support a 3.5 in diameter shaft. The bearing load will be 4000 lb at each bearing, and shaft speed may range up to 6000 rpm. Propose a bearing. Determine friction torque and power loss.

Design Decisions: Select a 3.5 in long journal bearing with 0.0025 in radial clearance. Lubricating oil with a viscosity of 24 cP at the operating temperature will be used. Use the Newton–Petrov theory. Plot friction torque, friction horsepower, and Sommerfeld number against shaft speed. (*Note:* The Newton–Petrov theory may not produce an accurate solution for the given conditions.)

18.21. Bearings are needed to support a 1 in diameter shaft. The bearing load will be 100 lb at each bearing, and shaft speed may range up to 6000 rpm. Propose a bearing. Determine friction torque, power loss, and coefficient of friction.

Design Decisions: Select a 1 in long journal bearing with 0.0015 in radial clearance. Lubricating oil with a viscosity of 13.8 cP at the operating temperature will be used. Use the Newton–Petrov theory. Plot friction torque, friction horsepower, and coefficient of friction against shaft speed. (*Note:* The Newton–Petrov theory will not be valid for low-speed operation.)

18.22 You are asked to select bearings to support a 2 in diameter shaft. The load will be 250 lb at each bearing, and shaft speed may range up to 6000 rpm. Propose a bearing. Determine friction torque, power loss, and coefficient of friction.

Design Decisions: Select a 1.5 in long journal bearing with 0.0015 in radial clearance. Lubricating oil with a viscosity of 27.6 cP at the operating temperature will be used. Use the Newton–Petrov theory. Plot friction torque, friction horsepower, and coefficient of friction against shaft speed. (*Note:* The Newton–Petrov theory will not be valid for low-speed operation.)

18.23. Bearings are required to support a shaft load of 3000 lb per bearing. The shaft is 3 inches in diameter and rotates at 550 rpm. Select bearings of this application, and determine eccentricity, friction loss, heat generated, and other relevant values.

Design Decisions: We will tentatively select a 3 in long journal bearing with 0.003 in radial clearance. We will use a lubricant with 25 cP viscosity at the expected operating temperature. Hydrodynamic lubrication theory for heavy loads (HDL–HL) will be used.

18.24. Select bearings to support a shaft load of 700 lb per bearing. The shaft is 1.5 inches in diameter and rotates at 1800 rpm. Determine eccentricity, friction loss, heat generated, and other relevant values.

Design Decisions: We will tentatively select a 1.5 in long journal bearing with 0.002 in radial clearance. We will use a lubricant with 20 cP viscosity at the expected operating temperature. Hydrodynamic lubrication theory for heavy loads (HDL–HL) will be used.

18.25. Select bearings to support a shaft load of 800 lb per bearing. The shaft is 1.75 inches in diameter and rotates at 1200 rpm. Determine eccentricity, friction torque, power loss, heat generated, and other relevant values.

Design Decisions: We will tentatively select a 1.75 in long journal bearing with 0.002 in radial clearance. We will use a lubricant with 24 cP viscosity at the expected operating temperature. Hydrodynamic lubrication theory for heavy loads (HDL–HL) will be used.

18.26. Make a tentative bearing selection for supporting 500 lb per bearing. Shaft diameter = 2 in. Plot eccentricity ratio, torque, and power loss against shaft speed.

Design Decisions: We will try a 2 in long bearing with 0.002 in radial clearance, and we will use oil with 20 cP viscosity at the expected operating temperature. Speeds up to 4000 rpm will be considered.

18.27. Make a tentative bearing selection for supporting 100 lb per bearing. Shaft diameter = 1 in. Plot eccentricity ratio, torque, and power loss against shaft speed.

Design Decisions: We will try a 1 in long bearing with 0.0015 in radial clearance, and we will use oil with a viscosity of 2 μreyns at the expected operating temperature. Speeds up to 6000 rpm will be considered. Note that low-speed results may be suspect.

18.28. Make a tentative bearing selection for supporting 200 lb per bearing. Shaft diameter = 1.5 in. Plot eccentricity ratio, torque, and power loss against shaft speed.

Design Decisions: We will try a 1.5 in long bearing with 0.002 in radial clearance, and we will use oil with 15 cP viscosity at the expected operating temperature. Speeds up to 6000 rpm will be considered. Note that low-speed results are suspect.

CHAPTER 19

Design for Manufacture

Symbols

C	cost per reject or failure	y_i	an individual measurement
D_{cum}	cumulative distribution function	y_{mean}	mean value of a set of measurements
P	probability	y_{var}	variance of a set of measurements
R	reliability	Y	target measurement
S_{YP}	yield point strength	SD	standard deviation

Units

cost	usually monetary but can be lost time, and so on
stress	psi or MPa
y_{mean}, Y, and SD	same as units selected for y_i (in, psi, mm, and so on)

The chapter title may seem redundant in that the purpose of machine design is to produce a product that *will* be manufactured. However, **design for manufacture** is a design review method that identifies optimal part design, materials choice, and assembly and fabrication operations to produce an efficient and cost-effective product. This chapter is intended to introduce some of those concepts.

CONCURRENT ENGINEERING

As noted in an earlier section, **concurrent engineering (CE)** is a team approach that considers all elements of a product life cycle. These elements include

- recognition of a need for a product
- formulation of specifications (with potential customer input)
- creative design
- documentation (computer-aided drafting, solid modeling, parts lists)
- analysis (finite element analysis for stress and deflection, kinematic analysis, analysis of product utility)
- manufacture (computer-integrated manufacture)
- testing
- marketing
- maintenance
- disposal and recycling

A team, with members representing all phases of the product life cycle, attempts to consider all of these elements in parallel. Potential benefits of concurrent engineering include

- low product cost
- high product quality and reliability
- shorter new-product introduction cycles
- quick "ramp up" to production volumes
- user-friendly products
- "green" (recyclable) products

MANUFACTURING PROCESSES

Manufacturing processes are treated in detail in a number of references, including DeGarmo, Black, and Kohser (1988); Doyle et al. (1985); and Neely and Kibbe (1987). Machine designers should be substantially familiar with common manufacturing processes and with the concepts treated in these and similar references. With such knowledge, they can help implement the concurrent engineering approach, integrating the design and manufacturing processes.

A few of the manufacturing processes of interest to designers include

- casting (expendable mold and reusable mold)
- hot-working (rolling, forging, drawing)
- cold-working (rolling, blanking, pressing, drawing, extruding, thread forming)

- powder metallurgy (hot-pressing, forging, injection molding)
- machining (milling, turning, shaping, planing, sawing, broaching, boring, drilling, reaming)
- abrasive machining (waterjet machining, grinding, ultrasonic machining)
- arc and flame cutting and machining
- electrodischarge and electrochemical discharge cutting and machining
- laser machining
- electron beam machining
- plasma beam machining
- plastics processing (blow molding, injection molding, extrusion)

Some joining processes include the use of

- welding (flame, arc, resistance, forge, laser, electron beam, ultrasonic)
- mechanical fasteners (bolts, nuts, screws, rivets, pins, retaining rings)
- adhesive bonding
- pressing (interference fits)
- crimping and seaming
- snap-in design (grooves and tabs in flexible parts)

Selection of a manufacturing process is often a local decision. The decision may depend on the equipment available in a given factory, the available hardware and software for numerically controlled and computer-controlled processes, and the skill of the employees. Purchases of new equipment must be justified by the expected production volume, particularly if the site has idle equipment that could serve almost as well as the proposed new equipment.

COMPUTER-AIDED MANUFACTURING

Numerical control (NC) and computer-aided design and manufacturing (CAD/CAM) systems are covered extensively in references by Ranky (1986), Groover and Zimmers (1984), and Hordeski (1986). Charles Babbage is often considered the father of the computer, based on his invention of the differential calculating machine in 1833. J. M. Jacquard (1757–1834) may be considered the father of computer-aided manufacturing (CAM). The Jacquard loom produced woven fabric in elaborate designs, utilizing a series of perforated cards to control the operation of the loom.

Numerical Control

Numerical control (NC) systems developed in the 1950s controlled machine tools with prerecorded coded directions. The early systems required the programmer to specify exact tool paths. They were followed by development of NC systems that included a dedicated computer with NC instruction stored in its memory. These systems, called **computer numerical control (CNC)** systems, make it easier to change the process.

Further advances in NC and CAM include

- the ability of the computer to provide feedback from the machine (tool status, production status, diagnostic information concerning the process)
- software to translate English-type statements into the required NC instructions

- menu-driven software that directs the user to select the appropriate machine, material (size and thickness), cutter (type and size), and machine operations (rough cut, finish cut)
- integrated CAD/CAM systems that extract the information from the design phase to generate the necessary manufacturing instructions (tool selection, tool path, etc.)

Flexible Manufacturing: Robots and Manipulators

Robots and manipulators allow for flexibility in production, the ability to change product lines rapidly. The **end-effector,** the final component at the free end of a robot arm, is critical to the operation. The end-effector holds the tool or, when used for material handling, grasps and releases a part. End-effectors may rely on one of the following methods for gripping:

- mechanical clamping, using a two- or three-finger gripper. The gripper can be operated with electric motors, solenoids, or pneumatic or hydraulic systems.
- magnetic attraction
- suction
- special tool-holder fittings

Robots are used for materials-handling in hostile environments, including the handling of radioactive materials and work in outer space. Manipulators are used to spray paint computers, automobiles, appliances, and furniture. They have the advantage of uniformity and of sparing humans exposure to hazardous environments.

Gas metal arc welding and spot welding are among the most cost-effective robot applications. Figure 19.1(a) shows gas metal arc welding (GMAW) of automotive exhaust components; Part (b) shows welding of display racks. The operator may be using the teach-pendant to adjust welding amperage or voltage, or wire feed speed. As noted in Chapter 16, "Welds and Adhesive Joints," the teach-pendant can also be used to program the robot.

The designer must decide between a flexible manufacturing system and a dedicated system. A dedicated system is a production system designed to perform only the specific operations required to produce a given product. The dedicated system may offer high precision and low unit cost for large production runs of a single product. However, flexible manufacturing systems can be reprogrammed for changes in product specifications, or even changes from one product to another.

Robots and manipulators can duplicate some assembly operations performed by humans. Feedback allows manipulators to sense forces. However, robots have primitive vision or none at all. Thus, currently available robots are not competitive with humans for most assembly operations. When robots and manipulators are to be used for assembly tasks, it is usually necessary to augment them with part feeders and orienters. Compliant manipulator designs are recommended for some assembly operations because incoming parts and partial assemblies may not be oriented precisely. Additional introductory information on robots is given by Roth (1983) and Wilson and Sadler (1993).

DESIGN FOR EASE OF MANUFACTURE AND ASSEMBLY

Costs attributed to assembly are an important part of manufacturing costs, whether the assembly process is manual or automated. Manufacturing costs are determined largely in the design phase. Much of the responsibility to minimize cost without reducing safety and utility

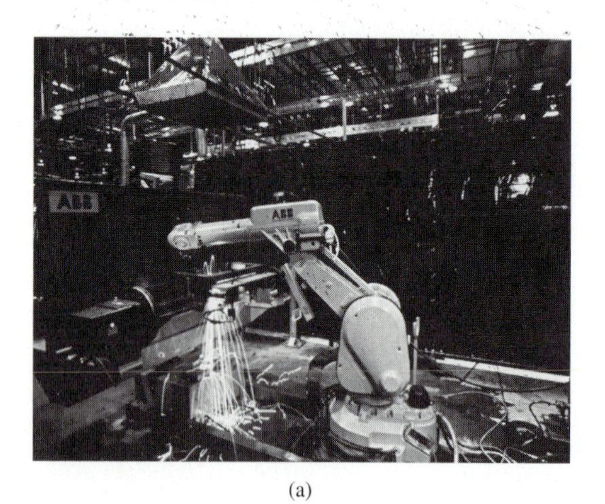

(a)

(b)

Figure 19.1 Flexible assembly. (a) Welding of 1996 Mustang exhaust systems; (b) welding of display racks. (Courtesy of ABB Flexible Automation, Inc.)

falls on the designer. A few assembly-cost-cutting methods worth considering are

- reducing the number of parts in a product
- making the assembly process easier
- standardization

Many of the suggestions listed below are discussed in detail by Shina (1991b).

Reducing the Number of Parts in a Product

Product cost is related to the number of parts contained in a product. In addition to direct part cost, we must add indirect costs such as the cost of maintaining part inventories. Reducing the number of parts in a product is also likely to increase the reliability of the product.

In an attempt to reduce the number of parts in a design, the designer should examine each part to determine its purpose. It may be possible to eliminate the part or to combine two or more parts into one. Factors to be considered include relative motion of parts, strength requirements, and the need for disassembly.

Examples of reduction of part numbers include the following:

- combining of functions (previously requiring two or more parts) into a single part and designing that part as a single casting
- automatic application of labels by including labels in the mold and/or combining information from two labels into one label

- use of snap fits to replace fasteners
- use of press fits to eliminate or reduce the number of fasteners
- use of adhesives or welds (unless this change would substantially increase assembly costs)

Standardization

Standardization involves the use of standard, off-the-shelf components when available. Standardization also refers to use of standard parts and standard modules that are common throughout a factory or within a product line. Standardization is encouraged by the development of databases that can identify and retrieve parts for designers. These practices should reduce tooling costs, assembly costs, and inventory costs.

It is usually cost-effective to specify standard fasteners and standard thicknesses of sheet metal when designing a product. Standardization also involves reducing the number of different types and sizes of fasteners across a product line, as well as reducing the number of different materials and thicknesses specified. Parts can sometimes be combined into subassemblies or modules. Where possible, the same subassemblies can be specified for a number of different products.

Symmetry

One way to make the assembly process easier is to design symmetrical parts. Symmetrical parts require less handling during manual assembly. If assembly is automated, part feeding and part orientation are simplified.

Symmetry is also an advantage when parts must be assembled in the field, particularly in cramped locations. Figure 19.2(a) shows an asymmetric cover plate design with only one correct assembly orientation. The redesigned part shown in Part (b) has four correct assembly orientations if the near-side and far-side of the plate are identical.

Asymmetry

Some parts will operate properly only if assembled with a particular orientation. In that case, we may build asymmetry into the design. If the asymmetry is subtle, the assembler may try to force the parts together with the wrong orientation, particularly if there is a critical, but hidden, feature. Exaggerated asymmetry assures the correct orientation for manual or machine assembly.

Figure 19.3(a) shows an almost symmetrical part that must be oriented in a particular way at assembly. It is unlikely that the redesigned part shown in Part (b) would be incorrectly assembled. When incorrect assembly would result in a safety hazard, the asymmetry should be obvious, and incorrect assembly should be virtually impossible. For example, to prevent interchanging of welding gases, right-hand thread fittings are used on one gas cylinder, and left-hand on the other.

Slotted Holes

Slotted holes and similar features can be used to accommodate variation in parts. For example, a motor base may have slotted holes to allow adjustment for belt stretching. The cover plate shown in Figure 19.4 is designed for easy field assembly. Machine screws are partially inserted in the box below the plate. The cover plate is then placed over the screws and rotated

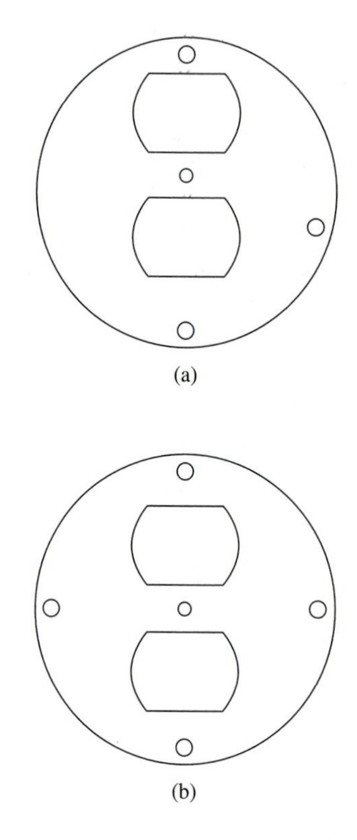

Figure 19.2 Redesign for symmetry.

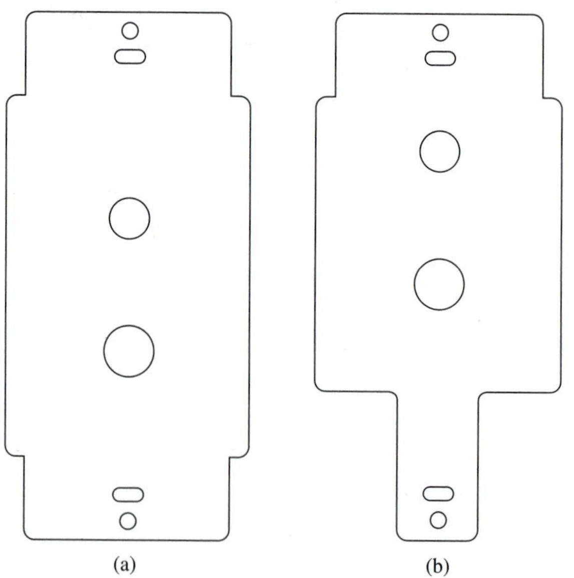

Figure 19.3 (a) Asymmetry; (b) exaggerated asymmetry.

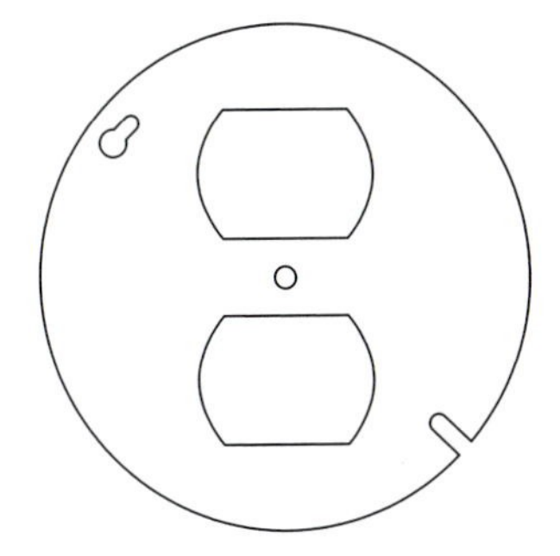

Figure 19.4 Slotted holes.

into place, and the screws are tightened. Although the cover plate may appear asymmetric, the location of the slot and the elongated hole can be interchanged.

DESIGN FOR THE ENVIRONMENT (DFE)

Discarded products often end up in landfills and incinerators or as litter on private and public lands and waterways. In **design for the environment (DFE),** in the early stages of the design process, the designer plans how to recycle products and reduce their negative impact on the environment.

The following suggestions are based principally on DFE guidelines given by Billatos and Nevrekar (1994):

1. Use recycled materials.
2. Use a single material, or only a few different materials, or materials that can be recycled together when the useful life of the product is over.
3. Avoid unnecessary finishes, toxic materials, and other materials that complicate recycling or disposal.
4. Find multiple or secondary uses for a product.
5. Design reusable containers for the product.
6. Keep the design simple.
7. Consider modular design to facilitate maintenance, repair, and recycling.
8. Design for long product life and easy maintenance.
9. Mold in logos to identify classes of materials for recycling.
10. Establish a tracking database for product recycling.
11. Look for product components that can be reused. Establish and publicize a buy-back system.
12. Look for production waste that can be turned into a useful by-product.
13. Design processes to use solvents, cleaning materials, and cutting fluids that are nonhazardous to workers and the environment.
14. Design the manufacturing process to dispose of all waste responsibly.
15. Reduce energy consumption in manufacturing.
16. Reduce product energy consumption.

17. Design the product for minimum air, water, and noise pollution.
18. Select materials that ensure a nonhazardous environment within the factory and the surrounding community.
19. Design production machinery and processes to avoid accidents and noise-induced hearing loss within the factory, and avoid contributing to community noise.

Some materials are more compatible with DFE than others. For example, thermoplastics can be melted down and remolded, but thermoset plastics cannot. DFE should be a part of CE. DFE includes design for disassembly (DFD), the practice of designing products for easy disassembly, component separation, and recycling.

PROTOTYPES

A prototype or mock-up is the first model of a product. Prototypes are used for

- design evaluation (including aesthetics or styling)
- fit-up (tolerance analysis and evaluation of assembly methods)
- selecting manufacturing options for full-scale production of the final product
- demonstration
- functional performance studies (including ergonomics)
- testing and analysis (including estimation of durability and reliability)

Computer Solid Models

Some of the prototype functions listed above can be performed by computer solid models. These computer models can supplement or replace actual physical prototypes. The solid model can be viewed from any angle and can be used to check for tolerances and interference. Moving parts of the model can be animated to view kinematic performance. Forces, moments, pressures, and thermal loads can be applied to a finite element representation of the model to determine deflections, stresses, and support reactions.

Computer simulation results uncover critical points in the product design, such as locations of high stress. At this design stage, it is still feasible to consider design alternatives, including dimension changes, material changes, and even gross changes in the product design.

Machined Prototypes

Physical prototypes can be produced in a variety of materials, from plastics to steel, using specialized machinery. The process of producing a machined prototype might include the following steps:

1. Generate a preliminary computer solid model of the proposed part.
2. Perform all required analysis for redesign.
3. Generate the redesigned computer solid model of the part.
4. Transfer the solid model file to a CNC tool-path generation file.
5. Machine the prototype with a CNC mill.

In some cases, the prototype will be almost identical to the final production-quantity part, except for the production process. The prototype can then be tested to confirm performance and strength requirements.

A desktop prototyping and manufacturing system is shown in Figure 19.5. The system includes a three-axis CNC mill with menu-driven software.

Stereolithography

Stereolithography is a rapid prototyping method particularly suited to parts with complex geometry. The prototype is generated in a vat of liquid photosensitive resin. The following steps are typical:

1. A computer solid model of the part is constructed.
2. A platform attached to an elevator mechanism is submerged a few thousandths of an inch below the surface of the resin.
3. A microcomputer-controlled laser beam draws a cross section of the proposed part on the surface of the platform, curing the resin and forming the first layer of the part.
4. The platform and the first layer of the part are lowered an additional few thousandths of an inch into the resin.
5. A second layer of the part is generated above the first.
6. The part is generated layer-by-layer.
7. The part is removed from the vat.
8. Surplus resin is stripped off, and the remaining uncured resin in the part is cured.

Figure 19.5 Desktop prototyping and manufacturing system. (Courtesy of Light Machines Corporation)

Advantages of stereolithography include its ability to generate complicated parts and its short turnaround time. It is sometimes practical to generate prototypes of two or more alternative designs for comparison purposes. The prototype can be used as a master pattern. Typical stereolithography resins can be machined, sanded, or tapped.

Testing of the stereolithography prototype is usually not practical because the resin is typically brittle and unable to withstand high stress or high temperature. Stereolithography methods can generate parts with undercuts and internal features that would be impractical or impossible to produce by machining. This can sometimes be a disadvantage, since most production runs will utilize conventional manufacturing processes.

Other Rapid Prototyping Methods

Additional methods of producing prototypes include the following:

- **Fused deposition modeling.** This rapid prototyping method is controlled by CAD files or NC code. A spooled filament of wax-filled plastic material is fed into a heated extruder. The extruder head lays down the softened material in thin layers as directed by the CAD or NC files. The material solidifies to form the model.
- **Laminated object manufacturing.** This method utilizes sheets of metal, polymer, paper, or other materials to produce a model. A laser cuts the appropriate cross-sectional outline of the part on each sheet as directed by CAD files. The process is repeated as sheets of material are piled on top of one another. Glue is used to bind the sheets together.
- **Selective laser sintering.** In this, another CAD-controlled process, a thin layer of fine wax or polymer powder is spread over a platform. The powder is heated by a laser in a pattern corresponding to the appropriate cross section. Another layer of powder is added to the first and is heated so that the layers weld (bond together). The heated area of each layer is the cross-sectional area of the part stored in the CAD file. This process is similar to stereolithography except that stereolithography uses liquids, and selective laser sintering uses powdered material.

ROBUST DESIGN AND VARIABILITY REDUCTION

The word *robust* makes us think of strength and endurance, possibly under adverse conditions. **Robust design (RD)** is a set of techniques used to optimize products and processes for performance to specifications. The product designer and the designer of the manufacturing process work together. The RD goals can be achieved if the following conditions are met:

1. The product designer can increase allowable tolerances to the maximum value consistent with proper functioning of the product.
2. The process designer can reduce product variability.
3. The process designer can center the process.

If a product design calls for close tolerances, then tool wear, temperature changes, and other effects may make the product unserviceable. In such cases, we should consider redesign of tolerance-sensitive parts to make the product more robust. **Product variability** refers to the variation of critical dimensions or other attributes. A manufacturing process is **centered** if critical dimensions or other product attributes are clustered about the target value. Shina (1991b) discusses the experiments and the analysis tools required to implement RD techniques.

EVALUATING QUALITY IMPROVEMENT

The consequences of product failure can range from annoying to disastrous. Even the mildest failure consequences involve scrapping of manufactured products and associated costs.

The Cost of Product Failures and Rejects

Significant human and economic loss may result from products that fail and products that are rejected by the manufacturer or customer. These costs may be related to

- scrapping of products containing defective parts
- product recalls
- injuries or deaths related to product defects
- delays in filling orders
- subsequent loss by the customer due to manufacturer delays
- loss of customer confidence in the manufacturer or loss of good will

It is sometimes possible to assign a cost to product rejects or failures when all relevant cost contributions are considered.

Target Measurements and Tolerances

Certain product attributes are critical. Examples of product attributes include the strength of a shaft, the capacity of a brake or clutch, and critical dimensions of a part in an assembly. A **target measurement** is the nominal value of that attribute, and the **tolerance** is the allowable variation from that value.

Tolerances should be realistic and based on actual needs. When unnecessary, close tolerances increase production costs without adding value to the product. In some cases, close tolerances may reduce product serviceability, for example, when a temperature change causes uneven expansion in moving parts in a machine. Examples of tolerances on a dimension could be

$$\pm 0.001 \quad \text{or} \quad +0.000/-0.002 \quad \text{or} \quad +0.0005/-0.0015$$

In the case of a favorable attribute, we might specify a lower limit only. For example, the strength of a material could be specified as

$$S_{YP} \geq 145{,}000 \text{ psi} \quad \text{or} \quad S_{YP} \geq 1000 \text{ MPa}$$

Product Tests and Measurements

If we could measure and test every part, statistical analysis would be unnecessary. However, extensive measurements are expensive and unnecessary, while destructive tests of 100% of manufactured products would result in zero output. Thus, we make a small number of measurements or tests of a critical product attribute, and then we compare the measurements with target measurements and tolerances.

Useful Statistical Descriptors

Suppose that we make n tests or take n measurements y_i where $i = 0, \ldots, n - 1$. Then the **mean** is given by

$$y_{\text{mean}} = \frac{1}{n}\sum_{i=0}^{n-1} y_i \tag{19.1}$$

The variance is

$$y_{\text{var}} = \frac{1}{n}\sum_{i=0}^{n-1} (y_i - y_{\text{mean}})^2 \tag{19.2}$$

and the standard deviation is

$$SD = \sqrt{y_{\text{var}}} \tag{19.3}$$

It is assumed that these descriptors, based on a few measurements, are representative of the total production process. We also assume that the distribution of values of a product attribute is a normal distribution about the mean. The validity of our analysis depends on how closely the assumptions model the actual manufacturing process.

Probability

The cumulative distribution function for a normal distribution is given by

$$D_{\text{cum}} = P(x < x_1) = \frac{1}{\sqrt{2\pi}}\int_{-\infty}^{x_1} e^{-x^2/2}\,dx \tag{19.4}$$

where $x = \dfrac{y - y_{\text{mean}}}{SD}$

$D_{\text{cum}} = P(x < x_1) =$ the cumulative distribution function $=$ the probability of $x < x_1$

Probability of a Reject

In practice, estimation of the probability of an out-of-specification measurement is not as difficult as it would appear from the above equations. The cumulative distribution function is tabulated in books on statistics. The mean, variance, standard deviation, and cumulative distribution function are available on mathematical software. We proceed as follows:

1. Select a critical attribute.
2. Make several measurements of that attribute.
3. Determine the mean and the standard deviation of the measurements. Use mathematical software or a preprogrammed calculator if available.
4. Establish a target measurement Y and acceptable limits Y_{MIN} and Y_{MAX}.
5. Calculate $x_1 = (Y_{\text{MIN}} - y_{\text{mean}})/SD$.

Figure 19.6 The normal curve.

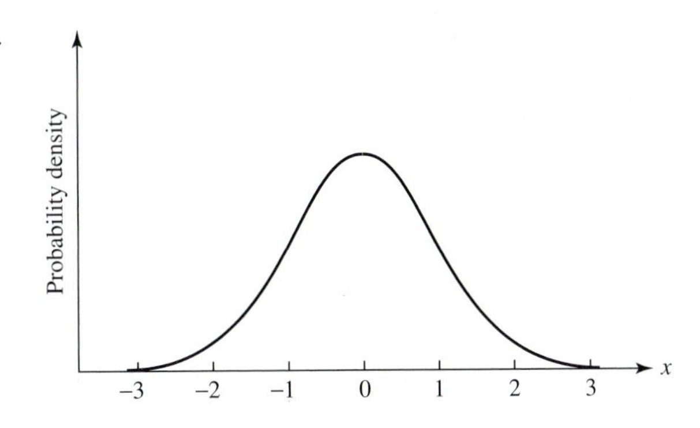

6. Determine $P_1 = D_{cum}(x_1) = P(x < x_1)$. Use mathematics software or statistical tables. This is the probability that a measurement will fall below Y_{MIN}.
7. Calculate $x_2 = (Y_{MAX} - y_{mean})/SD$.
8. Determine $P_2 = 1 - D_{cum}(x_2) = 1 - P(x < x_2)$. Again, use mathematics software or statistical tables. This is the probability that a measurement will fall above Y_{MAX}.
9. Calculate $P = P_1 + P_2$. This is the probability that a product will be out of specification.

The Normal Curve

We are using a normal or Gaussian distribution to approximate a chance happening, for example, the distribution of measurements about a mean value. Figure 19.6 shows a normal curve, a plot of probability distribution. The cumulative distribution function is given by

$$D_{cum}(x_1) = P(x < x_1) = \text{area under the curve from } x = -\infty \text{ to } x = x_1$$

Cost Calculation

The cost per product due to failure to meet the given specification is simply

$$C_1 = CP \qquad \textbf{(19.5)}$$

where C_1 = cost attributed to each product due to the probability of some out-of-specification products

C = cost per reject or failure

P = probability of failing to meet specifications

EXAMPLE PROBLEM Evaluating the Probability That a Product Will Not Meet Specifications and Calculating the Associated Costs

A manufacturer has received a large order for shafting. The shafts are to have a diameter of 2.000 ± 0.002 in. Ten shafts were produced, with the following diameters (in inches):

1.999 (2 samples) 2.000 (3 samples) 2.001 (2 samples) 2.002 (3 samples)

Find the probability of manufacturing a shaft that does not meet specifications. Find the cost due to product rejects.

Decisions: We estimate a cost of $90 per reject, including production costs and indirect costs related to loss of customer confidence. We will assume that shaft diameters (for the entire order) are normally distributed about a mean value.

Solution Summary: The 10 samples have a mean diameter of 2.0006 in with a standard deviation of 0.001114. All 10 samples are within the specified limits. However, using the cumulative normal distribution function, we predict that about 1% of the entire order will be rejected because the diameter is too small. Approximately 10% more will be rejected because the diameter is too large. The total probability of an out-of-specification shaft is 0.114, and the cost due to the probability of rejects is $10.27 per shaft.

The detailed solution follows.

Detailed Solution:

Cost of Reject ($/Reject):

$$C = 90$$

Number of Measurements:

$$n = 10 \qquad i = 0, \ldots, n - 1$$

Target Measurement:

$$Y = 2.000$$

Tolerance (+/−):

$$Y_t = 0.002$$

Acceptable Limits:

$$Y_{\text{MIN}} = Y - Y_t \qquad Y_{\text{MIN}} = 1.998$$
$$Y_{\text{MAX}} = Y + Y_t \qquad Y_{\text{MAX}} = 2.002$$

Measured Values:

y_i

2.002
2.000
2.001
2.000
2.001
1.999
2.000
2.002
2.002
1.999

Mean:

$$\text{mean}(y) = 2.0006$$

Variance:

$$\text{var}(y) = 1.24 \cdot 10^{-6}$$

Standard Deviation:

$$SD = \text{stdev}(y) \qquad SD = 0.001114$$

Probability of y < Y_{MIN}:

$$x_1 = \frac{Y_{MIN} - \text{mean}(y)}{SD} \qquad x_1 = -2.335$$

$$P_1 = c_{\text{norm}}(x_1) \qquad P_1 = 0.009775$$

Probability of y > Y_{MAX}:

$$x_2 = \frac{Y_{MAX} - \text{mean}(y)}{SD} \qquad x_2 = 1.257$$

$$P_2 = 1 - c_{\text{norm}}(x_2) \qquad P_2 = 0.104334$$

Probability of Out-of-Specification Product:

$$P = P_1 + P_2 \qquad P = 0.114$$

Cost per Unit ($/Unit):

$$C_1 = C \cdot P \qquad C_1 = 10.27$$

■■

DESIGN AND MANUFACTURING ALTERNATIVES

There are several possibilities for reducing the cost of out-of-specification products. They include

- controlling the manufacturing process to reduce variation in output
- increasing the tolerances by a design change
- centering the process

A reduction in product variation may be desirable, but it may be difficult and expensive in some cases. The second possibility is very attractive. For example, the close tolerances may be due to the necessity of fitting parts together. A different method of fastening or assembly may be possible, or we can choose to combine two parts into one casting. Alternatively, we can eliminate close tolerances by specifying elongated holes and locating pins in an assembly. The third possibility, centering the process, is discussed below.

Cost Saving by Centering the Process

If the mean of a dimension is not halfway between the upper and lower acceptable limits, then centering the process should reduce the number of rejects. In most manufacturing processes, it is easier to change the mean value of a dimension than it is to reduce variability. A change in the mean value may require only adjustment of a machine or process.

In the preceding example problem, the target diameter is 0.0006 in less than the mean diameter of the samples. As a result, there is a high probability of rejects of oversize shafts. Oversize shafts are represented by the area of the tail of the normal curve to the right of x_2. By centering the process, we shift the distribution, putting the mean measurement ($x = 0$ on the normal curve) on top of the target measurement, thus increasing the area of the curve representing in-specification parts.

EXAMPLE PROBLEM Centering the Process for Cost Reduction

Investigate the possible cost savings due to centering the process described in the preceding example problem.

Decision: We will attempt to adjust the manufacturing process so that the mean part dimension equals the target measurement.

Solution Summary: We assume that a new set of sample shafts would resemble the original set, except that each diameter would change by the amount of the adjustment. That is,

$$z_i = y_i + Y - y_{\text{mean}}$$

where z_i = diameter of each shaft in the new set

Following the same procedure as in the previous example, we obtain a lower probability of out-of-specification production and a lower cost per unit. The theoretical cost saving is about $3.75 per shaft. Note that we do not change the variance or the standard deviation by adjusting the mean. If we were doing hand calculations, we would simply use the values obtained above. In the detailed solution which follows, we let the computer redo this calculation.

Detailed Solution:

$$z = \text{projected (not actual) product measurements}$$
$$z_i = y_i + Y - \text{mean}(y)$$

Mean:

$$\text{mean}(z) = 2$$

Variance:

$$\text{var}(z) = 1.24 \cdot 10^{-6}$$

Standard Deviation:

$$SD = \text{stdev}(z) \qquad SD = 0.001114$$

Probability of $z < Y_{MIN}$:

$$x_1 = \frac{Y_{\text{MIN}} - \text{mean}(z)}{SD} \qquad x_1 = -1.796$$

$$P_1 = c_{\text{norm}}(x_1) \qquad P_1 = 0.036243$$

Probability of $z > Y_{MAX}$:

$$x_2 = \frac{Y_{\text{MAX}} - \text{mean}(z)}{SD} \qquad x_2 = 1.796$$

$$P_2 = 1 - c_{\text{norm}}(x_2) \qquad P_2 = 0.036243$$

Probability of Out-of-Specification Product:

$$P = P_1 + P_2 \qquad P = 0.072$$

Cost per Unit ($/Unit):

$$C_c = C \cdot P \qquad C_c = 6.524$$

Potential Cost Saving by Centering the Process ($/Unit):

$$C_s = C_1 - C_c \qquad C_s = 3.746$$

Reducing Output Variation and Increasing Tolerances

After the process is centered, it may be necessary to take further steps to reduce the probability of rejects. Either increasing tolerances or reducing part variation can have significant effect. For the centered process, we calculate

$$x_1 = \frac{-Y_t}{SD} \quad \textbf{(19.6)}$$

and

$$P = 2D_{\text{cum}}(x_1) = 2P(x < x_1) \quad \textbf{(19.7)}$$

where Y_t = tolerance
$Y \pm Y_t$ = critical part dimension
P = probability of a reject

Reliability is the probability of an in-specification product. It is given by

$$R = 1 - P \quad \textbf{(19.8)}$$

where R = process reliability = predicted fraction of satisfactory output

■■ EXAMPLE PROBLEM Reducing Rejects in a Centered Process by Decreasing Variation or Increasing Tolerance

In the centered process of the previous example, with $x_1 = -1.796$, the probability of an out-of-specification product was $P = 0.072$. How could this probability be reduced to 1/500?

Decision (a): We will try to control the manufacturing process more closely.

Solution (a): Using statistical tables or mathematics software and the procedure outlined above, we find that we must reduce the standard deviation to obtain $x_1 = -3.09$. The process must be improved to produce a standard deviation not greater than

$$SD = \frac{Y_t}{|x_1|} = \frac{0.002}{3.09} = 6.47 \cdot 10^{-4}$$

Decision (b): We will examine the design for the possibility of increasing the tolerance.

Solution (b): If the process is unchanged, then we must use the standard deviation obtained from the 10 samples. To yield the required reliability, the new tolerance must be no less than

$$Y_t = SD \cdot |x_1| = 0.001114 \cdot 3.09 = 0.00344 \text{ (say, 0.0035)}$$

■■

Problems Involving One Tail of the Normal Distribution Curve and Problems Involving Two Tails

Note that the shaft dimension in the preceding example problem must fall between an upper and a lower limit. A value of x in either tail of the normal distribution curve represents a reject. A requirement that $|x_1| = 3.09$ results in a prediction of 0.998 process reliability, or 1 reject in 500 products. In an earlier chapter, we considered shaft strength. The shaft must be strong enough, but there is no upper limit. Thus, only one tail of the normal distribution curve represents failure. In that case, requirement that $x_1 = -3.09$ represents a prediction of 0.999 reliability, or 1 failure in 1000 products.

Probability of a Reject for a Centered Process

Values of x_1 ranging from -6 to 0 are shown in Table 19.1. The probability of a reject for a centered process is shown to the right of each x_1 value. Of course, we cannot specify a tolerance of zero. Zero tolerance corresponds to $x_1 = 0$ and 100% rejects. At $x_1 = -6$, the reject probability is about $2 \cdot 10^{-9}$. At $x_1 = -6$, the tolerance is six times the standard deviation; this is called the **six sigma level** since the Greek letter σ (sigma) is often used for standard deviation.

Table 19.1 Probability of a reject if a process is centered.

$x_1 = -6, -5.9, \ldots, -3$		$x_1 = -3, -2.9, \ldots, 0$	
x_1	$2 \cdot c_{norm}(x_1)$	x_1	$2 \cdot c_{norm}(x_1)$
−6	0.00000000197	−3	0.0027
−5.9	0.00000000364	−2.9	0.00373
−5.8	0.00000000663	−2.8	0.00511
−5.7	0.00000001198	−2.7	0.00693
−5.6	0.00000002144	−2.6	0.00932
−5.5	0.00000003798	−2.5	0.01242
−5.4	0.00000006664	−2.4	0.0164
−5.3	0.0000001158	−2.3	0.02145
−5.2	0.00000019929	−2.2	0.02781
−5.1	0.00000033965	−2.1	0.03573
−5	0.0000005733	−2	0.0455
−4.9	0.00000095837	−1.9	0.05743
−4.8	0.00000158666	−1.8	0.07186
−4.7	0.00000260161	−1.7	0.08913
−4.6	0.00000422491	−1.6	0.1096
−4.5	0.00000679535	−1.5	0.13361
−4.4	0.00001082509	−1.4	0.16151
−4.3	0.00001707981	−1.3	0.1936
−4.2	0.0000266915	−1.2	0.23014
−4.1	0.00004131501	−1.1	0.27133
−4	0.00006334248	−1	0.31731
−3.9	0.00009619269	−0.9	0.36812
−3.8	0.00014469609	−0.8	0.42371
−3.7	0.00021559947	−0.7	0.48393
−3.6	0.00031821718	−0.6	0.54851
−3.5	0.00046525816	−0.5	0.61708
−3.4	0.00067385853	−0.4	0.68916
−3.3	0.00096684828	−0.3	0.76418
−3.2	0.00137427588	−0.2	0.84148
−3.1	0.00193520643	−0.1	0.92034
−3	0.00269979606	0	1

Although the values in Table 19.1 are calculated with many significant figures, they represent only a crude estimate of future manufacturing performance. Prediction of future production trends based on a few measurements is an inexact science. Furthermore, the normal distribution curve is only one possible model of a production process. The probability of rejects for a centered process is also given as a plot in Figure 19.7.

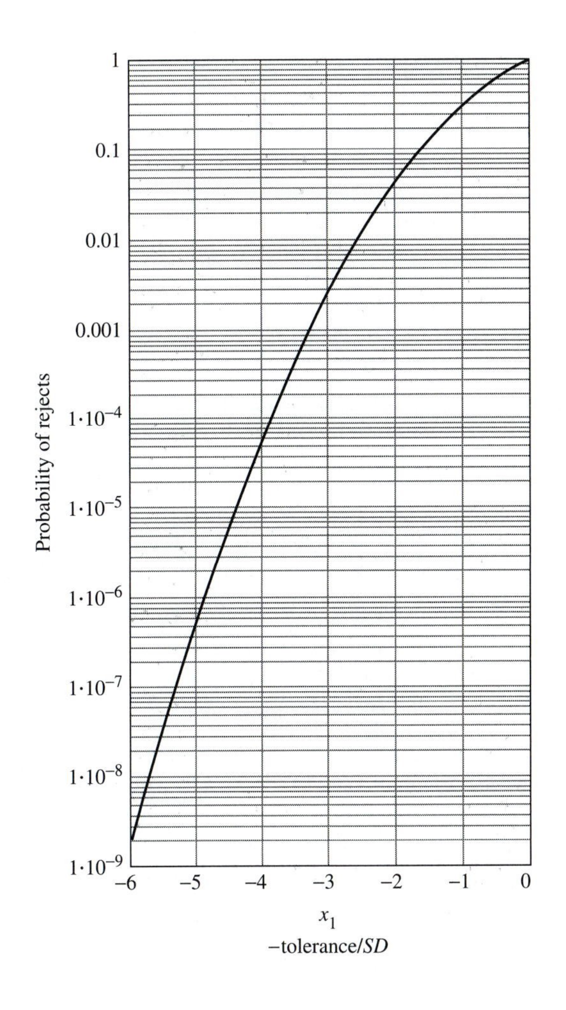

Figure 19.7 Probability of rejects for a centered process.

References and Bibliography

Billatos, S. B., and V. V. Nevrekar. "Challenges and Practical Solutions to Designing for the Environment." ASME National Design Engineering Conference. Chicago, March 1994.

Boothroyd, G., and P. Dewhurst. "Design for Assembly: Selecting the Right Method." *Machine Design* [November 10, 1983(a)]: 94–98.

Boothroyd, G., and P. Dewhurst. "Design for Assembly: Manual Assembly." *Machine Design* [December 8, 1983(b)]: 140–45.

Carter, D. E., and B. S. Baker. *Concurrent Engineering, The Product Development Environment for the 1990's.* Reading, MA: Addison-Wesley, 1992.

DeGarmo, E. P., J. T. Black, and R. A. Kohser. *Materials and Processes in Manufacturing.* 7th ed. New York: Macmillan, 1988.

Doyle, L. E., C. A. Keyser, J. L. Leach, G. F. Schrader, and M. B. Singer. *Manufacturing Processes and Materials for Engineers.* 3d ed. Englewood Cliffs, NJ: Prentice-Hall, 1985.

Groover, M. P., and E. W. Zimmers. *CAD/CAM: Computer-Aided Design and Manufacturing.* Englewood Cliffs, NJ: Prentice-Hall, 1984.

Hordeski, M. F. *CAD/CAM Techniques.* Reston, VA: Reston Publishing, 1986.

King, R. *Better Designs in Half the Time.* 3d ed. Methuen, MA: GOAL/QPC, 1989.

Marsh, S., J. W. Moran, S. Nakui, and G. Hoffherr. *Facilitating and Training in Quality Function Deployment.* Methuen, MA: GOAL/QPC, 1991.

McCormick, E. J., and M. S. Sanders. *Human Factors in Engineering and Design.* 5th ed. New York: McGraw-Hill, 1982.

Neely, J. E., and R. R. Kibbe. *Modern Materials and Manufacturing Processes.* New York: John Wiley, 1987.

Ranky, P. G. *Computer Integrated Manufacturing.* London: Prentice-Hall International, 1986.

Roth, B. "Introduction to Robotics." *Design and Application of Small Standardized Components.* Vol. 2, pp. 723–773, Mineola, NY: Stock Drive Products, 1983.

Shina, S. G. *Concurrent Engineering and Design for Manufacture.* Lowell, MA: University of Lowell, 1991a.

Shina, S. G., *Concurrent Engineering and Design for Manufacture of Electronics Products.* New York: Van Nostrand Reinhold, 1991b.

Wilson, C. E., and J. P. Sadler. *Kinematics and Dynamics of Machinery.* New York: HarperCollins, 1993.

Design Problems

19.1. A critical dimension must fall between 200 and 224 mm. Eight samples were examined, yielding the following measurements (mm):

201 221 200 210 210 209 211 205

Find the mean and the standard deviation of the measurements. Find the probability that a dimension will not meet the specifications. Estimate the cost of production that is attributable to rejected products.

Decisions: The cost per out-of-specification product is estimated to be \$1000/reject. We will assume that the critical dimension is described by a normal distribution.

19.2. Consider lowering the cost of production that is attributable to rejected products in the process described in Problem 19.1.

Decision: We will determine the possible cost saving due to centering the process.

19.3. The cost of each out-of-specification product is estimated to be $120. A critical dimension on that product must fall between 1.120 and 1.130 in. Ten samples were examined, yielding the following measurements (in):

1.124 1.121 1.125 1.120 1.129
1.121 1.126 1.122 1.127 1.123

Find the mean and the standard deviation of the measurements. Find the probability that a dimension will not meet specifications. Estimate the cost of production that is attributable to rejected products.

Design Decision: We will assume that the critical dimension is described by a normal distribution.

19.4. Consider lowering the cost of production that is attributable to rejected products in the process described in Problem 19.3.

Design Decision: We will determine the possible cost saving due to centering the process.

19.5. A critical dimension has a target value of 25 mm and a tolerance of ± 0.010 mm. Ten samples were examined, yielding the following measurements (mm):

24.999 24.990 25.000 25.002 25.002
24.991 25.009 24.995 24.996 24.994

Find the mean and the standard deviation of the measurements. Find the probability that a dimension will not meet the specifications. Estimate the cost of production that is attributable to rejected products.

Design Decisions: The cost per out-of-specification product is estimated to be $50/reject. We will assume that the critical dimension is described by a normal distribution.

19.6. Consider lowering the cost of production that is attributable to rejected products in the process described in Problem 19.5.

Design Decision: We will determine the possible cost saving due to centering the process.

19.7. The tolerance on a critical dimension is 0.003 in. A number of samples were measured, and the standard deviation of the measurements was found to be 0.0015. The target probability of rejects is 1 in 1000. If the current probability of rejects exceeds that value, consider changes to attain the target value.

Design Decision (a): We will consider tighter control over the manufacturing process. Find the new required standard deviation.

Design Decision (b): We will consider redesign of the product to allow a tolerance change for the critical dimension. The process related to the critical dimension will be unchanged. Determine the required tolerance.

19.8. The tolerance on a critical dimension is 0.0015 in. A number of samples were measured, and the standard deviation of the measurements was found to be 0.0019. The target probability of rejects is 0.01. If the current probability of rejects exceeds that value, consider changes to attain the target value.

Design Decision (a): We will consider tighter control over the manufacturing process. Find the new required standard deviation.

Design Decision (b): We will consider redesign of the product to allow a tolerance change for the critical dimension. The process related to the critical dimension will be unchanged. Determine the required tolerance.

19.9. The tolerance on a critical dimension is 0.005 in. The standard deviation of several part measurements was found to be 0.0015. The target probability of rejects is $2 \cdot 10^{-9}$. If the current probability of rejects exceeds that value, consider changes to attain the target value.

Design Decision (a): We will consider tighter control over the manufacturing process. Find the new required standard deviation.

Design Decision (b): We will consider redesign of the product to allow a tolerance change for the critical dimension. The process related to the critical dimension will be unchanged. Determine the required tolerance.

CHAPTER 20

Special Topics in Machine Design

Symbols

A	annual cost	lg	base 10 logarithm
C	cost function	n_y	number of times per year that interest is compounded
CL	criterion level	N	number of measurements
D	noise dose	P	original principal, present value
ER	exchange rate	P_R	investment cost to recover
i	index number, interest	P_S	salvage value
I	percent interest	P_Y	principal at end of term
i, j, k	unit vectors	T	time interval, exposure time
L	sound level	TWA	time-weighted average
L_{DN}	day–night sound level	Y	term
L_{eq}	equivalent sound level		

Units

cost function	property to be optimized (e.g., weight in pounds)
noise dose	percent
principal, annual cost	usually dollars
sound or noise level, criterion level, exchange rate, time-weighted average	decibels, usually A-weighted (dBA)

ENVIRONMENTAL CONCERNS

A number of product design suggestions are listed in the preceding chapter under the section "Design for the Environment (DFE)." These suggestions relate to materials conservation and to the environmental effects of product manufacture, use, and disposal. Additional topics related to design, energy conservation, and the environment are considered in this chapter.

Energy Use

According to the Energy Information Administration (1994), world production of primary energy was 343 quadrillion Btu in 1992, increasing at about 2.3% per year. One quadrillion = 10^{15}, and 3410 Btu = 1 kWh. Primary energy includes petroleum, natural gas, coal, hydropower, and nuclear fuel.

The United States is first among the world's major energy producers; U.S. primary energy production was $66.68 \cdot 10^{15}$ Btu in 1992. However, the United States is by far the largest primary energy user, consuming $82.19 \cdot 10^{15}$ Btu the same year. Thus, the U.S. energy deficit is about 15.5 quadrillion Btu/yr.

Total U.S. energy use and the energy deficit may have significant political and economic implications. High use of energy throughout the world, particularly in the industrially developed countries, has significant environmental consequences. Fossil fuel supplies are limited, and the extraction, transportation, and use of fossil fuels cause environmental degradation.

In the United States in 1980, there were 70 operable nuclear reactor units, and 2 in start-up. In the same year, there were 82 reactor construction permits effective, 12 more permits pending, and 3 reactors on order, for a total of 169 operable and planned nuclear reactor units with a design capacity of 163 million kW. However, nuclear energy has failed to live up to its promise as a safe, clean, and inexpensive replacement for fossil fuels. High costs and problems with safety, waste disposal, and decommissioning of existing reactors have led to cancellation of construction projects. As a result, the 1980 design capacity (including operable and planned plants) was not realized. In 1993, there were 109 operable reactor units and 7 permits effective, for a total of 116 units with a design capacity of 110 million kW.

Conservation of Energy and Other Resources

Resource conservation should be considered in design decisions. Design decisions based on efficiency may seem to have a small effect on worldwide energy use. However, conservation-based decisions have the potential to bring about significant change.

Energy Efficiency. **Efficiency** is defined as the output of a process divided by the input. Conversion of fuel to electric power is typically about 32% to 36% efficient. That is, the electric energy generated is about one-third of the heating value of the fuel. In cogeneration systems, low-grade heat, which would otherwise be wasted, is recovered for use in other industrial processes, increasing the overall efficiency.

If we define the **effective efficiency** of a device as the useful output divided by the input of primary energy, then very low efficiencies may apply. For example, only a small fraction of the electrical energy supplied to an incandescent light bulb is converted into light; the rest is converted into heat. Consider a room which is occupied intermittently, but in which the lights are not turned off when the room is unoccupied. The effective efficiency of the lighting is given by the efficiency of the generating plant (reduced by losses in transmitting power) times the efficiency of the bulb times the use factor for the room. Other systems that we design can be analyzed in a similar manner. Using the above definition, effective efficiencies may be as low as 1% to 3%. Thus, there is plenty of room for improvement in the design of typical machines and processes.

Some arguments in favor of design for energy and resource conservation include the following:

- Increases in efficiency have a local effect, reducing energy use and pollution.
- Efficient products serve as a model, showing that "it can be done."
- Environmentally minded and cost-conscious customers may select products on the basis of efficiency.
- Governments may include requirements for efficiency, use of recycled materials, and low environmental impact in their purchase specifications.
- Some utilities give rebates for energy-efficient appliances.

Global Warming and Air, Water, and Noise Pollution

Many of our environmental problems are the result of technological advances. It is reasonable to expect that technology should play a large part in remedying these problems as well.

Global Warming. It is generally assumed that increases in atmospheric carbon dioxide are disturbing the balance between the radiation energy received by the earth from the sun and the outward radiation from the earth. The result is an increase in the surface temperature of the earth. The warming effect is called the **greenhouse effect,** and carbon dioxide and the other gases involved are called **greenhouse gases.**

According to Houghton (1994), there is a consensus among the world's scientific community about the existence of global warming. However, there is substantial debate over the scale of the problem, and what to do about it. Houghton notes that the following actions can slow and stabilize climate change:

- reducing carbon dioxide emissions to the 1990 level
- phasing out the use of ozone-depleting chlorofluorocarbons (CFCs) and using CFC replacements
- reducing deforestation
- increasing reforestation
- reducing methane emissions
- increasing energy-saving and conservation measures

Machine designers can play an important part, particularly by designing products that save energy and conserve resources.

Air, Land, and Water Pollution. Air and water pollution and toxic waste sites are often thought of as problems related to waste disposal, transportation systems, power generation, and the chemical industry. However, designers of machines also have a part in resolving these problems. Examples of possible involvement include the following:

- designing improvements in motor vehicles to reduce pollution
- designing mass transportation innovations
- designing nonpolluting transportation systems such as improved bicycles and bicycle components
- utilizing suggestions found in the "Design for the Environment (DFE)" section of the previous chapter
- using general technology training to assess environmental issues and then working with other concerned citizens for environmental protection
- developing personal strategies for pollution prevention and resource conservation

The computer industry, and the electronics industry in general, are often viewed as clean and nonpolluting. Unfortunately, as noted by Brown et al. (1994), Silicon Valley has the largest concentration of hazardous waste sites in the United States. Instead of the "paperless office" promised by the computer industry, the use of computers and high-speed printers has resulted in the use of more paper than ever before.

Computers are also a valuable tool for improving the environment. Environmental monitoring and modeling are possible to a much greater extent than in precomputer times. The ability of computers to store and sort data and solve problems has permitted the creation of the toxins release inventory (TRI). TRI data are available to governmental agencies and citizen groups in computer-readable format. Remediation efforts are facilitated by the availability of TRI data. Tracking systems of this type have repeatedly exposed chronic polluters, putting pressure on them to reform.

Noise Pollution. **Noise** can be defined as "unwanted sound." Thus, the terms **sound** and **noise** are used interchangeably except when referring to wanted sound such as communication and music. Although noise pollution may seem a transient event, it reduces the quality of life for a large part of the population.

Equivalent Sound Level. Noise levels are commonly expressed in A-weighted decibels (dBA). The decibel is a logarithmic quantity, and A-weighting is an attempt to model human response to noise energy. Equivalent sound level is an energy average defined by

$$L_{eq} = 10 \lg\left[\frac{1}{T}\int_0^T 10^{L/10}\, dt\right] \qquad \textbf{(20.1)}$$

where L_{eq} = equivalent sound level (dBA)

lg = base 10 logarithm

T = time interval over which L_{eq} is defined

L = instantaneous sound level at time t

If the time interval is short, an integrating sound level meter can be used to produce L_{eq} automatically. If we have a series of N representative sound level measurements, equivalent sound level can be obtained from

$$L_{eq} = 10 \lg\left[\frac{1}{N} \sum_{i=1}^{N} 10^{L_i/10}\right] \tag{20.2}$$

where $\sum$ indicates summation.

EXAMPLE PROBLEM Equivalent Sound Level

The following five representative sound level measurements were taken during a 1 h interval (dBA):

60 60 88 90 89

Find the equivalent sound level for that hour. Determine the effect of design changes that would reduce the first two measurements.

Solution Summary: In the detailed solution which follows, the equivalent sound level is calculated as 86.9 dBA. The first two measurements are the lowest two sound level readings. If they are reduced from 60 to 30 dBA, the reduction in equivalent sound level is insignificant. It is necessary to reduce the highest noise contributions to make a significant reduction in L_{eq}.

Detailed Solution:

Number of Representative Measurements:

$$N = 5 \qquad i = 1, \ldots, N$$

Noise Measurements (dBA)

L_i

60
60
88
90
89

Equivalent Sound Level (dBA):

$$L_{eq} = 10 \cdot \log\left(\frac{1}{N} \cdot \sum_{i=1}^{N} 10^{L_i/10}\right) \qquad L_{eq} = 86.9$$

Effect of Lower Values on Equivalent Sound Level:

L_i

30
30
88
90
89

Equivalent Sound Level (dBA):

$$L_{eq} = 10 \cdot \log\left(\frac{1}{N} \cdot \sum_{i=1}^{N} 10^{L_i/10}\right) \qquad L_{eq} = 86.9$$

24 Hour Equivalent Sound Level. In some cases, our sound level measurements represent unequal time intervals. We can then weight the measurements by the length of the intervals. If the total averaging time is 24 h, we obtain

$$L_{eq24} = 10 \lg\left[\frac{1}{24} \sum_{i=1}^{N} (T_i \cdot 10^{L_i/10})\right] \quad \textbf{(20.3)}$$

where L_{eq24} = 24 h equivalent sound level
T_i = time interval (h) when the level was L_i

Equivalent Sound Levels for Protection of Public Health and Welfare. The Environmental Protection Agency (1978; Wilson, 1994) has identified values of sound levels that protect public health and welfare with a margin of safety. The values of equivalent sound level are given in Table 20.1. The EPA has defined L_{eq24} as an energy average sound level for a typical 24 h day. Thus, L_{eq24} should approximate the value that we would obtain by averaging over an entire year.

Day–Night Sound Level. Equivalent sound level L_{eq24} treats nighttime noise the same as daytime noise. Day–night sound level takes sleep interference into account by adding a 10 dBA penalty to nighttime (10:00 P.M.–7:00 A.M.) noise.

We first calculate daytime equivalent sound level

$$L_{eqD} = 10 \lg\left[\frac{1}{15} \sum_{i=1}^{N} (T_i \cdot 10^{L_i/10})\right] \quad \textbf{(20.4)}$$

based on noise levels measured during the 15 hours between 7:00 A.M. and 10:00 P.M.

Next we calculate nighttime equivalent sound level

$$L_{eqN} = 10 \lg\left[\frac{1}{9} \sum_{i=1}^{N} (T_i \cdot 10^{L_i/10})\right] \quad \textbf{(20.5)}$$

based on measurements during the 9 hours between 10:00 P.M. and 7:00 A.M.

Finally, we calculate day–night sound level for the 24 h day

$$L_{DN} = 10 \lg\left[\frac{1}{24}(15 \cdot 10^{L_{eqD}/10} + 90 \cdot 10^{L_{eqN}/10})\right] \quad \textbf{(20.6)}$$

Table 20.1 EPA-identified protective equivalent sound levels.

Effect	Sound Level	Area
Hearing	$L_{eq24} \leq 70$ dBA	Any area, measured at the ear
Outdoor activity interference and annoyance	$L_{eq24} \leq 55$ dBA	Outdoor, nonresidential areas where people spend limited time
Indoor activity interference and annoyance	$L_{eq24} \leq 45$ dBA	Indoor, nonresidential areas with human activities (such as schools)

Source: Environmental Protection Agency (1978).

Table 20.2 EPA-identified protective day–night sound levels.

Effect	Sound Level	Area
Outdoor activity interference and annoyance	$L_{DN} \leq 55$ dBA	Outdoors in residential areas
Indoor activity interference and annoyance	$L_{DN} \leq 45$ dBA	Indoor residential areas

or, if we wish to calculate L_{DN} in one step,

$$L_{DN} = 10 \lg\left\{\frac{1}{24} \cdot \left[\sum_{i=1}^{ND} (T_{Di} \cdot 10^{L_{Di}/10}) + 10 \sum_{i=1}^{NN} (T_{Ni} \cdot 10^{L_{Ni}/10})\right]\right\} \tag{20.7}$$

where L_{DN} = day–night sound level (dBA) which incorporates a penalty for nighttime noise
ND = number of daytime measurements (taken between 7:00 A.M. and 10:00 P.M.)
NN = number of nighttime measurements (taken between 10:00 P.M. and 7:00 A.M.)
L_i = ith measurement (dBA)
T_i = period for which that measurement was valid (h)
Subscripts D and N refer to day and night, respectively.

The Environmental Protection Agency (1978) has also identified protective day–night sound levels; they are listed in Table 20.2.

EXAMPLE PROBLEM Day–Night Sound Level

The following sound level pattern is typical outdoors in a certain residential area:

Sound Level (dBA)	*Time (h)*	*Day or Night*
57	7	Day
56	2	Day
43	2	Day
53	2	Day
44	2	Day
49	4	Night
35	1	Night
56	2	Night
42	2	Night

Find the day–night sound level, and evaluate the site for outdoor activity interference.

Solution Summary: The day–night sound level exceeds the protective value based on outdoor activity interference, as shown in the detailed solution which follows.

Detailed Solution:

Number of Day Measurements:

$$ND = 5 \qquad i = 1, \ldots, ND$$

Number of Night Measurements:

$$NN = 4 \qquad j = 1, \ldots, NN$$

Daytime:

Noise Measurements (dBA) L_i	*Time (h)* T_i
57	7
56	2
43	2
53	2
44	2

Equivalent Sound Level (dBA)

$$\sum_{i=1}^{ND} T_i = 15 \qquad L_{eqD} = 10 \cdot \log\left[\frac{1}{15} \cdot \left(\sum_{i=1}^{ND} T_i \cdot 10^{L_i/10}\right)\right]$$

$$L_{eqD} = 55$$

Nighttime:

LN_j	TN_j
49	4
35	1
56	2
41	2

Equivalent Sound Level (dBA)

$$\sum_{j=1}^{NN} TN_j = 9 \qquad L_{eqN} = 10 \cdot \log\left[\frac{1}{9} \cdot \left(\sum_{j=1}^{NN} TN_j \cdot 10^{LN_j/10}\right)\right] \qquad L_{eqN} = 51$$

24 Hour Equivalent Sound Level:

$$L_{eq24} = 10 \cdot \log\left[\frac{1}{24} \cdot (15 \cdot 10^{L_{eqD}/10} + 9 \cdot 10^{L_{eqN}/10})\right] \qquad L_{eq24} = 53.9$$

Day–Night Sound Level:

$$L_{DN} = 10 \cdot \log\left[\frac{1}{24} \cdot (15 \cdot 10^{L_{eqD}/10} + 90 \cdot 10^{L_{eqN}/10})\right] \qquad L_{DN} = 58.3$$

Addition of Noise Contributions. Several different sources add to the noise level at a given location. If there are N sources making individual contributions L_i to the noise at a given location, then the total noise level at that location is

$$L_{\text{total}} = 10 \lg\left[\sum_{i=1}^{N} 10^{L_i/10}\right] \qquad \textbf{(20.8)}$$

EXAMPLE PROBLEM Combining Noise Contributions

Five different noise sources contribute to the noise at a given location. Each one was measured in the absence of the other four. The individual contributions are (dBA)

70 70 68 67 50

Find the noise level at that location when all five sources are operating.

Solution Summary: The result is $L_{\text{total}} = 75$ dBA, as shown in the following detailed solution.

Detailed Solution:

Number of Measurements:

$$N = 5 \qquad i = 1, \ldots, N$$

Noise Measurements (dBA)

L_i

70

70

68 $\quad L_{\text{total}} = 10 \cdot \log\left(\sum_{i=1}^{N} 10^{L_i/10}\right) \qquad L_{\text{total}} = 75$

67

50

An Alternative Method for Combining Two Noise Contributions. Equation (20.8) can be used to add the contributions from any number of noise sources. If the contributions are to be added two at a time, then we can rearrange the equation (with some difficulty) to obtain

$$\text{Add} = 10 \cdot \lg[1 + 10^{-\text{Dif}/10}] \qquad \textbf{(20.9)}$$

where Add = number of dBA to be added to the larger noise level

Dif = difference in noise levels

Table 20.3 Combining two noise contributions.

Difference in Noise Levels (dBA)	Add to Higher Noise Level (dBA)
Dif	**Add (Dif)**
0	3.01
1	2.539
2	2.124
3	1.764
4	1.455
5	1.193
6	0.973
7	0.79
8	0.639
9	0.515
10	0.414
11	0.332
12	0.266
13	0.212
14	0.17
15	0.135
16	0.108
17	0.086
18	0.068
19	0.054
20	0.043

This alternative method is more difficult when we are using a computer, or even a calculator, to add noise levels. However, plotted or tabulated values from the Equation (20.9) may be preferred for quick estimates of noise levels. In addition, the equation gives us some insight into the importance of various noise contributions and the potential effect of product noise control.

EXAMPLE PROBLEM A Plot and a Table for Adding Two Noise Contributions

(a) Plot and tabulate Equation (20.9).
(b) Apply the plot or table to the previous example problem.

Solution (a): The tabulated and plotted values are given in Table 20.3 and Figure 20.1. We see that when two equal noise levels are combined, the total noise level is about 3 dBA higher than one of the levels. If one level is 6 dBA below another, combining the two produces a total noise level about 1 dBA greater than the higher noise level. If one level is 20 dBA lower than the other, the smaller noise level makes an insignificant contribution to the total.

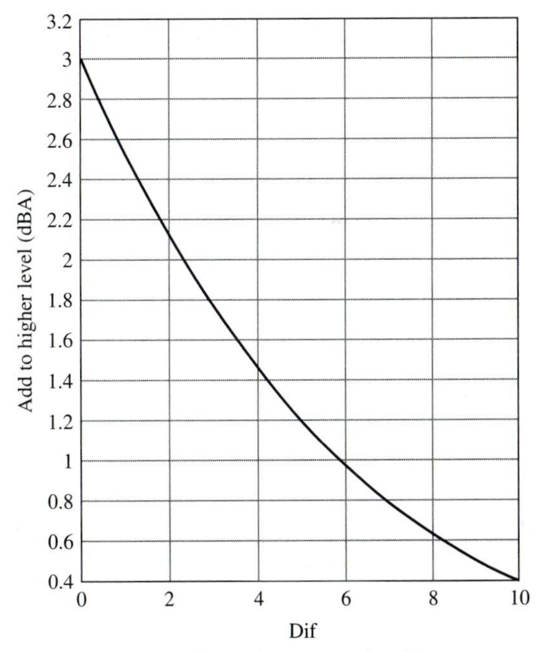

Figure 20.1 Combining two noise contributions.

Solution (b): We use the plot and/or the table as follows to add the noise levels given in the previous example problem:

- The two 70 dBA contributions are combined to produce 70 + 3 = 73 dBA.
- The 68 dBA contribution is 5 dBA less than 73 dBA, so we add about 1.2 dBA, to obtain 73 + 1.2 = 74.2 dBA.
- The 67 dBA contribution is 7.2 dBA less than 74.2 dBA, so we add about 0.7 dBA to obtain 74.2 + 0.7 = 74.9 dBA.
- The last contribution of 50 dBA is more than 20 dBA less than 74.9, so we add zero.
- Thus, L_{total} = 74.9 dBA. Because of rounding, this value varies about 0.1 dBA from the value that the computer gave us.

■■

Noise Control Implications. The noise contributions in the above examples can represent intrusive highway and aircraft noise as well as noise from domestic sources such as dishwashers, refrigerators, and plumbing and from heating, ventilating, and air conditioning (HVAC) systems.

Whenever possible, noise control should begin with reduction of noise at the source. Next, we should try interruption of sound transmission paths. Finally, hearing protection devices (HPDs) may be specified. HPDs are used in industrial settings and with powered garden and workshop equipment. Personal protection should be considered as a last resort, after maximum noise reduction has been obtained from all other methods.

Efforts at noise control must first deal with the source producing the greatest noise contribution. Note that in the preceding example problem, the 67 dBA contribution adds little to total noise when the 70 dBA contributions are present. The 50 dBA contribution to total noise level is negligible. Once the most prominent noise source is controlled, then control of other sources may be practical.

DESIGN FOR SAFETY IN MANUFACTURING

Workplace safety depends on the worker, the employer, and the designer of production machinery.

The Occupational Safety and Health Act and the Occupational Safety and Health Administration

In the United States, the Occupational Safety and Health Act (OSHA) was passed "to assure so far as possible every working man and woman in the Nation safe and healthful working conditions and to preserve our human resources."

According to OSHA (1992a), the Occupational Safety and Health Administration (which is also called OSHA) was created within the Department of Labor to

- encourage employers and employees to reduce workplace hazards and to implement new or improve existing safety and health programs;
- provide for research in occupational safety and health to develop innovative ways of dealing with occupational safety and health problems;
- establish "separate but dependent responsibilities and rights" for employers and employees for achievement of better safety and health conditions;

- maintain a reporting and record-keeping system to monitor job-related injuries and illnesses;
- establish training programs to increase the number and competence of occupational safety and health personnel;
- develop mandatory job safety and health standards and enforce them effectively; and
- provide for development, analysis, evaluation, and approval of state occupational safety and health programs.

Mechanical Hazards

Mechanical hazards occur at the following locations:

- **the point of operation:** the point where work is performed on the material (cutting, forming, boring)
- **power transmission apparatus:** components that transmit energy (belts, pulleys, connecting rods, cranks, chains, gears)
- **other moving parts:** reciprocating and rotating parts (feed mechanisms and auxiliary parts)

Mechanical hazards include

- rotating parts (Even smooth rotating parts can grip clothing or force a hand or an arm into a dangerous position.)
- projections on rotating parts
- nip points between a pair of gears or rollers, between a belt and a pulley, or between a chain and a sprocket
- cutting and shearing operations

Machine Guarding

Machine guarding may be the most reliable method of preventing industrial accidents. "Any machine part, function, or process which can cause injury must be safeguarded. When the operation of a machine or accidental contact with it can injure the operator or others in the vicinity, the hazards must be either eliminated or controlled" (OSHA, 1992b). The following minimum requirements for guards are based principally on that publication.

Machine Guards Must:

- prevent contact. A good guard system eliminates the possibility that an operator or other workers will place parts of their bodies near hazardous parts.
- be secure. Workers should not be easily able to disable or remove a guard.
- protect from falling objects. A tool or other object dropped into moving machinery can become a projectile.
- create no new hazards (such as sharp edges or shear points).
- create zero or minimal interference with work. A guard that impedes a worker is likely to be removed.
- allow for normal service and lubrication. If possible, oil reservoirs can be located outside the guard.

A manufacturing system using a robot for materials handling is shown in Figure 20.2. The area within reach of the robot and the machine tool defines a hazard space. The perimeter can be identified by markings on the floor, or a barrier can be used. A barrier with gate interlocks is used closer to the hazard space. A presence-sensing device is installed around the machine tool.

Two-hand control buttons, a common protective device, are shown in Figure 20.3. With two-hand control, both of the operator's hands must maintain pressure on the control buttons for the machine to operate. Both hands must be a safe distance from the danger area while the machine operates. Note that the buttons are recessed to prevent accidental actuation.

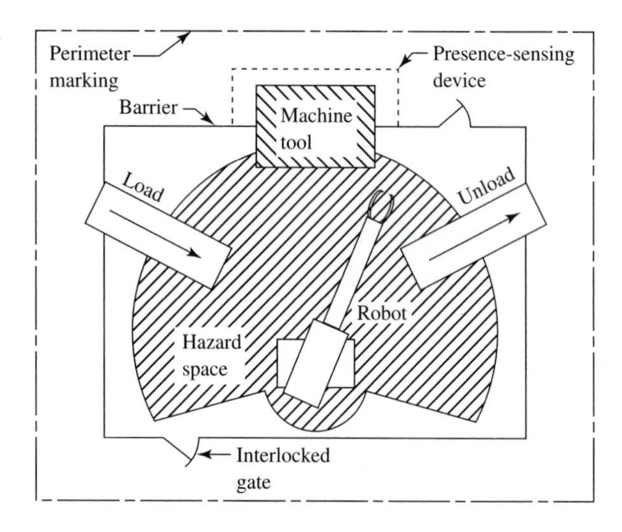

Figure 20.2 Guarding of a manufacturing cell that uses a robot for material handling.

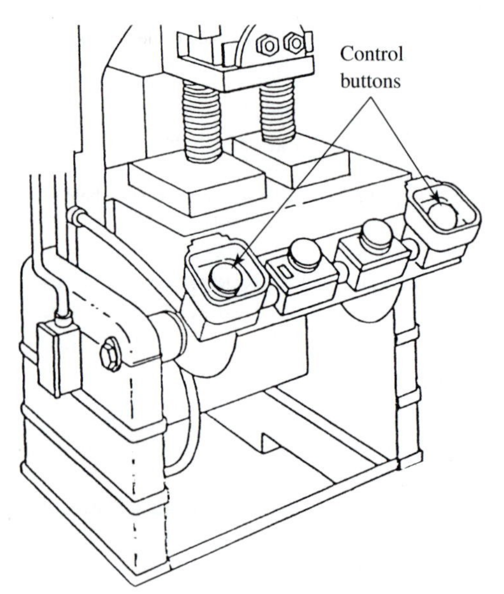

Figure 20.3 Two-hand control buttons.

Types of Machine Guards. OSHA (1992b) and the National Safety Council (1992) have compiled extensive information on machine guarding. Much of the information in Table 20.4 is adapted from those references.

Industrial Noise

Noise is one of the most pervasive occupational health problems. The Code of Federal Regulations hearing conservation amendment (OSHA, 1983) is an attempt to protect workers from material hearing impairment.

Table 20.4 Advantages and limitations of various machine guards and material handling methods.

Type	Action	Advantages	Limitations
Area guard; pressure-sensitive floormat or photoelectric device.	Detects intrusion into hazard area; activates alarm or stops machine or robot.	Senses presence anywhere in protected area.	Usually limited to automatic manufacturing.
Perimeter guard; physical barrier with interlocked gates or light curtain.	Opening of gate or interruption plane of invisible light detects intrusion.	Light curtain may be wrapped around a machine or work cell by use of mirrors.	Usually limited to automatic manufacturing.
Photoelectric or capacitance point-of-operation machine guard.	Activates brake when light curtain or capacitance field is interrupted by any part of operator's body.	Allows operator freedom and machine access.	Requires braking device.
Pullback.	Hands are pulled out of danger area when cycle begins.		Limits operator movement.
Restraint.	Holds operator's hands out of danger area.	Simplicity.	Limits operator movement.
Safety trip: tripwire or crash bar.	Stops machine when tripped.	Simplicity.	Requires machine brake.
Two-hand control.	Both hands must be held on control to activate cycle.	Operator's hands are in safe location.	Requires partial-cycle machine with brake; only operator is protected.
Automatic and semiautomatic feed.	Stock fed from rolls, hoppers, or chutes.	Operator not needed in hazard area.	Requires perimeter guards; may not be adaptable to stock variation.
Automatic and semiautomatic ejection.	Work ejected by air or mechanical means.	Operator not needed in hazard area.	Air ejection may cause a noise hazard; chips may be an eye hazard.
Robots.		Operator not needed in hazard area.	Requires perimeter guard; frequent maintenance required.

Source: Adapted from OSHA (1992b) and National Safety Council (1992).

Hearing Conservation Programs. Hearing conservation programs are required for all employees whose noise exposures equal or exceed an 8 h time-weighted average (TWA) of 85 dBA (OSHA, 1992c). Hearing conservation programs have the following components:

- noise monitoring to identify employees with exposures at or above 85 dBA
- audiometric testing, baseline audiograms, and annual audiograms
- audiogram evaluation
- hearing protector availability
- noise exposure record keeping

Allowable Exposure Time. Allowable daily exposure time for noise can be calculated from

$$T = 8 \cdot 2^{(CL-L)/ER} \tag{20.10}$$

where T = allowable exposure time (h/day) at level L

L = noise level (dBA measured with slow time constant setting)

CL = criterion level

ER = exchange rate

OSHA (1983) specifies the following values:

$$CL = 90 \text{ dBA} \quad \text{and} \quad ER = 5 \text{ dBA}$$

These equations apply to levels in the range 80 dBA $\leq L \leq$ 115 dBA. Levels below 80 dBA are not considered; levels above 115 dBA are in violation of the standard.

■■ **EXAMPLE PROBLEM Daily Exposure Limit**

Plot daily exposure limit versus noise level.

Decision: We will use the 90 dBA criterion level and the 5 dBA exchange rate, and we will consider noise in the range 85 dBA $\leq L \leq$ 115 dBA.

Solution Summary: As calculated in the following detailed solution, the daily exposure limits are 16 h at 85 dBA and 8 h at 90 dBA, halving for each 5 dBA noise-level increase to 1/4 h at 115 dBA. The plot is shown in Figure 20.4.

Detailed Solution:

Criterion Level (dBA):

$$CL = 90$$

Exchange Rate (dBA):

$$ER = 5$$

Noise Level (dBA):

$$L = 85, 86, \ldots, 115$$

Allowable Time (h) at Level L:

$$T(L) = 8 \cdot 2^{(CL-L)/ER}$$

See Figure 20.4. ■■

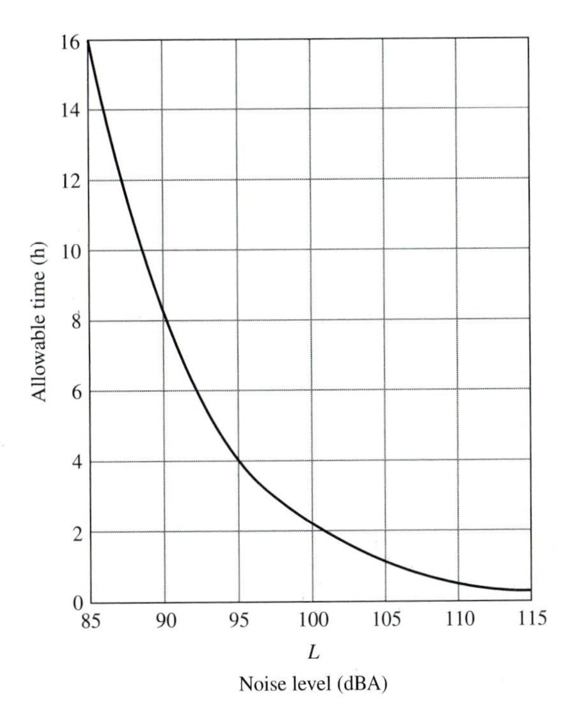

Figure 20.4 Occupational noise exposure limits. (Courtesy of Occupational Safety and Health Administration)

Daily exposure limit applies to each individual worker. The limits cannot be satisfied by averaging the exposure of workers in noisy environments with those in quieter environments. Note that Equation (20.10) produces a daily exposure limit of

$$T = 32 \text{ h} \quad \text{for} \quad L = 80 \text{ dBA}$$

a seemingly absurd result. This limit is not a mistake, however, but is used in determining percent noise dose and time-weighted average for time-varying noise.

Noise Dose. Note that the daily exposure limit is 4 h at a noise level of 95 dBA. If a worker is exposed to 1 h at that level, then the noise dose is 25%. For a worker exposed to different noise levels over the workday, the noise dose is given by

$$D = 100 \sum_{i=1}^{n} \frac{C_i}{T_i} \tag{20.11}$$

where D = daily noise dose (%)

n = number of intervals during a day

C_i = time (h) at noise level L_i

T_i = allowable exposure time (h) at level L_i

Daily noise dose should not exceed 100%.

Time-Weighted Average. Time-weighted average (TWA) is an indication of a steady noise level that would result in the same noise dose as the varying noise level. It is given by

$$TWA = CL + ER\left[\frac{\log(D/100)}{\log 2}\right] \tag{20.12}$$

where TWA = time-weighted average noise level (dBA)

Both numerator and denominator can use base 10 logarithms, or both can use natural logarithms.

EXAMPLE PROBLEM Noise Dose and Time-Weighted Average

A worker is exposed to the following daily noise exposure pattern:

Noise Level L (dBA)	*Exposure Time (h)*
113	0.5
99	3
98	2
80	2
89	0.5

Find noise dose and time-weighted average. Evaluate the results.

Decision: We will use a criterion level of 90 dBA and an exchange rate of 5 dBA.

Solution Summary: As shown in the following detailed solution, we find a total noise dose of almost 370% and a time-weighted average noise level of more than 99 dBA. These values compare with limits of 100% noise dose and 90 dBA time-weighted average. The situation requires remediation.

Detailed Solution:

Criterion Level (dBA):

$$CL = 90$$

Exchange Rate (dBA):

$$ER = 5$$

Number of Measurements:

$$n = 5 \qquad i = 1, \ldots, n$$

Noise Level (dBA) L_i	*Exposure Time (h)* C_i	*Allowable Time (h) at Level L* $T_i = 8 \cdot 2^{(CL-L_i)/ER}$ T_i	*Dose Contribution (%)* $\frac{C_i}{T_i} \cdot 100$
113	0.5	0.33	151.572
99	3	2.297	130.583
98	2	2.639	75.786
80	2	32	6.25
89	0.5	9.19	5.441

Noise Dose (%):

$$D = \sum_{i=1}^{n} 100 \cdot \frac{C_i}{T_i} \qquad D = 369.6$$

Time-Weighted Average (dBA):

$$TWA = CL + ER \cdot \left(\frac{\log\left(\frac{D}{100}\right)}{\log(2)} \right) \qquad TWA = 99.4$$

Exposure Limits:

L must not exceed 115 dBA.

D must not exceed 100%.

TWA must not exceed 90 dBA.

A hearing conservation program is required if *TWA* is equal to or greater than 85 dBA. ■■

Control of Industrial Noise Exposure. We should first try to control noise at the source by designing quieter machines. Then interruption on noise transmission paths should be considered. This remedy may include enclosures and other noise barriers as well as acoustical treatment to reduce reverberant sound. Administrative controls are next, including assignment of personnel to quieter areas whenever possible. Finally, personal hearing protective devices (HPDs) can be used.

■■ **EXAMPLE PROBLEM Control of Noise Exposure**

Attempt to reduce the noise exposure described in the preceding example problem to an acceptable value.

Decisions: Noise source control (design changes), enclosures, and other remedies should be investigated. In the meantime, we will specify HPDs.

Solution: We see that the three highest noise levels (113, 99, and 98 dBA) make dose contributions of about 152%, 130%, and 76%, respectively. Typical HPDs have noise reduction ratings ranging from 8 to 35 dB. We will assume a noise reduction of 11 dBA for field use. Furthermore, we will assume that the HPDs are used during the 5.5 h of maximum noise exposure. The calculation is repeated with the three highest levels each reduced by 11 dBA. The result is a noise dose of 89.6% and a time-weighted average of 89.2 dBA. ■■

Alternative Standards and Recommendations for Workplace Noise. If the EPA protective level of $L_{eq24} \leq 70$ dBA (based on hearing protection) is adjusted for 8 h/day for 5 days/week work patterns, then the criterion level becomes about 76.4 dBA. Equivalent sound level corresponds to an exchange rate of 3 dBA. Thus, a criterion level of 90 dBA and an exchange rate of 5 dBA are far less protective than the EPA protective level.

The International Institute of Noise Control Engineering (INCE) working party report on upper limits of noise levels in the workplace (Embleton, 1994) recommends an 85 dBA criterion level (for 8 h/day exposure) and a 3 dBA exchange rate. The report also lists 8 h

Table 20.5 Workplace noise standards.

Country	Criterion Level (dBA), 8 h Exposure	Exchange Rate (dBA)
Australia (varies)	85	3
Brazil	85	5
Canada	85–90	3–5
China	70–90	3
Finland	85	3
France	85	3
Germany	85	3
Hungary	85	3
Israel	85	5
Italy	85	3
Netherlands	80	3
New Zealand	85	3
Norway	85	3
Spain	85	3
United Kingdom	85	3
United States	90	5

Source: Based on Embleton (1994).

exposure limits, exchange rates, upper limits for continuous and peak noise, and levels for monitoring and controls in various countries. Table 20.5 is based on the INCE report.

PRODUCT SAFETY AND PRODUCT LIABILITY

The designer of a product has the greatest influence on product safety. The designer must use a reasonable safety factor and must follow applicable codes and standards. In addition, all foreseeable uses of the product should be considered. The last requirement is difficult, because the product may be used in unintended ways.

Guidance for Safe Design

The designer is usually prepared to design a product that will not fail owing to high stresses in normal use. Additional guidance and suggestions for safe design are found in publications of the Consumer Product Safety Commission (CPSC) and the American National Standards Institute (ANSI).

CPSC is an independent U.S. federal regulatory agency charged with protecting the public from dangerous products found in homes, schools, and public areas. These consumer products kill about 28,000 Americans annually and injure 33 million. CPSC works with industry to develop voluntary safety standards; enacts mandatory standards; and bans, or develops corrective action programs for hazardous products. CPSC also informs and educates consumers about product safety. CPSC contributions to product safety include development of a mandatory standard for bicycles, which sets safety requirements for reflectors, wheels, tires, chains, pedals, and braking and steering systems.

ANSI publishes consensus standards based on substantial agreement of affected interests. ANSI safety standards are developed to establish minimum safety requirements. They include safety precautions based on experience and testing by manufacturers, users, and regulatory agencies. The use of ANSI standards is voluntary; the standards do not preclude anyone from manufacturing, marketing, or using products or procedures that do not conform. However, if an applicable ANSI safety standard exists, designers are advised to comply with the standard.

Design Implications of Case Reports

An investigation of safety-related incidents often leads to improved design. Some of the following incidents may be worth considering.

Bicycle Incidents. One report (CPSC, 1994a) considered the role of mechanical design and performance in selected bicycle incidents. Incidents were randomly selected through the National Electronic Injury Surveillance System (NEISS). Incidents included

- cases in which the chain broke or fell off a sprocket, causing the rider to fall. In some, the chain jammed the rear wheel.
- cases involving serious finger injury to a rider attempting hand adjustments on a chain
- cases of brakes failing to stop the bicycle
- cases of brake components or a gear (sprocket) cable coming loose and jamming the front wheel or tangling with the pedal, causing the rider to be thrown over the handlebars
- cases of loose handlebars causing a rider to fall

Poor maintenance, component modification, and rider inexperience and behavior were contributing factors to many of the accidents. Design-related questions include the following:

- How can braking and speed-changing systems be made more reliable?
- Will design changes make maintenance easier?
- Can design changes make it unnecessary to adjust and maintain some components?

Special Hazards to Infants and Small Children. Products for use by infants and small children require special consideration. Products intended for adult use should also be examined for potential hazards to children, particularly if they are attractive to children. Product safety alerts (CPSC, 1994b) have identified the following incidents. Numbers of deaths and injuries are based on reports to CPSC over varying periods of time. It is likely that many incidents are not reported.

- strangulation deaths where a necklace, window-blind or drapery cord, ribbon, or toy guitar strap twisted around a child's neck.
- deaths resulting from tipovers of movable soccer goals.
- suffocation deaths of infants in drop-sided playpens.
- head entrapment in playground cargo nets.
- 150 reported deaths and over 1000 injuries resulting from children playing with cigarette lighters.

- deaths from aspirin and other oral medications. CPSC estimates that the lives of 700 children have been saved by mandatory child-resistant packaging.
- deaths from riding motorized all-terrain vehicles (ATVs). For drivers under 16, there is a 1-in-3 chance of injury during the life of the ATV.
- 45 children (between the ages of 2 and 14) trapped and killed under automatic garage doors which failed to stop and reverse direction after contacting the child. In addition many serious injuries were reported, including brain damage.
- deaths, loss of sight, and other serious injuries from lawn darts and pellet guns, including 25 deaths from pellet rifles or BB guns. CPSC knows of three children killed by lawn darts, and it estimates that 6700 lawn-dart injuries were treated in U.S. hospital emergency rooms from 1978 through 1987.

Note that CPSC is not charged with regulating firearms and motor vehicles. Publications of the Centers for Disease Control and Prevention (CDC) state that about 55% of all injury deaths are caused by motor vehicles and firearms. Fingerhut, Jones, and Makuc (1994) list 38,317 firearm deaths and 43,516 motor vehicle deaths in the United States for 1991. They observe that motor vehicle deaths have been declining, while firearms deaths have been increasing. This trend leads them to predict that firearm deaths will exceed vehicle deaths in the United States for the year 2000 or shortly thereafter. Both motor vehicles and firearms claim a substantial number of young children. Fingerhut et al. (1991) list 738 firearm deaths of children 1–14 years of age in the United States during 1988. Of these, 91 were children 1–4 years of age, and 124 were children 5–9 years of age.

It is obvious that children see medications, guns, and lighters as attractive toys. As a result, CPSC set a mandatory safety standard requiring child-resistant lighters. Effective 1988, lawn darts were banned from sale in the United States.

It is difficult to make specific recommendations with regard to design of items intended for use by infants and children, and even more difficult to deal with adult-use items that may fall into the hands of children. The following suggestions may be of use:

- Toys designed for very young children should avoid small parts, sharp edges, straps, and parts that could come loose. Small parts are a choking hazard when put in the mouth. Straps and cords have caused strangulation deaths.
- Anticipate adventurous behavior from children. When designing products that may be climbed (furniture, soccer goals, etc.), consider center-of-gravity and methods of support or anchoring. Guard against tipover hazard.
- Provide warnings for parents. Indicate appropriate age group for toys. Indicate the consequences of misuse and lack of supervision if an item is potentially dangerous (e.g., death by strangulation).
- Do not design inherently unsafe toys or other items for use by children.
- When designing items for adult use, consider possible use by children. Try to redesign the item to reduce the danger of misuse. For example, the space between the car gate and the shaftway door on some elevators can accommodate a child. Fatal accidents have resulted when a shaftway door closed and locked, trapping a child outside the car gate of a moving elevator. There are a number of redesign possibilities, the most obvious being a reduction of the space between the car gate and the shaftway door.

Guards and Special Safety Features

Some potentially dangerous products require special guards. Product hazards are often similar to the hazards discussed in the section "Design for Safety in Manufacturing." The principles of machine guarding discussed in that section may apply to nonindustrial hazards as well.

Powered gardening equipment and agricultural equipment present hazards involving inadvertent contact with blades and drive machinery, projectiles thrown by the blades, and high noise levels. Safety specifications are sponsored by the Outdoor Power Equipment Institute and are published by the American National Standards Institute (ANSI/OPEI, 1990). A few of the requirements, recommendations, and tests are listed below. Some may be adapted to other products as well.

- operator presence controls that require intentional, continuous actuation.
- a shutoff device. If stopping time exceeds 7 s, there should be audible or visible indications of movement.
- guarding and shielding to minimize inadvertent contact with machinery hazards during mounting, starting, operating, and dismounting of equipment.
- a guard or shield that remains effective when subject to the force of a 250 lb (113 kg) individual leaning or falling against it.
- finger-probe requirements. The finger probe is a 4 in (102 mm) long simulated finger used to test accessibility of hazard areas. The probe should not contact any hazard when inserted into any opening from the operator zone.
- foot-probe test. A test similar to the finger-probe test is used to check for possible exposure to cutting blades.
- thrown-object tests. Test projectiles (nails) are introduced into the blade area of a mower. Nails must not strike the operator target zone; a limited number of nails may strike a "bystander" target zone.
- material failure requirement. The components used in assembling the cutting elements should not become worn and should not fail in a hazardous manner before the elements themselves are worn out.

Chain saw kickback is a phenomenon in which the chain saw can be propelled upward and rearward at high speed toward the operator. Kickback accounts for about 20% of all chain saw injuries, including some of the most severe injuries. Chain saw hazards also include danger from falling trees, exposure to hand-transmitted vibration, noise exposure, and contact with the cutting chain due to causes other than kickback.

Chain saw safety standards (ANSI, 1991) include

- recommendations and requirements for guards
- sound and vibration test procedures
- chain brake test procedures
- standards for safety procedures labeling and owner's manual instructions
- a mathematical simulation of kickback including a flowchart and BASIC and FORTRAN program listings

Some gasoline-powered rotary brush cutters accommodate a string-trimmer head and also accommodate a sawlike blade (with four or more teeth) for cutting larger weeds and brush. When used with a blade, these tools are also subject to the problem of kickback and have caused amputation injuries. Possible solutions include

- improved guarding
- redesign of the handle, giving the operator more leverage to resist kickback
- a blade brake
- total redesign, utilizing a different cutting principle

Unforeseen Circumstances

Since no responsible person would intentionally design a faulty or unsafe product, unforeseen conditions account for most accidents and premature failures. It appears that the conditions leading to the product failures and accidents noted below were not foreseen by the designer.

A rack was designed to hold bakery products and mounted in a small truck. The truck was involved in an accident. Inertia forces caused the rack to break from its mounting and kill the driver of the truck. It was determined that the accident would not otherwise have killed the driver. The designer failed to consider the possibility of an (otherwise minor) accident and the resulting inertia effects.

A large industrial mixer was lashed to a flatbed truck for transportation from manufacturer to buyer. Inertia forces due to acceleration, deceleration, and turns in transit caused the top of the mixer to separate from the base. The bolts used in fabricating the mixer were adequate for forces that the mixer would encounter in use. However, the designer failed to consider inertia forces due to transportation.

The blade of a gasoline-powered brush cutter flew off the machine, causing severe injury to the operator. The blade had been held on the shaft of the cutter by a nut and lock washer. The shaft and nut had left-hand threads so that normal cutting forces would tend to tighten the nut. As in the above cases, the designer failed to consider inertia effects. Inertia torques arise from starting and stopping the machine and from the combustion cycle of a one-cylinder gasoline engine. A different fastening system could have prevented the accident.

The ball-joint component of a vehicle steering system separated, causing an accident. Ball-joints are periodically lubricated with a high-pressure grease system. The part originally had shallow grooves to relieve the pressure and disperse the lubricant. After a period of use, the part was worn to the depth of the grooves, so that lubricant pressure would no longer be relieved. It was determined that lubricant pressure contributed to the forces that caused the joint to separate. Wear and high-pressure lubrication should have been considered by the designer.

A 1994 failure of a 36 in diameter natural gas transmission pipeline resulted in an explosion that destroyed eight buildings of an apartment complex, injured 100 people, and left hundreds of people homeless. After the explosion, it took workers 3 hours to close pipeline valves manually. The National Transportation Safety Board (NTSB) investigated the explosion, and a few of their findings are listed below.

- The pipe was gouged, causing a crack that grew until the pipe burst. It is assumed that the gouge resulted from excavation activity subsequent to a 1986 inspection using a "smart pig" robot sensing device.

- The location of the underground pipeline was not marked in the explosion vicinity.
- Although excavation had taken place near the pipeline for years, aerial pipeline surveillance programs had failed to identify and report excavations in industrial areas through which the pipeline passed.
- A tougher, less brittle pipe may have sustained the gouges without failure or may have sustained a smaller failure opening that would have reduced the gas release rate.
- The lack of automatic or remote-control valves delayed shutoff and exacerbated damage to nearby property.

When the ship *Titanic* struck an iceberg and sank on its maiden voyage, 1500 people died. The accident happened in 1912 in 31°F water in the North Atlantic. The ship was traveling at 22 knots and sank only 2 h and 40 min after striking the iceberg. Recent expeditions utilizing mini-submarines and undersea robots have viewed the wreckage and have recovered samples. It has been determined that the steel plates that made up the ship's hull were brittle. As a result, the accident caused many cracks in the steel which propagated great distances. Even at the time, tougher, more ductile steels were available. If the designer had avoided the possibility of brittle fracture, the ship might have stayed afloat longer, and more lives could have been saved.

Product Recall

Good design is the best defense against product liability. However, faulty material and mistakes in design and manufacture sometimes happen. A product recall may be necessary to reduce the probability of accidents and exposure to product liability, to conform with the law, or simply to improve product performance.

In 1995, automakers agreed to repair or replace seat belts on more than 8 million cars and small trucks. Pieces of the plastic release button sometimes fall into and jam the latching mechanism of the belts made by one corporation. This prevents the seat belt from latching or unlatching properly. The problem resulted in 90 reported injuries. It is estimated that the cost of this product recall could exceed $1 billion.

When there is substantial risk of injury or damage because of faulty product, a recall should be considered. Costs can be reduced if there is an identification system in place so that only the faulty products are recalled. Jacobs (1995) discusses planning to reduce the problems associated with a product recall. He suggests that manufacturers establish full traceability of their products, both by identifying their products clearly and unambiguously and by maintaining a method of locating products in the field.

Product Liability

Until 1916, the responsibility of a manufacturer to a third party was very limited. However, a case tried that year changed the situation.

The facts in the case (*MacPherson* v. *Buick Motor Co.,* NY 382, 1916; McGoldrick and Smith, 1994) were as follows. Buick manufactured an automobile and sold it to a dealer who sold it to MacPherson, the plaintiff. The wheels that Buick installed on the car (without inspecting them) were purchased from a wheelmaker. One of the wheels collapsed, causing the car to go into a ditch and injuring MacPherson. In this landmark case, the court found the automobile manufacturer responsible for injuries to the plaintiff.

In general, manufacturers and sellers can be found liable for injuries and other losses when

- a product is unreasonably dangerous;
- a product is defective;
- negligence or lack of care can be proven in design, manufacture, inspection, or testing of a product;
- the manufacturer has failed to instruct and warn the user;
- the designer has failed to foresee reasonable use (and sometimes misuse) of the product.

Manufacturing Defects. Failure to control manufacturing processes, failure to inspect, failure to test, and material variations may result in a manufacturing defect. The probability of a manufacturing defect is reduced by quality control procedures.

Design Defects. When something is wrong with the basic design concept, the problem is called a **design defect.** If an injured party can identify a serious design defect in a product, the manufacturer may have to reconstitute a line of merchandise or terminate production completely. Design decisions and omissions that have been cited as design defects include the following:

- specifying an unsatisfactory material, for example, specifying a brittle steel when a tough, ductile steel would be safer
- designing a vehicle with gasoline tanks dangerously located
- failure to consider the consequences of normal wear on product safety
- failure to consider inertia forces and other loading that a product could encounter
- failure to guard moving parts
- failure to consider intended and unintended use of the product by children

Risk, Utility, and Consumer Expectations. One purpose of product liability law is to protect consumers from the dangers of complex products. In many cases, these dangers are unknown, unexpected, and undetectable by the average user of a product. In deciding cases involving product liability, courts assume that the manufacturer is in a better position than the consumer to assess and eliminate product dangers.

If a reasonable consumer would expect a particular product danger, then it can be argued that the product is not unreasonably dangerous. If a product danger is hidden, the consumer expectation is different, and the product may be considered unreasonably dangerous. The ratio of utility to risk is also a factor. If the product performs an important useful service, and there is no safer alternative, then some risk may be acceptable. The injured party may prevail in a design-defect case if

1. there is some basis to find that the risks in the product's use outweigh its utility, or
2. it can be shown that some safe and reasonably feasible alternative to the product exists.

Expert Testimony and Reports

When products fail or cause injury, experts may be called to determine the cause of the failure. Reports may be written by experts for the plaintiff (the injured party) and by experts for the defense (usually a manufacturer). The reports are usually furnished to the opposing side before the trial. Most cases are settled on the basis of the reports and other facts. A small percentage of product liability suits go to trial.

Expert Reports. Product designers may be called as expert witnesses or may be involved in product liability cases in other ways. An investigation of the incident is usually described in a report. If the investigation and report are done well, it is more likely that there will be an out-of-court settlement or that the case will be dropped, saving both sides the expense of a trial. The information required for a report could include the following:

- design evaluation: determination of how and why a product failed; the relation of the failure to an injury
- laboratory test results
- practice: Was the product designed according to standard practice? In accordance with applicable standards?
- accident reconstruction: a reconstruction of probable events before and during the accident
- computer simulations of the accident
- risk assessment: design features that contributed to the risk of injury
- an opinion as to the cause of the accident. The opinion should be founded on adequate knowledge, usually based on a substantial investigation and study.

Expert Testimony. Expert testimony is likely to be based on the items covered in the report. An expert witness possesses superior, applicable knowledge based on education and/or experience that relates to the case. The following suggestions may be useful to someone called as an expert witness:

- Tell the truth without exaggeration.
- Do not guess. If uncertain, say, "I don't know," or "I don't remember," or "I do not understand the question." An incorrect statement can discredit your entire testimony.
- If you realize an error in a previous response, indicate the mistake and give the correct answer.
- The opposing attorney may try to show that you are unqualified or that your testimony is incorrect. Keep your temper. Avoid sarcasm. Try not to appear argumentive.
- If possible, use ordinary language with a minimum of technical terms. Do not show off your technical knowledge. Give simple, clear explanations of complicated technical matters. Look at the jury.
- Wait for the complete question before answering. Pause to allow for any objection to the question. Answer *only* the question. Do not elaborate unless it is necessary. Note, however, that an expert witness may give an answer with an explanation if a simple answer would be misleading.

Depositions. Technical personnel are sometimes called to testify in a deposition, which may take place in an attorney's office. If there are several plaintiffs and/or several defendants in the case, lawyers representing each party may be present and may question the witness. During deposition, the witness testifies under oath, and the testimony is recorded by a court reporter. The deposition may be videotaped as well. Preparation for a deposition requires the same care as preparation for court testimony. Most of the above suggestions for testimony as an expert witness apply to both depositions and in-court testimony.

PATENTS AND TRADE SECRETS

A patent is a property right granted by a government to an inventor for a fixed time period. It excludes others from making, using, or selling an invention. Much of this section is based on publications of the U.S. Patent and Trademark Office (Patent and Trademark Office, 1993). Patent information, changes in the law, and copies of patent disclosures can be obtained from that office.

Types of Patents

There are three major patent categories:

1. **Utility patents** are granted to one who invents or discovers a new and useful process (method), machine, manufacture (product), or composition of matter, or an improvement thereof. Machine designers would be most interested in this patent category.
2. **Design patents** are granted to inventors of a new, original, or ornamental design for articles of manufacture. The appearance of the article is protected.
3. **Plant patents** are granted to those who invent and asexually reproduce a new variety of plant.

Inventor's Records

Clear, concise records of an invention should be kept in ink in a bound book that identifies the inventor or inventors. The record should

- describe the invention.
- contain sketches and/or photographs of the invention.
- include the date the invention was conceived and the date and signature of the inventor and a witness for each entry. The witness should be one who understands the invention but will not benefit from it.
- describe invention features that are new and useful.
- demonstrate that the concept can be reduced to practice, that is, the embodiment of the invention can perform as intended. (A working model is not required.)
- record meetings and conversations with others concerning the invention.

Patentability

Patents are granted only to the true inventor. The rules and conditions for patents vary from country to country. Recent changes in the patent law that resulted from the General Agreement on Tariffs and Trade (GATT) include the introduction of a "provisional" patent application that protects the invention for a 1 year period.

A patent **will not** be granted

- if the invention was known to others or used by others before invention by the applicant.
- if the invention was patented or described in a printed publication in the United States or a foreign country more than 1 year prior to the date of application for a U.S. patent. (*Caution:* Foreign patent applications should be filed before the invention is described in a publication.)
- if the invention was in public use or on sale in the United States more than 1 year before the applicant filed for a patent. Even the inventor's use and sale of the invention more than a year before the application is filed will bar the inventor's right to a patent.

Trade Secrets

Sometimes an invention will be maintained as a trade secret and will not be patented. However, there is a risk that the trade secret may be disclosed or that someone will independently invent and patent the product or process. Patent law generally supports inventors who obtain patents over inventors who conceal their inventions as trade secrets.

Consider an undisclosed invention (trade secret) that is later independently invented and patented by a second party in good faith. The patent holder may then claim infringement, if the "owner" of the trade secret continues to use the technology. A prior user's right is incorporated in proposed statutes. If the statutes are adopted, the patent would be valid, but the prior user would be free to use the technology. The patent laws of some countries include a prior user's right.

Ownership and Sale of Patent Rights

An inventor can sell all or part interest in a patent or patent application by properly worded assignment. Most corporations require each employee to sign an agreement that assigns to the corporation all rights to any invention created by the employee. Since companies present a united front on this practice, some individuals consider it a restraint of trade. Riley (1994) suggests that a law requiring a company to turn over a percentage of a patent's value to the inventor would lead to more inventions, improve national competitiveness, and result in more skilled jobs. He feels that companies would benefit from increased innovation.

ETHICS

Ethics involves the moral choices that an individual makes. Professional ethics describes the rules and standards governing the conduct of members of a profession. It could be argued that the discoveries of pure science are ethically neutral. However, technology, which puts science into practice, is not value-neutral; technology requires many ethical decisions.

Ethical Practice

Ethical practice requires responsible behavior—"doing the right thing." Some codes and standards of behavior define the "right thing" as

- promoting the safety and reliability of products
- holding paramount the safety, health, and welfare of the public
- using knowledge and skill for the advancement of human welfare; contributing professional advice to civic and charitable organizations as appropriate

- speaking out against abuses that affect the public interest or welfare
- acting as faithful agents and trustees for employers and clients (provided that such actions are not otherwise unethical)
- maintaining high standards of diligence, creativity, and productivity
- upholding the honor and dignity of the profession
- continuing to learn; keeping up-to-date on technology
- working *within* the limits of one's experience and training and accepting responsibility for one's actions
- analyzing designs and processes carefully; performing adequate tests; and reporting test results accurately and completely
- crediting the contributions of others; treating colleagues and co-workers fairly, regardless of race, religion, age, sex, or national origin; encouraging colleagues and co-workers to act ethically, and supporting them when they do

Designers of machines and other products are prepared to analyze their designs, to consider reliability, and to use adequate safety factors to avoid dangerous failures. It is clearly unethical to misrepresent and "doctor" test results. Although an employer may have no claim on an employee's time outside working hours, conflicts of interest, accepting favors from vendors, and unauthorized disclosure of trade secrets are unethical.

Ethical Questions

Other ethical decisions, including decisions involving social responsibility, are more difficult. Such decisions may involve short-term gains and profits versus public health and welfare. Many ethical questions can be answered only by an individual in possession of the relevant facts and conditions. For example, is it ethical to design

- products and processes that are wasteful of resources?
- products and processes that cause unnecessary pollution?
- machinery or processes to produce assault weapons for general use?
- machinery or processes to produce tobacco products?
- military weapons to be sold to the highest bidder?

Whistle Blowers

Whistle blowers are those who speak out against unsafe design, unsafe practices, and other abuse of the public trust. Anyone with direct knowledge of designs or practices that pose significant risk to public health, safety, or welfare has an ethical obligation to make that information known. Most whistle blowers act out of moral outrage, with no thought of compensation.

Whistle blowers in industry usually begin by advising their employers of the consequences of an unsafe design or practice. In one case, an individual went through an internal review process to complain about possible bonding problems with jet engines that his employer produced for the Air Force. After receiving no satisfaction, he and his attorney went to the Federal Bureau of Investigation, which arranged for him to secretly tape conversations related to engine standards. In a 1995 court filing, the Justice Department alleged that the manufacturer falsely certified the engine line. In this case, the individual could share in any judgment if the government prevails.

Technical personnel should be assured of the facts before making public allegations regarding unsafe designs and violation of the public trust. Unfounded claims made on the basis of rumor could do substantial harm. However, if the abuse is significant and cannot be corrected internally, then inaction would be unethical. In such cases, careful records should be kept, with the names of potential witnesses noted.

ECONOMIC DECISIONS

Cost plays an important part in many design and manufacturing decisions. Economic decisions may be based on the questions, "Will it 'pay' to make a given design change? Is it cost-effective to purchase a new machine to produce a part?"

The cost of money then becomes a component of these decisions. If money is to be borrowed to purchase equipment, then a bank can provide information about interest rates. The cost of using cash includes lost interest. The rate for lost interest can be based on interest rates of U.S. treasury bills, notes, and bonds of various maturities, which are listed in newspapers. Depreciation rules and "costing" options based on the tax laws, and inflation should be considered as well.

Compound Interest

Some investments in materials and equipment can be expected to "bear fruit" at a future date. In such cases, we can compare our expectations of future reward with the growth that we would get from fixed-income securities or deposits. If all interest is reinvested, or if we invest in zero-coupon bonds, the total principal at the end of a time period is given by

$$P_Y = P\left(1 + \frac{I}{100n_y}\right)^{Y \cdot n_y} \qquad \textbf{(20.13)}$$

where P_Y = principal (account value) at end of term (\$)

P = original principal (\$)

I = annual interest rate (%)

Y = term (yr)

n_y = number of times per year that interest is compounded (or paid and reinvested)

Bond interest is usually paid twice per year; interest on other investments can be compounded monthly or at other intervals.

Time Required to Increase Principal by a Given Percent

Some cost decisions are related to increasing an investment over a period of time. The compound interest equation can be solved for time for this purpose. Mathematics software with symbolic capabilities is helpful. The time to increase an investment by a certain amount is given by

$$Y = \frac{\ln r_p}{\ln\left(\frac{100n_y + I}{100n_y}\right) \cdot n_y} \qquad \textbf{(20.14)}$$

where Y = time to produce increase (yr)

ln = natural logarithm

r_p = ratio of principal at end of term to original principal

n_y = number of compounding periods per year

EXAMPLE PROBLEM Increasing Principal by a Given Amount

How long does it take to double principal at a 10% annual interest rate with

(a) semiannual compounding?
(b) monthly compounding?
(c) How long does it take to increase principal by one-third if the interest rate is 4.5%?

Solutions (a) and (b): Equation (20.14) is used with interest rate $I = 10$ and principal ratio $r_p = 2$. The results are (a) 7.1 years with semiannual compounding and (b) 6.69 years with monthly compounding.

Solution (c): With $I = 4.5$ and $r_p = 1.3333$, we find that it takes 6.46 and 6.4 years for semiannual and monthly compounding, respectively.

Rules of Thumb for Interest, Time, and Increase in Principal

Suppose that we want to find an easier way to relate interest and time to increase in principal. It obviously takes less time to reach a particular principal goal if interest rates go up. This could lead us to examine the product of the interest and time. It turns out that this product is nearly constant for a given percent principal increase. That is,

$$I \cdot Y \approx K \qquad \textbf{(20.15)}$$

where I = annual interest rate (%)

Y = time (yr)

K = an approximate constant for a given percent increase in principal

EXAMPLE PROBLEM A Rule of Thumb for Interest and Time

Write a simple equation relating percent interest and years of investment if principal is to increase by 50%.

Solution Summary: As shown in the following detailed solution, the time required for a 50% principal increase was determined for interest rates ranging from 2% to 10%. The product $I \cdot Y$ ranged from 40.95 to 42.54 based on annual compounding. Thus, our approximate equation is

$$I \cdot Y \approx K \quad \text{where } K = 42$$

For example, either 6% interest for 7 years or 7% interest for 6 years will yield about 50% increase in principal. For monthly compounding, $K = 40.6$ to 40.7.

Detailed Solution:

$$Y(I) = \frac{\ln(r_p)}{\left[\ln\left[\frac{1}{100} \cdot \frac{(100 \cdot n_y + I)}{n_y}\right] \cdot n_y\right]}$$

Annual Compounding: $n_y = 1$

Monthly Compounding: $n_y = 12$

Rate (%) I	*Years* $Y(I)$	*Product* $I \cdot Y(I)$	*Rate (%)* I	*Years* $Y(I)$	*Product* $I \cdot Y(I)$
2	20.48	40.95	2	20.29	40.58
2.5	16.42	41.05	2.5	16.24	40.59
3	13.72	41.15	3	13.53	40.6
3.5	11.79	41.25	3.5	11.6	40.61
4	10.34	41.35	4	10.15	40.61
4.5	9.21	41.45	4.5	9.03	40.62
5	8.31	41.55	5	8.13	40.63
5.5	7.57	41.65	5.5	7.39	40.64
6	6.96	41.75	6	6.77	40.65
6.5	6.44	41.85	6.5	6.25	40.66
7	5.99	41.95	7	5.81	40.66
7.5	5.61	42.05	7.5	5.42	40.67
8	5.27	42.15	8	5.09	40.68
8.5	4.97	42.25	8.5	4.79	40.69
9	4.7	42.34	9	4.52	40.7
9.5	4.47	42.44	9.5	4.28	40.71
10	4.25	42.54	10	4.07	40.72

For 100% increase (a doubling of principal), we have the more familiar approximation

$$I \cdot Y \approx 72$$

We obtain the constant for a 200% increase (tripling of principal) by adding the constant for doubling and the constant for 50% increase, yielding

$$I \cdot Y \approx 72 + 42 = 114$$

■■

Present Value

Present value is our estimate of the value (today) of an amount of future income or savings. We are simply looking at compound interest from a different viewpoint. Present value is given by

$$P = \frac{P_Y}{\left(1 + \frac{I}{100n_y}\right)^{Yn_y}} \qquad \textbf{(20.16)}$$

where P = present value (\$)

P_Y = expected future income or savings (\$)

Y = time we must wait for income or savings (yr)

n_y = number of times per year that interest is compounded

EXAMPLE PROBLEM Determining Present Value

We are considering purchasing a piece of land for a manufacturing facility that is to be constructed in about 25 years. In the meantime, the land can be leased for an amount that will just cover taxes. The land will be worth about \$1 million 25 years from now. What is it worth now?

Decisions: Since our intended use is quite far in the future, there are many uncertainties. Thus, we will make our decision on the basis of a 12% annual interest rate, compounded annually.

Solution Summary: As calculated in the following detailed solution, the present value of \$1 million 25 years from now is about \$58,800 based on annual compounding. We will offer \$58,800 for the land. If we based our decision on 12% interest compounded monthly, our offer would be about \$50,500.

Detailed Solution:

Value at End of Term:

$$P_Y = 10^6$$

Interest Rate per Year (%):

$$I = 12$$

Annual Compounding:

Compounding Periods per Year

$n_y = 1$

Term (Years of Compounding)

$Y = 25$

Present Value

$$P = \frac{P_Y}{\left(1 + \frac{I}{100 \cdot n_y}\right)^{Y \cdot n_y}} \qquad P = 58{,}823.31$$

Monthly Compounding:

Compounding Periods per Year

$n_y = 12$

Present Value

$$P = \frac{P_Y}{\left(1 + \frac{I}{100 \cdot n_y}\right)^{Y \cdot n_y}} \qquad P = 50{,}534.49$$

Annual Cost of an Investment

The annual cost concept is difficult, but very useful for making decisions about investments in production machinery. It is a level (constant) annual charge to cover interest and depreciation or amortization.

Salvage Value. Annual cost is similar to making mortgage payments, except that salvage value is included in the calculations. In some cases, there is negative salvage value. For example, nuclear reactors incur decommissioning costs which can be treated as negative salvage value.

Salvage value is converted to present value with the following equation:

$$P_{SO} = \frac{P_S}{(1 + i)^Y} \quad \textbf{(20.17)}$$

Where P_{SO} = present value of salvage ($) = the value today of the future salvage value
P_S = salvage value of equipment ($) after Y years
Y = useful life of equipment (yr)
i = annual interest (fraction) = $I/100$
I = annual interest (%)

We subtract present value of salvage to obtain

$$P_R = P_O - P_{SO} \quad \textbf{(20.18)}$$

where P_R = investment costs to recover ($)

Annual Cost. We then find annual cost as follows:

$$A = \frac{P_R \cdot i(1 + i)^Y}{(1 + i)^Y - 1} \quad \textbf{(20.19)}$$

where A = annual cost ($) = a constant for the useful life of the equipment

If production equipment is bought for cash, annual cost is useful in pricing of products. Annual cost also helps us decide whether to purchase components from a vendor or to purchase equipment to manufacture those components.

Some industrial equipment is financed by mortgaging the equipment. If there is no salvage value at the end of the useful life of the equipment, and no significant removal cost, then the annual cost is the same as the mortgage payment (if paid annually).

Costs on a Monthly or Other Basis. If equipment is financed by a mortgage, then monthly payments may be required. Even if we are using our calculations for pricing and decision-making purposes only, it may be convenient to find monthly or quarterly cost. To change from an annual basis, we make the following changes:

$$i \Rightarrow i/n_y = I/(100n_y)$$
$$Y \Rightarrow Y \cdot n_y$$

where n_y = number of intervals per year

For example, if the annual interest rate is $I = 9\%$, payments are monthly, and the useful life of the equipment is 6 years, then

$$n_y = 12$$
$$i \Rightarrow i/n_y = I/(100n_y) = 9/(100 \cdot 12) = 0.0075$$
$$Y \Rightarrow Y \cdot n_y = 6 \cdot 12 = 72$$

EXAMPLE PROBLEM Annual Cost of an Investment

We are considering the purchase of automated production machinery for $100,000. The useful life (for our operations) is estimated to be 8 years. At the end of that time, we could probably resell the equipment to another manufacturer for about 20% of its original cost. Estimate the annual cost of the investment.

Decision: After checking current interest rates, we have decided to base our calculations on an annual interest rate of 9% compounded annually.

Solution Summary: The estimated salvage value is 20% of $100,000 = $20,000 eight years from now. We find that the present value of the $20,000 is $10,040. The investment to be recovered is $100,000 − 10,040 = $89,960. We then calculate a cost of $16,254 for each of the 8 years to cover the investment. To this, we must add the cost of electricity, maintenance or a maintenance contract, taxes, insurance, factory space, and other fixed and production costs.

Check: We now check the calculations by "charging" the machine its annual cost for 8 years. For year zero, it owes us the entire $100,000 original investment. At the end of the year, it must pay 9% of $100,000 as interest. The rest of the $16,254 goes to amortization (reduction of the debt). In subsequent years, the debt principal is reduced by the amortization of the previous year, and the interest is paid on the reduced principal. The interest goes down every year, and the amortization goes up. The total annual cost is constant.

We could check the results with a calculator, but it would be tedious. Instead, we let the computer do the calculations for principal (debt), interest, and amortization (debt reduction). Note that we also check the sum of interest and amortization to see that it equals annual cost. We let the year y go from zero to eight, using subscript $y + 1$ for the subsequent year. At the beginning of year eight, we sell the machine for $20,000. Therefore, the figures in the interest, amortization, and annual cost columns for year eight are meaningless.

Mortgaged Machinery: We determined annual cost in the above calculations. The result is similar to a $100,000 mortgage with a "balloon payment" of $20,000 at the end of 8 years. Suppose that we had mortgaged the machinery, and the bank wanted us to pay off the entire loan in 8 years. Then we could have determined the mortgage payment by recalculating annual cost with the salvage value set at zero. If the payment was to be monthly or quarterly, then we would make substitutions similar to those indicated in the section "Costs on a Monthly or Other Basis."

The detailed solution follows.

Detailed Solution:

Original Investment ($):

$$P_O = 100{,}000$$

Interest Rate:

%	Fraction	
$I = 9$	$i = \frac{I}{100}$	$i = 0.09$

Useful Life (yr):

$$Y = 8$$

Salvage Value ($) after Y Years:

$$P_S = 20{,}000$$

Present Value of Salvage ($):

$$P_{SO} = \frac{P_S}{(1+i)^Y} \qquad P_{SO} = 1.004 \cdot 10^4$$

Investment to Recover ($):

$$P_R = P_O - P_{SO} \qquad P_R = 8.996 \cdot 10^4$$

Annual Cost ($):

$$A = \frac{P_R \cdot i \cdot (1+i)^Y}{(1+i)^Y - 1} \qquad A = 16{,}253.95$$

$$y = 0, \ldots, Y$$

$$\begin{matrix}\text{Principal:} \\ \text{Interest:} \\ \text{Amortization:}\end{matrix} \quad \begin{bmatrix}\mathrm{Pr}_0 \\ \mathrm{Int}_0 \\ \mathrm{Am}_0\end{bmatrix} = \begin{bmatrix}P_O \\ i \cdot P_O \\ A - i \cdot P_O\end{bmatrix} \qquad \begin{bmatrix}\mathrm{Pr}_{y+1} \\ \mathrm{Int}_{y+1} \\ \mathrm{Am}_{y+1}\end{bmatrix} = \begin{bmatrix}\mathrm{Pr}_y - \mathrm{Am}_y \\ i \cdot (\mathrm{Pr}_y - \mathrm{Am}_y) \\ A - i \cdot (\mathrm{Pr}_y - \mathrm{Am}_y)\end{bmatrix}$$

Year y	*Principal ($)* Pr_y	*Interest ($)* Int_y	*Amortization ($)* Am_y	*Annual cost ($)* $Int_y + Am_y$
0	100,000	9000	7253.95	16253.95
1	92,746.05	8347.14	7906.81	16253.95
2	84,839.24	7635.53	8618.42	16253.95
3	76,220.83	6859.87	9394.08	16253.95
4	66,826.75	6014.41	10,239.54	16253.95
5	56,587.21	5092.85	11,161.1	16253.95
6	45,426.11	4088.35	12,165.6	16253.95
7	33,260.5	2993.45	13,260.5	16253.95
8	20,000	1800	14,453.95	16253.95
		*	*	*

*Last payment does not apply.

AN INTRODUCTION TO OPTIMIZATION

Ideally, an optimum design is the best possible design. In real-world problems, however, possible design variations are uncountable, and the ideal optimum design is impossible.

Cost Function and Constraints

The cost function is what we want to minimize or maximize. It may even be difficult to decide on the cost function. Do we want minimum dollar cost? Or do we want minimum weight? Or should the cost function be some function of both weight and dollar cost, say, 10 times the weight in pounds plus 3 times the cost in dollars? There are often constraints on variables. For example, component size may be limited by packaging of the overall product.

Multivariable Optimization Problems

Most designers use their education and experience to reduce a complicated design assignment to a series of problems involving only one variable. Problems of this type were considered earlier. If we wish to consider many variables, we have a large number of methods to choose from. Multivariable optimization methods include the exhaustive search, the lattice method, and the gradient method (steepest ascent).

An exhaustive search is easy to program, but relatively inefficient. The lattice method and the gradient method require more programming effort in return for their better efficiency in finding an optimum solution. The more difficult mathematical concepts of the gradient method sometimes discourage its use. No optimization method works perfectly for all problems; the optimum solution may sometimes be missed.

Exhaustive Search. Figure 20.5 shows a contour map of a cost function. In this case, the cost function is made up of positive attributes rather than dollar cost or weight. We must maximize the cost function. The contour lines are lines of constant cost function. The cost func-

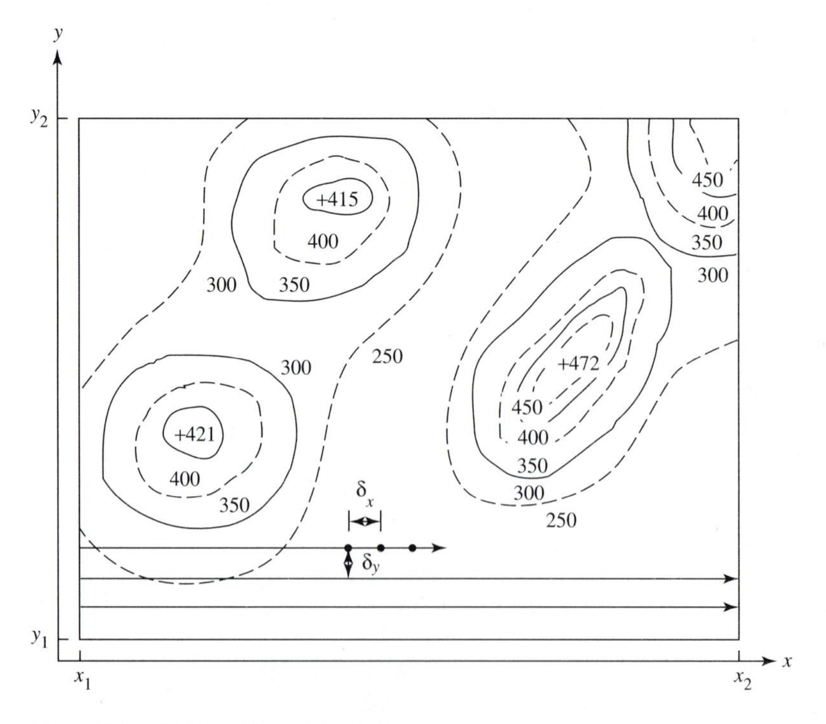

Figure 20.5 Multivariable optimization.

tion peaks at values of 415, 421, and 472 as indicated by the + marks. Acceptable design solutions are limited by constraints

$$x_1 \leq x \leq x_2 \quad \text{and} \quad y_1 \leq y \leq y_2$$

That is, we do not want to consider any solution outside that range.

Suppose that as a designer, you have made a mathematical formulation of the cost function, but you do not have the contour map. The problem is analogous to trying to find the highest point of land in a dense fog. For an exhaustive search, let the region of interest be covered with a grid of spacing δx and δy. The cost function can be calculated at every grid point as follows:

$$x_1, y_1; \quad x_1 + \delta x, y_1; \quad x_1 + 2\delta x, y_1, \ldots, x_2, y_1$$
$$x_1, y_1 + \delta y, x_1 + \delta x, y_1 + \delta y; \quad x_1 + 2\delta x, y_1 + \delta y, \ldots, x_2, y_1 + \delta y$$

and so on. Each value of cost function is compared with the highest previous value until the search reaches point x_2, y_2. This search method is easy to program. Considering the contour plot in Figure 20.5, an exhaustive search with a fine grid should not be fooled by the relative maxima at "hill 415" and "hill 421." This method should locate a point near the actual maximum cost function, which is either the summit of "hill 472" or the highest point on the boundary near point x_2, y_2. The disadvantage of this type of exhaustive search is that it takes considerable computer time and thus may not be practical for searching a region of three or more variables.

Lattice Method. This method uses a small grid or lattice, which is moved in the most promising direction toward the optimum. Consider Figure 20.6, a contour plot of cost function to be maximized. Again, since we do not have the contour plot, we cannot see the optimum value. A 5-point lattice like *ABCDE* in the figure or a 9-point lattice like *abcdefghi* can be used. Let the center of the lattice, point *e* or *E*, be placed arbitrarily at some location *x*, *y*. The other lattice points are located about point *e* or *E* as shown. We calculate the cost function at each of the lattice points and compare them. In the sketch, point *c* (or *C*) apparently has the highest value. The center of the lattice is then moved to the point formerly occupied

Figure 20.6 Lattice method.

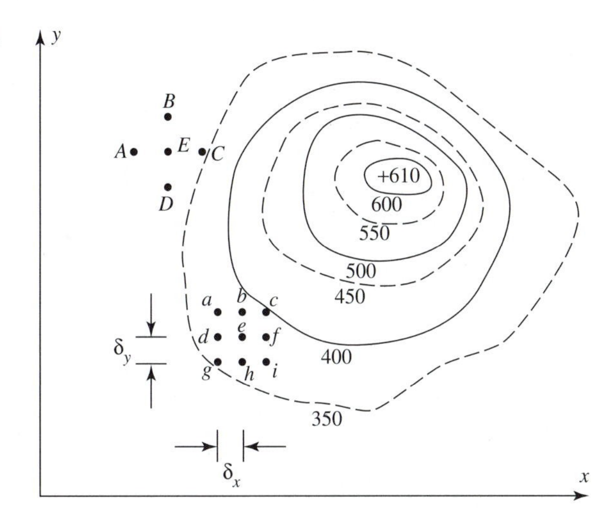

by point c or C, and the process is repeated. When the center of the lattice is at or near the summit, the value of cost function at the center of the lattice is better than at the other lattice points. We can then halve δx and δy to try to get closer to the summit. The process is stopped when further adjustments in lattice size and location produce no significant change in cost function.

For three independent variables, cost function optimization is analogous to finding the hottest point in a room. The two-dimensional lattice is replaced with a three-dimensional grid, a center point surrounded by 6 to 26 additional points in the x-, y-, and z-directions. It is difficult to imagine a center point surrounded by additional points displaced in the w-, x-, y-, and z-directions in four-dimensional space. We can, however, optimize a cost function made up of four or more independent variables. Although programming is more difficult, the optimization procedure for three or more variables is similar to that for two variables.

Gradient Method (Steepest Ascent).[1] The gradient method is analogous to climbing a hill by moving in the steepest direction—by taking each step in the direction of the gradient vector. This method can be applied to optimization problems involving two or more independent variables. We can avoid tedious, repetitive calculation by writing a computer program for this application.

For a two-independent-variable problem, the **gradient vector** is defined by

$$\nabla C = \mathbf{i}\, \partial C/\partial x + \mathbf{j}\, \partial C/\partial y$$

and for a three-independent-variable problem,

$$\nabla C = \mathbf{i}\, \partial C/\partial x + \mathbf{j}\, \partial C/\partial y + \mathbf{k}\, \partial C/\partial z$$

where **i**, **j**, and **k** = unit vectors in the x-, y-, and z-directions, respectively
C = cost function.

Formation of the gradient vector may appear complicated, but application of the gradient method is relatively simple. We begin by specifying the cost function. Consider the two-independent-variable case. We wish to take steps toward an optimum such that δx and δy, the x- and y-direction components of the steps, have the same ratio as the gradient vector components. Arbitrarily selecting a starting point x, y and initial x-increment δx, we calculate

$$\delta y = \delta x[\partial C/\partial y]/[\partial C/\partial x]$$

Referring to Figure 20.7, note that the first step takes us from location a (x, y) to location b $(x + \delta x, y + \delta y)$. If the cost function is improved, we continue with the same increment δx and compute a new increment δy based on conditions at location b. We continue the process in this way, trying to locate an optimum design.

If the cost function changes in the wrong direction, increment δx is halved and its sign is changed. If the cost function is well-behaved, after several iterations, cost function change will be less than a predetermined tolerance, and the calculation ends with the message

```
Search ended, cost function change < tolerance
```

---

[1]Some users of this text will omit the gradient method because they are not familiar with the underlying mathematical concept. Those who still wish to solve optimization problems will find the lattice method a useful alternative. Answers are essentially unaffected by the method chosen for solution.

Figure 20.7 Gradient method.

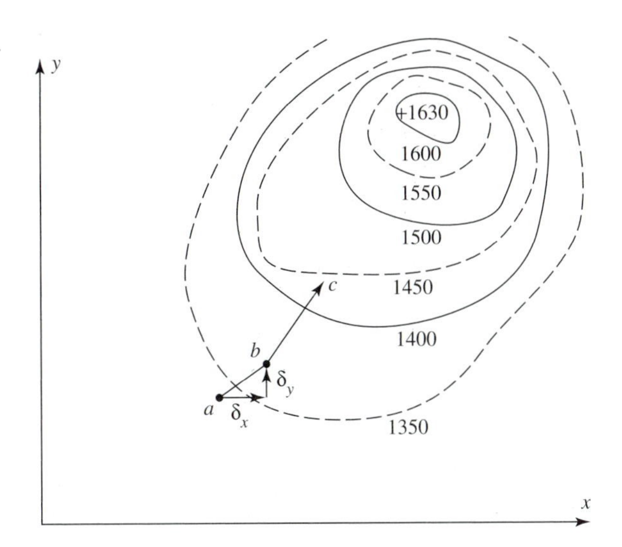

The gradient method is not foolproof; in some cases, it will not locate an optimum within the specified tolerance. A **do-loop** or a **for-next-loop** can be used to end an ineffective search after a given number of iterations. For example, if we allow 30 tries, an ineffective search could end with the message

```
Search ended after 30 iterations, out of tolerance
```

In some cases, this message will indicate a programming error.

Figure 20.8 shows a flowchart for a gradient-method optimization program where cost function C is to be **minimized.**

EXAMPLE PROBLEM Optimization Using the Gradient Method

Given the cost function

$$C = x^2 + y^2 - 11x - 8y + e^{xy/20} + y^{1.5} + 72$$

find the values of x and y for which C will be minimum.

Solution: We obtain the partial differential with respect to x by differentiating with respect to x, while holding y constant. Thus,

$$\partial C/\partial x = 2x - 11 + (y/20)e^{xy/20}$$

Similarly,

$$\partial C/\partial y = 2y - 8 + (x/20)e^{xy/20} + 1.5y^{0.5}$$

These equations are incorporated into a program based on the flowchart in Figure 20.8.

We can arbitrarily select $x = 0$, $y = 0$ as the starting point, with the x-component of the step $\delta x = 0.5$, and use a tolerance of 0.001. The program gives the value of the cost function as $C = 73$ at 0, 0. It then steps in the gradient direction to point 0.5, 0.364 (i.e., $x = 0.5$, $y = 0.364$) where $C = 65.2$

Figure 20.8 Flowchart for gradient method.

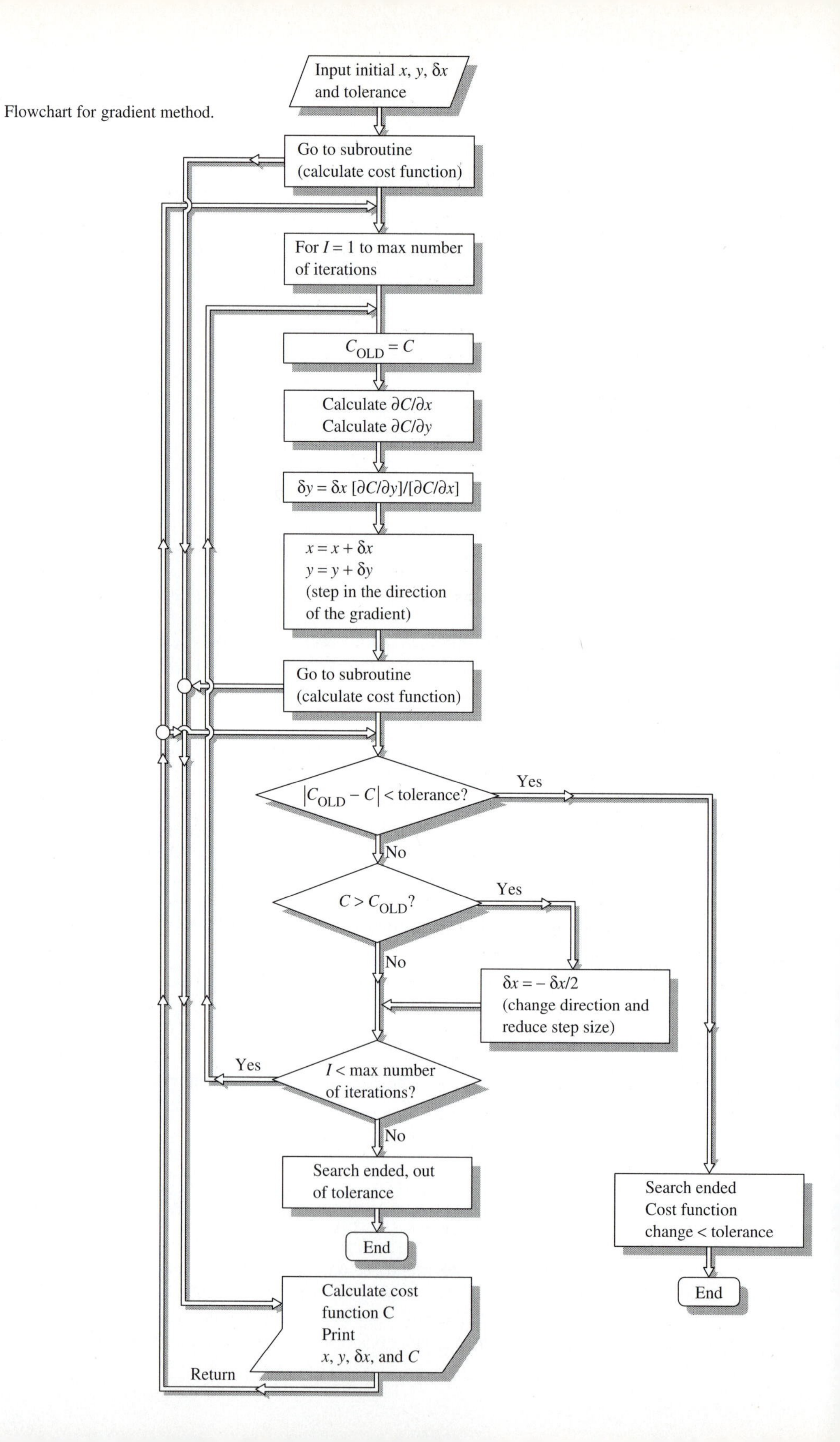

Table 20.6 Multivariable optimization by the gradient method (steepest ascent).

```
Cost function C is to be minimized
Initial x, y, x step and tolerance:      0      0      0.5      0.001
```

x	y	x step	cost function C
0	0	.5	73
.5	.3636364	.5	65.20156
1	.6813676	.5	58.61042
1.5	.9796111	.5	52.91856
2	1.261181	.5	48.05188
2.5	1.526728	.5	43.96378
3	1.775716	.5	40.61889
3.5	2.006458	.5	37.98702
4	2.215627	.5	36.03951
4.5	2.396795	.5	34.74566
5	2.535649	.5	34.06699
5.5	2.580918	.5	33.94359
6	2.830201	.5	34.46718
5.75	2.663922	-.25	34.09642
5.5	2.538539	-.25	33.94039
5.25	2.519209	-.25	33.94103

```
Search ended, cost function change < tolerance
```

(see the computer output shown in Table 20.6). Since we wish to minimize C, we are moving in the correct direction. Each step reduces C until we get to point 5.5, 2.58. The next step increases C, and the program responds by changing the sign of δx and halving it. Finally the search ends when the cost function changes less than the tolerance. It appears that the next-to-last point (5.5, 2.5385) represents the optimum values of x and y based on lowest cost function. ■■

Except in relatively simple situations, we cannot be sure that the result is the true optimum value. In some cases, we may take steps along a "ridge" or on a "plateau" of the cost function contours, or we may locate a relative extreme rather than the optimum. In an attempt to verify the result in the preceding problem, the optimum was approached from another direction, starting at point 10, 10 with an initial step of $\delta x = 0.5$ as before. The first step resulted in an increase in cost function, and the program set $\delta x = -0.25$ for the next step. The search ended after 30 iterations (the preset maximum) at a point near the point obtained above, and with a slightly better value of C.

Sometimes cost function C is made up of positive attributes, and the goal of the optimization is maximization of C. In that case, the test $C > C_{OLD}$? is replaced by $C < C_{OLD}$? If the answer is **yes**, then the program can change the sign of δx and halve it.

References and Bibliography

American National Standards Institute. *American National Standard for Commercial Turf Care Equipment*—Safety Specifications. ANSI/OPEI B71.4. 1990.

American National Standards Institute. *American National Standard for Power Tools—Gasoline Powered Chain Saws—Safety Requirements.* ANSI B175.1. 1991.

Brown, L. R., et al. *State of the World.* New York: W. W. Norton, 1994.

Consumer Product Safety Commission. *Bicycle Use and Hazard Patterns in the United States.* Washington, DC: CPSC, June 1994a.

Consumer Product Safety Commission. *Consumer Product Safety Alerts, 1985–93.* Washington, DC: CPSC, June 1994b.

Dieter, G. E. *Engineering Design, a Materials and Processing Approach.* 2d ed. New York: McGraw-Hill, 1991.

Embleton, T. F. W. "Upper Limits on Noise in the Workplace." *Noise/News International* 2 (no. 4): 227–37. Institute of Noise Control Engineering, December 1994.

Energy Information Administration. *Annual Energy Review 1993.* Washington, DC: U.S. Department of Energy, 1994.

Environmental Protection Agency. *Protective Noise Levels.* EPA 550/9-79-100. Washington, DC: EPA, 1978.

Famighetti, R., ed. *The World Almanac and Book of Facts, 1995.* Mahwah, NJ: Funk & Wagnalls, 1994.

Fingerhut, L. A., J. C. Kleinman, E. Godfrey, and H. Rosenberg. "Firearm Mortality among Children, Youth and Young Adults 1–34 Years of Age, Trends and Current Status: United States, 1979–88." *Monthly Vital Statistics Report* 39 (no. 11). Hyattsville, MD: Centers for Disease Control, March 14, 1991.

Fingerhut, L. A., and J. C. Kleinman. "Firearm Mortality among Children and Youth." *Advance Data* (no. 178). Hyattsville, MD: Centers for Disease Control, November 3, 1989.

Fingerhut, L. A., C. Jones, and D. M. Makuc. "Firearm and Motor Vehicle Injury Mortality—Variations by State, Race, and Ethnicity: United States, 1990–91." *Advance Data* (no. 242). Hyattsville, MD: Centers for Disease Control, January 27, 1994.

Houghton, J. *Global Warming.* Oxford: Lion Publishing, 1994.

Jacobs, R. M. "Product Recall—A Vendor/Vendee Nightmare." *Microelectronics Reliability* 37 (no. 7): 7–9. Elsevier Science Ltd., 1995.

McGoldrick, J. L., and F. T. Smith. *New Jersey Product Liability Law.* Newark: NJ Law Journal Books, 1994.

National Safety Council. *Safeguarding Concepts Illustrated.* 5th ed. Chicago: NSC, 1992.

Occupational Safety and Health Administration. *All about OSHA.* OSHA 2056 (revised). Washington, DC, 1992a.

Occupational Safety and Health Administration. *Concepts and Techniques of Machine Guarding.* OSHA 3067 (revised). Washington, DC, 1992b.

Occupational Safety and Health Administration. *Hearing Conservation.* OSHA 3074 (revised). Washington, DC, 1992c.

Occupational Safety and Health Administration. *Occupational Noise Exposure.* Code of Federal Regulations, 29 CFR 1910.95, 1981. Washington, DC, effective 1983.

Patent and Trademark Office. *Basic Facts about Patents.* Washington, DC: U.S. Department of Commerce, USPTO, October 1993.

Riley, R. "Inventors Deserve Their Fair Share." *Machine Design* 109. Cleveland, OH: Penton Publishing, March 21, 1994.

Wilson, C. E. *Noise Control, Measurement, Analysis and Control of Sound and Vibration.* Malabar, FL: Krieger Publishing, 1994.

Design Problems

20.1 The following representative sound level measurements were made during a 10 min period: 77, 65, 80, 70, and 89 dBA. Find equivalent sound level for that period.

20.2 The following five representative sound level measurements were made during a 1 h period: 105, 95, 98, 92, and 87 dBA. Find equivalent sound level for that period.

20.3 The following eight representative sound level measurements were made during a 1 h period: 60, 63, 70, 61, 60, 62, 67, and 80 dBA. Find equivalent sound level for that period.

20.4 An individual is exposed to the following noise levels and exposure times during a typical 24 h day:

Level (dBA)	*Time (h)*
70	3
59	2.5
69	5.5
81	4
85	5
60	1.5
79	2.5

Find 24 h equivalent sound level. Compare with a protective equivalent sound level based on hearing protection.

20.5 An individual is exposed to the following noise levels and exposure times during a typical 24 h day:

Level (dBA)	*Time (h)*
84	2.1
49	3.4
62	5.5
54	4
63	5
68	1.5
65	2.5

Find 24 h equivalent sound level. Compare with a protective equivalent sound level based on hearing protection.

20.6 An individual is exposed to the following noise levels and exposure times during a typical 24 h day:

Level (dBA)	*Time (h)*
55	2
49	3.5
67	5.5
68	4
67	5
59	1.5
60	2.5

Find 24 h equivalent sound level.

20.7 The following noise level measurements were made outdoors in a residential area:

Level (dBA)	*Hours (Day)*	*Level (dBA)*	*Hours (Night)*
32	5	43	2
55	4	35	2
49	2	40	3
40	3	39	2
42	1		

Find equivalent sound level and day–night sound level. Compare with protective levels for activity interference and hearing conservation.

20.8 The following noise level measurements were made outdoors in a residential area:

Level (dBA)	*Hours (Day)*	*Level (dBA)*	*Hours (Night)*
45	4	55	3
56	4	35	2
49	2	51	3
40	3	39	1
52	2		

Find equivalent sound level and day–night sound level. Compare with protective levels for activity interference and hearing conservation.

20.9 Find equivalent sound level and day–night sound level for the following measurements. Compare with protective levels for hearing conservation and activity interference in residential and other areas.

Level (dBA)	*Hours (Day)*	*Level (dBA)*	*Hours (Night)*
55	3	54	2
43	4	37	1
59	2	60	3
41	3	41	3
50	3		

20.10 The following noise measurements represent contributions to noise at a point from five individual sources: 100, 98, 86, 80, and 77 dBA. Find the noise level at that point if all the sources are operating at one time.

20.11 Four household appliances make individual noise contributions of 70, 70, 63, and 57 dBA. Find the noise level if all four operate at the same time.

20.12 Intrusive transportation noise, appliances, and HVAC sources make the following individual contributions: 49, 47, 40, 45, 41, and 42 dBA. Find the noise level for the combined contributions.

20.13 A worker is exposed to a daily noise pattern described as follows:

Noise Level (dBA)	*Exposure Time (h)*
85	3
87	2
90	1
81	1.5
83	0.5

Find percent noise dose and time-weighted average noise level. Compare results with exposure limits.

Decisions: Use a criterion level of 90 dBA and an exchange rate of 5 dBA.

20.14 A worker performs several operations, using different machines. He is exposed to a daily noise pattern described as follows:

Noise Level (dBA)	*Exposure Time (h)*
95	2
106	0.5
80	2
85	1.5
90	2

(a) Find percent noise dose and time-weighted average noise level. Compare results with exposure limits.

Decisions: Use a criterion level of 90 dBA and an exchange rate of 5 dBA.

(b) If the results are unacceptable, propose reducing the noise level of certain machines. Check to see that the proposed reductions are adequate. There are many possible solutions.

20.15 A worker operates several machines and is exposed to a daily noise pattern described as follows:

Noise Level (dBA)	*Exposure Time (h)*
85	3
87	2
90	1
81	1.5
83	0.5

(a) Find percent noise dose and time-weighted average noise level. Compare results with exposure limits.

Decisions: Use a criterion level of 90 dBA and an exchange rate of 5 dBA.

(b) If the results are unacceptable, propose reducing the noise level of one or more machines. Check to see that the proposed noise reduction is adequate. There are many possible solutions.

20.16 It is proposed to spend $25,000 on a project that will not yield results for about 7 years. For comparison purposes, determine the final account value if this sum is invested for the same period.

Assume interest at an annual rate of 5.25% with
(a) semiannual compounding.
(b) monthly compounding.

20.17 A proposed $1 million equipment investment is expected to yield results in 6.5 years. For comparison purposes, determine the final account value if this sum is invested for the same period. Assume interest at an annual rate of 8% with
(a) semiannual compounding.
(b) monthly compounding.

20.18 It is proposed to spend $1000 on a project that will not yield results for about 12 years. For comparison purposes, determine the final account value if this sum is invested for the same period. Assume interest at an annual rate of 6% with
(a) semiannual compounding.
(b) monthly compounding.

20.19 How long does it take to double principal at an annual interest rate of 8% if interest is compounded
(a) semiannually?
(b) monthly?

20.20 How long does it take to double principal at an annual interest rate of 5% if interest is compounded
(a) semiannually?
(b) monthly?

20.21 Based on an annual interest rate of 9%, compute the time for a sum to increase by 50%. Assume that interest is compounded
(a) semiannually.
(b) monthly.

20.22 Based on an annual interest rate of 6.5%, compute the time for a sum to increase by 25%. Assume that interest is compounded
(a) semiannually.
(b) monthly.

20.23 Calculate the time required to produce a 100% principal increase for interest rates ranging from 2% to 12%. Calculate the products of rate and time to verify the familiar "doubling-of-principal" approximation.

20.24 Calculate the time required to produce a 40% principal increase for interest rates ranging from 2% to 12%. Calculate the products of rate and time.

20.25 Calculate the time required to produce a 200% principal increase for interest rates ranging from 2% to 12%. Calculate the products of rate and time.

20.26 A certain purchase is expected to yield $1000 twelve years from now. What is its present value? Use an interest rate of 6% and
(a) annual compounding.
(b) monthly compounding.

20.27 We expect to receive $5000 in 5.5 years. What is the present value? Use an interest rate of 4.5% and
(a) annual compounding (prorating the half year).
(b) monthly compounding.

20.28 What is the present value of an investment that will produce $3000 in 8.5 years? Use an interest rate of 8.5% and
(a) annual compounding (prorating the half year).
(b) monthly compounding.

20.29 We are considering the purchase of equipment costing $70,000. The useful life is estimated to be 7 years. At the end of that time, the equipment will have no value, and we expect to incur cleanup costs totaling $3000. Estimate the annual cost of the investment.

Decision: We will base our calculations on an annual interest rate of 11% compounded annually.

20.30 We are considering the purchase of production machinery for $45,000. The useful life (for our operations) is estimated to be 6 years. At the end of that time, we could probably resell the equipment to another manufacturer for about 10% of its original cost. Estimate the annual cost of the investment.

Decision: After checking current interest rates, we have decided to base our calculations on an annual interest rate of 6.5% compounded annually.

20.31 We are considering the purchase of a robot for $750,000. The useful life is estimated to be 8 years. At the end of that time, we estimate that the value would be approximately zero. Estimate the annual cost of the investment.

Decision: After checking current interest rates, we have decided to base our calculations on an annual interest rate of 8% compounded annually.

20.32 (a) Write a computer program to optimize the cost function

$$C = x^2 + y^2 - 11x - 8y + e^{xy/20} + y^{1.5} + 72$$

Use any convenient language or software.

(b) Find the values of x and y for which C will be minimum. Use the starting point $x = 10$, $y = 10$.

20.33 Write a computer program for optimization of the cost function

$$C = x^2 + y^2 - 39x - 30y + 725 + 10^{-6}y^3$$

20.34 Find the values of x and y for which cost function C in Problem 20.33 will be minimum.

20.35 Write a computer program for optimization of the cost function

$$C = 1.9x^2 + 2.1y^2 - 80x - 60y + 1350 + e^{-y}$$

20.36 Find the values of x and y that will minimize cost function C in Problem 20.35.

APPENDIX

Machine Design Projects

Design projects can be used to expose students to real-world machine design problems that go beyond the scope of the text assignments and classroom lectures.

SOURCES OF PROJECT IDEAS

Periodicals such as *Machine Design* and *Design News* are good sources of interesting projects. Professors often obtain project ideas from their research, industrial experience, consulting assignments, or expert witness assignments. Design projects can also be proposed by industrial, government, or academic sponsors with a real need.

GENERAL CONSIDERATIONS

Project goals can include the fostering of group interaction, independent thinking, decision making, and creativity, as well as the development of planning and scheduling abilities. Some projects include several machine design topics as well as concepts learned throughout the curriculum. Project work can include collaborative learning, brainstorming, and use of libraries and other information resources. Note that problems faced by machine designers usually require far more time than is available in an academic course. It is suggested that the scope of each project be limited to that which can be accomplished in a professional manner within the allotted time. If sponsored projects are used, learning goals should take precedence over needs of the sponsor.

SUGGESTIONS FOR PROJECT PROPOSALS AND REPORTS

- Make a bar chart showing proposed tasks and starting and finishing dates for these tasks. Revise the proposal if the chart shows the project to be too ambitious. Update the bar chart when necessary.
- Keep a log of work attempted and completed.
- Perform a feasibility study of the proposed work.
- Make brief oral and written progress reports. Use simple visual aids. Solicit comments for design improvements. Incorporate design changes if comments are valid.

Where applicable, reports should include

- a table of contents (used as a cover sheet)
- explanations of design decisions
- problem-solving methodology, methods of analysis, and synthesis
- engineering principles used in the design
- selection of failure theories and design criteria (fatigue, elastic stability, etc.), design equations, and sample calculations
- safety aspects of the project and the designer's response to potential safety problems
- creative thinking; decision making
- optimization
- computer simulation
- identification of group work and individual work
- discussion and interpretation of results
- references

SUGGESTED PROJECT TOPICS

Human-Powered Vehicle

Design a human-powered vehicle (HPV). Concentrate on one or two HPV features, or redesign some HPV component. If the HPV rider is in a recumbent position, drag can be reduced significantly. However, there are visibility and safety implications. Consider cost and efficient use of energy. It is estimated that bicycling consumes about 35 cal/mi, compared with about 100 cal/mi for walking and about 1900 cal/mi for an automobile. Motorists spend about 4 h/day driving, maintaining, and paying for an automobile.

Reverted Gear Train

Write a program for designing gear trains. The program should include checking of contact ratio and interference, and design for strength and pitting resistance. Test the program by designing a two-speed reverted gear train with 100 hp capacity and 4000 rpm input speed. Approximate output speeds should be 2500 rpm and $1000 + x$ rpm where $x =$ the last three digits of your Social Security number. Design the shaft to the six sigma reliability level. Use the Soderberg–vonMises or the Soderberg–maximum shear stress criterion. Try to optimize the design.

Machine Guarding

Examine a machine. Design an innovative guard for that machine.

Lubrication

Design a system to test lubricants in discontinuous, non-hydrodynamic situations. Possibilities include the following:

1. **Oscillation.** A pendulumlike system in which a shaft oscillates within a bearing. Lubrication effectiveness is based on loss of energy. Initial energy and cumulative rotation angle through all cycles can be measured. Air resistance can be accounted for.

2. **Gear meshing.** Gears A and B mounted on shaft 1 mesh respectively with gears C and D on shaft 2. Gear B is mounted on a helical spline with a small spline angle. Gear D is mounted on a straight spline. Gears B and D are moved axially, twisting the shafts and producing a tooth load. The torque required to drive shaft 1 is related to the friction in the gear teeth.

Brake Noise

Design, construct, and utilize an experimental device for determining the conditions under which brake squeal occurs.

Laboratory Experiment

Design an experiment to teach some principle of physics or engineering.

Crash Protection for Side Impact

Design a system for occupant protection in highway crashes that involve side impact. The system can incorporate an air bag or other device.

Pedestrian Safety for Light-Rail Systems

Light-rail vehicles (LRVs) operate in shared right-of-way; that is, they operate on, adjacent to, or across city streets. Design a system to reduce injury due to interaction of LRVs with pedestrians and bicyclists. Assume that the LRV operates at low speed in the shared right-of-way.

Energy-Absorbing Structure for Trucks

Truck crashes account for about 5000 deaths annually in the United States. Design an energy-absorbing, crushable, frontal structure for trucks that would reduce injuries to automobile drivers in car–truck accidents.

Device to Aid Persons with Disabilities

Design a device to aid a person with a disability/disabilities in performing some task.

Pipeline Valve

A 36 in diameter pipeline carrying gas at 900 psi ruptured. The resulting explosion destroyed eight buildings and forced hundreds of people to flee for their lives. As flames shot into the air, it took hours to shut off the gas in the pipeline. Design an automatic shutoff system for gas pipelines.

Aircraft Door Latch

A crash investigation indicated the following probable scenario: The aircraft cargo door was not properly latched at takeoff. As the aircraft gained altitude, the pressurization of the cargo hold caused the door to open. Cabin pressure then exceeded cargo hold pressure, and the pressure difference distorted the cabin floor. Control cables located in the cabin floor were prevented from moving, causing the pilot to lose control.

Design a system to prevent a recurrence of accidents of this type. Options include (but are not limited to)

- a fail-safe latch
- a warning device when the latch is not closed
- control-cable protection

Seat-Belt Mechanism

The National Highway Traffic Safety Administration has received reports of injuries due to improperly functioning seat belts. It was determined that pieces of the plastic release button on one type of seat belt can fall into and jam the mechanism. Automakers have agreed to repair or replace the belts on over 8 million cars and small trucks. Design a jam-proof seat belt mechanism, or propose an innovative design to replace conventional seat belts.

Transportation for People with Disabilities

Under the Americans with Disabilities Act (ADA), public entities that provide fixed-route public transportation service must offer complementary paratransit service to people with disabilities unable to use fixed-route service. A major concern is the high cost of paratransit. To serve the most people cost-effectively, there is a need to shift paratransit passengers to fixed-route transit.

- Investigate barriers to the disabled, the elderly, and others who do not utilize fixed-route service.
- Design a system to make fixed-route transportation accessible to this group.

Consider vehicle design changes that make it more convenient for the disabled to enter. Consider also the development of a simple wheelchair restraint system for transit vehicles. The system should be operated by the wheelchair occupant and should protect the occupant and other riders in the event of an accident.

Database/Expert System for Selecting Materials

Design a simple database/expert system for selecting material for engineering designs.

Suggestions: Limit the scope of the system, considering time allotted to the project. For example, the information base could be obtained by library research into a particular type of material. The system should make if–then decisions based on required material properties (tensile strength, high-temperature performance, weight restrictions, etc.). Depending on the goals of the course,

(a) the program can be written in BASIC, FORTRAN, or other computer language, using logical IF statements; or

(b) a spreadsheet can be used; or

(c) mathematics software can be used; or

(d) a database management system or expert system shell software can be used.

Partial Answers to Selected Problems

Note: In some cases, results are reported to several decimal places so that they may be used to check calculations. In actual design practice, calculated dimensions may be rounded upward to standard sizes. Some problems have many acceptable solutions.

Chapter 1

1.1. Solid modeling software is used to represent the crankshaft.

1.3. The mass is 2.99 kg.

1.5. The mass is 0.221 kg.

1.7. Optimum cost function is about 25.683.

1.9. Optimum cost function is about 27.527.

1.11. Optimum cost function is about 20.677 corresponding to a radius of about 0.5 m.

Chapter 2

2.1c. Ultimate strength = 75,000 psi or 518 Mpa.

2.3. 0.362 in. diameter circular section.

2.5. Thickness = 1.025 in.

2.7. Thickness = 0.855 in.

2.9. 0.75 in. diameter long-fiber-reinforced thermoplastic.

2.11. Load/weight ratio = 280 for glass-reinforced thermoplastic polyester.

Chapter 3

3.1b Width $\geq 0.81(6\text{-}x)/6$. Width at end must be adequate to locate actual load and avoid transverse shear failure.

3.3a. Width = 1.6 in.

3.5. 0.706 in. outside diameter.

3.7. von Mises stress = 62,895 psi.

3.9. 1.158 in. diameter based on maximum shear theory.

3.11. 30.28 mm based on von Mises theory.

3.13. von Mises stress = 492.5 MPa.

3.15. Maximum shear stress = 18,564 psi.

Chapter 4

4.1. A reliability of 0.9999 corresponds to 3.72 standard deviations and a reliability factor of 0.628 for this material.

4.3. $h = 0.744$.

4.5. 0.325 in. deep by 0.976 in wide.

4.7. 0.493 in. deep by 1.726 in. wide.

4.9b. $R = 0.999998$. Actual reliability affected by system inspection frequency and maintenance.

4.11a. $R = 0.358$.

4.13. $A = 1.973$; $B = -0.189$.

Chapter 5

5.1. Bending moment is approximately $-30{,}000$ N · mm at $x = 40$ mm.

5.3. Bending moment is approximately -250 lb · in. at $x = 6$ in.

5.5. Bending moment is approximately 8000 N · mm at the center.

5.7. $I = 61$ in^4 (approx.).

5.9. Displacement = $1.045 \cdot 10^{-3}$ in. at the load.

5.11. Displacement = $7.77 \cdot 10^{-3}$ in. at the load.

5.13. Displacement = $12.5 \cdot 10^{-3}$ in. at the load.

5.15. Width = 33.8 mm.

5.17. Diameter = 21.4 mm.

5.19. Diameter = 0.906 in.

Chapter 6

6.1. The FEA model indicates a maximum stress of $1.28 \cdot 10^9$ Pa. The expected stress (based on a longer plate) is lower.

6.3. The maximum stress is about 31,300 psi, using the theoretical stress concentration factor.

6.5. Maximum tangential stress is 47,620 psi at the inner radius.

Chapter 7

7.1. Cross section: 0.25 by 0.19 in. It may be necessary to weld the ends due to the small dimensions.

7.3. 6 mm by 20 mm cross section.

7.5. $a = 0.235$; $b = 0.471$.

Chapter 8

8.1. $D = 0.694$ in. using von Mises theory.

8.3. $D = 1.071$ in. using maximum shear stress theory.

8.5. $OD = 28.32$ and $ID = 22.65$ mm based on von Mises theory.

8.7. $D = 1.202$ in.

8.9. $D = 1.086$ in.

8.11. The critical location is to the left of $x = 6$ in. $D = 0.998$ based on Soderberg-maximum shear stress theory.

8.13. $M = 1043$ at $x = 3.5$; $D = 0.782$ based on Soderberg-von Mises theory.

8.15. $M = 774$ at $x = 2$; $D = 0.84$ based on Soderberg-von Mises theory.

Chapter 9

9.1. Unsatisfactory due to interference.

9.3. $CR = 1.605$ (satisfactory).

9.5. No interference; $CR = 2.175$.

9.7. 14.5° pressure angle: unsatisfactory due to interference. 25° pressure angle: No interference. $CR = 1.431$.

9.9. Combined moment $= 1061$ at $x = 2.5$; $D = 0.993$ based on Soderberg-von Mises theory.

9.11. Face width based on pitting resistance governs. $F = 1.264$.

9.13. Face width based on pitting resistance governs. $F = 1.517$.

9.15. 20° bending strength geometry factor is approximated by $J = 0.311 \ln(N) + 0.15$. Other solutions possible.

9.17. $N_2 = 65$, $N_4 = 50$.

9.19. $N_P = 26$, $N_R = 92$.

9.21. Three 17-tooth planets can be used, or four 18-tooth planets. We will not chose 19-tooth planets because neither three nor four would result in a balanced train.

Chapter 10

10.1. One 5V belt may be used with a 12.5 in. *OD* input sheave.

10.3. $A = 369.7$; $B = -467.5$.

10.5. $A = 613.3$; $B = -1667.9$.

10.7. A 20 in. diameter input sheave results in the lowest value of maximum stress.

Chapter 11

11.1. 113 horsepower governed by roller-bushing fatigue.

11.3. The peak capacity is about 110 horsepower at 500 rpm.

11.5. The peak capacity is about 21 horsepower at 1200 rpm.

11.7. 1/2 inch pitch will be used.

11.9. Two strands will be used.

11.11. 2 in. wide, 1 in. pitch silent chain will be used.

11.13. 2.5 in. wide silent chain will be used.

Chapter 12

12.1. Let gears *A* and *B* on the input shaft mesh with gears *C* and *D*, which rotate freely on the output shaft. A clutch engages either gear *C* or gear *D* to the output shaft, producing ratios $-N_A/N_C$ or $-N_B/N_D$. Other configurations may be used.

12.3. Use gear *A* on the input shaft, gears *B*, *C*, and *E* on the countershaft, and gears *D* and *F* rotating freely on the output shaft. Gears *A* and *B* are always in mesh, as are *C* and *D*, and *E* and *F*. A clutch engages either *D* or *F* to the output shaft. Alternative configurations may be used.

12.5. Eight pairs are required at 150 rpm; six pairs at 200 rpm.
12.7. Maximum contact pressure is $4.974 \cdot 10^5$ Pa.
12.9. $r_o = 1.544$ in. for 5 horsepower.
12.11. $r_o = 59$ mm for 5 kw.
12.13. $r_o = 5.5$ in. if 10 pairs are used.

Chapter 13

13.1. $J = 1.367$ lb $\cdot$ s^2 $\cdot$ in
13.3. $J = 1.03$ kg $\cdot$ m^2
13.5. $T_B = 1420$ N $\cdot$ m
13.7. $T_B = 786$ N $\cdot$ m
13.9. $T_B = 7.29 \cdot 10^4$ in·lb
13.11. $T_B = 1.95 \cdot 10^4$ in·lb
13.13. 3968 N actuating force.
13.15. 340 N actuating force.
13.17. 36.5 mm wide shoe with 2113 N actuating force.
13.19. Band width = 1.02 in.
13.21. Band width = 1.24 in.
13.23. 4.3 cable turns.
13.25. 69.2 lb force.

Chapter 14

14.1. $f = 0.11$.
14.3. About 78% efficiency at 40° lead angle.
14.5. About 50% efficiency for 10° lead angle if $f = 0.16$.
14.7. 57–58% efficiency for both with 10° lead angle.
14.9. 15.5% system efficiency.
14.11. 7.4% system efficiency.
14.13. 3.324 horsepower input.

Chapter 15

15.1. Use 8 bolts based on pullout strength and stress at the root area.
15.3. Use 9 bolts based on pullout strength.
15.5. The part is always in compression if the preload is 900 N/bolt.
15.7a. Preload must exceed 833 lb/bolt.
15.9. Preload must exceed 800 lb/bolt.
15.11. Fastener diameter must not be less than 0.662 in.

Chapter 16

16.1. Weld leg dimension not less than 6.442 mm (to be rounded upward).
16.3. Weld leg dimension not less than 5.909 mm (to be rounded upward).

16.5. Weld leg dimension not less than 0.445 in. (to be rounded upward).

16.7. Weld leg dimension not less than 0.34 in. (to be rounded upward).

16.9. Weld leg dimension not less than 7.73 mm (to be rounded upward).

Chapter 17

17.1. Optimum diameter = 0.464 in.; length = 52.25 in. If 1/2 in. diameter steel is used, length = 70.6 in.

17.3. 17.5 mm diameter.

17.5. Wire diameter not less than 0.565 in.

17.7. Wire diameter not less than 0.678 in. Use 7.8 active coils if 3/4 in. wire diameter is used.

17.9. Wire diameter not less than 0.331 in. based on Soderberg criterion modified for springs.

17.11. Wire diameter not less than 7.918 mm based on Soderberg criterion modified for springs. Free height = 115 mm if 8 mm wire is used.

17.13. Maximum transmissibility is about 2.3 for a damping ratio of 0.25.

17.15. Use a frequency ratio of 2.57.

17.17. Spring rate less than 118 lb/in per spring.

17.19. Spring rate less than 35.65 lb/in per spring.

17.21. Spring rate less than 1883 lb/in per spring.

Chapter 18

18.1. Total stopping distance is about 480 ft at 70 mph.

18.3. Total stopping distance is about 440 ft at 80 mph for given conditions.

18.5b. Geometry factor approximated by $a(r_d - 0.2)^2 + b$ where a = -25,130; b = 4491.

18.7. The life expectancy is 69.5 weeks.

18.9. The life expectancy is 22.4 weeks.

18.11. Basic load rating should not be less than 1131 lb.

18.13. Basic load rating should not be less than 433 lb.

18.15. Basic load rating should not be less than 32,713 lb.

18.17. 19 microreyns.

18.19. 12.7 microreyns.

18.21. The friction loss is about 0.044 hp at 5000 rpm.

18.23. The friction loss is about 0.1 hp.

18.25. The friction loss is about 0.06 hp.

18.27. The eccentricity ratio is about 0.68 at 3000 rpm.

Chapter 19

19.1. $94 per unit.

19.3. $11.91 per unit.

19.5. $4.50 per unit.

19.7b. 0.0049 tolerance with process unchanged.

19.9b. 0.009 tolerance with process unchanged.

Chapter 20

20.1. 82.8 dBA.

20.3. 71.8 dBA.

20.5. 73.7 dBA.

20.7. $L_{DN} = 50$.

20.9. $L_{DN} = 62$.

20.11. 73.5 dBA.

20.13. 55.5% dose.

20.15. 92.8% dose.

20.17b. $1,679,128.

20.19a. 8.84 yrs.

20.21b. 4.52 yrs.

20.23. 8.04 years at 9% compounded annually.

20.25. 22.02 yrs. at 5% compounded monthly.

20.27b. $3906.

20.29. $15,162.

20.31. $128,045.

20.33. $\partial C/\partial x = 2x - 39$.

20.35. $\partial C/\partial x = 3.8x - 80$.

Index

G

H

I